Mathematical Statistics

Dieter Rasch
University of Natural Resources and Life Sciences,
Institute of Applied Statistics and Computing
Vienna, Austria

Dieter Schott
Faculty of Engineering, Hochschule Wismar, University of Applied
Sciences: Technology, Business and Design
Wismar, Germany

WILEY

This edition first published 2018

Registered Offices
John Wiley & Sons, Inc., 111 River Street, Hoboken, NJ 07030, USA
John Wiley & Sons Ltd, The Atrium, Southern Gate, Chichester, West Sussex, PO19 8SQ, UK

Editorial Office
9600 Garsington Road, Oxford, OX4 2DQ, UK

For details of our global editorial offices, customer services, and more information about Wiley products visit us at www.wiley.com.

Wiley also publishes its books in a variety of electronic formats and by print-on-demand. Some content that appears in standard print versions of this book may not be available in other formats.

Library of Congress Cataloging-in-Publication Data

Names: Rasch, Dieter, author. | Schott, Dieter, author.
Title: Mathematical statistics / by Dieter Rasch, Dieter Schott.
Description: Hoboken, NJ : John Wiley & Sons, 2018. | Includes
bibliographical references and index. |
Identifiers: LCCN 2017039506 (print) | LCCN 2017046479 (ebook) | ISBN
9781119385264 (pdf) | ISBN 9781119385233 (epub) | ISBN 9781119385288 (cloth)
Subjects: LCSH: Mathematical statistics.
Classification: LCC QA276 (ebook) | LCC QA276 .R362 2018 (print) | DDC
519.5–dc23
LC record available at https://lccn.loc.gov/2017039506

Cover design: Wiley
Cover image: (Background) © Xavier Antoinet/EyeEm/Getty Images;
(Graph) Density Functions of the Estimator of the Location Parameter μ
Depending on the Hypothesis Values $\mu = 0$ and $\mu = 2$, respectively.
Courtesy of Dieter Rasch and Dieter Schott

Set in 10/12pt Warnock by SPi Global, Pondicherry, India

Printed in Singapore by C.O.S. Printers Pte Ltd

10 9 8 7 6 5 4 3 2 1

Contents

Preface

'Mathematical statistics' never lost its attractiveness, both as a mathematical discipline and for its applications in nearly all parts of empirical research. During the last years it was found that not everything that is mathematically optimal is also practically recommendable if we are not sure whether the assumptions (for instance, normality) are valid.

As an example we consider the two-sample t-test that is an optimal (uniformly most powerful unbiased) test if all assumptions are fulfilled. In applications however, we are often not sure that both variances are equal. Then the approximate Welch test is preferable. Such results have been found by extensive simulation experiments that played a much greater role the last time (see the eight international conferences about this topic since 1994 under http://iws.boku.ac.at).

Therefore we wrote in 2016 a new book in German (Rasch and Schott, 2016) based on Rasch (1995) incorporating the developments of the last years.

We dropped the first part of the book from 1995 containing measure and probability theory because we have excellent books about this such as Billingsley (2012) and Kallenberg (2002).

Considering the positive resonance to this book in the community of statistics, we decided to present an English version of our book from 2016. We thank Alison Oliver for the reception into Wiley's publishing programme.

We assume from probability theory knowledge about exponential families as well as central and non-central t-, χ^2- and F-distributions. Because the definition of exponential families is basic for some chapters, it is repeated in this book.

Most of the authors of books about mathematical statistics assume that data already exist and must be analysed. But we think that the optimal design for collecting data is at least as important as the statistical analysis. Therefore, in addition to statistical analysis, we included the design of experiments. The optimal allocation is described in the chapters on regression analysis. Finally a chapter about experimental designs is added.

For practical calculations of data, we present and use in some parts of the book *IBM SPSS Statistics 24* for the statistical analysis, and we thank Dr. Johannes Gladitz (Berlin) for giving us access to it. Unfortunately, it is not possible to change within SPSS to British English – therefore, you find in the screens and in our command 'Analyze'.

The determination of sample sizes can be found together with the description of the method of analysis, and for the sample size determination and other design problems, we offer the package OPDOE (Optimal Design of Experiments) under R.

We heartily thank Prof. Dr. Rob Verdooren (Wageningen, Netherlands) for proving the correctness of statistics and Sandra Almgren (Kremmling, CO, USA) for improving the English text.

Rostock, December 2017 *Dieter Rasch and Dieter Schott*

References

Billingsley, P. (2012) *Probability and Measure*, John Wiley & Sons, Inc., New York.

Kallenberg, O. (2002) *Foundations of Modern Probability*, 2nd edition, Springer, New York.

Rasch, D. (1995) *Mathematische Statistik*, Johann Ambrosius Barth, Berlin, Heidelberg.

Rasch, D. and Schott, D. (2016) *Mathematische Statistik*, Wiley VCH, Weinheim.

1

Basic Ideas of Mathematical Statistics

Elementary statistical computations have been carried out for thousands of years. For example, the arithmetic mean from a number of measures or observation data has been known for a very long time.

First descriptive statistics arose starting with the collection of data, for example, at the national census or in registers of medical cards, and followed by compression of these data in the form of statistics or graphical representations (figures). Mathematical statistics developed on the fundament of probability theory from the end of 19th century on. At the beginning of the 20th century, Karl Pearson and Sir Ronald Aylmer Fisher were notable pioneers of this new discipline. Fisher's book (1925) was a milestone providing experimenters such basic concepts as his well-known maximum likelihood method and analysis of variance as well as notions of sufficiency and efficiency. An important information measure is still called the Fisher information (see Section 1.4).

Concerning historical development we do not want to go into detail. We refer interested readers to Stigler (1986, 1990). Instead we will describe the actual state of the theory. Nevertheless many stimuli come from real applications. Hence, from time to time we will include real examples.

Although the probability calculus is the fundament of mathematical statistics, many practical problems containing statements about random variables cannot be solved with this calculus alone. For example, we often look for statements about parameters of distribution functions although we do not partly or completely know these functions. Mathematical statistics is considered in many introductory textbooks as the theory of analysing experiments or samples; that is, it is assumed that a random sample (corresponding to Section 1.1) is given. Often it is not considered how to get such a random sample in an optimal way. This is treated later in design of experiments. But in concrete applications, the experiment first has to be planned, and after the experiment is finished, the analysis has to be carried out. But in theory it is appropriate to determine firstly the optimal evaluation, for example, the smallest sample size for a variance optimal estimator. Hence we proceed in such a way and start with the optimal evaluation, and after

Mathematical Statistics, First Edition. Dieter Rasch and Dieter Schott.

this we work out the design problems. An exception is made for sequential methods where planning and evaluation are realised together.

Mathematical statistics involves mainly the theory of point estimation, statistical selection theory, the theory of hypothesis testing and the theory of confidence estimation. In these areas theorems are proved, showing which procedures are the best ones under special assumptions.

We wish to make clear that the treatment of mathematical statistics on the one hand and its application to concrete data material on the other hand are totally different concepts. Although the same terms often occur, they need not be confused. Strictly speaking, the notions of the empirical sphere (hence of the real world) are related to corresponding models in theory.

If assumptions for deriving best methods are not fulfilled in practical applications, the question arises how good these best methods still are. Such questions are answered by a part of empirical statistics – by simulations. We often find that the assumption of a normal distribution occurring in many theorems is far from being a good model for many data in applications. In the last years simulation developed into its own branch in mathematics. This shows a series of international workshops on simulation. The first to sixth workshops took place in St. Petersburg (Russia) in 1994, 1996, 1998, 2001, 2005 and 2009. The seventh international workshop on simulation took place in Rimini (Italy) in 2013 and the eighth one in Vienna (Austria) in 2015.

Because the strength of assumptions has consequences mainly in hypothesis testing and confidence estimation, we discuss such problems first in Chapter 3, where we introduce the concept of robustness against the strength of assumptions.

1.1 Statistical Population and Samples

1.1.1 Concrete Samples and Statistical Populations

In the empirical sciences, one character or several characters simultaneously (character vector) are observed in certain objects (or individuals) of a population. The main task is to conclude from the sample of observed values to the whole set of character values of all objects of this population. The problem is that there are objective or economical points of view that do not admit the complete survey of all character values in the population. We give some examples:

- The costs to register all character values were out of all proportion to the value of the statement (for instance, measuring the height of all people worldwide older than 18 years).
- The registration of character values results in destruction of the objects (destructive materials testing such as resistance to tearing of ropes or stockings).

- The set of objects is of hypothetic nature, for example, because they partly do not exist at the moment of investigation (as all products of a machine).

We can neglect the few practical cases where all objects of a population can be observed and no more extensive population is demanded, because for them mathematical statistics is not needed. Therefore we assume that a certain part (subset) is chosen from the population to observe a character (or character vector) from which we want to draw conclusions to the whole population. We call such a part a (concrete) sample (of the objects). The set of character values measured for these objects is said to be a (concrete) sample of the character values. Each object of the population is to possess such a character value (independent of whether we register the value or not). The set of character values of all objects in the population is called the corresponding statistical population.

A concrete population as well as the (sought-after/relevant) character and therefore also the corresponding statistical population need to be determined uniquely. Populations have to be circumscribed in the first line in relation to space and time. In principle it must be clear for an arbitrary real object whether it belongs to the population or not. In the following we consider some examples:

Original population		**Statistical population**	
A	Heifer of a certain breed in a certain region in a certain year	A_1	Yearly yield of milk of these heifer
		A_2	Body mass of these heifer after 180 days
		A_3	Back height of these heifer
B	Inhabitants of a town at a certain day	B_1	Blood pressure of these inhabitants at 6.00 o'clock
		B_2	Age of these inhabitants

It is clear that applying conclusions from a sample to the whole population can be wrong. For example, if the children of a day nursery are chosen from the population B in the table above, then possibly the blood pressure B_1 but without doubt the age B_2 are not applicable to B. Generally we speak of characters, but if they can have a certain influence to the experimental results, they are also called factors. The (mostly only a few) character values are said to be factor levels, and the combinations of factor levels of several factors factor level combinations.

The sample should be representative with respect to all factors that can influence the character of a statistical population. That means the composition of the population should be mirrored in the sample of objects. But that is impossible for small samples and many factor level combinations. For example, there are already about 200 factor level combinations in population B concerning the factors age and sex, which cannot be representatively found in a sample of 100 inhabitants. Therefore we recommend avoiding the notion of 'representative sample' because it cannot be defined in a correct way.

Samples should not be assessed according to the elements included but according to the way these elements have been selected. This way of selecting a sample is called sampling procedure. It can be applied either to the objects as statistical units or to the population of character values (e.g. in a databank). In the latter case the sample of character values arises immediately. In the first case the character must be first registered at the selected objects. Both procedures (but not necessarily the created samples) are equivalent if the character value is registered for each registered object. This is assumed in this chapter. It is not the case in so-called censored samples where the character values could not be registered in all units of the experiment. For example, if the determination of lifespans of objects (as electronic components) is finished at a certain time, measured values of objects with longer lifespans (as time of determination) are missing.

In the following we do not differ between samples of objects and samples of character values; the definitions hold for both.

Definition 1.1 A sampling procedure is a rule of selecting a proper subset, named sample, from a well-defined finite basic set of objects (population, universe). It is said to be at random if each element of the basic set has the same probability p to come into the sample. A (concrete) sample is the result of a sampling procedure. Samples resulting from a random sampling procedure are said to be (concrete) random samples.

There are a lot of random sampling procedures in the theory of samples (see, e.g. Cochran and Boing, 1972; Kauermann and Küchenhoff, 2011; Quatember, 2014) that can be used in practice. Basic sets of objects are mostly called (statistical) populations or synonymously sometimes (statistical) universes in the following.

1.1.2 Sampling Procedures

Concerning random sampling procedures, we distinguish (among other cases)

- The simple sampling, where each element of the population has the same probability to come into the sample.
- The stratified sampling, where a random sampling is done within the before-defined (disjoint) subclasses (strata) of the population. This kind of sampling is only at random as a whole if the sampling probabilities within the classes are chosen proportionally to the cardinalities of the classes.
- The cluster sampling, where the population is divided again into disjoint subclasses (clusters), but the sampling of objects is done not among the objects of the population itself but among the clusters. In the selected clusters all objects are registered. This kind of selection is often used as area samples. It is only at

random corresponding to Definition 1.1 if the clusters contain the same number of objects.

- The multistage sampling, where at least two stages of sampling are taken. In the latter case the population is firstly decomposed into disjoint subsets (primary units). Then a random sampling is done in all primary units to get secondary units. A multistage sampling is favourable if the population has a hierarchical structure (e.g. country, province, towns in the province). It is at random corresponding to Definition 1.1 if the primary units contain the same number of secondary units.
- The (constantly) sequential sampling, where the sample size is not fixed at the beginning of the sampling procedure. At first a small sample is taken and analysed. Then it is decided whether the obtained information is sufficient, for example, to reject or to accept a given hypothesis (see Chapter 3), or if more information is needed by selecting a further unit.

Both a random sampling (procedure) and an arbitrary sampling (procedure) can result in the same concrete sample. Hence we cannot prove by inspecting the sample itself whether the sample is randomly chosen or not. We have to check the sampling procedure used.

For the pure random sampling, Definition 1.1 is applied directly: each object in the population of size N is drawn with the same probability p. The number of objects in a sample is called sample size, mostly denoted by n.

The most important case of a pure random sampling occurs if the objects drawn from a population are not put back. An example is a lottery, where n numbers are drawn from N given numbers (in Germany the well-known Lotto uses $N = 49$ and $n = 6$). Using an unconditioned sampling of size N, the number $M = \binom{N}{n}$ of all possible subsets have the same probability $p = \frac{1}{M}$ to come into the sample.

As mentioned before, the sample itself is only at random if a random sampling method was used. But persons become at once suspicious if the sample is extreme. If somebody gets the top prize buying one lot of 10.000 possible lots, then this case is possible although rather unlikely. It can happen at random, or in other words, it can be the result of (a correct) random sampling. But if this person gets the top prize buying one lot at three consecutive lotteries of the mentioned kind, and if it turns out additionally that the person is the brother of the lot seller, then doubts are justified that there was something fishy going on. We would refuse to accept such unlikely events and would suppose that something is wrong. In our lottery case, we would assume that the selection was not at random and that cheats were at work. Nevertheless, there is an extremely small possibility that this event is at random, namely, $p = 1/1.000.000.000.000$.

Incidentally, the strategy of statistical tests in Chapter 3 is to refuse models (facts) under which observed events possess a very small probability and instead to accept models where these events have a larger probability.

A pure sampling also occurs if the random sample is obtained by replacing the objects immediately after drawing and observing that each object has the same probability to come into the sample using this procedure. Hence, the population always has the same number of objects before a new object is taken. That is only possible if the observation of objects works without destroying or changing them (examples where that is impossible are tensile breaking tests, medical examinations of killed animals, felling of trees and harvesting of food). The discussed method is called simple random sampling with replacement.

If a population of N objects is given and n objects are selected, then it is $n < N$ for sampling without replacement, while objects that can multiply occur in the sample and $n > N$ is possible for sampling with replacement.

A method that can sometimes be realised more easily is the systematic sampling with random start. It is applicable if the objects of the finite sampling set are numbered from 1 to N and the sequence is not related to the character considered. If the quotient $m = N/n$ is a natural number, a natural number i between 1 and m is chosen at random, and the sample is collected from objects with numbers $i, m + i, 2m + i, \ldots, (n - 1)m + i$. Detailed information about this case and the case where the quotient m is not natural can be found in Rasch et al. (2008) in Method (1/31/1210).

The stratified sampling already mentioned is advantageous if the population of size N is decomposed in a content-relevant manner into s disjoint subpopulations of sizes $N_1, N_2, \ldots, N_s$. Of course, the population can sometimes be divided into such subpopulations following the levels of a supposed interfering factor. The subpopulations are denoted as strata. Drawing a sample of size n is to realise in such a population an unrestricted sampling procedure holds the danger that not all strata are considered in general or at least not in appropriate way. Therefore in this case a stratified random sampling procedure is favourable. Then partial samples of size n_i are collected from the ith stratum ($i = 1, 2, \ldots, s$) where pure random sampling procedures are used in each stratum. This leads to a random sampling procedure for the whole population if the numbers n_i/n are chosen proportional to the numbers N_i/N.

While for the stratified random sampling objects are selected from each subset, for the multistage sampling, subsets or objects are selected at random at each stage as described below. Let the population consist of k disjoint subsets of size N_0, the primary units, in the two-stage case. Further, it is supposed that the character values in the single primary units differ only at random, so that objects need not to be selected from all primary units. If the wished sample size is $n = r\,n_0$ with $r < k$, then, in the first step, r of the k given primary units are selected using a pure random sampling procedure. In the second step n_0 objects (secondary units) are chosen from each primary unit again applying a pure random sampling. The number of possible samples is $\binom{k}{r} \cdot \binom{N_0}{n_0}$, and each

Table 1.1 Possible samples using different sampling procedures.

Sampling	Number K of possible samples
Simple random sampling	$K = \binom{1000}{100} > 10^{140}$
Systematic sampling with random start $k = 10$	$K = 10$
Stratified random sampling $k = 20$, $N_i = 50$, $i = 1, \ldots, 20$	$K = \left[\binom{50}{5}\right]^{20} = 2.11876 \cdot 10^{120}$
Stratified random sampling $k = 10$, $N_i = 100$, $i = 1, \ldots, 10$	$K = \left[\binom{100}{10}\right]^{10} = 1.7310309 \cdot 10^{130}$
Stratified random sampling $k = 5$, $N_i = 200$, $i = 1, \ldots, 5$	$K = \left[\binom{200}{20}\right]^{5} = 1.6135878 \cdot 10^{135}$
Stratified random sampling $k = 2$, $N_1 = 400$, $N_2 = 600$	$K = \binom{400}{40} \cdot \binom{600}{60} = 5.4662414 \cdot 10^{138}$
Two-stage sampling $k = 20$, $N_0 = 50$, $r = 4$	$K = \binom{20}{4} \cdot \binom{50}{25} = 6.1245939 \cdot 10^{17}$
Two-stage sampling $k = 20$, $N_0 = 50$, $r = 5$	$K = \binom{20}{5} \cdot \binom{50}{20} = 7.3069131 \cdot 10^{17}$
Two-stage sampling $k = 10$, $N_0 = 100$, $r = 2$	$K = \binom{10}{2} \cdot \binom{100}{50} = 4.5401105 \cdot 10^{30}$
Two-stage sampling $k = 10$, $N_0 = 100$, $r = 4$	$K = \binom{10}{4} \cdot \binom{100}{25} = 5.0929047 \cdot 10^{25}$
Two-stage sampling $k = 5$, $N_0 = 200$, $r = 2$	$K = \binom{5}{2} \cdot \binom{200}{50} = 4.5385838 \cdot 10^{48}$
Two-stage sampling $k = 2$, $N_0 = 500$, $r = 1$	$K = \binom{2}{1} \cdot \binom{500}{100} > 10^{100}$

object of the population has the same probability $p = \frac{r}{k} \cdot \frac{n_0}{N_0}$ to reach the sample corresponding to Definition 1.1.

Example 1.1 A population has $N = 1000$ objects. A sample of size $n = 100$ should be drawn without replacement of objects. Table 1.1 lists the number of possible samples using the discussed sampling methods. The probability of selection for each object is $p = 0.1$.

1.2 Mathematical Models for Population and Sample

In mathematical statistics notions are defined that are used as models (generalisations) for the corresponding empirical notions. The population, which corresponds to a frequency distribution of the character values, is related to the model of probability distribution. The concrete sample selected by a random procedure is modelled by the realised (theoretical) random sampling. These model concepts are adequate, if the size N of the populations is very large compared with the size n of the sample.

Definition 1.2 An n-dimensional random variable

$$\boldsymbol{Y} = (\boldsymbol{y}_1, \boldsymbol{y}_2, \dots, \boldsymbol{y}_n)^T, n \ge 1$$

with components $\boldsymbol{y}_i$ is said to be a random sample, if

- All $\boldsymbol{y}_i$ have the same distribution characterised by the distribution function $F(y_i, \theta) = F(y, \theta)$ with the parameter (vector) $\theta \in \Omega \subseteq R^p$ and
- All $\boldsymbol{y}_i$ are stochastically independent from each other, that is, it holds for the distribution function $F(Y,\theta)$ of $\boldsymbol{Y}$ the factorisation

$$F(Y,\theta) = \prod_{i=1}^{n} F(y_i, \theta), \theta \in \Omega \subseteq R^p.$$

The values $Y = (y_1, y_2, \dots, y_n)^T$ of a random sample $\boldsymbol{Y}$ are called realisations. The set $\{Y\}$ of all possible realisations of $\boldsymbol{Y}$ is called sample space.

In this book the random variables are printed with bold characters, and the sample space $\{Y\}$ belongs always to an n-dimensional Euclidian space, that is, $\{Y\} \subset R^n$.

The function

$$L(Y,\theta) = \begin{cases} f(Y,\theta) = \dfrac{\partial F(Y,\theta)}{\partial Y}, & \text{for continuous } y \\ p(Y,\theta), & \text{for discrete } y \end{cases}$$

with the probability function $p(Y, \theta)$ and the density function $f(Y, \theta)$ correspondingly is said to be for given Y as function of θ the likelihood function (of the distribution).

Random sample can have two different meanings, namely:

- Random sample as random variable $\boldsymbol{Y}$ corresponding to Definition 1.2
- (Concrete) random sample as subclass of a population, which was selected by a random sample procedure.

The realizations Y of a random sample $\boldsymbol{Y}$ we call a realized random sample. The random sample $\boldsymbol{Y}$ is the mathematical model of the simple random sample procedure, where concrete random sample and realised random sample correspond to each other also in the symbolism.

We describe in this book the 'classical' philosophy, where $\boldsymbol{Y}$ is distributed by the distribution function $F(Y, \theta)$ with the fixed (not random) parameter $\theta \in \Omega \subseteq R^p$. Besides there is the philosophy of Bayes where a random $\boldsymbol{\theta}$ is supposed, which is distributed a priori with a parameter φ assumed to be known. In the empirical Bayesian method, the a priori distribution is estimated from the data collected.

1.3 Sufficiency and Completeness

A random variable involves certain information about the distribution and their parameters. Mainly for large n (say, $n > 100$), it is useful to condense the objects of a random sample in such a way that fewest possible new random variables contain as much as possible of this information. This vaguely formulated concept is to state more precisely stepwise up to the notion of minimal sufficient statistic. First, we repeat the definition of an exponential family.

The distribution of a random variable $\boldsymbol{y}$ with parameter vector $\theta = (\theta_1, \theta_2, \ldots, \theta_p)^T$ belongs to a k-parametric exponential family if its likelihood function can be written as

$$f(y, \theta) = h(y) e^{\sum_{i=1}^{k} \eta_i(\theta) \cdot T_i(y) - B(\theta)},$$

where the following conditions hold:

- η_i and B are real functions of θ and B does not depend on y.
- The function $h(y)$ is non-negative and does not depend on θ.

The exponential family is in canonical form with the so-called natural parameters η_i, if their elements can be written as

$$f(y, \eta) = h(y) e^{\sum_{i=1}^{k} \eta_i \cdot T_i(y) - A(\eta)} \text{ with } \eta = (\eta_1, \ldots, \eta_k)^T.$$

Let $(P_\theta, \theta \epsilon \Omega)$ be a family of distributions of random variables $\boldsymbol{y}$ with the distribution function $F(y, \theta)$, $\theta \in \Omega$. The realisations $Y = (y_1, \ldots, y_n)^T$ of the random sample

$$\boldsymbol{Y} = (\boldsymbol{y}_1, \boldsymbol{y}_2, \ldots, \boldsymbol{y}_n)^T,$$

where the components $\boldsymbol{y}_i$ are distributed as $\boldsymbol{y}$ itself lie in the sample space $\{Y\}$. According to Definition 1.2 the distribution function $F(Y, \theta)$ of a random sample $\boldsymbol{Y}$ is just as $F(y, \theta)$ uniquely determined.

Definition 1.3 A measurable mapping $\boldsymbol{M} = M(\boldsymbol{Y}) = [M_1(\boldsymbol{Y}), \dots, M_r(\boldsymbol{Y})]^T, r \le n$ of $\{Y\}$ on a space $\{M\}$, which does not depend on $\theta \in \Omega$, is called a statistic.

Definition 1.4 A statistic $\boldsymbol{M}$ is said to be sufficient relative to a distribution family $(P_\theta, \theta \in \Omega)$ or relative to $\theta \in \Omega$, respectively, if the conditional distribution of a random sample $\boldsymbol{Y}$ is independent of θ for given $\boldsymbol{M} = M(\boldsymbol{Y}) = M(Y)$.

Example 1.2 Let the components of a random sample $\boldsymbol{Y}$ satisfy a two-point distribution with the values 1 and 0. Further, let be $P(\boldsymbol{y}_i = 1) = p$ and $P(\boldsymbol{y}_i = 0) = 1 - p, 0 < p < 1$. Then $\boldsymbol{M} = M(\boldsymbol{Y}) = \sum_{i=1}^n \boldsymbol{y}_i$ is sufficient relative (corresponding) to $\theta \in (0, 1) = \Omega$. To show this, we have to prove that $P\left(\boldsymbol{Y} = Y \mid \sum_{i=1}^n \boldsymbol{y}_i = M\right)$ is independent of p. Now it is

$$P(\boldsymbol{Y} = Y \mid M) = \frac{P(\boldsymbol{Y} = Y, \boldsymbol{M} = M)}{P(\boldsymbol{M} = M)}, M = 0, 1, \dots, n.$$

We know from probability theory that $\boldsymbol{M} = M(\boldsymbol{Y}) = \sum_{i=1}^n \boldsymbol{y}_i$ is binomially distributed with the parameters n and p. Hence it follows that

$$P(\boldsymbol{M} = M) = \binom{n}{M} p^M (1-p)^{n-M}, M = 0, 1, \dots, n.$$

Further we get with $y_i = 0$ or $y_i = 1$ and $A(M) = \{Y \mid M(Y) = M\}$ the result

$$P[\boldsymbol{Y} = Y, M(\boldsymbol{Y}) = M] = P(\boldsymbol{y}_1 = y_1, \dots, \boldsymbol{y}_n = y_n) I_{A(M)}(Y)$$

$$= \prod_{i=1}^n \left(\binom{1}{y_i} p^{y_i} (1-p)^{1-y_i} \right) I_{A(M)}(Y)$$

$$= p^{\sum_{i=1}^n y_i} (1-p)^{n - \sum_{i=1}^n y_i} I_{A(M)}(Y)$$

$$= p^M (1-p)^{(n-M)}.$$

Consequently it is $P(\boldsymbol{Y} = Y \mid M) = \dfrac{1}{\binom{n}{M}}$, and this is independent of p.

This way of proving sufficiency is rather laborious, but we can also apply it for continuous distributions as the next example will show.

Example 1.3 Let the components $\boldsymbol{y}_i$ of a random sample $\boldsymbol{Y}$ of size n be distributed as $N(\mu,1)$ with expected value μ and variance $\sigma^2 = 1$. Then $\boldsymbol{M} = \sum \boldsymbol{y}_i$ is sufficient relative to $\mu \in R^1 = \Omega$. To show this we first remark that $\boldsymbol{Y}$ is distributed as $N(\mu e_n, I_n)$. Then we apply the one-to-one transformation

$$\boldsymbol{Z} = A\boldsymbol{Y} = \left(\boldsymbol{z}_1 = \sum \boldsymbol{y}_i, \boldsymbol{y}_2 - \boldsymbol{y}_1, \ldots, \boldsymbol{y}_n - \boldsymbol{y}_1\right) \text{ with } A = \begin{pmatrix} 1 & e_{n-1}^T \\ -e_{n-1} & I_{n-1} \end{pmatrix}$$

where $|A| = n$. We write $\boldsymbol{Z} = (\boldsymbol{z}_1, \boldsymbol{Z}_2) = (\sum \boldsymbol{y}_i, \boldsymbol{y}_2 - \boldsymbol{y}_1, \ldots, \boldsymbol{y}_n - \boldsymbol{y}_1)$ and recognise that

$$\text{cov}(\boldsymbol{Z}_2, \boldsymbol{z}_1) = \text{cov}(\boldsymbol{Z}_2, \boldsymbol{M}) = \text{cov}\left((-e_{n-1}, I_{n-1})\boldsymbol{Y}, e_n^T \boldsymbol{Y}\right) = 0_{n-1}.$$

Considering the assumption of normal distribution, the variables $\boldsymbol{M}$ and $\boldsymbol{Z}_2$ are stochastically independent. Consequently $\boldsymbol{Z}_2$ but also $\boldsymbol{Z}_2 \,|\boldsymbol{M}$ and $\boldsymbol{Z} \,|\boldsymbol{M}$ are independent of μ. Taking into account that the mapping $\boldsymbol{Z} = A\boldsymbol{Y}$ is biunique, also $\boldsymbol{Y}|\boldsymbol{M}$ is independent of μ. Hence $\boldsymbol{M} = \sum \boldsymbol{y}_i$ is sufficient relative to $\mu \in R^1$. With a sufficient $\boldsymbol{M} = \sum \boldsymbol{y}_i$ and a real number $c \neq 0$, then $c\,\boldsymbol{M}$ is also sufficient, that is, $\frac{1}{n}\sum \boldsymbol{y}_i = \bar{\boldsymbol{y}}$ is sufficient.

But sufficiency plays such a crucial part in mathematical statistics, and we need simpler methods for proving sufficiency and mainly for finding sufficient statistics. The following theorem is useful in this direction.

Theorem 1.1 (Decomposition Theorem)
Let a distribution family $(P_\theta, \theta \epsilon \Omega)$ of a random sample $\boldsymbol{Y}$ be given that is dominated by a finite measure ν. The statistic $M(\boldsymbol{Y})$ is sufficient relative to θ, if the Radon–Nikodym density f_θ of P_θ can be written corresponding to ν as

$$f_\theta(Y) = g_\theta[M(Y)]h(Y) \tag{1.1}$$

ν – almost everywhere. Then the following holds:

- The ν – integrable function g_θ is non-negative and measurable.
- h is non-negative and $h(Y) = 0$ is fulfilled only for a set of P_θ – measure 0.

The general proof came from Halmos and Savage (1949); it can be also found, for example, in Bahadur (1955) or Lehmann and Romano (2008).

In the present book we work only with discrete and continuous probability distributions satisfying the assumptions of Theorem 1.1. A proof of the theorem for such distributions is given in Rasch (1995). We do not want to repeat it here.

For discrete distributions Theorem 1.1 means that the probability function is of the form

$$p(Y, \theta) = g[M(Y), \theta]h(Y). \tag{1.2}$$

For continuous distributions the density function has the form

$$f(Y,\theta)=g[M(Y),\theta]h(Y). \tag{1.3}$$

Corollary 1.1 If the distribution family $(P^*(\theta), \theta\epsilon\Omega)$ of the random variable $\boldsymbol{y}$ is a k-parametric exponential family with the natural parameter η and the likelihood function

$$L^*(y,\eta)=h^*(y)e^{\sum_{i=1}^{k} M_j^*(y)-A(\eta)}, \tag{1.4}$$

then, denoting the random sample $\boldsymbol{Y}=(\boldsymbol{y}_1,\boldsymbol{y}_2,\ \dots,\boldsymbol{y}_n)^T$,

$$M(\boldsymbol{Y})=\left(\sum_{i=1}^{n}\boldsymbol{M}_1^*(\boldsymbol{y}_i),\dots,\sum_{i=1}^{n}\boldsymbol{M}_k^*(\boldsymbol{y}_i)\right)^T \tag{1.5}$$

is sufficient relative to θ.

Proof: It is

$$L(y,\eta)=\prod_{i=1}^{n}h^*(y_i)e^{\sum_{j=1}^{k}\eta_j\sum_{i=1}^{n}M_j^*(y_i)-nA(\eta)}, \tag{1.6}$$

which is of the form (1.2) and (1.3), respectively, where $h(Y)=\prod_{i=1}^{n}h^*(y_i)$ and $\theta = \eta$.

Definition 1.5 Two likelihood functions, $L_1(Y_1,\theta)$ and $L_2(Y_2,\theta)$, are said to be equivalent, denoted by $L_1 \sim L_2$, if

$$L_1(Y_1,\theta)=a(Y_1,Y_2)\,L_2(Y_2,\theta) \tag{1.7}$$

with a function $a(Y_1, Y_2)$ that is independent of θ.

Then it follows from Theorem 1.1

Corollary 1.2 $M(\boldsymbol{Y})$ is sufficient relative to θ if and only if (*iff*) the likelihood function $L_M(M,\theta)$ of $\boldsymbol{M}=M(\boldsymbol{Y})$ is equivalent to the likelihood function of a random sample $\boldsymbol{Y}$.

Proof: If $M(\boldsymbol{Y})$ is sufficient relative to θ, then because of

$$L_M(M,\theta)=a(Y)L(Y,\theta), a(Y)>0, \tag{1.8}$$

$L_M(M,\theta)$ has together with $L(Y,\eta)$ the form (1.1). Reversely, (1.8) implies that the conditional distribution of a random sample $\boldsymbol{Y}$ is for given $M(\boldsymbol{Y}) = M$ independent of θ.

Example 1.4 Let the components y_i of a random sample $Y = (y_1, y_2, \ldots, y_n)^T$ be distributed as $N(\mu, 1)$. Then it is

$$L(Y,\mu) = \frac{1}{\left(\sqrt{2\pi}\right)^n} e^{-\frac{1}{2}(Y-\mu e_n)^T(Y-\mu e_N)} = \frac{1}{\left(\sqrt{2\pi}\right)^n} e^{-\frac{1}{2}\sum_{i=1}^{n}(y_i-\bar{y})^2} e^{-\frac{n}{2}(\bar{y}-\mu)^2}. \tag{1.9}$$

Since $M(Y) = \bar{y}$ is distributed as $N\left(\mu, \frac{1}{n}\right)$, we get

$$L_M(\bar{y},\mu) = \frac{\sqrt{n}}{\sqrt{2\pi}} e^{-\frac{n}{2}(\bar{y}-\mu)^2}. \tag{1.10}$$

Hence $L_M(\bar{y},\mu) \sim L(Y,\mu)$ holds, and $\bar{y}$ is sufficient relative to μ.

Generally we immediately obtain from Definition 1.4 the

Corollary 1.3 If $c > 0$ is a real number chosen independently of θ and $M(Y)$ is sufficient relative to θ, then $c\,M(Y)$ is also sufficient relative to θ.

Hence, for example, by putting $M = \sum y_i$ and $c = \frac{1}{n}$, also $\frac{1}{n}\sum y_i = \bar{y}$ is sufficient.

The problem arises whether there exist under the statistics sufficient relative to the distribution family $(P^*(\theta), \theta \in \Omega)$ such, which are minimal in a certain sense, containing as few as possible components. The following example shows that this problem is no pure invention.

Example 1.5 Let $(P^*(\theta), \theta \in \Omega)$ be the family of $N(\mu,\sigma^2)$-normal distributions $(\sigma > 0)$. We consider the statistics

$$M_1(Y) = Y$$

$$M_2(Y) = \left(y_1^2, \ldots, y_n^2\right)^T$$

$$M_3(Y) = \left(\sum_{i=1}^{r} y_i^2, \sum_{i=r+1}^{n} y_i^2\right)^T, \quad r = 1, \ldots, n-1$$

$$M_4(Y) = \left(\sum_{i=1}^{n} y_i^2\right)$$

of a random sample Y of size n, which are all sufficient relative to σ^2. This can easily be shown using Corollary 1.1 of Theorem 1.1 (decomposition theorem). The likelihood functions of $M_1(Y)$ and Y are identical (and therefore equivalent). Since both the y_i and the y_i^2 are independent and $\frac{y_i^2}{\sigma^2} = \chi_i^2$ are distributed as $CS(1)$ (χ^2-distribution with 1 degree of freedom; see Appendix A: Symbolism), it follows after the transformation $y_i^2 = \sigma^2\chi_i^2$:

$$L_M\left(M_2(Y),\sigma^2\right) \sim L\left(Y,\sigma^2\right) = \frac{1}{\left(2\pi\sigma^2\right)^{\frac{n}{2}}} e^{-\frac{1}{2\sigma^2}\sum_{i=1}^{n} y_i^2}. \tag{1.11}$$

Analogously we proceed with $M_3(Y)$ and $M_4(Y)$.

Obviously, $M_4(Y)$ is the most extensive compression of the components of a random sample Y, and therefore it is preferable compared with other statistics.

Definition 1.6 A statistic $M^*(Y)$ sufficient relative to θ is said to be minimal sufficient relative to θ if it can be represented as a function of each other sufficient statistic $M(Y)$.

If we consider Example 1.5, there is

$$M_4(Y) = M_1^T(Y)M_1(Y) = e_n^T M_2 = (1\ 1)M_3, \quad r = 1,\ldots,n-1.$$

Hence, $M_4(Y)$ can be written as a function of each sufficient statistic of this example. This is not true for $M_1(Y)$, $M_2(Y)$ and $M_3(Y)$; they are not functions of $M_4(Y)$. $M_4(Y)$ is the only statistic of Example 1.5 that could be minimal sufficient relative to σ^2. We will see that it has indeed this property. But, how can we show minimal sufficiency? We recognise that the sample space can be decomposed with the help of the statistic $M(Y)$ in such a way into disjoint subsets that all Y for which $M(Y)$ supplies the same value M belong to the same subset. Vice versa, a given decomposition defines a statistic. Now we present a decomposition that is shown to generate a minimal sufficient statistic.

Definition 1.7 Let $Y_0 \in \{Y\}$ be a fixed point in the sample space (a certain value of a realised random sample), which contains the realisations of a random sample Y with components from a family $(P^*(\theta), \theta\epsilon\Omega)$ of probability distributions. The likelihood function $L(Y,\theta)$ generates by

$$M(Y_0) = \{Y : L(Y,\theta) \sim L(Y_0,\theta)\} \tag{1.12}$$

a subset in $\{Y\}$. If Y_0 runs through the whole sample space $\{Y\}$, then a certain decomposition is generated. This decomposition is called likelihood decomposition, and the corresponding statistic $M_L(Y)$ satisfying $M_L(Y)$ = const. for all $Y \epsilon M(Y_0)$ and each Y_0 is called likelihood statistic.

Before we construct minimal sufficient statistics for some examples by this method, we state

Theorem 1.2 The likelihood statistic $M_L(Y)$ is minimal sufficient relative to θ.

Proof: Considering the likelihood statistic $M_L(Y)$, it holds

$$M_L(Y_1) = M_L(Y_2)$$

for $Y_1, Y_2 \in \{Y\}$ iff $L(Y_1, \theta) \sim L(Y_2, \theta)$ is fulfilled. Hence, $L(Y, \theta)$ is a function of $M_L(\boldsymbol{Y})$ having the form

$$L(Y,\theta) = a(Y)g^*\left(M_L(Y),\theta\right). \tag{1.13}$$

Therefore $M_L(\boldsymbol{Y})$ is sufficient relative to θ taking Theorem 1.1 (decomposition theorem) into account. If $M(\boldsymbol{Y})$ is any other statistic that is sufficient relative to θ, if further for two points $Y_1, Y_2 \in \{Y\}$ the relation $M(\boldsymbol{Y}_1) = M(\boldsymbol{Y}_2)$ is satisfied, and finally, if $L(Y_i, \theta) > 0$ holds for $i = 1, 2$, then again Theorem 1.1 supplies

$$L(Y_1,\theta) = h(Y_1)g(M(Y_1),\theta) = h(Y_2)g(M(Y_2),\theta)$$

because of $M(\boldsymbol{Y}_1) = M(\boldsymbol{Y}_2)$ and

$$L(Y_2,\theta) = h(Y_2)g(M(Y_2),\theta) \text{ or equivalently } g(M(Y_2),\theta) = \frac{L(Y_2,\theta)}{h(Y_2)}.$$

Hence, we obtain

$$L(Y_1,\theta) = \frac{h(Y_1)}{h(Y_2)}L(Y_2,\theta), h(Y_2) > 0,$$

which means $L(Y_1, \theta) \sim L(Y_2, \theta)$. But this is just the condition for $M(\boldsymbol{Y}_1) = M(\boldsymbol{Y}_2)$. Consequently $M_L(\boldsymbol{Y})$ is a function of $M(\boldsymbol{Y})$, independent of how $M(\boldsymbol{Y})$ is chosen, that is, it is minimal sufficient.

We demonstrate the method giving two examples.

Example 1.6 Let the components $\boldsymbol{y}_i$ of a random sample $\boldsymbol{Y}$ fulfil a binomial distribution $B(N, p)$, N fixed and $0 < p < 1$. We look for a statistic that is minimal sufficient relative to p. The likelihood function is

$$L(Y,p) = \prod_{i=1}^{n} \binom{N}{y_i} p^{y_i}(1-p)^{N-y_i}, y_i = 0,1,\dots,N.$$

For all $Y_0 = (y_{01}, \dots, y_{0N})^T \in \{Y\}$ with $L(Y_0, p) > 0$, it is

$$\frac{L(Y,p)}{L(Y_0,p)} = \frac{\prod_{i=1}^{n}\binom{N}{y_i}}{\prod_{i=1}^{n}\binom{N}{y_{0i}}}\left(\frac{p}{1-p}\right)^{\sum_{i=1}^{n}(y_i - y_{0i})}.$$

Therefore $M(Y_0)$ is also defined by $M(Y_0) = \left\{Y : \sum_{i=1}^{n} y_i = \sum_{i=1}^{n} y_{0i}\right\}$, since just there $L(Y, p) \sim L(Y_0, p)$ holds. Hence $M(\boldsymbol{Y}) = \sum_{i=1}^{n} \boldsymbol{y}_i$ is a minimal sufficient statistic.

Example 1.7 Let the components $\boldsymbol{y}_i$ of a random sample $\boldsymbol{Y} = (\boldsymbol{y}_1, \boldsymbol{y}_2, \ldots, \boldsymbol{y}_n)^T$ be gamma distributed. Then for $y_i > 0$ we get the likelihood function

$$L(Y, a, k) = \frac{a^{nk}}{[\Gamma(k)]^n} e^{-a\sum_{i=1}^{n} y_i} \prod_{i=1}^{n} y_i^{k-1}.$$

For all $Y_0 = (y_{01}, \ldots, y_{0N})^T \epsilon \{Y\}$ with $L(Y_0, a, k) > 0$, it is

$$\frac{L(Y, a, k)}{L(Y_0, a, k)} = e^{-a\left(\sum_{i=1}^{n} y_i - \sum_{i=1}^{n} y_{0i}\right)} \frac{\prod_{i=1}^{n} y_i^{k-1}}{\prod_{i=1}^{n} y_{0i}^{k-1}}.$$

For given a the product $\prod_{i=1}^{n} \boldsymbol{y}_i$ is minimal sufficient relative to k. If k is known, then $\sum_{i=1}^{n} \boldsymbol{y}_i$ is minimal sufficient relative to a. If both a and k are unknown parameters, then $(\prod_{i=1}^{n} \boldsymbol{y}_i, \sum_{i=1}^{n} \boldsymbol{y}_i)$ is minimal sufficient relative to (a, k).

More generally the following statement holds:

Theorem 1.3 If $(P^*(\theta), \theta \epsilon \Omega)$ is a k-parametric exponential family with likelihood function in canonical form

$$L(y, \theta) = e^{\sum_{i=0}^{k} \eta_i M_i(y) - A(\eta)} h(y),$$

where the dimension of the parameter space is k (i.e. the $\eta_1, \ldots, \eta_k$ are linearly independent), then

$$M(\boldsymbol{Y}) = \left(\sum_{i=1}^{n} M_1(\boldsymbol{y}_i), \ldots, \sum_{i=1}^{n} M_k(\boldsymbol{y}_i) \right)^T$$

is minimal sufficient relative to $(P^*(\theta), \theta \in \Omega)$.

Proof: The sufficiency of $M(\boldsymbol{Y})$ follows from Corollary 1.1 of the decomposition theorem (Theorem 1.1), and the minimal sufficiency follows from the fact that $M(\boldsymbol{Y})$ is the likelihood statistic, because it is $L(Y, \theta) \sim L(Y_0, \theta)$ if

$$\sum_{j=1}^{k} \eta_j \sum_{i=1}^{n} \left[M_j(\boldsymbol{y}_i) - M_j(\boldsymbol{y}_{0i}) \right] = 0.$$

Regarding the linear independence of η_i, it is only the case if $M(\boldsymbol{Y}) = M(\boldsymbol{Y}_0)$ is fulfilled.

Example 1.8 Let $(P^*(\theta), \theta \in \Omega)$ the family of a two-dimensional normal distributions with the random variable $\begin{pmatrix} \boldsymbol{x} \\ \boldsymbol{y} \end{pmatrix}$, the expectation $\mu = \begin{pmatrix} \mu_x \\ \mu_y \end{pmatrix}$ and the covariance matrix $\Sigma = \begin{pmatrix} \sigma_x^2 & 0 \\ 0 & \sigma_y^2 \end{pmatrix}$. This is a four-parametric exponential family with the natural parameters

$$\eta_1 = \frac{\mu_x}{\sigma_x^2}, \eta_2 = \frac{\mu_y}{\sigma_y^2}, \eta_3 = -\frac{1}{2\,\sigma_x^2}, \eta_4 = -\frac{1}{2\,\sigma_y^2}$$

and the factors

$$M_1\left[\begin{pmatrix} \boldsymbol{x} \\ \boldsymbol{y} \end{pmatrix}\right] = \boldsymbol{x},\ M_2\left[\begin{pmatrix} \boldsymbol{x} \\ \boldsymbol{y} \end{pmatrix}\right] = \boldsymbol{y},$$

$$M_3\left[\begin{pmatrix} \boldsymbol{x} \\ \boldsymbol{y} \end{pmatrix}\right] = \boldsymbol{x}^2,\ M_4\left[\begin{pmatrix} \boldsymbol{x} \\ \boldsymbol{y} \end{pmatrix}\right] = \boldsymbol{y}^2,$$

$$A(\eta) = \frac{1}{2}\left(\frac{\mu_x^2}{\sigma_x^2} + \frac{\mu_y^2}{\sigma_y^2}\right).$$

If $\dim(\Omega) = 4$, then

$$\boldsymbol{M} = \left(\sum_{i=1}^{n} \boldsymbol{M}_{1i}, \sum_{i=1}^{n} \boldsymbol{M}_{2i}, \sum_{i=1}^{n} \boldsymbol{M}_{3i}, \sum_{i=1}^{n} \boldsymbol{M}_{4i}\right)^T$$

is minimal sufficient relative to $(P^*(\theta), \theta \epsilon \Omega)$. Assuming that $\left(\breve{P}^{**}(\theta), \theta \epsilon \Omega\right) \subseteq (P^*(\theta), \theta \epsilon \Omega)$ is the subfamily of $(P^*(\theta), \theta \epsilon \Omega)$ with $\sigma_x^2 = \sigma_y^2 = \sigma^2$, then $\dim(\Omega) = 3$ follows, and $\boldsymbol{M}$ is not minimal sufficient relative to $(\breve{P}^{**}(\theta), \theta \epsilon \Omega)$.

The natural parameters of $(\breve{P}^{**}(\theta), \theta \epsilon \Omega)$ are

$$\eta_1 = \frac{\mu_x}{\sigma^2}, \eta_2 = \frac{\mu_y}{\sigma^2}, \eta_3 = -\frac{1}{2\sigma^2}.$$

Further we have $A(\eta) = \frac{1}{2\sigma^2}\left(\mu_x^2 + \mu_y^2\right)$, and the factors of the η_i are

$$\breve{M}_1\left[\begin{pmatrix} \boldsymbol{x} \\ \boldsymbol{y} \end{pmatrix}\right] = \boldsymbol{x}, \breve{M}_2\left[\begin{pmatrix} \boldsymbol{x} \\ \boldsymbol{y} \end{pmatrix}\right] = \boldsymbol{y},\ \breve{M}_3\left[\begin{pmatrix} \boldsymbol{x} \\ \boldsymbol{y} \end{pmatrix}\right] = \boldsymbol{x}^2 + \boldsymbol{y}^2.$$

Relative to $\left(\breve{P}^{**}(\theta), \theta \epsilon \Omega\right)$,

$$\breve{\boldsymbol{M}} = \left(\sum_{i=1}^{n} \breve{\boldsymbol{M}}_{1i}, \sum_{i=1}^{n} \breve{\boldsymbol{M}}_{2i}, \sum_{i=1}^{n} \breve{\boldsymbol{M}}_{3i}\right)^T$$

is minimal sufficient.

As it will be shown in Chapter 6 for model II of analysis of variance, the result of Theorem 1.3 is suitable also in more sophisticated models to find minimal sufficient statistics.

Completeness and bounded completeness are further important properties for the theory of estimation. We want to introduce both together by the following definition.

Definition 1.8 A distribution family $P = (P_\theta, \theta \epsilon \Omega)$ with distribution function $F(y, \theta)$, $\theta \epsilon \Omega$ is said to be complete, if for each P-integrable function $h(\boldsymbol{y})$ of the random variable $\boldsymbol{y}$ the condition

$$E[h(\boldsymbol{y})] = \int h(y)dF(y) = 0 \text{ for all } \theta \epsilon \Omega \tag{1.14}$$

implies the relation

$$P_\theta[h(\boldsymbol{y}) = 0] = 1 \text{ for all } \theta \epsilon \Omega. \tag{1.15}$$

If this is true only for bounded functions $h(y)$, then $P = (P_\theta, \theta \epsilon \Omega)$ is called bounded complete.

We want to consider an example for a complete distribution family.

Example 1.9 Let P be the family $\{P_p\}$, $p \in (0,1)$ of binomial distributions with the probability function

$$p(y,p) = \binom{n}{y} p^y (1-p)^{n-y} = \binom{n}{y} \nu^y (1-p)^n, \quad 0 < p < 1,$$

$$y = 0, 1, \ldots, n, \ \nu = \frac{p}{1-p}.$$

Integrability of $h(y)$ means finiteness of $(1-p)^n \sum_{y=0}^{n} h(y) \binom{n}{y} \nu^y$, and (1.14) implies

$$\sum_{y=0}^{n} h(y) \binom{n}{y} \nu^y = 0 \text{ for all } p \in (0,1).$$

The left-hand side of the equation is a polynomial of nth degree in ν, which has at most n real zeros. To fulfil this equation for all $\nu \in R^+$, the factor $\binom{n}{y} h(y)$ must vanish for y = 0, 1, …, n, and because of $\binom{n}{y} > 0$, it follows

$$P_\theta[h(y) = 0)] = 1 \text{ for all } p \in (0,1).$$

Theorem 1.4 A k-parametric exponential family of the distribution of a sufficient statistic is complete under the assumptions of Theorem 1.3 (dim(Ω) = k).

The proof can be found in Lehmann and Romano (2008).

Definition 1.9 Let a random sample $\boldsymbol{Y} = (\boldsymbol{y}_1, \boldsymbol{y}_2, \dots, \boldsymbol{y}_n)^T$ be given whose components satisfy a distribution from the family

$$P^* = (P_\theta, \theta \in \Omega).$$

A statistic $M(\boldsymbol{Y})$, whose distribution is independent of θ, is called an ancillary statistic. If P is the family of distributions induced by the statistic $M(\boldsymbol{Y})$ in P^* and if P is complete and $M(\boldsymbol{Y})$ is sufficient relative to P^*, then $M(\boldsymbol{Y})$ is said to be complete sufficient.

Example 1.10 Let P^* be the family of normal distributions $N(\mu, 1)$ with expectation $\mu = \theta$ and variance 1, that is, it holds $\Omega = R^1$. This is a one-parametric exponential family with dim(Ω) = 1, which is complete by Theorem 1.4. If $\boldsymbol{Y} = (\boldsymbol{y}_1, \boldsymbol{y}_2, \dots, \boldsymbol{y}_n)^T$ is a random sample with components from P^*, then $M_1(\boldsymbol{Y}) = \bar{\boldsymbol{y}}$ is distributed as $N(\mu, \frac{1}{n})$. Consequently the family of distributions P^* is also complete. Because of Theorem 1.3, $\bar{\boldsymbol{y}}$ is minimal sufficient and therefore complete sufficient. The distribution family of $CS(n-1)$-distributions (χ^2-distributions with $n - 1$ degrees of freedom) induced by $(n-1)M_2(\boldsymbol{Y}) = \sum \boldsymbol{y}_i^2 - n\bar{\boldsymbol{y}}^2$ is independent of μ. Hence $\boldsymbol{s}^2 = \frac{1}{n-1}\sum_{i=1}^{n}(\boldsymbol{y}_i - \bar{\boldsymbol{y}})^2$ is an ancillary statistic relative to $\mu = \theta$.

We close this section with the following statement:

Theorem 1.5 Let $\boldsymbol{Y}$ be a random sample with components from $P = (P_\theta, \theta \in \Omega)$ and let $M_1(\boldsymbol{Y})$ be bounded complete sufficient relative to P. Further, if $M_2(\boldsymbol{Y})$ is a statistic with a distribution independent of θ, then $M_1(\boldsymbol{Y})$ and $M_2(\boldsymbol{Y})$ are (stochastically) independent.

Proof: Let $\{Y_0\} \subset \{Y\}$ be a subset of the sample space $\{Y\}$. Then $M_2(\boldsymbol{Y})$ maps $\{Y\}$ onto $\{M\}$ and $\{Y_0\}$ onto $\{M_0\}$. Since the distribution of $M_2(\boldsymbol{Y})$ is independent of θ, $P[M_2(\boldsymbol{Y}) \in \{M_0\}]$ is independent of θ. Moreover, observing the sufficiency of $M_1(\boldsymbol{Y})$ relative to θ, also $P[M_2(\boldsymbol{Y}) \in \{M_0\} | M_1(\boldsymbol{Y})]$ is independent of θ. We consider the statistic

$$h(M_1(\boldsymbol{Y})) = P[M_2(\boldsymbol{Y}) \in \{M_0\} | M_1(\boldsymbol{Y})] - P[M_2(\boldsymbol{Y}) \in \{M_0\}]$$

depending on $M_1(\boldsymbol{Y})$, such that analogously to (1.14)

$$E_\theta[h(M_1(Y))] = E_\theta[P[M_2(Y)\,\epsilon\,\{M_0\}]M_1(Y) - P[M_2(Y)\,\epsilon\,\{M_0\}]] = 0$$

follows for all $\theta \in \Omega$. Since $M_1(Y)$ is bounded complete,

$$P[M_2(Y)\epsilon\{M_0\}|M_1(Y)] - P[M_2(Y)\epsilon\{M_0\}] = 0$$

holds for all $\theta \in \Omega$ with probability 1, analogously to (1.15). But this means that $M_1(Y)$ and $M_2(Y)$ are independent.

1.4 The Notion of Information in Statistics

Concerning the heuristic introduction of sufficient statistics in Section 1.2, we emphasised that a statistic should exhaust the information of a sample to a large extent. Now we turn to the question what the information of a sample really means. The notion of information was introduced by R. A. Fisher in the field of statistics, and his definition is still today of great importance. We speak of the Fisher information in this connection. A further notion of information originates from Kullback and Leibler (1951), but we do not present this definition here. We restrict ourselves in this section at first to distribution families

$$P = (P_\theta, \theta \in \Omega), \Omega \subset R^1$$

with real parameters θ. We denote the likelihood function $(Y = y)$ of P by $L(y, \theta)$.

Definition 1.10 Let $\boldsymbol{y}$ be distributed as

$$P = (P_\theta, \theta \in \Omega), \Omega \subset R^1.$$

Further let the following assumption V1 be fulfilled:

1) Ω is an open interval.
2) For each $y \in \{Y\}$ and for each $\theta \in \Omega$, the derivative $\frac{\partial}{\partial\theta}L(y,\theta)$ exists and is finite. The set of points satisfying $L(y, \theta) = 0$ does not depend on θ.
3) For each $\theta \in \Omega$ there exist an $\varepsilon > 0$ and a positive P_θ-integrable function $k(y,\theta)$ such that for all θ_0 in an ε-neighbourhood of θ the inequality

   $$\left|\frac{L(y,\theta) - L(y,\theta_0)}{\theta - \theta_0}\right| \leqslant k(y,\theta_0)$$

   holds.
4) The derivative $\frac{\partial}{\partial\theta}L(y,\theta)$ is quadratic P_θ-integrable, and it holds for all $\theta \in \Omega$

   $$0 < E\left\{\left[\frac{\partial}{\partial\theta}\ln L(\boldsymbol{y},\theta)\right]^2\right\}.$$

Then the expectation

$$I(\theta) = E\left\{\left[\frac{\partial}{\partial\theta}\ln L(\boldsymbol{y},\theta)\right]^2\right\} \tag{1.16}$$

is said to be the Fisher information of the distribution P_θ and of the variable $\boldsymbol{y}$, respectively.

It follows from the third condition of V1 that P_θ-integration and differentiation by θ can be exchanged for $L(\mathrm{y}, \theta)$, and because of

$$\frac{\partial}{\partial\theta}\ln L(y,\theta) = \frac{\frac{\partial}{\partial\theta}L(y,\theta)}{L(\mathrm{y},\theta)},$$

we obtain

$$E_\theta\left[\frac{\partial}{\partial\theta}\ln L(\boldsymbol{y},\theta)\right] = \int_Y \frac{\partial}{\partial\theta}\ln L(y,\theta)L(y,\theta)dy = \frac{\partial}{\partial\theta}\int_Y L(y,\theta)dy = \frac{\partial}{\partial\theta}1 = 0$$

for all $\theta \in \Omega$. Hence, we have

$$I(\theta) = \mathrm{var}\left\{\frac{\partial}{\partial\theta}\ln L(\boldsymbol{y},\theta)\right\}. \tag{1.17}$$

Now let the second derivative of $\ln L(y, \theta)$ with respect to θ for all y and θ exist, and let $\int_Y L(\mathrm{y},\theta)dP_\theta$ be differentiable twice, where integration and double differentiation can be commuted. Then by considering

$$\frac{\partial^2}{\partial\theta^2}\ln L(y,\theta) = \frac{L(y,\theta)\frac{\partial^2}{\partial\theta^2}L(y,\theta) - \left(\frac{\partial}{\partial\theta}L(\boldsymbol{y},\theta)\right)^2}{(L(y,\theta))^2} = \frac{\frac{\partial^2}{\partial\theta^2}L(y,\theta)}{L(y,\theta)} - \left[\frac{\frac{\partial}{\partial\theta}L(\boldsymbol{y},\theta)}{L(y,\theta)}\right]^2$$

and

$$0 = \frac{\partial^2}{\partial\theta^2}\int \ln L(y,\theta)dP_\theta = \int \frac{\partial^2}{\partial\theta^2}\ln L(y,\theta)dP_\theta,$$

the relation

$$E_\theta\left[\frac{\partial^2}{\partial\theta^2}\ln L(\boldsymbol{y},\theta)\right] = -E_\theta\left\{\left[\frac{\partial}{\partial\theta}\ln L(\boldsymbol{y},\theta)\right]^2\right\} = -I(\theta)$$

follows and therefore also

$$I(\theta) = -E_\theta\left[\frac{\partial^2}{\partial\theta^2}\ln L(\boldsymbol{y},\theta)\right]. \tag{1.18}$$

We present an example determining the Fisher information for both a discrete and a continuous distribution.

Example 1.11 Let P be the family of binomial distributions for given n and $\Omega = (0,1)$. The likelihood function is

$$L(y,p) = \binom{n}{y} p^y (1-p)^{n-y}.$$

The assumption V1 is satisfied, namely, the square of

$$\frac{\partial}{\partial p} \ln L(y,p) = \frac{y}{p} - \frac{n-y}{1-p}$$

after replacing y by random $\boldsymbol{y}$ has the finite expectation

$$I(p) = E_p\left\{\left[\frac{\partial}{\partial p} \ln L(\boldsymbol{y},p)\right]^2\right\} = \sum_{y=0}^{n} \left(\frac{y}{p} - \frac{n-y}{1-p}\right)^2 \binom{n}{y} p^y (1-p)^{n-y},$$

and this means

$$I(p) = \frac{n}{p(1-p)}, 0 < p < 1.$$

Example 1.12 Let P be the family of normal distributions $N(\mu, \sigma^2)$ with known σ^2. It is $\Omega = R^1$, and the likelihood function has the form

$$L(y,\mu) = \frac{1}{\sigma\sqrt{2\pi}} e^{-\frac{1}{2\sigma^2}(y-\mu)^2}.$$

For these distributions assumption V1 is fulfilled, too. We obtain

$$\frac{\partial}{\partial \mu} \ln L(y,\mu) = \frac{1}{\sigma^2}(y-\mu)$$

and

$$I(\mu) = \frac{1}{\sigma^4} E(\boldsymbol{y}-\mu)^2 = \frac{1}{\sigma^4} \operatorname{var}(\boldsymbol{y}) = \frac{1}{\sigma^2}.$$

Now we show the additivity of the Fisher information.

Theorem 1.6 If the Fisher information $I(\theta) = I_1(\theta)$ exists for a family P of probability distributions with $\Omega = R^1$ and if $\boldsymbol{Y} = (\boldsymbol{y}_1, \boldsymbol{y}_2, \ldots, \boldsymbol{y}_n)^T$ is a random sample with components $\boldsymbol{y}_i$ ($i = 1, \ldots, n$) all distributed as $P_i \in P$, then the Fisher information $I_n(\theta)$ of the distribution corresponding to $\boldsymbol{Y}$ is given by

$$I_n(\theta) = n I_1(\theta). \tag{1.19}$$

Proof: It follows from Definition 1.2 that the likelihood function $L_n(Y, \theta)$ of a random sample $\boldsymbol{Y}$ is

$$L_n(Y,\theta) = \prod_{i=1}^{n} L(y_i,\theta).$$

Consequently we get

$$\ln L_n(Y,\theta) = \sum_{i=1}^{n} \ln L(y_i,\theta)$$

and

$$\frac{\partial}{\partial\theta}\ln L_n(Y,\theta) = \sum_{i=1}^{n} \frac{\partial}{\partial\theta}\ln L(y_i,\theta).$$

Observing (1.17) we finally arrive at

$$I_n(\theta) = \mathrm{var}\left\{\frac{\partial}{\partial\theta}\ln L_n(\boldsymbol{Y},\theta)\right\} = \sum_{i=1}^{n} \mathrm{var}\left\{\frac{\partial}{\partial\theta}\ln L(\boldsymbol{y}_i,\theta)\right\} = nI_1(\theta).$$

Theorem 1.7 Let $M(\boldsymbol{Y})$ be a sufficient statistic with respect to the distribution $P_\theta \in P, \Omega \subseteq R^1$ of the components of the random sample $\boldsymbol{Y} = (\boldsymbol{y}_1, \boldsymbol{y}_2, \ldots, \boldsymbol{y}_n)^T$. Let the distribution P_θ fulfil the condition V1 of Definition 1.10. Then the Fisher information

$$I_M(\theta) = E\left\{\left[\frac{\partial}{\partial\theta}\ln L_M(\boldsymbol{M},\theta)\right]^2\right\} \tag{1.20}$$

of $\boldsymbol{M} = M(\boldsymbol{Y})$ exists where $L_M(M, \theta)$ is the likelihood function of $\boldsymbol{M}$ and

$$I_n(\theta) = I_M(\theta). \tag{1.21}$$

Proof: Considering (1.2) and (1.3), respectively, we have

$$L(Y,\theta) = h(Y)g(M(Y),\theta)$$

and therefore

$$\frac{\partial}{\partial\theta}\ln L(Y,\theta) = \frac{\partial}{\partial\theta}\ln g(M(Y),\theta)$$

since h (Y) is by assumption independent of θ. Taking Corollary 1.1 of Theorem 1.1 into account, the likelihood function $L_M(M, \theta)$ of $\boldsymbol{M}$ satisfies also condition V1 of Definition 1.10, and therefore $I_M(\theta)$ in (1.20) exists. Observing the equivalence of $L_M(M, \theta)$ and $L(Y, \theta)$, the assertion follows because of $\frac{\partial}{\partial\theta}\ln L_M(M,\theta) = \frac{\partial}{\partial\theta}\ln g(M,\theta)$.

Consequently the Fisher information of a sufficient statistic is the Fisher information of the corresponding random sample.

Now we consider parameters $\theta \in \Omega \subseteq R^p$.

Definition 1.11 Let $\boldsymbol{y}$ be distributed as $P_\theta \in P, \Omega \subseteq R^p, \theta = (\theta_1, \ldots, \theta_p)^T$. Let the conditions 2, 3 and 4 of V1 in Definition 1.10 be fulfilled for each component θ_i $(i = 1, \ldots, p)$. Let Ω be an open interval in R^p. Further, assume that the expectation of $\frac{\partial}{\partial \theta_i} \ln L(\boldsymbol{y}, \theta) \frac{\partial}{\partial \theta_j} \ln L(\boldsymbol{y}, \theta)$ exists for all θ and all $i, j = 1, \ldots, p$. Then the quadratic matrix

$$I(\theta) = (I_{i,j}(\theta)),\ \ i, j = 1, \ldots, p$$

of order p given by

$$I(\theta) = E\left\{\frac{\partial}{\partial \theta_i} \ln L(\boldsymbol{y}, \theta) \frac{\partial}{\partial \theta_j} \ln L(\boldsymbol{y}, \theta)\right\}$$

is said to be the (Fisher) information matrix with respect to P_θ.

Example 1.13 Let the random variable $\boldsymbol{y}$ be distributed as $N(\mu, \sigma^2)$ where $\theta = (\mu, \sigma^2)^T \in R^1 \times R^+ = \Omega$. Then

$$\ln L(y, \theta) = -\ln\sqrt{2\pi} - \frac{1}{2}\ln\sigma^2 - \frac{1}{2\sigma^2}(y-\mu)^2$$

holds, and the assumption of Definition 1.11 is fulfilled with $\theta_1 = \mu$ and $\theta_2 = \sigma^2$. Further we have

$$\frac{\partial}{\partial \mu} \ln L(y, \theta) = \frac{y-\mu}{\sigma^2} \text{ and } \frac{\partial}{\partial \sigma^2} \ln L(y, \theta) = -\frac{1}{2\sigma^2} + \frac{1}{2\sigma^4}(y-\mu)^2.$$

Because $E[(\boldsymbol{y} - \mu)^2] = \mathrm{var}(\boldsymbol{y}) = \sigma^2$, it follows $I_{11}(\theta) = \frac{1}{\sigma^2}$. Since the skewness $\gamma_1 = 0$ and since $E(\boldsymbol{y} - \mu) = 0$, we get $I_{12}(\theta) = I_{21}(\theta) = 0$. Further

$$\left[\frac{\partial}{\partial \sigma^2} \ln L(y, \theta)\right]^2 = \frac{1}{4\sigma^4} - \frac{2}{4\sigma^6}(y-\mu)^2 + \frac{1}{4\sigma^8}(y-\mu)^4.$$

Moreover, considering $\gamma_2 = 0$ and $E[(\boldsymbol{y} - \mu)^4] = 3\sigma^4$, it follows

$$I_{22} = E\left\{\left[\frac{\partial}{\partial \sigma^2} \ln L(\boldsymbol{y}, \theta)\right]^2\right\} = \frac{1}{\sigma^4}\left[\frac{1}{4} - \frac{1}{2} + \frac{3}{4}\right] = \frac{1}{2\sigma^4},$$

and we obtain

$$I(\theta) = \begin{pmatrix} \frac{1}{\sigma^2} & 0 \\ 0 & \frac{1}{2\sigma^4} \end{pmatrix}.$$

On the other hand, if we put $\theta_1 = \mu$ and $\theta_2 = \sigma$, then we have

$$\frac{\partial}{\partial \sigma} \ln L(y,\theta) = -\frac{1}{\sigma} + \frac{1}{\sigma^2}(y-\mu)^2.$$

While I_{11}, I_{12} and I_{21} remain unchanged, we find now

$$I_{22} = E\left\{ \left[\frac{\partial}{\partial \sigma} \ln L(\mathbf{y},\theta) \right]^2 \right\} = \frac{1}{\sigma^2}[1-2+3] = \frac{2}{\sigma^2}$$

and therefore

$$I(\theta) = \begin{pmatrix} \frac{1}{\sigma^2} & 0 \\ 0 & \frac{2}{\sigma^2} \end{pmatrix}.$$

This example shows that the Fisher information is not invariant with respect to parameter transformations. Using the chain rule of differential calculus, the following general statement arises.

Theorem 1.8 Let $\psi = h(\theta)$ be a monotone in $\Omega \subseteq R^1$, and with respect to θ differentiable function, let h map Ω onto Π. Then with respect to ψ, the differentiable inverse function $\theta = g(\psi)$ exists. Under the assumptions of Definition 1.10, let $I(\theta)$ be the Fisher information of the distribution $P_o \in \Pi$. Then the Fisher information $I^*(\psi)$ of the distribution P_ψ (i.e. P_θ written with the transformed parameter) is

$$I^*(\psi) = I(\theta)\left(\frac{d}{d\psi} g(\psi)\right)^2. \tag{1.22}$$

Considering Example 1.13 we set (for fixed μ) $\theta = \sigma^2$, $\psi = \sqrt{\theta} = \sigma$ and $\frac{d\theta}{d\psi} = 2\psi = 2\sigma$. By Theorem 1.8 we get with $I(\sigma^2) = \frac{1}{\sigma^2}$ the information

$$I^*(\psi) = I(\sigma^2)\, 4\sigma^2 = \frac{2}{\sigma^2}.$$

In Chapter 2 we need the following statement.

Theorem 1.9 (Inequality of Rao and Cramér)
Let assumption V1 of Definition 1.10 hold for all components of the random sample Y possessing the likelihood function $L(Y, \theta)$. Let the set $\{Y_0\} = \{Y \epsilon \{Y\} : L(Y, \theta) = 0\}$ of the points in the sample space satisfying $L(Y, \theta) = 0$ do not depend on θ. Let $P_\theta \in P = (P_\theta, \theta \in \Omega)$, $\Omega \subseteq R^1$ be the distribution of the components, and let $M(\boldsymbol{Y})$ be a statistic with expectation $E[M(\boldsymbol{Y})]$ and variance $\text{var}[M(\boldsymbol{Y})]$ mapping the sample space $\{Y\}$ into Ω. Then the inequality of Rao and Cramér

$$\text{var}[M(\boldsymbol{Y})] \geq \frac{\left(\frac{dE[M(\boldsymbol{Y})]}{d\theta}\right)^2}{nI(\theta)} \tag{1.23}$$

is fulfilled.

Proof: With the notation $M(\boldsymbol{Y}) = \boldsymbol{M}$, we get $E = E[\boldsymbol{M} - E(\boldsymbol{M})] = 0$. Hence

$$\frac{dE}{d\theta} = -\int_{\{Y\}} \frac{dE[M(\boldsymbol{Y})]}{d\theta} dP_\theta + \int_{\{Y\}} (M - E(M)) \frac{d}{d\theta} L(Y,\theta) dP_\theta = 0$$

and

$$\frac{dE}{d\theta} = -\frac{dE[M(\boldsymbol{Y})]}{d\theta} \int_{\{Y\}} dP_\theta + \int_{\{Y\}} (M - E(M)) \frac{d}{d\theta} \ln L(Y,\theta) dY = 0$$

hold, respectively. Then

$$E\left\{(\boldsymbol{M} - E(\boldsymbol{M})) \frac{d}{d\theta} \ln L(Y,\theta)\right\} = \frac{dE(\boldsymbol{M})}{d\theta}$$

follows. Taking Schwarz's inequality into account, we arrive at

$$\left\{\frac{dE(\boldsymbol{M})}{d\theta}\right\}^2 = \left\{E[(\boldsymbol{M} - E(\boldsymbol{M}))] \frac{d}{d\theta} \ln L(Y,\theta)\right\}^2$$
$$\leq E\left[(\boldsymbol{M} - E(\boldsymbol{M}))^2 E\left\{\left[\frac{d}{d\theta} \ln L(\boldsymbol{Y},\theta)\right]^2\right\}\right].$$

Considering (1.16) completes the proof.

When choosing $E(\boldsymbol{M}) = \theta$, the inequality of Rao and Cramér takes the form

$$\text{var}[M(\boldsymbol{Y})] \geq \frac{1}{nI(\theta)}. \tag{1.24}$$

Theorem 1.10 If y is distributed as a one-parametric exponential family and if $g(\theta) = \eta = E(\boldsymbol{M})$, then

$$I^*(\eta) = \frac{1}{\operatorname{var}(\boldsymbol{M})}. \tag{1.25}$$

Proof: Since the assumption V1 of Definition 1.10 is fulfilled, $I(\eta)$ exists. Observing

$$\frac{d}{d\eta} \ln L(Y,\eta) = M(Y) - \frac{d}{d\eta} A(\eta)$$

and Theorem 1.8, we obtain

$$\operatorname{var}(\boldsymbol{M}) = I^*(\eta) = I(\theta)[\operatorname{var}(\boldsymbol{M})]^2.$$

Hence, the assertion is true.

Considering Schwarz's inequality for the second moments of statistics $M(\boldsymbol{Y})$ with finite second moment and an arbitrary function $h(\boldsymbol{Y},\theta)$ with existing second moment, then the inequality

$$\operatorname{var}(\boldsymbol{M}) \geq \frac{\operatorname{cov}^2[\boldsymbol{M}, h(\boldsymbol{Y},\theta)]}{\operatorname{var}[h(\boldsymbol{Y},\theta)]}$$

follows.

Theorem 1.11 Let $M(\boldsymbol{Y})$ be a statistic with expectation $g(\theta)$ and existing second moment, and let $\boldsymbol{h}_j = h_j(\boldsymbol{Y}, \theta), j = 1, \ldots, r$ be functions with existing second moments. Then with the notations

$$c_j = \operatorname{cov}(M(\boldsymbol{Y}), \boldsymbol{h}_j), \sigma_{ij} = \operatorname{cov}(\boldsymbol{h}_i, \boldsymbol{h}_j), c^T = (c_1, \ldots, c_r)$$

and $\Sigma = (\sigma_{ij})$, $(|\Sigma| \neq 0)$, the inequality

$$\operatorname{var}(\mathbf{M}) \geq c^T \Sigma^{-1} c \tag{1.26}$$

is fulfilled.

Proof: The assertion follows from $\dfrac{c^T \Sigma^{-1} c}{\operatorname{var}(\boldsymbol{y})} \leq 1.$

With the help of (1.26), the inequality (1.24) of Rao and Cramér can be generalised to the p-dimensional case.

Theorem 1.12 Let the components of a random sample $\boldsymbol{Y} = (\boldsymbol{y}_1, \boldsymbol{y}_2, \ldots, \boldsymbol{y}_n)^T$ be distributed as

$$P_\theta \in P = (P_\theta, \theta \in \Omega), \Omega \subseteq R^p, \theta^T = (\theta_1, \ldots, \theta_p), p > 1.$$

Let $L(Y,\theta)$ be the likelihood function of $\boldsymbol{Y}$. Further, let the assumptions of Definition 1.10 be fulfilled. Additionally, let the set of points in $\{Y\}$ satisfying $L(Y,\theta) = 0$ do not depend on θ. Finally, let $M(\boldsymbol{Y})$ be a statistic, whose expectation $E[M(\boldsymbol{Y})] = w(\theta)$ exists and is differentiable with respect to all θ_i. Then the inequality

$$\text{var}[M(\boldsymbol{Y})] \geq a^T I^{-1} a$$

holds, where I^{-1} is the inverse of $I(\theta)$ and a is the vector of the derivatives of $w(\theta)$ with respect to the θ_i.

Proof: As $I(\theta)$ is positive definite and therefore I^{-1} exists, the assertion follows with $h_j = \frac{d}{d\theta_j} L(Y,\theta)$ from (1.26) considering Definition 1.11.

1.5 Statistical Decision Theory

First, we formulate the general statistical decision problem. Let us start from the assumption that there is a set of random variables $\{\boldsymbol{y}_t\}$, $t \epsilon R^1$ whose distribution $P_\theta \in P = (P_\theta, \theta \in \Omega)$, $\dim \{\Omega\} = p$ is at least partly unknown.

Here we restrict ourselves to the case that only statements about $\psi = g(\theta)$ are demanded, where Ω is mapped by g onto Z and $\dim(Z) = s$. This set Z is called state space.

The statistician has for statements about ψ a set $\{E\}$ of decisions at his disposal. $\{E\}$ is called decision space. Let $Y_{t_i} = \left(y_{t_i 1}, \ldots, y_{t_i n_i}\right)^T$ for each fixed t_i be a random sample of size n_i.

Let the set of results of an experiment for which a decision has to be made (i.e. to select from $\{E\}$) be with

$$N = \sum_{i=1}^{k} n_i,\ A_k = (Y_{t_1}, \ldots, Y_{t_k}) \in \prod_{i=1}^{k} \{Y_{t_i}\} = \{Y_{k,N}\},$$

the realisation of a random variable $\boldsymbol{A}_k = (\boldsymbol{Y}_{t_1}, \ldots, \boldsymbol{Y}_{t_k})$. Now let $d \in D$ be a measurable mapping from $\{Y_{k,N}\}$ onto E, which relates each A_k to a decision $d(A_k)$. Then d is called a decision function, and D is the set of admissible decision functions. A_k will depend on the distribution of $\boldsymbol{A}_k$, the support $\mathfrak{S}_k = (t_1, \ldots, t_k)$ of the experiment and its allocation vector $\mathfrak{N}_k = (n_1, \ldots, n_k)$. We denote the concrete design by

$$\begin{pmatrix} \mathfrak{S}_k \\ \mathfrak{N}_k \end{pmatrix} = \begin{pmatrix} t_1, \ldots, t_k \\ n_1, \ldots, n_k \end{pmatrix} \in V_N$$

belonging to a set V of admissible designs. Additionally, let a loss function L be given as measurable mapping from $E \times Z \times R^1$ into R^m (its definition and therefore that of m is a problem outside of mathematics), which means

$$L[d(A_k), \psi, f(M)], d(A_k) \in E, \psi \in Z \tag{1.27}$$

with a non-negative real function $f(M)$ where $M = (d, \mathfrak{S}_k, \mathfrak{N}_k)$.

The function Z registers the loss occurring if $d(A_k)$ is chosen, and ψ is the value in the transformed parameter space, while $f(M)$ corresponds to the costs for realising M. The task of statistics consists in providing methods for selection of triples $M = (d, \mathfrak{S}_k, \mathfrak{R}_k)$, which minimise a functional $R(d, \mathfrak{S}_k, \mathfrak{R}_k, \psi, f(M))$ of random loss called risk function R. We will denote by d either a decision function (for fixed n) or a sequence of decision functions of the same structure, whose elements differ only with respect to the sample size n. We assume that

$$R(d, \mathfrak{S}_k, \mathfrak{R}_k, \psi, f(M)) = F(d, \mathfrak{S}_k, \mathfrak{R}_k, \psi) + f(\mathfrak{S}_k, \mathfrak{R}_k), \tag{1.28}$$

where f does not depend on d and d^* is the decision function (sequence of decision functions) for which

$$\min_{d \in D} (d, \mathfrak{S}_k, \mathfrak{R}_k, \psi) = F(d^*, \mathfrak{S}_k, \mathfrak{R}_k, \psi) \tag{1.29}$$

is satisfied. Then the risk R can be minimised in two steps. First, d^* is determined in such a way that (1.29) is fulfilled, and second, $(\mathfrak{S}_k^*, \mathfrak{R}_k^*)$ is chosen to fulfil

$$R(d^*, \mathfrak{S}_k^*, \mathfrak{R}_k^*, \psi, f(\mathfrak{S}_k^*, \mathfrak{R}_k^*)) = \min_{\binom{\mathfrak{S}_k}{\mathfrak{R}_k} \in V} R(d^*, \mathfrak{S}_k, \mathfrak{R}_k, \psi, f(\mathfrak{S}_k, \mathfrak{R}_k)).$$

Definition 1.12 A triple $T^* \in V \times D$ is said to be locally R-optimal at the point $\psi_0 \in Z$ relative to $V \times D$ if for all $T \in V \times D$ the inequality

$$R[T^*, \psi_0, f(M^*)] \leq R[T, \psi_0, f(M)]$$

holds. If M^* is for all $\psi_0 \in Z$ locally R-optimal, then M^* is called global R-optimal.

Example 1.14 Let $k = 1$ and $\boldsymbol{y}_{t_1} = \boldsymbol{y}$ be distributed as $N(\mu, \sigma^2)$. Then $\theta = \begin{pmatrix} \mu \\ \sigma^2 \end{pmatrix} \in \Omega = R^1 \times R^+$ and $\boldsymbol{A}_1 = \boldsymbol{Y}$. Further, let $\psi = g(\theta) = \mu \in R^1$ and $d(\boldsymbol{Y}) = \hat{\boldsymbol{\mu}}$ be a statistic with realisations in $R^1 = \{E\}$. Finally, let D be the class of statistics with finite second moment and realisations in R^1. Now we choose the loss function

$$L[\hat{\mu}, \mu, f(T)] = c_1(\hat{\mu} - \mu)^2 + c_2 nK, c_1, c_2, K > 0,$$

where K represents the costs of an experiment (a measurement). Besides we define as risk R the expected random loss

$$R(\hat{\boldsymbol{\mu}}, n, \mu, Kn) = E\left[c_1(\mu - \hat{\boldsymbol{\mu}})^2 + c_2 nK\right] = c_2 nK + c_1\left[\operatorname{var}(\hat{\boldsymbol{\mu}}) + B(\hat{\boldsymbol{\mu}})^2\right],$$

where $B(\hat{\boldsymbol{\mu}}) = E(\hat{\boldsymbol{\mu}}) - \mu$. In the class D the choice $\hat{\mu} = \psi_0$ is together with $n = 0$ locally R-optimal for the decision, and for (ψ_0, n) the risk R is equal to 0. The class D can be restricted to exclude this unsatisfactory trivial case. We denote by $D_E \subseteq D$ the subset in D with $B(\hat{\boldsymbol{\mu}}) = 0$. Then we obtain

$$R(\hat{\boldsymbol{\mu}}, n, \mu, Kn) = c_2 nK + c_1 \operatorname{var}(\hat{\boldsymbol{\mu}}), \hat{\boldsymbol{\mu}} \in D_E$$

in the form (1.28). We will see in Chapter 2 that $\operatorname{var}(\hat{\boldsymbol{\mu}})$ becomes minimal for $\hat{\boldsymbol{\mu}}^* = \bar{\boldsymbol{y}}$.

Since for a random sample of n elements we have $\operatorname{var}(\bar{\boldsymbol{y}}) = \frac{\sigma^2}{n}$, the first step of minimising R leads to

$$\min_{d \in D_E} c_1 \operatorname{var}(\hat{\boldsymbol{\mu}}) = \frac{c_1}{n}\sigma^2$$

and

$$R(\hat{\boldsymbol{\mu}}, n, \mu, Kn) = c_2 Kn + \frac{c_1}{n}\sigma^2.$$

If we derive the right-hand side of the equation with respect to n and put the derivative equal to 0, then we get $n^* = \sigma\sqrt{\frac{c_1}{Kc_2}}$, and this as well as $\bar{\boldsymbol{y}}$ does not depend on $\psi = \mu$. The convexity of the considered function shows that we have indeed found a (global) minimum. Hence, the R-optimal solution of the decision problem in $E \times Z$ (which is in Z global, but in Ω only local because of the dependence on σ) is given by

$$M^* = \left(\bar{\boldsymbol{y}}, n^* = \sigma\sqrt{\frac{c_1}{Kc_2}}\right).$$

If we choose $\psi = g(\theta) = \sigma^2 > 0$, then we obtain $E = R^+$, $k = 1$, $\boldsymbol{A}_1 = \boldsymbol{Y}$ and $N = n$. Let the loss function be

$$L\left[d(\boldsymbol{Y}), \sigma^2, f(M)\right] = c_1\left(\sigma^2 - d\right)^2 + c_2 nK, c_i > 0, K > 0.$$

If we take again

$$R\left(d(\boldsymbol{Y}), n, \sigma^2, Kn\right) = \mathrm{R} = \mathrm{E}(\boldsymbol{L}) = c_1 E\left\{\left(\sigma^2 - d(\boldsymbol{Y})\right)^2\right\} + c_2 nK$$

as risk function, it is of the form (1.28). If we restrict ourselves to $d \in D_E$ by analogous causes as in the previous case such that $E[d(\boldsymbol{Y})] = \sigma^2$ holds, then the first summand of R is minimal for

$$d(\boldsymbol{Y}) = \boldsymbol{s}^2 = \frac{1}{n-1}\sum_{i=1}^{n}(\boldsymbol{y}_i - \bar{\boldsymbol{y}})^2,$$

which will be seen in Chapter 2.

Since $\frac{s^2}{\sigma^2}(n-1)$ is distributed as CS (n – 1) and has therefore the variance 2(n – l), we find

$$\operatorname{var}\left(s^2\right) = \frac{2\sigma^4}{n-1}.$$

The first step of optimisation supplies

$$R\left(s^2, n, \sigma^2, Kn\right) = c_1 \frac{2\sigma^4}{n-1} + c_2 nK.$$

The R-optimal n is given by

$$n^* = 1 + \sigma^2 \sqrt{\frac{2c_1}{Kc_2}}.$$

The locally R-optimal solution of the decision problem is

$$M^* = \left(s^2, n^* = 1 + \sigma^2 \sqrt{\frac{2c_1}{Kc_2}}\right).$$

We consider more detailed theory and further applications in the next chapters where the selection of minimal sample sizes is discussed. We want to assume that d has to be chosen for fixed $\mathfrak{S}_k$ and $\mathfrak{N}_k$ R-optimal relative to a certain risk function. Concerning the optimal choice of $\begin{pmatrix} \mathfrak{S}_k \\ \mathfrak{N}_k \end{pmatrix}$, we refer to Chapters 8 and Chapter 9 treating regression analysis. We write therefore with τ from Definition 1.13

$$R(\boldsymbol{d}, \psi) = E\{L[d(\boldsymbol{Y}), \psi]\} = r(d, \tau). \tag{1.30}$$

In Example 1.14 a restriction to a subset $D_E \subseteq D$ was carried out to avoid trivial locally R-optimal decision functions d. Now two other general procedures are introduced to overcome such problems.

Definition 1.13 Let $\boldsymbol{\theta}$ be a random variable with realisations $\theta \in \Omega$ and the probability distribution $P_\tau, \tau \in \mathfrak{T}$. Let the expectation

$$\int_\Omega R(d, \psi) d\Pi_\tau = r(d, \tau) \tag{1.31}$$

of (1.30) exist relative to P_τ, which is called Bayesian risk relative to the a priori distribution P_τ.

A decision function $d_0(\boldsymbol{Y})$ that fulfils

$$r(d_0, \tau) = \inf_{d \in D}[r(d, \tau)],$$

is called Bayesian decision function relative to the a priori distribution P_τ.

Definition 1.14 A decision function $d_0 \in D$ is said to be minimax decision function if

$$\max_{\theta \in \Omega} R(d_0,\psi) = \min_{d \in D} \max_{\theta \in \Omega} R(d,\psi). \tag{1.32}$$

Definition 1.15 Let $d_1, d_2 \in D$ be two decision functions for a certain decision problem with the risk function $R(d, \psi)$ where $\psi = g(\theta), \theta \in \Omega$. Then d_1 is said not to be worse than d_2, if $R(d_1, \psi) \le R(d_2, \psi)$ holds for all $\theta \in \Omega$. The function d_1 is said to be better than the function d_2 if apart from $R(d_1, \psi) \le R(d_2, \psi)$ for all $\theta \in \Omega$ at least for one $\theta^* \in \Omega$ the strong inequality $R(d_1, \psi^*) < R(d_2, \psi^*)$ holds where $\psi^*=g(\theta^*)$. A decision function d is called admissible in D if there is no decision function in D better than d. If a decision function is not admissible, it is called inadmissible.

In this chapter it is not necessary to develop the decision theory further. In Chapter 2 we will consider the theory of point estimation, where $d(Y) = S(Y)$ is the decision function. Regarding the theory of testing in Chapter 3, the probability for the rejection of a null hypothesis and in the confidence interval estimation a domain in Ω covering the value θ of the distribution P_θ with a given probability is the decision function $d(Y)$. Selection rules and multiple comparison methods are further special cases of decision functions.

1.6 Exercises

1.1 For estimating the average income of the inhabitants of a city, the income of owners of each 20th private line in a telephone directory is determined. Is this sample a random sample with respect to the whole city population?

1.2 A set with elements 1, 2, 3 is considered. Selecting elements with replacement, there are $3^4 = 81$ different samples of size $n = 4$. Write down all possible samples, calculate $\bar{y}$ and s^2 and present the frequency distribution of $\bar{y}$ and s^2 graphically as a bar chart. (You may use a program package.)

1.3 Prove that the statistic $M(Y)$ is sufficient relative to θ, where $\boldsymbol{Y} = (\boldsymbol{y}_1, \boldsymbol{y}_2, \dots, \boldsymbol{y}_n)^T$, $n \ge 1$ is a random sample from a population with distribution $P_\theta, \theta \in \Omega$, by determining the conditional distribution of $\boldsymbol{Y}$ for given $M(Y)$.

a) $M(\boldsymbol{Y}) = \sum_{i=1}^{n} \boldsymbol{y}_i$ and P_θ is the Poisson distribution with the parameter $\theta \in \Omega \subset R^+$.

b) $M(Y) = (\boldsymbol{y}_{(1)}, \boldsymbol{y}_{(n)})^T$ and P_θ is the uniform distribution in the interval $(\theta, \theta + 1)$ with $\theta \in \Omega \subset R^1$.

c) $M(Y) = \boldsymbol{y}_{(n)}$ and P_θ is the uniform distribution in the interval $(0, \theta)$ with $\theta \in \Omega = R^+$.

d) $M(\boldsymbol{Y}) = \sum_{i=1}^{n} \boldsymbol{y}_i$ and P_θ is the exponential distribution with the parameter $\theta \in \Omega = R^+$.

1.4 Let $Y = (y_1, y_2, \ldots, y_n)^T$, $n \geq 1$ be a random sample from a population with the distribution P_θ, $\theta \in \Omega$. Determine a sufficient statistic with respect to θ using Corollary 1.1 of the decomposition theorem if P_θ, $\theta \in \Omega$ is the density function

a) $(y, \theta) = \theta y^{\theta - 1}, 0 < y < 1; \theta \in \Omega = R^+$,
b) Of the Weibull distribution

$$f(y,\theta) = \theta a(\theta y)^{a-1} e^{-(\theta y)^a}, y \geq 0, \theta \in \Omega = R^+, a > 0 \text{ known}$$

c) Of the Pareto distribution

$$f(y,\theta) = \frac{\theta a^\theta}{y^{\theta+1}}, y > a > 0, \theta \in \Omega = R^+ \text{ known}$$

1.5 Determine a minimal sufficient statistic $M(Y)$ for the parameter θ, if $Y = (y_1, y_2, \ldots, y_n)^T$, $n \geq 1$ is a random sample from a population with the following distribution P_θ:

a) Geometric distribution with the probability function

$$p(y,p) = p(1-p)^{y-1}, \; y = 1,2,\ldots, 0 < p < 1$$

b) Hypergeometric distribution with the probability function

$$p(y,M,N,n) = \frac{\binom{M}{y}\binom{N-M}{n-y}}{\binom{N}{n}}, n \in \{1,\ldots,N\}, y \in \{0,\ldots,N\}; M \leq N \text{ integer},$$

c) Negative binomial distribution with the probability function

$$\mathrm{p}(y,p,r) = \binom{y-1}{r-1} p^r (1-p)^{y-r}, 0 < p < 1, y \geq r \text{ integer}, r \in \{0,1,\ldots\}$$

and
i) $\theta = p$ and r known;
ii) $\theta^T = (p, r)$,

d) Beta distribution with the density function

$$f(y,\theta) = \frac{1}{B(a,b)} y^{a-1}(1-y)^{b-1}, 0 < y < 1, 0 < a, b < \infty$$

and
i) $\theta = a$ and b known;
ii) $\theta = b$ and a known

1.6 Prove that the following distribution families $\{P_\theta, \theta \in \Omega\}$ are complete:

a) P_θ is the Poisson distribution with the parameter $\theta \in \Omega = R^+$.
b) P_θ is the uniform distribution in the interval $(0, \theta)$, $\theta \in \Omega = R^+$.

1.7 Let $\boldsymbol{Y} = (\boldsymbol{y}_1, \boldsymbol{y}_2, \ldots, \boldsymbol{y}_n)^T, n \geq 1$ be a random sample, whose components are uniformly distributed in the interval $(0, \theta)$, $\Omega = R^+$. Show that $M(\boldsymbol{Y}) = \boldsymbol{y}_{(n)}$ is complete sufficient.

1.8 Let the variable $\boldsymbol{y}$ have the discrete distribution P_θ with the probability function

$$p(y,\theta) = P(\boldsymbol{y} = y) = \begin{cases} \theta & \text{for } y = -1 \\ (1-\theta)^2 \theta^y & \text{for } y = 0, 1, 2, \ldots \end{cases}.$$

Show that the corresponding distribution family with $\theta \in (0, 1)$ is bounded complete, but not complete.

1.9 Let a one-parametric exponential family with the density or probability function

$$f(y,\theta) = h(y)\, \mathrm{e}^{\eta(\theta)M(y) - B(\theta)}, \theta \in \Omega$$

be given.

a) Express the Fisher information of this distribution by using the functions $\eta(\theta)$ and $B(\theta)$.
b) Use the result of a) to calculate $I(\theta)$ for the
   i) Binomial distribution with the parameter $\theta = p$
   ii) Poisson distribution with the parameter $\theta = \lambda$
   iii) Exponential distribution with the parameter θ
   iv) Normal distribution $N(\mu, \sigma^2)$ with $\theta = \sigma$ and μ fixed

1.10 Let the assumptions of Definition 1.11 be fulfilled. Besides, the second partial derivatives $\frac{\partial^2}{\partial\theta_i \partial\theta_j} L(y,\theta)$ are to exist for all $i, j = 1, \ldots, p$ and $y \in \{Y\}$ as well as their expectations for random $\boldsymbol{y}$. Moreover, let $\int_{\{Y\}} L(y, \theta)\, dy$ be twice differentiable, where integration and differentiation are commutative.

Prove that in this case the elements of the information matrix in Definition 1.11 have the form

a) $I_{i,j}(\theta) = \mathrm{cov}\left[\frac{\partial}{\partial\theta_i} \ln L(\boldsymbol{y},\theta), \frac{\partial}{\partial\theta_j} \ln L(\boldsymbol{y},\theta)\right]$,

b) $I_{i,j}(\theta) = -E\left[\frac{\partial^2}{\partial\theta_i \partial\theta_j} \ln L(\boldsymbol{y},\theta)\right]$,

respectively.

1.11 Let $Y = (y_1, y_2, \ldots, y_n)^T$ be a random sample from a population with the distribution $P_\theta, \theta \in \Omega$ and $M(Y)$ a given statistic.

Calculate $E[M(Y)]$, $\mathrm{var}[M(Y)]$, the Fisher information $I(\theta)$ of the distribution and the Rao–Cramér bound for $\mathrm{var}[M(Y)]$.

Does equality hold in the inequality of Rao and Cramér under the following assumptions?

a) P_θ is the Poisson distribution with the parameter $\theta \in R^+$ and

$$\mathrm{M}(\mathbf{Y}) = \begin{cases} 1 & \text{for } y = 0 \\ 0 & \text{else} \end{cases},$$

(here we have $n = 1$, i.e. $\mathbf{y} = \mathbf{Y}$).

b) P_θ is the Poisson distribution with the parameter $\theta \in R^+$ and $M(Y) = \left(1 - \frac{1}{n}\right)^{n\bar{y}}$, (generalisation of a) for the case $n > 1$).

c) P_θ is the distribution with the density function

$$f(y, \theta) = \theta y^{\theta - 1}, 0 < y < 1, \theta \in R^+$$

and $M(Y) = -\frac{1}{n}\sum_{i=1}^{n} \ln y_i$.

1.12 In a certain region it is intended to drill for oil. The owner of drilling rights has to decide between strategies from $\{E_1, E_2, E_3\}$.

The following notations are introduced:

E_1 – The drilling is carried out under its own direction.
E_2 – The drilling rights are sold.
E_3 – A part of drilling rights are alienated.

It is not known so far if there really is an oil deposit in the region. Further let be $\Omega = \{\theta_1, \theta_2\}$ with the following meanings:

$\theta = \theta_1$ – Oil occurs in the region.
$\theta = \theta_2$ – Oil does not occur in the region.

The loss function $L(d, \theta)$ has for the decisions $d = E_i, i = 1, 2, 3$ and $\theta = \theta_j$, $j = 1, 2$ the form

	E_1	E_2	E_3
θ_1	0	10	5
θ_2	12	1	6

The decision is made considering expert's reports related to the geological situation in the region. We denote the result of the reports by $y \in \{0, 1\}$.

Let $p_\theta(y)$ be the probability function of the random variable $\boldsymbol{y}$ – in dependence on θ – with values

	y = 0	y = 1
θ_1	0.3	0.7
θ_2	0.6	0.4

Therefore the variable $\boldsymbol{y}$ states the information obtained by the 'random experiment' of geological reports about existing ($y = 1$) or missing ($y = 0$) deposits of oil in the region. Let the set D of decision functions $d(\boldsymbol{y})$ contain all possible 3^2 discrete functions:

	1	2	3	4	5	6	7	8	9
$d_i(0)$	E_1	E_1	E_1	E_2	E_2	E_2	E_3	E_3	E_3
$d_i(1)$	E_1	E_2	E_3	E_1	E_2	E_3	E_1	E_2	E_3

a) Determine the risk $R(d(\boldsymbol{y}), \theta) = E_\theta[L\{d(\boldsymbol{y})\}, \theta]$ for all 18 above given cases.
b) Determine the minimax decision function.
c) Following the opinion of experts in the field of drilling technology, the probability of finding oil after drilling in this region is approximately 0.2. Then $\boldsymbol{\theta}$ can be considered as random variable with the probability function

$\boldsymbol{\theta}$	θ_1	θ_2
$\pi(\theta)$	0.2	0.8

Determine for each decision function the Bayesian risk $r(d_i, \pi)$ and then the Bayesian decision function.

1.13 The strategies of treatment using two different drugs M_1 and M_2 are to be assessed. Three strategies are at the disposal of doctors:

E_1 – Treatment with the drug M_1 increasing blood pressure
E_2 – Treatment without using drugs
E_3 – Treatment with the drug M_2 decreasing blood pressure

Let the variable θ characterise the (suitable transformed) blood pressure of a patient such that $\theta < 0$ indicates too low blood pressure, $\theta = 0$ normal blood pressure and $\theta > 0$ too high blood pressure. The loss function is defined as follows:

	E_1	E_2	E_3
$\theta < 0$	0	c	b + c
$\theta = 0$	b	0	b
$\theta > 0$	b + c	c	0

The blood pressure of a patient is measured. Let the measurement $\boldsymbol{y}$ be distributed as $N(\theta, 1)$ and let it happen n-times independently from each other: $\boldsymbol{Y} = (\boldsymbol{y}_1, \boldsymbol{y}_2, \dots, \boldsymbol{y}_n)^T$. Based on this sample the decision function

$$d_{r,s} = \begin{cases} E_1, & \text{if } \bar{y} < r \\ E_2, & \text{if } r \le \bar{y} \le s \\ E_3, & \text{if } \bar{y} > s \end{cases}$$

is defined.

a) Determine the risk $R(d_{r,s}(\bar{\boldsymbol{y}}), \theta) = E\{L[d_{r.s}(\bar{\boldsymbol{y}}), \theta]\}$.
b) Sketch the risk function in the case $b = c = 1$, $n = 1$ for
   i) $r = -s = -1$;
   ii) $r = -\frac{1}{2} s = -1$.

For which values of θ the decision function $d_{-1,1}(y)$ should be preferred to the function $d_{-1,2}(y)$?

References

Bahadur, R. R. (1955) Statistics and subfields. *Ann. Math. Stat.*, **26**, 490–497.

Cochran, W. G. and Boing, W. (1972) *Stichprobenverfahren*, De Gruyter, Berlin, New York.

Fisher, R. A. (1925) *Statistical Methods for Research Workers*, Oliver & Boyd, Edinburgh.

Halmos, P. R. and Savage, L. J. (1949) Application of the Radon-Nikodym theorem to the theory of sufficient statistics. *Ann. Math. Stat.*, **20**, 225–241.

Kauermann, G. and Küchenhoff, H. (2011) *Stichproben und praktische Umsetzung mit R*, Springer, Heidelberg.

Kullback, S. and Leibler, R. A. (1951) On information and sufficiency. *Ann. Math. Stat.*, **22**, 79–86.

Lehmann, E. L. and Romano J. P. (2008) *Testing Statistical Hypothesis*, Springer, Heidelberg.

Quatember, A. (2014) *Datenqualität in Stichprobenerhebungen*, Springer, Berlin.
Rasch, D. (1995) *Mathematische Statistik*, Johann Ambrosius Barth, Berlin, Heidelberg.
Rasch, D., Herrendörfer, G., Bock, J., Victor, N. and Guiard, V. (Eds.) (2008) *Verfahrensbibliothek Versuchsplanung und - auswertung*, 2. verbesserte Auflage in einem Band mit CD, R. Oldenbourg Verlag, München, Wien.
Stigler, S. M. (1986, 1990) *The History of Statistics: The Measurement of Uncertainty Before 1900*, Harvard University Press, Cambridge.

2

Point Estimation

In this chapter we consider so-called point estimations. The problem can be described as follows. Let the distribution P_θ of a random variable $\boldsymbol{y}$ belong to a family $P = (P_\theta, \theta \in \Omega)$, $\Omega \subseteq R^p$, $p \geq 1$. With the help of a realisation Y of a random sample $\boldsymbol{Y} = (\boldsymbol{y}_1, \boldsymbol{y}_2, \ldots, \boldsymbol{y}_n)^T$, $n \geq 1$, a statement is to be given concerning the value of a prescribed real function $\psi = g(\theta) \in Z$. Often $g(\theta) = \theta$. Obviously the statement about $g(\theta)$ should be as precise as possible. What this really does mean depends on the choice of the loss function defined in Section 1.5. We define a statistic $M(\boldsymbol{Y})$ taking the value $M(Y)$ for $\boldsymbol{Y} = Y$ where $M(Y)$ is called the estimate of $\psi = g(\theta)$.

The notation 'point estimation' reflects the fact that each realisation $M(Y)$ of $M(\boldsymbol{Y})$ defines a point in the space Z of possible values of $g(\theta)$.

The problem of interval estimators is discussed in Chapter 3 following the theory of testing.

By $L[g(\theta), M(Y)] = L(\psi, M)$, we denote a loss function taking the value $L(\psi_0, M)$ if ψ takes the value ψ_0 and $\boldsymbol{Y}$ the value Y (i.e. $\boldsymbol{M} = M(\boldsymbol{Y})$ takes the value $M = M(Y)$).

Although many statements in this chapter can be generalised to arbitrary convex loss functions, we want to use mainly the most convenient loss function, the quadratic loss function without costs. If it is not explicitly stated in another way, our loss function

$$L(\psi,M) = \|\psi - M\|^2, \psi \in Z, M \in D \tag{2.1}$$

is the square of the L_2-norm of the vector $\psi - M$ supposing that it is P_θ-integrable. Then we define the risk function as expectation

$$R(\psi,M) = E\left(\|\psi - \boldsymbol{M}\|^2\right) = \int_{\{Y\}} \|\psi - M(Y)\|^2 dP_\theta \tag{2.2}$$

of the random loss. Here $R(\psi, M)$ is the risk (the expected or mean loss) occurring if the statistic $M(\boldsymbol{Y}) \in D$ is used to estimate $\psi = g(\theta) \in Z$. We will come back

Mathematical Statistics, First Edition. Dieter Rasch and Dieter Schott.

later to the problem of finding a suitable set D of statistics. First we want to assure by the following definition that the difference $\psi - M$ makes sense.

Definition 2.1 Let $\boldsymbol{Y} = (\boldsymbol{y}_1, \boldsymbol{y}_2, \dots, \boldsymbol{y}_n)^T$ be a random sample of size $n \geq 1$ with components $\boldsymbol{y}_i$ whose distribution P_θ is from the family $P = (P_\theta, \theta \in \Omega)$. A statistic is said to be an estimator (in the stronger sense) or also estimation $\boldsymbol{S} = S(\boldsymbol{Y})$ with respect to the real function $g(\theta) = \psi$ with $\psi \in Z = g(\Omega)$, if $\boldsymbol{S}$ maps the sample space into a subset of Z. By D we denote the set of all estimators with respect to $g(\theta)$ based on samples of size n.

Two remarks should be made concerning Definition 2.1.

First, if we look for optional estimators, we always suppose that n is fixed and not itself a variable of the optimisation problem. Therefore we assume that both n and $\boldsymbol{S} \in D$ can be chosen separately optimal considering the total optimisation process according to Section 1.5. Hence, if we speak about 'the estimator', we mean the estimator for a fixed n. For example, the arithmetic mean

$$\bar{\boldsymbol{y}} = \frac{1}{n}\sum_{i=1}^{n} \boldsymbol{y}_i$$

is an estimator for each n. But we want to give statements about the asymptotic behaviour, for example, referring to n in the case of the arithmetic mean. Then we consider the sequence $\{S(\boldsymbol{Y}_n)\}$ of estimators $S(\boldsymbol{Y}_n)$ with $n = 1,2, \dots$, for example, the sequence $\left\{\bar{\mathbf{y}} = \frac{1}{n}\sum_{i=1}^{n} \mathbf{y}_i\right\}$ of the arithmetic means. For short we keep to the common speech that 'the arithmetic mean is consistent' instead of the more precise expression that 'the sequence of the arithmetic means is consistent'.

Second, demanding that $\boldsymbol{S}$ is only an estimator, if $\boldsymbol{S}$ maps the space $\{Y\}$ measurably into a space $\{M(Y)\} \subset Z$, is sometimes too restrictive. In older publications also such statistics, $\boldsymbol{M}$ are admitted as estimators if $Z \subseteq \{M(Y)\}$, $\dim(Z) = \dim(\{M(Y)\})$ is fulfilled. Often such cases occurred in model II of analysis of variance (ANOVA) (Chapter 6). Variance components estimated by the ANOVA method can also take negative values. In this book we will call such procedures not as estimators and remain with Definition 2.1.

In non-linear regression we also speak of estimators if the corresponding mapping is not measurable. We call such statistics estimators in a weaker sense.

It could be suggested to declare the aim of the theory of estimation as finding such estimators that are $R(\psi, S)$-optimal (i.e. that minimise the value of $R(\psi, S)$ under all $S(\boldsymbol{Y}) \in D$). But, since $R(\psi, S)$ is minimal, namely, equal to 0 for $\psi = \psi_0$, if we put $S(Y) = \psi_0$ for all $Y \in \{Y\}$, a problem stated in this way has no solution, which is uniformly R-optimal (i.e. for all $\psi \in Z$). This dilemma can be eliminated, as already described in Section 1.5, either by restricting to a subset $D_0 \subset D$ and looking for R-optimal estimators in this subset D_0 or, analogously to use the Bayesian approach, by minimising a weighted risk, the so-called Bayesian risk

$$R_B(\psi) = \int_Z R(\psi,S)dP_\lambda \tag{2.3}$$

with respect to a measure P_λ standardised to 1, where P_λ ($\lambda \in K$) is chosen as a weight function, which has an existing integral according to (2.3) and which moreover measures the 'importance' of single θ-values. For random $\boldsymbol{\theta}$ the weight P_λ is the probability measure of the random variable $g(\boldsymbol{\theta})$, that is, the a priori distribution of $\boldsymbol{\psi} = g(\boldsymbol{\theta})$.

Finally there is a third approach that is often used. Here we look for a minimax estimator $S(\boldsymbol{Y})$ satisfying

$$R(\tilde{\psi},\tilde{S}) = \min_{S \in D} \max_{\psi \in Z} R(\psi,S). \tag{2.4}$$

We use the first approach in this book, as already indicated in Section 1.5. We consider in Section 2.1 the subset $D'_E = D_0 \subseteq D$ of unbiased estimators; further we restrict ourselves to linear (D_L), linear unbiased (D_{LE}), quadratic (D_Q) or quadratic unbiased (D_{QE}) estimators.

2.1 Optimal Unbiased Estimators

We suppose that all estimators $\boldsymbol{S}$ used in this chapter are P_θ-integrable, which means that for each $P_\theta \in P = (P_\theta, \theta \in \Omega)$ and for each $\boldsymbol{S}$ the expectation

$$E[S(\boldsymbol{Y})] = \int_{\{Y\}} S(Y)dF_\theta(Y) \tag{2.5}$$

exists. Here $F_\theta(Y)$ is the distribution function of the random sample $\boldsymbol{Y} = (\boldsymbol{y}_1, \boldsymbol{y}_2, \dots, \boldsymbol{y}_n)^T$ (and therefore the distribution function of the product measure of the distributions P_θ belonging to $\boldsymbol{y}_i$).

Definition 2.2 An estimator $S(\boldsymbol{Y})$ based on a random sample $\boldsymbol{Y} = (\boldsymbol{y}_1, \boldsymbol{y}_2, \dots, \boldsymbol{y}_n)^T$ of size $n \geq 1$ is said to be unbiased with respect to $\psi = g(\theta)$ if

$$E[S(\boldsymbol{Y})] = g(\theta) \tag{2.6}$$

holds for all $\theta \in \Omega$. We denote the class of unbiased estimators of an estimation problem by D'_E. The difference $v_n(\theta) = E[S(\boldsymbol{Y})] - g(\theta)$ is called the bias of $S(\boldsymbol{Y})$.

A statistic $U(\boldsymbol{Y})$ is said to be unbiased with respect to 0 if

$$E[U(\boldsymbol{Y})] = 0 \tag{2.7}$$

for all $\theta \in \Omega$.

Naturally the expression 'for all $\theta \in \Omega$' in the definitions and theorems always means 'for all $P_\theta \in P$', that is, more precisely for all measures P_θ with existing integral in (2.5). First we show by an example that there are problems of estimation with a non-empty class D_E'.

Example 2.1 Let the components y_i of the random sample $Y = (y_1, y_2, \dots, y_n)^T$ be distributed as $N(\mu, \sigma^2)$. Then $\theta = (\mu, \sigma^2)^T$. Let $\psi_1 = g_1(\theta) = (1\ 0)^T\theta = \mu$ and $\psi_2 = g_2(\theta) = (0\ 1)^T\theta = \sigma^2$. We consider

$$S_1(Y) = \bar{y} \text{ and } S_2(Y) = \frac{1}{n-1}\sum_{i=1}^{n}(y_i - \bar{y})^2 = s^2.$$

We know that $\bar{y}$ is distributed as $N\left(\mu, \frac{\sigma^2}{n}\right)$ and $X^2 = \frac{(n-1)s^2}{\sigma^2}$ as $CS(n-1)$. Consequently we have $E(\bar{y}) = \mu$ (for all θ) and $E(s^2) = \sigma^2$ because $E(X^2) = n - 1$. Hence, $\bar{y}$ is unbiased with respect to μ and s^2 is unbiased with respect to σ^2.

However, there are problems of estimation possessing no unbiased estimators. This is shown in the next example.

Example 2.2 Let the random variable $Y = y$ be distributed as $B(n, p)$ with $0 < p < 1$. Let n be known and $\psi = g(p) = 1/p$. The sample space is $\{Y\} = \{0, 1, \dots, n\}$. Assuming that there is an unbiased estimator $S(y)$ with respect to $1/p$, the expectation

$$E[S(y)] = \sum_{y=0}^{n}\binom{n}{y}p^y(1-p)^{n-y}S(y)$$

would be $1/p$. But this is not possible, because $E[S(y)]$ tends to $S(0)$ for $p \to 0$ while $1/p$ tends to infinity for $p \to 0$.

The following statement is obvious.

Theorem 2.1 If $S_0(Y)$ is an unbiased estimator with respect to $\psi = g(\theta)$, then each other unbiased estimator $S(Y)$ with respect to ψ has the form

$$S(Y) = S_0(Y) - U(Y) \tag{2.8}$$

where $U(Y)$ is an unbiased statistic with respect to 0.

We want to use this theorem to find $R(\psi, S)$-optimal estimators $S(Y) \in D$. First we see that

$$R(\psi, S) = \text{var}(S)$$

is true for $S(Y) \in D'_E$, that is, the variance-optimal unbiased estimator has to be found. Assuming that $\boldsymbol{S}_0$ is an unbiased estimator with respect to g and that $\boldsymbol{S}_0$, $\boldsymbol{S}$ and $\boldsymbol{U}$ have a finite variance, then

$$\operatorname{var}(\boldsymbol{S}) = \operatorname{var}(\boldsymbol{S}_0 - \boldsymbol{U}) = E\left[(\boldsymbol{S}_0 - \boldsymbol{U})^2\right] - \psi^2. \tag{2.9}$$

Hence, we can find the variance-optimal estimator by minimising $E(\boldsymbol{S}_0 - \boldsymbol{U})$. We want to demonstrate this approach in the next example.

Example 2.3 Let $Y = \boldsymbol{y}$ where $\boldsymbol{y}$ take the values −1, 0, 1, … with the probabilities

$$P(\boldsymbol{y} = -1) = p, P(\boldsymbol{y} = y) = p^y(1-p)^2 \text{ for } y = 0, 1, \ldots$$

where $0 < p = \theta < 1$. Since

$$\sum_{k=0}^{\infty} x^k = \frac{1}{1-x} \text{ for } |x| < 1,$$

a distribution is defined observing

$$p + \sum_{k=0}^{\infty} p^k (1-p)^2 = 1.$$

If $U(y) = -y\,U(-1)$ for $y = 0, 1, \ldots$ and $U(-1) \in R^1$, then $U(\boldsymbol{y})$ is unbiased with respect to 0.

This can be seen from

$$\sum_{k-1}^{\infty} kx^{k-1} = \frac{1}{(1-x)^2} \text{ for } |x| < 1$$

and

$$E[U(\boldsymbol{y})] = U(-1)\left[p + 0 - (1-p)^2 p \sum_{y=1}^{\infty} y p^{y-1}\right] = 0.$$

However, $E[U(\boldsymbol{y})] = 0$ implies $U(y) = -y\,U(-1)$ for $y = 0, 1, \ldots$, namely, in

$$pU(-1) + (1-p)^2 U(0) + (1-p)^2 p \sum_{y=1}^{\infty} U(y) p^{y-1} = 0,$$

the series converges for $U(y) = y \cdot \text{const}$. Hence, the solution is $U(0) = 0$, $U(y) = -y\,U(-1)$.

Otherwise the series does not converge or converges in dependence on p. Now we consider two special cases:

a) Let $\psi = g(p) = p$. Then, for example, $S_0(y)$ with

$$S_0(y) = \begin{cases} 1 & \text{for } y = -1 \\ 0 & \text{else} \end{cases}$$

is unbiased with respect to p, and $S(y)$ in (2.8) is a variance-optimal unbiased estimator, since it minimises

$$Q = \sum_{y=-1}^{\infty} P(y = y)[S_0(y) + yU(-1)]^2$$

because of (2.9). Fixing $p = p_0$ we get for Q

$$Q_0 = p_0\,[1 - U(-1)]^2 + 0 + \sum_{y=-1}^{\infty} [yU(-1)]^2 p_0^y (1 - p_0)^2.$$

By differentiating Q_0 for $U(-1)$ and putting the derivative equal to 0, we get as variance-optimal value (the second derivative for $U(-1)$ is positive) the minimum at

$$U_0(-1) = \frac{1 - p_0}{2},$$

that is, there is only one variance-optimal unbiased estimator dependent on the parameter value p_0.

The situation is favourably disposed if we consider another function $g(p)$.

b) Let $\psi = g(p) = (1 - p)^2$. Therefore we have to estimate $(1 - p)^2$ (and not p itself) unbiasedly. An unbiased estimator is, for example, $S_0(y)$ with

$$S_0(y) = \begin{cases} 1 & \text{for } y = 0 \\ 0 & \text{else} \end{cases}.$$

Naturally as unbiased estimation of 0, $U(y)$ is the same for all functions g, and analogous to the case (a), we want to determine the minimum of

$$Q_0 = p_0[U(-1)]^2 + (1 - p_0)^2 1^2 + (1 - p_0)\sum_{y=-1}^{\infty} [yU(-1)]^2 p_0^y.$$

Again the second derivative of Q_0 with respect to p_0 is positive. Now the minimum is at $U(-1) = 0$. Consequently $S(y) = S_0(y)$ is the variance-optimal unbiased estimator for $(1 - p)^2$ with respect to each $p_0 \in (0, 1)$.

We want especially to emphasise the property of the estimator in case (b) of Examples 2.3.

Definition 2.3 Let $\boldsymbol{Y} = (y_1, y_2, \ldots, y_n)^T$ be a random sample with components distributed as $P_\theta \in P = (P_\theta, \theta \in \Omega)$, and let $\tilde{S}(\boldsymbol{Y})$ be an unbiased estimator with respect to $g(\theta) = \psi$ with finite variance. Besides, let $D_{\mathrm{E}} \subseteq D'_E$ be the class of all unbiased estimators with finite positive variance and D'_E the class of all unbiased estimators. If

$$\operatorname{var}\left[\tilde{S}(\boldsymbol{Y})\right] = \min_{S(\boldsymbol{Y}) \in D_E} \operatorname{var}_{\theta_0}[S(\boldsymbol{Y})], \theta_0 \in \Omega, \tag{2.10}$$

then $\tilde{S}(\boldsymbol{Y})$ is said to be a locally variance-optimal unbiased estimator (LVUE) at $\theta = \theta_0$.

Definition 2.4 If (2.10) is satisfied for all $\theta_0 \in \Omega$, then $\tilde{S}(\boldsymbol{Y})$ is said to be a uniformly variance-optimal unbiased estimator (UVUE).

The class D_{E} introduced in Definition 2.3 is used in the same sense also in the following. The next theorem contains a necessary and sufficient condition for an estimator to be a UVUE.

Theorem 2.2 Let the components of the random sample $\boldsymbol{Y} = (y_1, y_2, \ldots, y_n)^T$ be distributed as $P_\theta \in P = (P_\theta, \theta \in \Omega)$, and (let be) $S(\boldsymbol{Y}) \in D_E$. Further, let D_E^0 be the class of unbiased estimators with respect to 0 with finite second moment. Then the condition

$$E[S(\boldsymbol{Y})U(\boldsymbol{Y})] = 0 \text{ for all } U(\boldsymbol{Y}) \in D_E^0 \text{ and all } \theta \in \Omega \tag{2.11}$$

is necessary and sufficient for $S(\boldsymbol{Y})$ to be a UVUE with respect to $g(\theta)$.

Proof: If $S(\boldsymbol{Y})$ is a UVUE with respect to $g(\theta)$, then $S^*(\boldsymbol{Y}) = S(\boldsymbol{Y}) + \lambda\, U(\boldsymbol{Y})$ is unbiased with respect to $g(\theta)$ for $U(\boldsymbol{Y}) \in D_E^0, \theta_0 \in \Omega$ and $\lambda \in R^1$. Moreover

$$\operatorname{var}_{\theta_0}[S^*(\boldsymbol{Y})] = \operatorname{var}_{\theta_0}[S(\boldsymbol{Y}) + \lambda U(\boldsymbol{Y})] \geq \operatorname{var}[S^*(\boldsymbol{Y})] \text{ for all } \lambda \in R^1$$

is fulfilled. But then

$$\lambda^2 \operatorname{var}_{\theta_0}[U(\boldsymbol{Y})] + 2\lambda \operatorname{cov}_{\theta_0}[S(\boldsymbol{Y}), U(\boldsymbol{Y})] \geq 0 \text{ for all } \lambda \in R^1$$

follows. Assuming equality, the quadratic equation in λ has the two solutions:

$$\lambda_1 = 0, \lambda_2 = -\frac{2\operatorname{cov}_{\theta_0}[S(\boldsymbol{Y}), U(\boldsymbol{Y})]}{\operatorname{var}_{\theta_0}[U(\boldsymbol{Y})]}.$$

But the expression on the left-hand side of the inequality is only non-negative for arbitrary λ if the condition

$$\operatorname{cov}_{\theta_0}[S(\boldsymbol{Y}), U(\boldsymbol{Y})] = E_{\theta_0}[S(\boldsymbol{Y})U(\boldsymbol{Y})] = 0$$

is satisfied. This derivation is independent of the special parameter value θ_0. Therefore it is true everywhere in Ω.

Reversely, assume that

$$E[S(\boldsymbol{Y})U(\boldsymbol{Y})] = 0$$

is fulfilled for all $U(\boldsymbol{Y}) \in D_E^0$. Besides, let $S'(\boldsymbol{Y})$ be another unbiased estimator with respect to $g(\theta)$. If $S'(\boldsymbol{Y})$ is not in D_E, that is, if it is in $D'_E \setminus D_E$, then trivially $\text{var}[S(\boldsymbol{Y})] < \text{var}[S'(\boldsymbol{Y})]$ holds. Therefore let $S'(\boldsymbol{Y}) \in D_E$. But then $S(\boldsymbol{Y}) - S'(\boldsymbol{Y}) \in D_E^0$ follows since the finite variances of $S(\boldsymbol{Y}) \in D_E$ and $S'(\boldsymbol{Y}) \in D_E$ imply also the finite variance of the difference $S(\boldsymbol{Y}) - S'(\boldsymbol{Y})$ by considering

$$\text{var}[S(\boldsymbol{Y}) - S'(\boldsymbol{Y})] = \text{var}[S(\boldsymbol{Y})] + \text{var}[S'(\boldsymbol{Y})] - 2\,\text{cov}[S(\boldsymbol{Y}), S'(\boldsymbol{Y})].$$

Namely, with $\text{var}[S(\boldsymbol{Y})]$ and $\text{var}[S'(\boldsymbol{Y})]$,the right-hand side of the equation is finite such that the assertion $S(\boldsymbol{Y}) - S'(\boldsymbol{Y}) \in D_E^0$ follows. Moreover, the assumption implies

$$E\{S(\boldsymbol{Y})[S(\boldsymbol{Y}) - S'(\boldsymbol{Y})]\} = E\{\boldsymbol{S}[\boldsymbol{S} - \boldsymbol{S}']\} = 0$$

and

$$E(\boldsymbol{S}^2) = E(\boldsymbol{S}\boldsymbol{S}'),$$

respectively. Now

$$\text{cov}(\boldsymbol{S}, \boldsymbol{S}') = E\{[\boldsymbol{S} - g(\theta)][\boldsymbol{S}' - g(\theta)]\} = E(\boldsymbol{S}\boldsymbol{S}') - g(\theta)^2 = E(\boldsymbol{S}^2) - \psi^2 = \text{var}(\boldsymbol{S}).$$

Observing the inequality of Schwarz, we get

$$[\text{var}(\boldsymbol{S})]^2 = \text{cov}(\boldsymbol{S}, \boldsymbol{S}')^2 \le \text{var}(\boldsymbol{S})\,\text{var}(\boldsymbol{S}')$$

and therefore as asserted

$$\text{var}(\boldsymbol{S}) \le \text{var}(\boldsymbol{S}').$$

We want to demonstrate the consequences of this theorem by returning to Example 2.3.

Example 2.3 (continuation)

Our aim is to determine all UVUE of $g(p)$. Since D_E^0 contains only elements of the form $U(y) = -\boldsymbol{y}\,U(-1)$ and since (2.11) holds, it is necessary and sufficient to be an UVUE that under the assumption $U(-1) \neq 0$ the equality $E_p(S(\boldsymbol{y})) = 0$ is fulfilled for all $p \in (0, 1)$, that is, then $S(\boldsymbol{y})$ belongs to D_E^0 and satisfies therefore the relations $U(\boldsymbol{y}) = S(\boldsymbol{y})\,\boldsymbol{y} = -\boldsymbol{y}\,U(-1) = \boldsymbol{y}\,S(-1)$.

These relations hold if $S(0)$ is an arbitrary real value and $S(y) = S(-1)$ for $y = 1,2, \ldots$.

If we put $S(-1) = a$, $S(0) = b$ (a, b real), then we obtain

$$E_p(\boldsymbol{S}) = pS(-1) + (1-p)^2 S(0) + \sum_{y=1}^{\infty} p^y (1-p)^2 S(-1)$$
$$= pa + b(1-p)^2 + (1-p)^2 a\frac{p}{1-p} = b(1-p)^2 + a\left[1-(1-p)^2\right]$$
$$= a + (b-a)(1-p)^2.$$

Hence, $g(p)$ must be of the form $a + (b - a)(1 - p)^2$, if it is to possess an UVUE, but $g(p) = p$ is not of this form. Therefore it is impossible to find a UVUE.

The following statement is of fundamental significance for estimators belonging to D_E.

Theorem 2.3 (Rao, 1945; Blackwell, 1947; Lehmann and Scheffé, 1950) Let the components of the random sample $\boldsymbol{Y} = (\boldsymbol{y}_1, \boldsymbol{y}_2, \dots, \boldsymbol{y}_n)^T$ be distributed as $P_\theta \in P = (P_\theta, \theta \in \Omega)$, and let $S(\boldsymbol{Y}) \in D_E$ be unbiased with respect to $g(\theta) = \psi$. If there is a sufficient statistic $M(\boldsymbol{Y})$ with respect to P_θ, then the following exists:

$$\hat{\psi}(\boldsymbol{Y}) = E[S(\boldsymbol{Y})|M(\boldsymbol{Y})] = h[M(\boldsymbol{Y})] \tag{2.12}$$

and is unbiased with respect to ψ and

$$\text{var}[\hat{\psi}(\boldsymbol{Y})] \leq \text{var}[S(\boldsymbol{Y})] \text{ for all } \theta \in \Omega.$$

If $M(\boldsymbol{Y})$ is complete (and) minimal sufficient, then $\hat{\psi}(\boldsymbol{Y})$ with probability 1 is the uniquely determined unbiased estimator of $g(\theta)$ with minimal variance for each $\theta \in \Omega$.

Proof: Considering the sufficiency of $M(\boldsymbol{Y})$, the expectation in (2.12) does not depend on θ and consequently is an estimator. Observing that $S(\boldsymbol{Y})$ is unbiased, it follows via

$$E[\hat{\psi}(\boldsymbol{Y})] = E\{E[S(\boldsymbol{Y})|M(\boldsymbol{Y})]\} = E[S(\boldsymbol{Y})] = \psi$$

that $\hat{\psi}(\boldsymbol{Y})$ is unbiased, too. Further we get

$$\text{var}[S(\boldsymbol{Y})] = E\{\text{var}[S(\boldsymbol{Y})|M(\boldsymbol{Y})] + \text{var}\{E[S(\boldsymbol{Y})|M(\boldsymbol{Y})]\}\}.$$

The second summand on the right-hand side of the equation is equal to $\text{var}[\hat{\psi}(\boldsymbol{Y})]$, and the first summand is non-negative. This implies the second part of the assertion.

Now let $M(\boldsymbol{Y})$ be additionally complete and $\hat{\psi}(\boldsymbol{Y}) = h[M(\boldsymbol{Y})]$. Further, let $M^*(\boldsymbol{Y})$ be an arbitrary estimator from D_E dependent on $M(\boldsymbol{Y})$ such that $M^*(\boldsymbol{Y}) = t[M(\boldsymbol{Y})]$. Then for all $\theta \in \Omega$ the statement

$$E[\hat{\psi}(\boldsymbol{Y})] = E[M^*(\boldsymbol{Y})] \text{ and } E\{h[M(\boldsymbol{Y})] - t[M(\boldsymbol{Y})]\} = 0,$$

respectively, holds. As $M(Y)$ is complete, this implies $h = t$ (with probability l). This completes the proof.

Under the assumption that P_θ is a k-parametric exponential family of full rank, it follows from Section 1.3 that it suffices to find an estimator $\boldsymbol{S} \in D_E$ and a vector

$$M(\boldsymbol{Y}) = \{M_1(\boldsymbol{Y}),\dots,M_k(\boldsymbol{Y})\}.^T$$

Via (2.12) the UVUE $\hat{\boldsymbol{\psi}}$ with probability 1 is unique.

We want to demonstrate the applicability of this theorem by examples.

Example 2.4 Let $\boldsymbol{Y} = (\boldsymbol{y}_1, \boldsymbol{y}_2, \dots, \boldsymbol{y}_n)^T$ be a random sample.

a) Let the components of $\boldsymbol{Y}$ be distributed as $N(\mu, 1)$, that is, it is $\theta = \mu$. If $g(\theta) = \mu$, then $\bar{\boldsymbol{y}} \in D_E$. Since $\bar{\boldsymbol{y}}$ is complete minimal sufficient with respect to the $N(\mu, 1)$ family, $\bar{\boldsymbol{y}}$ is with probability 1 the only UVUE with respect to μ.
b) Let the components of $\boldsymbol{Y}$ be distributed as $N(0,\sigma^2)$. Then $\sum_{i=1}^n \boldsymbol{y}_i^2 = \boldsymbol{SQ}_y$ with respect to this family is complete minimal sufficient. It is $\theta = \sigma^2$ and we choose $g(\theta) = \sigma^2$. As $\dfrac{\boldsymbol{SQ}_y}{\sigma^2}$ is distributed as $CS(n)$, $\dfrac{\boldsymbol{SQ}_y}{n}$ is with probability 1 the only UVUE with respect to σ^2.
c) Let the components of $\boldsymbol{Y}$ be distributed as $N(\mu, \sigma^2)$. With $\theta = \begin{pmatrix}\mu\\ \sigma^2\end{pmatrix}$ we put $g(\theta) = \begin{pmatrix}\mu\\ \sigma^2\end{pmatrix} = \theta$. Then $\boldsymbol{H}^T = \left(\sum_{i=1}^n \boldsymbol{y}_i, \sum_{i=1}^n \boldsymbol{y}_i^2\right)$ is complete minimal sufficient with respect to θ. The statistic

$$M = \left[\bar{\boldsymbol{y}}, \sum_{i=1}^{n} (\boldsymbol{y}_i - \bar{\boldsymbol{y}})^2\right]^T$$

is equivalent to $\boldsymbol{H}^T$, meaning that $H(Y_1) = H(Y_2)$ iff $M(Y_1) = M(Y_2)$. This is clear if $\sum_{i=1}^n (\boldsymbol{y}_i - \bar{\boldsymbol{y}})^2 = \sum_{i=1}^n \boldsymbol{y}_i^2 - n\bar{\boldsymbol{y}}^2$ is considered. Therefore $M(Y)$ is also complete minimal sufficient with respect to θ. As $\frac{1}{\sigma^2}\sum_{i=1}^n (\boldsymbol{y}_i - \bar{\boldsymbol{y}})^2$ is distributed as $CS(n-1)$, it follows that $(\bar{\boldsymbol{y}}, \boldsymbol{s}^2)$ with $\boldsymbol{s}^2 = \frac{1}{n-1}\sum_{i=1}^n (\boldsymbol{y}_i - \bar{\boldsymbol{y}})^2$ with probability 1 is the only UVUE with respect to (μ, σ^2).

Example 2.5 Let the components of a random sample $\boldsymbol{Y} = (\boldsymbol{y}_1, \boldsymbol{y}_2, \dots, \boldsymbol{y}_n)^T$ be two-point distributed. W.l.o.g. we assume $P(\boldsymbol{y} = 0) = 1 - p$, $P(\boldsymbol{y} = 1) = p$; $0 < p < 1$. The likelihood function

$$L(Y,p) = p^{\sum_{i=1}^n y_i}(1-p)^{n-\sum_{i=1}^n y_i}$$

shows that the distribution of $\boldsymbol{Y}$ belongs to a one-parametric exponential family and $M(Y) = \sum_{i=1}^n \boldsymbol{y}_i$ is complete sufficient. Because of $E(\boldsymbol{y}_i) = p$, we have

$E[M(\boldsymbol{Y})] = np$, and $\dfrac{M(\boldsymbol{Y})}{n} = \bar{y}$ is UVUE with respect to $p = g(\theta)$. Observing that $\boldsymbol{M} = M(\boldsymbol{Y})$ is distributed as $B(n, p)$ and assuming $g(\theta) = \text{var}(\boldsymbol{y}_i) = p(1-p)$, the estimator $S(\boldsymbol{Y}) = \dfrac{1}{n-1}(1-p)\boldsymbol{M}$ because of

$$E\left(\bar{y}-\bar{y}^2\right) = E(\bar{y}) - E\left(\bar{y}^2\right) = p - \left[\text{var}(\bar{y}) + p^2\right]$$

is unbiased with respect to $p(1-p)$ and therefore UVUE. Considering $\text{var}(\bar{y}) = \dfrac{p(1-p)}{n}$ and consequently

$$E\left[\frac{1}{n-1}(1-\bar{y})\boldsymbol{M}\right] = \frac{n}{n-1}E[\bar{y}(1-\bar{y})] = \frac{n}{n-1}\left[\frac{n-1}{n}\left(p-p^2\right)\right] = p(1-p).$$

$S(\boldsymbol{Y})$ with probability 1 is the uniquely determined UVUE.

Example 2.6 Let the components of the random sample $\boldsymbol{Y} = (y_1, y_2, \ldots, y_n)^T$ be distributed as $N(\mu, \sigma^2)$, where μ is known. Further, put $g(\theta) = \sigma^t$ $(t = 1,2, \ldots)$. The estimator $S(\boldsymbol{Y}) = \sum_{i=1}^{n}(y_i-\mu)^2$ is complete minimal sufficient, and $\boldsymbol{X}^2 = \dfrac{1}{\sigma^2}\boldsymbol{S}(\boldsymbol{Y})$ is $CS(n)$-distributed. The components and moments of $\boldsymbol{X}^2$ are only dependent on n, that is, we have

$$E\left(\boldsymbol{X}^{2r}\right) = c(n, 2r) \text{ and } E[S^r(\boldsymbol{Y})] = \sigma^{2r}c(n, 2r),$$

respectively. Hence, $[S(\boldsymbol{Y})]^{\frac{t}{2}}\dfrac{1}{c(n,t)}$ is UVUE with respect to σ^t.

The factor $c(n, 2r)$ is known from probability theory, which is

$$c(n, 2r) = \frac{\Gamma\left(\frac{n}{2}+r\right)2^r}{\Gamma\left(\frac{n}{2}\right)}. \tag{2.13}$$

For $t = 1$ and $r = ½$, respectively, the UVUE with respect to σ is obtained from

$$\frac{\sqrt{S(\boldsymbol{Y})}}{c(n,1)} = \frac{\Gamma\left(\frac{n}{2}\right)\sqrt{\sum_{i=1}^{n}(y_i-\mu)^2}}{\sqrt{2}\Gamma\left(\frac{n+1}{2}\right)}. \tag{2.14}$$

For $t = 2$ and $r = 1$, respectively, the UVUE with respect to σ^2 results from

$$\frac{S(\boldsymbol{Y})}{c(n,2)} = \frac{1}{n}\sum_{i=1}^{n}(y_i-\mu)^2.$$

However, if μ is unknown, then $(\bar{\boldsymbol{y}}, (\boldsymbol{n}-1)\boldsymbol{s}^2)$ is complete minimal sufficient with respect to $\theta^T = (\mu, \sigma^2)$ according to Example 2.4, and $\bar{\boldsymbol{y}}$ is UVUE with respect to μ. Since $\frac{(n-1)s^2}{\sigma^2}$ is distributed as $CS(n-1)$, the UVUE with respect to σ^2 is obtained by

$$E\left(\boldsymbol{s}^{2r}\right) = \frac{2^r\Gamma\left(\frac{n-1}{2}+r\right)}{\Gamma\left(\frac{n-1}{2}\right)(n-1)^r}\sigma^{2r}. \tag{2.15}$$

For $r = \frac{1}{2}$ this implies

$$E(\boldsymbol{s}) = \frac{\sqrt{2}\Gamma\left(\frac{n}{2}\right)}{\Gamma\left(\frac{n-1}{2}\right)\sqrt{n-1}}\sigma,$$

such that

$$\boldsymbol{s}\frac{\Gamma\left(\frac{n-1}{2}\right)\sqrt{n-1}}{\sqrt{2}\Gamma\left(\frac{n}{2}\right)} \tag{2.16}$$

is UVUE with respect to σ. For $r = 1$ the estimator $\boldsymbol{s}^2$ is UVUE with respect to σ^2.

Example 2.7 Let the component of a random sample $\boldsymbol{Y} = (\boldsymbol{y}_1, \boldsymbol{y}_2, \ldots, \boldsymbol{y}_n)^T$ be distributed as $P(\lambda)$, $0 < \lambda < \infty$. Now we want to estimate $g(\lambda) = \frac{e^{-\lambda}\lambda^k}{k!}$ ($k = 0, 1, 2, \ldots$), a value of the probability function for a given k. An unbiased estimator based on the first element $\boldsymbol{y}_1$ of a random sample $\boldsymbol{Y}$ with $\boldsymbol{I} = \{\boldsymbol{Y}, \boldsymbol{y}_1 = \boldsymbol{k}\}$ is given by

$$S_1(\boldsymbol{Y}) = g(\boldsymbol{y}_1)\,I(\boldsymbol{Y})\ (k = 0,1,2,\ldots),$$

that is, $S_1(\boldsymbol{Y})$ is equal to $\frac{e^{-\lambda}\lambda^k}{k!}$ for $y_1 = k$ and 0 else. As $M(\boldsymbol{Y}) = \sum_{i=1}^n \boldsymbol{y}_i$ is complete minimal sufficient with respect to λ, we can determine a UVUE $S_2(\boldsymbol{Y})$ according to

$$S_2(\boldsymbol{Y}) = E[S_1(\boldsymbol{Y})|M(\boldsymbol{Y})] = P[\boldsymbol{y}_1 = k|M(\boldsymbol{Y})].$$

For all $M = M(\boldsymbol{Y})$ the conditional distribution of $\boldsymbol{y}_1$ for a given value of M is a binomial distribution with $n = M$ and $p = 1/n$. This is not difficult to see. Since the y_i are independent and identically distributed, for a fixed sum M, each $\boldsymbol{y}_i$ takes the value a ($a = 0, \ldots, M$) with probability

$$\binom{M}{a}\left(\frac{1}{n}\right)^{a}\left(1-\frac{1}{n}\right)^{n-a}.$$

Therefore the UVUE is equal to

$$S_2(\boldsymbol{Y})=\begin{cases} 0 \ \text{for}\, M(Y)=0, k>0 \\ 1 \ \text{for}\, M(Y)=0, k=0 \\ \binom{M}{k}\left(\frac{1}{n}\right)^{k}\left(\frac{n-1}{n}\right)^{n-k} \ \text{for}\, M(Y)>0, k=0,\dots,M \end{cases}.$$

Theorem 2.4 If a random sample $\boldsymbol{Y}=(\boldsymbol{y}_1,\boldsymbol{y}_2,\dots,\boldsymbol{y}_n)^T$ is $N(\mu,\Sigma)$-distributed with $\mu=(\mu_1,\dots,\mu_k)^{\mathrm{T}}\in R^k$, $\mathrm{rk}(\Sigma)=k$ and $|\Sigma|>0$, then the UVUE of the $\frac{k(k+3)}{2}$-dimensional parameter vector

$$\theta=\left(\mu_1,\dots,\mu_k,\sigma_1^2,\dots,\sigma_k^2,\sigma_{i,j}\right)^T, i<j, j=2,\dots,k$$

based on a random sample $\boldsymbol{X}=(\boldsymbol{Y}_1,\dots,\boldsymbol{Y}_{\boldsymbol{n}})^T$ with components $\boldsymbol{Y}_i$ distributed as $\boldsymbol{Y}$ is given by

$$\boldsymbol{\theta}=\left(\boldsymbol{y}_{1.},\dots,\bar{\boldsymbol{y}}_{k.},\boldsymbol{s}_1^2,\dots,\boldsymbol{s}_k^2,\boldsymbol{s}_{i,j}\right)^T, i<j, j=2,\dots,k$$

where

$$\bar{\boldsymbol{y}}_{i.}=\frac{1}{n}\sum_{j=1}^{n}\boldsymbol{y}_{ij}, \boldsymbol{s}_i^2=\frac{1}{n-1}\sum_{j=1}^{\boldsymbol{n}}\left(\boldsymbol{y}_{\boldsymbol{ij}}-\bar{\boldsymbol{y}}_{i.}\right)^2,$$

$$\boldsymbol{s}_{jk}=\frac{1}{n-1}\sum_{\boldsymbol{i}=1}^{\boldsymbol{n}}\left(\boldsymbol{y}_{\boldsymbol{ij}}-\bar{\boldsymbol{y}}_{j.}\right)(\boldsymbol{y}_{\boldsymbol{ik}}-\bar{\boldsymbol{y}}_{k.}).$$

Proof: If $k=2$, then the family of two-dimensional normal distributions with positive definite covariance matrix is a five-parametric exponential family, and

$$M_2(\boldsymbol{X})=(\bar{\boldsymbol{y}}_{1.},\bar{\boldsymbol{y}}_{2.},\boldsymbol{SS}_1,\boldsymbol{SS}_2,\boldsymbol{SP}_{12})$$

is a complete minimal sufficient statistic with respect to this family, where

$$\boldsymbol{SS}_{\boldsymbol{i}}=\sum_{\boldsymbol{j}=1}^{\boldsymbol{n}}\left(\boldsymbol{y}_{\boldsymbol{ij}}-\bar{\boldsymbol{y}}_{i.}\right)^2, \ i=1,2$$

and

$$\boldsymbol{SP}_{1,2}=\sum_{\boldsymbol{j}=1}^{\boldsymbol{n}}\left(\boldsymbol{y}_{1j}-\bar{\boldsymbol{y}}_{1.}\right)\left(\boldsymbol{y}_{2j}-\bar{\boldsymbol{y}}_{2.}\right).$$

The marginal distributions of y_{ij} , $i = 1 , 2$ are $N(\mu_i, \sigma_i^2)$-distributed with the UVUE $\bar{y}_{i.}$ and

$$s_i^2 = \frac{1}{n-1} SS_i.$$

If we can show that $E(SP_{1,2}) = (n-1)\sigma_{1,2}$, then the proof of the theorem is completed for $k = 2$, because the five parameters are then estimated by an unbiased estimator only depending on the sufficient statistic $M_2(X)$.

But by definition it is now

$$\sigma_{1,2} = E[(y_1 - \mu_1)(y_2 - \mu_2)] = E(y_1 y_2) - \mu_1 \mu_2.$$

Then, we have

$$SP_{1,2} = \sum_{j=1}^{n} y_{1j} y_{2j} - n \bar{y}_{1.} \bar{y}_{2.}$$

and

$$E(SP_{1,2}) = \sum_{j=1}^{n} E\left(y_{1j} y_{2j}\right) - n E(\bar{y}_{1.} \bar{y}_{2.})$$

$$= n(\sigma_{12} + \mu_1 \mu_2) - \frac{1}{n}[n(\sigma_{12} + \mu_1 \mu_2) + n(n-1)\mu_1 \mu_2] = \sigma_{12}(n-1),$$

since

$$E\left(y_{1j} y_{2j}\right) = \mu_1 \mu_2$$

taking into account that the y_{1j} and y_{2j} are independent for $i \neq j$.

Now we consider the case $k > 2$ where X follows a $\frac{k(k+3)}{2}$-parametric exponential family with an analogously to the case $k = 2$ defined complete minimal sufficient statistic

$$M_k(X) = (\bar{y}_1, \ldots, \bar{y}_k, SS_1, \ldots, SS_k, SP_{12}, \ldots, SP_{k-1,k})^T.$$

All $\frac{k(k+3)}{2}$ parameters can be estimated from two-dimensional marginal distributions unbiased and only depending on $M_k(X)$. This finishes the proof.

Sometimes it is indicated to compare the variance of any estimator from D_E to the variance of the UVUE. This leads to a new definition.

Definition 2.5 Let $S_0(Y)$ and $S(Y)$ be estimators from D_E with respect to $g(\theta)$ and let $S_0(Y)$ be a UVUE. Then the ratio

$$E_0 = \frac{\operatorname{var}(S_0(Y))}{\operatorname{var}(S(Y))}$$

is called the relative efficiency of $S(Y)$. All UVUE are called efficient estimators.

If there is no UVUE, then we often look for the best linear or the best quadratic estimators, meaning that a statistic has to be minimised in the class D_L of linear, in the class D_{LE} of linear unbiased, in the class D_Q of quadratic or in the class D_{QE} of quadratic unbiased estimators, respectively. The best linear or quadratic estimators and the best linear predictions are treated in the chapters about linear models. Linear estimators are used to estimate fixed effects in linear models. However, quadratic estimators are suitable to estimate variance components of random effects in linear models.

2.2 Variance-Invariant Estimation

In applications of statistics, measurements are carried out within a certain scale, which is sometimes chosen arbitrarily. In the biological testing for active agents, for instance, concentrations of solutions are registered directly using a logarithmic scale. Temperatures are given in degrees with respect to the Celsius, Fahrenheit, Réaumur or Kelvin scales. Angles are measured in degrees or radians. Assume now that two methods of measuring differ only by an additive constant c, such that $y_i^* = y_i + c$ holds for the realisations of random samples $\boldsymbol{Y}^*$ and $\boldsymbol{Y}$, respectively. If components $\boldsymbol{y}_i^*$ and $\boldsymbol{y}_i$ of these random samples are distributed as P_{θ^*} and P_θ, respectively, and if $\theta^* = \theta + c$, then the relations

$$S(\boldsymbol{Y}^*) = S(\boldsymbol{Y}) + c$$

and

$$\text{var}[S(\boldsymbol{Y}^*)] = \text{var}[S(\boldsymbol{Y})]$$

are fulfilled. The variances of the estimators are equal in both problems, and we say that the problem of estimation is variance-invariant with respect to translations.

Definition 2.6 Let a random variable $\boldsymbol{y}$ be distributed as $P_\theta \in P = (P_\theta, \theta \in \Omega)$ and take values $y \in \{Y\}$ in the sample space $\{Y\}$. Further, let h be a measurable one-to-one mapping of $\{Y\}$ onto $\{Y\}$ such that for each $\theta \in \Omega$ the distribution P_{θ^*} of $h(\boldsymbol{y}) = \boldsymbol{z}$ is in $P = (P_\theta, \theta \in \Omega)$, too, where $\tilde{h}(\theta) = \theta^*$ covers with θ the whole parameter space. Then we say that $P_\theta \in P = (P_\theta, \theta \in \Omega)$ is invariant relative to h, where $\tilde{h}$ is the mapping from Ω onto itself induced by h. If $\{T\}$ is a class of transformations such that $P_\theta \in P = (P_\theta, \theta \in \Omega)$ is invariant relative to the whole class, and if $H(\{T\}) = H$ is the set of all transformations, arising by taking all finite products of transformations in $\{T\}$ and their inverses, then $P_\theta \in P = (P_\theta, \theta \in \Omega)$ is invariant relative to H, and H is the group induced by $\{T\}$.

W.l.o.g. we will therefore assume in the following that a class of transformations is a group, where the operation is the product (or concatenation, the stepwise application of transformations). In the following let H be a group of one-to-one mappings of $\{T\}$ onto itself. If $P = (P_\theta, \theta \in \Omega)$ is invariant relative to H, then we obtain

$$E_\theta[h(y)] = E_{\tilde{h}(\theta)}[y] \tag{2.17}$$

because

$$E_\theta[h(y)] = \int_{\{Y\}} h(y)dP_\theta = \int_{\{Y\}} ydP_{\tilde{h}(\theta)} = E_{\tilde{h}(\theta)}[y].$$

Example 2.8 The family of normal distributions $N(\mu,\sigma^2)$, $\theta^T = (\mu, \sigma^2)$ is invariant relative to the group of real affine transformations. Namely, if $\boldsymbol{z} = h(\boldsymbol{y}) = a + b\boldsymbol{y}$ $(a, b \in R^1)$ and if $\boldsymbol{y}$ is distributed as $N(\mu,\sigma^2)$, then it is known that $\boldsymbol{z}$ is distributed as $N(\mu^*,\sigma^{*2})$ with $\mu^* = a + b\mu$, $\sigma^{*2} = b^2\sigma^2$.

Moreover, $\theta^{*\mathrm{T}} = (\mu^*,\sigma^{*2})$ covers with θ^{T} for fixed a and b the whole set Ω.

Definition 2.7 Let $Y = (\boldsymbol{y}_1, \boldsymbol{y}_2, \dots, \boldsymbol{y}_n)^T$ be a random sample with realised components $y_i \in \{Y\}$ and let H be a group of transformations as described before. Further, let $\boldsymbol{y}_i$ be distributed as $P_\theta \in P = (P_\theta, \theta \in \Omega)$ and assume that S is invariant relative to H. Then a statistic $M(Y) = M(\boldsymbol{y}_1, \dots, \boldsymbol{y}_n)$ is said to be invariant relative to H if

$$\tilde{h}[M(\boldsymbol{Y})] = M(\boldsymbol{y}_1,\dots,\boldsymbol{y}_n) \tag{2.18}$$

holds with $\tilde{h}(M) = M[h(y_1),\dots,h(y_n)]$ for all $h \in H$. If $M(Y)$ is an estimator and

$$h[M(\boldsymbol{Y})] = \tilde{h}[M(\boldsymbol{Y})] \tag{2.19}$$

is fulfilled, then $M(Y)$ is said to be equivariant (relative to H).

The induced transformations $\tilde{h}$ from Ω onto Ω introduced in Definition 2.6 constitute a group $\tilde{H}$ if h runs through the whole group H.

Let the components of the random sample $\boldsymbol{Y}$ be distributed as $N(\mu,\sigma^2)$, and let H be the group of real affine transformations introduced in Example 2.8. The (minimal sufficient and complete) estimator $S^T(Y) = (\bar{\boldsymbol{y}}_., \boldsymbol{s}^2)$ is equivariant, because, according to Example 2.8, we have

$$\tilde{h}(\theta^T) = \theta^{*T} = (a + b\mu, b^2\sigma^2) \text{ and } \tilde{h}[S^T(\boldsymbol{Y})] = (a + b\bar{\boldsymbol{y}}, b^2\boldsymbol{s}^2).$$

Definition 2.8 If $g(\theta) = \psi \in Z$ is to be estimated in a problem of estimation, if besides $g(\theta_1) = g(\theta_2)$ implies $g\left[\tilde{h}(\theta_1)\right] = g\left[\tilde{h}(\theta_2)\right]$ for all $h \in H$ and if finally

$$\left\|g\left[\tilde{h}(\theta)\right] - \tilde{h}[S(Y)]\right\|^2 = \|g(\theta) - S(Y)\|^2$$

holds for each estimator $S(\boldsymbol{Y}) \in D_E$ with respect to $g(\theta)$ and for all $h \in H, \theta \in \Omega$, then the problem of estimation is said to be invariant relative to H (with respect to the quadratic loss).

Theorem 2.5 If $S(\boldsymbol{Y})$ is an equivariant estimator with finite variance in a problem of estimation, which is invariant relative to a group H of transformations, then

$$\text{var}_{\tilde{h}(\theta)}[S(\boldsymbol{Y})] = \text{var}_\theta[S(\boldsymbol{Y})]. \tag{2.20}$$

Proof: Observing

$$\text{var}_{\tilde{h}(\theta)}[S(\boldsymbol{Y})] = \int_{\{Y\}} \left\|g\left[\tilde{h}(\theta)\right] - \tilde{h}[S(Y)]\right\|^2 dP_{\tilde{h}(\theta)} = E_{\tilde{h}(\theta)}\left\{\left\|g\left[\tilde{h}(\theta)\right] - \tilde{h}[S(\boldsymbol{Y})]\right\|^2\right\}$$

from (2.17) and the invariance of the estimation problem, the assertion follows that

$$\text{var}_{\tilde{h}(\theta)}[S(\boldsymbol{Y})] = E_\theta\left\{\|g(\theta) - S(\boldsymbol{Y})\|^2\right\} = \text{var}_\theta[S(\boldsymbol{Y})].$$

Corollary 2.1 Under the assumptions of Theorem 2.5, the variance of all with respect to H equivariant estimators in Ω is constant (i.e. independent of θ), if the group H is transitive over Ω.

Proof: The transitivity of a transformation group H over Ω means that to any pair $(\theta_1, \theta_2) \in \Omega$, there exists a transformation $\tilde{h} \in H$, which transfers θ_1 into θ_2. Then it follows by Theorem 2.5 for each such pair

$$\text{var}_{\theta_1}[S(\boldsymbol{Y})] = \text{var}_{\theta_2}[S(\boldsymbol{Y})] = \text{const}.$$

Let $P = (P_\theta, \theta \in \Omega)$ be a group family of distributions, which is invariant relative to a group H of transformations for which $\tilde{H}$ acts transitive over Ω and additionally where $g(\theta_1) \neq g(\theta_2)$ always implies $\tilde{h}[g(\theta_1)] \neq \tilde{h}[g(\theta_2)]$.

If the distribution of $\boldsymbol{y}$ is given by P_{θ_0} with arbitrary $\theta_0 \in \Omega$, then the set of distributions induced by $h(y)$ with $h \in H$ is just the group family $(P_\theta, \theta \in \Omega)$. Therefore a group family is invariant relative to the group of transformations defining this family. Hence, especially the location families are invariant relative to translations.

Now we look for equivariant estimators with minimal mean square deviation. If D_A is the class of equivariant estimators with existing second moment and if for R in (2.2) the mean square deviation $MSD[S(Y)]$ is chosen, then an estimator $S_0(Y) \in D_A$ satisfying

$$MSD[S_0(Y)] = \inf_{S(Y) \in D_A} MSD[S(Y)]$$

is called equivariant estimator with minimal *MSD*.

Example 2.9 Let the components of a random sample Y be distributed as $N(\mu,\sigma^2)$ where $\theta^T = (\mu, \sigma^2) \in \Omega$. As in Example 2.8 let H be the group of affine transformations. The statistic $M^T(Y) = (\bar{y}, SS_y)$ with $SS_y = \sum_{i=1}^n (y_i - \bar{y})^2$ is minimal sufficient with respect to θ. Let $\hat{\mu}(\bar{y}, SS_y)$ be equivariant for $g_1(\theta) = \mu$, and let $\hat{\sigma}^2(\bar{y}, SS_y)$ be equivariant for $g_2(\theta) = \sigma^2$, that is, for all $h \in H$ and $\tilde{h} \in H$ the relations

$$h[\hat{\mu}(\bar{y}, SS_y)] = a + b\hat{\mu}(\bar{y}, SS_y) = \hat{\mu}(a + b\bar{y}, b^2 SS_y) = \tilde{h}[\hat{\mu}(\bar{y}, SS_y)]$$

and analogously

$$\tilde{h}[\hat{\sigma}^2(\bar{y}, SS_y)] = b^2\hat{\sigma}^2(\bar{y}, SS_y)$$

are fulfilled. If we put $a = -\bar{y}$, $b = 1$, then $\tilde{h}[M^T(Y)] = (0, SS_y)$ arises, and we can write all equivariant estimators with respect to μ as $\hat{\mu}(\bar{y}, SS_y) = \bar{y} + w(SS_y)$ and all equivariant estimators with respect to σ^2 as $\hat{\sigma}^2(\bar{y}, SS_y) = \alpha SS_y$ with suitably chosen w and α. Since $\bar{y}$ and SS_y are independent,

$$MSD[\bar{y} + w(SS_y)] = \text{var}(\bar{y}) + w^2 E[SS_y]$$

holds. This becomes minimal for $SS_y = 0$, such that $\bar{y}$ is the equivariant estimator with minimal *MSD* with respect to μ. Nonetheless it is

$$MSD[\alpha SS_y] = E\left[(\alpha SS_y - \sigma^2)^2\right] = \alpha^2 E(SS_y{}^2) - 2\alpha\sigma^2 E(SS_y) + \sigma^4.$$

Since $\frac{SS_y}{\sigma^2}$ is distributed as $CS(n-1)$, we get $E(SS_y) = (n-1)\sigma^2$, $\text{var}(SS_y) = 2(n-1)\sigma^4$ and consequently

$$MSD[\alpha SS_y] = \alpha^2\sigma^4(n-1)(n+1) - 2\alpha\sigma^4(n-1) + \sigma^4.$$

This expression becomes minimal for $\alpha = \frac{1}{n+1}$. Therefore

$$\breve{\sigma}^2 = \frac{SS_y}{n+1}$$

is the equivariant estimator with minimal *MSD* with respect to σ^2.

2.3 Methods for Construction and Improvement of Estimators

In Sections 2.1 and 2.2 we checked estimators whether they fulfilled certain optimality criteria. But first we need one or more estimators at our disposal for use in applications. In the following we want to consider methods for constructing estimators.

2.3.1 Maximum Likelihood Method

We assume now that the likelihood function $L(Y, \theta)$ for all $Y \in \{Y\}$ has a uniquely determined supremum with respect to $\theta \in \Omega$. The reader should be not confused by the double meaning of $L(.,.)$, namely, both for the loss function and the likelihood function – but this is common use in the statistical community.

Definition 2.9 Fisher (1925)
Let the components of the random sample $\boldsymbol{Y}$ be distributed as $P_\theta \in P = (P_\theta, \theta \in \Omega)$ and let $L(Y,\theta)$ be the corresponding likelihood function.

An estimator $S_{ML}(\boldsymbol{Y})$ is said to be the maximum likelihood estimator or shortly *ML* estimator (MLE) with respect to $g(\theta) = \psi \in Z$, if its realisation is defined for each realisation Y of $\boldsymbol{Y}$ by

$$L\{Y, g[S_{ML}(Y)]\} = \max_{\psi \in Z} L(Y,\psi) = \max_{\theta \in \Omega} L[Y, g(\theta)]. \tag{2.21}$$

It is obvious that equivalent likelihood functions imply the same set of MLE. Many standard distributions possess, as generally supposed in this section, exactly one MLE. Sometimes their determination causes considerable numerical problems. For exponential families the calculations can be simplified by looking for the supremum of $\ln L(Y,\theta)$ instead of $L(Y,\theta)$. Since the logarithmic function is strictly monotone increasing, this implies the same extremal θ.

If the distribution of the components of a random sample $\boldsymbol{Y}$ follows a k-parametric exponential family with a natural parameter η, then

$$\ln L(Y,\eta) = \sum_{i=1}^{k} \eta_j M_j(Y) - nA(\eta) + \ln h(Y)$$

with $M_j(Y) = \sum_{i=1}^{n} M_j(y_i)$.

For $\psi = g(\theta) = \eta$ with $\eta = (\eta_1, \dots, \eta_k)^T$, the MLE of η is obtained by solving the simultaneous equations

$$\frac{\partial}{\partial \eta_j} A(\eta) = \frac{1}{n} \sum_{i=1}^{n} M_j(y_i), j = 1, \dots, k \tag{2.22}$$

if $A(\eta)$ is partially differentiable relative to η_j. If the expectations of $M_j(\boldsymbol{Y})$ exist for a random sample $\boldsymbol{Y}$, then

$$\frac{\partial}{\partial \eta_j} A(\eta) = E\left[M_j(\boldsymbol{y}_i)\right]$$

and

$$E\left\{\frac{\partial}{\partial \eta_j} A[S_{ML}(\boldsymbol{Y})]\right\} = \frac{\partial}{\partial \eta_j} A(\eta),\ j = 1,\ldots,k$$

follow. Moreover, if $A(\eta)$ is twice partially differentiable relative to the coordinates of η and if the matrix

$$\left(\frac{\partial^2 A(\eta)}{\partial \eta_j \partial \eta_l}\right)$$

of these partial derivatives is positive definite at $\eta = S_{ML}(Y)$ (for all $\eta = \psi \in Z$), then (2.22) has the unique solution $S_{ML}(Y)$, which is minimal sufficient.

Example 2.10 Let $\boldsymbol{Y} = (\boldsymbol{y}_1, \boldsymbol{y}_2, \ldots, \boldsymbol{y}_n)^T$ be a random sample with components $\boldsymbol{y}_i$ satisfying a two-point distribution, where $\boldsymbol{y}_i$ ($i = 1, \ldots, n$) takes the value 1 with the probability p and the value 0 with the probability $1 - p$ ($\Omega = (0;1), \theta = p$). Then by putting $\boldsymbol{y} = \sum_{i=1}^{n} \boldsymbol{y}_i$ and $g(p) = p$, we get

$$L(Y,p) = \prod_{i=1}^{n} p^{y_i}(1-p)^{1-y_i} = p^y(1-p)^{n-y},\ y = 0,1,\ldots,n.$$

This likelihood function is equivalent to the likelihood function of a random sample Y of size 1 distributed as $B(n, p)$. By setting the derivative

$$\frac{\partial \ln L(Y,p)}{\partial p} = \frac{y}{p} - \frac{n-y}{1-p}$$

equal to 0, we get the solution $p = \frac{y}{n}$, which supplies a maximum of L as the second derivative of $\ln L$ relative to p is negative. Therefore the uniquely determined ML estimator is

$$S_{ML}(\boldsymbol{Y}) = \frac{\boldsymbol{y}}{n} = \hat{\boldsymbol{p}}.$$

Example 2.11 Let the components of a random sample $\boldsymbol{Y}$ be distributed as $N(\mu, \sigma^2)$ and let $\theta = (\mu, \sigma^2)^T \in \Omega$. Then we obtain

$$\ln L(Y,\theta) = -\frac{n}{2}\ln 2\pi - \frac{n}{2}\ln\sigma^2 - \frac{1}{2\sigma^2}\sum_{i=1}^{n}(y_i - \mu)^2.$$

We consider two cases.

a) Let $g(\theta) = \theta$. By partial differentiation we get

$$\frac{\partial \ln L(Y,\theta)}{\partial \mu} = \frac{1}{\sigma^2}\sum_{i=1}^{n}(y_i-\mu),$$

$$\frac{\partial \ln L(Y,\theta)}{\partial \sigma^2} = -\frac{n}{2\sigma^2} + \frac{1}{2\sigma^4}\sum_{i=1}^{n}(y_i-\mu)^2.$$

After putting both right-hand sides of these equations equal to 0, we arrive at the unique solution

$$S(\boldsymbol{Y}) = \left[\bar{y}, \frac{1}{n}\sum_{i=1}^{n}(y_i-\bar{y})^2\right]^T$$

of this system. Since the matrix of second partial derivatives of ln L is negative definite at this point, we have the ML estimator

$$S_{ML}(\boldsymbol{Y}) = \left[\bar{y}, \frac{1}{n}\sum_{i=1}^{n}(y_i-\bar{y})^2\right]^T = \left(\tilde{\mu}, \tilde{\sigma^2}\right)^T.$$

b) Let $g(\theta) = (\mu,\sigma)^T$. By partial derivation we obtain instead of

$$\frac{\partial \ln L(Y,\theta)}{\partial \sigma^2} = -\frac{n}{2\sigma^2} + \frac{1}{2\sigma^4}\sum_{i=1}^{n}(y_i-\mu)^2$$

now as second equation

$$\frac{\partial \ln L(Y,\theta)}{\partial \sigma} = -\frac{n}{\sigma} + \frac{1}{\sigma^3}\sum_{i=1}^{n}(y_i-\mu)^2.$$

Setting again the partial derivatives to 0 and solving the system, we find the ML estimator

$$S_{ML}(\boldsymbol{Y}) = \left[\bar{y}, \sqrt{\frac{1}{n}\sum_{i=1}^{n}(y_i-\bar{y})^2}\right]^T = (\tilde{\mu}, \tilde{\sigma})^T.$$

Since the $N(\mu,\sigma^2)$ distributions form an exponential family with the natural parameters

$$\eta_1 = \frac{\mu}{\sigma^2},\ \eta_2 = -\frac{1}{2\sigma^2},$$

the function ln $L(Y,\theta)$ can be written as

$$\ln L(Y,\theta) = \ln L^*(Y,\eta) = -\frac{n}{2}\ln 2\pi + \eta_1 M_1 + \eta_2 M_2 - nA(\eta),$$

where

$$M_1 = \sum_{i=1}^{n} y_i, M_2 = \sum_{i=1}^{n} y_i^2 \text{ and } A(\eta) = -\frac{\eta_1^2}{4\eta_2} + \frac{1}{2}\ln\left(-\frac{1}{2\eta_2}\right).$$

If we partially differentiate $\ln L^*(Y,\eta)$ relative to η_1 and η_2, then we get with $S_{ML}^T(Y) = (\tilde{\eta}_1, \tilde{\eta}_2)$ after putting the partial derivatives to 0 and solving the system the expressions

$$\frac{1\tilde{\eta}_1}{2\tilde{\eta}_2} = -\frac{M_1}{n}, -\frac{\widetilde{\eta_1\eta}^2}{4\tilde{\eta}_2{}^2} = \frac{1}{2\tilde{\eta}_2} = -\frac{M_2}{n}$$

and finally the solutions

$$\tilde{\eta}_1 = \frac{\tilde{\mu}}{\tilde{\sigma}^2}, \tilde{\eta}_2 = -\frac{1}{2\tilde{\sigma}^2}.$$

Since the matrix $\left(\frac{\partial^2 A(\eta)}{\partial\eta_1\partial\eta_2}\right)$ is positive definite, the estimators $(\tilde{\eta}_1, \tilde{\eta}_2)$ and $(\tilde{\mu}, \tilde{\sigma}^2)$, respectively, are minimal sufficient.

Often numerical problems occur if the equations that have to be solved are non-linear or if even the function $L(Y,\theta)$ cannot be differentiated with respect to θ.

As a consequence of the decomposition theorem (Theorem 1.1) the following statement is given.

Theorem 2.6 If the statistic $M(\boldsymbol{Y})$ is sufficient with respect to P_θ under the conditions of Definition 2.9, then a *ML* estimator $S_{ML}(\boldsymbol{Y})$ with respect to θ only depends on $M(\boldsymbol{Y})$.

2.3.2 Least Squares Method

If the form of a distribution function from the family $P = (P_\theta, \theta \in \Omega)$ is unknown or (as in the case of non-parametric families) not sufficiently specified, then the maximum likelihood method does not apply. Concerning the following method, we need a model for the components $\boldsymbol{y}_i$ of the random sample. Then the problem of estimation consists in the estimation of the model parameters. We now write for the components

$$\boldsymbol{y}_i = E(\boldsymbol{y}_i) + \boldsymbol{e}_i = f(\theta) + \boldsymbol{e}_i \tag{2.23}$$

with an unknown real function f and with random 'errors' $\boldsymbol{e}_i$. Hence, we suppose that we know a parametric model $f(\theta)$ for the expectations $E(\boldsymbol{y}_i)$ of $\boldsymbol{y}_i$. So we have to estimate the model parameter θ and, if necessary, the distribution parameters of $\boldsymbol{e}_i$.

Since $\boldsymbol{Y} = (\boldsymbol{y}_1, \boldsymbol{y}_2, \dots, \boldsymbol{y}_n)^T$ is a random sample, all $\boldsymbol{e}_i$ have the same distribution and are (stochastically) independent, that is, $\boldsymbol{e} = (\boldsymbol{e}_1, \dots, \boldsymbol{e}_n)^T$ is a vector of identically and independently distributed components.

Besides we assume that $E(\boldsymbol{e}_i) = 0$. We restrict ourselves to estimate only θ and $\text{var}(\boldsymbol{e}_i) = \sigma^2$. The model class (2.23) originates from theory of errors. If an object is measured n times and if the measuring method includes errors, then the measured values y_i differ by an experimental error e_i from the real value μ (in this case we have $f(\theta) = \mu$).

The question arises how to get a statement about μ by the n single measurements y_i. Gauss, but earlier also Legendre, proposed the least squares method (*LSM*) that determines a value $\theta \in \Omega$ with minimal sum of squared errors $\sum_{i=1}^{n} e_i^2$.

Definition 2.10 A measurable statistic $S_Q(\boldsymbol{Y})$ whose realisation $S_Q(Y)$ fulfils the condition

$$\sum_{i=1}^{n} \{y_i - f[S_Q(Y)]\}^2 = \min_{\theta \in \Omega} \sum_{i=1}^{n} \{y_i - f(\theta)\}^2 \tag{2.24}$$

is said to be estimator according to the least squares method with respect to $\theta \in \Omega$ or shortly *LSM* estimator of $\theta \in \Omega$.

Usually the variance $\sigma^2 = \text{var}(\boldsymbol{e}_i)$ is estimated by

$$\boldsymbol{s}^2 = \frac{\sum_{i=1}^{n} \{\boldsymbol{y}_i - f[S_Q(\boldsymbol{Y})]\}^2}{n - \dim(\Omega)}. \tag{2.25}$$

if $\dim(\Omega) < n$ holds.

The *LSM* estimator is mainly used in the theory of linear (and also non-linear) models. In these models $\boldsymbol{Y}$ is not a random sample, since the components of $\boldsymbol{Y}$ have different expectations. For example, for a simple linear model, we have

$$\boldsymbol{y}_i = \beta_0 + \beta_1 x_i + \boldsymbol{e}_i \ (i = 1, \dots, n),$$

and if $\beta_1 \neq 0$, the expectation $E(\boldsymbol{y}_i)$ is dependent on x_i. Hence, the vector $\boldsymbol{Y} = (\boldsymbol{y}_1, \dots, \boldsymbol{y}_n)^T$ is not a random sample. Nevertheless the parameters of the model can be estimated by the *LSM*. We refer to Chapters 4 and 8 where parameter estimation in linear models is investigated. The *LSM* can be generalised also for dependent $\boldsymbol{e}_i$ with arbitrary, but known positive definite covariance matrix.

2.3.3 Minimum Chi-Squared Method

The minimum chi-squared method (or minimum χ^2 method; this notation is used in the following) is applicable if the observed values are frequencies of observations, which belong to a finite number of mutually disjoint subsets whose union represents the totality of possible realisations of a component of a random sample $\boldsymbol{Y}$. It is unimportant whether these classes are possible realisations of a discrete random variable (natural classes) or subsets of values of continuous random variables. In each case let $\boldsymbol{n}_1, \ldots, \boldsymbol{n}_k$ be the number of components of a random sample $\boldsymbol{Y}$, which fall into k classes. Because of

$$\sum\nolimits_{i=1}^{k} n_i = n,$$

the variables $\boldsymbol{n}_i$ are dependent. Let $\psi_1 = g_1(\theta), \ldots, \psi_k = g_k(\theta)$ be the corresponding probabilities, determined by the distribution P_θ, for which an element of a random sample $\boldsymbol{Y}$ belongs to one of the k classes.

Definition 2.11 An estimator $S_0(\boldsymbol{Y})$ whose realisations fulfil

$$X^2 = \sum_{i=1}^{k} \frac{\{\boldsymbol{n}_i - n g_i[S_0(\boldsymbol{Y})]\}^2}{n g_i[S_0(\boldsymbol{Y})]} = \min_{\theta \epsilon \Omega} \sum_{i=1}^{k} \frac{\{\boldsymbol{n}_i - n \boldsymbol{g}_i(\theta)\}^2}{n \boldsymbol{g}_i(\theta)} \tag{2.26}$$

is said to be minimum χ^2 estimator.

The notation minimum χ^2-estimator originates from the fact that $\boldsymbol{X}^2$ is asymptotically distributed as $CS(n-k)$. If the functions $g_i(\theta)$ are differentiable relative to θ, then a minimum of the convex function X^2 in (2.26) is obtained if the partial derivatives of X^2 relative to the components of θ are put equal to 0 and the simultaneous equations are solved. This leads to

$$\sum_{i=1}^{k} \left\langle \frac{\{n_i - n g_i[S_0(Y)]\}}{g_i[S_0(Y)]} + \frac{\{n_i - n g_i[S_0(Y)]\}^2}{2n\{g_i[S_0(Y)]\}^2} \right\rangle \frac{\partial g_i(\theta)}{\partial \theta_i}\bigg|_{\theta = S_0(Y)} = 0. \tag{2.27}$$

Unfortunately it is difficult to solve (2.27). But often the second part in this sum can be neglected without severe consequences. In these cases (2.27) is replaced by the simpler equation

$$\sum_{i=1}^{k} \frac{\{n_i - n g_i[S_0(Y)]\}}{g_i[S_0(Y)]} \frac{\partial g_i(\theta)}{\partial \theta_i}\bigg|_{\theta = S_0(Y)} = 0. \tag{2.28}$$

This approach is also called modified minimum χ^2 method.

2.3.4 Method of Moments

If just p product moments of the distribution $P_\theta \in P = (P_\theta, \theta \in \Omega)$, $\dim(\Omega) = p$ controlling the components of a random sample $\boldsymbol{Y}$ are known as explicit functions of θ, then the method of moments can be used.

Definition 2.12 If $n \geq p$, then an estimator $S_M(\boldsymbol{Y})$ whose realisation $S_M(Y)$ solves the simultaneous equations

$$m'_r = \mu'_r[S_M(Y)] \tag{2.29}$$

is said to be an estimator according to the method of moments. In (2.29) μ'_r is the usual rth moment. Observe that

$$m'_r = \frac{1}{n}\sum_{i=1}^{n} y_i^r.$$

Example 2.12 Let $\boldsymbol{Y}$ be a random sample from a non-central $CS(\nu,\lambda)$ distribution where ν and λ are assumed to be unknown. Then we get

$$E(\boldsymbol{y}_i) = \nu + \lambda,\ \mathrm{var}(\boldsymbol{y}_i) = 2(\nu + 2\lambda),\ \mathrm{var}(\boldsymbol{y}_i) = E\left(\boldsymbol{y}_i^2\right) - [E(\boldsymbol{y}_i)]^2,\ i = 1,\ldots,n.$$

For $p = 2$ relation (2.29) implies with $r = 1$ and $r = 2$ the system

$$\bar{y} = \hat{\nu} + \hat{\lambda} \text{ and } \frac{1}{n}\sum_{i=1}^{n} y_i^2 = 2\left(\hat{\nu} + 2\hat{\lambda}\right) + \left(\hat{\nu} + \hat{\lambda}\right)^2,$$

with the solution $\boldsymbol{S}_M^T = \left(\hat{\nu},\hat{\lambda}\right)$ with

$$\hat{\nu} = 2\bar{y} - \frac{1}{2}\left[\frac{1}{n}\sum_{i=1}^{n} y_i^2 - \bar{y}^2\right], \hat{\lambda} = \frac{1}{2}\left[\frac{1}{n}\sum_{i=1}^{n} y_i^2 - \bar{y}^2\right] - \bar{y}.$$

2.3.5 Jackknife Estimators

It is supposed for this method that an estimator $S(\boldsymbol{Y})$ for the problem of estimation is already known. The aim is now to improve the given estimator. Here we restrict ourselves to such cases, where $S(\boldsymbol{Y})$ is biased with respect to $g(\theta) = \psi$. We look for possibilities to reduce the bias $v_n(\theta) = E[S(\boldsymbol{Y})] - g(\theta)$.

Definition 2.13 Let $S_n(\boldsymbol{Y})$ be an estimator with respect to $g(\theta)$ including all n (>1) elements of a random sample $\boldsymbol{Y}$. Besides, let $S_{n-1}(\boldsymbol{Y}^{(i)})$ be an element of the same sequence of estimators based on

$$\boldsymbol{Y}^{(i)} = (\boldsymbol{y}_1,\ldots,\boldsymbol{y}_{i-1},\boldsymbol{y}_{i+1},\ldots,\boldsymbol{y}_n)^T.$$

Then

$$J[S(\boldsymbol{Y})] = nS_n(\boldsymbol{Y}) - \frac{n-1}{n}\sum_{i=1}^{n} S_{n-1}\left(\boldsymbol{Y}^{(i)}\right) \tag{2.30}$$

is said to be the jackknife estimator of first order with respect to $g(\theta)$ based on $S_n(\boldsymbol{Y})$.

If $S_n(Y)$ and $S_{n-1}(Y^{(i)})$ have finite expectations and if the bias of $S_n(Y)$ has the form

$$v_n(\theta) = \sum_{l=1}^{\infty} \frac{a_l(\theta)}{n^l},$$

then

$$E\{J[S(\boldsymbol{Y})] - g(\theta)\} = nv_n(\theta) - (n-1)v_{n-1}(\theta) = \sum_{l=1}^{\infty} \frac{a_l(\theta)}{n^{l-1}} - \sum_{l=1}^{\infty} \frac{a_l(\theta)}{(n-1)^{l-1}}$$

$$= -\frac{a_2(\theta)}{n(n-1)} - \sum_{l=2}^{\infty} a_{l+1}(\theta)\left[\frac{1}{(n-1)^l} - \frac{1}{n^l}\right]$$

follows such that the order of bias is reduced from $O\left(\frac{1}{n}\right)$ to $O\left(\frac{1}{n^2}\right)$.

Example 2.13 Let Y be a random sample whose components have the expectation μ. Further, let $g(\theta) = \mu$ and $S_n(Y) = \bar{y}_n$. Then the jackknife estimator based on $\bar{\boldsymbol{y}}_n$ is given by $J(\bar{\boldsymbol{y}}_{n.}) = \bar{\boldsymbol{y}}_{n.}$

Indeed, it is

$$J(\bar{\boldsymbol{y}}_{n.}) = n\,\bar{\boldsymbol{y}}_{n.} - \frac{n-1}{n} \sum_{i=1}^{n} \bar{\boldsymbol{y}}_n^{(i)} = \bar{\boldsymbol{y}}_{n.}$$

2.3.6 Estimators Based on Order Statistics

The estimators of this subsection are to estimate mainly location parameters. First we want to introduce statistics that are important for certain problems of estimation (but also of testing).

2.3.6.1 Order and Rank Statistics

Definition 2.14 Let Y be a random sample of size $n > 1$ from a certain distribution family.

If we arrange the elements of the realisation Y according to their magnitude, and if we denote the jth element of this ordered set by $y_{(j)}$ such that $y_{(1)} \leq \ldots \leq y_{(n)}$ holds, then

$$Y_{(.)} = \left(y_{(1)},\ldots,y_{(n)}\right)^T$$

is a function of the realisation of $\boldsymbol{Y}$, and $S^*(\boldsymbol{Y}) = \boldsymbol{Y}_{(.)} = (\boldsymbol{y}_{(1)}, \ldots, \boldsymbol{y}_{(n)})^T$ is said to be the order statistic vector, the component $\boldsymbol{y}_{(i)}$ is called the ith order statistic, and $\boldsymbol{y}_{(n)} - \boldsymbol{y}_{(1)} = \boldsymbol{w}$ is called the range of $\boldsymbol{Y}$.

Since $Y \in \{Y\}$ implies also $Y_{(.)} \in \{Y\}$, the sample space $\{Y\}$ is mapped by $S^*(\boldsymbol{Y})$ into itself.

Theorem 2.7 Let Y be a random sample with continuous components possessing the distribution function $F(y)$ and the density function $f(y)$. Then the density function $h(Y_{(.)})$ is given by

$$h\left(Y_{(.)}\right) = n!\prod_{i=1}^{n} f\left(y_{(i)}\right). \tag{2.31}$$

If $1 \le k \le n$ and if $\boldsymbol{R}_k = \left(\boldsymbol{y}_{(i_1)},\ldots,y_{(i_k)}\right)^T$ is the vector of a subset with k elements of $Y_{(.)}$, then the density function $h(R_k)$ of $\boldsymbol{R}_k$ is given by

$$h(R_k) = \frac{n!}{\prod_{j=1}^{k+1}\left(i_j - i_{j-1} - 1\right)} \prod_{j=1}^{k+1}\left[F\left(y_{(i_j)}\right) - F\left(y_{(i_{j-1})}\right)\right]^{i_j - i_{j-1} - 1} \prod_{j=1}^{k} f\left(y_{(i_j)}\right), \tag{2.32}$$

where we put $i_0 = 0$, $i_{k+1} = k + 1$, $y_{(i_0)} = -\infty$ and $y_{(k+1)} = +\infty$ and observe

$$y_{(i_1)} \le \ldots \le y_{(i_k)}.$$

We sketch only the basic idea of the proof. Let $B_{i_j} = \left(y_{(i_{j-1})}, y_{(i_j)}\right)$ and E the following event: considering the components of a random sample Y (and $Y_{(.)}$, respectively) lie $i_1 - 1$ in B_{i_1}, $i_2 - i_1 - 1$ in $B_{i_2}, \ldots, k - i_k$ in B_k.

If P_{i_j} is the probability for $\boldsymbol{y} \in B_{i_j}$, then $P_{i_j} = \int_{B_{i_j}} f(y)dy = F\left(y_{(i_j)}\right) - F\left(y_{(i_{j-1})}\right)$ holds. Since

$$P(E) = n!\frac{P_{i_1}^{i_1-1} P_{i_2}^{i_2-i_2-1} \ldots P_{i_k}^{k-i_k}}{(i_1-1)!(i_2-i_1-1)!\ldots(k-i_k)!},$$

we obtain (2.32) and for $k = n$ also (2.31).

Corollary 2.2 The density function of the ith order statistic is

$$h\left(y_{(i)}\right) = \frac{n!}{(i-1)!(n-i)!}\left[F\left(y_{(i)}\right)\right]^{i-1}\left[1 - F\left(y_{(i)}\right)\right]^{n-1} f\left(y_{(i)}\right). \tag{2.33}$$

Especially significant are

$$h\left(y_{(1)}\right) = n\left[1 - F\left(y_{(1)}\right)\right]^{n-1} f\left(y_{(1)}\right) \tag{2.34}$$

and

$$h\left(y_{(n)}\right) = n\left[F\left(y_{(n)}\right)\right]^{n-1} f\left(y_{(n)}\right). \tag{2.35}$$

Definition 2.15 Taking the notations of Definition 2.14 into account, let the n positive integers $r_i = r(y_{(i)})$ defined by $y_i = y_{(r_i)}$. The numbers r_i are called the rank numbers or simply ranks of y_i ($i = 1, \ldots, n$). The vector $\boldsymbol{R} = (\boldsymbol{r}_1, \ldots, \boldsymbol{r}_n)^T = [r(\boldsymbol{y}_1), \ldots, r(\boldsymbol{y}_n)]^T$ is called rank statistic vector of the random sample $\boldsymbol{Y}$, and the components $r(\boldsymbol{y}_i)$ are called rank statistics.

2.3.6.2 *L*-Estimators

L-estimators are weighted means of order statistics (where L stands for linear combination).

Definition 2.16 If $\boldsymbol{Y}$ is a random sample and $\boldsymbol{Y}_{(.)}$ the corresponding order statistic vector, then

$$L(\boldsymbol{Y}) = S_L(\boldsymbol{Y}) = \sum_{i=1}^{n} c_i \boldsymbol{y}_{(i)};\, c_i \geq 0, \sum_{i=1}^{n} c_i = 1 \tag{2.36}$$

is said to be an L-estimator.

It has to be indicated with respect to which parameters $L(\boldsymbol{Y})$ is to be an estimator. In the most cases we have to do with location parameters. The main causes for this are the conditions $c_i \geq 0, \sum_{i=1}^{n} c_i = 1$. Linear combinations within order statistics without these restrictions can also be used to estimate other parameters, but often they are not called L-estimators.

Thus with $c_1 = -1$, $c_2 = \cdots = c_{n-1} = 0$, $c_n = 1$, we get the range $S(\boldsymbol{Y}) = \boldsymbol{w} = \boldsymbol{y}_{(n)} - \boldsymbol{y}_{(1)}$, which is an estimator with respect to $\sigma = \sqrt{\operatorname{var}(\boldsymbol{y})}$ in distributions with existing second moment.

Example 2.14 Trimmed mean value

If we put

$$c_1 = \ldots = c_t = c_{n-t+1} = \ldots = c_n = 0 \text{ and } c_{t+1} = \ldots = c_{n-t} = \frac{1}{n-2t}$$

in (2.36) with $t < \frac{n}{2}$, then the so-called $\frac{t}{n}$-trimmed mean

$$L_T(\boldsymbol{Y}) = \frac{1}{n-2t} \sum_{i=t+1}^{n-t} \boldsymbol{y}_{(i)} \tag{2.37}$$

arises. It is used if some measured values of the realised sample can be strongly influenced by observation errors (so-called outliers). For $n = 2t + 1$ the $\frac{t}{n}$-trimmed mean is the sample median

$$L_M(\boldsymbol{Y}) = \boldsymbol{y}_{(t+1)} = \boldsymbol{y}_{(n-t)}. \tag{2.38}$$

Example 2.15 Winsorized mean

If we do not suppress, as in Example 2.14, the t smallest and the t largest observations, but concentrate them in the value $y_{(t+1)}$ and $y_{(n-t)}$, respectively, then we get the so-called $\frac{t}{n}$ -winsorized mean

$$L_W(\boldsymbol{Y}) = \frac{1}{n}\left[\sum_{i=t+1}^{n-t} \boldsymbol{y}_{(i)} + t\boldsymbol{y}_{(t+1)} + t\boldsymbol{y}_{(n-t)}\right] \tag{2.39}$$

$$c_1 = \ldots = c_t = c_{n-t+1} = \ldots = c_n = 0 \text{ and } c_{t+1} = \ldots = c_{n-t} = \frac{1}{n}.$$

The median in samples even of size $n = 2t$ can be defined as ½-winsorized mean

$$L_W(\boldsymbol{Y}) = \frac{1}{2}\left(\boldsymbol{y}_{(t+1)} + \boldsymbol{y}_{(n-t)}\right). \tag{2.40}$$

Definition 2.17 The median $\boldsymbol{y}_{\text{med}}$ of a random sample of size $n \geq 2$ is defined by

$$\boldsymbol{y}_{\text{med}} = \text{Med}(\boldsymbol{Y}) = \begin{cases} \boldsymbol{y}_{(n-t)} & \text{for } n = 2t+1 \\ \frac{1}{2}\left(\boldsymbol{y}_{(t+1)} + \boldsymbol{y}_{(n-t)}\right) & \text{for } n = 2t \end{cases}.$$

For $n = 2t + 1$ and $t = \frac{n-1}{2}$, respectively, it is $\text{Med}(\boldsymbol{Y}) = L_T(\boldsymbol{Y})$.

2.3.6.3 *M*-Estimators

Definition 2.18 An estimator $S(\boldsymbol{Y}) = M(\boldsymbol{Y})$, which minimises for each realisation Y of a random sample $\boldsymbol{Y}$ the expression

$$\sum_{i=1}^{n} \rho(y_i - S(Y)), \tag{2.41}$$

where

$$\rho(t) = \begin{cases} \frac{1}{2}t^2 & \text{for } |t| \leq k \\ k|t| - \frac{1}{2}k^2 & \text{for } |t| > k \end{cases} \tag{2.42}$$

holds for suitable chosen k, is said to be an M-estimator.

Huber (1964) introduced M-estimators for the case that the distributions of the components $\boldsymbol{y}_i$ of a random sample $\boldsymbol{Y}$ have the form

$$F(y) = (1-\varepsilon)\,G(y) + \varepsilon\,H(y),$$

where $0 < \varepsilon < 1$ and G and H are known distributions. If $0 < \varepsilon < ½$, then F can be considered as distribution G contaminated by H.

2.3.6.4 R-Estimators

Definition 2.19 Let Y be a random sample and $Y_{(.)}$ the corresponding order statistic vector of Y. For $1 \le j \le k \le n$ we denote

$$m_{jk} = \frac{1}{2}\left(y_{(j)} + y_{(k)}\right).$$

Further, let $d_1, \ldots, d_n$ be n given non-negative numbers. The numbers

$$w_{jk} = \frac{d_{n-(k-j)}}{\sum_{i=1}^{n} i d_i}, 1 \le j \le k \le n$$

define the probabilities of a $\frac{1}{2}n(n+1)$-point distribution, that is, of a discrete distribution with $\frac{1}{2}n(n+1)$ possible values m_{jk} (constituting the support), which occur with the positive probabilities w_{jk}. If $R(Y)$ is the median of this distribution, then $R(\boldsymbol{Y})$ is said to be an R-estimator after transition to a random variable.

It is easy to see that the values w_{jk} define a probability distribution. Namely, these w_{jk} are non-negative and also not greater than 1, since the numerators in the defining term are not greater than the common denominator. Finally, considering the $\frac{1}{2}n(n+1)$ pairs (j, k), the numerator d_n occurs n-times, the numerator d_{n-1} occurs $(n - 1)$-times and so on, up to the numerator d_1 that occurs once (viz. for the pair $j = 1$, $k = n$).

Example 2.16 Hodges–Lehmann estimator
Assume that $d_1 = \cdots = d_n = 1$. Then $R(\boldsymbol{Y})$ is the median of the $\boldsymbol{m}_{jk}$. This estimator is called Hodges–Lehmann estimator.

2.4 Properties of Estimators

If we construct R-optimal estimators as in Section 2.1, we know in the case of global R-optimality that the obtained estimator is the best one in the sense of R-optimisation. Sometimes it is interesting to know how these optimal estimators behave according to other criteria. But it is more important to validate estimators constructed by methods, which were described in Section 2.3. Is it possible to state properties of these estimators?

What can be done if R-optimal solutions do not exist as it was shown in Example 2.2? Are there estimators that have at least asymptotically (i.e. for $n \to \infty$) certain desired properties? We will present some results for such problems.

2.4.1 Small Samples

The first question should be to define the meaning of 'small' in this connection. This verbal expression has become a technical term of statistics. The focus is then on samples of such a size, which needs exact methods and excludes the approximate use of asymptotic results. This holds mainly for samples of size $n < 50$. For larger samples, partly asymptotic results can be used supplying good approximation for sequences of samples with $n \to \infty$. It depends on the problem from which n on this is possible. We will see in Chapter 9 about non-linear regression that in special cases asymptotic results can be exploited already for $n = 4$. But this is the exception. Unfortunately in most cases, it is not known where the limit of applicability really lies.

In this section we describe properties that hold for each $n > 1$. Such essential properties are to be unbiased (Definition 2.2) or to be in Ω global variance optimal unbiased (Definition 2.3). If no local variance-optimal unbiased estimator exists, then the relative efficiency in Definition 2.5 can be extended to arbitrary estimators in D_E fulfilling condition V1 in Definition 1.10, and the variance in Definition 2.5 can be replaced by the lower bound given in the inequality of Rao and Cramér.

All random samples and estimators of this section may be assumed to satisfy the assumption V1 of Definition 1.10. Let the components of a random sample $\boldsymbol{Y}$ be distributed as $P_\theta \in P = (P_\theta, \theta \in \Omega)$ with $\dim(\Omega) = 1$. We start with the generalised concept of relative efficiency.

Definition 2.20 Let $\boldsymbol{S}_1 = S_1(\boldsymbol{Y})$ and $\boldsymbol{S}_2 = S_2(\boldsymbol{Y})$ be two unbiased estimators based on the random sample $\boldsymbol{Y}$ with respect to $g(\theta)$. Then

$$e(\boldsymbol{S}_1, \boldsymbol{S}_2) = \frac{\operatorname{var}[S_1(\boldsymbol{Y})]}{\operatorname{var}[S_2(\boldsymbol{Y})]} \tag{2.43}$$

is said to be the relative efficiency of $\boldsymbol{S}_2$ with respect to $\boldsymbol{S}_1$. For each unbiased estimator $\boldsymbol{S} = S(\boldsymbol{Y})$ with respect to $g(\theta)$, the quotient

$$e(\boldsymbol{S}) = \frac{\left(\frac{\partial g(\theta)}{\partial \theta}\right)^2}{I_n(\theta)\operatorname{var}(S(\boldsymbol{Y}))} \tag{2.44}$$

is called efficiency function, where $I_n(\theta)$ denotes the Fisher information (see (1.16)).

The concepts of efficiency just introduced in (2.43) and (2.44) are not related to the existence of a UVUE; they need weaker assumptions as, for example, the existence of the second moments of $\boldsymbol{S}_2$ and $\boldsymbol{S}_1$ in (2.43) or the assumptions of Theorem 1.8 with respect to $g(\theta)$ in (2.44). The equation (2.44) measures the variance of all $S(Y) \in D_E$ at the lower bound of the inequality of Rao and Cramér for $dE[M(Y)] = d\theta$. Sometimes it is interesting to compare estimators with different bias according to the risk (2.2), which is based on the quadratic loss in (2.1).

Definition 2.21 If $S(Y)$ is an estimator with respect to $g(\theta) = \psi$ with the bias $v_n = v_n(\theta)$ according to Definition 2.2 and if the second moment of $S(Y)$ exists, then

$$MSD[S(\boldsymbol{Y})] = \mathrm{E}\{[\psi - S(\boldsymbol{Y})]^2\} = \mathrm{var}[S(\boldsymbol{Y})] + v_n^2 \tag{2.45}$$

is said to be the mean square deviation of $S(Y)$. For two estimators with existing second moments, the quotient

$$r(\boldsymbol{S}_1, \boldsymbol{S}_2) = \frac{MSD[S_1(\boldsymbol{Y})]}{MSD[S_2(\boldsymbol{Y})]} \tag{2.46}$$

is called relative mean square deviation of $S_2(Y)$ with respect to $S_1(Y)$.

The following example shows that there exist estimators outside of D_E with a mean square deviation smaller than that of the UVUE.

Example 2.17 If the components of a random sample $\boldsymbol{Y} = (y_1, y_2, \dots, y_n)^T$ $(n > 1)$ are distributed as $N(\mu, \sigma^2)$ and if $g(\theta) = \sigma^2$, then $\boldsymbol{s}^2$ is a UVUE with respect to σ^2 (see Example 2.4c). The formula for the variance of the χ^2 distribution implies that $\mathrm{var}(\boldsymbol{s}^2) = \dfrac{2\sigma^4}{n-1}$ holds. The maximum likelihood estimator

$$\tilde{\sigma}^2 = \frac{n-1}{n}\boldsymbol{s}^2$$

has the bias $v_n(\boldsymbol{\sigma}^2) = -\frac{\sigma^2}{n}$ and the variance

$$\mathrm{var}(\tilde{\boldsymbol{\sigma}}^2) = \frac{(n-1)^2}{n^2}\mathrm{var}(\boldsymbol{s}^2) = \frac{2(n-1)}{n^2}\sigma^4.$$

Therefore it is

$$MSD(\boldsymbol{s}^2) = \mathrm{var}(\boldsymbol{s}^2) = \frac{2\sigma^4}{n-1} \text{ and } MSD(\tilde{\boldsymbol{\sigma}}^2) = \mathrm{var}(\tilde{\boldsymbol{\sigma}}^2) + v_n^2(\tilde{\boldsymbol{\sigma}}^2) = \frac{2n-1}{n^2}\sigma^4.$$

We get

$$r(\tilde{\boldsymbol{\sigma}}^2, \boldsymbol{s}^2) = \frac{(2n-1)(n-1)}{2n^2} = \frac{2n^2-3n+1}{2n^2} < 1,$$

That is, $MSD(\tilde{\boldsymbol{\sigma}}^2)$ is always smaller than $MSD(\boldsymbol{s}^2)$. Therefore $\tilde{\boldsymbol{\sigma}}^2$, with respect to the risk function, $R(\psi,S)$ is uniformly better in Ω than $\boldsymbol{s}^2$ (and $\boldsymbol{s}^2$ is not admissible in the sense of Definition 1.15). Nevertheless $\boldsymbol{s}^2$ is used in applications with only a few exceptions. The equivariant estimator with minimal *MSD* is according to Example 2.9

$$\breve{\boldsymbol{\sigma}}^2 = \frac{n-1}{n+1}\boldsymbol{s}^2$$

with the bias $-\frac{2}{n+1}\sigma^2$. Consequently we obtain

$$MSD\left(\breve{\boldsymbol{\sigma}}^2\right) = \operatorname{var}\left(\breve{\boldsymbol{\sigma}}^2\right) + \frac{4}{(n+1)^2}\sigma^4 = \frac{2}{n+1}\sigma^4$$

and because of $n > 1$

$$r\left(\breve{\boldsymbol{\sigma}}^2,\tilde{\boldsymbol{\sigma}}^2\right) = \frac{2n^2}{(2n-1)(n+1)} = \frac{2n^2}{2n^2+n-1} < 1\,.$$

Among the three estimators, $\breve{\boldsymbol{\sigma}}^2$ has the largest bias, but the smallest *MSD*.

Definition 2.22 If $S_1(\boldsymbol{Y})$ is an unbiased estimator of $\psi_1 = g_1(\theta)$ and $S_2(\boldsymbol{Y})$ an unbiased estimator of $\psi_2 = g_2(\theta)$, then

$$e_{\psi_1\psi_2}(\boldsymbol{S}_1,\boldsymbol{S}_2) = \left(\frac{\frac{d\,g_1(\theta)}{d\theta}}{\frac{d\,g_2(\theta)}{d\theta}}\right)^2 \frac{\operatorname{var}[S_1(\boldsymbol{Y})]}{\operatorname{var}[S_2(\boldsymbol{Y})]} \tag{2.47}$$

is said to be Pitman efficiency of $S_2(\boldsymbol{Y})$ with respect to $S_1(\boldsymbol{Y})$ (Pitman 1979). Here the existence of the derivatives of g_1 and g_2 and of the second moments of the estimators is supposed.

For $g_1 = g_2$ the efficiency (2.47) is reduced to (2.43).

2.4.2 Asymptotic Properties

Sometimes it is useful to investigate the limit behaviour of estimator sequences for $n \to \infty$.

Briefly, an estimator possesses a certain asymptotic property, if the corresponding estimator sequence possesses this property. In each case we suppose a sequence $\boldsymbol{Y}_1, \boldsymbol{Y}_2, \ldots$ of random samples $\boldsymbol{Y} = (\boldsymbol{y}_1, \boldsymbol{y}_2, \ldots, \boldsymbol{y}_n)^T$ with $n = 1,2, \ldots$.

Definition 2.23 Let $\boldsymbol{S}_1, \boldsymbol{S}_2, \ldots$ be a sequence $\{\boldsymbol{S}_n\}$ of estimators with respect to $g(\theta)$, where $\boldsymbol{S}_n = S(\boldsymbol{Y}_n)$. Then $\{\boldsymbol{S}_n\}$ is said to be consistent, if this sequence stochastically converges to $g(\theta)$ for all $\theta \in \Omega$, that is, if for all $\varepsilon > 0$

$$\lim_{n\to\infty} P\{\|S_n - g(\theta)\| \ge \varepsilon\} = 0$$

holds. Further, the sequence $\{S_n\}$ is called asymptotically unbiased, if the sequence $v_n(\theta) = E(S_n) - g(\theta)$ of bias tends to zero for all $\theta \in \Omega$, that is,

$$\lim_{n\to\infty} v_n(\theta) = 0.$$

The concept of consistency is not really suitable to evaluate competing estimators. Thus the estimators s^2, $\tilde{\sigma}^2$ and $\breve{\sigma}^2$ of Section 2.4.1 are consistent (in the family of normal distributions) with respect to σ^2; all three are also asymptotically unbiased. But we have

$$MSD(\breve{\sigma}^2) < MSD(\tilde{\sigma}^2) < MSD(s^2).$$

Definition 2.24 Let $\{S_{1,n}\}$ be a sequence of estimators with respect to $g(\theta)$ and let $\sqrt{n}[S_{1,n} - g(\theta)]$ be distributed asymptotically as $N(0, \sigma_1^2)$. Additionally, let $\{S_{2,n}\}$ be another sequence of estimators with respect to $g(\theta)$ so that $\sqrt{n}[S_{2,n} - g(\theta)]$ is distributed asymptotically as $N(0, \sigma_2^2)$. Then the quotient

$$e_A(S_1, S_2) = \frac{\sigma_1^2}{\sigma_2^2} \tag{2.48}$$

is said to be the asymptotic relative efficiency of $\{S_{2,n}\}$ with respect to $\{S_{1,n}\}$. Here σ_i^2 is called the asymptotic variance of $\{S_{i,n}\}$ $(i = 1,2)$.

A general definition of the asymptotic relative efficiency of two sequences of estimators can be given also for the case that the limit distributions are no normal distributions.

Example 2.18 We consider the asymptotic relative efficiency of the sample median with respect to the arithmetic mean based on location families of distributions P_θ. If $F(y - \theta)$ is the distribution function and $L(y,\theta) = f(y)$ the density function of the components of the random sample $Y = (y_1, y_2, \ldots, y_n)^T$, then θ is for $F(0) = 1/2$ and $f(0) > 0$ the median of the distribution P_θ. Now let

$$S_{2,n} = \tilde{y}_n = \begin{cases} y_{(m+1)} & \text{for } n = 2m+1 \\ \frac{1}{2}\left[y_{(m)} + y_{(m+1)}\right] & \text{for } n = 2m \end{cases}$$

be the median of Y_n.

We show that $\sqrt{n}(\tilde{y}_n - \theta)$ is distributed asymptotically as $N\left(0, \frac{1}{4f^2(0)}\right)$. First let $n = 2m + 1$. Since the distribution of $\tilde{y}_n - \theta$ is independent of θ,

$$P_\theta\{\sqrt{n}(\tilde{y}_n-\theta)\le c\}=P_0\{\sqrt{n}\tilde{y}_n\le c\}=P_0\left\{\tilde{y}_n\le\frac{c}{\sqrt{n}}\right\}.$$

holds for real c. If w_n is the number of realisations y_i greater than $\frac{c}{\sqrt{n}}$, then $\tilde{y}_n\le\frac{c}{\sqrt{n}}$ is satisfied iff $w_n\le m=\frac{n-1}{2}$. Observing that $\boldsymbol{w_n}$ is distributed as $B(n, p_n)$ with $p_n = 1 - F\left(\frac{c}{\sqrt{n}}\right)$, the relation

$$P_\theta\{\sqrt{n}(\tilde{y}_n-\theta)\le c\}=P_0\left\{\boldsymbol{w}_n\le\frac{n-1}{2}\right\}=P_0\left\{\frac{\boldsymbol{w}_n-np_n}{\sqrt{np_n(1-p_n)}}\le\frac{\frac{1}{2}(n-1)-np_n}{\sqrt{np_n(1-p_n)}}\right\}$$

is fulfilled. If we apply the inequality of Berry and Esseen [Berry (1941), Esseen (1944), see also Lehmann and Romano (2008)] (taking into account that the third moment exists for the binomial distribution), it follows that the difference

$$P_0\left\{\boldsymbol{w}_n\le\frac{n-1}{2}\right\}-\Phi[u_n],u_n=\frac{\frac{1}{2}(n-1)-np_n}{\sqrt{np_n(1-p_n)}}$$

tends to 0 for $n\to\infty$ (Φ distribution function of $N(0, 1)$ distribution). It is

$$\lim_{n\to\infty}u_n=\lim_{n\to\infty}\frac{1}{\sqrt{p_n(1-p_n)}}\left[\sqrt{n}\left(\frac{1}{2}-p_n\right)-\frac{1}{2\sqrt{n}}\right].$$

For $n\to\infty$ the sequence $F\left(\frac{c}{\sqrt{n}}\right)$ converges to $F(0) = ½$ and therefore $p_n(1-p_n)$ to $\frac{1}{4}$. Hence,

$$\lim_{n\to\infty}u_n=2\lim_{n\to\infty}\left[\sqrt{n}\left(\frac{1}{2}-p_n\right)\right]=2c\lim_{n\to\infty}\frac{F\left(\frac{c}{\sqrt{n}}\right)-F(0)}{\frac{c}{\sqrt{n}}}.$$

But the limit of the right-hand side of the equation is just the first derivative of $F(y)$ at $y = 0$, that is,

$$\lim_{n\to\infty}u_n=2cf(0).$$

Consequently $P\left\{\sqrt{n}\left(\boldsymbol{y}_{(m)}-\theta\right)\le c\right\}$ tends to $\Phi[2cf(0)]$ for $n\to\infty$. If $\boldsymbol{y}$ is distributed as $N(0, \sigma^2)$, then we get

$$\Phi\left(\frac{c}{\sigma}\right)=P\left(\frac{\boldsymbol{y}}{\sigma}\le\frac{c}{\sigma}\right)=P(\boldsymbol{y}<c),$$

and vice versa. $P(y < c) = \Phi(\frac{c}{\sigma})$ implies that y is distributed as $N(0, \sigma^2)$. Therefore $\sqrt{n}\left(y_{(m)} - \theta\right)$ is distributed asymptotically as $N\left(0, \frac{1}{4f^2(0)}\right)$. It can be shown [see Lehmann and Romano (2008)] that this is also the case for even n, which finally means for arbitrary n.

Now we consider on the other hand the arithmetical mean

$$S_{1n} = \bar{y} = \frac{1}{n}\sum_{i=1}^{n} y_i.$$

It is well known that $\bar{y}$ is distributed with expectation θ and variance $\frac{\sigma^2}{n}$, which means that $\sqrt{n}(\bar{y} - \theta)$ has expectation 0 and variance σ^2. Hence the distribution of $\sqrt{n}(\bar{y} - \theta)$ converges to the $N(0,\sigma^2)$ distribution. By (2.48) we obtain

$$e_A(\bar{y}, \tilde{y}) = 4\sigma^2 f^2(0).$$

If y is distributed as $N(\mu, 1)$, then $f(0) = \frac{1}{\sqrt{2\pi}}$ and

$$e_A(\bar{y}, \tilde{y}) = \frac{4}{2\pi} \approx 0.6366.$$

Bahadur (1964) showed the following result under certain regularity conditions, which are omitted here. If $\sqrt{n}[S_n(y) - \theta]$ is distributed asymptotically as $N(0,\sigma^2)$ for estimators $S_n(y)$ with respect to θ, then

$$\sigma^2(\theta) \geq \frac{1}{I(\theta)} \tag{2.49}$$

is true, where $I(\theta)$ denotes the Fisher information with respect to P_θ.

Definition 2.25 Let $S_n(y)$ be an estimator with respect to $\theta \in \Omega$ and let us assume that the Fisher information with respect to P_θ exists. Further, let $\sqrt{n}[S_n(y) - \theta]$ be asymptotically distributed as $N(0, \sigma^2(\theta))$. If the equality holds for $\sigma^2(\theta)$ in (2.49), then $S_n(y)$ is said to be a best asymptotically normally distributed estimator or simply *BAN* estimator.

Let $\theta^T = (\theta_1, \ldots, \theta_p)$ and let the information matrix $I(\theta)$ defined in Section 1.4 exist and be positive definite. Then a (vectorial) estimator S_n with respect to θ is called *BAN* estimator, if $\sqrt{n}[S_n - \theta]$ is asymptotically distributed as $N[0_n, I^{-1}(\theta)]$.

The following theorem is given without proof.

Theorem 2.8 Let $L(y,\theta)$ be the likelihood function of the components in the sequence $\{Y_n\}$ of random samples and assume that $\ln L(y,\theta)$ has second partial derivatives according to all components of θ. For sufficient small $\varepsilon > 0$ and for all $\theta_0 \in \Omega$ with $|\theta_0 - \theta| < \varepsilon$, let the supremum of

$$\left|\frac{\partial^2}{\partial\theta_i\partial\theta_j}\ln L(y,\theta_0) - \frac{\partial^2}{\partial\theta_i\partial\theta_j}\ln L(y,\theta)\right|$$

be bounded with respect to y by a function, which is integrable relative to the components of θ. Let the sequence $\{\tilde{\boldsymbol{\theta}}_n\}$ of maximum likelihood estimators be consistent. Finally assume that the information matrix $I(\theta)$ exists and is positive definite. Then $\{\tilde{\boldsymbol{\theta}}_n\}$ is a BAN estimator with respect to θ.

Generally BAN estimators are not unique. For example, the estimators $\boldsymbol{s}^2$, $\breve{\boldsymbol{\sigma}}^2$ and $\tilde{\boldsymbol{\sigma}}^2$ given in Section 2.4.1 are BAN estimators.

2.5 Exercises

2.1 Let $\boldsymbol{y}$ be a random variable whose values −1, 0, 1, 2, 3 occur with the probabilities

$$P(\boldsymbol{y} = -1) = 2p(1-p), P(\boldsymbol{y} = k) = p^k(1-p)^{3-k}, 0 < p < 1, k = 0,1,2,3.$$

a) Show that this defines a probability distribution for $\boldsymbol{y}$.
b) Give the general form of all functions $U(\boldsymbol{y})$, which are unbiased with respect to 0.
c) Determine locally variance-optimal unbiased estimators for p and for $p(1 - p)$ on the basis of Theorem 2.3.
d) Are the LVUE obtained in c) also UVUE? Check the necessary and sufficient condition (2.11).

2.2 Let $Y = (\boldsymbol{y}_1, \boldsymbol{y}_2, \ldots, \boldsymbol{y}_n)^T, n \ge 1$ be a random sample from a binomial distributed population with parameters n and p, $0 < p < 1$, n fixed. Determine the uniformly variance-optimal unbiased estimator for p and for $p(1 - p)$.

2.3 Let $Y = (\boldsymbol{y}_1, \boldsymbol{y}_2, \ldots, \boldsymbol{y}_n)^T, n \ge 1$ be a random sample, where the second moments for the components exist and are equal, that is, $\text{var}(\boldsymbol{y}) = \sigma^2 < \infty$.

a) Show that

$$S(\boldsymbol{Y}) = \frac{1}{n-1}\sum_{i=1}^{n}(\boldsymbol{y}_i - \bar{\boldsymbol{y}})^2$$

is unbiased with respect to σ^2.

b) Suppose that the random variables $\boldsymbol{y}_i$ take the values 0 and 1 with the probabilities $P(\boldsymbol{y}_i = 0) = 1 - p$ and $P(\boldsymbol{y}_i = 1) = p$, $0 < p < 1$, respectively.

Prove that in this case $S(\boldsymbol{Y})$ is a uniformly variance-optimal unbiased estimator with respect to $p(1 - p)$.

2.4 Let $\boldsymbol{Y} = (\boldsymbol{y}_1, \boldsymbol{y}_2, \dots, \boldsymbol{y}_n)^T$, $n \geq 1$ be a random sample whose components have the distribution P_θ.

Calculate the maximum likelihood estimator with respect to θ as well as the estimator according to the method of moments using the first usual moments of P_θ, where P_θ is the uniform distribution in the interval (a) $(0, \theta)$, (b) $(\theta, 2\theta)$ and (c) $(\theta, \theta+1)$.

2.5 Let $\boldsymbol{Y} = (\boldsymbol{y}_1, \boldsymbol{y}_2, \dots, \boldsymbol{y}_n)^T$, $n \geq 1$ be a random sample from a population uniformly distributed in the interval $(0, \theta)$, $\theta \in R^+$. Let $S_{ML}(\boldsymbol{Y})$ and $S_M(\boldsymbol{Y})$ be the estimators described in Exercise 2.4 (a).

a) Are the estimators $S_{ML}(\boldsymbol{Y})$ and $S_M(\boldsymbol{Y})$ unbiased with respect to θ? If not, then change them in such a way that unbiased estimators $\check{S}_{ML}(\boldsymbol{Y})$ and $\tilde{S}_M(\boldsymbol{Y})$ are created.
b) Determine the UVUE with respect to θ and the relative efficiency of $\check{S}_{ML}(\boldsymbol{Y})$ and $\tilde{S}_M(\boldsymbol{Y})$.

2.6 Consider three stochastically independent random samples

$$\boldsymbol{X} = (\boldsymbol{x}_1, \boldsymbol{x}_2, \dots, \boldsymbol{x}_n)^T, \boldsymbol{Y} = (\boldsymbol{y}_1, \boldsymbol{y}_2, \dots, \boldsymbol{y}_n)^T \text{ and } \boldsymbol{Z} = (\boldsymbol{z}_1, \boldsymbol{z}_2, \dots, \boldsymbol{z}_n)^T.$$

Let the random variables $\boldsymbol{x}_i$, $\boldsymbol{y}_i$, $\boldsymbol{z}_i$ be distributed as $N(a, \sigma_a^2)$, $N(b, \sigma_b^2)$ and $N(c, \sigma_c^2)$, respectively. Further, we suppose that σ_a^2, σ_b^2, σ_c^2 are known and that $c = a + b$ holds.

a) Determine ML estimators for a, b and c, where only the sample from the population is used for estimation whose expectation is to be estimated.
b) Calculate estimators for a, b and c applying the maximum likelihood method, if the united sample and $c = a + b$ are used for estimation.
c) Determine the expectations and variances of the ML estimators from (a) and (b).

2.7 The task is to estimate the parameter θ in the model

$$\boldsymbol{y}_i = f_i(x_i, \theta) + \boldsymbol{e}_i, i = 1, \dots, n.$$

Further, assume that the random variables $\boldsymbol{e}_i$ are distributed as $N(0, \sigma^2)$ and are stochastically independent. Show that the maximum likelihood method and the least squares method are equivalent under these conditions.

2.8 a) Let $Y = (y_1, y_2, \dots, y_n)^T, n \geq 1$ be a random sample with $E(y_i) = \theta < \infty$ $(i = 1, \dots, n)$. Determine the *LSM* estimator of the expectation θ.
b) Estimate the parameters α and β of the linear model

$$y_i = \alpha + \beta x_i + e_i, i = 1, \dots, n$$

according to the least squares method, where $x_i \neq x_j$ holds for at least one pair (i, j) of the indices.

2.9 Let $Y = (y_1, y_2, \dots, y_n)^T, n \geq 1$ be a random sample whose components are uniformly distributed in $(0, \theta)$, and let $S(Y) = y_{(n)}$ be the maximum likelihood estimator with respect to θ. Calculate the bias of this estimator.

2.10 Let $y_1, y_2, \dots, y_n$ be independently and identically distributed and positive random variables with $E(y_i) = \mu > 0$, $\text{var}(y_i) = \sigma^2 < \infty$ and $x_1, x_2, \dots, x_n$ independently and identically distributed random variables with $E(x_i) = \eta > 0$, $\text{var}(x_i) = \tau^2 < \infty$. Further, let

$$\text{cov}\left(x_i, y_j\right) = \begin{cases} \rho\sigma\tau \text{ for } i = j \\ 0 \quad \text{for } i \neq j \end{cases}, \; i,j = 1, \dots, n, \; |\rho| < 1.$$

First estimate $g(\theta) = \frac{\eta}{\mu}$. Then show that the estimator $\bar{x}/\bar{y}$ and its jackknife estimator with respect to $g(\theta)$ have biases of order $O(1/n)$ and $O(1/n^2)$, respectively.

2.11 Let $Y = (y_1, y_2, \dots, y_n)^T, n \geq 1$ be a random sample whose components are uniformly distributed in the interval $[\mu - \alpha; \mu + \alpha]$.

a) Determine the expectation of the ith order statistic $(i = 1, \dots, n)$.
b) Show that the median of the sample (see Definition 2.17) is in this case an unbiased estimator with respect to μ.

2.12 Let the random sample $Y = (y_1, y_2, \dots, y_n)^T$ of size $n > 2$ be from a population exponentially distributed with the parameter $\alpha > 0$.

a) Give the efficiency function for the estimators that are unbiased with respect to α.
b) Starting with the *ML* estimator for α, determine an unbiased estimator and calculate its relative efficiency.

2.13 Show that the *ML* estimator of Exercise 2.12 (b) is asymptotically unbiased and consistent.

2.14 Let $y_1, y_2, \dots, y_n$ be independently and identically $N(\theta, 2\theta)$-distributed random variables. Determine the ML estimator of the parameter $\theta > 0$ and check its consistency.

References

Bahadur, R.R. (1964) On Fisher's bound for asymptotic variances. *Ann. Math. Stat.*, **35**, 1545–1552.

Berry, A.C. (1941) The accuracy of the Gaussian approximation to the sum of independent variables. *Trans. Amer. Math. Soc.*, **49**, 122–136.

Blackwell, D. (1947) Conditional expectations and unbiased sequential estimation. *Ann. Math. Stat.*, **18**, 105–110.

Esseen, C.-G. (1944) Fourier analysis of distribution functions. A mathematical study of the Laplace-Gaussian law. Dissertation. *Acta Math.*, **77**, 1–125.

Fisher, R.A. (1925) *Statistical Methods for Research Workers*, Oliver & Boyd, Edinburgh.

Huber, P.J. (1964) Robust estimation of a location parameter. *Ann. Math. Stat.*, **35**, 73–101.

Lehmann, E.L. and Romano, J.P. (2008) *Testing Statistical Hypothesis*, Springer, Heidelberg.

Lehmann, E.L. and Scheffé, H. (1950) Completeness, similar regions and unbiased estimation. *Sankhya*, **10**, 305–340.

Pitmann, E.J.G. (1979) *Some Basic Theory for Statistical Inference*, Chapman and Hall, London.

Rao, C.R. (1945) Information and accuracy attainable in estimation of statistical parameters. *Bull. Calc. Math. Soc.*, **37**, 81–91.

3

Statistical Tests and Confidence Estimations

3.1 Basic Ideas of Test Theory

Sometimes the aim of investigation is neither to determine certain statistics (to estimate parameters) nor to select something, but to test or to examine carefully considered hypotheses (assumptions, suppositions) and often also wishful notions on the basis of practical material. Also in this case a mathematical model is established where the hypothesis can be formulated in the form of model parameters. We want to start with an example.

Table potatoes are examined beside other things, whether they are infected by so-called brown foulness. Since the potato under examination is cut for that reason, it is impossible to examine the whole production. Hence, the examiner takes at random a certain number n of potatoes from the produced amount of potatoes and decides to award the rating 'table potatoes' if the number r of low-quality potatoes is less or equal to a certain number c, and otherwise he declines to do so. (For example, we can suppose that a quantity of potatoes is classified as table potatoes, if the portion p of damaged or bad potatoes is smaller than or equal to 3%.) This is a typical statistical problem, because it concludes from a random sample (the n examined potatoes) to a population (the whole amount of potatoes of a certain producer in a certain year).

The above described situation is a bit more complicated than that for estimation and selection problems, because evidently two wrong decisions can appear with different effect. We call the probability to make an error of the first kind or type I error (e.g. by classifying table potatoes wrongly as fodder potatoes) the risk of the first kind α and correspondingly the probability to make an error of the second kind or type II error (e.g. by classifying fodder potatoes wrongly as table potatoes) the risk of the second kind β.

Both errors have different consequences. Assuming that table potatoes are more expensive than fodder potatoes, the error of the first kind implies that the producer is not rewarded for his effort to supply good quality; therefore

Mathematical Statistics, First Edition. Dieter Rasch and Dieter Schott.

the risk of the first kind is called also producer's risk. However, the error of the second kind implies that the consumers get bad quality for their money; therefore the risk of the second kind is called also consumer's risk. The choices of numbers α and β depend on n and c, and reversely, n and c have to be chosen suitably for given α and β.

Generally a statistical test is a procedure to allow a decision for accepting or rejecting a hypothesis about the unknown parameter occurring in the distribution of a random variable. We shall suppose in the following that two hypotheses are possible. The first (or main) hypothesis is called null hypothesis H_0, and the other one alternative hypothesis H_A. The hypothesis $H = H_0$ is right, if H_A is wrong, and vice versa. Hypotheses can be composite or simple. A simple hypothesis prescribes the parameter value θ uniquely, for example, the hypothesis $H_0 : \theta = \theta_0$ is simple. A composite hypothesis admits that the parameter θ can have several values.

Examples for composite null hypotheses are:

$$H_0 : \theta = \theta_0 \text{ or } \theta = \theta_1$$

$$H_0 : \theta < \theta_1,$$

$$H_0 : \theta \neq \theta_0.$$

Let $\boldsymbol{Y}$ be a random sample of size n, and let the distribution of their components belong to a family $P = \{P_\theta, \theta \in \Omega\}$ of distributions. We suppose the null hypothesis $H_0 : \theta \in \omega = \Omega_0 \subset \Omega$ and the alternative hypothesis $H_A : \theta \in \Omega_A = \Omega \backslash \omega \subset \Omega$. We denote the acceptance of H_0 by d_0 and the rejection of H_0 by d_A. A non-randomised statistical test has the property that it is fixed for each possible realisation Y of the random sample $\boldsymbol{Y}$ in the sample space $\{Y\}$ whether the decision has to be d_0 or d_A. According to this test, the sample space $\{Y\}$ is decomposed into two disjoint subsets $\{Y_0\}$ and $\{Y_A\}$ ($\{Y_0\} \cap \{Y_A\} = \emptyset$, $\{Y_0\} \cup \{Y_A\} = \{Y\}$), defining the decision function

$$d(Y) = \begin{cases} d_0 & \text{for } Y \in \{Y_0\} \\ d_A & \text{for } Y \in \{Y_A\} \end{cases}$$

The set $\{Y_0\}$ is called acceptance region, and the set $\{Y_A\}$ critical region or rejection region. We consider a simple case for illustration. Let θ be a one-dimensional parameter in $\Omega = (-\infty, \infty)$. We suppose that the random variable $\boldsymbol{y}$ has the distribution P_θ. With respect to the parameter, two simple hypotheses are established, the null hypothesis $H_0 : \theta = \theta_0$ and the alternative hypothesis $H_A : \theta = \theta_1$ where $\theta_0 < \theta_1$. Based on the realisation of a random sample $\boldsymbol{Y} = (\boldsymbol{y}_1, \dots, \boldsymbol{y}_n)^T$, we have to decide between both hypotheses. We calculate an estimator $\hat{\boldsymbol{\theta}}$ from the sample whose distribution function $G(\hat{\theta}, \theta)$ is known. Let $\hat{\boldsymbol{\theta}}$ be a continuous variable with the density function $g(\hat{\theta}, \theta)$, which evidently depends on the true value of the parameter. Consequently $\hat{\boldsymbol{\theta}}$ has under the null

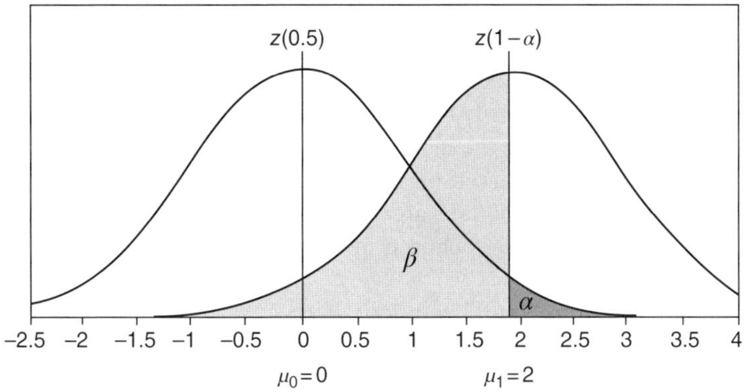

Figure 3.1 Density functions of the estimator of the location parameter μ depending on the hypothesis values $\mu = 0$ and $\mu = 2$, respectively.

hypothesis the density function $g(\hat{\theta}, \theta_0)$ and under the alternative hypothesis the density function $g(\hat{\theta}, \theta_1)$. Further, we assume for the sake of simplicity that $\boldsymbol{y}$ is normally distributed with unknown expectation $\theta = \mu$ and known variance σ^2. Then the densities $g(\hat{\theta}, \theta_0)$ and $g(\hat{\theta}, \theta_1)$ are of the same type. Their graphs have the same shape and are only mutually shifted along the θ-axis, as Figure 3.1 shows. Here we put $\hat{\boldsymbol{\theta}} = \bar{\boldsymbol{y}}$.

Both hypotheses are simple hypotheses. In this case usually the test statistic

$$\mathbf{z} = \frac{\bar{\mathbf{y}} - \mu_0}{\sigma}\sqrt{\boldsymbol{n}}$$

is applied where $\bar{\boldsymbol{y}}$ is the mean taken from the random sample $\boldsymbol{Y}$ of size n.

Starting with the (random) sample mean $\bar{\boldsymbol{y}}$, first the value μ_0 of the null hypothesis is subtracted, and then this difference is divided by the standard deviation $\frac{\sigma}{\sqrt{n}}$ of $\bar{\boldsymbol{y}}$.

Therefore $\mathbf{z} = \frac{\bar{\mathbf{y}} - \mu_0}{\sigma}\sqrt{\boldsymbol{n}}$ has the variance 1 and under the null hypothesis the expectation 0. Under the alternative hypothesis, the expectation is $E(\mathbf{z}) = \frac{\mu_1 - \mu_0}{\sigma}\sqrt{\boldsymbol{n}}$. The corresponding number $\lambda = \frac{\mu_1 - \mu_0}{\sigma}\sqrt{n}$ is called non-centrality parameter.

Having in mind a test decision to define on the base of the realisation z of $\boldsymbol{z}$, we determine for the chosen α with $0 < \alpha < 1$ the $(1 - \alpha)$-quantile $(z(1 - \alpha) = z_{1-\alpha})$ of the standard normal distribution that can be found in Table D.2 and for special values of α in the last line of Table D.3 (see Appendix). Then the decision is as follows: reject H_0, if $z > z(1 - \alpha)$, the so-called critical value generally denoted by θ_k, and otherwise accept the null hypothesis (i.e. for $z \leq z(1 - \alpha)$).

This decision rule is illustrated in Figure 3.1. In the coordinate system, for example, the values $\mu_0 = 0$ and $\mu_1 = 2$ are marked on the z-axis (where z represents certain realisations of $\boldsymbol{z}$). Taking each of the two values as corresponding expectation, the curves of the density functions can be plotted (shifted standard normal distributions). Regarding the left curve belonging to $\mu_0 = 0$ the critical value, the quantile $z(1-\alpha)$ is marked on the z-axis; besides a vertical straight line through $z(1-\alpha)$ separates certain area parts under both curves (shaded in Figure 3.1). Besides $\alpha = 0.025$ is chosen in Figure 3.1 supplying $z(1-\alpha) = z(0.975) = 1.96$.

The decision of rejecting the null hypothesis, if z > 1.96 (where z is obtained from the realisations of the random sample or in other words is calculated from the measurements), can be wrong, because such a value z is also possible for $\mu > 0$ (e.g. for $\mu = 2$).

Let us come back to the general case. The probability to get an estimate of $\theta > \theta_k$ for valid null hypothesis is equal to

$$\int_{\theta_k}^{\infty} g\left(\hat{\theta}, \theta_0\right) d\hat{\theta} = \alpha.$$

The value of α is represented by the darker shaded area under the curve of $g\left(\hat{\theta}, \theta_0\right)$ in Figure 3.1. If the null hypothesis is rejected although it is right, then an error is made, called error of the first kind. The maximal probability of wrongly rejecting the null hypothesis in a test is said to be the risk α of the first kind or significance level. However, the alternative hypothesis is said to have a significance of $(1-\alpha) \cdot 100\%$.

The better a test seems, the smaller its risk of the first kind. Considering practical investigations a risk of the first kind $\alpha = 0.05$ seems to be only just acceptable in the most cases. Users may ask why the test is not designed in such a way that α has a very small value, say, $\alpha = 0.00001$. Figure 3.1 clearly illustrates that the further to the right the bound θ_k (in this case $z(1-\alpha)$) between both regions is shifted, the smaller α (i.e. the area under the curve of $g\left(\hat{\theta}, \theta_0\right)$) is to be chosen on the right-hand side of $z(1-\alpha)$. But then the probability to make another error increases. Namely, if we calculate an estimate $\hat{\theta} < \theta_k$ from the realisation of the sample, then the null hypothesis is accepted, although this value would be also possible in the case that the alternative hypothesis is right and consequently the null hypothesis is wrong.

If we accept the null hypothesis although it is wrong, then another error is made, called the error of the second kind. The probability β of wrongly accepting the null hypothesis, that is, the probability to make an error of the second kind, is said to be the risk of the second kind. In Figure 3.1 this risk is represented by the lightly shaded area under the curve of $g\left(\hat{\theta}, \theta_1\right)$ at the left-hand side of θ_k. Its value is obtained by integrating the density function $g(\hat{\theta}, \theta_1)$ from $-\infty$ up to θ_k, that is,

$$\int_{-\infty}^{\theta_k} g(\hat{\theta}, \theta_1) d\hat{\theta} = \beta.$$

Figure 3.1 makes clear that α can only be reduced for a certain test and a fixed sample size if a larger β is accepted. Hence, the risks of the first and the second kind cannot simultaneously be made arbitrarily small for a fixed sample size. Applying statistical tests, it is wrong but common to focus mainly on the risk of the first kind while the risk of the second kind is neglected. There are a lot of examples where the wrong acceptance of the null hypothesis can produce serious consequences (consider 'genetic corn has no damaging side effects' or 'nuclear power stations are absolutely safe'). Therefore, it is advisable to control both risks, which is always possible by suitably chosen sample size. In the following scheme the decisions performing a statistical test with respect to the true facts (H_0 null hypothesis, H_A alternative hypothesis) are shown.

True fact	Decision	Result of the decision	Probability of the result
H_0 right (H_A wrong)	H_0 accepted (H_A rejected)	Right decision	Acceptance (or confidence) probability $1 - \alpha$
	H_0 rejected (H_A accepted)	Error of the first kind	Significance (or error) level, risk α of the first kind
H_0 wrong (H_A right)	H_0 accepted (H_A rejected)	Error of the second kind	Risk β of the second kind
	H_0 rejected (H_A accepted)	Right decision	Power $1 - \beta$

To be on the safe side, it is recommended to declare that hypothesis as null hypothesis, which causes the more serious consequences in the case of wrong rejection.

A generalisation of the situation just described is given if, after knowing the experimental results, that is, the realisation $\hat{\theta}$ of the statistic $\hat{\boldsymbol{\theta}}$, it is not instantly decided which of the two hypotheses is accepted. But instead a random procedure (a kind of tossing a coin) is used accepting the null hypothesis with probability $1 - k(Y) = 1 - k(\hat{\theta})$ and the alternative hypothesis with probability $k(\hat{\theta})$, if $Y \in \{Y\}$ was observed (or $\hat{\theta}$ calculated). Although the user of statistical methods will hardly agree, leaving it after carefully planned and often cost intensive experiments to leave it to chance which of the two hypotheses should be accepted, the theory of testing is firstly based on the concept of such randomised tests. The significance of the Neyman–Pearson lemma in Section 3.2 is just to recognise that non-randomised tests are sufficient for continuous distributions.

Definition 3.1 Let Y be a random sample with $Y \in \{Y\}$ whose components are distributed as $P_\theta \in P = \{P_\theta, \theta \in \Omega\}$. Further, let $k(Y)$ be a measurable mapping of the sample space $\{Y\}$ onto the interval (0, 1). It is called critical function. If $k(Y)$ states the probability for rejecting $H_0 : \theta \in \omega$ (i.e. for accepting $H_A : \theta \in \Omega \backslash \omega$) in the case that Y takes the value $Y \in \{Y\}$, then the critical function defines a statistical test for the pair (H_0, H_A) of hypotheses. Then $k(Y)$ is shortly called a test. The test $k(Y)$ is said to be randomised if it does not take with probability 1 only the values 0 or 1.

Now we want to define the risks of the first and the second kind for such general tests $k(Y)$. In this chapter we consider only such functions $k(Y)$ whose expectation exists for all $\theta \in \Omega$. The notation $E[k(Y)|\theta]$ means that the expectation is taken with respect to the distribution $P_\theta \in \Omega$.

Definition 3.2 If $k(Y)$ is a statistical test for a pair (H_0, H_A) of hypotheses according to Definition 3.1, then

$$E[k(\boldsymbol{Y}) | \theta \in \omega] = \int_{\{Y\}} k(Y) dP_\theta = \alpha(\theta), P_\theta \in P, \theta \in \omega \tag{3.1}$$

is said to be the risk function of the first kind and

$$1 - E[k(\boldsymbol{Y}) | \theta \in \Omega \backslash \omega] = \beta(\theta) \tag{3.2}$$

the risk function of the second kind. The function

$$\pi(\theta) = \int_{\{Y\}} k(\boldsymbol{Y}) dP_\theta, P_\theta \in P, \theta \in \Omega$$

is said to be the power function of the test. Further

$$\max_{\theta \in \Omega} \alpha(\theta) = \alpha$$

is called the significance level of the test $k(Y)$. A test with the significance level α is briefly called an α-test (alpha-test).

If $\alpha(\theta) = \alpha$ for all $\theta \in \omega$, then the test $k(Y)$ is said to be α-similar or simply similar.

If $\bar{\omega}$ and $\overline{\Omega \backslash \omega}$, respectively, are the closures of ω and $\Omega \backslash \omega$, respectively, and if $\bar{\omega} \cap \overline{\Omega \backslash \omega} = \Omega^*$ is the common boundary of both subsets, then $k(Y)$ is called α-similar on the boundary, if $E[k(\boldsymbol{Y}) | \theta \in \Omega^*] = \alpha$ is fulfilled with P_θ - probability 1.

Definition 3.3 Considering tests $\omega = \{\theta_0\}$ and $\Omega \setminus \omega = \{\theta_A\}$ with simple null and alternative hypotheses, then $k^*(Y)$ is said to be most powerful α-test, if for all α in the interval (0, 1)

$$\max_{k(\boldsymbol{Y}) \in K_\alpha} E[k(\boldsymbol{Y}) \,|\theta_A] = E[k^*(\boldsymbol{Y}) \,|\theta_A] = \alpha,$$

where K_α is the class of all α-tests.

According to Definition 3.2, an α-test $k(Y) \in K_\alpha$ for the pair of hypotheses

$$H_0 : \theta = \theta_0, \quad H_A : \theta = \theta_A$$

considered in Definition 3.3 is a test with

$$E[k(\boldsymbol{Y})|\theta_0] = \alpha. \tag{3.3}$$

Definition 3.4 An α-test $k^*(Y)$ for the pair $H_0 : \theta \in \omega$ against $H_A : \theta \in \Omega \setminus \omega$ is said to be uniformly best α-test, if

$$E[k^*(\boldsymbol{Y})|\,\theta \in \Omega \setminus \omega] \geq E[k(\boldsymbol{Y})|\,\theta \in \Omega \setminus \omega] \tag{3.4}$$

for each other test $k(Y)$ with a significance level not larger than α and for all $\alpha \in (0, 1)$. The test $k^*(Y)$ is also briefly called a uniformly most powerful test (UMP-test).

Definition 3.5 If $k(Y)$ is with respect to the pair $H_0 : \theta \in \omega\,, H_A : \theta \in \Omega \setminus \omega$ an α-test and if for its power function $\pi(\theta) \geq \alpha$ holds for all $\theta \in \Omega \setminus \omega$ and for all $\alpha \in (0, 1)$, then $k(Y)$ is said to be an unbiased α-test. If $K_{u\alpha}$ is the class of all unbiased α-tests and if

$$\max_{k(\boldsymbol{Y}) \in K_{u\alpha}} E[k(\boldsymbol{Y})|\,\theta_A] = E[k^{**}(\boldsymbol{Y})|\,\theta_A] \quad \text{for all} \quad \theta_A \in \Omega \setminus \omega,$$

then $k^{**}(Y)$ is said to be a uniformly most powerful unbiased α-test (UMPU-test).

We need the following statement in the next sections.

Lemma 3.1 Let (H_0, H_A) be a pair of hypotheses $H_0 : \theta \in \omega, H_A : \theta \in \Omega \setminus \omega$ concerning the parameter θ of the distribution family $P = \{P_\theta, \theta \in \Omega\}$, assuming that each test has in θ a continuous power function $\pi(\theta)$. If $k(Y)$ is with respect to (H_0, H_A) in the class $\{K_{\Omega^*}\}$ of all on the boundary α-similar tests the uniformly most powerful α-test, then it is also a uniformly most powerful unbiased α-test.

The proof uses the facts that the class K_{Ω^*} contains the class of unbiased α-tests, taking the continuity of $\pi(\theta)$ into account, and considering that $k(Y)$ fulfils the inequality (3.4) for all $k^*(\boldsymbol{Y}) \in K_{\Omega^*}$, it all the more fulfils this inequality for all $k^*(Y) \in K_{ua}$. Moreover, $k(Y)$ is in $K_{u\alpha}$, since it fulfils as uniformly most powerful α-test in K_{Ω^*} also inequality (3.4) for $k^*(\boldsymbol{Y}) \in K_{\Omega^*}$. Hence, its power function in $\Omega \setminus \omega$ cannot lie under that of $k^*(Y)$, which is just α.

Example 3.1 Let $\boldsymbol{Y} = (\boldsymbol{y}_1, \boldsymbol{y}_2, \ldots, \boldsymbol{y}_n)^T$ be a random sample of size $n > 1$ from a $N(\mu, \sigma^2)$-distribution, where $\mu \in R^1 = \Omega$, σ^2 known. Let $\omega = (-\infty, a]$ and therefore $\Omega \backslash \omega = (a, \infty)$. Then $\boldsymbol{z} = \frac{\bar{\boldsymbol{y}} - a}{\sigma}\sqrt{n}$ is distributed as $N\left[(\mu - a)\frac{\sqrt{n}}{\sigma}, 1\right]$. We consider the test $k(\boldsymbol{Y})$ with

$$k(Y) = \begin{cases} 0 & \text{for } z \le z_{0.95} \\ 1 & \text{for } z > z_{0.95} \end{cases}.$$

Here we have $z_{0.95} = 1.6449$ and $\Phi(z_{0.95}) = 0.95$. This is a 0.05-test because

$$P\{\boldsymbol{z} > z_{0.95} | \mu \le a\} \le 0.05.$$

Regarding $P\{\boldsymbol{z} > z_{0.95} | \mu > a\} > 0.05$, this is an unbiased 0.05-test. For each other α in the interval (0, 1)

$$k(Y) = \begin{cases} 0 & \text{for } z \le z_{1-\alpha} \\ 1 & \text{for } z > z_{1-\alpha} \end{cases}$$

is an unbiased α-test which can easily be seen.

Let $\mu = a + \delta$ ($\delta \ge 0$). Then the power function for $\alpha = 0.05$ is

$$\pi(\delta) = P\left\{\boldsymbol{z} > 1.6449 - \frac{\sqrt{n}\delta}{\sigma}\right\}.$$

Table 3.1 lists $\pi(\delta)$ for special δ and n.

In the applications δ is chosen as practically interesting minimum difference to the value of the null hypothesis (also called effect size). If we want to avoid such a difference with at most probability β, that is, to discover it with probability $1 - \beta$, we have to prescribe a corresponding sample size. Again we consider the general case that $\boldsymbol{Y}$ is a random sample of size n taken from an $N(\mu,\sigma^2)$-distribution.

Putting $\mu = a + \delta$ ($\delta > 0$), assuming $\alpha = 0.05$ and after that fixing $\beta = 0.1$, then the difference

$$1.6449 - \frac{\sqrt{n}\,\delta}{\sigma}$$

has to be the 0.1-quantile of the standard normal distribution, namely, −1.2816. Therefore

$$1.6449 - \frac{\sqrt{n}\delta}{\sigma} = -1.2816$$

is satisfied. This equation has to be solved for n.

Table 3.1 Values of the power function in Example 3.1 for n = 9, 16, 25, σ = 1 and special δ.

δ	$\pi(\delta)$, n = 9	$\pi(\delta)$, n = 16	$\pi(\delta)$, n = 25
0	0.05	0.05	0.05
0.1	0.0893	0.1066	0.1261
0.2	0.1480	0.1991	0.2595
0.3	0.2282	0.3282	0.4424
0.4	0.3282	0.4821	0.6387
0.5	0.4424	0.6387	0.8038
0.6	0.5616	0.7749	0.9123
0.7	0.6755	0.8760	0.9682
0.8	0.7749	0.9400	0.9907
0.9	0.8543	0.9747	0.9978
1.0	0.9123	0.9907	0.9996
1.1	0.9510	0.9971	0.9999
1.2	0.9747	0.9992	1.0000

Choosing $\delta = \sigma$, we find

$$1.6449 - \sqrt{n} = -1.2816$$

and n = 8.56. But we have to look for the smallest integer n, which is larger or equal to the calculated value (using rounding up function CEIL(x) denoted here by $\lceil x \rceil$). Hence, we get $n = \lceil (1.6449 + 1.2816)^2 \rceil = \lceil 8.56 \rceil$ and that is 9. Generally, we get the sample size for given α, β,σ and δ by the formula

$$n = \left\lceil \left(z_{1-\alpha} + z_{1-\beta}\right)^2 \frac{\sigma^2}{\delta^2} \right\rceil.$$

We call $\frac{\delta}{\sigma}$ the relative effect size.

3.2 The Neyman–Pearson Lemma

The authors Neyman and Pearson as a lemma introduced the following very important theorem.

Theorem 3.1 Neyman–Pearson Lemma (Neyman and Pearson, 1933)
Let $L(Y, \theta)$ be the likelihood function of the random sample $\boldsymbol{Y} = (y_1, \dots, y_n)^T$ with $Y \in \{Y\}$ and $\theta \in \Omega = \{\theta_0, \theta_A\}$ with $\theta_0 \neq \theta_A$. Further, the null hypothesis $H_0: \theta = \theta_0$ is to be tested against the alternative hypothesis $H_A: \theta = \theta_A$. Then with a constant $c \geq 0$, the following holds:

1) Each test $k(\boldsymbol{Y})$ of the form

$$k(Y) = \begin{cases} 1 & \text{for } L(Y, \theta_A) > c\,L(Y, \theta_0) \\ \gamma(Y) & \text{for } L(Y, \theta_A) = c\,L(Y, \theta_0) \\ 0 & \text{for } L(Y, \theta_A) < c\,L(Y, \theta_0) \end{cases} \tag{3.5}$$

with $0 \leq \gamma(Y) \leq 1$ is for a certain $\alpha = \alpha[c, \gamma(Y)]$ a most powerful α-test $(0 \leq \alpha \leq 1)$.
The test $k(\boldsymbol{Y})$ with

$$k(Y) = \begin{cases} 1 & \text{for } L(Y, \theta_0) = 0 \\ 0 & \text{for } L(Y, \theta_0) > 0 \end{cases} \tag{3.6}$$

is a best 0-test, and the test $k(\boldsymbol{Y})$ with

$$k(Y) = \begin{cases} 1 & \text{for } L(Y, \theta_A) > 0 \\ 0 & \text{for } L(Y, \theta_A) = 0 \end{cases} \tag{3.7}$$

is a best 1-test.

2) For testing H_0 against H_A, there exist for each $\alpha \in (0, 1)$ constants $c = c_\alpha, \gamma = \gamma_\alpha$ so that the corresponding test $k(\boldsymbol{Y})$ in the form (3.5) is a best α-test.
3) If $k(\boldsymbol{Y})$ is a best α-test with $\alpha \in (0, 1)$, then it is with probability 1 of the form (3.5) (apart from the set $\{Y: L(Y, \theta_A) = c\,L(Y, \theta_0)\}$ of P_θ-measure 0) if there is no α_0-test $k^*(\boldsymbol{Y})$ with $\alpha_0 < \alpha$ and $E[k^*(\boldsymbol{Y}) | \theta_A] = 1$.

Proof:
Assertion (1)
If $\alpha = 0$, then $k(\boldsymbol{Y})$ satisfies the relation (3.6). If $k'(Y)$ is another α-test, then

$$E[k'(\boldsymbol{Y}) | \theta_0] = \int_{B_0} k'(Y)\, dP_{\theta_0} = 0$$

for $B_0 = \{Y: L(Y, \theta_0) > 0\}$. If $L(Y, 0) > 0$ holds, then $k'(\boldsymbol{Y})$ has to be equal to 0 with probability 1. Putting $B_A = \{Y\} \setminus B_0$, we get

$$E[k(\boldsymbol{Y})|\theta_A] - E[k'(\boldsymbol{Y})]|\theta_A] = \int_{\{Y\}} [k(\boldsymbol{Y}) - k'(\boldsymbol{Y})]\, dP_{\theta_A}$$

$$= \int_{B_0 \cup B_A} [k(\boldsymbol{Y}) - k'(\boldsymbol{Y})]\, dP_{\theta_A}$$

$$= \int_{B_A} [k(\boldsymbol{Y}) - k'(\boldsymbol{Y})]\, dP_{\theta_A} = \int_{B_A} [1 - k'(\boldsymbol{Y})]\, dP_{\theta_A} \geq 0$$

and therefore the assertion (1) for $\alpha = 0$ from (3.6). Analogously the assertion (1) follows for $\alpha = 1$ from (3.7). Hence, we can now consider α-tests with $0 < \alpha < 1$. We show that they are most powerful α-tests if they fulfil (3.5).

Let $k(\boldsymbol{Y})$ be an α-test of the form (3.5), that is, besides (3.5) assume also

$$E[k(\boldsymbol{Y})|\,\theta_0] = \alpha. \tag{3.8}$$

If $k'(\boldsymbol{Y})$ is an arbitrary test with a significance level not larger than α, then we have to show

$$E[k(\boldsymbol{Y})|\,\theta_A] \geq E[k'(\boldsymbol{Y})|\,\theta_A]. \tag{3.9}$$

For $L(Y,\theta_A) > c\,L(Y,\theta_0)$ it is $1 = k(Y) \geq k'(Y)$, and for $L(Y,\theta_A) < c\ L(Y,\theta_0)$, it is $0 = k(Y) \leq k'(Y)$. That means

$$[L(Y,\theta_A) - c\,L(Y,\theta_0)][k(Y) - k'(Y)] \geq 0,$$

and therefore

$$[k(Y) - k'(Y)][dP_{\theta_A} - c\,dP_{\theta_0}] \geq 0,$$

and further

$$E[k(\boldsymbol{Y})|\,\theta_A] - E[k'(\boldsymbol{Y})]|\,\theta_A] \geq c\{E[k(\boldsymbol{Y})|\,\theta_0] - E[k'(\boldsymbol{Y})]|\,\theta_0]\} \geq 0.$$

Finally, this implies (3.9).

Assertion (2)

For $\alpha = 0$ and $\alpha = 1$ the formulae (3.6) and (3.7), respectively, have the form (3.5) putting $c_0 = \infty$ (where $0 \cdot \infty = 0$), $\gamma_0 = 0$ and $c_1 = 0$, $\gamma_1 = 0$, respectively. Therefore we can restrict ourselves to $0 < \alpha < 1$.

If we put $\gamma(Y) = \gamma$ in (3.5), then the constants c_0 and γ_0 are to be determined so that

$$\alpha = E[k(\boldsymbol{Y})|\theta_0] = 1 \cdot P[L(\boldsymbol{Y},\theta_A) > c_\alpha L(\boldsymbol{Y},\theta_0)] + \gamma_\alpha P[L(\boldsymbol{Y},\theta_A) = c_\alpha L(\boldsymbol{Y},\theta_0)],$$

and with the notation

$$q = \frac{L(Y, \theta_A)}{L(Y, \theta_0)},$$

then

$$\alpha = 1 - P[q \le c_\alpha | \theta_0] + \gamma_\alpha P[q = c_\alpha | \theta_0]$$

holds. In the continuous case we choose the (1 – α)-quantile of the distribution of q for c_α and $\gamma_\alpha = 0$. If q is discrete, then a constant c_α exists in such a way that

$$P[q < c_\alpha | \theta_0] \le 1 - \alpha \le P[q \le c_\alpha | \theta_0] \tag{3.10}$$

holds. We put

$$\gamma_\alpha = \frac{P[q \le c_\alpha | \theta_0] - (1 - \alpha)}{P[q = c_\alpha | \theta_0]}, \tag{3.11}$$

if the equality does not hold twice (= for both ≤) in (3.10) (i.e. if $P[q = c_\alpha | \theta_0] > 0$). Otherwise (for vanishing denominator), we proceed just as in the continuous case and write $k(Y)$ in the form (3.5).

Assertion (3)

Since $0 < \alpha < 1$ is supposed, we take $k(Y)$ as a most powerful α-test of the form (3.5) with $c = c_\alpha$ and $\gamma(Y) = \gamma_\alpha$ from (3.10) and (3.11), respectively, and choose the α-quantile c_α of q and $\gamma_\alpha = 0$. Let $k'(Y)$ be an arbitrary most powerful α-test. Then both

$$E[k(Y) | \theta_0] = E[k'(Y) | \theta_0] = \alpha$$

and

$$E[k(Y) | \theta_A] = E[k'(Y) | \theta_A]$$

must hold, which means

$$\int_{\{Y\}} [k(Y) - k'(Y)]\, dP_{\theta_A} = 0, \quad \theta \in \{\theta_0, \theta_A\}$$

as well as

$$\int_{\{Y\}} [k(Y) - k'(Y)][dP_{\theta_A} - c_\alpha\, dP_{\theta_0}] = 0.$$

This implies the assertion. If there is an α_0-test $k^*(Y)$ with $\alpha_0 < \alpha$ and $E[k^*(Y) | \theta_A] = 1$, then this conclusion is not possible.

The Neyman–Pearson lemma has some interesting consequences.

Corollary 3.1 Let the general assumptions of Theorem 3.1 be fulfilled and put $\beta = E[k(\boldsymbol{Y}) \mid \theta_A]$. Then always $\alpha < \beta$ for the most powerful a-test $k(\boldsymbol{Y})$ if $L(Y, \theta_0) \neq L(Y, \theta_A)$.

Proof: Since $E[k^*(\boldsymbol{Y})|\theta_A] = \alpha$ holds for the special α-test $k^*(Y) \equiv \alpha$, it follows $\alpha \le \beta$ for the most powerful α-test $k(\boldsymbol{Y})$. However, $\alpha = \beta$ is not possible. Otherwise $k^*(\boldsymbol{Y}) \equiv \alpha$ would be a most powerful α-test and would have with probability 1 the form (3.5) considering (3) in Theorem 3.1. Nevertheless, both would only hold, if $L\ (Y, \theta_0)$ is with probability 1 equal to $L(Y, \theta_A)$. This contradicts the assumption of the corollary.

One of the best books of test theory was that of Lehmann (1959), and we also cite the revised edition of Lehmann and Romano (2008).

Theorem 3.1 can be generalised (for the proof, see Lehmann, 1959, pp. 84–87).

Corollary 3.2 Let K be the set of all critical functions $k(\boldsymbol{Y})$ of a random sample $\boldsymbol{Y}$ with respect to a distribution $P_\theta \in P = \{P_\theta,\ \theta \in \Omega\}$. Further, let $g_1, \ldots, g_m$ and g_0 be in R^n defined real P_θ-integrable functions. Additionally, for given real constants $c_1, \ldots, c_m$, let $k(\boldsymbol{Y})$ exist so that

$$\int_{\{Y\}} k(Y) g_i(Y)\, dP_\theta = c_i, \quad i = 1, \ldots, m.$$

We denote the class of functions $k(\boldsymbol{Y}) \in K$ satisfying this equation by K_c.

1) There is a function $k^*(\boldsymbol{Y})$ in K_c with the property

$$\int_{\{Y\}} k^*(Y) g_0(Y)\, dP_\theta = \max_{k(Y) \in K_c} \int_{\{Y\}} k(Y) g_0(Y)\, dP_\theta.$$

2) Let real constants $k_1, \ldots, k_m$ and a function $\gamma(Y)$ with $0 < \gamma(Y) < 1$ exist so that

$$k^*(Y) = \begin{cases} 1 & \text{for } g_0 > \sum_{i=1}^m k_i g_i \\ \gamma(Y) & \text{for } g_0 = \sum_{i=1}^m k_i g_i \\ 0 & \text{for } g_0 < \sum_{i=1}^m k_i g_i \end{cases}$$

for all $Y \in \{Y\}$. Then (1) is satisfied using this $k^*(\boldsymbol{Y})$.

3) If $k^*(\boldsymbol{Y}) \in K_c$ fulfils the sufficient condition in (2) with non-negative k_i, then

$$\int_{\{Y\}} k^*(Y) g_0(Y)\, dP_\theta = \max_{k(Y) \in K_c^*} \int_{\{Y\}} k(Y) g_0(Y)\, dP_\theta$$

follows, where $K_c^* \subset K_c$ is the set of critical functions $k(Y)$ with

$$\int_{\{Y\}} k(Y) g_i(Y)\, dP_\theta \le c_i, \quad i = 1, \dots, m.$$

4) The set $M \subset R^m$ of the points

$$\left\{ \int_{\{Y\}} k(Y) g_1(Y)\, dP_\theta, \dots, \int_{\{Y\}} k(Y) g_m(Y)\, dP_\theta \right\}$$

generated by the functions g_i for any $k(\boldsymbol{Y}) \in K_c$ is convex and closed. If $c = (c_1, \dots, c_m)^T$ is an inner point of M, then there exist m constants $k_1, \dots, k_m$ and any $k^*(\boldsymbol{Y}) \in K_c$ so that the condition in (2) is true.

The condition that $k^*(\boldsymbol{Y})$ with probability 1 has the form in (2) is necessary for a $k^*(\boldsymbol{Y}) \in K_c$ to fulfil the equation in (3).

If we put $m = 1$ in Corollary 3.2, we get the statements of Theorem 3.1.

Example 3.2 Let $\boldsymbol{Y}$ be a random sample of size n taken from a $N(\mu, \sigma^2)$-distribution, where σ^2 is known. Besides we assume that $\mu \in \{a, b\}$, that is, μ, can be either equal to a or equal to $b \neq a$. We want to test $H_0 : \mu = a$ against $H_A : \mu = b$. Since the components of $\boldsymbol{Y}$ are continuously distributed, a most powerful α-test for this pair of hypotheses has according to Theorem 3.1 the form (3.5) with $\gamma_\alpha = 0$ ($0 < \alpha < 1$). Besides, we have

$$L(Y, \theta_0) = L(Y, a) = \frac{1}{(2\pi\sigma^2)^{\frac{n}{2}}} e^{-\frac{1}{2\sigma^2}\left(\sum_{i=1}^n y_i^2 - 2a\sum_{i=1}^n y_i + na^2\right)}$$

and

$$L(Y, \theta_A) = L(Y, b) = \frac{1}{(2\pi\sigma^2)^{\frac{n}{2}}} e^{-\frac{1}{2\sigma^2}\left(\sum_{i=1}^n y_i^2 - 2b\sum_{i=1}^n y_i + nb^2\right)}$$

as well as

$$q = \frac{L(Y, b)}{L(Y, a)} = e^{\frac{1}{\sigma^2}\left[n\bar{y}(b-a) - \frac{n}{2}(a+b)(b-a)\right]}.$$

The quantity $c = c_\alpha$ in (3.5) has to be chosen so that $1 - \alpha = P(\boldsymbol{q} < c_\alpha) = P(\ln \boldsymbol{q} < \ln c_\alpha)$.

Considering

$$\ln q = \frac{1}{\sigma^2}\left[n\bar{y}(b-a) - \frac{n}{2}(a+b)(b-a)\right],$$

the relation $\ln q < \ln c_\alpha$ is equivalent to

$$\bar{y}\begin{cases} < \dfrac{\sigma^2 \ln c_\alpha}{n(b-a)} + \dfrac{(a+b)}{2} & \text{for } a<b \\ > \dfrac{\sigma^2 \ln c_\alpha}{n(b-a)} + \dfrac{(a+b)}{2} & \text{for } a>b \end{cases}.$$

Since

$$z = \frac{\bar{y}-\mu}{\sigma}\sqrt{n}. \tag{3.12}$$

is $N(0, 1)$-distributed, it holds with the $(1 - \alpha)$-quantile $z_{1-\alpha}$ of the standard normal distribution under the null hypothesis H_0

$$P\left(\frac{\bar{y}-a}{\sigma}\sqrt{n} < z_{1-\alpha}\right) = 1-\alpha$$

and

$$P\left(\bar{y} < \frac{\sigma}{\sqrt{n}}z_{1-\alpha} + a\right) = 1-\alpha,$$

respectively. Regarding the case $a < b$, it follows under H_0

$$\frac{\sigma}{\sqrt{n}}z_{1-\alpha} + a = \frac{\sigma^2 \ln c_\alpha}{n(b-a)} + \frac{(a+b)}{2}$$

and

$$c_\alpha = e^{\frac{1}{\sigma}z_{1-\alpha}\sqrt{n}(b-a) - \frac{n}{2\sigma^2}(a-b)^2},$$

respectively. Analogously we get for $a > b$

$$c_\alpha = e^{\frac{1}{\sigma}z_\alpha\sqrt{n}(b-a) - \frac{n}{2\sigma^2}(a-b)^2}.$$

This leads to an important statement.

Theorem 3.2 Let the random sample $\boldsymbol{Y} = (\boldsymbol{y}_1, \boldsymbol{y}_2, \ldots, \boldsymbol{y}_n)^T$ be for known $\sigma^2 > 0$ distributed as $N(\mu 1_n, \sigma^2 I_n)$. Assume that μ can only have the values a and b (with $a \neq b$). If $H_0: \mu = a$ is tested against $H_A: \mu = b$, then a most powerful α-test $k(\boldsymbol{Y})$ is given in form of (3.5) with $\gamma_\alpha = 0$ and

$$c_\alpha = e^{\frac{1}{\sigma}z_{1-\alpha}\sqrt{n}(b-a) - \frac{n}{2\sigma^2}(a-b)^2},$$

which can be written with z in (3.12) also in the form

$$k(Y) = \begin{cases} 1 & \text{for } |z| > z_{1-\alpha} \\ 0 & \text{else} \end{cases},$$

that is, H_0 is rejected for $|z| > z_{1-\alpha}$.

This test is one-sided, since it is known whether $b > a$ or $b < a$ holds. The test corresponds to the heuristic-derived test in Example 3.1. Hence, the sample size given there is always the smallest possible.

Now we turn to discrete random variables.

Example 3.3 Let the random variables $\boldsymbol{y}_i$ with the values 0 and 1 be independent from each other as $B(1, p)$ two-point distributed, where $p = P(\boldsymbol{y}_i = 1)$ and $1 - p = P(\boldsymbol{y}_i = 0)$ with $p \in \{p_0, p_A\}$; $i = 1, \dots, n$. We want to test the null hypothesis H_0: $p = p_0$ against H_A: $p = p_A$. Then $\boldsymbol{y} = \sum_{i=1}^{n} \boldsymbol{y}_i$ is $B(n, p)$ binomial distributed, and for $\boldsymbol{Y} = (\boldsymbol{y}_1, \dots, \boldsymbol{y}_n)^T$

$$L(Y,p) = \binom{n}{y} p^y (1-p)^{n-y}.$$

According to Theorem 3.1, there exists a most powerful α-test of the form (3.5). We determine now γ_α and c_α. Regarding

$$q = \frac{L(Y,p_A)}{L(Y,p_0)} = \left(\frac{p_A}{p_0}\right)^y \left(\frac{1-p_A}{1-p_0}\right)^{n-y} \tag{3.13}$$

the following equation is satisfied:

$$\ln q = y[\ln p_A - \ln(1-p_A) - \ln p_0 + \ln(1-p_0)] + n[\ln(1-p_A) - \ln(1-p_0)].$$

Case A:

For the chosen α there exists a y^* so that the distribution function of $B(n,p_0)$ has at y^* the value $F(y^*, p_0) = 1 - \alpha$. In (3.5) we put $\gamma_\alpha = 0$ and calculate c_α, obtaining

$$c_\alpha = \left(\frac{p_A}{p_0}\right)^{y^*} \left(\frac{1-p_A}{1-p_0}\right)^{n-y^*}. \tag{3.14}$$

Case B:

For the chosen α there does not exist such a value y^* considered in case A. But, assuming $p_A > p_0$, there is a value y^* so that $F(y^*, p_0) < 1 - \alpha \le F(y^* + 1, p_0)$. Then we choose according to (3.11)

$$\gamma_\alpha = \frac{F(y^* + 1, p_0) - (1-\alpha)}{\binom{n}{y^*} p^{y^*} (1-p)^{n-y^*}}, \tag{3.15}$$

and calculate c_α again by (3.14).

If $p_A < p_0$, then a value y^* exists with $F(y^*, p_0) \le \alpha < F(y^* + 1, p_0)$. Then we choose

$$\gamma_\alpha = \frac{\alpha - F(y^*, p_0)}{\binom{n}{y^*} p^{y^*}(1-p)^{n-y^*}}, \tag{3.16}$$

where c_α is calculated again according to (3.14). Therefore, we can formulate the test also directly with y.

For $p_A > p_0$ it is

$$k(y) = \begin{cases} 1 & \text{for } y > y^* \\ \gamma_\alpha & \text{for } y = y^* \text{ (with } \gamma_\alpha \text{ from (3.15))}. \\ 0 & \text{for } y < y^* \end{cases}$$

For $p_A < p_0$ it is

$$k(y) = \begin{cases} 1 & \text{for } y < y^* \\ \gamma_\alpha & \text{for } y = y^* \text{ (with } \gamma_\alpha \text{ from (3.16))}. \\ 0 & \text{for } y > y^* \end{cases}$$

Now we turn to special data. If $n = 10$ and $H_0: p = 0.5$ is to be tested against $H_A: p = 0.1$, then the value $y^* = 3$ follows by (3.16) because of $0.1 < 0.5$, and for $\alpha = 0.1$ we get

$$\gamma_{0.1} = \frac{0.1 - 0.05469}{0.11719} = 0.3866.$$

Then $k(\boldsymbol{Y})$ has the form

$$k(y) = \begin{cases} 1 & \text{for } y < 3 \\ 0.3866 & \text{for } y = 3, \\ 0 & \text{for } y > 3 \end{cases}$$

that is, for $y < 3$, the hypothesis H_0: $p = 0.5$ is rejected; H_0 is rejected for $y = 3$ with the probability 0.3866; and H_0 is accepted for $y > 3$. The random trial in the case $y = 3$ can be simulated on a computer. Using a generator of random numbers supplying uniformly distributed pseudorandom numbers in the interval (0, 1), a value v is obtained. For $v < 0.3866$ the hypothesis H_0 is rejected and otherwise accepted. This test is a most powerful 0.1-test.

Now our considerations can be summarised. The proof is analogous to that in Example 3.3.

Theorem 3.3 If y is distributed as $B(n, p)$, then a most powerful α-test for $H_0: p = p_0$ against $H_A: p = p_A < p_0$ is given by

$$k^-(y) = \begin{cases} 1 & \text{for } y < y^- \\ \gamma_\alpha^- & \text{for } y = y^- \quad \left(\text{with } \gamma_\alpha^- \text{ from } (3.16)\right). \\ 0 & \text{for } y > y^- \end{cases} \tag{3.17}$$

and for $H_0: p = p_0$ against $H_A: p = p_A > p_0$ by

$$k^+(y) = \begin{cases} 1 & \text{for } y > y^+ \\ \gamma_\alpha^+ & \text{for } y = y^+ \quad \left(\text{with } \gamma_\alpha^+ \text{ from } (3.15)\right) \\ 0 & \text{for } y < y^+ \end{cases} \tag{3.18}$$

where y^- is determined by

$$F(y^-, p_0) \le \alpha < F(y^- + 1, p_0)$$

and y^+ by

$$F(y^+, p_0) < 1 - \alpha \le F(y^+ + 1, p_0).$$

Here $F(y, p)$ is the distribution function of $B(n, p)$.

If possible, randomised tests are avoided. As mentioned earlier, users do not really accept that the decisions after well-planned experiments depend on randomness.

3.3 Tests for Composite Alternative Hypotheses and One-Parametric Distribution Families

Theorem 3.1 allows finding most powerful tests for one-sided null and alternative hypotheses. In this section we will clarify the way to transfer this theorem to the case of composite hypotheses.

3.3.1 Distributions with Monotone Likelihood Ratio and Uniformly Most Powerful Tests for One-Sided Hypotheses

It is supposed in the Neyman–Pearson lemma that the null hypothesis as well as the alternative hypothesis is simple and the parameter space consists of only two points. However, such prerequisites are rather artificial and do not meet practical requirements. We intend to decrease these restrictions systematically; however, we have to accept that the domain of validity of such extended statements is reduced. First we consider the case $\Omega \subset R^1$ and one-sided (one-tailed) hypotheses. We demonstrate the new situation in the next example.

Example 3.4 Let the components of the random sample $\boldsymbol{Y} = (y_1, y_2, \ldots, y_n)^T$ be distributed as $N(\mu, \sigma^2)$, where $\sigma^2 > 0$ is known. The hypothesis $H_0 : \mu \in (-\infty, a]$ is to be tested against $H_A : \mu \in (a, \infty)$. Looking for an α-test the condition

$$\max_{-\infty < \mu \leq a} E[k(\boldsymbol{Y}) | \mu] = \alpha$$

must hold. Regarding the pair of hypotheses

$$H_0^* : \mu = a; \quad H_A^* : \mu = b > a,$$

a most powerful α-test is defined (see Theorem 3.2). Since $k(\boldsymbol{Y})$ is a most powerful test for each $b \in (a, \infty)$, $k(\boldsymbol{Y})$ is a uniformly most powerful α-test for $H_0^* : \mu = a$ against $H_A : \mu \in (a, \infty)$ and for H_0 against H_A in the class K_α of all α-tests, respectively. If

$$z = \frac{\sqrt{n}}{\sigma}(\bar{y} - a)$$

is distributed as $N\left[\frac{\sqrt{n}}{\sigma}(\mu - a), 1\right]$ and

$$E[k(\boldsymbol{Y}) | \mu] = P\left[\frac{\sqrt{n}}{\sigma}(\bar{y} - a) > z_{1-\alpha}\right],$$

then $E[k(\boldsymbol{Y})| \mu]$ increases monotone in μ and has for $\mu = a$ the value α and for $\mu \leq a$ a value $\nu \leq \alpha$. Since $k(\boldsymbol{Y})$ is a uniformly most powerful test in the class K_α, it is a uniformly most powerful test for the pair $H_0 : \mu \in (-\infty, a]$, $\sigma^2 > 0$ against $H_A : \mu \in (a, \infty)$, $\sigma^2 > 0$, because the class of tests satisfying $E[k(\boldsymbol{Y})|\mu] \leq \alpha$ for all $\mu \in (-\infty, a]$ is a subset of K_α.

The results are summarised in the following statements.

Theorem 3.4 Under the assumptions of Theorem 3.2, $H_0 : \mu \leq a$ is to be tested against $H_A : \mu > a$. Then

$$k(\boldsymbol{Y}) = \begin{cases} 1 & \text{for } \frac{\bar{y} - a}{\sigma}\sqrt{n} \geq z_{1-\alpha} \\ 0 & \text{else} \end{cases} \tag{3.19}$$

is a uniformly most powerful α-test. Analogously

$$k(\boldsymbol{Y}) = \begin{cases} 1 & \text{for } \frac{\bar{y} - a}{\sigma}\sqrt{n} < z_{\alpha} \\ 0 & \text{else} \end{cases}$$

is a uniformly most powerful α-test for $H_0 : \mu \geq a$ against $H_A : \mu < a$.

Now we consider normal distributed random samples with known expectation.

Example 3.5 Let the components of the random sample $\boldsymbol{Y} = (\boldsymbol{y}_1, \boldsymbol{y}_2, \ldots, \boldsymbol{y}_n)^T$ be distributed as $N(\mu, \sigma^2)$, where μ is known. It is $H_0 : \sigma^2 \leq \sigma_0^2$ to be tested against $H_A : \sigma^2 = \sigma_A^2 > \sigma_0^2$. Then

$$\boldsymbol{q}(n) = \frac{1}{\sigma^2}\boldsymbol{Q}(n) = \frac{1}{\sigma^2}\sum_{i=1}^{n}(\boldsymbol{y}_i - \mu)^2$$

is as $CS(n)$ centrally χ^2-distributed with n degrees of freedom. Considering the pair $\{H_0^* : \sigma^2 = \sigma_0^2, H_A : \sigma^2 = \sigma_A^2 > \sigma_0^2\}$, the test

$$k^*(Y) = \begin{cases} 1 & \text{for}\, Q(n) \geq \sigma_0^2 CS(n|1-\alpha) \\ 0 & \text{else} \end{cases}$$

is according to Theorem 3.1 a most powerful α-test, where $CS(n|1-\alpha)$ is the $(1-\alpha)$-quantile of the $CS(n)$-distribution.

Since this holds for arbitrary $\sigma_A^2 > \sigma_0^2$, $k^*(\boldsymbol{Y})$ is a uniformly most powerful α-test for the pair $\{H_0^*, H_A\}$. Observe that

$$E\left[k^*(\boldsymbol{Y})|\sigma^2\right] = P\left\{\sigma^2\boldsymbol{q}(n) \geq \sigma_0^2 CS(n|1-\alpha)\right\} \leq \alpha$$

holds for all $\sigma^2 \leq \sigma_0^2$. Besides it is

$$\left\{q(n) \geq \frac{\sigma_0^2}{\sigma_1^2}CS(n|1-\alpha)\right\} \subset \left\{q(n) \geq \frac{\sigma_0^2}{\sigma_2^2}CS(n|1-\alpha)\right\}$$

for $0 < \sigma_1^2 < \sigma_2^2 \leq \sigma_0^2$. This implies

$$E\left[k^*(\boldsymbol{Y})|\sigma_1^2\right] \leq E\left[k^*(\boldsymbol{Y})|\sigma_2^2\right] \leq E\left[k^*(\boldsymbol{Y})|\sigma_0^2\right]$$

and

$$\max_{\sigma^2 \leq \sigma_0^2} E\left[k^*(\boldsymbol{Y})|\sigma^2\right] = \alpha, \tag{3.20}$$

respectively. Hence, $k^*(\boldsymbol{Y})$ is a uniformly most powerful α-test for the pair $\{H_0, H_A\}$.

The results of this example can be stated in a theorem.

Theorem 3.5 Let the components of the random sample $\boldsymbol{Y} = (\boldsymbol{y}_1, \boldsymbol{y}_2, \ldots, \boldsymbol{y}_n)^T$ be distributed as $N(\mu, \sigma^2)$, where $\sigma^2 > 0$ and μ is known. Considering the pairs of hypotheses

a) $H_0 : \sigma^2 \leq \sigma_0^2;\ H_A : \sigma^2 = \sigma_A^2 > \sigma_0^2$
b) $H_0 : \sigma^2 \geq \sigma_0^2;\ H_A : \sigma^2 = \sigma_A^2 < \sigma_0^2$,

a uniformly most powerful α-test is given by

$$\text{a) } k^+(Y) = \begin{cases} 1 & \text{for } Q(n) \geq \sigma_0^2 CS(n|1-\alpha) \\ 0 & \text{else} \end{cases} \tag{3.21}$$

and

b) $k^-(Y) = \begin{cases} 1 & \text{for } Q(n) \le \sigma_0^2 CS(n|\alpha) \\ 0 & \text{else} \end{cases}$ (3.22)

respectively, where

$$\boldsymbol{q}(n) = \frac{1}{\sigma^2}\boldsymbol{Q}(n) = \frac{1}{\sigma^2}\sum_{i=1}^{n}(\boldsymbol{y}_i - \mu)^2.$$

The proof of this theorem essentially exploits the fact that the ratio

$$\frac{L(Y,\sigma_A^2)}{L(Y,\sigma_0^2)} = \left(\frac{\sigma_0^2}{\sigma_A^2}\right)^{\frac{n}{2}} e^{-\frac{Q(n)}{2}\left(\frac{1}{\sigma_A^2}-\frac{1}{\sigma_0^2}\right)}$$

is monotone increasing in Q for $\sigma_0^2 < \sigma_A^2$ and monotone decreasing in Q for $\sigma_0^2 < \sigma_A^2$.

Such a property is generally significant to get α-tests for one-sided hypotheses concerning real parameters.

Definition 3.6 A distribution family $P = \{P_\theta,\ \theta \in \Omega \subset R^1\}$ is said to possess a monotone likelihood ratio, if the quotient

$$\frac{L(y,\theta_2)}{L(y,\theta_1)} = LR(y|\,\theta_1,\theta_2);\ \theta_1 < \theta_2$$

at the positions y, where at least one of the two likelihood functions $L(y,\theta_1)$ and $L(y,\theta_2)$ is positive, is monotone non-decreasing (isotone) or monotone non-increasing (antitone) in y. Observe that $LR(y|\theta_1,\theta_2)$ is defined as ∞ for $L(y,\theta_1) = 0$.

Theorem 3.6 Let P be a one-parametric exponential family in canonical form with respect to the parameter $\theta \in \Omega \subset R^1$. Then P has a monotone likelihood ratio, provided that concerning the exponent of the likelihood function, the factor $T(y)$ is monotone in y and the factor $\eta(\theta)$ monotone in θ.

Proof: W.l.o.g. we assume $\theta_1 < \theta_2$. Then the assertion can evidently be seen regarding

$$LR(y|\theta_1,\theta_2) = r(y)e^{T(y)[\eta(\theta_2)-\eta(\theta_1)]} \text{ with } r(y) \ge 0.$$

Now we are ready to design uniformly most powerful α-tests for one-parametric exponential families and for one-sided hypotheses if we still refer to the next statements.

Theorem 3.7 Karlin (1957)
Let $P = \{P_\theta,\ \theta \in \Omega \subset R^1\}$ be a family with isotone and antitone likelihood ratio LR, respectively. If $g(y)$ is P_θ-integrable and isotone (antitone) in $y \in \{Y\}$, then $E[g(\boldsymbol{y})|\ \theta]$ is isotone (antitone) and antitone (isotone) in θ, respectively. For the distribution function $F(y, \theta)$ of $\boldsymbol{y}$ in the case of isotone LR and g for all $\theta < \theta'$ and $y \in \{Y\}$, we have

$$F(y,\theta) \geq F(y,\theta')$$

and in the case of antitone LR and g for all $\theta < \theta'$ and $y \in \{Y\}$

$$F(y,\theta) \leq F(y,\theta').$$

Proof:
Without loss of generality the assertion is shown in the case of isotone.

First we suppose $\theta < \theta'$. Further, let M^+ and M^- be two sets from the sample space $\{Y\}$ defined by

$$M^+ = \{y : L(y,\theta') > L(y,\theta)\},\quad M^- = \{y : L(y,\theta') < L(y,\theta)\}.$$

Since $LR(y|\theta_1, \theta_2)$ is isotone in y, we obtain for $y \in M^-$, $y' \in M^+$ the relation $y < y'$. Then the isotone of $g(y)$ implies

$$a = \max_{y \in M^-} g(y) \leq \max_{y \in M^+} g(y) = b.$$

Therefore it is

$$\begin{aligned} D &= E[g(\boldsymbol{y})|\theta'] - E[g(\boldsymbol{y})|\theta] \\ &= \int_{\{Y\}} g(y)(dP_{\theta'} - dP_\theta) = \int_{M^-} g(y)(dP_{\theta'} - dP_\theta) + \int_{M^+} g(y)(dP_{\theta'} dP_\theta) \\ &\geq a \int_{M^-} (dP_{\theta'} - dP_\theta) + b \int_{M^+} (dP_{\theta'} - dP_\theta). \end{aligned} \tag{3.23}$$

Evidently we have for each $\theta^* \in \Omega$ the relations

$$\int_{\{Y\}} dP_{\theta^*} = \int_{M^-} dP_{\theta^*} + \int_{M^+} dP_{\theta^*} = 1 - P\{L(\boldsymbol{y}|\theta') = L(\boldsymbol{y}|\theta)|\theta = \theta'\}.$$

It follows for $\theta^* = \theta'$

$$\int_{M^-} dP_{\theta'} = -\int_{M^+} dP_{\theta'} + 1 - P\{L(\boldsymbol{y}|\theta') = L(\boldsymbol{y}|\theta)|\,\theta = \theta'\}$$

and correspondingly for $\theta^* = \theta$

$$-\int_{M^-} dP_\theta = \int_{M^+} dP_\theta - 1 + P\{L(\mathbf{y}|\theta') = L(\mathbf{y}|\theta) | \theta = \theta'\}.$$

This means that

$$\int_{M^-} (dP_{\theta'} - dP_\theta) = -\int_{M^+} (dP_{\theta'} - dP_\theta).$$

If this is inserted in (3.23), we finally arrive at

$$D \geq (b-a) \int_{M^+} (dP_{\theta'} - dP_\theta) \geq 0,$$

considering $b > a$ and the definition of M^+. Observing the assumption $\theta' > \theta$, this shows that $E[g(\mathbf{y})|\ \theta]$ is isotone in θ.

Now we choose $g(y) = \varphi_t(y)$, $t \in R^1$ and

$$\varphi_t(y) = \begin{cases} 1 & \text{for } y > t \\ 0 & \text{else} \end{cases}.$$

Since the function $\varphi_t(y)$ is isotone in y, we get

$$E[\varphi_t(\mathbf{y})|\theta] \leq E[\varphi_t(\mathbf{y})|\theta']$$

using the first part of the proof. Because of

$$E[\varphi_t(\mathbf{y})|\theta] = P(\mathbf{y} > t) = 1 - F(t,\theta),$$

also the last part of the assertion is shown.

Theorem 3.8 Let $P = \{P_\theta,\ \theta \in \Omega \subset R^1\}$ be a distribution family of the components $\mathbf{y}_1, \ldots, \mathbf{y}_n$ of a random sample $\mathbf{Y}$ and $\mathbf{M} = M(\mathbf{Y})$ be a sufficient statistic with respect to P.

Further, let the distribution family P^M of $\mathbf{M}$ possess an isotone likelihood ratio. Denoting the $(1-\alpha)$-quantile $M_{1-\alpha}$ of the distribution belonging to $\mathbf{M}$, the function

$$k(\mathbf{Y}) = \begin{cases} 1 & \text{for } M > M_{1-\alpha} \\ \gamma_\alpha & \text{for } M = M_{1-\alpha} \\ 0 & \text{for } M < M_{1-\alpha} \end{cases} \tag{3.24}$$

is a test with the following properties:

1) $k(\mathbf{Y})$ is an UMP-test for $H_0: \theta \leq \theta_0$ against $H_A: \theta = \theta_A > \theta_0$ and $0 < \alpha < 1$. (Analogously a test for $H_0: \theta \geq \theta_0$ against $H_A: \theta = \theta_A < \theta_0$ can be formulated.)

2) For all $\alpha \in (0, 1)$, there exist M_0^α and γ_α with $-\infty \le M_0^\alpha \le \infty$, $0 \le \gamma_\alpha \le 1$ and M_0^α satisfying

$$P\{\boldsymbol{M} < M_0^\alpha | \theta_0\} \le 1-\alpha \le P\{\boldsymbol{M} \le M_0^\alpha | \theta_0\},$$

so that the corresponding test $k(\boldsymbol{Y})$ in (3.24) is with γ_α and $M_0^\alpha = M_{1-\alpha}$, an UMP-$\alpha$-test for H_0 against H_A.
3) The power function $E[k(\boldsymbol{Y})|\theta]$ is isotone in $\theta \in \Omega$.

Proof: According to Theorem 3.1, a most powerful α-test for $\mathrm{H}_0^*: \theta = \theta_0$ against H_A^*: $\theta = \theta_A$ has the form

$$k(M) = \begin{cases} 1 & \text{for } c_\alpha L_M(M,\theta_0) < L_M(M,\theta_A) \\ \gamma(M) & \text{for } c_\alpha L_M(M,\theta_0) = L_M(M,\theta_A), \\ 0 & \text{for } c_\alpha L_M(M,\theta_0) > L_M(M,\theta_A) \end{cases}$$

where $L_M(M, \theta)$ is the likelihood function of $\boldsymbol{M}$. Since $M > M_0$ implies

$$LR_M(M|\theta_0,\theta_A) \ge LR_{M_0}(M|\theta_0,\theta_A)$$

because LR_M is isotone,

$$LR_M(M|\theta_0,\theta_A) = \frac{L_M(M,\theta_A)}{L_M(M,\theta_0)} \begin{cases} > \\ = \\ < \end{cases} c_\alpha \frac{L_M(M_0^\alpha,\theta_A)}{L_M(M_0^\alpha,\theta_0)}$$

implies

$$M \begin{cases} > \\ = \\ < \end{cases} M_0^\alpha.$$

Hence $k(\boldsymbol{M})$ is the same as $k^*(\boldsymbol{M})$ and therefore a most powerful α-test for (H_0^*, H_A^*), where $M_0^\alpha = M_{1-\alpha}$ and γ_α has to be determined according to the proof of Theorem 3.1. Since $k^*(M)$ is isotone in M, the power function $E[k^*(\boldsymbol{M})|\theta]$ is by Theorem 3.7 isotone in θ. Hence, assertion (3) is true.

Further we have

$$\max_{\theta \le \theta_0} E[k^*(\boldsymbol{M})|\theta] = \alpha,$$

and $k(\boldsymbol{Y}) = k^*(\boldsymbol{M})$ in (3.24) is an UMP-α-test. Therefore, assertion (1) holds.

If we put $M_0^\alpha = M_{1-\alpha}$ for a fixed $\alpha \in (0, 1)$ taking the $(1-\alpha)$-quantile $M_{1-\alpha}$ of the distribution of $\boldsymbol{M}$ for θ_0, then we get $\gamma_\alpha = 0$ for $0 = P(\boldsymbol{M} = M_{1-\alpha}| \theta_0)$ (and, e.g. for all continuous distributions), which shows assertion (2) for this case. Otherwise we choose M_0^α analogously to (3.10) so that

$$P(\boldsymbol{M} < M_0^\alpha | \theta_0) < 1-\alpha \le P(\boldsymbol{M} \le M_0^\alpha | \theta_0).$$

Put $M_{1-\alpha} = M_0^{\alpha}$ in (3.24) and determine γ_α analogously to (3.11) as

$$\gamma_\alpha = \frac{P(\boldsymbol{M} \le M_{1-\alpha} | \theta_0) - (1-\alpha)}{P(\boldsymbol{M} = M_{1-\alpha} | \theta_0)}.$$

Hence, assertion (2) generally follows.

Corollary 3.3 If the components of a random sample $\boldsymbol{Y} = (\boldsymbol{y}_1, \boldsymbol{y}_2, \ldots, \boldsymbol{y}_n)^T$ satisfy a distribution of a one-parametric exponential family with $\theta \epsilon \Omega \subset R^1$ and if the natural parameter $\eta(\theta)$ is monotone increasing, then $H_0 : \theta \le \theta_0$ can be tested against $H_A : \theta = \theta_A > \theta_0$ for each $\alpha \in (0, 1)$ using an UMP-test $k(Y)$ of the form (3.24).

It is easy to see that the tests $H_0 : \theta \le \theta_0$ against $H_A : \theta = \theta_A > \theta_0$ can be analogously designed for antitone $\eta(\theta)$.

Example 3.6 Let the random variable $\boldsymbol{y}$ be $B(n, p)$-distributed. We want to test case A: $H_0 : p \le p_0$ against $H_A : p = p_A, p_A > p_0$ and case B: $H_0 : p \ge p_0$ against $H_A : p = p_A, p_A < p_0$ using a sample of size 1. Instead we can take $\boldsymbol{Y} = (\boldsymbol{y}_1, \boldsymbol{y}_2, \ldots, \boldsymbol{y}_n)^T$, where $\boldsymbol{y}_i$ is distributed as B$(1, p)$, because $\sum_{i=1}^{n} \boldsymbol{y}_i$ is sufficient. The distribution belongs for fixed n to a one-parametric exponential family with the natural parameter

$$\eta = \eta(p) = \ln\left(\frac{p}{1-p}\right),$$

which is isotone in p. The likelihood function is

$$L(y, \eta) = \binom{n}{p} e^{y\eta(p) - n\ln(1-p)},$$

and we have $\boldsymbol{T} = \boldsymbol{M} = \boldsymbol{y}$. According to Theorem 3.6, the random variable $\boldsymbol{y}$ has a monotone likelihood ratio. Therefore we have to choose $k^*(\boldsymbol{y})$ in case A according to (3.18) with γ_α^+ from (3.15) (putting $y^* = y^+$) and in case B according to (3.17) with γ_α^- from (3.16) (putting $y^* = y^{--}$). These tests are UMP-tests for the corresponding α.

If θ is a vector and if we intend to test one-sided hypotheses relating to the components of this vector for unknown values of the remaining components, then UMP-tests exist only in exceptional cases. The same holds already in case $\theta \in R^1$ for simple null hypotheses and two-sided alternative hypotheses. In Section 3.3.2 the latter case is considered, while in Section 3.4 tests are developed for multi-parametric distribution families.

But there are UMP-tests for a composite alternative hypothesis and a two-sided null hypothesis, where two-sided has a special meaning here. This is shown in the next theorem.

Theorem 3.9 We consider for the parameter θ of the distribution family $P = \{P_\theta,\ \theta \in \Omega \subset R^1\}$ the pair

$$H_0 : \theta \le \theta_1 \text{ or } \theta \ge \theta_2, \theta_1 < \theta_2; \theta_1,\ \theta_2 \in \Omega$$

$$H_A : \theta_1 < \theta < \theta_2; \theta_1 < \theta_2; \theta_1,\ \theta_2 \in \Omega$$

of hypotheses. If P is an exponential family and if θ is the natural parameter (until now η), then also the distribution of the random sample $\boldsymbol{Y} = (y_1, y_2, \dots, y_n)^T$ belongs to an exponential family with the sufficient statistic $\boldsymbol{T} = T(\boldsymbol{Y})$ and the natural parameter $\theta = \eta$. Then the following statements hold:

1) There is a uniformly most powerful α-test for $\{H_0, H_A\}$ of the form

$$h(T) = k(Y) = \begin{cases} 1 & \text{for } c_{1\alpha} < T < c_{2\alpha}; c_{1\alpha} < c_{2\alpha} \\ \gamma_{i\alpha} & \text{for } T = c_{i\alpha};\ i = 1,2 \\ 0 & \text{else} \end{cases} \tag{3.25}$$

   where $c_{i\alpha}$ and $\gamma_{i\alpha}$ have to be chosen so that

$$E[h(\boldsymbol{T})|\theta_1] = E[h(\boldsymbol{T})|\theta_2] = \alpha. \tag{3.26}$$

   (Then we say that $h(\boldsymbol{T}) \in K_\alpha$.)
2) The test $h(\boldsymbol{T})$ from (1) has the property that $E[h(\boldsymbol{T})|\ \theta]$ for all $\theta < \theta_1$ and $\theta > \theta_2$ is minimal in the class K_α of all tests fulfilling (3.26).
3) For $0 < \alpha < 1$ there is a point θ_0 in the interval (θ_1, θ_2) so that the power function $\pi(\theta)$ of $k(\boldsymbol{Y})$ given in (1) takes its maximum at this point and is monotone decreasing in $|\theta - \theta_0|$, provided that there is no pair $\{T_1, T_2\}$ fulfilling

$$P(\boldsymbol{T} = T_1|\ \theta) + P(\boldsymbol{T} = T_2|\ \theta) = 1$$

for all $\theta \in \Omega$.

The proof of this theorem can be found in the book of Lehmann (1959, pp. 102–103). It is based on Corollary 3.2 with $m = 2$. This theorem is hardly important in practical testing.

If null and alternative hypotheses are exchanged in Theorem 3.9, or, more precisely, if we consider under the assumptions of Theorem 3.9 the pair

$$H_0 : \theta_1 \le \theta \le \theta_2; \theta_1, \theta_2 \in \Omega \subset R^1$$

$$H_A : \theta < \theta_1 \text{ or } \theta > \theta_2; \theta_1 < \theta_2; \theta_1, \theta_2 \in \Omega \subset R^1$$

of hypotheses, there is no uniformly most powerful (UMP-) test, but a uniformly most powerful unbiased (UMPU-) test. This will be shown in the next Section 3.3.2.

3.3.2 UMPU-Tests for Two-Sided Alternative Hypotheses

Let the assumptions of Theorem 3.9 hold for the just defined pair $\{H_0, H_A\}$ of hypotheses. We will show that

$$h(T) = k(Y) = \begin{cases} 1 & \text{for } c_{1\alpha} < T \text{ or } T > c_{2\alpha};\ c_{1\alpha} < c_{2\alpha} \\ \gamma'_{i\alpha} & \text{for } T = c_{i\alpha};\ \ i = 1,2 \\ 0 & \text{else} \end{cases} \tag{3.27}$$

is a uniformly most powerful unbiased (UMPU)-test for this pair, if $c_{i\alpha}$ and $\gamma'_{i\alpha}$ are chosen so that (3.26) holds. Since $k(\boldsymbol{Y})$ is a bounded and measurable function, $E[k(\boldsymbol{Y}) \mid \theta]$ is continuous in θ, and therefore differentiation and integration (related to expectation) with respect to θ can be commuted. Regarding continuity all assumptions of Lemma 3.1 in Section 3.1 are satisfied, where $\Omega^* = \{\theta_1, \theta_2\}$. We have to maximise $E[k(\boldsymbol{Y})|\theta']$ for all $k(\boldsymbol{Y}) \in K_\alpha$ and any θ' outside of $[\theta_1, \theta_2]$ and to minimise $E[\lambda(\boldsymbol{Y})|\theta']$ with $\lambda(\boldsymbol{Y}) = 1 - k(\boldsymbol{Y})$ outside of $[\theta_1, \theta_2]$, respectively, where $\lambda(\boldsymbol{Y})$ lies in the class $K_{1-\alpha}$ of tests fulfilling

$$E[\lambda(\boldsymbol{Y})|\theta_1] = E[\lambda(\boldsymbol{Y})|\theta_2] = 1 - \alpha.$$

Theorem 3.9 implies that $\lambda(\boldsymbol{Y})$ has the form (3.25), and therefore $k(\boldsymbol{Y}) = 1 - \lambda(\boldsymbol{Y})$ has the form (3.27), where all $\gamma'_{i\alpha}$ in (3.27) have to be put equal to $1 - \gamma_{i\alpha}$ in (3.25).

Consequently, the test (3.27) is a UMP-α-test in K_α and because of Lemma 3.1 also a UMPU-α-test. These results are summarised in the next theorem.

Theorem 3.10 If $P = \{P_\theta, \theta \in \Omega \subset R^1\}$ is an exponential family with the sufficient statistic $T(\boldsymbol{Y})$ and $k(\boldsymbol{Y})$ is a test of the form (3.27) for the pair

$$H_0: \theta_1 \le \theta \le \theta_2;\ \ \theta_1 < \theta_2;\ \ \theta_1, \theta_2 \in \Omega \subset R^1$$

$$H_A: \theta < \theta_1 \quad \text{or} \quad \theta > \theta_2;\ \ \theta_1, \theta_2 \in \Omega \subset R^1$$

of hypotheses, then $k(\boldsymbol{Y})$ is a UMPU-α-test.

In the applications, a pair $\{H_0, H_A\}$ is often tested with the simple null hypothesis H_0: $\theta = \theta_o$ and the alternative hypothesis $H_A: \theta \neq \theta_o$. This case is now considered.

Theorem 3.11 If under the assumptions of Theorem 3.10 the pair

$$H_0: \theta = \theta_0, \theta_0 \in \Omega \subset R^1$$

$$H_A:\ \theta \neq \theta_0, \theta_0 \in \Omega \subset R^1$$

of hypotheses is tested using $k(\boldsymbol{Y})$ in the form (3.27), where all $c_{i\alpha}$ and $\gamma'_{i\alpha}$ so that

$$E[k(\boldsymbol{Y})|\theta_0] = \alpha \tag{3.28}$$

and

$$E[T(\boldsymbol{Y})k(\boldsymbol{Y})|\theta_0] = \alpha E[T(\boldsymbol{Y})|\theta_0] \tag{3.29}$$

hold, then $k(\boldsymbol{Y})$ is a UMPU-α-test.

Proof: The condition (3.28) ensures that $k(\boldsymbol{Y})$ is an α-test. To get an unbiased test $k(\boldsymbol{Y})$, the expectation $E[k(\boldsymbol{Y})|\theta]$ has to be minimal at θ_0. Therefore

$$D(\theta) = \frac{\partial}{\partial\theta}E[k(\boldsymbol{Y})|\theta] = \int_{Y}\frac{\partial}{\partial\theta}k(Y)dP_\theta$$

necessarily has to be 0 at $\theta = \theta_0$. Since $L(Y,\theta) = C(\theta)e^{\theta T}h(Y) \sim L_T(T,\theta)$ by assumption, with the notation $C' = \frac{\partial C}{\partial\theta}$, we get

$$\frac{\partial}{\partial\theta}L_T(T,\theta) = \frac{C'(\theta)}{C(\theta)}L_T(T,\theta) + TL_T(T,\theta).$$

and therefore

$$D = D(\theta) = \frac{C'(\theta)}{C(\theta)}E[k(\boldsymbol{Y})|\theta] + E[T(\boldsymbol{Y})k(\boldsymbol{Y})|\theta].$$

Regarding

$$0 = \frac{\partial}{\partial\theta}\int_{\{Y\}} dP_\theta = \frac{C'(\theta)}{C(\theta)}\int_{\{Y\}} dP_\theta + E[T(\boldsymbol{Y})|\theta],$$

it follows

$$\frac{C'(\theta)}{C(\theta)} = -E[T(\boldsymbol{Y})|\theta].$$

Because of (3.28) we get

$$0 = -\alpha E[T(\boldsymbol{Y})|\theta_0] + E[T(\boldsymbol{Y})k(\boldsymbol{Y})|\theta_0]$$

by putting $\theta = \theta_0$ and therefore (3.29). This shows that (3.29) is true because the test is unbiased.

Now let M be the set of the points $\{E[k(\boldsymbol{Y})|\theta_0], E[T(\boldsymbol{Y})k(\boldsymbol{Y})|\theta_0]\}$ taking all critical functions $k(Y)$ on $\{Y\}$. Then M is convex and contains for $0 < z < 1$ all points $\{z, zE[T(\boldsymbol{Y})|\theta_0]\}$ as well as all points (α, x_2) with $x_2 > \alpha E[T(\boldsymbol{Y})|\theta_0]$. This follows because there are tests with $E[k(\boldsymbol{Y})|\theta_0] = \alpha$, where $D(\theta_0) > 0$. Analogously we get that M contains also points (α, x_1) with $x_1 > \alpha E[T(\boldsymbol{Y})|\theta_0]$. But this means that $(\alpha, \alpha E[T(\boldsymbol{Y})|\theta_0])$ is an inner point of M. Taking Corollary 3.2, part (4), in Section 3.2, into account, there exist two constants k_1, k_2 and a test $k(\boldsymbol{Y})$ fulfilling (3.28) and (3.29) and supplying $k(Y) = 1$ iff

$$C(\theta_0)(k_1 + k_2T)\,e^{\theta_0 T} < C(\theta')\,e^{\theta' T}.$$

The T-values satisfying this inequality lie either below or above a real constant, respectively, or outside an interval $[c_{1\alpha}, c_{2\alpha}]$. However, the test can have neither the form (3.24) given in Theorem 3.8 for the isotone case nor the corresponding form for the antitone case, because the statement (3) of Theorem 3.8 contradicts (3.29). This shows that the UMPU-test has the form (3.27).

Example 3.7 Let P be the family of Poisson distributions with $\boldsymbol{Y} = (\boldsymbol{y}_1, \ldots, \boldsymbol{y}_n)^T$ and the likelihood function

$$L(Y,\lambda) = \prod_{i=1}^{n} \frac{1}{y_i!} e^{\left[\ln \lambda \sum_{i=1}^{n} y_i - \lambda n\right]}, \quad y_i = 0,1,2,\ldots; \ \lambda \in R^+$$

with the natural parameter $\theta = \ln \lambda$. We want to test the pair $H_0 : \lambda = \lambda_0, H_A : \lambda \neq \lambda_0$ of hypotheses. The likelihood function of the sufficient statistic $T = \sum_{i=1}^{n} \boldsymbol{y}_i$ is with $\theta = \ln \lambda$

$$L_T(T,\theta) = \frac{1}{T!} e^{\theta T - A(\theta)}.$$

It defines also a distribution from a one-parametric exponential family with $\theta = \ln (n\lambda)$ and

$$H_0 : \theta = \theta_0 \text{ with } \theta_0 = \ln(n\lambda_0); \ H_A : \theta \neq \theta_0.$$

Hence, all assumptions of Theorem 3.11 are fulfilled. Therefore (3.27) is a UMPU-α-test for $\{H_0, H_A\}$, if $c_{i\alpha}$ and $\gamma'_{i\alpha}$ $(i = 1, 2)$ are determined so that (3.28) and (3.29) hold. Considering

$$T L_T(T,\lambda) = n\lambda L_T(T-1,\lambda) \quad (T = 1,2,\ldots),$$
$$E(\boldsymbol{T}\theta_0) = \lambda_0 = e^{\theta_0},$$

and putting now w.l.o.g. $n = 1$, the simultaneous equations

$$\alpha = P(\boldsymbol{T} < c_{1\alpha}|\theta_0) + P(\boldsymbol{T} > c_{2\alpha}|\theta_0) + \gamma'_{1\alpha} L_T(c_{1\alpha}, \theta_0) + \gamma'_{2\alpha} L_T(c_{2\alpha}, \theta_0),$$
$$\alpha = P(\boldsymbol{T} - 1 < c_{1\alpha}|\theta_0) + P(\boldsymbol{T} - 1 > c_{2\alpha}|\theta_0) + \gamma'_{1\alpha} L_T(c_{1\alpha} - 1, \theta_0) + \gamma'_{2\alpha} L_T(c_{2\alpha} - 1, \theta_0) \tag{3.30}$$

have to be solved.

The results of this example supply the following statements.

Theorem 3.12 If $\boldsymbol{y}$ is distributed as $P(\lambda)$, then a UMPU-α-test for the pair

$$H_0 : \lambda = \lambda_0, H_A : \lambda \neq \lambda_0, \quad \lambda_0 \in R^+$$

of hypotheses has the form (3.27), where constants $c_{i\alpha}$ and $\gamma'_{i\alpha}$ are solutions of (3.30) with natural $c_{i\alpha}$ and $0 \leq \gamma'_{i\alpha} \leq 1$.

Example 3.7 (continuation)
It needs some time of calculation to find the constants $c_{i\alpha}$ and $\gamma'_{i\alpha}$. We give a numerical example to illustrate the solution procedure. We test $H_0: \lambda = 10$ against $H_A: \lambda \neq 10$. The values of the probability function and the likelihood function can be determined by statistical software, for instance, by SPSS or R. We choose $\alpha = 0.1$ and look for possible pairs (c_1, c_2).

For $c_1 = 4$, $c_2 = 15$, we obtain equations

$$0.006206 = 0.018917\gamma'_1 + 0.034718\gamma'_2$$
$$0.013773 = 0.007567\gamma'_1 + 0.052077\gamma'_2$$

from (3.30) supplying the improper solutions $\gamma'_1 = -0.215$, $\gamma'_2 = 0.296$. The pairs (4, 16) and (5, 15) lead to improper values (γ'_1, γ'_2), too. Finally, we recognise that only the values $c_1 = 5$, $c_2 = 16$ and $\gamma'_1 = 0.697$, $\gamma'_2 = 0.799$ solve the problem. Hence, (3.27) has the form

$$k(y) = \begin{cases} 1 & \text{for } y < 4 \text{ or } y > 15 \\ 0.697 & \text{for } y = 4 \\ 0.799 & \text{for } y = 15 \\ 0 & \text{else} \end{cases},$$

and $k(y)$ is the uniformly most powerful unbiased 0.1-test.

Example 3.8 Let $\boldsymbol{y}$ be distributed as $B(n, p)$. Knowing one observation $\boldsymbol{y} = \boldsymbol{Y}$, we want to test H_0: $p = p_0$ against $H_A: p \neq p_0$, $p_0 \in (0, 1)$. The natural parameter is $\eta = \ln \frac{p}{1-p}$, and $\boldsymbol{y}$ is sufficient with respect to the family of binomial distributions. Therefore the UMPU-α-test is given by (3.27), where $c_{i\alpha}$ and $\gamma'_{i\alpha}$ $(i = 1, 2)$ have to be determined from (3.28) and (3.29). With

$$L_n(y|p) = \binom{n}{y} p^y (1-p)^{n-y},$$

Equation (3.28) has the form

$$\sum_{y=0}^{c_{1\alpha}-1} L_n(y|p_0) + \sum_{y=c_{2\alpha}+1}^{n} L_n(y|p_0) + \gamma'_{1\alpha} L_n(c_{1\alpha}|p_0) + \gamma'_{2\alpha} L_n(c_{2\alpha}|p_0) = \alpha. \tag{3.31}$$

Regarding

$$y L_n(y|p) = np\, L_{n-1}(y-1|p) \text{ and } E(\boldsymbol{y}|p_0) = np_0,$$

the relation (3.29) leads to

$$\sum_{y=0}^{c_{1\alpha}-1} L_{n-1}(y-1|p_0) + \sum_{y=c_{2\alpha}+1}^{n} L_{n-1}(y-1|p_0) + \gamma'_{1\alpha} L_{n-1}(c_{1\alpha}-1|p_0) + \gamma'_{2\alpha} L_{n-1}(c_{2\alpha}-1|p_0) = \alpha. \tag{3.32}$$

The solution of these two simultaneous equations can be obtained by statistical software, for example, by R. Further results can be found in the book of Fleiss et al. (2003).

Example 3.9 If $\boldsymbol{Y} = (\boldsymbol{y}_1, \ldots, \boldsymbol{y}_n)^{\mathrm{T}}$ is a random sample with components distributed as $N(0, \sigma^2)$, then the natural parameter is $\eta = -\frac{1}{2\sigma^2}$, and $\sum_{i=1}^{n} \boldsymbol{y}_i^2 = T(\boldsymbol{Y}) = \boldsymbol{T}$ is sufficient with respect to the family of $N(0, \sigma^2)$-distributions. The variable $\boldsymbol{T}$ is distributed with the density function $\frac{1}{\sigma^2} g_n\left(\frac{T}{\sigma^2}\right)$, where $g_n(x)$ is the density function of a $CS(n)$-distribution. Therefore, the distribution family of $\boldsymbol{T}$ is a one-parametric exponential family (depending on σ^2). Then

$$h(T) = k(Y) = \begin{cases} 1 & \text{for } T < c_{1\alpha}\sigma_0^2 \text{ or } T > c_{2\alpha}\sigma_0^2 \\ 0 & \text{else} \end{cases}$$

is a UMPU-α-test for the hypotheses $H_0: \sigma^2 = \sigma_0^2 \left(0 < \sigma_0^2 < \infty\right)$ against $H_A: \sigma^2 \neq \sigma_0^2$, where constants $c_{i\alpha}$, $i = 1,2$ are non-negative and satisfy

$$\int_{c_{1\alpha}}^{c_{2\alpha}} g_n(x)dx = 1 - \alpha \tag{3.33}$$

$$\int_{c_{1\alpha}}^{c_{2\alpha}} x g_n(x)dx = (1-\alpha)E\left[\frac{\boldsymbol{T}}{\sigma_0^2} | \sigma_0^2\right] = n(1-\alpha)$$

A symmetric formulation supplies under the conditions of the next corollary of Theorem 3.11 a UMPU-test (naturally these conditions are not fulfilled for Example 3.9).

Corollary 3.4 Let the distribution of the sufficient statistic $\boldsymbol{T} = T(\boldsymbol{Y})$ be for $\theta = \theta_0$ symmetric with respect to a constant m, and let the assumptions of Theorem 3.11 be fulfilled. Then a UMPU-α-test is given by (3.27), $c_{2\alpha} = 2m - c_{1\alpha}$, $\gamma_\alpha = \gamma'_{1\alpha} = \gamma'_{2\alpha}$ and

$$P\{T(\boldsymbol{Y}) < c_{1\alpha}|\theta_0\} + \gamma_\alpha P\{T(\boldsymbol{Y}) = c_{1\alpha}|\theta_0\} = \frac{\alpha}{2}. \tag{3.34}$$

Proof: Regarding

$$P\{T(\boldsymbol{Y}) < m - x\} = P\{T(\boldsymbol{Y}) > m + x\}.$$

Equation (3.28) is satisfied for $x = m - c_{1\alpha}$, where $c_{1\alpha}$ is the $\alpha/2$-quantile of the $\boldsymbol{T}$-distribution. Then we have

$$E\{T(\boldsymbol{Y})k(\boldsymbol{Y})|\theta_0\} = E\{[T(\boldsymbol{Y}) - m]k(\boldsymbol{Y})|\theta_0\} + mE\{k(\boldsymbol{Y})|\theta_0\}.$$

Since the first summand on the right-hand side vanishes for a $k(\boldsymbol{Y})$, which fulfils the assumptions above (symmetry), it follows $mE\{k(\boldsymbol{Y})|\theta_0\} = m\alpha$ and because of $E\{T(\boldsymbol{Y})|\theta_0\} = m$ also (3.29).

Example 3.10 If $\boldsymbol{Y} = (\boldsymbol{y}_1, \boldsymbol{y}_2, \dots, \boldsymbol{y}_n)^T$ is a random sample with components distributed as $N(\mu, \sigma^2)$ and σ^2 is known, then $\boldsymbol{T} = T(\boldsymbol{Y}) = \sum_{i=1}^{n} \boldsymbol{y}_i$ is sufficient with respect to μ. The statistic $\frac{1}{n}\boldsymbol{T} = \bar{\boldsymbol{y}}$ is distributed as $N\left(\mu, \frac{\sigma^2}{n}\right)$, that is, it is symmetrically distributed with respect to μ. Regarding the hypotheses $H_0: \mu = \mu_0, H_A: \mu \neq \mu_0$, a UMPU-$\alpha$-test is given by

$$k(Y) = \begin{cases} 1 & \text{for } z < z_{\frac{\alpha}{2}} \text{ or } z > z_{1-\frac{\alpha}{2}} \\ 0 & \text{else} \end{cases}, \tag{3.35}$$

if z_P is the P-quantile of the standard normal distribution and $\boldsymbol{z} = \frac{\bar{\boldsymbol{y}} - \mu_0}{\sigma}\sqrt{\boldsymbol{n}}$. In the description of (3.27), we obtain

$$c_{1\alpha} = \mu_0 + z_{\frac{\alpha}{2}}\frac{\sigma}{\sqrt{n}} \text{ and } c_{2\alpha} = \mu_0 - z_{\frac{\alpha}{2}}\frac{\sigma}{\sqrt{n}} = \mu_0 + z_{1-\frac{\alpha}{2}}\frac{\sigma}{\sqrt{n}}.$$

3.4 Tests for Multi-Parametric Distribution Families

In several examples we supposed normal distributed components $\boldsymbol{y}_i$, where either μ or σ^2 were assumed to be known. However, in the applications in the most cases both parameters are unknown. When we test a hypothesis with respect to one parameter the other unknown parameter is a disturbance parameter. Here we mainly describe a procedure for designing α-tests that are on the common boundary of the closed subsets of Ω belonging to both hypotheses independent of a sufficient statistic with respect to the noisy parameters. At the end of this section, we briefly discuss a further possibility. Then we need the concept of α-similar tests and especially of α-similar tests on the common boundary Ω^* of ω and $\Omega \setminus \omega$ given in Definition 3.2. We start with an example.

Example 3.11 Let the vector $\boldsymbol{Y} = (\boldsymbol{y}_1, \boldsymbol{y}_2, \ldots, \boldsymbol{y}_n)^T$ be a random sample whose components are distributed as $N(\mu, \sigma^2)$. We test the null hypothesis H_0: $\mu = \mu_0$, against $H_A : \mu \neq \mu_0$ for arbitrary σ^2. The statistic

$$\boldsymbol{t}(\mu) = \frac{\bar{\boldsymbol{y}} - \boldsymbol{\mu}}{\boldsymbol{s}}\sqrt{n}$$

is a function of the sufficient statistic $\boldsymbol{M} = (\bar{\boldsymbol{y}}, \boldsymbol{s}^2)^T$. It is centrally distributed as $t(\mathrm{n} - 1)$. Here and in the following examples,

$$\boldsymbol{s}^2 = \frac{1}{n-1}\sum_{i=1}^{n}(\boldsymbol{y}_i - \bar{\boldsymbol{y}})^2$$

is the sample variance, the unbiased estimator of the variance σ^2. The statistic $\boldsymbol{t}(\mu_0)$ is non-centrally distributed as $t\left(n-1; \frac{\mu - \mu_0}{\sigma}\sqrt{n}\right)$. Therefore the test

$$k(Y) = \begin{cases} 1 & \text{for } |t(\mu_0)| > t\left(n-1 \middle| 1 - \frac{\alpha}{2}\right) \\ 0 & \text{else} \end{cases} \quad (3.36)$$

is an α-test, where $t\left(n-1 \middle| 1-\frac{\alpha}{2}\right)$ is the α-quantile of the central t-distribution with $n - 1$ degrees of freedom. These quantiles are shown in Table D.3. Since Ω^* is in this case, the straight line in the positive (μ, σ^2)-half-plane ($\sigma^2 > 0$) defined by $\mu = \mu_0$ and $P\{k(\boldsymbol{Y}) = 1 | \mu_0\} = \alpha$ for all σ^2, $k(\boldsymbol{Y})$ is an α-similar test on Ω^*.

3.4.1 General Theory

Definition 3.7 We consider a random sample $\boldsymbol{Y} = (\boldsymbol{y}_1, \boldsymbol{y}_2, \ldots, \boldsymbol{y}_n)^T$ from a family $P = \{P_\theta, \theta \in \Omega\}$ of distributions P_θ and write $\Omega_0 = \omega$ and $\Omega_A = \Omega \setminus \omega$ for the subsets in Ω defined by the null and alternative hypothesis, respectively. The set $\Omega^* = \bar{\Omega}_0 \cap \bar{\Omega}_A$ denotes the common boundary of the closed sets $\bar{\Omega}_0$ and $\bar{\Omega}_A$. Let $P^* \subset P$ be the subfamily $P^* = \{P_\theta, \ \theta \in \Omega^* \subset \Omega\}$ on this common boundary. We assume that there is a (non-trivial) sufficient statistic $T(\boldsymbol{Y})$ with respect to Ω^* so that $E[k(\boldsymbol{Y}) | T(\boldsymbol{Y})]$ is independent of $\theta \in \Omega^*$, that is, $k(\boldsymbol{Y})$ is α-similar on Ω^* with

$$\alpha = E[k(\boldsymbol{Y}) | T(\boldsymbol{Y}), \theta \in \Omega^*]. \quad (3.37)$$

A test $k(\boldsymbol{Y})$ satisfying (3.37) is said to be an α-test with Neyman structure.

Hence, tests with Neyman structure are always α-similar on Ω^*. Moreover they have the property that α can be calculated by (3.37) as conditional expectation of the sufficient statistic $T(\boldsymbol{Y})$ for the given value $T(Y)$. Since the conditional expectation in (3.37) for each surface is defined by $T(\boldsymbol{Y}) = T(Y) = T$ independent of $\theta \in \Omega^*$, the tests of this section can be reduced for each single

T-value to such of preceding sections (provided that (3.37) holds). Therefore we will look for UMP-tests or UMPU-tests in the set of all tests with Neyman structure by trying to find a sufficient statistic with respect to P^*. But first of all we want to know whether tests with Neyman structure exist. The next theorem states conditions for it.

Theorem 3.13 If the statistic $T(\boldsymbol{Y})$ with the notations of Definition 3.7 is sufficient with respect to P^*, then a test $k(\boldsymbol{Y})$ that is α-similar on the boundary has with probability 1 a Neyman structure with respect to $T(\boldsymbol{Y})$ iff the family P_T of the distributions of $T(\boldsymbol{Y})$ is bounded complete.

Proof:

a) Let P_T be bounded complete, and let $k^*(\boldsymbol{Y})$ be α-similar on the boundary. Then the equation $E[k(\boldsymbol{Y}) - \alpha|\theta \epsilon \Omega^*] = 0$ is fulfilled. Now consider

$$d(\boldsymbol{Y}) = k(\boldsymbol{Y}) - \alpha = E[k^*(\boldsymbol{Y}) - \alpha|T(\boldsymbol{Y}), \theta \epsilon \Omega^*].$$

Since $T(\boldsymbol{Y})$ is sufficient, we get $E[d(\boldsymbol{Y}) \mid P_T] = 0$. However, critical functions $k(\boldsymbol{Y})$ are bounded by definition. Thus the assertion follows from the bounded completeness.

b) If P_T is not bounded complete, then there exist a function f and a real $C > 0$ so that $|f[T(Y)]| \le C$ and $E\{f[T(\boldsymbol{Y})]| \theta \in \Omega^*\} = 0$, but that $f[T(Y)] \neq 0$ holds with positive probability for at least one element of P_T. Putting $\frac{1}{C}$ min$(\alpha, 1 - \alpha) = K$, the function

$$k(\boldsymbol{Y}) = h[T(\boldsymbol{Y})] = Kf[T(\boldsymbol{Y})] + \alpha$$

because of $0 \le k(Y) \le 1$ for all $Y \in \{Y\}$ is a test and because of

$$E[k(\boldsymbol{Y})|\theta \epsilon \Omega^*] = K\, E\{f[T(\boldsymbol{Y})]|\theta \epsilon \Omega^*\} + \alpha = \alpha$$

α-similar on the boundary Ω^*. But it holds $k(Y) \neq \alpha$ for elements of P_T with $f(T) \neq 0$. Therefore the test has no Neyman structure.

Using Theorems 1.3 and 1.4 as well as Lemma 3.1, the problems of this section can be solved for k-parametric exponential families. Solutions can also be found for other distribution families. But we do not want to deal with them here.

Theorem 3.14 Let us choose $\theta = (\lambda, \theta_2, \dots, \theta_k)^T, \lambda \in R^1$ in Definition 3.7 and consider each of the hypotheses $H_0 : \lambda \in \widetilde{\Omega}_0 \subset R^1$ and $H_A : \lambda \notin \widetilde{\Omega}_0 \subset R^1$ with arbitrary $\theta_2, \dots, \theta_k$. Besides, let P be a k-parametric exponential family with natural parameters $\eta_1, \dots, \eta_k$, where we put $\eta_1 = \lambda$ and $T_1(\boldsymbol{Y}) = S(\boldsymbol{Y}) = \boldsymbol{S}$. Then UMPU-$\alpha$-tests exist for $\{H_0, H_A\}$, namely, we get for $\widetilde{\Omega}_0 = (-\infty, \lambda_0]$ the form

$$k(Y) = h(S|T^*) = \begin{cases} 1 & \text{for } S > c_\alpha(T^*) \\ \gamma_\alpha(T^*) & \text{for } S = c_\alpha(T^*), \\ 0 & \text{else} \end{cases} \tag{3.38}$$

for $\widetilde{\Omega}_0 = [\lambda_0, \infty)$ the form

$$k(Y) = h(S|T^*) = \begin{cases} 1 & \text{for } S < c_\alpha(T^*) \\ \gamma_\alpha(T^*) & \text{for } S = c_\alpha(T^*), \\ 0 & \text{else} \end{cases} \tag{3.39}$$

and for $\widetilde{\Omega}_0 = [\lambda_1, \lambda_2]$ the form

$$k(Y) = h(S|T^*) = \begin{cases} 1 & \text{for } S < c_{1\alpha}(T^*) \text{ or } S > c_{2\alpha}(T^*) \\ \gamma_{i\alpha}(T^*)\ (i = 1{,}2) & \text{for } S = c_{i\alpha}(T^*) \\ 0 & \text{else} \end{cases}, \tag{3.40}$$

The constants in (3.38) and (3.39) have to determined so that

$$E\left[h(\boldsymbol{S}|\boldsymbol{T}^*)|\boldsymbol{T}^* = T^*, \theta \in \widetilde{\Omega_0}\right] = \alpha$$

for all T^*. Further, the constants in (3.40) must fulfil the equation

$$E\left[h(\boldsymbol{S}|\boldsymbol{T}^*)|\boldsymbol{T}^* = T^*, \theta \in \widetilde{\Omega_0}\right] = \alpha$$

and, in the special case $\lambda_1 = \lambda_2 = \lambda_0$, both equations

$$E[h(\boldsymbol{S}|\boldsymbol{T}^*)|\boldsymbol{T}^* = T^*, \lambda = \lambda_0] = \alpha,$$
$$E[\boldsymbol{S}h(\boldsymbol{S}|\boldsymbol{T}^*)|\boldsymbol{T}^* = T^*, \lambda = \lambda_0] = \alpha E[\boldsymbol{S}|\boldsymbol{T}^* = T^*, \lambda = \lambda_0]$$

with probability 1, respectively (analogous to (3.28) and (3.29)).

Proof: Regarding the null hypothesis, we have

$$\Omega^* = \{\theta : \lambda = \lambda_0; \eta_2, \dots, \eta_k \text{ arbitrary}\}$$

if $\widetilde{\Omega_0} = (-\infty, \lambda_0]$, $\widetilde{\Omega_0} = [\lambda_0, \infty)$ or $\widetilde{\Omega_0} = \{\lambda_0\}$ and

$$\Omega^* = \{\theta : \lambda = \lambda_1 \text{ or } \lambda = \lambda_2; \eta_2, \dots, \eta_k \text{ arbitrary}\}$$

if $\widetilde{\Omega_0} = [\lambda_1, \lambda_2]$, $\lambda_1 \neq \lambda_2$. Because of Theorems 1.3 and 1.4, $\boldsymbol{T}$ is complete (and therefore bounded complete) as well as sufficient with respect to P and therefore also with respect to P^*. The conditional distribution of $\boldsymbol{S}$ for $\boldsymbol{T}^* = T^*$ belongs to a one-parametric exponential family with the parameter space $\Omega \cap R^1 = \widetilde{\Omega_0}$. In the case of one-sided hypotheses, the test $k(\boldsymbol{Y})$ in (3.38) and (3.39), respectively, designed analogously to (3.24), is a UMP-α-test for known $\eta_2, \dots, \eta_k$ by Corollary 3.3 of Theorem 3.8 for suitable choice of the constants. Taking the sufficiency of $T(\boldsymbol{Y})$ into account, these constants can be determined independent of $\eta_2, \dots, \eta_k$. Hence, $k(\boldsymbol{Y})$ in (3.38) and (3.39), respectively, have Neyman structure by Theorem 3.14. Therefore, both are UMPU-α-tests because of Lemma 3.1. The assertion for the two-sided case follows analogously by applying Theorems 3.10 and 3.11.

Example 3.12 Let $\begin{pmatrix} \boldsymbol{x} \\ \boldsymbol{y} \end{pmatrix}$ be bivariate distributed with independent components. Let the marginal distributions of $\boldsymbol{x}$ be $P(\lambda_x)$ and of $\boldsymbol{y}$ be $P(\lambda_y)$, respectively ($0 < \lambda_x, \lambda_y < \infty$). We want to test H_0: $\lambda_x = \lambda_y$ against $H_A : \lambda_x \neq \lambda_y$. We take a sample of size n and put $\boldsymbol{T} = \boldsymbol{x} + \boldsymbol{y}$. The conditional distribution of $\boldsymbol{x}$ for $\boldsymbol{T}^* = T^*$ is a $B(T^*, p)$-distribution with $p = \frac{\lambda_x}{\lambda_x + \lambda_y}$, where $\boldsymbol{T}^*$ is distributed as $P(\lambda_x + \lambda_y)$. Hence, the probability function of the two-dimensional random variable $(\boldsymbol{x}, \boldsymbol{T}^*)$ is

$$P(x, T^* \mid \theta, \eta_2) = \binom{T^*}{x} \frac{1}{T^*!} \mathrm{e}^{\theta x + \eta_2 T^* - \lambda_x - \lambda_y}$$

which has the form of an exponential family for $\theta = \ln \frac{\lambda_x}{\lambda_y}$, $\eta_2 = \ln \lambda_y$ and $A(\eta) = e^{\eta_2}\left(1 + e^{\theta}\right)$. Now the pair $\{H_0, H_A\}$ can be written as H_0: $\theta = 0, \eta_2$ arbitrary and $H_A : \theta \neq 0$, η_2 arbitrary. Therefore the optimal UMPU-α-test for $\{H_0, H_A\}$ has the form (3.40).

Assume H_0 (i.e. for p = ½) the conditional distribution of $\boldsymbol{x}$ under the condition $\boldsymbol{T}^* = T^*$ is symmetric with respect to $\frac{1}{2} T^*$. By Corollary 3.4 of Theorem 3.11, the constants in (3.28) and (3.29), respectively, have to be calculated from

$$c_{1\alpha}(T^*) = c_{\alpha}, \qquad c_{2\alpha}(T^*) = T^* - c_{\alpha}$$

$$\gamma'_{1\alpha}(T^*) = \gamma'_{2\alpha}(T^*) = \frac{\frac{\alpha}{2} - F\left(c_{\alpha} \middle| T^*, p = \frac{1}{2}\right)}{P\left(c_{\alpha} \middle| T^*, p = \frac{1}{2}\right)} \tag{3.41}$$

where c_{α} is the largest integer for which the distribution function $F\left(x_{\alpha} \middle| T^*, p = \frac{1}{2}\right)$ of the $B\left(T^*, \frac{1}{2}\right)$-distribution is less or equal to $\frac{\alpha}{2}$. Further, $P\left(x_{\alpha} \middle| T^*, p = \frac{1}{2}\right)$ is the probability function of the $B\left(T^*, \frac{1}{2}\right)$-distribution.

The results of the example imply the following statements.

Theorem 3.15 If $\boldsymbol{x}$ and $\boldsymbol{y}$ are independent from each other distributed as $P(\lambda_x)$ and $P(\lambda_y)$, respectively, and if H_0: $\lambda_x = \lambda_y$ is tested against $H_A : \lambda_x \neq \lambda_y$, then a UMPU-$\alpha$-test is given by (3.40), where the constants are determined with the notations of Example 3.12 by (3.41).

The following theorem allows the simple construction of further tests. The present theory does not supply the t-test of (3.36) in Example 3.11, which is often used in applications.

Theorem 3.16 Let the assumptions of Theorem 3.14 be fulfilled. If moreover there exists a function $g(S,T^*)$, which is isotone in S for all T^*, and if $\boldsymbol{g} = g(\boldsymbol{S},\boldsymbol{T}^*)$ is under H_0 independent of $\boldsymbol{T}^*$, then the statements of Theorem 3.14 hold for the tests

$$k(Y) = r(g) = \begin{cases} 1 & \text{for } g > c_\alpha \\ \gamma_\alpha & \text{for } g = c_\alpha, \\ 0 & \text{else} \end{cases} \tag{3.42}$$

in the case $\widetilde{\Omega_0} = (-\infty, \lambda_0]$,

$$k(Y) = r(g) = \begin{cases} 1 & \text{for } g < c_\alpha \\ \gamma_\alpha & \text{for } g = c_\alpha \\ 0 & \text{else} \end{cases} \tag{3.43}$$

in the case $\widetilde{\Omega_0} = [\lambda_0, \infty)$, and

$$k(Y) = r(g) = \begin{cases} 1 & \text{for } g < c_{1\alpha} \text{ or } g > c_{2\alpha} \\ \gamma_{i\alpha} & \text{for } g = c_{i\alpha}\ (i = 1,2) \\ 0 & \text{else} \end{cases} \tag{3.44}$$

in the case $\widetilde{\Omega_0} = [\lambda_1, \lambda_2]$, if c_α and γ_α are determined in (3.42) and (3.43), respectively, that $k(\boldsymbol{Y})$ is an α-test and for (3.44) conditions analogous to both of the last equations of Theorem 3.14 are fulfilled.

Proof: The prescriptions for determining the constants imply $E[r(\boldsymbol{g})|H_0] = \alpha$, that is, for instance, in the case of the test in (3.42)

$$P(\boldsymbol{g} > c_\alpha) + \gamma_\alpha P(\boldsymbol{g} = c_\alpha) = \alpha.$$

Since $\boldsymbol{g}$ is independent of $\boldsymbol{T}^*$ for $\lambda = \lambda_0$, c_α and γ_α are independent of $\boldsymbol{T}^*$.

Further, since $g(S, T^*)$ is isotone in S for each T^*, the tests in (3.42) and in (3.38) as well as analogously the tests in (3.43) and in (3.39) are equivalent (i.e. there rejection regions in the sample space $\{Y\}$ are identical). The same conclusion can be made in the two-sided case with respect to the tests in (3.44) and (3.40) if only the last equations of Theorem 3.14 are replaced by the equivalent conditions

$$E[r(\boldsymbol{g})|\, T^*, \lambda_0] = \alpha,$$
$$E[\boldsymbol{g}\, r(\boldsymbol{g})|\, T^*, \lambda_0] = \alpha\, E[\boldsymbol{g}|\, T^*, \lambda_0].$$

We will use this theorem for showing that the t-test in Example 3.11 is a UMPU-test.

Example 3.11 (continuation)

We know from Chapter 1 that $\left(\sum_{i=1}^{n} y_i, \sum_{i=1}^{n} y_i^2\right)^T = \boldsymbol{T}$ is minimal sufficient with respect to the family of $N(\mu, \sigma^2)$-distributions. With the notations of Theorem 3.14, we put

$$\boldsymbol{S} = \bar{\boldsymbol{y}} = \frac{1}{n}\sum_{i=1}^{n} \boldsymbol{y}_i \text{ and } \boldsymbol{T}^* = \sum_{i=1}^{n} \boldsymbol{y}_i^2,$$

where $\boldsymbol{T}^*$ is complete sufficient with respect to P^* (e.g. with respect to the family of $N(\mu_0, \sigma^2)$-distributions). Now we consider

$$\boldsymbol{t} = \boldsymbol{g} = g(\boldsymbol{S}, \boldsymbol{T}^*) = \frac{\sqrt{n}\,(\boldsymbol{S} - \mu_0)}{\sqrt{\frac{1}{n-1}\left(\boldsymbol{T}^* - n\boldsymbol{S}^2\right)}} = \frac{\bar{\boldsymbol{y}} - \mu_0}{\boldsymbol{s}}\sqrt{n}. \tag{3.45}$$

We know that $\boldsymbol{g}$ is for $\mu = \mu_0$ distributed as $t(\mathrm{n} - 1)$ and to be precise independent of $\sigma^2 \in R^+$. However, for known $\mu = \mu_0$ the statistic $\frac{1}{\sigma^2}\boldsymbol{T}^*$ is distributed as $CS(n)$. Therefore Theorem 1.5 implies that $\boldsymbol{g}$ and $\boldsymbol{T}^*$ are independent for all $\theta \in \Omega^*$ (i.e. for $\mu = \mu_0$). Thus the assumptions of Theorem 3.16 are fulfilled because g is isotone in S for each T^*. Consequently the t-test is a UMPU-α-test.

This leads to the test of W.S. Gosset published in 1908 under the pseudonym Student (Student, 1908).

Theorem 3.17 Student (1908)

If $n > 1$ components of a random sample $\boldsymbol{Y} = (\boldsymbol{y}_1, \boldsymbol{y}_2, \ldots, \boldsymbol{y}_n)^T$ are distributed as $N(\mu, \sigma^2)$, then the so-called t-test (Student's test) for testing H_0: $\mu = \mu_0$, σ^2 arbitrary, of the form

$$k(Y) = \begin{cases} 1 & \text{for } t > t(n-1|1-\alpha) \\ 0 & \text{else} \end{cases},$$

for H_A: $\mu > \mu_0$, σ^2 arbitrary, of the form

$$k(Y) = \begin{cases} 1 & \text{for } t < -t(n-1|1-\alpha) \\ 0 & \text{else} \end{cases},$$

for H_A: $\mu < \mu_0$, σ^2 arbitrary, of the form

$$k(Y) = \begin{cases} 1 & \text{for } |t| > t\left(n-1\middle|1-\frac{\alpha}{2}\right) \\ 0 & \text{else} \end{cases}$$

and for $H_A: \mu \neq \mu_0, \sigma^2$ arbitrary, respectively, is a UMPU-α-test, where $t(\mathrm{n}-1|P)$ is the P-quantile of the central t-distribution with $n-1$ degrees of freedom.

First we will show how the sample size can be determined appropriately corresponding to Example 3.1. We want to calculate the sample size so that for given risks of the first and the second kind, a fixed difference of practical relevance related to the value of the null hypothesis can be recognised. We suppose that for each $n > 1$ $\boldsymbol{Y} = (\boldsymbol{y}_1, \boldsymbol{y}_2, \ldots, \boldsymbol{y}_n)^T$ is a random sample with components distributed as $N(\mu, \sigma^2)$.

We test the null hypothesis H_0: $\mu = \mu_0$, σ^2 arbitrary, against one of the alternatives:

a) $H_A: \mu > \mu_0, \sigma^2$ arbitrary,
b) $H_A: \mu < \mu_0, \sigma^2$ arbitrary,
c) $H_A: \mu \neq \mu_0, \sigma^2$ arbitrary.

The test statistic

$$\boldsymbol{t}(\mu) = \frac{\overline{\boldsymbol{y}} - \mu}{\boldsymbol{s}}\sqrt{n}$$

in (3.45) is under H_0 centrally t-distributed; in general it is non-centrally t-distributed with the non-centrality parameter $\lambda = \frac{\mu - \mu_0}{\sigma}\sqrt{n}$.

Actually, each difference of the parameters under the null hypothesis (μ_0) on the one hand and under the alternative hypothesis (μ_1) on the other hand can become significant as soon as the sample size is large enough. Hence, a significant result alone is not yet meaningful. Basically it expresses nothing, because the difference could also be very small, for instance, $|\mu_1 - \mu_0| = 0.00001$. Therefore, investigations have to be planned by fixing the difference to the parameter value of the null hypothesis (μ_0) to be practically relevant. For explaining the risk β of the second kind, we pretended the alternative hypothesis would consist only of one single value μ_1. But in most applications μ_1 can take all values apart from μ_0 for two-sided test problems and all values smaller than or larger than μ_0 for one-sided test problems. The matter is that each value of μ_1 causes another value for the risk β of the second kind. More precisely the smaller β is, the larger the difference $\mu_1 - \mu_0$. The quantity $E = (\mu_1 - \mu_0)/\sigma$, that is, the relative or standardised practically relevant difference, is called (relative) effect size.

Therefore the fixing of the practically interesting minimal difference $\delta = \mu_1 - \mu_0$ is an essential step for planning an investigation. Namely, if δ is determined and if certain risks α of the first kind and β of the second kind are chosen, then the necessary sample size can be calculated. The fixing of α, β and δ is called the precision requirement. The crucial point is that differences $\mu_1 - \mu_0$ equal or larger than the prescribed δ should not be overlooked insofar as it is possible. To say it more precisely, it is to happen only with a probability less or equal to β that such differences are not recognised.

The sample size that fulfils the posed precision requirement can be obtained by the power function of the test. This function states the power for given sample size and for all possible values of δ, that is, the probability for rejecting the null hypothesis if indeed the alternative hypothesis holds. If the null hypothesis is true, the power function has the value α. It would not be fair to compare the power of a test with $\alpha = 0.01$ with that of a test with $\alpha = 0.05$, because a larger α also means that the power is larger for all arguments referring to the alternative hypothesis. Hence only tests with the same α are compared with each other.

For calculating the required sample size, we first look for all power functions related to all possible sample sizes that have the probability α for μ_0, that is, the parameter value under the null hypothesis. Now we look up the point of the minimum difference δ. Then we choose under all power functions the one that has the probability $1 - \beta$ at this point, that is, the probability for the justified rejection of the null hypothesis; hence, at this point the probability of unjustified accepting, that is, of making an error of the second kind, is β. Finally, we have to choose the size n corresponding to this power function. For two-sided test problems, the points $-\delta$ and $+\delta$ have to be fixed. Figure 3.2 illustrates that deviations larger than δ are overlooked with still lower probability. A practical method is as follows: divide the expected range of the investigated character, that is, the difference between the imaginably maximal and minimal realisation of the character, by 6 (assuming a normal distribution approximately 99% of the realisations lie between $\mu_0 - 3\sigma$ and $\mu_0 + 3\sigma$) and use the result as estimation for σ.

For unknown variance σ^2 we can use the sample variance of a prior sample of size n between 10 and 30.

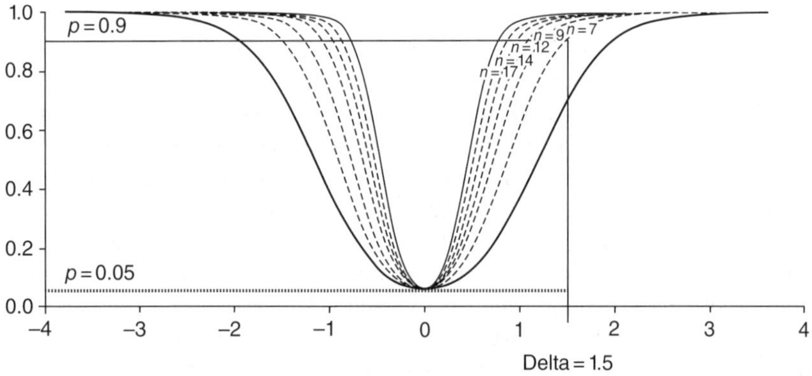

Figure 3.2 The power functions of the *t*-test testing the null hypothesis $H_0: \mu = \mu_0$ against H_A: $\mu \neq \mu_0$ for a risk $\alpha = 0.05$ of the first kind and a sample size $n = 5$ (bold-plotted curve below) as well as other values of n (broken-lined curves) up to $n = 20$ (bold-plotted curve above).

For example, assuming a power of 0.9, the relative effect can be read on the abscissa, and it is approximately 1.5 for $n = 7$.

Hints Referring to the Statistical Software Package R

In practical investigations professional statistical software is used to determine appropriate sample sizes for given values of α, β and δ; in this book we apply mainly `R`. The software package `R` is an adaptation of the programming language `S`, which has been developed since 1976 by *John Chambers* and colleagues in the *Bell Laboratories*. The functionality of `R` can be extended by everybody without any restrictions using free software tools; moreover it is possible to implement also special statistical methods as well as certain procedures of C and FORTRAN. Such tools are offered in the Internet in standardised archives. The most popular archive is probably CRAN (*Comprehensive* `R` *Archive Network*), a server net that is supervised by the `R` *Development Core Team*. This net also offers the package OPDOE (**O**ptimal **D**esign **of** **E**xperiments), which was thoroughly described in the book of Rasch et al. (2011b).

Apart from only a few exceptions, `R` contains implementations for all statistical methods concerning analysis, evaluation and planning.

The software package `R` is available free of charge under http://cran.r-project.org/ for the operating systems Linux, Mac OS X and Windows. The installation under Microsoft Windows takes place via 'Windows'. Choosing their 'base', the installation platform is reached. With 'Download R 2.X.X for Windows' (X stands for the required version number), the setup file can be downloaded. After this file is started, the setup assistant is running through the single installation steps. Concerning this book, all standard settings can be adopted. The interested reader will find more information about `R` under http://www.r-project.org/

After starting `R` the input window will be opened presenting the red-coloured input request: '>'. Here commands can be written up and carried out by the enter button. The output is given directly below the command line. But the user can also realise line changes as well as line indents for increasing clarity. All this does not influence the functional procedure. Needing a line change the next line has to be continued with '+'. A sequence of commands is read, for instance, as follows:

```
> cbind(u1_t1.tab, u1_t1.pro, u1_t1.cum)
```

The `Workspace` is a special working environment in `R`. There certain *objects* can be stored, which were obtained during the current work with `R`. Such objects contain not only results of computations but also data sets. A `Workspace` is loaded using the menu

```
File - Load Workspace...
```

Now we turn to the calculation of sample sizes. We describe the procedure for calculations by hand and list a corresponding file in R.

The test statistic (3.45) is non-centrally t-distributed with $n - 1$ degrees of freedom and the non-centrality parameter $\lambda = \frac{\mu - \mu_0}{\sigma}\sqrt{n}$. Under the null hypothesis $\mu = \mu_0$ is $\lambda = 0$. Taking the $(1-\alpha)$-quantile $t(n-1 \mid 1-\alpha)$ of the central t-distribution with $n - 1$ degrees of freedom and the β-quantile of the corresponding non-central t-distribution $t(n-1, \lambda \mid \beta)$, we obtain in the one-sided case the condition

$$t(n-1 \mid 1-\alpha) = t(n-1, \lambda \mid \beta)$$

because of the requirement $1 - \pi(\mu) = P(\boldsymbol{t} < t(n-1,\ \lambda \mid 1-\alpha)) = \beta$. This means that the $(1 - \alpha)$-quantile of the central t-distribution (the distribution under the null hypothesis) has to be equal to the β-quantile of the non-central t-distribution with non-centrality parameter λ, where λ depends on the minimum difference δ. We illustrate these facts by Figure 3.3.

We apply an approximation that is sufficiently precise for the calculation of sample sizes by hand, namely,

$$t(n-1, \lambda|\beta) \approx t(n-1|\beta) + \lambda = -t(n-1|1-\beta) + \frac{\delta}{\sigma}\sqrt{n}.$$

Analogous to Example 3.1 the minimum sample size n is therefore obtained by

$$n = \left\lceil [t(n-1|1-\alpha) + t(n-1|1-\beta)]^2 \frac{\sigma^2}{\delta^2} \right\rceil,$$

where $\lceil x \rceil$ again denotes the round-off function.

After fixing α, β, δ and σ, the sample size n can be iteratively calculated by this formula. Now we put $\delta = \sigma$, that is, deviations of at least the standard

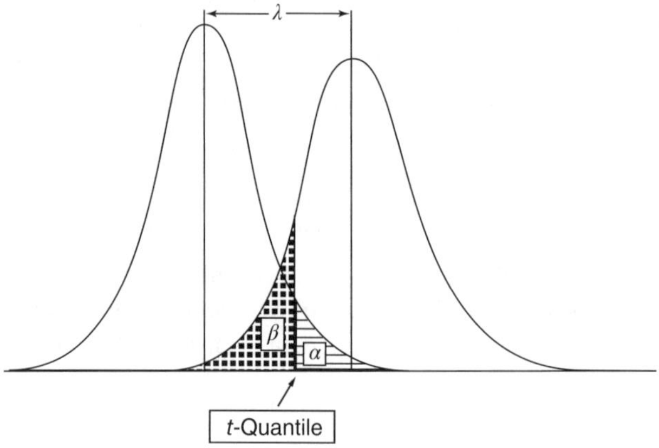

Figure 3.3 Graphical illustration of the risks α and β.

deviation are to be overlooked at most with the probability β. For $\alpha = 0.05$ and $\beta = 0.2$, we start iterations with $n_{(0)} = \infty$ and get

$$t(\infty|0.95) = 1.6449,\ t(\infty|0.8) = 0.8416,$$

followed by

$$n_{(1)} = \lceil [1.6449 + 0.8416]^2 \rceil = \lceil 6.18 \rceil = 7;$$
$$t(6|0.95) = 1.9432, t(6|0.8) = 0.957;$$
$$n_{(2)} = \lceil [1.9432 + 0.9057]^2 \rceil = \lceil 8.11 \rceil = 9$$
$$t(8|0.95) = 1.8595,\ t(8|0.8) = 0.8889$$
$$n_{(3)} = \lceil [1.9432 + 0.9057]^2 \rceil = \lceil 7.56 \rceil = 8$$
$$t(7|0.95) = 1.8946,\ t(7|0.8) = 0.896;$$
$$n_{(4)} = \lceil [1.8946 + 0.896]^2 \rceil = \lceil 7.78 \rceil = 8.$$

Hence, $n = 8$ is the minimum sample size. In the case of a two-sided alternative, we calculate $n = 10$ using R (see Table 3.2). Here $1 - \alpha$ has to be replaced in the t-quantile by $1 - \alpha/2$.

Table 3.2 lists the sample sizes in the just considered case for a two-sided alternative with $\alpha = 0.05$, $\beta = 0.2$ and some δ computed with the software package OPDOE (according to the exact formula). The extract of commands is

```
>size.t.test(delta=1, sd=1, sig.level=0.05, power =
+0.8, +type:"one.sample", alternative = "two.sided")
```

where sd = σ, sig.level = α and power = $1 - \beta$. Remember that a new command line needs the sign '+' at the beginning.

Exploiting the previous results the reader can prove for many of the customary tests used in applications that they are UMPU-α-tests.

Example 3.13 Supposing the conditions of Example 3.11, we want to test $H_0: \sigma^2 = \sigma_0^2$, μ arbitrary, against a one- or two-sided alternative hypothesis. Here we restrict ourselves to the alternative $H_A: \sigma^2 \neq \sigma_0^2$, μ arbitrary. Put

$$\lambda = -\frac{1}{2\sigma^2},\quad \eta_2 = \frac{n}{\sigma^2},\quad \boldsymbol{S} = \sum_{i=1}^{n} \boldsymbol{y}_i^2, T^* = \bar{\boldsymbol{y}}$$

Table 3.2 Values of n depending on $\delta = c{\cdot}\sigma$ for $\alpha = 0.05$, $\beta = 0.20$ and a two-sided alternative (i.e. $P = 1 - \alpha/2$).

δ	1/25 σ	1/10 σ	1/5 σ	1/4 σ	1/3 σ	1/2 σ	1 σ
n	4908	787	199	128	73	34	10

and

$$\boldsymbol{g} = g(\boldsymbol{S}, \boldsymbol{T}^*) = \frac{1}{\sigma_0^2}\left[\boldsymbol{S} - n\,\boldsymbol{T}^{*2}\right] = \frac{1}{\sigma_0^2}\sum_{i=1}^{n}(\boldsymbol{y}_i - \bar{\boldsymbol{y}})^2.$$

The function g is for each T^* isotone in $\boldsymbol{S}$. Besides $\bar{\boldsymbol{y}}$ is complete sufficient with respect to the family $N(\mu, \sigma_0^2)$ (i.e. with respect to P^*). Since $\boldsymbol{g}$ is distributed as $CS(n-1)$, the mapping

$$k(Y) = \begin{cases} 1 & \text{for } g < c_{1\alpha} \text{ or } g > c_{2\alpha} \\ 0 & \text{else} \end{cases}$$

is a UMPU-α-test if $c_{1\alpha}$ and $c_{2\alpha}$ are determined according to (3.32) and (3.33) with $n-1$ instead of n.

Now we discuss whether it is always favourable to look for UMPU-tests. The exclusion of noisy factors $\eta_2, \ldots, \eta_k$, as it was described in this section, is only one of several possibilities. We can also design tests so that the condition

$$\max_{\theta \in \Omega_0} E[k(\boldsymbol{Y})|\theta] = \alpha$$

is fulfilled. We want to consider both possibilities in the following case. Let the random variables $\boldsymbol{x}$ and $\boldsymbol{y}$ be mutually independently distributed as $B(1, p_x)$ and $B(1, p_y)$ correspondingly (satisfying each a two-point distribution), where

$$P(\boldsymbol{x}=0) = p_x, P(\boldsymbol{x}=1) = 1 - p_x, 0 < p_x < 1,$$
$$P(\boldsymbol{y}=0) = p_y, P(\boldsymbol{y}=1) = 1 - p_y, 0 < p_y < 1.$$

Further, the null hypothesis H_0: $p_x = p_y = p$, p arbitrary in $\{0, 1\}$, is to be tested in $\Omega^* = (0, 1)$ against H_A: $p_x < p_y$; p_x, p_y arbitrary in $\{0, 1\}$ with a risk α $(0 < \alpha < 0.25)$ of the first kind. The set of possible realisations of $(\boldsymbol{x}, \boldsymbol{y})$ is

$$\{Y\} = \{(x, y) : x = 0, 1; y = 0, 1\}.$$

The boundary Ω^* is the set of possible p-values, namely, $\Omega^* = \{0, 1\}$, the diagonal in the discrete square $\{0, 1\} \times \{0, 1\}$.

First we design a test that fulfils the above given maximum condition. Because of $\Omega_0 = \Omega^*$, the condition

$$\max_{p \in \{0,1\}} E[k(\boldsymbol{Y})|p] = \alpha$$

has to hold, which supplies for $\alpha < 0.25$ (with $c_\alpha = 0$)

$$k_1(Y) = \begin{cases} 4\alpha & \text{for } (x, y) = (0, 1) \\ 0 & \text{else} \end{cases},$$

taking

$$E[k_1(\boldsymbol{Y})] = 4\alpha P(\boldsymbol{x}=0, \boldsymbol{y}=1) = 4\alpha p_x(1-p_y).$$

into account. Under the null hypothesis the expectation is equal to 4 $\alpha p(1-p)$. This functional expression takes its maximum α for $p = 1/2$. Therefore $k_1(\boldsymbol{Y})$ is an α-test.

Now we design a UMPU-α-test. Evidently we have

$$p_{xy} = P(\boldsymbol{x}=x, \boldsymbol{y}=y) = \binom{1}{x} p_x^x (1-p_x)^{1-x} \binom{1}{y} p_y^y (1-p_y)^{1-y}$$
$$= p_x^x p_y^y (1-p_x)^{1-x} (1-p_y)^{1-y}.$$

Putting $\boldsymbol{T}^* = \boldsymbol{x} + \boldsymbol{y}$, $\boldsymbol{S} = \boldsymbol{x}$, we see from (3.37) analogous to Example 3.12 that

$$P(\boldsymbol{x}=x, \boldsymbol{T}^* = T^*, p_x = p_y) = \frac{1}{\binom{2}{T^*}}$$

is true under H_0. Then

$$k_2(Y) = \begin{cases} 2\alpha & \text{for } x=0, y=1 \\ \alpha & \text{for } x=y \\ 0 & \text{for } x=1, y=0 \end{cases}$$

is with $c_\alpha = 0$ the realisation of a UMPU-α-test, since

$$E[k_2(\boldsymbol{Y})] = 2\alpha p_y(1-p_x) + \alpha\left[p_x p_y + (1-p_x)(1-p_y)\right] + 0 = \alpha(1+p_y-p_x),$$

which is equal to α under the null hypothesis. If the power functions $\pi_1(p_x, p_y)$ and $\pi_2(p_x, p_y)$ of both tests, namely,

$$\pi_1(p_x, p_y) = 4\alpha p_x(1-p_y);\ \pi_2(p_x, p_y) = \alpha(1-p_x+p_y),$$

are compared, then we get (here 'more powerful' means a larger power)

$$k_2(\boldsymbol{Y}) \text{ is more powerful than } k_1(\boldsymbol{Y}), \text{if } 4p_x(1-p_y) > 1+p_y-p_x,$$
$$k_1(\boldsymbol{Y}) \text{ is biased, if } 4p_x(1-p_y) < 1.$$

The parameter space is determined by $p_x \le p_y$. It is easy to see that the biased test $k_1(\boldsymbol{Y})$ is in a considerable part of the parameter space more powerful than the unbiased test $k_2(\boldsymbol{Y})$. If an a priori information is available that the differences between p_x and p_y are rather great or that only rather great differences are of interest, then $k_1(\boldsymbol{Y})$ should be preferred to $k_2(\boldsymbol{Y})$.

3.4.2 The Two-Sample Problem: Properties of Various Tests and Robustness

The following examples for UMPU-α-tests are of such great practical importance that we dedicate an entire section to them. Moreover, as representatives of all test problems in this chapter, these tests are to be compared with tests not belonging to the UMPU-class where also consequences of violated or modified assumptions concerning the underlying distributions are also pointed out. We consider two independent random samples $\boldsymbol{Y}_1 = \left(\boldsymbol{y}_{11},\ldots,\boldsymbol{y}_{1n_1}\right)^T$, $\boldsymbol{Y}_2 = \left(\boldsymbol{y}_{21},\ldots,\boldsymbol{y}_{2n_2}\right)^T$, where components $\boldsymbol{y}_{ij}$ are supposed to be distributed as $N\left(\mu_i,\sigma_i^2\right)$. We intend to test the null hypothesis

$$H_0: \mu_1 = \mu_2 = \mu, \sigma_1^2, \sigma_2^2 \text{ arbitrary}$$

against

$$H_A: \mu_1 \neq \mu_2, \sigma_1^2, \sigma_2^2 \text{ arbitrary}.$$

The UMPU-α-tests for one-sided alternatives with $\sigma_1^2 = \sigma_2^2$ can be designed analogously. This work is left to the reader.

The second class of tests we consider concerns the pair

$$H_0: \sigma_1^2 = \sigma_2^2 = \sigma^2, \mu_1, \mu_2 \text{ arbitrary}$$
$$H_A: \sigma_1^2 \neq \sigma_2^2, \mu_1, \mu_2 \text{ arbitrary}.$$

of hypotheses. Since we use two random samples belonging to different distributions, it is called a two-sample problem. Regarding each pair (i,j), $1 \le i \le n_1$, $1 \le j \le n_2$ the vector variable $\begin{pmatrix} \boldsymbol{y}_{1i} \\ \boldsymbol{y}_{2j} \end{pmatrix}$ belongs to a two-dimensional (or bivariate) normal distribution with the expectation vector $\begin{pmatrix} \mu_1 \\ \mu_2 \end{pmatrix}$ and the covariance matrix $\begin{pmatrix} \sigma_1^2 & 0 \\ 0 & \sigma_2^2 \end{pmatrix}$ representing a four-parametric exponential family. Therefore, the random vector $\boldsymbol{Y} = \begin{pmatrix} \boldsymbol{Y}_1 \\ \boldsymbol{Y}_2 \end{pmatrix}$ has also a distribution from a four-parametric exponential family with the natural parameters

$$\eta_k = \frac{n_k \mu_k}{\sigma_k^2} (k = 1,2); \quad \eta_3 = -\frac{1}{2\sigma_1^2}, \quad \eta_4 = -\frac{1}{2\sigma_2^2}$$

and the complete sufficient statistics

$$T_i(\boldsymbol{Y}) = \bar{\boldsymbol{y}}_i \ (i = 1,2); \ T_3(\boldsymbol{Y}) = \sum_{i=1}^{n_1} \boldsymbol{y}_{1i}^2; \ T_4(\boldsymbol{Y}) = \sum_{j=1}^{n_2} \boldsymbol{y}_{2j}^2.$$

3.4.2.1 Comparison of Two Expectations

Considering the pair of hypotheses with respect to the expectations, we cannot design a UMPU-α-test in general. We are only successful for the special case $\sigma_1^2 = \sigma_2^2 = \sigma^2$ (variance homogeneity).

3.4.2.1.1 A UMPU-α-Test for Normal Distributions in the Case of Variance Homogeneity

We want to design a test for the pair

$$H_0: \mu_1 = \mu_2 = \mu,\ \sigma_1^2 = \sigma_2^2 = \sigma^2 \text{ arbitrary}$$
$$H_A: \mu_1 \neq \mu_2,\ \sigma_1^2 = \sigma_2^2 = \sigma^2 \text{ arbitrary}$$

of hypotheses. Then the common distribution of a random variable $\boldsymbol{Y} = \begin{pmatrix} \boldsymbol{Y}_1 \\ \boldsymbol{Y}_2 \end{pmatrix}$ is an element of a three-parametric exponential family, which can be written with the natural parameters

$$\eta_1 = \lambda = \frac{\mu_1 - \mu_2}{\left(\frac{1}{n_1} + \frac{1}{n_2}\right)\sigma^2},\quad \eta_2 = \frac{n_1\mu_1 + n_2\mu_2}{(n_1 + n_2)\sigma^2},\quad \eta_3 = -\frac{1}{2\sigma^2}$$

and the corresponding statistics

$$\boldsymbol{S} = \bar{\boldsymbol{y}}_1 - \bar{\boldsymbol{y}}_2;\ \boldsymbol{T}_1^* = n_1\bar{\boldsymbol{y}}_1 + n_2\bar{\boldsymbol{y}}_2;\ \boldsymbol{T}_2^* = \sum_{i=1}^{n_1} \boldsymbol{y}_{1i}^2 + \sum_{j=1}^{n_2} \boldsymbol{y}_{2j}^2.$$

Besides, $(\boldsymbol{T}_1^*, \boldsymbol{T}_2^*) = \boldsymbol{T}^*$ is complete sufficient with respect to P^* (i.e. for the case $\mu_1 = \mu_2 = 0$, where P^* is a two-parametric exponential family). According to Theorem 3.14, there is a UMPU-a-test for our problem. We consider

$$\boldsymbol{g} = g(\boldsymbol{S}, \boldsymbol{T}^*) = \frac{\boldsymbol{S}}{\sqrt{\boldsymbol{T}_2^* - \frac{1}{n_1 + n_2}\boldsymbol{T}_1^* - \frac{n_1 n_2}{n_1 + n_2}\boldsymbol{S}^2}} = \frac{\bar{\boldsymbol{y}}_1 - \bar{\boldsymbol{y}}_2}{\boldsymbol{s}\sqrt{n_1 + n_2 - 2}}$$
$$= \frac{\frac{1}{\sigma}(\bar{\boldsymbol{y}}_1 - \bar{\boldsymbol{y}}_2)}{\frac{\boldsymbol{s}}{\sigma}\sqrt{n_1 + n_2 - 2}} \tag{3.46}$$

with

$$\boldsymbol{s}^2 = \frac{\sum_{i=1}^{n_1}(\boldsymbol{y}_{1i} - \bar{\boldsymbol{y}}_1)^2 + \sum_{j=1}^{n_2}(\boldsymbol{y}_{2i} - \bar{\boldsymbol{y}}_2)^2}{n_1 + n_2 - 2}.$$

The distribution of $\boldsymbol{g}$ under H_0 depends neither on the value $\mu = \mu_1 = \mu_2$ nor on σ^2, since the nominator of $\boldsymbol{g}$ is distributed as $N\left(0, \frac{n_1+n_2}{n_1 n_2}\right)$ and the square of the denominator independent of it distributed as $CS(n_1 + n_2 - 2)$ referring to the first quotient representation in (3.46). Hence, by Theorem 1.5 the random variable $\boldsymbol{g}$ is independent of $\boldsymbol{T}$. The test statistic

$$\boldsymbol{t} = \frac{\bar{\boldsymbol{y}}_1 - \bar{\boldsymbol{y}}_2}{\boldsymbol{s}}\sqrt{\frac{n_1 n_2}{n_1+n_2}} \tag{3.47}$$

is distributed as $t\left[n_1+n_2-2; \frac{\mu_1-\mu_2}{\sigma}\sqrt{\frac{n_1 n_2}{n_1+n_2}}\right]$. Therefore the UMPU-$\alpha$-test for H_0 against H_A given in (3.47) has the form

$$k(Y) = \begin{cases} 1 & \text{for } |t| > t\left(n_1+n_2-2 \middle| 1-\frac{\alpha}{2}\right) \\ 0 & \text{else} \end{cases}.$$

This test is called the two-sample-t-test.

Example 3.14 (Optimal Sample Size)

We want to calculate the optimal (i.e. the minimal) total size of both samples so that the precision requirements $\alpha = 0.05$, $\beta = 0.1$ and $\sigma = \delta = \mu_1 - \mu_2$ hold. For given total size $N = n_1 + n_2$ the factor $\sqrt{\frac{n_1 n_2}{n_1+n_2}}$ in (3.47) becomes maximal if $n_1 = n_2 = n$. We take this choice. Observing the mentioned precision requirements, the non-centrality parameter of the t-distribution is

$$\lambda = \frac{\mu_1-\mu_2}{\sigma}\sqrt{\frac{n_1 n_2}{n_1+n_2}} = \frac{\delta}{\sigma}\sqrt{\frac{n}{2}}.$$

Analogous to the one-sample case, the condition

$$t\left[2(n-1); \sqrt{\frac{n}{2}} \,|\, \beta\right] = t\left[2(n-1); \sqrt{\frac{n}{2}} \,|\, P\right]$$

has to be realised. Using OPDOE in CRAN – R, the size n of a random sample can be determined. Again we choose for one-sided alternatives $P = 1 - \alpha$ and for two-sided alternatives $P = 1 - \alpha/2$.

The commands in R have to be modified only slightly compared with the one-sample problem as you can see below:

```
>size.t.test(delta=1, sd=1, sig.level=0.05, power = 0.8,
+type="two.sample", alternative = "two.sided")
```

If the calculation is made by hand, we can again use the formula

$$n = \left\lceil [t(2(n-1)|P) + t(2(n-1)|1-\beta)]^2 \frac{2\sigma^2}{\delta^2} \right\rceil$$

obtained by approximation.

Warning: It should be explicitly mentioned here that the two-sample t-test is not really suitable for practical applications. This is the consequence of an article published by Rasch et al. (2011a). Some comments can also be found at the end of this section concerning robustness. We urgently recommend using the Welch test instead of the two-sample-t-test. This test is described now.

3.4.2.1.2 Welch Test

We previously assumed that the unknown variances of the populations from which both samples are taken are equal. But often this is not fulfilled or not reliably known. Then we advise for practical purposes applying an approximate t-test, namely, a test whose test statistic is nearly t-distributed. Such a test is sufficiently precise concerning practical investigations. Moreover, it is a so-called conservative test – meaning a test guaranteeing a risk of the first kind not larger than the prescribed α.

The distribution of the test statistic

$$\boldsymbol{t}^* = \frac{\bar{\boldsymbol{y}}_1 - \bar{\boldsymbol{y}}_2 - (\mu_1 - \mu_2)}{\sqrt{\frac{\boldsymbol{s}_1^2}{n_1} + \frac{\boldsymbol{s}_2^2}{n_2}}}, \quad \boldsymbol{s}_k^2 = \frac{1}{n_k - 1} \sum_{i=1}^{n_k} (\boldsymbol{y}_{ik} - \bar{\boldsymbol{y}}_k)^2, \quad k = 1, 2$$

for unknown variances was derived by Welch (1947). The result is given in the next theorem.

Theorem 3.18 (Welch)

Let $\boldsymbol{Y}_1 = (\boldsymbol{y}_{11}, \ldots, \boldsymbol{y}_{1n_1})^T, \boldsymbol{Y}_2 = (\boldsymbol{y}_{21}, \ldots, \boldsymbol{y}_{2n_2})^T$ be two independent random samples with components $\boldsymbol{y}_{ij}$ distributed as $N(\mu_i, \sigma_i^2)$. Introducing the notations

$$\gamma = \frac{\frac{\sigma_1^2}{n_1}}{\frac{\sigma_1^2}{n_1} + \frac{\sigma_2^2}{n_2}}, \quad \boldsymbol{b} = \frac{(n_1 - 1)\frac{\boldsymbol{s}_1^2}{\sigma_1^2}}{(n_1 - 1)\frac{\boldsymbol{s}_1^2}{\sigma_1^2} + (n_2 - 1)\frac{\boldsymbol{s}_2^2}{\sigma_2^2}}$$

and

$$p(b) = \frac{1}{B\left(\frac{n_1 - 1}{2}, \frac{n_2 - 1}{2}\right)} b^{\frac{n_1 - 1}{2} - 1} (1 - b)^{\frac{n_2 - 1}{2} - 1},$$

with values $B\left(\frac{n_1-1}{2}, \frac{n_2-1}{2}\right)$ of the beta function, the distribution function of $\boldsymbol{t}^*$ in the case $\mu_1 = \mu_2$ is given by

$$F(t^*) = \int_0^1 H_{n_1+n_2-2}\left\{\sqrt{n_1+n_2-2}\,\frac{\gamma b}{n_1-1} + \frac{(1-\gamma)(1-b)}{n_2-1}\right\} p(b)db,$$

where $H_{n_1+n_2-2}$ is the distribution function of the central t-distribution with n_1+n_2-2 degrees of freedom.

The proof of the theorem is contained, for example, in Welch (1947) or in Trickett and Welch (1954). The critical value t_P^* can only be iteratively determined. An iterative method is presented in Trickett and Welch (1954), Trickett et al. (1956) and Pearson and Hartley (1970). Tables listing critical values are given in Aspin (1949).

If the pair

$$H_0: \mu_1 = \mu_2 = \mu,\ \sigma_1^2, \sigma_2^2 \text{ arbitrary}$$
$$H_A: \mu_1 \neq \mu_2, \sigma_1^2, \sigma_2^2 \text{ arbitrary}$$

of hypotheses is to be tested often, the approximate test statistic

$$\boldsymbol{t}^* = \frac{\bar{\boldsymbol{y}}_1 - \bar{\boldsymbol{y}}_2}{\sqrt{\frac{\boldsymbol{s}_1^2}{n_1} + \frac{\boldsymbol{s}_2^2}{n_2}}}$$

is used. H_0 is rejected if $|t^*|$ is larger than the corresponding quantile of the central t-distribution with

$$f^* = \frac{\left(\frac{s_1^2}{n_1} + \frac{s_2^2}{n_2}\right)^2}{\frac{s_1^4}{n_1^2(n_1-1)} + \frac{s_2^4}{n_2^2(n_2-1)}}$$

degrees of freedom.

Example 3.14 Optimum Sample Size (continuation)

We want to determine the size of both samples so that the precision requirements $\alpha = 0.05$; $\beta = 0.1$; $\sigma_x = C\sigma_y$ with known C and $\delta = \mu_1 - \mu_2 = 0.9\sigma_y$ are fulfilled. Using these data the software package OPDOE in CRAN – `R` is ready for calculating the sizes of both samples. As before we have to put $P = 1 - \alpha$ for one-sided alternatives and $P = 1 - \alpha/2$ for two-sided alternatives. The sequence of commands in `R` has to be only slightly changed compared with the one for the t-test.

Concerning calculations by hand, we use $n_y = \frac{\sigma_y}{\sigma_x} n_x$ and again the approximation

$$n_x \approx \left[t(f^* | P) + t(f^* | 1 - \beta)\right]^2 \frac{\sigma_x\left(\sigma_x + \sigma_y\right)}{\delta^2}.$$

The data in this example supply the sizes $n_x = 105$ and $n_y = 27$ for $\sigma_x = 4\sigma_y(C = 4)$.

Hints for Program Packages

At this point, we give an introduction to the package IBM SPSS 24 Statistics (SPSS in short). When we open the package, we find a data matrix (which is empty at the beginning) into which we will put our data.

Clicking on variables you can give names for the characters you wish to enter as shown in Figure 3.4.

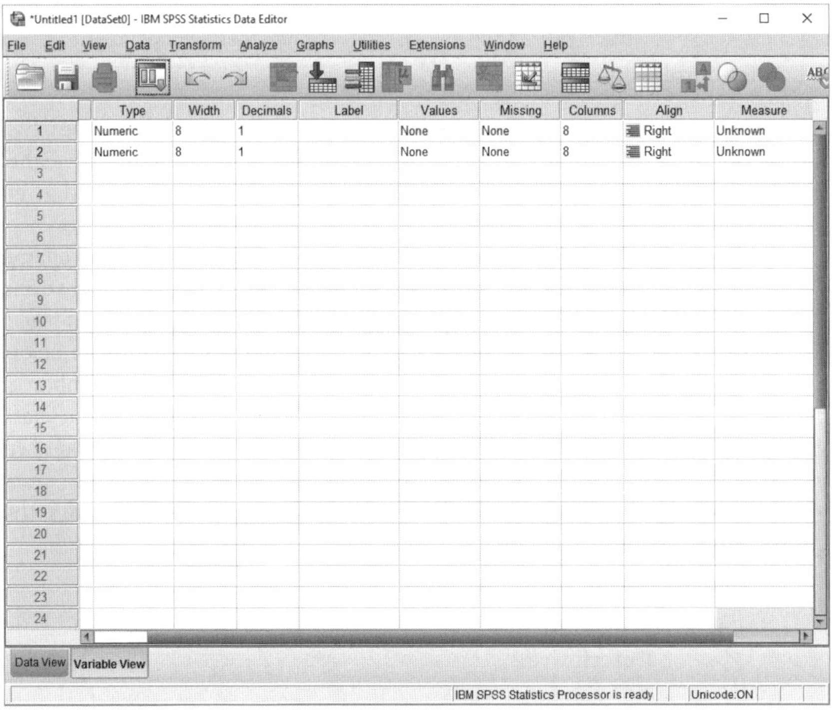

Figure 3.4 SPSS data file in variable view. *Source:* Reproduced with permission of IBM.

Table 3.3 The litter weights of mice (in g) of Example 3.15.

i	x_i	y_i
1	7.6	7.8
2	13.2	11.1
3	9.1	16.4
4	10.6	13.7
5	8.7	10.7
6	10.6	12.3
7	6.8	14.0
8	9.9	11.9
9	7.3	8.8
10	10.4	7.7
11	13.3	8.9
12	10.0	16.4
13	9.5	10.2

Let us consider an example.

Example 3.15 Two independent samples of 13 mice are drawn from two mouse populations (26 mice in all). The x- and y-values are the litter weights of the first litters of mice in populations 1 (x_i) and 2 (y_i), respectively, and given in Table 3.3.

We will now create this data as an SPSS data file. First we need to rename var in the first column as x and var in the second column as y as already done in Figure 3.5. Then we need three digits in each column and one decimal place. To do these we change from Data View to Variable View (see Figure 3.5 below left). Now we can change the variable names to x and y and the number of decimal places to 1. Having returned to Data View, we now enter the data values. We save the file under the name mice-data.sav. The SPSS file is shown in Figure 3.5.

SPSS allows us at first to calculate some descriptive statistics from the observations via

Analyze
Descriptive statistics
Descriptive

and then we choose Options as shown in Figure 3.6

Here we select what we like and receive the output in Figure 3.7.

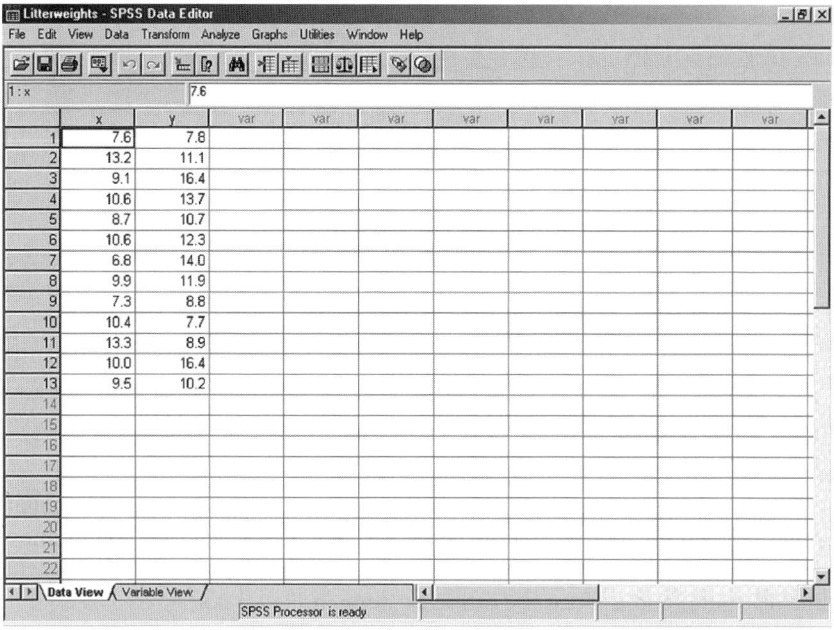

Figure 3.5 SPSS data file for Example 3.15. *Source:* Reproduced with permission of IBM.

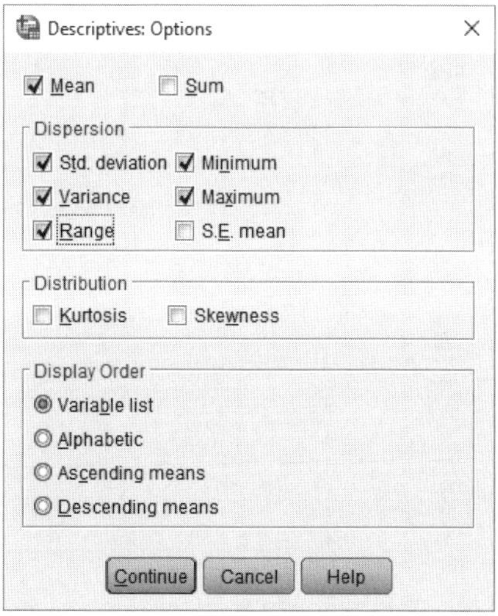

Figure 3.6 Options in descriptive statistics. *Source:* Reproduced with permission of IBM.

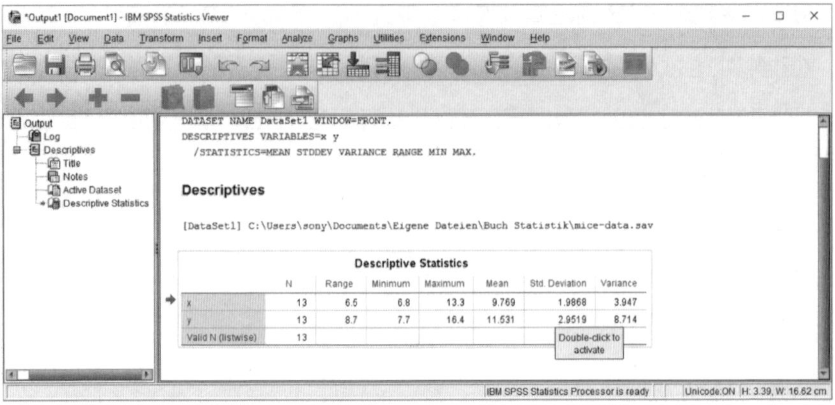

Figure 3.7 SPSS output of Example 3.15. *Source:* Reproduced with permission of IBM.

To test the hypothesis, that the expectation of both variables are equal against a two-sided alternative, we first have to rearrange the data in one column now proceed named 'weight' and a second column group where we put a '1' for the first 13 values and a '2' for the others as done in Figure 3.8.

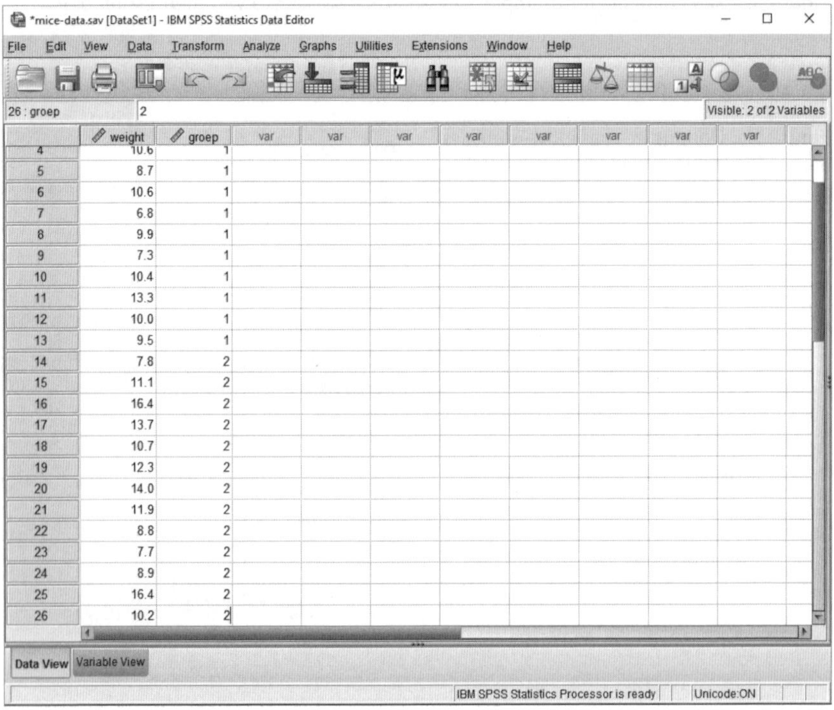

Figure 3.8 Rearranged data of Example 3.15. *Source:* Reproduced with permission of IBM.

Now we proceed with

Analyze

  Compare means

    independent samples t-test

and receive Figure 3.9.

In the upper row we find the result for the t-test and below that for the Welch test. As we mentioned above, we always use only the Welch test output. The decision about the rejection of the null hypothesis is as follows. If the first kind risk chosen in advance is larger than the value significance in the output, we reject the null hypothesis. In our example it must be accepted if $\alpha = 0.05$ (because it is below 0.089 in the output).

Confidence intervals can be found in the corresponding test output right from the test results.

3.4.2.1.3 *Wilcoxon–Mann–Whitney Test*

Assume that we do not know whether the sample components of a two-sample problem are normally distributed, but the distributions are continuous, all moments exist and at most the expectations of the distributions are different. Then the pair

$$H_0 : \mu_1 = \mu_2 = \mu, \text{ all higher moments equal, but arbitrary}$$

$$H_A : \mu_1 \neq \mu_2, \text{ all higher moments equal, but arbitrary}$$

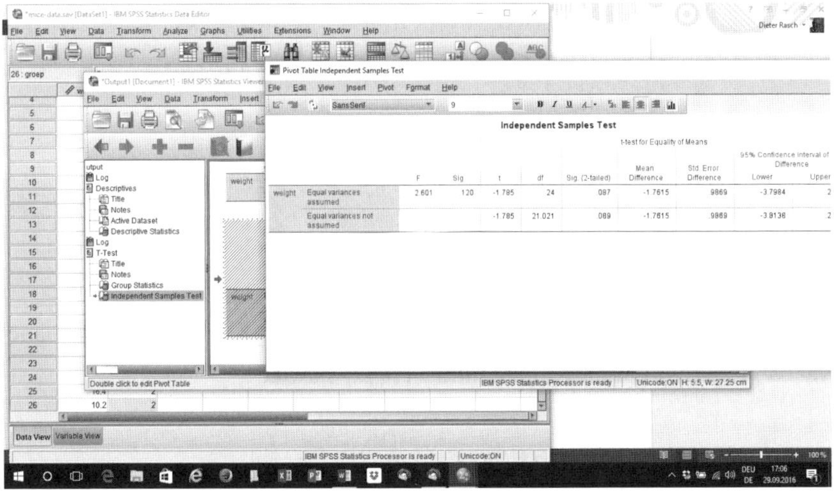

Figure 3.9 SPSS output for comparing means. *Source:* Reproduced with permission of IBM.

of hypotheses can also be written as

$$H_0 : f_1(y_1) = f_2(y_2),$$

where $f_1(y_1), f_2(y_2)$ are the densities of both distributions. If higher moments of the distributions are different (e.g. $\sigma_1^2 \neq \sigma_2^2$ or skewness and excess of both distributions are different, respectively), then the rejection of the null hypothesis does not say anything about the expectations. However, if the equality of all kth moments ($k \geq 2$) of both distributions is guaranteed, then non-parametric tests can be used for the hypotheses. Such tests are generally not treated in this book (see Bagdonavicius et al., 2011; Rasch et al., 2011c). We only want to describe a special representative, the Wilcoxon test, also called Mann–Whitney test (see Wilcoxon, 1945; Mann and Whitney, 1947).

For $i = 1, \ldots, n_1$; $j = 1, \ldots, n_2$, we consider

$$d_{ij} = \begin{cases} 1 & \text{for } y_{2j} < y_{1i} \\ 0 & \text{for } y_{2j} > y_{1i} \end{cases}.$$

The equality occurs for continuous random variables with probability 0.

In Rasch et al. (2011c) it is described how to proceed in practical cases if equality happens (ties).

The test statistic is

$$\boldsymbol{U} = \sum_{i=1}^{n_1} \sum_{j=1}^{n_2} \boldsymbol{d}_{ij}.$$

If $F_i\,(y_i)$ are the distribution functions of $\boldsymbol{y}_{ij}$ ($i = 1, 2$) and if

$$p = P(\boldsymbol{y}_2 < \boldsymbol{y}_1) = \int_{-\infty}^{\infty} \int_{-\infty}^{y_1} f_2(y_2) f_1(y_1)\, dy_2\, dy_1 = \int_{-\infty}^{\infty} F_2(t) f_1(t)\, dt,$$

then $H_0 : f_1(y_1) = f_2(y_2)$ implies

$$p = \int_{-\infty}^{\infty} F_1(t) f_1(t)\, dt = \frac{1}{2}.$$

The $n_1 n_2$ random variables $\boldsymbol{d}_{ij}$ are distributed as $B(1, p)$, where $E(\boldsymbol{d}_{ij}) = p$ and $\operatorname{var}\left(\boldsymbol{d}_{ij}^2\right) = p(1-p)$. Mann and Whitney (1947) showed

$$E(\boldsymbol{U} | H_0) = \frac{n_1 n_2}{2}, \quad \operatorname{var}(\boldsymbol{U} | H_0) = \frac{n_1 n_2 (n_1 + n_2 + 1)}{12}.$$

Further, the distribution of $\boldsymbol{U}$ is under H_0 symmetric with respect to $\frac{n_1 n_2}{2}$. With the notation $\boldsymbol{U}' = n_1 n_2 - \boldsymbol{U}$ the function

$$k_U(Y) = \begin{cases} 1 & \text{for } U < c_{\alpha/2} \text{ or } U' < c_{\alpha/2} \\ 0 & \text{else} \end{cases}$$

is an α-test, provided that $c_{\alpha/2}$ is determined by $P(\boldsymbol{U} < c_{\alpha/2} | H_0) = \alpha/2$. The random variable

$$\boldsymbol{W} = \boldsymbol{U} + \frac{n_1(n_1+1)}{2}$$

is equal to the sum of the ranks of the n_1 random variables $\boldsymbol{y}_{1i}$ in the vector of the ranks of the composed random vector $\boldsymbol{Y} = \begin{pmatrix} \boldsymbol{Y}_1 \\ \boldsymbol{Y}_2 \end{pmatrix}$, representing the test statistic of the Wilcoxon test. Therefore $k_U(Y)$ is equivalent to the test

$$k_W(Y) = \begin{cases} 1 & \text{for } W < W_{U\,\alpha/2} \text{ or } W > W_{O\,\alpha/2} \\ 0 & \text{else} \end{cases}$$

The quantiles $W_{U\,\alpha/2}$ and $W_{O\,\alpha/2}$ of this test can for $n_i > 20$ be replaced by the quantiles of the standard normal distribution; for smaller n these quantiles should be calculated with the help of R.

Example 3.15 (continued)

For the data in Figure 3.8, we now calculate the Mann–Whitney test by SPSS.

We use

Analyze
  Nonparametric Tests
    Independent Samples

and use the entry **Fields** putting weight as **Test Fields** and groups as **Groups**. Then we use **Run** and obtain Figure 3.10.

3.4.2.1.4 Robustness

All statistical tests in this chapter are proved to be α-tests and to have other wished properties if some distributional assumptions are fulfilled. An experimenter looking for a proper statistical test often does not know whether or not these assumptions are fulfilled, or he knows that they are not fulfilled. How can we help him? Certainly not by deriving some theorems about this topic.

Hypothesis Test Summary

	Null Hypothesis	Test	Sig.	Decision
1	The distribution of weight is the same across categories of group.	Independent-Samples Mann-Whitney U Test	.091[1]	Retain the null hypothesis.

Asymptotic significances are displayed. The significance level is .05.

[1] Exact significance is displayed for this test.

Figure 3.10 SPSS output of the Mann–Whitney test.

We give here an introduction to methods of empirical statistics (see Chapter 1) via simulations and methods, which will be used later not only in this chapter but also in other chapters (especially in Chapter 11).

General problems concerning robustness are not thoroughly discussed in this book. We restrict ourselves to such comments, which are necessary for understanding the tests presented above (and later). The robustness of a statistical method means that the essential properties of this method are relatively insensitive to variations of the assumptions. We especially want to investigate the robustness of the methods in Section 3.4.2.1 with respect to violating normality or variance equality. Problems of robustness are discussed in detail in a paper of Rasch and Guiard (2004).

Definition 3.8 Let $\boldsymbol{k}_\alpha$ be an α-test ($0 < \alpha < 1$) for the pair $\{H_0, H_A\}$ of hypotheses in the class G_1 of distributions of the random sample $\boldsymbol{Y}$ with size n. And let G_2 be a class of distributions containing G_1 and at least one distribution, which does not fulfil all assumptions for guaranteeing $\boldsymbol{k}_\alpha$ to be an α-test.

Finally, let $\alpha(g)$ be the risk of the first kind for $\boldsymbol{k}_\alpha$ concerning the element g of G_2 (estimated by simulation). Here and in the sequel, we write $\alpha(g) = \alpha_{act}$, the actual α and the α fixed the nominal α written as α_{nom}.

Then $\boldsymbol{k}_\alpha$ is said to be $(1 - \varepsilon)$-robust in the class G_2 if

$$\max_{g \in G_2} |\alpha_{act} - \alpha_{nom}| \leq \varepsilon.$$

We call a statistical test acceptable if $100\,(1 - \varepsilon)\,\% \geq 80\%$.

For example, elements of the set difference $G_2 \setminus G_1$ are distributions with $\sigma_1^2 \neq \sigma_2^2$ for the two-sample t-test and the Wilcoxon test as well as distributions being not normal for the t-test and the Welch test. Rasch and Guiard (2004) report about extensive simulation experiments investigating the robustness of the t-test in a set of 87 distributions of the Fleishman system (Fleishman, 1978) as well as the robustness of the two-sample t-test and the Wilcoxon test for unequal variances. The results showed that both the one-sample t-test and the two-sample t-test (and also the corresponding confidence intervals given in Section 3.5) are extremely robust with respect to deviations from the normal distribution. So, Rasch et al. (2011a) conclude that the two-sample t-test cannot be recommended, and also it makes no sense to check in a pretest whether the variances of both random samples are equal or not. In most cases the Wilcoxon test yields unsatisfactory results, too. Only the Welch test works well. Its power is nearly that one of the two-sample t-test if both variances are equal. Moreover, for unequal variances, this test obeys the given risks in the sense of 80% robustness even for non-normal distributions with a skewness $|\gamma_1| < 3$.

3.4.3 Comparison of Two Variances

A* UMPU-α-*Test

A UMPU-α-test exists for the pair

$$H_0: \sigma_1^2 = \sigma_2^2, \quad \mu_1, \mu_2 \text{ arbitrary}$$
$$H_A: \sigma_1^2 \neq \sigma_2^2, \quad \mu_1, \mu_2 \text{ arbitrary}$$

of hypotheses and the random samples $Y_1 = \left(y_{11}, \dots, y_{1n_1}\right)^T$, $Y_2 = \left(y_{21}, \dots, y_{2n_2}\right)^T$, where components y_{ij} are distributed as $N\left(\mu_i, \sigma_i^2\right)$. The random vector $Y = \begin{pmatrix} Y_1 \\ Y_2 \end{pmatrix}$ has a distribution from a four-parametric exponential family with the natural parameters

$$\eta_1 = \lambda = -\frac{1}{2}\left(\frac{1}{\sigma_1^2} - \frac{1}{\sigma_2^2}\right), \eta_2, \eta_3, \eta_4$$

and the sufficient statistics S, $T* = (T_1^*, T_2^*, T_3^*)^T$ given by

$$S = \sum_{j=1}^{n_2} y_{2j}^2, \; T_1^* \sum_{i=1}^{n_1} y_{1i}^2 + \frac{\sigma_1^2}{\sigma_2^2} \sum_{j=1}^{n_2} y_{2j}^2, T_2^* = \bar{y}_1, T_2^* = \bar{y}_2.$$

Under H_0 we have $\frac{\sigma_1^2}{\sigma_2^2} = 1$, and the random variable

$$F = \frac{\sum_{i=1}^{n_1} \left(y_{1i} - \bar{y}_1\right)^2 \frac{n_2 - 1}{\sigma_1^2}}{\sum_{i=1}^{n_2} \left(y_{2i} - \bar{y}_2\right)^2 \frac{n_1 - 1}{\sigma_2^2}}$$

does not depend on μ_1, μ_2 and $\sigma_1^2 = \sigma_2^2 = \sigma^2$. Hence, $\boldsymbol{F}$ is independent of $\boldsymbol{T}^*$. Therefore Theorem 3.16 can be used. The random variable $\boldsymbol{F}$ is centrally distributed as $F(n_1 - 1, n_2 - 1)$ under H_0. The function

$$k(Y) = \begin{cases} 1 & \text{if } F < F\left(n_1 - 1, n_2 - 1 \middle| \frac{\alpha}{2}\right) \\ & \text{or } F > F\left(n_1 - 1, n_2 - 1 \middle| 1 - \frac{\alpha}{2}\right) \\ 0 & \text{else} \end{cases}$$

defines a UMPU-α-test, where $F(n_1 - 1, n_2 - 1|\ P)$ is the P-quantile of the F-distribution with $n_1 - 1$ and $n_2 - 1$ degrees of freedom. These quantiles for $\alpha = 0.05$ can be found in Table D.5. This test is very sensitive to deviations from the normal distribution. Therefore the following Levene test should be used instead of it in the applications.

Levene Test

Box (1953) already mentioned the extreme non-robustness of the F-test comparing two variances (introduced at the beginning of this Section 3.4.2.2). Rasch and Guiard (2004) report on extensive simulation experiments devoted to this problem. The results of Box show that non-robustness has to be expected already for relatively small deviations from the normal distribution. Hence, we generally suggest applying the test of Levene (1960), which is described now.

For $j = 1, 2$ we put

$$\boldsymbol{z}_{ij} = \left(\boldsymbol{y}_{ij} - \bar{\boldsymbol{y}}_{.j}\right)^2; i = 1, \ldots, n_j$$

and

$$\boldsymbol{SS}_{\text{between}} = \sum_{j=1}^{2}\sum_{i=1}^{n_j}\left(\bar{\boldsymbol{z}}_{.j} - \bar{\boldsymbol{z}}_{..}\right)^2, \quad \boldsymbol{SS}_{\text{within}} = \sum_{j=1}^{2}\sum_{i=1}^{n_j}\left(\boldsymbol{z}_{ij} - \bar{\boldsymbol{z}}_{.j}\right)^2$$

where $\bar{\boldsymbol{z}}_{.j} = \frac{1}{n_j}\sum_{i=1}^{n_j}\boldsymbol{z}_{ij}, \bar{\boldsymbol{z}}_{..} = \frac{1}{n_1 + n_2}\sum_{j=1}^{2}\sum_{i=1}^{n_j}\boldsymbol{z}_{ij}$.

The null hypothesis H_0 is rejected if

$$F^* = \frac{SS_{\text{between}}}{SS_{\text{within}}}(n_1 + n_2 - 2) > F\left(1, n_1 + n_2 - 2 \middle| 1 - \frac{\alpha}{2}\right).$$

3.4.4 Table for Sample Sizes

We present in Table 3.4 an overview listing formulae to determine the sample sizes for testing hypotheses.

Table 3.4 Approximate sample sizes for testing hypotheses with given risks α, β and given minimum difference δ (apply $P = 1 - \alpha$ for one-sided tests and $P = 1 - \alpha/2$ for two-sided tests).

Parameter	Sample size
μ	$n \approx \left\lceil \left[\{t(n-1;P) + t(n-1;1-\beta)\}\frac{\sigma}{\delta}\right]^2 \right\rceil$
$\mu_x - \mu_y$ paired observations	$n \approx \left\lceil \left[\{t(n-1;P) + t(n-1;1-\beta)\}\frac{\sigma_\Delta}{\delta}\right]^2 \right\rceil$
$\mu_x - \mu_y$ independent samples, equal variances	$n \approx \left\lceil 2\left[\{t(n-1;P) + t(n-1;1-\beta)\}\frac{\sigma}{\delta}\right]^2 \right\rceil$
$\mu_x - \mu_y$ independent samples, unequal variances	$n_x \approx \left\lceil \frac{\sigma_x(\sigma_x + \sigma_y)}{\delta^2}[\{t(f^*;P) + t(f^*,1-\beta)\}]^2 \right\rceil$
Probability p	$n = \left\lceil \frac{\left[z_{1-\alpha}\sqrt{p_0(1-p_0)} + z_{1-\beta}\sqrt{p_1(1-p_1)}\right]^2}{(p_1 - p_0)^2} \right\rceil$
Probabilities p_1 and p, $H_0 : p_1 = p_2$	$n = \left\lceil \frac{1}{\delta^2}\left[z(P)\sqrt{(p_1+p_2)\left(1-\frac{1}{2}(p_1+p_2)\right)} + z(1-\beta)\sqrt{p_1(1-p_1)+p_2(1-p_2)} \right]^2 \right\rceil$

3.5 Confidence Estimation

In applications, the user seldom contents oneself with point estimations for unknown parameters. On the contrary, he often tries to calculate or estimate the variance of the estimation. If this variance is sufficiently small, then there is no cause to distrust the estimated value.

Definition 3.9 Let $\boldsymbol{Y} = (\boldsymbol{y}_1, \boldsymbol{y}_2, \dots, \boldsymbol{y}_n)^T$ be a random sample with realisations $Y \in \{Y\}$, whose components are distributed as $P_\theta \in P = \{P_\theta : \theta \in \Omega\}$. Let $S(\boldsymbol{Y})$ be a measurable mapping of the sample space onto the parameter space and $K(\boldsymbol{Y})$ be a random set with realisations $K(Y)$ in Ω. Further, let P^S be the probability measure induced by $S(\boldsymbol{Y})$. Then $K(\boldsymbol{Y})$ is said to be a confidence region for θ with the corresponding confidence coefficient (confidence level) $1 - \alpha$ if

$$P^S[\theta \in K(\boldsymbol{Y})] = P[\theta \in K(\boldsymbol{Y})] \geq 1 - \alpha \text{ for all } \theta \in \Omega. \tag{3.48}$$

In a condensed form, $K(\boldsymbol{Y})$ is also said to be a $(1 - \alpha)$ confidence region. If $\Omega \subset R^1$ and $K(Y)$ is a connected set for all $Y \in \{Y\}$, then $K(\boldsymbol{Y})$ is called a confidence interval. The realisation $K(Y)$ of a confidence region is called a realised confidence region.

The interval estimation includes the construction of confidence intervals. It stands beside the point estimation. Nevertheless, we will see that there are analogies to the test theory concerning the optimality of confidence intervals that can be exploited to simplify many considerations. That is the cause for treating this subject in the chapter about tests.

Example 3.16 Let the $n > 1$ components of a random sample $\boldsymbol{Y} = (\boldsymbol{y}_1, \boldsymbol{y}_2, \ldots, \boldsymbol{y}_n)^T$ be distributed as $N(\mu,\sigma^2)$, where σ^2 is known. We consider the measurable mapping $S(\boldsymbol{Y}) = \bar{\boldsymbol{y}}$ from $\{Y\}$ onto $\Omega = R^1$. The mean $\bar{\boldsymbol{y}}$ follows a $N\left(\mu, \frac{\sigma^2}{n}\right)$-distribution. A $(1 - \alpha)$ confidence region $K(Y)$ with respect to μ has to satisfy $P[\mu \in K(Y)] = 1 - \alpha$ (here we write P for P^S). We suppose that $K(Y)$ is a connected set, that is, an interval $K(\boldsymbol{Y}) = [\hat{\boldsymbol{\mu}}_u, \hat{\boldsymbol{\mu}}_o]$. This means

$$P(\hat{\boldsymbol{\mu}}_u \le \mu \le \hat{\boldsymbol{\mu}}_o) = 1 - \alpha.$$

Since $\bar{\boldsymbol{y}}$ is distributed as $N\left(\mu, \frac{\sigma^2}{n}\right)$, it holds

$$P\left\{ z_{\alpha_1} \le \frac{\bar{\boldsymbol{y}} - \mu}{\sigma}\sqrt{n} \le z_{1-\alpha_2} \right\} = 1 - \alpha_1 - \alpha_2 = 1 - \alpha.$$

for $\alpha_1 + \alpha_2 = \alpha$, $\alpha_1 \ge 0$, $\alpha_2 \ge 0$. Consequently, we have

$$P\left\{ \bar{\boldsymbol{y}} - \frac{\sigma}{\sqrt{n}} z_{1-\alpha_2} \le \mu \le \bar{\boldsymbol{y}} - \frac{\sigma}{\sqrt{n}} z_{\alpha_1} \right\} = 1 - \alpha$$

so that $\hat{\boldsymbol{\mu}}_u = \bar{\boldsymbol{y}} - \frac{\sigma}{\sqrt{n}} z_{1-\alpha_2}$ and $\hat{\boldsymbol{\mu}}_o = \bar{\boldsymbol{y}} - \frac{\sigma}{\sqrt{n}} z_{\alpha_1}$ are fulfilled. For $1 - \alpha$ there are infinitely many confidence intervals according to the choice of α_1 and $\alpha_2 = \alpha - \alpha_1$. If $\alpha_1 = 0$ and $\alpha_2 = 0$, respectively, then the confidence intervals are one-sided (i.e. only one interval end is random). The more the values α_1 and α_2 differ from each other, the larger is the expected width $E(\hat{\boldsymbol{\mu}}_o - \hat{\boldsymbol{\mu}}_u) = \frac{\sigma}{\sqrt{n}}(z_{1-\alpha_2} - z_{\alpha_1})$. For example, the width becomes infinite for $\alpha_1 = 0$ or for $\alpha_2 = 0$. Finite confidence intervals result for $\alpha_1 > 0$, $\alpha_2 > 0$.

Now we set conditions helping to select suitable confidence intervals from the huge number of possible ones. First $K(Y)$ ought to be connected and finite with probability 1. Additionally, for fixed α we prefer confidence intervals possessing the smallest width or the smallest expected width with respect to all $\theta \in \Omega$.

3.5.1 One-Sided Confidence Intervals in One-Parametric Distribution Families

Definition 3.10 Let the components of a random sample $\boldsymbol{Y} = (\boldsymbol{y}_1, \boldsymbol{y}_2, \ldots, \boldsymbol{y}_n)^T$ be distributed as $P_\theta \in P = \{P_\theta, \theta \in \Omega$, where $\Omega = \{\theta_1, \theta_2\}$ and the improper values $-\infty$ for θ_1 and $+\infty$ for θ_2 are admitted. Then

$$\boldsymbol{K}_L = K_L(\boldsymbol{Y}) = [\theta_u(\boldsymbol{Y}), \theta_2)$$

and

$$\boldsymbol{K}_R = K_R(\boldsymbol{Y}) = (\theta_1, \theta_o(\boldsymbol{Y})],$$

respectively, are said to be one-sided confidence intervals for θ with the confidence coefficient $1 - \alpha$ if

$$P_\theta\{\theta \in \boldsymbol{K}_L\} \geq 1 - \alpha \text{ and } P_\theta\{\theta \in \boldsymbol{K}_R\} \geq 1 - \alpha, \tag{3.49}$$

respectively. $\boldsymbol{K}_L$ is called a left-sided and $\boldsymbol{K}_R$ a right-sided confidence interval. A left-sided (right-sided) confidence interval with coefficient $1 - \alpha$ is said to be a uniformly most powerful confidence interval (UMP $(1 - \alpha)$-interval), if for each $\theta^* < \theta$ $[\theta^* > \theta]$, $\theta^* \in \Omega$ the probability

$$P_\theta\{\theta_u(\boldsymbol{Y}) \leq \theta^*\} \quad [P_\theta\{\theta_o(\boldsymbol{Y}) \geq \theta^*\}]$$

becomes minimal under the condition (3.49).

A two-sided confidence interval $K(\boldsymbol{Y})$ satisfying (3.48) is called uniformly most powerful confidence interval (UMP $(1 - \alpha)$-interval), if for each $\theta^* \neq \theta, \theta^* \in \Omega$ the probability $P_\theta\{\theta^* \in K(\boldsymbol{Y})\}$ becomes minimal.

As we will see there is a close relation between UMP-α-tests and UMP $(1 - \alpha)$-intervals. At first we more generally state the relation between α-tests and confidence intervals with the coefficient $1 - \alpha$.

Theorem 3.19 Let the components of a random sample $\boldsymbol{Y} = (\boldsymbol{y}_1, \boldsymbol{y}_2, \ldots, \boldsymbol{y}_n)^T$ be distributed as $P_\theta \in P = \{P_\theta : \theta \in \Omega\}$. For each $\theta_o \in \Omega \subset R^1$, let $\{Y_0\} \subset \{Y\}$ be the region of the sample space $\{Y\}$, where the null hypothesis H_0: $\theta = \theta_0$ is accepted. Let $K(Y)$ be for each $Y \in \{Y\}$ the subset

$$K(Y) = \{\theta \in \Omega : Y \in \{Y_0\}\}. \tag{3.50}$$

of the parameter space Ω. Then $K(\boldsymbol{Y})$ is a $(1 - \alpha)$-confidence interval, if a test with a risk of the first kind not larger than α is defined by $\{Y_0\}$. If moreover $\{Y_0\}$ defines a UMP-α-test, then $K((\boldsymbol{Y})$ is a UMP $(1-\alpha)$-interval.

Proof: Since $\theta \in K(Y)$ iff $Y \in \{Y_0\}$, it follows

$$P_\theta\{\theta \in K(\boldsymbol{Y})\} = P_\theta\{\boldsymbol{Y} \in Y_0\}\} \geq 1 - \alpha.$$

If $K^*(\boldsymbol{Y})$ is another $(1 - \alpha)$-confidence interval for θ and if $\{Y_0^*\} = \{Y : \theta \in K^*(Y)\}$, then we analogously get

$$P_\theta\{\theta \in K^*(\boldsymbol{Y})\} = P_\theta\{\boldsymbol{Y} \in \{Y_0^*\}\} \geq 1 - \alpha,$$

that is, $\{Y_0^*\}$ defines another test with maximal risk α of the first kind. Since $\{Y_0\}$ generates a UMP-test, we obtain

$$P_\theta\{\theta \in K^*(\boldsymbol{Y}) \mid \theta_0\} \geq P_\theta\{\theta \in K(\boldsymbol{Y}) \mid \theta_0\}$$

for all $\theta \neq \theta_0 \in \Omega$ and therefore

$$P_\theta\{\theta \in K^*(\boldsymbol{Y})\} \geq P_\theta\{\theta \in K(\boldsymbol{Y})\} \quad \text{for all } \theta \neq \theta_0.$$

The equivalence given in the above theorem means that a realised confidence interval with coefficient $1 - \alpha$ contains a subset ω of Ω so that $H_o : \theta = \theta_0$ would be accepted for all $\theta_o \in \omega$ if Y is a realisation of $\boldsymbol{Y}$.

The next theorem is a consequence of Theorems 3.19 and 3.8 and its Corollary 3.3, respectively.

Theorem 3.20 If P^* is under the assumptions of Theorem 3.8 a family of continuous distributions with distribution functions $F_\theta(T)$, then there exists for each α with $0 \leq \alpha \leq 1$ a UMP $(1-\alpha)$-interval $K_L(\boldsymbol{Y})$ according to Definition 3.10. If the equation $F_\theta(T) = P_\theta\{T(\boldsymbol{Y}) < T\} = 1 - \alpha$ has a solution $\widehat{\theta} \in \Omega$, then it is unique. Further $\theta_u(\boldsymbol{Y}) = \widehat{\boldsymbol{\theta}}$.

Proof: The elements of P^* are continuous distributions. Hence, to each θ_0 there exists a number $T_{1-\alpha} = T_{1-\alpha}(\theta_0)$ so that $P^*_\theta\{\, T(\boldsymbol{Y}) > T_{1-\alpha}\} = \alpha$. Taking (3.24) into account, $Y_A(\theta_0) = \{T : T > T_{1-\alpha}(\theta_o)\}$ is the rejection region of a UMP-α-tests for H_o: $\theta = \theta_0$ against H_A: $\theta = \theta_A$. Then $Y_0(\theta_0) = \{\mathrm{T} : \mathrm{T} \leq T_{1-\alpha}(\theta_o)\}$ is the corresponding acceptance region. Now let $K(Y)$ be given by (3.50). Since $T_{1-\alpha}(\theta_o)$ is strictly monotone in θ_o (the test is unbiased), $K(Y)$ consists of all $\theta \in \Omega$ with $\theta_u(Y) \leq \theta$, where $\theta_u(Y) = \min_{\theta \in \Omega}\{\theta, T(Y)\} \leq T_{1-\alpha}(\theta_0)\}$. This implies the first assertion in Theorem 3.20.

It follows from Corollary 3.1 of Theorem 3.1 that $F_\theta(T)$ is strictly antitone in θ for each fixed T, provided that $0 < F_\theta(T) < 1$ holds for this T. Therefore the equation $F_\theta(T) = 1 - \alpha$ has at most one solution. Let $\widehat{\theta}$ be such a solution, that is, let $F_{\widehat{\theta}}(T) = 1 - \alpha$. Then $T_{1-\alpha}\left(\widehat{\theta}\right) = T$ follows, and the inequalities $T \leq T_{1-\alpha}(\theta_o)$ and $T_{1-\alpha}\left(\widehat{\theta}\right) \leq T_{1-\alpha}(\theta_o)$ or $\widehat{\theta} \leq \theta$ are equivalent. But this means $\theta_u(Y) = \widehat{\theta}$. Hence, $\theta_u(Y)$ is obtained by solving the equation $T(Y) = T_{1-\alpha}(\theta)$ in θ.

Example 3.17 Under the conditions of Example 3.4, we look for a UMP $(1-\alpha)$-interval for μ. Now $T(\boldsymbol{Y}) = \bar{\boldsymbol{y}}$ is distributed as $N\left(\mu, \dfrac{\sigma^2}{n}\right)$. Therefore $T_{1-\alpha}(\mu)$ is the $(1-\alpha)$-quantile of a $N\left(\mu, \dfrac{\sigma^2}{n}\right)$-distribution. Considering $\theta = \mu$ we must first solve the equation $F_\mu[T(Y)] = 1 - \alpha$. Because of $T_{1-\alpha}(\mu) = \mu + \dfrac{\sigma}{\sqrt{n}} z_{1-\alpha}$, the wished UMP $(1-\alpha)$-interval for μ has the form $\left[\bar{\boldsymbol{y}} - \dfrac{\sigma}{\sqrt{n}} z_{1-\alpha}, +\infty\right)$

Example 3.18 Starting with the random sample of Examples 3.5, we want to construct a one-sided confidence interval with coefficient $1-\alpha$ for σ^2 (where μ is known). Using the sufficient statistic $T(\boldsymbol{Y}) = \sum_{i=1}^{n} (\boldsymbol{y}_i - \mu)^2$, the region Y_0 for accepting $H_0 : \sigma^2 = \sigma_0^2$ is given by the inequality $T(Y) < \sigma_0^2 \, CS(n|1-\alpha)$. Here $CS(n|1-\alpha)$ is the $(1-\alpha)$-quantile of the chi-squared distribution. The quantiles are shown in Table D.4. Now $K(\boldsymbol{Y})$ written as $\sigma_u^2(\boldsymbol{Y})$ is determined by

$$\sigma_u^2(\boldsymbol{Y}) = \min_{\sigma^2 \in \Omega} \left\{ \sigma^2, T(Y) < \sigma_0^2 CS(n|1-\alpha) \right\}.$$

Hence, $\left[\sigma_u^2(\boldsymbol{Y}), +\infty\right)$ with the left end

$$\sigma_u^2(\boldsymbol{Y}) = \frac{\sum_{i=1}^{n} (\boldsymbol{y}_i - \mu)^2}{CS(n|1-\alpha)}$$

is for each α $(0 < \alpha < 1)$ a UMP $(1-\alpha)$-interval for σ^2.

Analogously the reader can as an exercise transform other UMP-α-tests into corresponding UMP $(1-\alpha)$-intervals.

If under the assumptions of Theorem 3.8 the distribution of $T(\boldsymbol{Y})$ is discrete, then the tests are randomised. Thus the corresponding confidence intervals are also called randomised. But in general we do not want to deal with such confidence intervals. However, in practical applications, they often are needed concerning the parameter p of the binomial distribution representing a probability p. Here we refer to Fleiss et al. (2003) and to the case of two-sided intervals in Section 3.5.2.

3.5.2 Two-Sided Confidence Intervals in One-Parametric and Confidence Intervals in Multi-Parametric Distribution Families

Definition 3.11 A two-sided confidence interval $K(\boldsymbol{Y})$ with coefficient $1-\alpha$ is said to be a uniformly most powerful interval, if $K(\boldsymbol{Y})$ is in the class

$$K_\alpha = \{K(\boldsymbol{Y}), P_\theta[\theta \in K(\boldsymbol{Y})] \geq 1-\alpha \ \text{ for all } \ \theta \in \Omega\} \tag{3.51}$$

and fulfils the condition

$$P_\theta[\theta^* \in K(\boldsymbol{Y})] = \min_{K^*(Y) \in K_\alpha} P_\theta\{\theta^* \in K^*(\boldsymbol{Y})\} \ \text{ for all } \ \theta^* \neq \theta \in \Omega. \tag{3.52}$$

Analogous to Section 3.5.1 for continuous distributions, we can construct two-sided uniformly most powerful $(1-\alpha)$-intervals on the base of UMP-α-tests for $H_0 : \theta = \theta_0$ against $H_A : \theta \neq \theta_0$. But generally such tests do not exist for all α, and therefore we introduce a weaker optimality condition analogous to the UMPU-tests.

Definition 3.12 A $(1-\alpha)$-confidence interval $K(Y) = [\boldsymbol{l}, \boldsymbol{u}]$ is said to be an unbiased $U(1-\alpha)$-interval, if it lies in K_α and satisfies

$$P_\theta[\theta^* \in K(\boldsymbol{Y})] \leq 1-\alpha \text{ for all } \theta^* \neq \theta \in \Omega. \tag{3.53}$$

Then we note briefly that $K(Y)$ is a U-$(1-\alpha)$-interval. $K(Y)$ is said to be a uniformly most powerful unbiased $(1-\alpha)$-confidence interval, if it fulfils the conditions (3.51) and (3.53) as well as a condition analogous to (3.52), where the minimum is taken within the class $\widetilde{K_\alpha} \subset K_\alpha$ of such $K(Y)$ satisfying both (3.51) and (3.53). We denote uniformly most powerful unbiased $(1-\alpha)$-confidence intervals shortly as UMPU $(1-\alpha)$-intervals.

If $\theta = (\lambda, \eta_2, \dots, \eta_k)^T$ is a parameter vector and if a confidence interval with respect to the real component λ is to be designed, then we can generalise with $\eta^* = (\eta_2, \dots, \eta_k)^T$ the Definitions 3.9 and 3.3 by replacing the demand 'for all θ "by the demand" for all λ and η^*'. If a UMPU-α-test exists, then it is easy to see that the procedure described in Section 3.5.1 can be used to construct a UMPU $(1-\alpha)$-interval. We want to demonstrate this by presenting some examples. But first we must mention the fact that UMPU $(1-\alpha)$-intervals satisfy for continuous distributions the condition

$$P_\theta[\theta \in K(\boldsymbol{Y})] = 1-\alpha.$$

Example 3.19 Under the conditions of Example 3.9, we want to construct a UMPU $(1-\alpha)$-interval for σ^2. For this purpose we use the sufficient statistic $T(\boldsymbol{Y}) = \sum_{i=1}^n \boldsymbol{y}_i^2$ and introduce

$$\{Y_0\} = A(\sigma^2) = \left\{\sigma^2,\ c_{1\alpha} \leq \frac{1}{\sigma^2} T(Y) \leq c_{2\alpha}\right\},$$

where $c_{1\alpha}$ and $c_{2\alpha}$ fulfil (3.33) and (3.34). Observing

$$A(\sigma^2) = \left\{\sigma^2,\ \frac{1}{c_{2\alpha}} \leq \frac{\sigma^2}{T(Y)} \leq \frac{1}{c_{1\alpha}}\right\}$$

and passing to random variables shows that

$$K(\boldsymbol{Y}) = \left[\frac{1}{c_{2\alpha}} \sum_{i=1}^n \boldsymbol{y}_i^2, \frac{1}{c_{1\alpha}} \sum_{i=1}^n \boldsymbol{y}_i^2\right]$$

is a two-sided UMPU $(1-\alpha)$-interval for σ^2.

Example 3.20 On the basis of Example 3.11, a UMPU $(1-\alpha)$-interval is to be constructed for the expectation μ of a normal distribution with unknown variance. Because of (3.36) we get

$$\{Y_0\} = A(\mu) = \left\{-t\left(n-1\middle|1-\frac{\alpha}{2}\right) \leq \frac{\bar{y}-\mu}{s}\sqrt{n} \leq t\left(n-1\middle|1-\frac{\alpha}{2}\right)\right\},$$

and therefore

$$K(\boldsymbol{Y}) = \left[\bar{\boldsymbol{y}} - t\left(n-1\middle|1-\frac{\alpha}{2}\right)\frac{\boldsymbol{s}}{\sqrt{n}}; \bar{\boldsymbol{y}} + t\left(n-1\middle|1-\frac{\alpha}{2}\right)\frac{\boldsymbol{s}}{\sqrt{n}}\right],$$

is a UMPU (1 – α)-interval for μ.

Example 3.21 On the base of Example 3.15, a UMPU (1 – α)-interval for the difference $\mu_1 - \mu_2$ is to be constructed. It follows from (3.46) and the form of the UMPU-α-test $K(\boldsymbol{Y})$ given afterwards, if in the numerator of (3.46) the expression $\mu_1 - \mu_2$ is inserted (which is 0 under H_o), that

$$K(\boldsymbol{Y}) = \left[\bar{\boldsymbol{y}}_1 - \bar{\boldsymbol{y}}_2 - t\left(n_1+n_2-2\middle|1-\frac{\alpha}{2}\right)\boldsymbol{s}\sqrt{\frac{n_1+n_2}{n_1 n_2}}, \bar{\boldsymbol{y}}_1 - \bar{\boldsymbol{y}}_2 \right.$$
$$\left. + t\left(n_1+n_2-2\middle|1-\frac{\alpha}{2}\right)\boldsymbol{s}\sqrt{\frac{n_1+n_2}{n_1 n_2}}\right]$$

is a UMPU (1 – α)-interval. In this case we also propose to use instead confidence intervals that are based on the Welch test.

If the distribution modelling the character is discrete as, for example, in the case of the binomial distribution, then exact tests are for all α always randomised tests. If the demand is slightly weakened by looking for a confidence interval that covers the parameter p at least with probability $1-\alpha$, then an exact interval $K(\boldsymbol{Y}) = [\boldsymbol{l}, \boldsymbol{u}]$ can be constructed according to Clopper and Pearson (1934) as follows. If $[l, u]$ is a realised confidence interval and y the observed value of the random variable $\boldsymbol{y}$ distributed as $B(n, p)$, then the endpoints l and u can be determined so that

$$\sum_{i=y}^{n} \binom{n}{i} l^i (1-l)^{n-i} = \alpha_1$$

and

$$\sum_{i=0}^{y} \binom{n}{i} u^i (1-u)^{n-i} = \alpha_2$$

hold, where $\alpha_1 + \alpha_2 = \alpha$ for given α_1 and α_2 is independent of y.

We put $l = 0$ and $u = 1 - \left(\frac{\alpha}{2}\right)^{\frac{1}{n}}$ for $y = 0$ as well as $l = \left(\frac{\alpha}{2}\right)^{\frac{1}{n}}$ and $u = 1$ for $y = n$.

The other values can be calculated according to Stevens (1950) with the help of the probability function p_{beta} of the beta distribution with the parameters x and $n-x-1$, for instance, with R using the commands

```
l< -qbeta(alfa/2,X,n-X+1)
```

and

```
u< -qbeta(1-alfa/2,X+1,n-X),
```

respectively. The Clopper–Pearson intervals can be calculated with the R command binom.test. In SPSS confidence intervals cannot be found in the menu bar.

The minimal covering probability is for $n \geq 10$ and for all p at least $1 - (\alpha_1 - \alpha_2) - 0.005$, but in the most cases larger than $1-\alpha$, that is, conservative. This was shown by Pires and Amado (2008). Both authors compared 20 construction methods for two-sided confidence intervals regarding the covering probability and the expected interval width using extensive simulation experiments. The study found that a method of Agresti and Coull (1998) had slight advantages in comparison with the Clopper–Pearson intervals. But we do not want to go into the matter here.

The needed sample size can be obtained in R with the command size.prop. confint by calculating confidence intervals via normal approximation (see Rasch et al., 2011a, p. 31).

3.5.3 Table for Sample Sizes

We present in Table 3.5 a list of formulae for determining suitable sample sizes of confidence estimations. It should be observed that for location parameters, either the width or, if it is random, the expected width of the interval has to be given before lying under a reasonable bound 2δ.

Table 3.5 Sample size for the construction of two-sided $(1 - \alpha)$-confidence intervals with half expected width δ.

Parameter	Sample size
μ	$n = \left\lceil t^2\left(n-1; 1-\frac{\alpha}{2}\right)\frac{2\cdot\Gamma^2\left(\frac{n}{2}\right)\cdot}{\Gamma^2\left(\frac{n-1}{2}\right)(n-1)}\frac{\sigma^2}{\delta^2}\right\rceil$
P	With R via size.prop.confint
$\mu_x - \mu_y$ paired observations	$n = \left\lceil t^2\left(n-1; 1-\frac{\alpha}{2}\right)\frac{2\cdot\Gamma^2\left(\frac{n}{2}\right)\cdot}{\Gamma^2\left(\frac{n-1}{2}\right)(n-1)}\frac{\sigma_\Delta^2}{\delta^2}\right\rceil$
$\mu_x - \mu_y$ independent samples, equal variances	$n = \left\lceil 2\sigma^2 \frac{t^2\left(2n-2; 1-\frac{\alpha}{2}\right)}{\delta^2(2n-2)} \frac{2\Gamma^2\left(\frac{2n-1}{2}\right)}{\Gamma^2(n-1)}\right\rceil$
$\mu_x - \mu_y$ independent samples, unequal variances	$n_x = \left\lceil \frac{\sigma_x(\sigma_x+\sigma_y)}{\delta^2} t^2\left(f^*; 1-\frac{\alpha}{2}\right)\right\rceil$; $n_y = \left\lceil \frac{\sigma_y}{\sigma_x} n_x \right\rceil$

3.6 Sequential Tests

3.6.1 Introduction

Until now a sample of fixed size n was given. The task of statistical design of experiments is to determine n so that the test satisfies certain precision requirements of the user. We have demonstrated this procedure in the previous sections.

For testing the null hypothesis that the expectation of a normal distribution with unknown variance takes a particular value against a one-sided alternative hypothesis, the sample size has to be determined after fixing the risks α, β and the minimum difference δ as

$$n = \left\lceil [t(n-1|1-\alpha) + t(n-1|1-\beta)]^2 \frac{\sigma^2}{\delta^2} \right\rceil \tag{3.54}$$

according to Section 3.4.1. Apart from the fact that (3.54) can only be iteratively solved, it needs also a priori information about σ^2. Therefore Stein (1945) proposed a method of realising a two-stage experiment. In the first stage a sample of size $n_0 > 1$ is drawn to estimate σ^2 by the variance s_0^2 of this sample and to calculate the sample size n of the method using (3.54). In the second stage $n - n_0$, further measurements are taken. Following the original method of Stein in the second stage, at least one further measurement is necessary from a theoretical point of view. In this subsection we simplify this method by introducing the condition that no further measurements are to be taken for $n - n_0 \leq 0$. Nevertheless, this supplies an α-test of acceptable power.

Since both parts of the experiment are carried out one after the other, such experiments are called sequential. Sometimes it is even tenable to make all measurements step by step, where each measurement is followed by calculating a new test statistic. A sequential testing of this kind can be used, if the observations of a random variable in an experiment take place successive in time. Typical examples are series of single experiments in a laboratory, psychological diagnostics in single sessions and medical treatments of patients in hospitals, consultations of clients of certain institutions and certain procedures of statistical quality control, where the sequential approach was used the first time (Dodge and Romig, 1929). The basic idea is to utilise the observations already made before the next are at hand.

For example, testing the hypothesis H_0: $\mu = \mu_0$ against H_A: $\mu > \mu_0$ there are three possibilities in each step of evaluation, namely,

1) Accept H_0.
2) Reject H_0.
3) Continue the investigation.

Comparing sequential tests with tests of fixed size, the advantage of the former is that on the average fewer experimental units are needed considering great series of investigations. But a decision between the abovementioned three cases is only possible for a priori given values of α, β and δ. Unfortunately, this a priori information is not compelling for testing with fixed size.

Nevertheless, we will only briefly deal with the theory of sequential tests for two good reasons. Firstly the up-to-now unsurpassed textbook of Wald (1947) has since been reprinted and is therefore generally available (Wald, 1947/2004), and new results can be found in books of Ghosh and Sen (1991) as well as DeGroot (2005). Secondly we do not recommend the application of this general theory, but we recommend closed plans, which end after finite steps with certainty (and not only with probability 1).

We start with some concepts.

Definition 3.13 Let a sequence $\boldsymbol{S} = \{\mathbf{y}_1, \mathbf{y}_2, \ldots\}$ of random variables (a stochastic process) be given, where the components are identically and stochastically independently distributed as $P_\theta \in P = \{P_\theta \in \Omega\}$. Let the parameter space Ω consist of two different elements, θ_0 and θ_A. Besides, let $y_i \in \{Y\} \subset R^1$. Concerning testing of the hypotheses H_0: $\theta = \theta_0$; H_A: $\theta = \theta_A$, we suppose that for each n in the above sequence a decomposition $\{M_0^n, M_A^n, M_F^n,\}$ of

$$\{Y^n\} = \{y_1\} \times \{y_2\} \times \cdots \times \{y_n\} \subset R^n$$

with

$$M_0^n \cup M_A^n \cup M_F^n = M_F^{n-1} \times \{y_n\} \subset R^n$$

exists. Then the sets M_0^n, M_A^n, M_F^n (n = 1, 2, ...) define together with the prescription

$$(y_1, \ldots, y_n) \in \begin{cases} M_0^n & \text{acceptation of } H_0 : \theta = \theta_0 \\ M_A^n & \text{rejection of } H_0 : \theta = \theta_0 \\ M_F^n & \text{continuation, observe } y_{n+1} \end{cases}$$

a sequential test with respect to the pair H_0: $\theta = \theta_0$; H_A: $\theta = \theta_A$. M_0^n and M_A^n are called final decisions. The pair (α, β) of risks is called the strength of a sequential test.

Definition 3.14 Let a sequence $\boldsymbol{S} = \{\boldsymbol{y}_1, \boldsymbol{y}_2, \ldots\}$ of random variables be given where the components are identically and stochastically independently distributed as $P_\theta \in P = \{P_\theta \in \Omega\}$. Let the parameter space Ω consist of the two different elements θ_0 and θ_A. A sequential test for H_0: $\theta = \theta_0$ against H_A: $\theta = \theta_A$, based on the ratio

$$\boldsymbol{LR}_n = \frac{L\left(\boldsymbol{Y}^{(n)} | \theta_A\right)}{L\left(\boldsymbol{Y}^{(n)} | \theta_0\right)}.$$

of the likelihood functions $L(\boldsymbol{Y}^{(n)}|\theta)$ of both parameter values and on the first n elements $\boldsymbol{Y}^{(n)}=\{\boldsymbol{y}_1, \boldsymbol{y}_2, \ldots, \boldsymbol{y}_n\}$ of the sequence $\boldsymbol{S}$ is said to be sequential likelihood ratio test (SLRT), if for certain numbers A and B with $0 < B < 1 < A$ the decomposition of $\{Y^{(n)}\}$ reads

$$M_0^n = \left\{Y^{(n)} : LR_n \le B\right\},\ M_A^n = \left\{Y^{(n)} : LR_n \ge A\right\},\ M_F^n = \left\{Y^{(n)} : B < LR_n < A\right\}.$$

Theorem 3.21 A sequential likelihood ratio test (SLRT) that leads with probability 1 to a final decision with the strength (α, β) fulfils with the numbers A and B from Definition 3.14 the conditions

$$A \le \frac{1-\beta}{\alpha}, \tag{3.55}$$

$$B \ge \frac{\beta}{1-\alpha}. \tag{3.56}$$

In the applications the equalities are often used in (3.56) and (3.57) to calculate approximately the bounds A and B. Such tests are called approximate tests.

It follows from the theory that SLRT can hardly be recommended, since they end under certain assumptions only with probability 1. So far they are the most powerful tests for a given strength as the expectation of the sample size – the average sample number (ASN) – for such tests is minimal and smaller than the size for tests where the size is fixed. Since it is unknown for which maximal sample size the SLRT ends with certainty, it belongs to the class of open sequential tests. In comparison there are also closed sequential tests, that is, tests with a secure maximal sample size, but this advantage is won by a bit larger ASN.

3.6.2 Wald's Sequential Likelihood Ratio Test for One-Parametric Exponential Families

All results are presented without proofs. The interested reader can find proofs in the book of Wald (1947). Some results come from an unpublished manuscript of B. Schneider (1992).

We thank him for the permission to use the results for our book.

Let a sequence $\boldsymbol{S} = \{\boldsymbol{y}_1, \boldsymbol{y}_2, \ldots\}$ of identically and independently distributed random variables be given, which are distributed as $\boldsymbol{y}$ with the same likelihood function $f(y, \theta)$. We test the null hypothesis

$$H_0 : \theta = \theta_0 [f(y,\theta) = f(y,\theta_0)]$$

against the alternative

$$H_A : \theta = \theta_1\ [f(y,\theta) = f(y,\theta_1)]$$

where $\theta_0 \neq \theta_1$ and $\theta_0, \theta_1 \in \Omega \subset R^1$.

The realised likelihood ratio after n observations is

$$\boldsymbol{LR}_n = \prod_{i=1}^{n} \frac{f(\boldsymbol{y}_i;\theta_1)}{f(\boldsymbol{y}_i;\theta_0)}; \quad n > 1. \tag{3.57}$$

The subsequent questions arise:

- How do we choose the numbers A and B in (3.55) and (3.56)?
- What is the mean size $E(\boldsymbol{n} \mid \theta)$ of the sequence $\{\boldsymbol{y}_1, \boldsymbol{y}_2, \ldots\}$?

Wald used the following approximations for A, B. If the nominal risks of the first and the second kind are given by α_{nom} and β_{nom}, then the real (actual) risks α_{act} and β_{act} satisfy

$$\alpha_{act} \leq \frac{1}{\boldsymbol{A}} = \alpha_{nom}; \quad \beta_{act} \leq \boldsymbol{B} = \beta_{nom}.$$

Hence, the approximate test introduced in the preceding subsection is conservative. This supplies the relations in (3.55) and (3.56). The corresponding bounds are called Wald bounds.

Example 3.22 Assuming that the nominal risks of the first and the second kind are 0.05 and 0.1, respectively, the relations (3.55) and (3.56) lead (in the equality case) to the values A = 18 and B = 0.10536. Therefore we have to continue the process up to the step where $0.10356 < LR_n < 18$ is fulfilled. In a system of coordinates with n on the abscissa and LR_n on the ordinate, the zone of continuation lies between two parallel lines.

The (approximate) power function of the SLRT is

$$\pi(\theta) = \frac{\left(\frac{1-\beta}{\alpha}\right)^{h(\theta)} - 1}{\left(\frac{1-\beta}{\alpha}\right)^{h(\theta)} - \left(\frac{\beta}{1-\alpha}\right)^{h(\theta)}} \quad \text{for } h(\theta) \neq 0. \tag{3.58}$$

The function $h(\theta)$ in (3.58) is uniquely defined in the continuous case by the equation

$$\int \left(\frac{f(y,\theta_1)}{f(y,\theta_0)}\right)^{h(\theta)} \cdot f(y,\theta)dy = 1$$

and in the discrete case by the equation

$$\sum_{\forall y_i} \left(\frac{f(y_i,\theta_1)}{f(y_i,\theta_0)}\right)^{h(\theta)} \cdot f(y_i,\theta) = 1.$$

Wald showed that for sequential likelihood ratio tests, the expected (average) sample size ASN is minimal under all sequential tests with risks not larger than α_{nom} and β_{nom} provided that one of the two hypotheses is true.

With the notations

$$\mathbf{z} = \ln\frac{f(\mathbf{y},\theta_1)}{f(\mathbf{y},\theta_0)},\quad \mathbf{z}_i = \ln\frac{f(\mathbf{y}_i,\theta_1)}{f(\mathbf{y}_i,\theta_0)} \tag{3.59}$$

we get $\ln\mathbf{LR}_{\boldsymbol{n}} = \sum \mathbf{z}_i$. For $E(|\mathbf{z}|) < \infty$ Wald showed also

$$E(\boldsymbol{n}|\theta) = \frac{\pi(\theta)\ln A + [1-\pi(\theta)]\ln B}{E(\mathbf{z}|\theta)},\quad \text{if } E(\mathbf{z}|\theta) \neq 0. \tag{3.60}$$

The experiment ends if in the current step at least one inequality (sign) becomes an equality (sign) in

$$\alpha_{act} \le \frac{1}{A} = \alpha_{nom},\quad \beta_{act} \le B = \beta_{nom}.$$

Wijsman (1991) presented an approximation for $E(\boldsymbol{n}|\theta)$ that reads in the special case $\theta = \theta_0$

$$E(\boldsymbol{n}|\theta_0) \approx \frac{1}{E(\mathbf{z}/\theta_0)}\left[\frac{A-1}{A-B}\ln B + \frac{1-B}{A-B}\ln A\right] \tag{3.61}$$

and in the general case

$$E(\boldsymbol{n}|\theta) \approx \frac{1}{E(\mathbf{z}|\theta)}\left[\frac{(A-1)\cdot B}{A-B}\ln B + \frac{A\cdot(1-B)}{A-B}\ln A\right]. \tag{3.62}$$

In an exponential family the first derivative of $A(\theta)$ supplies the expectation of $\boldsymbol{y}$ and the second derivative the variance of $\boldsymbol{y}$.

Wald (1947) proved in the continuous case that there exists a θ^* with $E(\boldsymbol{z}|\theta^*) = 0$ that fulfils $h(\theta^*) = 0$ in (3.58) and moreover that

$$E(\boldsymbol{n}|\theta^*) \approx \frac{|\ln A|\cdot|\ln B|}{E(\mathbf{z}^2|\theta^*)},\quad \text{if } h(\theta^*) = 0 \tag{3.63}$$

holds.

Example 3.23 We consider a one-parametric exponential family with density function $f(y, \theta) = h(y)e^{y\eta - A(\eta)}$.

We want to test

$$H_0: \theta = \theta_0 \ (\eta = \eta_0)$$

against

$$H_A: \theta = \theta_1 \ (\eta = \eta_1),\quad \theta_0 < \theta_1 (\eta_0 < \eta_1)$$

with $\eta_i = \eta(\theta_i)$; $i = 0, 1$. For $\theta_0 > \theta_1$ we interchange the hypotheses.

The variables z_i can be written (in a realised form) as $z_i = (\eta_1 - \eta_0)y_i - [A(\eta_1) - A(\eta_0)]$. We continue while

$$\ln B < (\eta_1 - \eta_0)\sum y_i - n[A(\eta_1) - A(\eta_0)] < \ln A$$

and because of $\eta_1 - \eta_0 > 0$ while

$$b_u^n = \frac{\ln B + n[A(\eta_1) - A(\eta_0)]}{(\eta_1 - \eta_0)} < \sum y_i < \frac{\ln A + n[A(\eta_1) - A(\eta_0)]}{(\eta_1 - \eta_0)} = b_o^n \quad (3.64)$$

is satisfied.

For $\eta_1 - \eta_0 < 0$ we continue if

$$b_u^n > \sum y_i > b_o^n$$

holds with the bounds b_u^n and b_o^n from (3.64). W.l.o.g. we restrict ourselves to the case $\eta_1 - \eta_0 > 0$.

In the discrete case the distribution function of the random process is between parallel lines a step function. Therefore it cannot be guaranteed that the Wald bound is met in the last step exactly. In such cases an algorithm of Young (1994) is useful, which is described now.

Suppose that the test ends with the nth observation. The probability for obtaining a value t_n of the variable $\mathbf{t}_n = \sum \mathbf{y}_i$ after n units were observed is the sum of the probability sequence that fulfils the conditions $b_u^i \le \mathbf{t}_i \le b_o^i; i = 1, 2, \ldots, n-1$ and $\boldsymbol{t}_n = t_n$. We write this probability as

$$P(\mathbf{t}_n = t) = \sum_{j=b_u^{n-1}}^{b_o^{n-1}} P(\mathbf{t}_n = t_n | \mathbf{t}_{n-1} = j) \cdot P(\mathbf{t}_{n-1} = j) = \sum_{j=b_u^{n-1}}^{b_o^{n-1}} f(t_n - j; \theta) \cdot P(\mathbf{t}_{n-1} = j).$$

We start with $P(t_0 = 0) = 1$ and determine all further probabilities by recursion.

For fixed n the probability for accepting H_A at the nth observation is

$$\begin{aligned} P(\mathbf{t}_n > b_u^n) &= \sum_{j=b_u^{n-1}}^{b_o^{n-1}} \sum_{k=b_u^n - j + 1}^{\infty} f(k; \theta) \cdot P(\mathbf{t}_{n-1} = j) \\ &= \sum_{j=b_u^{n-1}}^{b_o^{n-1}} \left[1 - F(b_o^n - j; \theta)\right] \cdot P(\mathbf{t}_{n-1} = j), \end{aligned} \quad (3.65)$$

where F is the distribution function.

For fixed n the probability for accepting H_0 at the nth observation is

$$\begin{aligned} P(\mathbf{t}_n < b_l^n) = k &= \sum_{j=b_u^{n-1}}^{b_o^{n-1}} \sum_{k=0}^{b_u^n - j - 1} f(k; \theta) \cdot P(\mathbf{t}_{n-1} = j) \\ &= \sum_{j=b_u^{n-1}}^{b_o^{n-1}} \left[F(b_0^n - j - 1; \theta)\right] \cdot P(\mathbf{t}_{n-1} = j). \end{aligned} \quad (3.66)$$

The power function is given by $\sum_{i=1}^{n} P\left(\mathbf{t}_i < b_u^i\right)$ if the procedure ends with step n, and the probability for this event is equal to $P\left(\mathbf{t}_n < b_u^n\right) + P\left(\mathbf{t}_n > b_o^n\right)$.

In the following example we use one-sided hypotheses with $\alpha = 0.05$, $\beta = 0.1$ and $\delta = \theta_1 - \theta_0 = 0.1$.

Example 3.24 Normal Distribution with Known Variance

If $\boldsymbol{y}$ is distributed as $N(\mu; \sigma^2)$ with known σ^2, then we get

$$\boldsymbol{z} = \ln \frac{1}{2\sigma^2}\left[2\boldsymbol{y}(\mu_1 - \mu_0) + \mu_0^2 - \mu_1^2\right],$$

$$E(z|\mu) = \ln \frac{1}{2\sigma^2}\left[2\mu(\mu_1 - \mu_0) + \mu_0^2 - \mu_1^2\right].$$

We test

$$H_0 : \mu = \mu_0$$

against the alternative

$$H_A : \mu = \mu_1; \ \ \mu_0 \neq \mu_1; \ \ \mu \in R^1.$$

For $\mu_0 - \mu_1 = \sigma$ and $\theta = \mu$, we obtain in (3.58) with $h(\theta) = h(\mu)$

$$\int \left(e^{-\frac{1}{2\sigma^2}\left[2y\sigma - \sigma^2\right]}\right)^{h(\mu)} \cdot \frac{1}{\sigma\sqrt{2\pi}} e^{-\frac{1}{2\sigma^2}(y-\mu)^2} dy = 1.$$

Finally an R-routine in OPDOE supplies $E(\boldsymbol{n}|\mu)$ and $\pi(\mu)$ as functions of μ.

3.6.3 Test about Mean Values for Unknown Variances

Now we deal with a two-parametric exponential family. We have to adapt the method from Section 3.6.1 to this case. We have a nuisance parameter, that is, the method cannot be used directly. The parameter vector of an exponential family is $\theta = (\theta_1, \theta_2)^T$. For $\varphi_0 \neq \varphi_1$ we have to test

$$H_0 : \varphi(\theta) \leq \varphi_0; \varphi \in R^1 \ \text{ against } \ H_A : \varphi(\theta) \geq \varphi_1; \varphi \in R^1$$

or

$$H_0 : \varphi(\theta) = \varphi_0; \varphi \in R^1 \ \text{ against } \ H_A : \varphi(\theta) \neq \varphi_1; \varphi \in R^1.$$

In this book, we consider the one-dimensional normal distribution; the corresponding test is called sequential t-test.

The Sequential t-Test

The normal distribution of a random variable $\boldsymbol{y}$ is a two-parametric exponential family with the parameter vector $\theta = (\mu; \sigma^2)^T$ and the log-likelihood function (natural logarithm of the likelihood function)

$$l\left(\mu; \sigma^2\right) = -\ln\sqrt{2\pi} - \ln\sigma - \frac{1}{2\sigma^2}(y - \mu)^2.$$

We put $\varphi(\theta) = \frac{\mu}{\sigma}$ and test

$$H_0: \frac{\mu}{\sigma} \leq \varphi_0 \text{ against } H_A: \frac{\mu}{\sigma} \geq \varphi_1$$

or

$$H_0: \frac{\mu}{\sigma} = \varphi_0 \text{ against } H_A: \frac{\mu}{\sigma} \neq \varphi_1.$$

If we replace the noisy parameter, as in the case of fixed sample size in Section 3.6.2, by its estimation, then LR_n in Definition 3.14 is no likelihood ratio.

We consider a sequence

$$\mathbf{z}_1 = z_1(\mathbf{y}_1); \mathbf{z}_2 = z_2(\mathbf{y}_1; \mathbf{y}_2), \ldots$$

so that for each $n > 1$ the conditional likelihood function $f_u^{(n)}(\mathbf{z}_1, \mathbf{z}_2, \ldots, \mathbf{z}_n; \varphi)$ of $(\mathbf{z}_1, \mathbf{z}_2, \ldots, \mathbf{z}_n)$ depends only via $\varphi(\theta)$ on θ.

Then we apply the theory of Section 3.6.1 with

$$LR_n^* = \prod_{i=1}^{n} \frac{f_u^{(n)}(z_i, \varphi_1)}{f_u^{(n)}(z_i, \varphi_0)}, n > 1 \tag{3.67}$$

instead of

$$\lambda_n = \prod_{i=1}^{n} \frac{f(y_i; \theta_1)}{f(y_i; \theta_0)}; n > 1.$$

The choice of the sequence $\mathbf{z}_1 = z_1(\mathbf{y}_1); \mathbf{z}_2 = z_2(\mathbf{y}_1; \mathbf{y}_2), \ldots$ is explained in the following.

Lehmann (1959) formulated the principle of invariant tests. If we multiply μ and σ with a positive real number c, then the hypotheses $H_0: \frac{\mu}{\sigma} \leq \varphi_0$ and $H_A: \frac{\mu}{\sigma} \geq \varphi_1$ remain unchanged since they are invariant with respect to affine transformations. The random variables $\mathbf{y}_i^* = c\mathbf{y}_i$ are normally distributed with expectation $c\mu$ and standard deviation $c\sigma$. Therefore, the family of the distributions of $\mathbf{y}_1, \mathbf{y}_2, \ldots, \mathbf{y}_n$ is for each $n \geq 1$ the same as that of $c\mathbf{y}_1, c\mathbf{y}_2, \ldots, c\mathbf{y}_n$. Summarised we see that both the hypotheses and the family of the distributions are invariant with respect to affine transformations. Now the sequential t-test can be implemented according to Eisenberg and Ghosh (1991) as follows:

- Specialise LR_n^* in (3.67) for a normal distribution, that is,

$$LR_n^* = e^{-\frac{1}{2}\left(n - v_n^2\right)\left(\varphi_1^2 - \varphi_0^2\right)} \frac{\int_0^\infty t^{n-1} e^{-\frac{1}{2}(t - v_n\varphi_1)^2} dt}{\int_0^\infty t^{n-1} e^{-\frac{1}{2}(t - v_n\varphi_0)^2} dt} \tag{3.68}$$

- Solve $LR_n^* = \frac{\beta}{1-\alpha}$ and $LR_n^* = \frac{1-\beta}{\alpha}$ with respect to v_n. We denote the solutions by v_u^n and v_o^n, respectively.

- Calculate

$$v_n = \frac{\sum_{i=1}^{n} y_i}{\sqrt{\sum_{i=1}^{n} y_i^2}} \tag{3.69}$$

  and continue while $v_u^n < v_n < v_o^n$ holds.
- Accept H_0 if $v_n \leq v_u^n$ and reject H_0 if $v_n \geq v_o^n$.

Approximation of the Likelihood Function for Constructing an Approximate *t*-Test

Now we want to use certain functions $\boldsymbol{z}$ and $\boldsymbol{v}$ related to the Taylor expansion of the log-likelihood function to construct simple sequential tests.

Let the sequence $(\boldsymbol{y}_1, \boldsymbol{y}_2, \ldots)$ of identically distributed and independent random variables be distributed as $\boldsymbol{y}$ with the likelihood function $f(y; \theta)$. We expand $l(y; \theta) = \ln f(y; \theta)$ according to Taylor with respect to θ at $\theta = 0$ up to the second order with third-order error term:

$$l(y;\theta) = l(y;0) + \theta \cdot l_\theta(y;0) + \frac{1}{2}\theta^2 l_{\theta\theta}(y;0) + O(\theta^3). \tag{3.70}$$

Now we put

$$z = l_\theta(y;0) = \frac{\partial \ln(y,\theta)}{\partial\theta}\Big|_{\theta=0}, \tag{3.71}$$

$$-v = l_{\theta\theta}(y;0) = \frac{\partial^2 \ln(y,\theta)}{\partial\theta^2}\Big|_{\theta=0}. \tag{3.72}$$

If we neglect the error term $O(\theta^3)$, we get a quadratic approximation around $\theta = 0$:

$$l(y;\theta) = \text{const.} + \theta \cdot z - \frac{1}{2}\theta^2 v. \tag{3.73}$$

In the case where the likelihood function depends also on a vector, $\tau = (\tau_1, \ldots, \tau_k)^T$ of noisy parameters. Whitehead (1997) proposed to replace this vector by the vector of the corresponding maximum likelihood estimators.

Then the likelihood function reads $f(y; \theta, \tau)$ and has the logarithm $l(y; \theta, \tau) = \ln f(y; \theta, \tau)$.

We denote the maximum likelihood estimator of τ by $\tilde{\tau}(\theta)$, which supplies $\tilde{\tau} = \tilde{\tau}(0)$ for $\theta = 0$. The maximum likelihood estimator of τ is a solution of the simultaneous equations

$$\frac{\partial}{\partial\tau} l(y;\theta,\tau) = 0$$

and under natural assumptions also the unique solution, since $l(y; \theta, \tau)$ is then concave (convex from above) in τ and has therefore only one maximum. We expand $\tilde{\tau}(\theta)$ according to Taylor at $\theta = 0$ taking a quadratic error term:

$$\tilde{\tau}(\theta) = \tilde{\tau} + \theta \cdot \frac{\partial}{\partial\theta}\tilde{\tau}(\theta)\big|_{\theta=0} + O(\theta^2). \tag{3.74}$$

The vector $\widetilde{\tau}_\theta = \frac{\partial}{\partial\theta}\widetilde{\tau}(\theta)$ is the first derivative of $\widetilde{\tau}(\theta)$ with respect to θ. The matrix of the second partial derivatives (called also Hessian matrix) of $\ln f(y;\theta,\tau)$ with respect to τ_i and τ_j for $\tau = \widetilde{\tau}$ is denoted by $M_{\tau\tau}(y,\theta,\widetilde{\tau}(\theta))$.

After some rearrangements (see Whitehead, 1997), we can write z and v with the notations

$$l_\theta(y;0,\widetilde{\tau}) = \frac{\partial \ln(y,\theta,\widetilde{\tau})}{\partial\theta}\Big|_{\theta=0}$$

$$l_{\theta\theta}(y;0,\widetilde{\tau}) = \frac{\partial^2 \ln(y,\theta,\widetilde{\tau})}{\partial\theta^2}\Big|_{\theta=0};\ l_{\theta\tau}(y;0,\widetilde{\tau}) = \frac{\partial^2 \ln(y,\theta,\tau)}{\partial\theta\cdot\partial\tau}\Big|_{\theta=0;\tau=\widetilde{\tau}}$$

in the following form:

$$z = l_\theta(y;0,\widetilde{\tau}), \tag{3.75}$$

$$v = -l_{\theta\theta}(y;0,\widetilde{\tau}) - l_{\theta\tau}(y;0,\widetilde{\tau})^T \cdot M_{\tau\tau}(y,\theta,\widetilde{\tau}(\theta)) \cdot l_{\theta\tau}(y;0,\widetilde{\tau}). \tag{3.76}$$

Using z- and v-values in (3.71) and (3.72) in the case without noisy parameters or in (3.75) and (3.77) in the case with noisy parameter(s), unique approximate sequential likelihood ratio tests (SLRT) can be constructed.

After observing n elements $\boldsymbol{y}_1, \boldsymbol{y}_2, \dots, \boldsymbol{y}_n$ of the sequence, which are distributed as $\boldsymbol{y}$ with the log-likelihood function $l(y;\theta,\tau) = \ln f(y;\theta,\tau)$, we write the z-function in (3.71) and (3.75), respectively, as

$$z_n = \sum_{i=1}^{n} z_i = \sum_{i=1}^{n} l_\theta(y_i;0,\widetilde{\tau}),$$

where the estimator of the noisy parameter is put 0, if it is missing. The number z_n represents the efficient value and characterises the deviation from the null hypothesis.

The v-function is connected with the Fisher information matrix

$$I(\theta) = -\sum_{i=1}^{n} E_\theta\left\{\frac{\partial^2 l(\mathbf{y}_i,\theta,\widetilde{\tau})}{\partial\theta^2}\right\} = n\cdot i(\theta),$$

where

$$\boldsymbol{i}(\theta) = -\frac{\partial^2 l(\mathbf{y},\theta,\widetilde{\tau})}{\partial\theta^2}$$

is the information of an observation with $i(\theta) = E_\theta(\boldsymbol{i}(\theta))$.

Now we put $v = nE[\boldsymbol{i}(\theta)]|_{\theta=0}$. Since likelihood estimations are asymptotically normally distributed, the variable

$$\mathbf{z}_n = \sum_{i=1}^{n} \mathbf{z}_i = \sum_{i=1}^{n} l_\theta(\mathbf{y}_i;0,\widetilde{\tau})$$

is approximately asymptotically normally distributed with the expectation $\theta\ \nu$ and the variance ν.

After observing n elements $\boldsymbol{y}_1, \boldsymbol{y}_2, \ldots, \boldsymbol{y}_n$ distributed as $\boldsymbol{y}$, we write the ν-functions (3.72) and (3.76) in the form

$$\nu_n = \sum_{i=1}^{n} \left\{ -l_{\theta\theta}(y;0,\tilde{\tau}) - l_{\theta\tau}(y;0,\tilde{\tau})^T \cdot M_{\tau\tau}(y,\theta,\tilde{\tau}(\theta)) \cdot l_{\theta\tau}(y;0,\tilde{\tau}) \right\}$$

For testing the null hypothesis

$$H_0: \theta = \theta_0;\ \ \theta_0 \in \Omega \subset R^1 \ \text{ against } \ H_A: \theta = \theta_1 \neq \theta_0;\ \ \theta_1 \in \Omega \subset R^1,$$

we use the approximate SLRT as follows:

- Continue in taking observation values, while
$$a_u = \frac{1}{\theta_1} \ln\frac{\beta}{1-\alpha} < z_n - b\nu_n < a_o = \frac{1}{\theta_1} \ln\frac{1-\beta}{\alpha} \text{ with } b = \frac{1}{2}\theta_1.$$
- Accept $H_A: \theta = \theta_1 > 0$, if $z_n - b\nu_n > a_o = \frac{1}{\theta_1}\ln\frac{1-\beta}{\alpha}$, and accept $H_A: \theta = \theta_1 < 0$, if $z_n - b\nu_n < a_u = \frac{1}{\theta_1}\ln\frac{\beta}{1-\alpha}$, respectively; otherwise accept H_0.

The power function of the test is

$$\pi(\theta) \approx \frac{1 - \left(\frac{\beta}{1-\alpha}\right)^{1-2\frac{\theta}{\theta_1}}}{\left(\frac{1-\beta}{\alpha}\right)^{1-2\frac{\theta}{\theta_1}} - \left(\frac{\beta}{1-\alpha}\right)^{1-2\frac{\theta}{\theta_1}}} \quad \text{for } \theta \neq 0.5\theta_1$$

and

$$\pi(\theta) = \frac{\ln\left(\frac{1-\alpha}{\beta}\right)}{\ln\left(\frac{1-\beta}{\alpha}\right) + \ln\left(\frac{1-\alpha}{\beta}\right)} \quad \text{for } \theta = 0.5\theta_1.$$

The expected sample size is given by

$$E(\boldsymbol{n}|\theta) = \frac{\ln\frac{1-\beta}{\alpha}\left\{1 - \left(\frac{\beta}{1-\alpha}\right)^{1-2\frac{\theta}{\theta_1}}\right\} - \ln\frac{\beta}{1-\alpha}\left\{1 - \left(\frac{1-\beta}{\alpha}\right)^{1-2\frac{\theta}{\theta_1}}\right\}}{\left[\theta_1(\theta - 0.5\theta_1)\left\{\left(\frac{1-\beta}{\alpha}\right)^{1-2\frac{\theta}{\theta_1}} - \left(\frac{\beta}{1-\alpha}\right)^{1-2\frac{\theta}{\theta_1}}\right\}\right]} \quad \text{for } \theta \neq 0.5\theta_1$$

and

$$E(\boldsymbol{n}|\theta) = \frac{1}{\theta_1^2} \ln\frac{1-\beta}{\alpha} \cdot \ln\frac{\beta}{1-\alpha} \quad \text{for } \theta = 0.5\theta_1.$$

The null hypothesis $H_0:\theta = 0$ can be used. W.l.o.g., as it is obtained if in the general hypothesis $H_0:\mu = \mu_0 \neq 0$, the value μ_0 is subtracted from all observation values.

Now we consider the normal distribution with unknown variance by using the approximation given in this section. We test

$$H_0:\mu = 0;\sigma^2 \text{ arbitrary against } H_A:\mu = \mu_1,\mu_1 \neq 0_1;\mu_1 \in \Omega \subset R^1.$$

We put $\tau = \sigma^2$ and $\theta = \dfrac{\mu}{\sigma}$. Then the log-likelihood function reads

$$\begin{aligned} l_\theta(y_1;y_2;\ldots;y_n;\theta) &= -\frac{1}{2}n\cdot\ln(2\pi\sigma) - \frac{1}{2\sigma^2}\sum_{i=1}^{n}(y_i-\mu)^2 \\ &= -\frac{1}{2}n\cdot\ln(2\pi\sigma) - \frac{1}{2}\sum_{i=1}^{n}\left(\frac{y_i}{\sigma}-\theta\right)^2. \end{aligned} \tag{3.77}$$

The efficient value is

$$z_n = \frac{\sum_{i=1}^{n} y_i}{\sqrt{\dfrac{1}{n}\sum_{i=1}^{n} y_i^2}}. \tag{3.78}$$

The v-function is

$$v_n = n - \frac{z_n^2}{2n}. \tag{3.79}$$

The formulae for z_n and v_n are listed for some distributions in Table 3.6.

3.6.4 Approximate Tests for the Two-Sample Problem

We have two distributions with parameters θ_1, θ_2 and a common noisy parameter ψ. Two random samples $(\boldsymbol{y}_{i1},\ldots,\boldsymbol{y}_{ini})$ of size n_i with $i = 1, 2$ are sequentially drawn to test the null hypothesis

Table 3.6 Values z_n and v_n for special distributions.

Distribution	Log-likelihood	Hypotheses	z_n	v_n
Normal, σ known	$-\frac{n}{2}\ln(2\pi\sigma)-\frac{1}{2}\sum_{i=1}^{n}\left[\frac{y_i}{\sigma}-\mu\right]^2$	$H_0:\mu=0$ $H_A:\mu=\mu_1$	$z_n = \frac{\sum_{i=1}^{n} y_i}{\sigma}$	$v_n = n$
Normal, σ unknown	$-\frac{n}{2}\ln(2\pi\sigma)-\frac{1}{2}\sum_{i=1}^{n}\left[\frac{y_i}{\sigma}-\mu\right]^2$	$H_0:\mu=0$ $H_A:\mu=\mu_1$	$z_n = \frac{\sum_{i=1}^{n} y_i}{\sqrt{\frac{1}{n}\sum_{i=1}^{n} y_i^2}}$	$v_n = n - \frac{z_n^2}{2n}$
Bernoulli	$n\ln\frac{p}{1-p}+n\ln(1-p)$	$H_0:p=p_0$	$z_n = \frac{y-np_0}{p_0(1-p_0)}$	$v_n = \frac{n}{p_0(1-p_0)}$

$$H_0 : \theta_1 = \theta_2$$

against one of the following alternative hypotheses:

a) $H_A : \theta_1 > \theta_{2.}$
b) $H_A : \theta_1 < \theta_{2.}$
c) $H_A : \theta_1 \neq \theta_{2.}$

A suitable reparametrisation is

$$\theta = \frac{1}{2}(\theta_1 - \theta_2); \varphi = \frac{1}{2}(\theta_1 + \theta_2).$$

The log-likelihood function of both samples reads

$$l(y_1, y_2; q, \varphi, \psi) = l^{(y_1)}(y_1; \theta_1, y) + l^{(y_2)}(y_2; \theta_2, \psi).$$

For simplification we omit now the arguments of the functions. Then the first and second partial derivatives of l are

$$l_\theta = l_{\theta_1}^{(y_1)} - l_{\theta_2}^{(y_2)}$$

$$l_\varphi = l_{\theta_1}^{(y_1)} + l_{\theta_2}^{(y_2)}$$

$$l_\psi = l_\psi^{(y_1)} + l_\psi^{(y_2)}$$

$$l_{\theta\theta} = l_{\theta_1\theta_1}^{(y_1)} + l_{\theta_2\theta_2}^{(y_2)}$$

$$l_{\theta\varphi} = l_{\theta_1\theta_1}^{(y_1)} l_{\theta_2\theta_2}^{(y_2)}$$

$$l_{\theta\psi} = l_{\theta_1\psi}^{(y_1)} l_{\theta_2\psi}^{(y_2)}$$

$$l_{\varphi\psi} = l_{\theta_1\psi}^{(y_1)} + l_{\theta_2\psi}^{(y_2)}$$

$$l_{\psi\psi} = l_{\psi\psi}^{(y_1)} + l_{\psi\psi}^{(y_2)}$$

The expectations $\widehat{\varphi}; \widehat{\psi}$ are solutions of

$$l_{\theta_1}^{(y_1)}(\widehat{\varphi}, \widehat{\psi}) + l_{\theta_2}^{(y_2)}(\widehat{\varphi}, \widehat{\psi}) = 0,$$

$$l_{\psi_1}^{(y_1)}(\widehat{\varphi}, \widehat{\psi}) + l_{\psi_2}^{(y_2)}(\widehat{\varphi}, \widehat{\psi}) = 0.$$

Now we can continue as described in Section 3.6.2. We do not want to go into detail here, since we are more interested in focusing on triangular tests (see the following Section 3.6.5).

Naturally, it is wrong to prefer sequential tests in each case to non-sequential ones. At most, it holds for the mean sample size. Namely, in a sequential test the actual n in a final decision could be larger than the one for a test with a priori given size. Besides, for these kinds of experiments, the necessary time interval has to be taken into account. A sequential experiment lasts at least n-times as long as it has a fixed size n. Sequential evaluations are beneficial (compared with other methods) in such cases where the data arise anyway sequentially (e.g. in medical tests or treatments for patients who are rarely ill).

3.6.5 Sequential Triangular Tests

In this section we turn to special closed tests, the triangular tests.

The values of the decision statistics of the triangular tests correspond to those of the approximate tests in the previous section. In a suitable coordinate system, the sequence of these values (ordinates) generates as a function of successive points in time and sample sizes (abscissas), respectively, a sequential path. The population is here the set of all possible paths. Within the plane coordinate system, a triangular 'zone of continuation' is defined, which contains the origin of the time axis. While the path runs in this zone, the process of drawing samples is continued. If the path meets or crosses the boundary of the zone, then the data collection is finished. The decision whether the null hypothesis is accepted or rejected depends on the point of boundary crossing. The separation of the boundary into two parts has to fulfil the condition that for true H_0 the part of rejecting H_0 is reached at most with probability α and for true H_A the part of accepting H_0 is reached at most with probability β. We want to point out that it is not necessary to make an evaluation after each newly taken sample value and to wait for the evaluation result before the sampling is continued. The sequence path is independent of the decision whether an evaluation is made or not. Hence, it is definitely possible to restrict the evaluations to certain time points or sample sizes fixed before or ad hoc.

Regarding sequential triangular tests for a parameter θ, we want to use the standardised null hypothesis $\theta = 0$. Otherwise a reparametrisation of θ has to be chosen so that θ becomes for the corresponding reference value θ_0 of the null hypothesis the value 0.

The two variables z and v create the sequence path. They are derived from the likelihood function $L(\theta)$ as described in Section 3.6.3. More precisely, z and v are introduced by using derivatives of $l(\theta) = \ln L(\theta)$ with respect to θ. Namely, z is the first derivative of $l(\theta)$, and v is the negative second derivative of $l(\theta)$ with respect to θ in place 0. If we replace the sample values in the likelihood function by corresponding random variables (whose realisations represent the sample), then $\boldsymbol{z}$ becomes a random variable itself, which is for not too small sample sizes and not too great absolute values of θ nearly normally distributed with the expectation θ and the variance v. Therefore $\boldsymbol{z}$ can be considered as a measure for the deviation of the parameters θ from the value 0 of the null hypothesis. The variable $\boldsymbol{v}$ characterises the amount of information in the sample with respect to the parameters θ. This amount increases with increasing sample size, that is, v is a monotone increasing function of the sample size.

For triangular tests the continuation zone is a closed triangular set. These tests are based on the asymptotic tests of the previous section.

We consider the one-sample problem and test

$$H_0 : \theta = \theta_0 \text{ against } H_A : \theta = \theta_1.$$

The continuation zone is given by

$$-a + 3bv_n < z_n < a + bv_n \quad \text{for} \quad \theta_1 > \theta_0,$$
$$-a + bv_n < z_n < a + 3bv_n \quad \text{for} \quad \theta_1 < \theta_0$$

where the sequence $(z_n; v_n)$ is defined in (3.75) and (3.76).

The hypothesis $H_0 : \theta = \theta_0$ is accepted, if

$$z_n \geq a + bv_n \quad \text{for} \quad \theta_1 > 0$$

and if

$$z_n \leq -a + bv_n \quad \text{for} \quad \theta_1 < 0.$$

If z_n leaves the continuation zone or meets its boundary, then $H_A : \theta = \theta_1$ is accepted.

The constants a and b are determined by

$$a = \left(1 + \frac{z_{1-\beta}}{z_{1-\alpha}}\right) \frac{\ln \frac{\alpha}{2}}{\theta_1}, \tag{3.80}$$

$$b = \frac{\theta_1}{\left(1 + \frac{z_{1-\beta}}{z_{1-\alpha}}\right)}. \tag{3.81}$$

Both straight lines on the boundary meet in the point

$$(v_{\max}; z_{\max}) = \left(\frac{a}{b}; 2a\right).$$

If this point is reached, we accept $H_A : \theta = \theta_1$. The point of intersection corresponds to the maximal sample size. This size is larger than that of experiments with fixed size for equally prescribed precision, but the latter is larger than the average sample size (ASN) of the triangular test.

Now some special cases follow, which can be solved using the software OPDOE in R.

First we consider the problem of Example 3.11. Let $\boldsymbol{S} = (\boldsymbol{y}_1, \boldsymbol{y}_2, \ldots)$ be a sequence with components distributed together with $\boldsymbol{y}$ as $N(\mu, \sigma^2)$. We want to test

$$H_0 : \mu = \mu_0, \sigma^2 \text{ arbitrary, against } H_A : \mu = \mu_1, \sigma^2 \text{ arbitrary.}$$

Then we get

$$z_n = \frac{\sum_{i=1}^{n} y_i}{\sqrt{\frac{\sum_{i=1}^{n} y_i^2}{n}}}, \quad v_n = n - \frac{z_n^2}{2n}.$$

The boundary lines of the triangle follow from (3.80) and (3.81) putting $\theta_1 = \frac{\mu_1 - \mu_0}{\sigma}$.

Regarding the two-sample problem, we test analogously to Section 3.4.2.1

$$H_0 : \mu_1 = \mu_2 = \mu, \quad \sigma_1^2 = \sigma_2^2 = \sigma^2 \text{ arbitrary}$$

against

$$H_A : \mu_1 \neq \mu_2, \quad \sigma_1^2 = \sigma_2^2 = \sigma^2 \text{ arbitrary}.$$

We put $\theta_1 = \frac{\mu_1 - \mu_2}{\sigma}$ and calculate from the n_1 and n_2, respectively, observations the maximum likelihood estimator

$$\tilde{\sigma}_n^2 = \frac{\sum_{i=1}^{n_1} (y_{1i} - \bar{y}_1)^2 + \sum_{i=1}^{n_2} (y_{2i} - \bar{y}_2)^2}{n_1 + n_2}.$$

Then we introduce

$$z_n = \frac{n_1 n_2}{n_1 + n_2} \frac{\bar{y}_1 - \bar{y}_2}{\tilde{\sigma}_n}, \quad v_n = \frac{n_1 n_2}{n_1 + n_2} \frac{z_n^2}{2(n_1 + n_2)}.$$

The constants a and b result again from (3.80) and (3.82). Analogously many tests can be derived from this general theory. More details on the R-files and examples using concrete data and including the accompanying triangles are presented in Rasch at al. (2011b). We want to clarify only one special case, since it stands out from the usual frame. This case was just recently investigated by Schneider et al. (2014).

3.6.6 A Sequential Triangular Test for the Correlation Coefficient

We suppose that the distribution $F(x,y)$ of a two-dimensional continuous random vector $(\boldsymbol{x}, \boldsymbol{y})$ has finite second moments σ_x^2, σ_y^2 and σ_{xy}. Then the correlation coefficient $\rho = \sigma_{xy}/(\sigma_x \sigma_y)$ of the distribution exists and can be calculated. We want to test the null hypothesis H_0: $\rho \le \rho_0$ (or $\rho \ge \rho_0$) against the alternative H_A: $\rho > \rho_0$ (or $\rho < \rho_0$). The probability for rejecting H_0 although $\rho \le \rho_0$ (or $\rho \ge \rho_0$) is to be less or equal to α, and the probability for rejecting H_A although $\rho = \rho_1 > \rho_0$ (or $\rho = \rho_1 < \rho_0$) is to be less or equal to β.

The empirical correlation coefficient $r = s_{xy}/(s_x s_y)$, which is determined from k data pairs (x_i, y_i) with $i = 1,\ldots,k$ as realisations from $(\boldsymbol{x}, \boldsymbol{y})$, is an estimate for the parameter ρ (where s_{xy}, s_x^2, s_y^2 are the empirical covariance and the empirical variances, respectively). Naturally $\boldsymbol{r}$ can be used as test statistic for ρ. Fisher

(1915) derived the distribution of $\boldsymbol{r}$ assuming a two-dimensional (bivariate) normal distribution. He showed that this distribution only depends on n and ρ. Later Fisher (1921) introduced the transformed variable

$$\boldsymbol{z} = \frac{1}{2}\ln\left(\frac{1+\boldsymbol{r}}{1-\boldsymbol{r}}\right) \tag{3.82}$$

as a test statistic. Moreover, he proved that for a small k the distribution of this statistic can already be quite well approximated by a normal distribution. Following Cramér (1946) it suffices already $k = 10$ to get for the interval $-0.8 \le \rho \le +0.8$ a very good adaptation to a normal distribution. We propose here for technical causes a further transformation, namely, $\boldsymbol{u} = 2\boldsymbol{z}$. This statistic has approximately the following expectation and variance:

$$E(\boldsymbol{u}) = \zeta(\rho) = \ln\frac{1+\rho}{1-\rho} + \frac{\rho}{k-1};\quad \mathrm{var}(\boldsymbol{u}) = \frac{4}{k-3} \tag{3.83}$$

If we look for a usable triangular test for hypotheses about the correlation coefficient ρ, then the sequence of data pairs (x_i, y_i) is unsuited, since their likelihood function depends not only on ρ but also on the expectations and variances of the two variables $\boldsymbol{x}$ and $\boldsymbol{y}$ (altogether five parameter), which cannot be estimated by one data pair alone. We need at least three data pairs. This suggests the idea with the sequence of the data pairs (x_i, y_i) to generate at first successive partial samples of arbitrarily chosen size k and to calculate with the data of each partial sample j a test statistic possessing a known distribution, which depends on the parameter ρ. A hot candidate for this is the already introduced $\boldsymbol{z}$-statistic of Fisher (used here with $\boldsymbol{u} = 2\boldsymbol{z}$ instead of $\boldsymbol{z}$). As mentioned above this statistic is for not too small sample sizes k approximately normally distributed with the expectation $\zeta(\rho)$ and the variance $4/(k-3)$ (see (3.84)). As we supposed for the triangular test to use for the null hypothesis the standardised parameter value 0, we transform for testing the hypothesis $\rho = \rho_0$ the u-values into u^*-values so that they have for $\rho = \rho_0$ the expectation 0 and the variance 1. Hence, our triangular test will use the sequence

$$\boldsymbol{u}_j^* = \left(\boldsymbol{u}_j - \ln\frac{1+\rho_0}{1-\rho_0} - \frac{\rho_0}{k-1}\right)\sqrt{\frac{k-3}{4}}$$

for $j = 1, 2, \ldots$. The expectation of $\boldsymbol{u}_j^*$ is the tested parameter θ:

$$\theta = E\left(\boldsymbol{u}_j^*\right) = \left(\ln\frac{1+\rho}{1-\rho} - \ln\frac{1+\rho_0}{1-\rho_0} + \frac{\rho-\rho_0}{k-1}\right)\frac{\sqrt{k-3}}{2}. \tag{3.84}$$

For $\rho = \rho_0$ we get the wanted standard $\theta = 0$. The value for $\rho = \rho_1$ is denoted by θ_1.

The numbers $\boldsymbol{u}_j^*$ that are calculated from the consecutively drawn partial samples j with the empirical correlation coefficients r_j (implicitly contained in u_j) are realisations of independent (approximately) normally distributed random variables with the expectation θ and the variance 1. If m consecutive values u_j^* are available, then the log-likelihood function involving these m values reads as

$$l(\theta) = \text{const.} - \frac{1}{2}\sum_{j=1}^{m}\left(u_j^* - \theta\right)^2. \tag{3.85}$$

Now we write again z for u to be in accordance with the usual notation for triangular tests. This rewriting leads to the test statistics

$$z_m = \frac{dl(\theta)}{d\theta} = \sum_{j=1}^{m} u_j^*, \quad v_m = -\frac{d^2 l(\theta)}{d\theta^2} = m. \tag{3.86}$$

Using a (z, v)-coordinate system, the sequence path is generated by the points (z_m, v_m) obtained by the evaluation steps $m = 1, 2, 3, \ldots$.

The continuation zone is a triangle whose sides are determined by two variables a and c depending on the risks α, β, the sample size k and the value of the alternative hypothesis θ_1:

$$a = \frac{\left(1 + \frac{z_{1-\beta}}{z_{1-\alpha}}\right)\ln\left(\frac{1}{2\alpha}\right)}{\theta_1}, c = \frac{\theta_1}{2\left(1 + \frac{z_{1-\beta}}{z_{1-\alpha}}\right)} \tag{3.87}$$

Here z_P denotes the P-quantile of the standardised normal distribution. One side of the triangle lies on the z-axis extending from a to $-a$. Both sides are created by the straight lines

$$G_1 : z = a + cv \text{ and } G_2 : z = -a + 3cv, \tag{3.88}$$

which meet in the point with the coordinates

$$v_{\max} = \frac{a}{c} \quad z_{max} = 2a. \tag{3.89}$$

For $\theta = \theta_1 > 0$ we have $a > 0$ and $c > 0$. The upper side of the triangular starting from a on the z-axis has the ascent c, while the lower side starting from $-a$ on the z-axis has the ascent $3c$ with respect to the v-axis. Moreover, for $\theta_1 < 0$ it is $a < 0$ and $c < 0$. Now the upper side starting from -a has the ascent $3c$, and the lower side starting from a has the ascent c with respect to the v-axis.

The decision rule is as follows: Continue making observations up to the step where z_m reaches the value $a + cv_m$ or goes under it, and accept H_0, if

$$-a + 3cv_m < z_m < a + cv_m \quad \text{for} \quad \theta_1 > 0, \text{ or} \tag{3.90}$$

$$-a + 3cv_m > z_m > a + cv_m \quad \text{for} \quad \theta_1 < 0.$$

In the case $\theta_1 > 0$ the alternative H_A has to be accepted, if z_m reaches at v_m the straight line $z = a + cv_m$ or goes over it, and H_0, if z_m reaches at v_m the value $z = a + 3cv_m$ or goes under it. In the case $\theta_1 < 0$ the alternative H_A has to be accepted, if z_m reaches the value $z = a + 3cv_m$ or goes over it. If the top of the triangular is hit exactly, then H_A has to be accepted.

Example 3.25 We want to test the null hypothesis $\rho \le 0.6$ against the alternative hypothesis $\rho > 0.6$ with prescribing the risks $\alpha = 0.05$, $\beta = 0.2$ and the minimum deviation $\rho_1 - \rho_0 = 0.1$. Further we choose $k = 12$. This means we calculate one correlation coefficient from samples of each 12 elements.

Then we get for $\rho_1 = 0.7$ and for $\rho_0 = 0.6$ the values

$$\varsigma(0.7) = \ln\left(\frac{1+0.7}{1-0.7}\right) + \frac{0.7}{11} = 1.798$$

and

$$\varsigma(0.6) = \ln\left(\frac{1+0.6}{1-0.6}\right) + \frac{0.6}{11} = 1.441.$$

Because of $\sqrt{k-3} = 3$ the Formula (3.84) supplies

$$\theta_1 = \frac{3}{2}(1.798 - 1.4444) = 0.5355.$$

Taking $z_{0.8} = 0.8416$ and $z_{0.95} = 1.6449$ into account, the sides of the triangular result by (3.87) from

$$a = \frac{\left(1 + \frac{0.8416}{1.6449}\right)\ln\left(\frac{1}{0.1}\right)}{0.5355} = 6.50$$

and

$$c = \frac{0.5355}{2\left(1 + \frac{0.8416}{1.6449}\right)} = 0.1771$$

(see Figure 3.11). Further, (3.89) supplies

$$v_{\max} = \frac{6.50}{0.1771} = 36.7, \quad z_{\max} = 13.$$

The number n_{fix} of observations can be calculated for a test with fixed sample size under corresponding precision requirements using the software **R** or the iterative procedure

$$n_i = \left\lceil 3 + ,4\left(\frac{z_{1-\alpha} + z_{1-\beta}}{\ln\left(\frac{1+\rho_1}{1-\rho_1}\right) - \ln\left(\frac{1+\rho_0}{1-\rho_0}\right) + \frac{\rho_1 - \rho_0}{n_{i-1} - 1}}\right)^2\right\rceil,$$

where the result at the end of the iteration is denoted by n_{fix}.

Schneider et al. (2014) investigated the approximation quality of such tests as well as Rasch and Yanagida (2015).

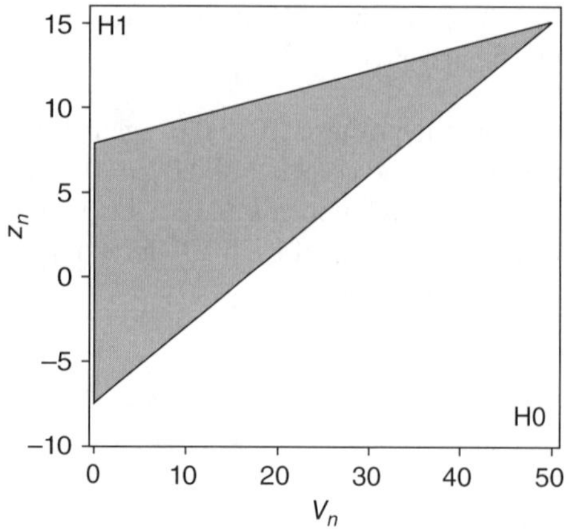

Figure 3.11 Graph of the triangle obtained in Example 3.25.

Simulations were carried out with 10 000 repetitions (calculations of the test statistics) using different sample sizes k, two-dimensional normally distributed random numbers $\boldsymbol{x}$, $\boldsymbol{y}$ with $\mu_x = \mu_y = 0$, $\sigma_x^2 = \sigma_y^2 = 1$ and a correlation coefficient $\sigma_{xy} = \rho$, nominal risks $\alpha_{nom} = 0.05$, $\beta_{nom} = 0.1$ and 0.2 as well as some values of ρ_0, ρ_1. Criteria for the quality of the tests were as follows:

a) The relative frequency of rejecting H_0 wrongly, if $\rho = \rho_0$. This is an estimator of the actual risk of the first kind, α_{act}.
b) The relative frequency of rejecting H_A if $\rho = \rho_1$. This is an estimator of the actual risk of the second kind, β_{act}.
c) The mean number of partial samples for determining r and z up to the stop for ρ_0 and ρ_1.
d) The mean number of pairs (x, y), that is, the ASN, for ρ_0 and ρ_1 taken over all 10 000 repetitions.

In a special case the sample size can be over or under the value of ASN. Some results are presented in Table 3.7. There are two values of k listed, for which α_{act} lies just under or just over 0.05 with exception of one case, where $\alpha_{act} = 0.05$ is exactly met. Table 3.8 lists the k-values that obey α_{act} and β_{act}. The ASN strongly depends on the value of ρ. This is demonstrated in the next example.

Example 3.26 In Table 3.7 we consider the case with $\alpha_{act} = 0.05$, $\rho_0 = 0.6$, $\rho_1 = 0.75$, $\alpha = 0.05$, $\beta = 0.1$ and $k = 20$. The following values of ρ were simulated:

$$\rho = 0.05; 0.1; 0.15; 0.2; 0.25; 0.3; 0.35; 0.4; 0.45;$$

$$0.5; 0.55; 0.65; 0.7; 0.8; 0.85; 0.9; 0.95.$$

Using 10 000 repetitions the ASN and the relative frequency of rejecting H_0 are plotted in Figure 3.12. The ASN is shown in Table 3.9, its graph tends for $\rho \to 0$ to 30 and for $\rho \to 1$ to 20. The maximum lies between $\rho = 0.6$ and $\rho = 0.75$.

Table 3.7 Simulation results for α =0.05.

	$\rho_0 = 0.5, \rho_1 = 0.7$				$\rho_0 = 0.6, \rho_1 = 0.75$			$\rho_0 = 0.6, \rho_1 = 0.8$				$\rho_0 = 0.7, \rho_1 = 0.8$			
	$\beta = 0.1$		$\beta = 0.2$		$\beta = 0.1$	$\beta = 0.2$		$\beta = 0.1$		$\beta = 0.2$		$\beta = 0.1$		$\beta = 0.2$	
k	*12*	*16*	*12*	*16*	*20*	*12*	*16*	*12*	*16*	*12*	*16*	*20*	*50*	*16*	*20*
α_{act}	0.060	0.049	0.053	0.042	0.050	0.063	0.052	0.052	0.041	0.048	0.038	0.064	0.036	0.064	0.057
β_{act}	0.043	0.053	0.114	0.130	0.049	0.103	0.112	0.040	0.047	0.109	0.117	0.044	0.053	0.102	0.112
$ASN\|\rho_0$	74.2	71.5	55.7	54.5	90.0	71.4	67.9	49	47	37.1	37.0	128.7	131.8	98.2	96.1
$ASN\|\rho_1$	72.2	72.3	62.1	62.3	90.0	77.0	76.1	48.2	49.1	41.6	42.8	124.3	137.0	104.9	105.5
n_{fix}	88	88	65	65	113	82	82	56	56	41	41	164	164	119	119

	$\rho_0 = 0.7, \rho_1 = 0.9$				$\rho_0 = 0.8, \rho_1 = 0.9$				$\rho_0 = 0.9, \rho_1 = 0.95$			
	$\beta = 0.1$		$\beta = 0.2$		$\beta = 0.1$		$\beta = 0.2$		$\beta = 0.1$		$\beta = 0.2$	
k	*8*	*12*	*6*	*8*	*16*	*20*	*12*	*16*	*16*	*20*	*16*	*20*
α_{act}	0.058	0.041	0.066	0.047	0.054	0.045	0.059	0.047	0.058	0.048	0.051	0.041
β_{act}	0.029	0.039	0.059	0.085	0.038	0.046	0.094	0.110	0.039	0.040	0.106	0.108
$ASN\|\rho_0$	28.2	25.8	24.9	21.2	56.9	56.2	44.7	43.3	61.6	60.8	46.4	46.9
$ASN\|\rho_1$	26.3	25.9	25.2	23.1	55.4	56.4	47.8	48.4	58.8	60.3	51.6	52.7
n_{fix}	27	27	20	20	65	65	48	48	70	70	51	51

Table 3.8 Admissible results for the simulated δ, ρ_0 and β for $\alpha = 0.05$.

$\beta = 0.1$			$\beta = 0.2$		
ρ_0	δ	k	ρ_0	δ	k
0.5	0.1	50	0.5	0.1	$20 < k < 50$
0.5	0.15	$20 < k < 50$	0.5	0.15	20
0.5	0.2	$12 < k < 16$	0.5	0.2	$12 < k < 16$
0.6	0.1	$20 < k < 50$	0.6	0.1	$20 < k < 50$
0.6	0.15	20	0.6	0.15	$12 < k < 16$
0.6	0.2	$12 < k < 16$	0.6	0.2	$12 < k < 16$
0.7	0.1	$20 < k < 50$	0.7	0.1	$16 < k < 20$
0.7	0.15	$12 < k < 16$	0.7	0.15	$8 < k < 12$
0.7	0.2	$8 < k < 12$	0.7	0.2	$6 < k < 8$
0.8	0.05	50	0.8	0.05	$20 < k < 50$
0.8	0.1	$16 < k < 20$	0.8	0.1	$12 < k < 16$
0.8	0.15	8	0.8	0.15	$6 < k < 8$
0.9	0.05	$16 < k < 20$	0.9	0.05	$16 < k < 20$

Table 3.9 Empirical ASN in dependence on ρ in Example 3.26.

ρ	0.60	0.65	0.70	0.75
ASN(ρ)	89.98	110.094	115.01	89.98

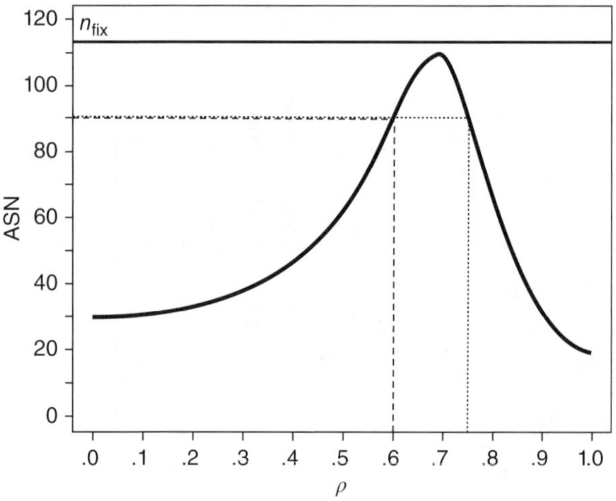

Figure 3.12 ASN graph of Example 3.26.

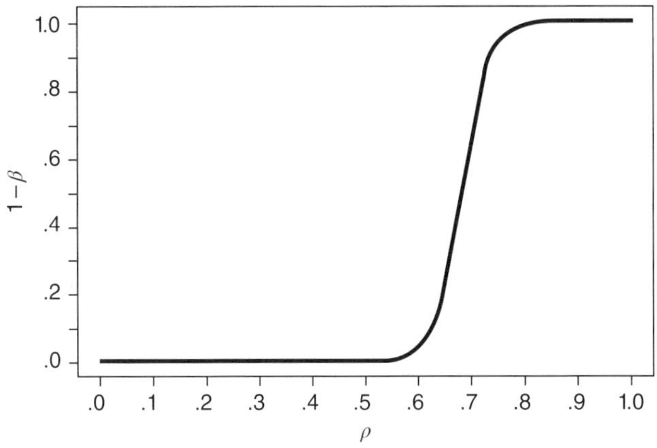

Figure 3.13 Empirical power function of Example 3.26.

As it is shown in Figure 3.12, the maximum of the empirical ASN function lies between the values of the hypotheses, but it is smaller than n_{fix}. The empirical power function is plotted in Figure 3.13.

The examples show that the optimal k-values can be found in relatively large regions and the risks of the second kind are conservative. Therefore, Rasch and Yanagida (2015) have developed tables where the user can see which value k has to be chosen and how the nominal risk of the second kind has to be increased to obtain the wished risk of the second kind as actual risk so that the ASN becomes minimal.

3.7 Remarks about Interpretation

At the end of statistical tests, we decide on one of two possibilities, namely, for accepting or for rejecting the null hypothesis. A confidence interval $K(\boldsymbol{Y})$ covers the unknown parameter of a distribution with a certain probability, and the tests $k(\boldsymbol{Y})$ are connected with risks, with probabilities for wrong rejection (risk of the first kind) or wrong acceptation (risk of the second kind) of the null hypothesis. Concerning the mathematical theory no questions arise. But practical applications need some clarification. What can be stated about a realised confidence interval $K(Y)$, and how we assess the value $k(Y)$ of a critical function $k(\boldsymbol{Y})$ that leads in the non-randomised case either to the acceptance or to the rejection of H_0?

Probability statements can never be made for a realisation of $K(\boldsymbol{Y})$ or after accepting H_0. Such probabilities relate to the method of constructing confidence intervals and tests, but not to their realisations.

It would be nonsense to say that a realised interval $K(Y) = (4.756; 29.560)$ contains the parameter σ^2 with a probability 0.95. As we know the realisation of this parameter is an unknown but although fixed non-negative real number. Hence, this parameter value either lies in $K(Y)$ or not. For example, no serious scientist would claim that the number 2 with probability 0.95 is in the interval (4.756; 29.560). Nevertheless, there are some books about applied statistics, but above all printouts of some statistical software tools, where you can read that a realised interval contains an unknown parameter with a calculated probability. Therefore it is no miracle that some students and user repeat this nonsense.

Analogously it is completely wrong to state after rejecting a null hypothesis based on a random sample that this decision is wrong with probability α. Evidently this decision is either right or wrong. However, it is correct to say that the decision is based on a procedure that supplies wrong rejections with probability α.

Let us turn to a further example. If a single die is thrown, then an even number is obtained with probability 0.5. Assume that the number 3 was thrown. It is nonsense to claim that the number 3 would be even with probability 0.5. Perhaps this simple example is helpful to realise that probability statements concerning realised test results or confidence intervals make no sense.

Therefore the user is recommended to choose α (and β, respectively), small enough that a rejection (or acceptation) of H_o or a realisation $K(Y)$ let the user behave with a clear conscience as H_o would be wrong (or H_A right) or as would θ lie in $K(Y)$. But there is also an important statistical consequence: if the user has to conclude during his/her investigations a lot of such decisions, then he/she will wrongly decide in about 100 α (and 100 β, respectively) percent of the cases. This is a realistic point of view that can be essentially confirmed by experience. If we move in traffic, we should realise the risk of one's own and other people's incorrect actions (observe that in this case α is considerably smaller then 0.05), but we must participate, just as a researcher must derive a conclusion from a random experiment, although he knows that it can be wrong. Moreover, it is very important to control risks. Concerning the risk of the second kind, this is only possible if the sample size is determined before the experiment or if it is sequentially tested during the experiment.

The user should take care not to transfer probability statements to single cases.

3.8 Exercises

3.1 Let P_0 and P_1 be the rectangular distributions acting over the intervals (0,1) and (1,2), respectively. Test the hypothesis H_0 that the distribution P_0 occurs against the hypothesis H_A that the distribution P_1 occurs taking one observation y. The following tests are proposed for given $\alpha \in (0, 1)$:

$$k_1(y) = \begin{cases} \alpha & \text{for } y \in (0,1) \\ 1 & \text{for } y \in (1,2) \end{cases}, \quad k_2(y) = \begin{cases} 0 & \text{for } y \in (\alpha,1) \\ 1 & \text{for } y \in (0,\alpha) \cup (1,2) \end{cases}$$

a) Show that these tests are most powerful α-tests.
b) Present these tests – if possible – in the form (3.5). Does the result contradict the statement (3) of the Neyman–Pearson lemma (Theorem 3.1)? Is one of these tests randomised?

3.2 Test for the geometric distribution with the probability function

$$P(y = k) = p^{k-1}(1-p),\ k = 1,2,\dots;0 < p < 1$$

the hypothesis $H_0: p = p_0$ against $H_A: p = p_1$ $(p_0 \neq p_1)$ based on a random sample $Y = (y_1, y_2, \dots, y_n)^T$.

a) Formulate the most powerful α-test using the test statistic $\bar{y}$.
b) Determine the numbers $c_\alpha, \gamma(Y)$ and β for this test assuming that $n = 1, \alpha = 0.05, p_0 = 0.5, p_1 = 0.1$ (see Section 3.2).

3.3 Test for the Poisson distribution with a random sample $Y = (y_1, y_2, \dots, y_n)^T$ of size $n = 10$ the hypothesis $H_0: \lambda = \lambda_0 = 0.1$ against the hypothesis $H_A: \lambda = \lambda_1 = 1$. Determine the most powerful α-test for $\alpha = 0.01$ and calculate for this test the risk β of the second kind.

3.4 Let the lifetime y of certain industrial instruments be exponentially distributed with the density function $f(y) = \lambda e^{-\lambda y}, y > 0$. Based on a random sample $Y = (y_1, y_2, \dots, y_n)^T$ the hypothesis $H_0: \lambda = \lambda_0$ is to be tested against the hypothesis $H_A: \lambda = \lambda_1,\ \lambda_0 \neq \lambda_1$. Determine the most powerful α-test.

3.5 Formulate and prove a modification of Theorem 3.8 concerning the UMP-test $H_0: \theta \le \theta_0$ against $H_A: \theta = \theta_A > \theta_0$, supposing an antitone likelihood ratio (instead of an isotone one) of the distribution family belonging to the sufficient statistic $M = M(Y)$.

3.6 Determine under the assumptions of Exercise 3.4

a) The UMP-α-test of $H_0: \lambda \le \lambda_0$ against $H_A: \lambda = \lambda_1 > \lambda_0$,
b) The power function of this test
c) In the case $\lambda_0 = 0.01$, $\alpha = 0.05$ the test result based on the sample

170.8; 211.7; 73.5; 52.1; 11.8; 22.1; 167.6; 26.7; 77.5; 17.3

3.7 Let $Y = (y_1, y_2, \dots, y_n)^T$ be a random sample whose components are uniformly distributed in the interval $(0, \theta)$. With the notation $y_{(n)} = \max(y_1, \dots, y_n)$, let

$$k_c(Y) = \begin{cases} 1 & \text{for } y_{(n)} \ge c \\ 0 & \text{else} \end{cases}, \quad (c > 0)$$

be the critical function of a test for the pair $\{H_0, H_A\}$ of hypotheses.

a) Determine the power function $\pi(\theta)$ for $k_c(Y)$ and show that $\pi(\theta)$ is monotone non-increasing in θ.
b) Test $H_0: \theta \le \frac{1}{2}$ against $H_A: \theta > \frac{1}{2}$ with the significance level $\alpha = 0.05$. For which value c is $k_c(Y)$ an α-test for $\{H_0, H_A\}$?
c) Sketch the power function of the test in (b) using $n = 20$. Is this test unbiased?
d) Which number n has to be chosen, so that the test in (b) has for $\theta = 0.6$ a risk 0.02 of the second kind?

3.8 Let the components of the random sample $Y = (y_1, y_2, \dots, y_n)^T$ satisfy a Rayleigh distribution with the density function

$$f(y,\theta) = \frac{y}{\theta^2} e^{-\frac{y^2}{2\theta^2}}, y > 0,\ \theta > 0.$$

The hypothesis H_0: $\theta \le \theta_0 = 1$ is to be tested against H_A: $\theta > \theta_0$.

a) Show that there is a UMP-α-test for $\{H_0, H_A\}$, and determine for great n with the help of the central limit theorem approximately the critical function of this test.
b) Determine for great n approximately the power function of this test.

3.9 Let the assumptions of Exercise 3.4 be fulfilled.

a) Show that there is a UMPU-α-test for the hypotheses $H_0: \lambda = \lambda_0$, $H_A: \lambda \ne \lambda_0$.
b) Determine for $n = 1$ the simultaneous equations whose solutions $c_{i\alpha}$ $(i = 1, 2)$ are necessary to describe the critical function of this test.
c) Show, for example, in the case $\lambda_0 = 10$, $\alpha = 0.05$, $n = 1$, that the corresponding test for a symmetric partition of α is biased by calculating the power function of this tests at $\lambda = 10.1$.

3.10 Let p be the probability that the event A happens. Based on a large sample of size n, where this event was h_n-times observed, the hypothesis $H_0: p = p_0$ is to be tested against the hypothesis $H_A: p \ne p_0$.

a) Construct an approximate UMPU-α-test for these hypotheses by applying the limit theorem of Moivre–Laplace.
b) A coin with head and tail on its faces was tossed 10 000 times, where the tail appeared 5280 times. Check with the help of (a) if it is justified to assume that the coin is not fair (that head and tail do not appear with the same probability). Choose $\alpha = 0.001$.
c) A dice is tossed 200 times, where the (side with) number 6 occurs 40 times. Is it justified to claim (with a significance level of 0.05) that this dice shows the number 6 with the probability $p = \frac{1}{6}$?

3.11 The milk-fat content of 280 randomly chosen young cows of a cattle breed was determined. The average value in the sample was $\bar{y} = 3.61\%$. We suppose that the random variable $\boldsymbol{y}$ modelling this fat content is normally distributed.

a) Let the variance of the fat content $\boldsymbol{y}$ be given, say, $\sigma^2 = 0.09$. Test the hypothesis that the average milk-fat content of young cows of this race is $\mu_0 = 3.5\%$ against the alternative that it is larger than 3.5%. Choose $\alpha = 0.01$.
b) Determine the probability that deviations of the population mean μ of 0.05% fat content from $\mu_0 = 3.5\%$ imply the rejection of the null hypothesis in (a).
c) Which deviations δ between the value μ and the reference value $\mu_0 = 3.5\%$ imply in the test from (a) that the null hypothesis is rejected with a probability larger than 0.9?
d) Let the variance σ^2 of the fat content $\boldsymbol{y}$ be unknown. From a sample of size 49 an estimator $s^2 = 0.076673$ for the variance was calculated. Test the hypotheses in (a) using the significance level $\alpha = 0.01$.

3.12 The producer of a certain car model declares that the fuel consumption for this model is in the city traffic approximately normally distributed with the expectation $\mu = 7.5$ l/100 km and the variance $\sigma^2 = (2.5 \text{ l}/100 \text{ km})^2$. These declarations are to be tested to satisfy the interests of the car buyers. Therefore the fuel consumption was measured for 25 (randomly chosen) cars of this model moving in the city traffic (of randomly chosen cities worldwide). Here are the results:

Average fuel consumption: 7.9 l/100 km
Sample variance $(3.2 \text{ l}/100 \text{ km})^2$

Test the statements of the car producer separately for both parameters choosing $\alpha = 0.05$.

3.13 The milk-fat content of Jersey cows is in general considerably higher than the one of black-coloured cows. It is to be tested whether the variability of the fat content is for both breeds equal or not. A random sample of $n_1 = 25$ Jersey cows supplied the estimator $s_1^2 = 0.128$, while an independent random sample of $n_2 = 31$ black-coloured cows led to the estimator $s_2^2 = 0.072$. The fat content is supposed to be in both breeds normally distributed. Test for $\alpha = 0.05$ the hypothesis $H_0 : \sigma_1^2 = \sigma_2^2$ against the alternative

a) $H_A : \sigma_1^2 > \sigma_2^2$,
b) $H_A : \sigma_1^2 \neq \sigma_2^2$.

3.14 Consider a random sample $\boldsymbol{Y} = (\boldsymbol{y}_1, \boldsymbol{y}_2, \ldots, \boldsymbol{y}_n)^T$ whose components are uniformly distributed in the interval $(0, \theta)$, $\theta \in R^+$ Determine the sample size n so that the random interval $(y_{(1)}, y_{(n)})$ of order statistics covers the parameter $\frac{\theta}{2}$ with the probability 0.999.

3.15 Let $Y = (y_1, y_2, \ldots, y_n)^T$ be a random sample whose components are uniformly distributed in the interval $(0, \theta_0)$ with unknown θ_0. Confidence intervals $K(Y)$ are to be constructed with respect to θ and with the confidence coefficient $1-\alpha$. They are to be of the form

$$K(Y) = \left[y_{(n)}c_1(\alpha_1), y_{(n)}c_2(\alpha_2)\right],$$

where $\alpha = \alpha_1 + \alpha_2; 0 \le \alpha_1, \alpha_2 < \frac{1}{2}$ holds and $c_1(\alpha_1), c_2(\alpha_2)$ are suitable constants.

a) Construct three confidence intervals $K_1(Y)$ for $\alpha_1, \alpha_2 < \frac{1}{2}$ arbitrary; $K_2(Y)$ for $\alpha_1 = 0, \alpha_2 = \alpha$; and $K_3(Y)$ for $\alpha_1 = \alpha,\ \alpha_2 = 0$.
b) Calculate the expected length $2\delta_i$ of the confidence intervals $K_i(Y)$ with i = 1, 2, 3 from (a). Which interval has the smallest expected length?
c) $W(\theta, \theta_0) = P(\theta \in K(Y) \mid \theta_0)$ is called the characteristic function of the confidence estimation $K(Y)$. Calculate the corresponding functions $W_i(\theta, \theta_0)$ of the intervals $K_i(Y)$ with i = 1, 2, 3 given in (a), and sketch these functions for $\theta_0 = 10$, $n = 16$, $\alpha = 0.06$ and $\alpha_1 = 0.04$ in the case of the interval $K_1(Y)$. Which confidence intervals are unbiased?

3.16 a) Determine the one-sided UMP-(1 – α) confidence intervals with respect to λ supposing the conditions of Exercise 3.4.
b) Determine the realisations of these confidence intervals based on the sample from Exercise 3.6 (c) using α = 0.05.

3.17 Let the assumptions of Section 3.4.2 be satisfied.
Determine das UMPU-(1 – α) confidence interval for the quotient $\frac{\sigma_1^2}{\sigma_2^2}$ of variances.

3.18 In a factory certain pieces are produced in large series. The probability p $(0 < p < 1)$ that a peace in a series is defect is unknown. The hypothesis $H_0: p = p_0$ is to be tested against the alternative $H_A: p = p_1$ where $p_0 \neq p_1$. We want to use the following sequential test. Let n_0 be a fixed natural number. We successively select independent pieces for the sample. If the kth piece $(k \le n_0)$ is defect, then H_0 is rejected. But if all n_0 pieces are intact, then H_0 is accepted.

a) Determine the power function of this test.
b) Calculate the average sample size $\mathrm{E}(\boldsymbol{n} \mid \mathrm{p})$.

c) Calculate for $p_0 = 0.01, p_1 = 0.1, n_0 = 10$ the risks α, β and the expectations $E(\boldsymbol{n} \mid p_i)$ with $i = 1,2$.

3.19 Let the components of the random vector $Y = (y_1, y_2, \ldots)^T$ be mutually independently distributed as $N(\mu, \sigma^2)$. The null hypothesis $H_0: \mu = \mu_0$ is to be tested by a 0.05-t-test. Which minimal sample size has to be chosen for a risk 0.1 of the second kind if

a) The alternative H_A is one-sided and the practically relevant minimum difference is $\delta = \frac{1}{4}\sigma$?
b) The alternative H_A is two-sided and the practically relevant minimum difference is $\delta = \frac{1}{2}\sigma$?

Hint: Use the approximate formula.

3.20 Let the components of the random vectors $Y_i = (y_{i1}, y_{i2}, \ldots)^T; i = 1, 2$ be independently distributed as $N(\mu_i, \sigma_i^2)$. It is unknown whether $\sigma_1^2 = \sigma_2^2$ holds or not. The null hypothesis $H_0: \mu_1 = \mu_0$ is to be tested with a 0.05-t-test.

b) Which test statistic should be used?
c) Which minimal sample size has to be chosen for a risk 0.1 of the second kind if
   i) The alternative H_A is one-sided and the practically relevant minimum difference is $\delta = \frac{1}{4}\sigma$?
   ii) The alternative H_A is two-sided and the practically relevant minimum difference is $\delta = \frac{1}{2}\sigma$?

Hint: Use the approximate formula.

3.21 We consider two independent random samples $Y_1 = (y_{11}, \ldots, y_{1n_1})^T$, $Y_2 = (y_{21}, \ldots, y_{2n_2})^T$, where components y_{ij} are supposed to be distributed as $N(\mu_i, \sigma_i^2)$ with $i = 1, 2$. The null hypothesis

$$H_0: \mu_1 = \mu_2 = \mu,\ \sigma_1^2, \sigma_2^2 \text{ arbitrary}$$

is to be tested against

$$H_A: \mu_1 \neq \mu_2,\ \sigma_1^2, \sigma_2^2 \text{ arbitrary.}$$

Construct a UMPU-α-test for one-sided alternatives in the case $\sigma_1^2 = \sigma_2^2$.

References

Agresti, A. and Coull, B. A. (1998) The analysis of contingency tables under inequality constraints. *J. Stat. Plan. Inf.*, **107**, 45–73.

Aspin, A. A. (1949) Tables for use in comparisons whose accuracy involves two variances separately estimated. *Biometrika*, **36**, 290–296.

Bagdonavicius, V., Kruopis J. and Nikulin, M. S. (2011) *Non-parametric Tests for Complete Data*, ISTE and John Wiley & Sons, Inc., London and Hoboken.

Box, G. E. P. (1953) Non-normality and tests on variances. *Biometrika*, **40**, 318–335.

Clopper, C. J. and Pearson, E. S. (1934) The use of confidence or fiducial limits illustrated in the case of the binomial. *Biometrika*, **26**, 404–413.

Cramér, H. (1946) *Mathematical Methods of Statistics*, Princeton Press, Princeton.

DeGroot, M. H. (2005) *Sequential Sampling in Optimal Statistical Decisions*, John Wiley & Sons, Inc., New York.

Dodge, H. F. and Romig, H. G. (1929) A method of sampling inspections. *Bell Syst. Tech. J.*, **8**, 613–631.

Eisenberg, B. and Ghosh, B. K. (1991) The sequential probability ratio test, in Ghosh, B. K and Sen, P. K. (Eds.), *Handbook of Sequential Analysis*, Marcel Dekker, New York.

Fisher, R. A. (1915) Frequency distribution of the values of the correlation coefficient in samples from an indefinitely large population. *Biometrika*, **10**, 507–521.

Fisher, R. A. (1921) On the "probable error" of a coefficient of correlation deduced from a small sample. *Metron*, **1**, 3–32.

Fleishman, A. J. (1978) A method for simulating non-normal distributions. *Psychometrika*, **43**, 521–532.

Fleiss, J. L., Levin, B. and Paik, M. C. (2003) *Statistical Methods for Rates and Proportions*, 3rd edition, John Wiley & Sons, Inc., Hoboken.

Ghosh, B. K and Sen, P. K (1991) *Handbook of Sequential Analysis*, Marcel Dekker, New York.

Karlin, S. (1957) Polyå type distributions II. *Ann. Math. Stat.*, **28**, 281–308.

Lehmann, E. L. (1959) *Testing Statistical Hypothesis*, John Wiley & Sons, Inc., Hoboken.

Lehmann, E. L. and Romano J. P. (2008) *Testing Statistical Hypothesis*, Springer, Heidelberg.

Levene, H. (1960) Robust tests for equality of variances, in Olkin, I., Ghurye, S., Hoeffding, W., Madow, W. and Mann H. (Eds.), *Contributions to Probability and Statistics*, Stanford University Press, Stanford, 278–292.

Mann, H. H. and Whitney, D. R. (1947) On a test whether one of two random variables is stochastically larger than the other. *Ann. Math. Stat.*, **18**, 50–60.

Neyman, J. and Pearson, E. S. (1933) On the problem of the most efficient tests of statistical hypothesis. *Philos. Trans. R. Soc. Lond. A*, **231**, 289–337.

Pearson, E. S. and Hartley, H. O. (1970) *Biometrika Tables for Statisticians Volume I*, Table 11, Cambridge University Press, Cambridge.

Pires, A. M. and Amado, C. (2008) Interval estimators for a binomial proportion: comparison of twenty methods. *REVSTAT – Stat. J.*, **6** (2), 165–197.

Rasch, D. and Guiard, V. (2004) The robustness of parametric statistical methods. *Psychol. Sci.*, **46**, 175–208.

Rasch, D. and Yanagida, T. (2015) A modified sequential triangular test of a correlation coefficient's null-hypothesis, Proceedings of the 8th International Workshop on Simulation, Vienna, 21–25 September, 2015.

Rasch, D., Kubinger, K. D. and Moder, K. (2011a) The two-sample *t* test: pre-testing its assumptions does not pay-off. *Stat. Pap.*, **52**, 219–231.

Rasch, D., Pilz, J., Verdooren, R. L. and Gebhardt, A. (2011b) *Optimal Experimental Design with R*, Chapman and Hall, Boca Raton.

Rasch, D., Kubinger, K. D. and Yanagida, T. (2011c) *Statistics in Psychology Using R and SPSS*, John Wiley & Sons, Inc., Hoboken.

Schneider, B. (1992) An interactive computer program for design and monitoring of sequential clinical trials, *Proceedings of the XVIth International Biometric Conference* (pp. 237–250), University of Waikato, Hamilton, New Zealand, December 7–11, 1992.

Schneider, B., Rasch, D., Kubinger, K. D. and Yanagida, T. (2014) A sequential triangular test of a correlation coefficient's null-hypothesis: $0 < \rho \leq \rho_0$. *Stat. Pap.*, **56**, 689–699.

Stein, C. (1945) A two sample test for a linear hypothesis whose power is independent of the variance. *Ann. Math. Stat.*, **16**, 243–258.

Stevens, W. L. (1950) Fiducial limits of the parameter of a discontinuous distribution. *Biometrika*, **37**, 117–129.

Student (Gosset, W. S.) (1908) The probable error of a mean. *Biometrika*, **6**, 1–25.

Trickett, W. H. and Welch, B. L. (1954) On the comparison of two means: further discussion of iterative methods for calculating tables. *Biometrika*, **41**, 361–374.

Trickett, W. H., Welch, B. L. and James, G. S. (1956) Further critical vales for the two-means problems. *Biometrika*, **43**, 203–205.

Wald, A. (1947) *Sequential Analysis*, Dover Publications, New York (Reprint: 2004 John Wiley & Sons, Inc., New York).

Welch, B. L. (1947) The generalization of students problem when several different population variances are involved. *Biometrika*, **34**, 28–35.

Whitehead, J. (1997) *The Design and Analysis of Sequential Clinical Trials*, 2nd revised edition. (Reprint: 2004 John Wiley & Sons, Inc., New York).

Wijsman, R. A. (1991) Stopping times: termination, moments, distribution, in Ghosh, B. K and Sen, P. K (Eds.), *Handbook of Sequential Analysis*, Marcel Dekker, New York.

Wilcoxon, F. (1945) Individual comparisons by ranking methods. *Biom. Bull.*, **1**, 80–82.

Young, L. Y. (1994) Computation of some exact properties of Wald's SPRT when sampling from a class of discrete distributions. *Biom. J.*, **36**, 627–637.

4

Linear Models – General Theory

4.1 Linear Models with Fixed Effects

The theory of linear statistical models plays an important role in the applications. Mainly the standard methods of analysis of variance and regression analysis have become firmly established in evaluating biological and technological experiments.

In this chapter we introduce the general theory concerning methods of analysis of variance and regression analysis with fixed effects. In the following $\Omega \subset R^n$ denotes a p-dimensional linear subspace with $p < n$ called parameter space, and $\theta \in \Omega$ denotes a parameter vector with n coordinates $\theta_i (i = 1, \ldots, n)$.

Further, let $\boldsymbol{Y}$ be an n-dimensional random variable (a random vector) with components $\boldsymbol{y}_i (i = 1, \ldots, n)$ and realisations Y from the n-dimensional sample space R^n. Finally, let $\boldsymbol{e}$ be an n-dimensional random variable with $E(\boldsymbol{e}) = 0_n$, $\text{var}(\boldsymbol{e}) = \sigma^2 V$, where V is a symmetric and positive definite matrix of size (n, n) and rank n. For constructing tests and confidence intervals, we will later suppose that $\boldsymbol{e}$ (and hence also $\boldsymbol{Y}$) are n-dimensional normally distributed (satisfy n-variate normal distributions).

Definition 4.1 The equation

$$\boldsymbol{Y} = \theta + \boldsymbol{e} \tag{4.1}$$

including the constraints $\theta \in \Omega$, $E(\boldsymbol{e}) = 0_n$, $\text{var}(\boldsymbol{e}) = \sigma^2 V$ is said to be a general linear model (with fixed effects). If $\omega \subset \Omega$ is a linear subspace of Ω, then the hypothesis $H_0 : \theta \in \omega$ is called linear.

The definition of a linear hypothesis obviously implies that $H_A : \theta \notin \omega$ is no linear hypothesis, since $\Omega \setminus \omega$ is no linear subspace of Ω. Namely, linear combinations of elements in this set can, for example, belong to ω. W.l.o.g. we assume $V = I_n$. This is indeed no restriction of generality if V is known as we will see. Since V is symmetric and positive definite, there is a non-singular matrix P with

Mathematical Statistics, First Edition. Dieter Rasch and Dieter Schott.

$V = P^T P$. We introduce the new variable $\boldsymbol{Z} = (P^{\mathrm{T}})^{-1}\boldsymbol{Y}$ that has the expectation $\lambda = E(\boldsymbol{Z}) = (P^{\mathrm{T}})^{-1}E(\boldsymbol{Y}) = (P^{\mathrm{T}})^{-1}\,\theta$ and the variance

$$\operatorname{var}(\boldsymbol{Z}) = \left(P^{\mathrm{T}}\right)^{-1} E\left[(\boldsymbol{Y}-\theta)(\boldsymbol{Y}-\theta)^{\mathrm{T}}\right] P^{-1} = \left(P^{\mathrm{T}}\right)^{-1} \operatorname{var}(\boldsymbol{Y}) P^{-1} = \sigma^2 I_n.$$

The model

$$\boldsymbol{Z} = \lambda + \boldsymbol{e}^*$$

$$\boldsymbol{e}^* = \left(P^{\mathrm{T}}\right)^{-1}\boldsymbol{e} \quad (\lambda \in \Omega^*), \quad E(\boldsymbol{e}^*) = 0_n, \quad \operatorname{var}(\boldsymbol{e}^*) = \sigma^2 I_n$$

has therefore the form (4.1) including corresponding constraints. Since Ω is mapped by $\lambda = (P^{\mathrm{T}})^{-1}\,\theta$ onto a set Ω^* with $\dim(\Omega) = \dim(\Omega^*)$, Ω^* is again a p-dimensional linear subspace. Analogously the matrix $(P^{\mathrm{T}})^{-1}$ maps ω onto ω^* where $\dim(\omega) = \dim(\omega^*)$ and $\omega^* \subset \Omega^*$ so that the linearity of the hypothesis is also conserved. Hence we will use $V = I_n$ for linear models.

4.1.1 Least Squares Method

First we want to estimate the parameter vector using the least squares method (LSM) (compare Section 2.3.2). An estimator $\hat{\boldsymbol{\theta}}$ for θ by the LSM is an estimator where its realisations $\hat{\theta}$ fulfil

$$\| \hat{e} \|^2 = \| Y - \hat{\theta} \|^2 = \inf_{\theta \in \Omega} \| Y - \theta \|^2. \tag{4.2}$$

The following theorem is known from approximation theory.

Theorem 4.1 A realisation $\hat{\theta}$ of the LSM $\hat{\boldsymbol{\theta}}$ satisfying (4.2) is the orthogonal projection of Y (the realisation of $\boldsymbol{Y}$) onto Ω.

Proof: Let $c_1, \ldots, c_p$ be an orthonormal (vector) basis of Ω. Introducing numbers (scalars) $k_i = Y^{\mathrm{T}} c_i$, the realisation Y can be written in the form

$$Y = \sum_{i=1}^{p} k_i c_i + Y - \sum_{i=1}^{p} k_i c_i = c + b, \quad c = \sum_{i=1}^{p} k_i c_i.$$

Because of $c_j^{\mathrm{T}} b = 0$ the representation $Y = c + b$ supplies a decomposition of Y in the sum of two orthogonal vectors $c \in \Omega$, $b \in \Omega^{\perp}$.

This decomposition is unique. Assuming that there is another decomposition $Y = c^* + b^*$, we get $c + b = c^* + b^*$ or $c - c^* = b^* - b$. Since $c - c^* \in \Omega$ and $b^* - b \in \Omega^{\perp}$, it follows $c - c^* = b^* - b = 0$. Hence, the uniquely determined vector c is the orthogonal projection of Y onto Ω.

Finally we have to show that $c = \hat{\theta}$. Taking $Y - \hat{\theta} = Y - c + c - \hat{\theta}$ into account, it follows

$$\left\| Y - \hat{\theta} \right\|^2 = \| Y - c \|^2 + \left\| c - \hat{\theta} \right\|^2 + 2(Y - c)^T \left(c - \hat{\theta}\right). \tag{4.3}$$

Since $c - \hat{\theta} \in \Omega$ and $b = Y - c \in \Omega^{\perp}$, the third summand on the right-hand side of (4.3) vanishes, and this side attains its minimum for $c = \hat{\theta}$.

Theorem 4.2 The LSM vector $\hat{\theta}$ satisfying (4.2) can be obtained from a realisation Y of the random vector $\boldsymbol{Y}$ by the linear transformation

$$\hat{\theta} = AY \tag{4.4}$$

with a (symmetric) idempotent matrix A of rank p.[1]

On the other hand, if A is an idempotent matrix of size (n,n) with rank p, then the linear transformation AY with $Y \in R^n$ realises the orthogonal projection of R^n onto a p-dimensional vector space.

Proof: First we show (4.4), where A is supposed to be idempotent with rank p ($\text{rk}(A) = p$). Considering the proof of Theorem 4.1, it is

$$\hat{\theta} = \sum_{i=1}^{p} k_i c_i = \sum_{i=1}^{p} c_i Y^{\mathrm{T}} c_i. \tag{4.5}$$

Because of $Y^{\mathrm{T}} c_i = c_i^{\mathrm{T}} Y$ (transposition rule), we get

$$\hat{\theta} = (c_1, \ldots, c_p)(c_1, \ldots, c_p)^{\mathrm{T}} Y.$$

With the notations $C = (c_1, \ldots, c_p)$ and $A = CC^{\mathrm{T}}$, the vector $\hat{\theta}$ becomes the form (4.4). Observing $A^{\mathrm{T}} = (CC^{\mathrm{T}})^{\mathrm{T}} = CC^{\mathrm{T}} = A$ and remembering the vectors c_i to be orthonormal (compactly written as $C^{\mathrm{T}}C = I_p$), it follows $A^{\mathrm{T}}A = CC^{\mathrm{T}}\, CC^{\mathrm{T}} = A^2 = A$, that is, A is idempotent.

Moreover it is $\text{rk}(A) = \text{rk}(C) = p$.

Now we prove the second part of the theorem. Let A be an idempotent matrix of size (n,n) and with rank p. For each such matrix there is an orthogonal matrix C so that $C^{\mathrm{T}}AC = I_p \otimes O_{n-p,n-p}$. Therefore A can be written as $A = (c_1, \ldots, c_p)(c_1, \ldots, c_p)^{\mathrm{T}}$ where the column vectors c_i of C ($i = 1, \ldots, p$) represent a basis of a p-dimensional subspace of R^n.

If we intend to estimate parameters in linear models for unknown distribution, then the LSM is usually applied. A justification for this approach is presented in the next theorem.

Theorem 4.3 Gauss–Markov Theorem

Let $L = a^{\mathrm{T}}\theta$ be a linear form in the parameter vector $\theta \in \Omega$ of model (4.1) with range R^1. Then there exists for $a^{\mathrm{T}}\theta$ in the class of all linear estimators with bounded mean square deviation $E(\boldsymbol{SD})$ a uniquely determined estimator

[1] In the following, we omit the attribute 'symmetric', since all idempotent matrices in this book will be symmetric.

with minimal $E(\boldsymbol{SD})$; this estimator is the LSM that has the form $a^{\mathrm{T}}AY$ with A described in Theorem 4.2 (see (4.4)).

Proof: Let $t^{\mathrm{T}}\boldsymbol{Y}$ be a linear estimator for $a^{\mathrm{T}}\theta = L$. We consider the mean square deviation (MSD)

$$E(\boldsymbol{SD}) = E\left(t^{\mathrm{T}}\boldsymbol{Y} - a^{\mathrm{T}}\theta\right)^2 = E\left(t^{\mathrm{T}}\boldsymbol{Y} - t^{\mathrm{T}}\theta + t^{\mathrm{T}}\theta - a^{\mathrm{T}}\theta\right)^2,$$

which can be written with $E(\boldsymbol{Y}) = \theta$ in the form

$$E(\boldsymbol{SD}) = \mathrm{var}\left(t^{\mathrm{T}}\boldsymbol{Y}\right) + \left(t^{\mathrm{T}}\theta - a^{\mathrm{T}}\theta\right)^2.$$

$E(\boldsymbol{SD})$ is only then bounded for all $\theta \in \Omega$, if $t^{\mathrm{T}}\theta - a^{\mathrm{T}}\theta = 0$ for all these θ. Hence, the class of all linear estimators for $a^{\mathrm{T}}\theta$ with bounded $E(\boldsymbol{SD})$ is described by the equation $t^{\mathrm{T}}\theta - a^{\mathrm{T}}\theta = 0$. The matrix A in (4.4) realises the orthogonal projection of R^n onto Ω. Therefore it is $A\theta = \theta$, and the class of linear estimators with bounded $E(\boldsymbol{SD})$ is characterised by $At = Aa$ if $(t^{\mathrm{T}} - a^{\mathrm{T}})\, A\theta = 0$ for all θ, and consequently $(t^{\mathrm{T}} - a^{\mathrm{T}})A = 0_n^{\mathrm{T}}$ is taken into account. This class of estimators satisfies

$$E(\boldsymbol{SD}) = \mathrm{var}\left(t^{\mathrm{T}}\boldsymbol{Y}\right) + t^{\mathrm{T}}t\sigma^2.$$

Now we have to determine t so that $E(\boldsymbol{SD})$ is minimised under the condition $At = Aa$ with A from (4.4). We write

$$t^{\mathrm{T}}t = (t + At - At)^{\mathrm{T}}(t + At - At).$$

Since A is idempotent, we get

$$t^{\mathrm{T}}t = (At)^{\mathrm{T}}(At) + [(I_n - A)t]^{\mathrm{T}}(I_n - A)t$$

and because of $At = Aa$ also

$$t^{\mathrm{T}}t = (Aa)^{\mathrm{T}}(Aa) + [(I_n - A)t]^{\mathrm{T}}(I_n - A)t. \tag{4.6}$$

The functional $t^{\mathrm{T}}t$ in (4.6) and consequently $E(\boldsymbol{SD})$ are minimised if the second summand of the right-hand side in (4.6) vanishes, that is, if $t = At = Aa$. This supplies the uniquely determined estimator $a^{\mathrm{T}}A\boldsymbol{Y}$ for $a^{\mathrm{T}}\theta$.

Many variants of the Gauss–Markov theorem are based on the class of linear unbiased estimators for $a^{\mathrm{T}}\theta$. The LSM under all these estimators is the one with minimal variance.

Example 4.1 Let $p = 1$ in the model equation (4.1) so that (4.1) can be written as

$$\boldsymbol{Y} = 1_n\theta_1 + \boldsymbol{e} \quad \left(\boldsymbol{y}_i = \theta_1 + \boldsymbol{e}_i,\ i = 1,\ldots,n\right).$$

Here 1_n is the vector whose coordinates are all 1 (see also Appendix A). Assuming $-\infty < \theta_1 < \infty$ the parameter space Ω has the dimension 1. The estimator of

θ_1 according to the LSM is obtained by putting the derivative of $e^{\mathrm{T}}e = f(\theta_1)$ to 0 as unique solution $\hat{\theta}_1 = \theta_1 = \bar{y}$ of the equation

$$D = \frac{\partial (Y - 1_n\theta_1)^{\mathrm{T}}(Y - 1_n\theta_1)}{\partial \theta_1} = -2\sum_{i=1}^{n} y_i + 2n\theta_1 = 0$$

(which produces a minimum, since the second derivative of $f(\theta_1)$ is positive). Therefore the parameter vector is estimated by $\hat{\theta} = 1_n\hat{\theta}_1$. The parameter space Ω has the orthonormal basis $c_1 = \left(\frac{1}{\sqrt{n}},\ldots,\frac{1}{\sqrt{n}}\right)^{\mathrm{T}}$. Using the notations of the proofs to Theorems 4.1 and 4.2, we get

$$C = c_1, \quad CC^{\mathrm{T}} = c_1^{\mathrm{T}}c_1 = A = \begin{pmatrix} \frac{1}{n} & \cdots & \frac{1}{n} \\ \vdots & & \vdots \\ \frac{1}{n} & \cdots & \frac{1}{n} \end{pmatrix}.$$

The orthogonal decomposition of Y given in the proof of Theorem 4.1 is realised with $k_1 = \frac{\Sigma y_i}{\sqrt{n}}$ and $c = \frac{\Sigma y_i}{\sqrt{n}}c_1$. Additionally in this special case, the general statement $c = \hat{\theta} = AY$ holds. The variance of $\hat{\boldsymbol{\theta}} = 1_n^{\mathrm{T}}\bar{\boldsymbol{y}}$ is

$$\operatorname{var}(\hat{\theta}) = A\sigma^2 = \begin{pmatrix} \frac{\sigma^2}{n} & \cdots & \frac{\sigma^2}{n} \\ \vdots & & \vdots \\ \frac{\sigma^2}{n} & \cdots & \frac{\sigma^2}{n} \end{pmatrix}.$$

It is easy to show that A is idempotent and has rank 1 (see also Exercise 4.5).

Theorem 4.4 If $\hat{\boldsymbol{\theta}}$ is the LSM for θ in (4.1) with $\operatorname{var}\left(\hat{\boldsymbol{\theta}}\right) = \sigma^2 I_n$, then

$$\boldsymbol{s}^2 = \frac{1}{n-p}\|\boldsymbol{Y} - \hat{\boldsymbol{\theta}}\|^2 = \frac{1}{n-p}\boldsymbol{Y}^{\mathrm{T}}(I_n - A)\boldsymbol{Y} \tag{4.7}$$

is an unbiased estimator for σ^2.

Proof: We have to show

$$E\left[\|\boldsymbol{Y} - \hat{\boldsymbol{\theta}}\|^2\right] = \sigma^2(n-p). \tag{4.8}$$

Regarding $\hat{\boldsymbol{\theta}} = A\boldsymbol{Y}$ and the idempotence of both A and $I_n - A$, we obtain

$$E\left[\| \boldsymbol{Y} - \hat{\boldsymbol{\theta}} \|^2\right] = E\left[\boldsymbol{Y}^{\mathrm{T}}(I_n - A)\boldsymbol{Y}\right] = E\left(\boldsymbol{Y}^{\mathrm{T}} I_n \boldsymbol{Y}\right) - E\left(\boldsymbol{Y}^{\mathrm{T}} A \boldsymbol{Y}\right). \tag{4.9}$$

Now

$$E\left(\boldsymbol{Y}^{\mathrm{T}} B \boldsymbol{Y}\right) = \mathrm{tr}(B\Sigma) + \mu^{\mathrm{T}} B \mu$$

if $E(\boldsymbol{Y}) = \theta = \mu$ and $\mathrm{var}(\boldsymbol{Y}) = \Sigma = \sigma^2 I_n$. This implies with $B = I_n$,

$$E\left(\boldsymbol{Y}^{\mathrm{T}} I_n \boldsymbol{Y}\right) = \sigma^2 \mathrm{tr}(I_n) + \theta^{\mathrm{T}}\theta = \sigma^2 n + \theta^{\mathrm{T}}\theta$$

and with $B = A$

$$E\left(\boldsymbol{Y}^{\mathrm{T}} A \boldsymbol{Y}\right) = \sigma^2 \mathrm{tr}(A) + \theta^{\mathrm{T}} A \theta = \sigma^2 p + \theta^{\mathrm{T}}\theta.$$

The difference of both equations leads to (4.8) that finishes the proof.

4.1.2 Maximum Likelihood Method

In this section in addition to the linear model conditions, that is, (4.1) and the constraints, we suppose that the random vector $\boldsymbol{e}$ in (4.1) follows an n-dimensional normal distribution $N(0_n, \sigma^2 I_n)$. Then $\boldsymbol{Y}$ is distributed as $N(\theta, \sigma^2 I_n)$. Now we look for a MLE, an estimator for θ according to the maximum likelihood method. The likelihood function has the form

$$L = L(\theta, \sigma^2 | Y) = (2\pi\sigma^2)^{-\frac{n}{2}} \exp\left(-\frac{(Y-\theta)^{\mathrm{T}}(Y-\theta)}{2\sigma^2}\right)$$
$$\theta \in \Omega, \quad \left(\theta^{\mathrm{T}}, \sigma^2\right)^{\mathrm{T}} \in \Omega^*, \quad \Omega^* = \Omega \times (0, \infty). \tag{4.10}$$

According to this method we get MLE for θ and σ^2. We start with determining the log-likelihood function in (4.10):

$$\ln L = -\frac{n}{2}\ln 2\pi - \frac{n}{2}\ln\sigma^2 - \frac{1}{2\sigma^2}(Y-\theta)^{\mathrm{T}}(Y-\theta). \tag{4.11}$$

Now we want to maximise ln L under the constraint $A\theta = \theta$ (i.e. $\theta \in \Omega$), where A is the matrix of orthogonal projection of R^n onto Ω. We denote the values maximising L and ln L, respectively, by $\tilde{\theta}$ and $\tilde{\sigma}^2$. We use the Lagrange method. Introducing the Lagrange multiplicator λ for $A\theta = \theta$ and deriving the modified function (4.11) partially according to λ, θ and σ^2 we get after putting the derivatives to zero and replacing the variable θ and σ^2 by $\tilde{\theta}$ and $\tilde{\sigma}^2$ the simultaneous equations

$$\left.\begin{array}{l} -\dfrac{n}{2\tilde{\sigma}^2}+\dfrac{1}{2\tilde{\sigma}^4}\left(Y-\widetilde{\theta}\right)^{\mathrm{T}}\left(Y-\widetilde{\theta}\right)=0 \\ \dfrac{1}{\tilde{\sigma}^2}\left(Y-\widetilde{\theta}\right)-(I_n-A)\lambda=0 \\ (I_n-A)\widetilde{\theta}=0 \end{array}\right\} \quad (4.12)$$

which have unique solutions supplying the maximum, since the matrix of second partial derivatives is negative definite. If we use the random variable $\boldsymbol{Y}$ in the solutions instead of its realisation Y, we find the MLE

$$\tilde{\sigma}^2=\frac{1}{n}\|\boldsymbol{Y}-\widetilde{\boldsymbol{\theta}}\|^2, \quad (4.13)$$

$$\widetilde{\boldsymbol{\theta}}=A\boldsymbol{Y}=\hat{\boldsymbol{\theta}}. \quad (4.14)$$

The MLE $\widetilde{\boldsymbol{\theta}}$ is identical with LSM $\hat{\boldsymbol{\theta}}$. The MLE $\tilde{\boldsymbol{\sigma}}^2$ is biased, but consistent.

Theorem 4.5 If $\boldsymbol{Y}$ is distributed with the likelihood function (4.10) and $\dim(\Omega)=p$, then $c^{\mathrm{T}}\widetilde{\boldsymbol{\theta}}=c^{\mathrm{T}}\hat{\boldsymbol{\theta}}$ is for each vector $c=(c_1,\ldots,c_n)^{\mathrm{T}}$, with real numbers c_i the uniformly variance-optimal unbiased estimator (UVUE) for $c^{\mathrm{T}}\theta$, and $\boldsymbol{s}^2$ in (4.7) is a UVUE for σ^2 (compare Definition 2.3).

Proof: The assertion follows from Theorem 2.4 in relation to Example 2.4, because $E\left(c^{\mathrm{T}}\hat{\boldsymbol{\theta}}\right)=c^{\mathrm{T}}\theta$ and $E(\boldsymbol{s}^2)=\sigma^2$ as well as $c^{\mathrm{T}}\hat{\boldsymbol{\theta}}$ and $\boldsymbol{s}^2$ are complete sufficient statistics.

4.1.3 Tests of Hypotheses

The linear hypothesis $H_0:\theta\in\omega$ with the $(p-q)$-dimensional linear subspace $\omega\subset\Omega$ is to be tested against the alternative hypothesis $\theta\notin\omega$. We design a likelihood quotient test by introducing

$$Q=\frac{\sup_{\theta\in\omega}L(\theta,\sigma^2\,|\,Y)}{\sup_{\theta\in\Omega}L(\theta,\sigma^2\,|\,Y)}, \quad (4.15)$$

where $\boldsymbol{Y}$ is again supposed to be distributed as $N(\theta,\sigma^2 I_n)$. After passing to random variable $\boldsymbol{Q}$ itself or a monotone function of $\boldsymbol{Q}$ considered as a function of $\boldsymbol{Y}$ are to be used as test statistic. We denote such values of σ^2 and θ that maximise the function L from (4.15) given in (4.10) on ω by $\breve{\sigma}^2$ and $\breve{\theta}$. Additionally, let B be the idempotent matrix, which orthogonally projects R^n onto ω. After passing from the realisations to the random variables analogously to (4.13) and (4.14). we get

$$\breve{\boldsymbol{\sigma}}^2=\frac{1}{n}\|\boldsymbol{Y}-\breve{\boldsymbol{\theta}}\|^2, \quad (4.16)$$

$$\breve{\boldsymbol{\theta}} = BY. \tag{4.17}$$

Regarding

$$\sup_{\theta \in \omega} L(\theta, \sigma^2 | Y) = (2\pi\breve{\sigma}^2)^{-\frac{n}{2}} \exp\left[-\frac{1}{2} \frac{\|Y - \breve{\theta}\|^2}{\frac{1}{n}\|Y - \breve{\theta}\|^2} \right] = (2\pi\breve{\sigma}^2)^{-\frac{n}{2}} \mathrm{e}^{-\frac{n}{2}}$$

and

$$\sup_{\theta \in \Omega} L(\theta, \sigma^2 | Y) = (2\pi\tilde{\sigma}^2)^{-\frac{n}{2}} \mathrm{e}^{-\frac{n}{2}}$$

after passing to random variables, the likelihood ratio (4.15) becomes

$$\boldsymbol{Q} = \left(\frac{\tilde{\boldsymbol{\sigma}}^2}{\breve{\boldsymbol{\sigma}}^2}\right)^{\frac{n}{2}} = \left[\frac{\|\boldsymbol{Y} - A\boldsymbol{Y}\|^2}{\|\boldsymbol{Y} - B\boldsymbol{Y}\|^2}\right]^{\frac{n}{2}} = \left[\frac{\|\boldsymbol{Y} - A\boldsymbol{Y}\|}{\|\boldsymbol{Y} - B\boldsymbol{Y}\|}\right]^{n}. \tag{4.18}$$

We consider a monotone function $F = F(Q)$ of Q, namely (in the random form),

$$\boldsymbol{F} = \left(\boldsymbol{Q}^{-\frac{2}{n}} - 1\right)\frac{n-p}{q} = \frac{\boldsymbol{Y}^{\mathrm{T}}(A-B)\boldsymbol{Y}}{\boldsymbol{Y}^{\mathrm{T}}(I_n - A)\boldsymbol{Y}} \frac{n-p}{q}, \tag{4.19}$$

where $q = \mathrm{rk}(A - B)$ to allow calculations on the base of tabulated distributions. Theorem 4.7 clarifies the distribution behind (4.19).

We repeat without proof a theorem from probability theory, which is needed here and also later.

Theorem 4.6 Theorem of Cochran (1934)
If $\boldsymbol{Y}$ is distributed as $N(1_n \mu, I_n)$, then the positive semi-definite quadratic forms $\boldsymbol{Y}^{\mathrm{T}} A_i \boldsymbol{Y}$ $(i = 1, 2, \ldots, k)$ of rank n_i are independently of each other distributed as $CS(n_i, \lambda_i)$ with the non-centrality parameters $\lambda_i = (1_n \mu)^{\mathrm{T}} A_i (1_n \mu)$ if and only if at least two of the three following conditions are fulfilled:

1) All A_i are idempotent.
2) $\sum_{i=1}^{k} A_i$ is idempotent.
3) $A_i A_j = 0$ for all $i \neq j$.

Corollary 4.1 If $\boldsymbol{Y}$ is distributed as $N(1_n \mu, I_n)$ and if $\boldsymbol{Y}^{\mathrm{T}}\boldsymbol{Y} = \sum_{i=1}^{k} \boldsymbol{Y}^{\mathrm{T}} A_i \boldsymbol{Y}$, then the quadratic forms $\boldsymbol{Y}^{\mathrm{T}} A_i \boldsymbol{Y}(i = 1, 2, \ldots, k)$ of rank n_i are mutually independent distributed as $CS(n_i, \lambda_i)$ with $n_i = \mathrm{rk}(A_i)$ and the non-centrality parameters $\lambda_i = (1_n \mu)^{\mathrm{T}} A_i (1_n \mu)$ if and only if either

- all A_i are idempotent

or

- $A_i A_j = 0$ for all $i \neq j$

or

- $\sum_{i=1}^{k} \text{rk}(A_i) = \text{rk}\left(\sum_{i=1}^{k} A_i\right) = n$

Corollary 4.2 If $\boldsymbol{Y}$ is distributed as $N(1_n\, \mu, \sigma^2 I_n)$ and if $\boldsymbol{Y}^T\boldsymbol{Y} = \sum_{i=1}^{k} \boldsymbol{Y}^T A_i \boldsymbol{Y}$ with $n_i = \text{rk}(A_i)$, then each of the three conditions of Corollary 4.1 is necessary and sufficient for the fact that the quadratic forms $(1/\sigma^2)\, \boldsymbol{Y}^T A_i \boldsymbol{Y}$ $(i = 1, 2, \ldots, k)$ are distributed independently of each other as $CS(n_i, \lambda_i)$ with the non-centrality parameters $\lambda_i = (1/\sigma^2)\, (1_n\, \mu)^T A_i\, (1_n\, \mu)$.

Now we come to the announced statement about the distribution of $\boldsymbol{F}$ in (4.19).

Theorem 4.7 Let $\boldsymbol{Y}$ be distributed as $N(\theta, \sigma^2 I_n)$ and let A and B be idempotent matrices that project R^n orthogonally onto Ω and onto $\omega \subset \Omega$, respectively (where $\text{rk}(A) = p$, $\text{rk}(B) = p - q$). Then $\boldsymbol{F}$ in (4.19) is distributed as $F(q, n-p, \lambda)$ with non-centrality parameter $\lambda = (1/\sigma^2)\theta^T(A - B)\theta$ and the degrees of freedom q and $n - p$.

Proof: Since A is the orthogonal projector of R^n onto the p-dimensional subspace Ω and B the orthogonal projector onto the $(p - q)$-dimensional subspace $\omega \subset \Omega$, we get $AB = BA = B$. Hence, $I_n - A$ and $A - B$ are idempotent. With the notations $A_1 = I_n - A$, $A_2 = A - B$ and $A_3 = B$, the conditions of Theorem 4.6 are satisfied. Regarding Corollary 4.2 to this theorem, $(1/\sigma^2)\boldsymbol{Y}^T(I_n - A)\boldsymbol{Y}$ and $(1/\sigma^2)$ $\boldsymbol{Y}^T(A - B)\boldsymbol{Y}$ are mutually independent distributed as $CS(n - p)$ and $CS(q, \lambda)$, respectively, with the non-centrality parameter $\lambda = (1/\sigma^2)\theta^T(A - B)\theta$, which supplies the assertion.

Using results in Section 4.1.1 we can show

$$E\left[\boldsymbol{Y}^T(I_n - A)\boldsymbol{Y}\right] = \sigma^2(n-p), \quad E\left[\boldsymbol{Y}^T(A-B)\boldsymbol{Y}\right] = \sigma^2 q + \sigma^2\lambda.$$

If the interim results for calculating F are to be represented in a clear way, an analysis of variance table often is used (see Table 4.1).

If H_0 is true, the non-centrality parameter becomes $\lambda = 0$, and $\boldsymbol{F}$ is centrally F-distributed with degrees of freedom q and $n - p$. H_0 is rejected, if

$$F > F_{1-\alpha}(q, n-p) = F(q, n-p \mid 1-\alpha),$$

where the quantile $F_{1-\alpha}(q, n-p)$ is chosen so that

$$\max P\{\boldsymbol{F} > F_{1-\alpha}(q, n-p) \mid \theta \in \omega\} = \alpha \tag{4.20}$$

is the significance level of the test. The power function is

$$\beta(\theta, \lambda) = P\left\{\frac{q\boldsymbol{F}}{q\boldsymbol{F} + n - p} > \frac{qF_{1-\alpha}(q, n-p)}{qF_{1-\alpha}(q, n-p) + n - p}\right\}. \tag{4.21}$$

Table 4.1 Analysis of variance table calculating the test statistic for the test of the hypothesis H_0: $\theta \in \omega \subset \Omega$.

Source of variation	Sum of squares SS	Degrees of freedom df	Mean square deviation $MS = \frac{SS}{df}$	$E(MS)$	F
Total	Y^TY	n			
Null hypothesis $\theta \in \omega$	$Y^T(A - B)Y$	q	$\frac{1}{q}Y^T(A-B)Y$	$\sigma^2\left(1+\frac{\lambda}{q}\right)$	$F = \frac{n-p}{q}\frac{Y^T(A-B)Y}{Y^T(I_n-A)Y}$
Residual	$Y^T(I_n - A)Y$	$n - p$	$\frac{1}{n-p}Y^T(I_n-A)Y$	σ^2	
Alternative hypothesis $\theta \notin \omega$	Y^TBY	$p - q$			

It can be shown (see Witting and Nölle, 1970, p. 37) that this test is invariant with respect to the group of affine transformations in R^n, thus the test problem is also invariant. The F-test is under all invariant tests with respect to these transformations a uniformly most powerful α-test.

Each linear hypothesis can be written in a basic form by using a suitable transformation of the sample space.

Definition 4.2 A linear hypothesis $\theta^* \in \omega$ according to Definition 4.1 is said to be in canonical form if

$$\theta^* \in \Omega \text{ means } \theta^*_{p+1} = \cdots = \theta^*_n = 0$$

and

$$\theta^* \in \omega \text{ means } \theta^*_1 = \cdots = \theta^*_q = \theta^*_{p+1} = \cdots = \theta^*_n = 0.$$

Theorem 4.8 Each linear hypothesis H_0: $\theta \in \omega$ can be transformed by orthogonal projection of the model equation (4.1) into canonical form so that

$$\boldsymbol{Y}^T(A-B)\boldsymbol{Y} = \boldsymbol{z}_1^2 + \cdots + \boldsymbol{z}_q^2, \quad \boldsymbol{Y}^T(I_n-A)\boldsymbol{Y} = \boldsymbol{z}_{p+1}^2 + \cdots + \boldsymbol{z}_n^2$$

is satisfied, and the distribution in (4.19) remains unchanged.

Proof: Let P be an orthogonal matrix of size (n,n). We put $\boldsymbol{Y} = P\boldsymbol{Z}$ and $\theta = P\theta^*$. W.l.o.g. we choose P so that

$$P^T(A-B)P = \begin{pmatrix} I_q & O \\ O & O \end{pmatrix}, \quad P^TBP = \begin{pmatrix} O & O & O \\ O & I_{p-q} & O \\ O & O & O \end{pmatrix}$$

and

$$P^{\mathrm{T}}(I_n - A)P = \begin{pmatrix} O & O \\ O & I_{n-p} \end{pmatrix}$$

which is always possible. For simplicity the sizes of the null matrices are omitted here. Then (4.1) is transferred into

$$\boldsymbol{Z} = \theta^* + \boldsymbol{e}^* \text{ where } \boldsymbol{Z} \in R^n, \theta^* \in \Omega, \boldsymbol{e}^* = P^T \boldsymbol{e}, \boldsymbol{Z}^* = (\boldsymbol{z}_1, \ldots, \boldsymbol{z}_n)^T.$$

As B is also $P^{\mathrm{T}}BP$ the orthogonal projector of R^n onto a $(p - q)$-dimensional subspace ω^*, and that means

$$H_0 : \theta_i^* = \begin{cases} 0 & \text{for } i = 1, \ldots, q, p+1, \ldots, n \\ \text{arbitrary} & \text{for } i = q+1, \ldots, p. \end{cases}.$$

Besides we find

$$\boldsymbol{Y}^{\mathrm{T}}(A - B)\boldsymbol{Y} = \boldsymbol{Z}^{\mathrm{T}}P^{\mathrm{T}}(A - B)P\boldsymbol{Z} = \boldsymbol{z}_1^2 + \cdots + \boldsymbol{z}_q^2$$

and

$$\boldsymbol{Y}^{\mathrm{T}}(I_n - A)\boldsymbol{Y} = \boldsymbol{Z}^{\mathrm{T}}P^{\mathrm{T}}(I_n - A)P\boldsymbol{Z} = \boldsymbol{z}_{p+1}^2 + \cdots + \boldsymbol{z}_n^2.$$

The non-centrality parameter of the numerator in (4.19) is

$$\lambda = \frac{1}{\sigma^2}\left(\theta_1^{*2} + \cdots + \theta_q^{*2}\right).$$

It is equal to 0 if and only if $\theta_1{}^* = \cdots = \theta_q{}^* = 0$, that is, if and only if H_0 is true. According to this Theorem (4.8) can also be applied for testing linear hypotheses in canonical form.

Definition 4.3 We understand as linear contrast of the parameter vector θ a linear functional $c^{\mathrm{T}}\,\theta$ with $c = (c_1, \ldots, c_n)^{\mathrm{T}}$ and $\sum_{i=1}^{n} c_i = 0$. Two linear contrasts $c_1^{\mathrm{T}}\theta$ and $c_2^{\mathrm{T}}\theta$ are said to be orthogonal (linear contrasts) if $c_1^{\mathrm{T}}c_2 = 0$.

Now we are able to express the null hypothesis $\theta \in \Omega$ by orthogonal contrasts. Let $n - p$ pairwise orthogonal contrasts $c_i^{\mathrm{T}}\theta \quad (i = 1, \ldots, n-p)$ be given that are equal to 0. Under this condition the hypothesis H_0, in which q further pairwise and to the given $c_i^{\mathrm{T}}\theta$ orthogonal contrasts $t_j^{\mathrm{T}}\theta \quad (j = 1, \ldots, q)$ are also 0, is to be tested against the alternative hypothesis that at least one of the additional contrasts $t_i^{\mathrm{T}}\theta$ is different from 0. We put $C = (c_1, \ldots, c_{n-p})$ and $T = (t_1, \ldots, t_q)$. Now the condition $C^{\mathrm{T}}\theta = 0_{n-p}$ defines the p-dimensional null space Ω, that is, $C^{\mathrm{T}}\theta = 0_{n-p}$ is equivalent to $\theta \in \Omega$. Correspondingly the hypothesis $H_0 : C^{\mathrm{T}}\theta = 0_{n-p} \wedge T^{\mathrm{T}}\theta = 0_q$ is equivalent to $\theta \in \omega$. Hence, the hypothesis containing the contrasts can be tested with $\boldsymbol{F}$ from (4.19). This test statistic can be rewritten in another form as it is shown in the next theorem.

Theorem 4.9 We consider $n-p+q$ orthogonal contrasts $c_i^{\mathrm{T}}\theta$ $(i=1,\dots,n-p)$ and $t_j^{\mathrm{T}}\theta$ $(j=1,\dots,q)$. Then the notations $C=(c_1,\dots,c_{n-p})$ and $T=(t_1,\dots,t_q)$ imply $C^{\mathrm{T}}T=O$. Besides $C^{\mathrm{T}}C=D_1$ and $T^{\mathrm{T}}T=D_2$ are diagonal matrices.

Now let $c_i^{\mathrm{T}}\theta=0$ $(i=1,\dots,n-p)$ and $\boldsymbol{Y}$ be distributed as $N(\theta,\sigma^2 I_n)$. Then the test statistic of the linear hypothesis $H_0: t_j^{\mathrm{T}}\theta=0$ for all $j=1,\dots,q$ $(\theta\in\omega)$ can be written with the estimator $\hat{\boldsymbol{\theta}}$ as

$$\boldsymbol{F}=\frac{n-p}{q}\frac{\sum_{j=1}^{q}\frac{1}{\|t_j\|^2}\left(t_j^{\mathrm{T}}\hat{\boldsymbol{\theta}}\right)^2}{\boldsymbol{Y}^{\mathrm{T}}(I_n-A)\boldsymbol{Y}}, \tag{4.22}$$

where A is again the orthogonal projector onto Ω.

Proof: The first assertions are evident. Finally we have to show that the term $\boldsymbol{Y}^{\mathrm{T}}(A-B)\boldsymbol{Y}$ in the numerator of (4.19) has the form

$$\boldsymbol{Y}^{\mathrm{T}}(A-B)\boldsymbol{Y}=\sum_{j=1}^{q}\frac{1}{\|t_j\|^2}\left(t_j\hat{\boldsymbol{\theta}}\right)^2,$$

where the matrices A and B are the projectors onto Ω and ω, respectively. Then the difference $A-B$ is the orthogonal projector of R^n onto the subspace $\omega^{\perp}\cap\Omega$, and we get

$$\theta=B\theta+(A-B)\theta$$

for $\theta\in\Omega$. The columns of T form a basis of $\omega^{\perp}\cap\Omega$, and the columns of P in $A-B=PP^{\mathrm{T}}$ also form an orthonormal basis of $\omega^{\perp}\cap\Omega$. Consequently a - non-singular matrix H exists so that $T=PH$ and $P=TH^{-1}$, respectively, as well as $A-B=PP^{\mathrm{T}}=T(H^{\mathrm{T}}H)^{-1}T^{\mathrm{T}}$ hold. Since $A-B$ is idempotent, it follows $A-B=T(T^{\mathrm{T}}T)^{-1}T^{\mathrm{T}}$ and

$$\boldsymbol{Y}^{\mathrm{T}}(A-B)\boldsymbol{Y}=\boldsymbol{Y}^{\mathrm{T}}A(A-B)A\boldsymbol{Y}=\hat{\boldsymbol{\theta}}^{\mathrm{T}}T\left(T^{\mathrm{T}}T\right)^{-1}T^{\mathrm{T}}\hat{\boldsymbol{\theta}}Y. \tag{4.23}$$

This implies the assertion because $T^{\mathrm{T}}T$ is a diagonal matrix.

4.1.4 Construction of Confidence Regions

As in the previous subsections, we assume that $\boldsymbol{Y}$ is distributed as $N(\theta,\sigma^2 I_n)$. In this subsection, methods are presented that can be used to construct confidence regions for linear combinations. The condition $\theta\in\Omega$ is also written as $C^{\mathrm{T}}\theta=0$, $C^{\mathrm{T}}C=D_1$.

Theorem 4.10 Let Y be distributed as $N(\theta, \sigma^2 I_n)$ so that the condition $C^T\theta = 0$ $(\theta \in \Omega)$ of the linear model (4.1) is fulfilled. If $C^T T = 0$ also holds, then a confidence region for $T^T\theta$ with the coefficient $1 - \alpha$ is given by

$$\frac{1}{qs^2}\left(\hat{\boldsymbol{\theta}}^T T - \theta^T T\right)\left(T^T A T\right)^{-1}\left(T^T\hat{\boldsymbol{\theta}} - T^T\theta\right) \le F_{1-\alpha}(q, n-p). \tag{4.24}$$

In (4.24) the matrix A is the projector of R^n onto Ω, $\boldsymbol{s}^2$ is the estimator for σ^2 according to (4.7) and q is the rank of T.

Proof: Regarding (4.19), (4.23) and (4.7) Theorem 4.7, the assumptions above imply that the statistic

$$\boldsymbol{F} = \frac{1}{qs^2}\left(\hat{\boldsymbol{\theta}}^T T - \theta^T T\right)\left(T^T A T\right)^{-1}\left(T^T\hat{\boldsymbol{\theta}} - T^T\theta\right)$$

is centrally F-distributed as $F(q, n - p)$, if $E\left(T^T\hat{\boldsymbol{\theta}}\right) = T^T\theta$ is taken into account. Hence, the assertion is true.

Example 4.2 Let $T = t$ be a $(n \times 1)$-vector (i.e. $q = 1$). Then it follows from the Gauss–Markov theorem (Theorem 4.3) that the LSM $\hat{\boldsymbol{L}}$ of $L = t^T\theta$ is equal to $\hat{\boldsymbol{L}} = t^T\hat{\boldsymbol{\theta}} = t^T A\boldsymbol{Y}$. We put $t^T A = a$. Because of $T^T A T = T^T A A T = a^T a$, we get as a special case of (4.24) with focus on L

$$\frac{1}{\|a\|^2 \boldsymbol{s}^2}\left(\hat{\boldsymbol{L}} - L\right)^2 \le F_{1-\alpha}(1, n-p) = t^2\left(n-p, 1-\frac{\alpha}{2}\right). \tag{4.25}$$

This supplies for L the $(1 - \alpha)$-confidence interval

$$\left[\hat{\boldsymbol{L}} - \boldsymbol{s}\|a\|t\left(n-p, 1-\frac{\alpha}{2}\right), \hat{\boldsymbol{L}} + \boldsymbol{s}\|a\|t\left(n-p, 1-\frac{\alpha}{2}\right)\right]. \tag{4.26}$$

4.1.5 Special Linear Models

Example 4.3 Regression Analysis

Let X be a $(n \times p)$-matrix of rank $p < n$ so that Ω is in (4.1) the rank space of X $(\mathrm{R}[X] = \Omega)$; that is, for a certain $\beta \in R^p$, we have

$$\theta = X\beta. \tag{4.27}$$

Since both X and $X^T X$ have the same rank p, the inverse matrix $(X^T X)^{-1}$ exists. Then (4.27) implies $\beta = (X^T X)^{-1} X^T \theta$. According to the Gauss–Markov theorem (Theorem 4.3) from (4.4), we get the estimator

$$\hat{\boldsymbol{\beta}} = \left(X^T X\right)^{-1} X^T A\boldsymbol{Y}, \tag{4.28}$$

where A is again the orthogonal projector of R^n onto Ω. Consequently there exists a matrix P whose columns form an orthonormal basis of Ω so that $A = PP^{\mathrm{T}}$.

Since Ω is the rank space of X, the columns of X also form a basis of Ω. Hence, there is a non-singular matrix H with $P = XH^{-1}$. As $A = XH^{-1}(H^{\mathrm{T}})^{-1}X^{\mathrm{T}}$ is idempotent, it has to be $A = X(X^{\mathrm{T}}X)^{-1}X^{\mathrm{T}}$. If this representation of A is used in (4.28), we obtain

$$\hat{\boldsymbol{\beta}} = \left(X^{\mathrm{T}}X\right)^{-1}X^{\mathrm{T}}\boldsymbol{Y}. \tag{4.29}$$

With this form of A, the formula for $\boldsymbol{s}^2$ in (4.7) becomes

$$\boldsymbol{s}^2 = \frac{1}{n-p}\|\boldsymbol{Y} - X\left(X^{\mathrm{T}}X\right)^{-1}X^{\mathrm{T}}\boldsymbol{Y}\|^2 = \frac{1}{n-p}\boldsymbol{Y}^{\mathrm{T}}\left(I_n - X\left(X^{\mathrm{T}}X\right)^{-1}X^{\mathrm{T}}\right)\boldsymbol{Y}.$$

Now we want to test the hypothesis

$$K^{\mathrm{T}}\beta = a \tag{4.30}$$

under the assumption that $\boldsymbol{Y}$ is distributed as $N(X\beta, \sigma^2 I_n)$, where K^{T} is a $(q \times p)$-matrix of rank q and a is a $(q \times 1)$-vector. The hypothesis (4.30) is according to Definition 4.1 in the case $a \neq 0_q$ no linear hypothesis. But (4.30) can be linearised as follows. We put

$$Z = Y - Xc, \quad \theta^* = \theta - Xc, \quad \gamma = \beta - c,$$

where c is chosen so that $K^{\mathrm{T}}c = a$. Considering the linear model

$$\boldsymbol{Z} = \theta^* + \boldsymbol{e} \tag{4.31}$$

with $\theta^* = \theta - Xc = X\beta - Xc = X\gamma$, the hypothesis $H_0 : K^{\mathrm{T}}\beta = a$ becomes the linear hypothesis

$$H_0 : K^{\mathrm{T}}\gamma = K^{\mathrm{T}}\beta - K^{\mathrm{T}}c = 0_q.$$

Now H_0: $K^{\mathrm{T}}\gamma = 0$ can be tested for the model equation (4.31) using the test statistic (4.19) with inclusion of formula (4.23). The test statistic has the form

$$\boldsymbol{F} = \frac{\boldsymbol{Z}^{\mathrm{T}}T(T^{\mathrm{T}}T)^{-1}T^{\mathrm{T}}\boldsymbol{Z}}{\boldsymbol{Z}^{\mathrm{T}}(I_n - A)\boldsymbol{Z}} \cdot \frac{n-p}{q},$$

where T^{T} is as in Section 4.1.3 the matrix occurring in the hypothesis

$$H_0 : \theta^* \in \omega \;\left(C^{\mathrm{T}}\theta^* = 0 \wedge T^{\mathrm{T}}\theta^* = 0\right).$$

The matrix T can be expressed by K^{T} and X.

Because of $\theta^* = X\gamma$ we get $\gamma = (X^{\mathrm{T}}X)^{-1}X^{\mathrm{T}}\theta^*$ and $K^{\mathrm{T}}\gamma = K^{\mathrm{T}}(X^{\mathrm{T}}X)^{-1}X^{\mathrm{T}}\theta^*$. Therefore it is $T^{\mathrm{T}} = K^{\mathrm{T}}(X^{\mathrm{T}}X)^{-1}X^{\mathrm{T}}$. The equation $K^{\mathrm{T}}c = a$ implies $c = K\,(K^{\mathrm{T}}\,K)^{-1}a$. If besides $Z = Y - Xc = Y - XK(K^{\mathrm{T}}K)^{-1}a$ is used, the test statistic reads

$$F = \frac{n-p}{q}\,\frac{\left(\boldsymbol{Y}-XK(K^{\mathrm{T}}K)^{-1}a\right)^{\mathrm{T}} X(X^{\mathrm{T}}X)^{-1}K\left[K^{\mathrm{T}}(X^{\mathrm{T}}X)^{-1}K\right]^{-1} K^{\mathrm{T}}(X^{\mathrm{T}}X)^{-1}X^{\mathrm{T}}\left(\boldsymbol{Y}-XK(K^{\mathrm{T}}K)^{-1}a\right)}{(\boldsymbol{Y}-XK(K^{\mathrm{T}}K)^{-1}a)^{\mathrm{T}}(I_n-X(X^{\mathrm{T}}X)^{-1}X^{\mathrm{T}})(\boldsymbol{Y}-XK(K^{\mathrm{T}}K)^{-1}a)}$$

$$= \frac{\left(K^{\mathrm{T}}\hat{\boldsymbol{\beta}}-a\right)^{\mathrm{T}}\left[K^{\mathrm{T}}(X^{\mathrm{T}}X)^{-1}K\right]^{-1}\left(K^{\mathrm{T}}\hat{\boldsymbol{\beta}}-a\right)^{\mathrm{T}}}{\boldsymbol{Y}^{\mathrm{T}}(I_n-X(X^{\mathrm{T}}X)^{-1}X^{\mathrm{T}})\boldsymbol{Y}}\,\frac{n-p}{q} \tag{4.32}$$

since $X^{\mathrm{T}}[I_n - X(X^{\mathrm{T}}X)^{-1}X^{\mathrm{T}}] = 0$.

The hypothesis $K^{\mathrm{T}}\beta = a$ that can be tested by (4.32) is very general. From Theorem 4.7 follows that $\boldsymbol{F}$ in (4.32) is non-centrally F-distributed as $F_{1-\alpha}(q, n-p, \lambda)$. The non-centrality parameter is

$$\lambda = \frac{(K^{\mathrm{T}}\beta-a)^{\mathrm{T}}\left[K^{\mathrm{T}}(X^{\mathrm{T}}X)^{-1}K\right]^{-1}(K^{\mathrm{T}}\beta-a)}{\sigma^2}.$$

It vanishes, if the null hypothesis is true.

Example 4.4 Analysis of Variance

As in Example 4.3 let X be a $(n \times p)$-matrix, but now of rank $r < p$. Using (4.27) the model equation (4.1) becomes

$$\boldsymbol{Y} = X\beta + \boldsymbol{e}.$$

Since the rank of X is smaller than p, the inverse matrix $(X^{\mathrm{T}}X)^{-1}$ does not exist. Consequently β cannot be uniquely determined from θ. The quantities $\beta = \beta*$ that minimise

$$S = \|\boldsymbol{Y}-X\beta\|^2 = (\boldsymbol{Y}-X\beta)^{\mathrm{T}}(\boldsymbol{Y}-X\beta)$$

are the solutions of the Gaussian normal equations

$$X^{\mathrm{T}}X\beta = X^{\mathrm{T}}Y \tag{4.33}$$

for $X\beta = Y$. These equations arises also if the derivative

$$\frac{\partial S}{\partial \beta} = 2X^{T}X\beta - 2X^{T}Y$$

is put to 0. (A minimum is reached for $\beta = \beta^*$, since the matrix of second derivatives is positive definite.)

Let G be a generalised inverse (or also inner inverse) of $X^{\mathrm{T}}X$ defined by the relation $X^{\mathrm{T}}XGX^{\mathrm{T}}X = X^{\mathrm{T}}X$. Then a solution of (4.33) can be written as

$$\beta^* = GX^{\mathrm{T}}Y.$$

We will see later that it makes no sense to call $\boldsymbol{\beta}^*$ an estimator for β. Naturally $X\boldsymbol{\beta}^* = \hat{\boldsymbol{\theta}}$ is an estimator for θ because XGX^{T} is in $\hat{\boldsymbol{\theta}} = XGX^{\mathrm{T}}\boldsymbol{\beta}^*$ independent of choosing G. Some further considerations need the concept of an estimable function.

Definition 4.4 A linear function $q^T\beta$ of the parameter vector β is said to be estimable if it is equal to at least one linear function $p^T E(\boldsymbol{Y})$ of the expectation vector of the random variable $\boldsymbol{Y}$ in the model equation

$$\boldsymbol{Y} = X\beta + \boldsymbol{e}.$$

Theorem 4.11 Let a random variable $\boldsymbol{Y}$ be given that satisfies the model equation $\boldsymbol{Y} = X\beta + \boldsymbol{e}$ with a $(n \times p)$-matrix X. Then it follows:

a) The expectations of all components of $\boldsymbol{Y}$ are estimable.
b) If all $q_j^T\beta$ $(j = 1,\dots,k)$ are estimable functions, then the linear combination $L = \sum_{j=1}^{k} c_j q_j^T \beta$ (c_j real) is also an estimable function.
c) The function $q^T\beta$ is estimable iff q^T can be written in the form $q^T = p^T X$ with a certain vector p.
d) If $q^T\beta$ is estimable, then $q^T\beta^*$ is independent of the special solution β^* of the normal equations (4.33).
e) The best linear unbiased estimator (BLUE) of an estimable function $q^T\beta$ is $\widehat{\boldsymbol{q}^T\boldsymbol{\beta}} = \boldsymbol{q}^T\boldsymbol{\beta}^*$ where $\boldsymbol{\beta}^*$ is the random variable of solutions of (4.33).

Proof:

a) If the i-th (coordinate) unit vector is chosen for p in $p^T E(\boldsymbol{Y})$, then $E(\boldsymbol{y}_i) = p^T E(\boldsymbol{Y})$ arises that is estimable.
b) $q_j^T\beta = p_j^T E(\boldsymbol{Y})$ implies $L = \sum_{j=1}^{k} c_j p_j^T E(\boldsymbol{Y}) = p^T E(\boldsymbol{Y})$ with $p = L = \sum_{j=1}^{k} c_j p_j^T$.
c) Starting with $E(\boldsymbol{Y}) = X\beta$ and $q^T\beta = p^T E(\boldsymbol{Y})$, it follows $q^T\beta = p^T X\beta$. Since the estimability is a property that does not depend on β, the latter relation must hold for all β. Hence, $q^T = p^T X$. On the other hand, if $q^T = \mathrm{p}^T X$, then $q^T\beta$ is obviously estimable.
d) We have $q^T\beta^* = p^T X\beta^* = p^T XGX^T Y$, where G is a generalised inverse of X. Since XGX^T is independent of the special choice of G, $q^T\beta^*$ does not depend on the special choice of a solution β^* in (4.33).
e) The Equation (4.33) implies that $q^T\beta^*$ is linear in Y and also that

$$E\left(q^T\boldsymbol{\beta}^*\right) = q^T E\left(GX^T\boldsymbol{Y}\right) = q^T GX^T E(\boldsymbol{Y})$$

is fulfilled. Since $\boldsymbol{Y} = X\beta + \boldsymbol{e}$ leads to $E(\boldsymbol{Y}) = X\beta$ we obtain

$$E\left(q^T\boldsymbol{\beta}^*\right) = q^T GX^T X\beta.$$

Because of (c) we can put $q^T = p^T X$ so that $E(q^T\boldsymbol{\beta}^*) = p^T XGX^T\beta$. Regarding $XGX^T X = X$ this supplies that $q^T\boldsymbol{\beta}^*$ is unbiased. We need the equation

$q^{\mathrm{T}}GX^{\mathrm{T}}X = q^{\mathrm{T}}$ again in the equivalent form $q = X^{\mathrm{T}}XG^{\mathrm{T}}q$. The variance $\mathrm{var}(\boldsymbol{\beta}^*)$ can be written as

$$\mathrm{var}(\boldsymbol{\beta}^*) = \mathrm{var}\left(GX^{\mathrm{T}}\boldsymbol{Y}\right) = GX^{\mathrm{T}}\mathrm{var}(\boldsymbol{Y})XG^{\mathrm{T}} = GX^{\mathrm{T}}XG^{\mathrm{T}}\sigma^2.$$

Therefore we get

$$\mathrm{var}\left(q^{\mathrm{T}}\boldsymbol{\beta}^*\right) = q^{\mathrm{T}}GX^{\mathrm{T}}XG^{\mathrm{T}}q\sigma^2 = \underbrace{q^{\mathrm{T}}GX^{\mathrm{T}}X}_{q^{\mathrm{T}}}\,\underbrace{GX^{\mathrm{T}}XG^{\mathrm{T}}q}_{q}\,\sigma^2 = q^{\mathrm{T}}Gq\sigma^2.$$

We have to show that the variance of arbitrary linear combinations $c^{\mathrm{T}}\boldsymbol{Y}$ of $\boldsymbol{Y}$ with $E(c^{\mathrm{T}}\boldsymbol{Y}) = q^{\mathrm{T}}\beta$ cannot fall under the just obtained variance above. The unbiasedness has the consequence $c^{\mathrm{T}}X = q^{\mathrm{T}}$ if $c^{\mathrm{T}}E(\boldsymbol{Y}) = c^{\mathrm{T}}X\beta$ is taken into account. Now we get

$$\mathrm{cov}\left(q^{\mathrm{T}}\boldsymbol{\beta}^*, c^{\mathrm{T}}\boldsymbol{Y}\right) = q^{\mathrm{T}}GX^{\mathrm{T}}XG^{\mathrm{T}}q\sigma^2 = q^{\mathrm{T}}Gq\sigma^2$$

and

$$\begin{aligned}\mathrm{var}\left(q^{\mathrm{T}}\boldsymbol{\beta}^* - c^{\mathrm{T}}\boldsymbol{Y}\right) &= \mathrm{var}\left(q^{\mathrm{T}}\boldsymbol{\beta}^*\right) + \mathrm{var}\left(c^{\mathrm{T}}\boldsymbol{Y}\right) - 2\,\mathrm{cov}\left(q^{\mathrm{T}}\boldsymbol{\beta}^*, c^{\mathrm{T}}\boldsymbol{Y}\right)\\ &= \mathrm{var}\left(c^{\mathrm{T}}\boldsymbol{Y}\right) - q^{\mathrm{T}}Gq\sigma^2 = \mathrm{var}\left(c^{\mathrm{T}}\boldsymbol{Y}\right) - \mathrm{var}\left(q^{\mathrm{T}}\boldsymbol{\beta}^*\right).\end{aligned}$$

Since $\mathrm{var}(q^{\mathrm{T}}\boldsymbol{\beta}^* - c^{\mathrm{T}}\boldsymbol{Y})$ is non-negative, $\mathrm{var}(c^{\mathrm{T}}\boldsymbol{Y}) \geq \mathrm{var}(q^{\mathrm{T}}\boldsymbol{\beta}^*)$ follows. Therefore the estimation of $q^{\mathrm{T}}\beta$ is a BLUE.

The estimability of a linear combination of θ is connected with the testability of a hypothesis that is introduced next.

Definition 4.5 A hypothesis $H: K^{\mathrm{T}}\beta = a$ with β from the model $\boldsymbol{Y} = X\beta + \boldsymbol{e}$ is said to be testable if the functions $k_i^{\mathrm{T}}\beta$ are estimable for all i ($i = 1,\ldots, q$), where k_i are the columns of K, that is, if K^{T} can be written as $P^{\mathrm{T}}X$ with a certain $(n \times q)$-matrix P.

In Definition 4.5 we can also write $K = (k_1, \ldots, k_q)$, $K^{\mathrm{T}}\beta = \left(k_i^{\mathrm{T}}\beta\right)$ and $P = (p_1, \ldots, p_q)$. If the hypothesis H is testable, then $K^{\mathrm{T}}\beta^* = a$ does not depend on the choice of the solution β^* in (4.33).

Now we want to find a test statistic for a testable null hypothesis $H_0: K^{\mathrm{T}}\beta = a$. We know that $K^{\mathrm{T}}\beta^*$ is an estimator for $K^{\mathrm{T}}\beta$ that is invariant with respect to β^*. It is also unbiased, since (because of $X = XGX^{\mathrm{T}}X$)

$$\begin{aligned}E\left(K^{\mathrm{T}}\boldsymbol{\beta}^*\right) &= K^{\mathrm{T}}E(\boldsymbol{\beta}^*) = K^{\mathrm{T}}GX^{\mathrm{T}}E(\boldsymbol{Y}) = K^{\mathrm{T}}GX^{\mathrm{T}}X\beta\\ &= P^{\mathrm{T}}XGX^{\mathrm{T}}X\beta = P^{\mathrm{T}}X\beta = K^{\mathrm{T}}\beta.\end{aligned}$$

We can derive a test statistic for the hypothesis $K^{\mathrm{T}}\beta = a$ similar to Example 4.3, where $\boldsymbol{Y}$ is again supposed to be distributed as $N(X\beta, \sigma^2 I_N)$. All conversions leading to (4.32) can be overtaken; only $(X^TX)^{-1}$ has to be replaced by the generalised inverse G of X. Hence, instead of (4.32) we get

$$\boldsymbol{F} = \frac{\left(\boldsymbol{Y}-XK(K^TK)^{-1}a\right)^T XGK[K^TGK]^{-1}K^TGX^T\left(\boldsymbol{Y}-XK(K^TK)^{-1}a\right)}{\left(\boldsymbol{Y}-XK(K^TK)^{-1}a\right)^T(I_n-XGX^T)\left(\boldsymbol{Y}-XK(K^TK)^{-1}a\right)}\cdot\frac{n-p}{q}.$$

We have only to show that $T^TT = K^TGX^TXG^TK = K^TGK$. Regarding $K^T = P^TX$ and $X = XGX^TX$ or $X^T = X^TXG^TX^T$, we find indeed

$$K^TGX^TXG^TK = P^TXGX^TXG^TX^TP = P^TXGX^TP = K^TGK.$$

The numerator of $\boldsymbol{F}$ (ignoring the scalar factor at the end) can be rewritten as $(K^T\boldsymbol{\beta}^* - a)^T(K^TGK)^{-1}\,(K^T\boldsymbol{\beta}^* - a)$ if

$$a = K^TK\left(K^TK\right)^{-1}a = P^TXK\left(K^TK\right)^{-1}a = P^TXGX^TXK\left(K^TK\right)^{-1}a$$
$$= K^TGX^TXK\left(K^TK\right)^{-1}a$$

is considered. Therefore the test statistic of the testable hypothesis $K^T\beta = a$ reads

$$\boldsymbol{F} = \frac{(K^T\boldsymbol{\beta}^*-a)^T(K^TGK)^{-1}(K^T\boldsymbol{\beta}^*-a)}{\boldsymbol{Y}^T(I_n-XGX^T)\boldsymbol{Y}}\cdot\frac{n-p}{p}, \tag{4.34}$$

since $X^T(I_n - XGX^T) = 0$.

According to Theorem 4.7 the statistic $\boldsymbol{F}$ in (4.34) is non-centrally F-distributed as $F(q, n - p, \lambda)$ with degrees of freedom q and $n - p$ and the non-centrality parameter

$$\lambda = \frac{1}{\sigma^2}\left(K^T\beta-a\right)^T\left(K^TGK\right)^{-1}\left(K^T\beta-a\right).$$

If H_0: $K^T\beta = a$ is true, then $\lambda = 0$ follows.

Example 4.5 Covariance Analysis

Often it happens that the matrix X in Example 4.4 contains some linear independent columns. This suggests to represent X in the form $X = (W, Z)$, where W is a $(n \times s)$-matrix of rank $r < s$ and Z is a $(n \times k)$-matrix of rank k (with linear independent columns). Obviously it is $s + k = p$. Now it is natural to split also $\beta = \begin{pmatrix}\alpha\\ \gamma\end{pmatrix}$ so that (4.1) obtains the form

$$\boldsymbol{Y} = W\alpha + Z\gamma + \boldsymbol{e}. \tag{4.35}$$

The parameter space Ω is the rank space of X, that is, $\Omega = R[X]$. If $R[W] \cap R[Z] = \{0\}$, then Ω is (equal to) the direct sum $R[W] \oplus R[Z]$ of these two rank spaces.

In the following, it is supposed on the one hand that the columns in Z are linear independent and on the other hand that the columns of W do not linearly depend on columns of Z.

The model equation (4.35) can be considered not only as a mixture of the model equations used in Example 4.3 and in Example 4.4 but also as special case of the model equation in Example 4.4. We obtain from (4.33)

$$X^{\mathrm{T}}X\beta^* = \begin{pmatrix} W^{\mathrm{T}}W & W^{\mathrm{T}}Z \\ Z^{\mathrm{T}}W & Z^{\mathrm{T}}Z \end{pmatrix}\begin{pmatrix} \alpha^* \\ \gamma^* \end{pmatrix} = \begin{pmatrix} W^{\mathrm{T}}Y \\ Z^{\mathrm{T}}Y \end{pmatrix}. \tag{4.36}$$

If G_W denotes a generalised inverse of $W^{\mathrm{T}}W$ and G a generalised inverse of $Z^{\mathrm{T}}(E_n - WG_WW^{\mathrm{T}})Z$ (in the sense we used it before), then α^* and γ^* can be determined in (4.36) as

$$\alpha^* = G_w\left(W^{\mathrm{T}}Y - W^{\mathrm{T}}Z\gamma^*\right) = G_wW^{\mathrm{T}}Y - G_wW^{\mathrm{T}}Z\gamma^* = \alpha_0^* - G_wW^{\mathrm{T}}Z\gamma^*$$

and

$$\gamma^* = GZ^{\mathrm{T}}\left(I_n - WG_WW^{\mathrm{T}}\right)Y.$$

Here α_0^* denotes a solution of (4.36) in the case $\gamma^* = 0$.

Since $S = I_n - WG_WW^{\mathrm{T}}$ is idempotent, the matrices SZ and $Z^{\mathrm{T}}SZ = Z^{\mathrm{T}}SSZ$ have the same rank. Because of rk(SZ) = rk(Z) (the columns of W are by assumption no linear combinations of columns in Z), the inverse $(Z^{\mathrm{T}}SZ)^{-1}$ exists and we get

$$\gamma^* = \left(Z^{\mathrm{T}}SZ\right)^{-1}Z^{\mathrm{T}}SY = \hat{\gamma}.$$

Therefore $\gamma^* = \hat{\gamma}$ is (together with the corresponding α^*) not only a special solution of (4.36) but also the unique one. Hence $\hat{\boldsymbol{\gamma}}$ is an estimator for γ. As we see, γ is estimable. Besides $q^{\mathrm{T}}\alpha$ is always estimable if it is estimable in a model with $\gamma = 0$. The representation $\hat{\boldsymbol{\gamma}} = (Z^{\mathrm{T}}SSZ)^{-1}Z^{\mathrm{T}}S\boldsymbol{Y}$ implies that $\hat{\boldsymbol{\gamma}}$ is the LSM of γ in the model $\boldsymbol{Y} = SZ\gamma + \boldsymbol{e}$.

We want to derive a test statistic for the hypothesis H_0: $\gamma = 0$. If we put $\theta = W\alpha + Z\gamma$, then Ω in (4.1) is a parameter space of dimension

$$p = \mathrm{rk}(W) + \mathrm{rk}(Z) = r + k.$$

The linear hypothesis H_0: $\gamma = 0$ corresponds to the parameter space ω, whose dimension is $p - q = \mathrm{rk}(W) = r$. Hence the hypothesis H_0: $\gamma = 0$ can be tested using the statistic (4.19). Let A again denote the orthogonal projector of R^n onto Ω and B the orthogonal projector of R^n onto ω. Regarding $\Omega \cap \omega^{\perp} = R[(I_n - B)Z]$ and $R[Z] \cap \omega = \{0\}$, we get

$$A - B = (I_n - B)Z\left(Z^{\mathrm{T}}(I_n - B)Z\right)^{-1}Z^{\mathrm{T}}(I_n - B) = SZ\left(Z^{\mathrm{T}}SZ\right)^{-1}Z^{\mathrm{T}}S.$$

Therefore $\boldsymbol{Y}^{\mathrm{T}}(A-B)\boldsymbol{Y}=\hat{\gamma}^{\mathrm{T}}Z^{\mathrm{T}}S\boldsymbol{Y}$ and $\boldsymbol{Y}^{\mathrm{T}}(I_n-A)\boldsymbol{Y}=\boldsymbol{Y}^{\mathrm{T}}(I_n-B)\boldsymbol{Y}-\hat{\gamma}^{\mathrm{T}}Z^{\mathrm{T}}S\boldsymbol{Y}$. Hence, the hypothesis $H_0 : \gamma = 0$ can be tested with

$$\boldsymbol{F}=\frac{\hat{\gamma}^{\mathrm{T}}Z^{\mathrm{T}}S\boldsymbol{Y}}{\boldsymbol{Y}^{\mathrm{T}}(I_n-B)\boldsymbol{Y}-\hat{\gamma}^{\mathrm{T}}Z^{\mathrm{T}}S\boldsymbol{Y}}\cdot\frac{n-r-k}{k}. \tag{4.37}$$

Moreover, $\boldsymbol{F}$ is centrally F-distributed with k and $n - r - k$ degrees of freedom if H_0 is true.

If the hypothesis $K\alpha^{\mathrm{T}} = a$ is to be tested with the estimable function $K^{\mathrm{T}}\alpha$, then the test statistic $\boldsymbol{F}$ is applied as in Example 4.4.

4.1.6 The Generalised Least Squares Method (GLSM)

Now we again want to consider the case where $V = \mathrm{var}(\boldsymbol{e}) \neq I_n$ with a positive definite matrix V. Although it was shown after Definition 4.1 that $V = I_n$ can be taken by transforming the model, it is sometimes useful to get estimators for arbitrary positive definite matrices in a direct way (without transformation). We apply the same notations as in the special case (see the passages after Definition 4.1).

If we use the LSM relation (4.2) with the notations

$$V=P^TP, Z=\left(P^T\right)^{-1}Y,\ \lambda=\left(P^T\right)^{-1}\theta,\ \hat{\lambda}=\left(P^T\right)^{-1}\hat{\theta},\ \Omega^*=\left(P^T\right)^{-1}\Omega,$$

then we get

$$\|Z-\hat{\lambda}\|^2=\inf_{\lambda\in\Omega^*}\|Z-\lambda\|^2$$

and

$$\begin{aligned}\|Z-\hat{\lambda}\|^2&=\left(\left(P^{\mathrm{T}}\right)^{-1}Y-\left(P^{\mathrm{T}}\right)^{-1}\hat{\theta}\right)^{\mathrm{T}}\left(\left(P^{\mathrm{T}}\right)^{-1}Y-\left(P^{\mathrm{T}}\right)^{-1}\hat{\theta}\right)\\&=\left(Y-\hat{\theta}\right)^{\mathrm{T}}P^{-1}\left(P^{\mathrm{T}}\right)^{-1}\left(Y-\hat{\theta}\right)=\left(Y-\hat{\theta}\right)^{\mathrm{T}}V^{-1}\left(Y-\hat{\theta}\right).\end{aligned}$$

Analogously to the transformation (4.4), we have $\hat{\lambda}=BZ$ with an idempotent matrix B of rank p. It follows

$$\hat{\theta}=P^{\mathrm{T}}B\left(P^{\mathrm{T}}\right)^{-1}Y \tag{4.38}$$

from $(P^{\mathrm{T}})^{-1}\theta = B(P^{\mathrm{T}})^{-1}Y$ after multiplying both sides of the equation by P^{T}. This corresponds with θ in (4.4) putting $A = P^{\mathrm{T}}B(P^{\mathrm{T}})^{-1}$.

Regarding the case in Example 4.3 $\left(\hat{\theta}=X\beta,\ \mathrm{rk}(X)=p\right)$, we have $\lambda = (P^{\mathrm{T}})^{-1}X\beta = X^*\beta$. Further, analogously as in Example 4.3 we find

$$B=X^*\left(X^{*\mathrm{T}}X^*\right)^{-1}X^{*\mathrm{T}}=\left(P^{\mathrm{T}}\right)^{-1}X\left(X^{\mathrm{T}}P^{-1}\left(P^{\mathrm{T}}\right)^{-1}X\right)^{-1}X^{\mathrm{T}}P^{-1}$$

and finally after introducing random variables and the UVUE of β

$$\hat{\boldsymbol{\beta}} = \left(X^{\mathrm{T}}V^{-1}X\right)^{-1}X^{\mathrm{T}}V^{-1}\boldsymbol{Y}. \tag{4.39}$$

If V is unknown, then (4.39) is often used with the estimator $\hat{\boldsymbol{V}}$ instead of V, that is, we estimate β by the quasi-UVUE

$$\hat{\boldsymbol{\beta}} = \left(X^{\mathrm{T}}\hat{\boldsymbol{V}}^{-1}X\right)^{-1}X^{\mathrm{T}}\hat{\boldsymbol{V}}^{-1}\boldsymbol{Y}. \tag{4.40}$$

$\hat{V}$ is the estimated covariance matrix of $\boldsymbol{Y}$. If the structure of X it permits (multiple measurements at single measuring points), V is estimated from observation values used for the estimation of β. In (4.40) the estimator $\hat{\boldsymbol{\beta}}$ is neither linear nor unbiased.

4.2 Linear Models with Random Effects: Mixed Models

If in model equation (4.1) at least one component of θ is random and at least one component is an unknown fixed parameter, then the corresponding linear model is called a mixed model. Up to now the theory of mixed models could not be developed in as much as the unified and complete theory of linear models with fixed effects. Further it is up to the diversity of models. If we arrange θ in such an order that $\boldsymbol{\theta}^T = \left(\theta_1^T, \boldsymbol{\theta}_2^T\right)$ is written with an unknown parameter vector θ_1 and a random vector $\boldsymbol{\theta}_2$, then we can split up the matrix X and the vector β in (4.27) analogously. Then we find with $X = (X_1, X_2)$, $\beta^{\mathrm{T}} = \left(\beta_1^{\mathrm{T}}, \beta_2^{\mathrm{T}}\right)$, the $(n \times p_1)$-matrix X_1, the $(n \times p_2)$-matrix X_2 and $p_1 + p_2 = p$ the following model variants:

$$\boldsymbol{Y} = X_1\beta_1 + \boldsymbol{X}_2\beta_2 + \boldsymbol{e}, \tag{4.41}$$

$$\boldsymbol{Y} = X_1\beta_1 + X_2\boldsymbol{\beta}_2 + \boldsymbol{e}, \tag{4.42}$$

$$\boldsymbol{Y} = X_1\beta_1 + \boldsymbol{X}_2\boldsymbol{\beta}_2 + \boldsymbol{e}. \tag{4.43}$$

All three models contain the linear model of Section 4.1 for $p_2 = 0$ as a special case. If $X_1\,\beta_1 = \mu 1_N$ (μ real), then each of the models (4.41) up to (4.43) is called model II. The other models with $p_2 > 0$ are called mixed models (in a stronger sense).

The special models in Section 4.1.5 are usually denoted in the following way (after the model name the chapter number is given, where this model is treated, and the model specification is recorded; the models of covariance analysis are omitted here to guarantee a certain clarity):

- Model I of regression analysis (8): (4.41) with $\mathrm{rk}(X_1) = p_1 = p$, $p_2 = 0$.
- Model II of regression analysis (8): (4.41) with $X_1\beta_1 = \beta_0 1_N$ (β_0 real), $\left(\boldsymbol{Y}_i, \boldsymbol{X}_{i,p_1+1}, \ldots, \boldsymbol{X}_{ip}\right)$ non-singular (p_2+1)-dimensional distributed with $p_2 \geq 1$.

- Mixed model of regression analysis (8): (4.41) with $\mathrm{rk}(X_1) = p_1 > 1$, $\left(\boldsymbol{Y}_i, X_{i,p_1+1}, \ldots, X_{ip}\right)$ non-singular (p_2+1)-dimensional distributed with $p_2 \geq 1$.
- Regression model I with random regressors (8): (4.42) with $\beta_1 = 0$ ($p_1 = 0$), $\mathrm{rk}(X_2) = p$ or with $X_1\beta_1 = \beta_0 e_N$ (β_0 real), $\mathrm{rk}(X_2) = p - 1$.
- Model I of analysis of variance (5): (4.42) with $X_2 = 0$ (i.e. with $p = p_1$), $\mathrm{rk}(X_1) < p$.
- Model II of analysis of variance (6): (4.42) with $X_1\beta_1 = \mu 1_N$ (μ real), $\mathrm{rk}(X_2) < p - 1$.
- Mixed model of analysis of variance (7): (4.42) with $p_1 > 1$, $\mathrm{rk}(X_1) < p_1, p_2 \geq 1$, $\mathrm{rk}(X_2) < p_2$.

This list does not contain all possible models, but is focused on the ones described in the literature under the above given name.

In the mixed models some problems arise, which are only briefly or even not treated in the preceding chapters. This concerns the estimation of variance components and the optimal prediction of random variables. The following problems occur in the mixed models (4.41) and (4.42):

- Estimation of β_1
- Prediction of $\boldsymbol{X}_2$ and $\boldsymbol{\beta}_2$, respectively
- Estimation of $\mathrm{var}(\boldsymbol{\beta}_2)$

The estimation of β_1 can principally done with methods described in Section 4.1 – but there are also methods of interest estimating β_1 and $\mathrm{var}(\boldsymbol{\beta}_2)$ together in an optimal way, based on a combined loss function. Prediction methods are briefly discussed in Section 4.2.1. Methods for estimating variance matrices $\mathrm{var}(\boldsymbol{\beta}_2)$ of special structure are dealt with in Section 4.2.2.

4.2.1 Best Linear Unbiased Prediction (BLUP)

We introduce a new concept, that of prediction.

Definition 4.6 Model equation (4.42) is considered with $E(\boldsymbol{e}) = 0_N$. Further, let $V = \mathrm{var}(\boldsymbol{Y}|\,\boldsymbol{\beta}_2)$ be positive definite, $E(\boldsymbol{\beta}_2) = b_2$, $\mathrm{var}(\boldsymbol{\beta}_2) = B$ be positive definite, β_1 be known and $\mathrm{cov}(\boldsymbol{e},\boldsymbol{\beta}_2) = O_{N,p_2}$. A linear function in $\boldsymbol{Y}$ of the form

$$\boldsymbol{L} = a^{\mathrm{T}}(\boldsymbol{Y} - X_1\beta_1) \quad \left(a = (a_1,\ldots,a_N)^{\mathrm{T}},\ a_i \text{ real}\right) \tag{4.44}$$

is said to be an unbiased prediction or briefly $\boldsymbol{L}$ from the set of unbiased predictions D_{UP} if

$$E[\boldsymbol{K} - \boldsymbol{L}] = 0, \tag{4.45}$$

and it is said to be a best linear unbiased prediction (BLUP) of $\boldsymbol{K} = c^{\mathrm{T}}\boldsymbol{\beta}_2,\ c^{\mathrm{T}} = \left(c_1,\ldots,c_{p_2}\right)$ if $\boldsymbol{L}$ is from D_{UP} and

$$\text{var}[\boldsymbol{K}-\boldsymbol{L}] = \min_{L^* \in D_{UP}} \text{var}[\boldsymbol{K}-\boldsymbol{L}^*] \tag{4.46}$$

is fulfilled for all V, b_2, B and $X_1\beta_1$.

Analogously BLUPs can be defined for linear combinations of elements from X_2 in model equation (4.41). Here we restrict ourselves to the case of Definition 4.6, since it is representative for all models.

Theorem 4.12 The BLUP of $\boldsymbol{c}^{\text{T}}\boldsymbol{\beta}_2 = \boldsymbol{K}$ (for unknown b_2) is given under the conditions of Definition 4.6 by

$$\boldsymbol{L} = a^{\text{T}}(\boldsymbol{Y} - X_1\beta_1)$$

where

$$a = V^{-1}X_2\left(X_2^{\text{T}}V^{-1}X_2\right)^{-1}c, \tag{4.47}$$

provided that D_{UP} has at least one element and $X_2^{\text{T}}V^{-1}X_2$ is positive definite. Then

$$\text{var}(\boldsymbol{K}-\boldsymbol{L}) = c^{\text{T}}\left(X_2^{\text{T}}V^{-1}X_2\right)^{-1}c. \tag{4.48}$$

Proof: First we show $\boldsymbol{L} \in D_{UP}$, that is, (4.45). Namely, we have

$$E[\boldsymbol{K}-\boldsymbol{L}] = c^{\text{T}}E(\boldsymbol{\beta}_2) - a^{\text{T}}E(\boldsymbol{Y}-X_1\beta_1) = c^{\text{T}}b_2 - a^{\text{T}}X_2b_2$$
$$= c^{\text{T}}b_2 - c^{\text{T}}\left(X_2^{\text{T}}V^{-1}X_2\right)^{-1}X_2^{\text{T}}V^{-1}X_2b_2 = 0$$

Now let $\boldsymbol{L}^* = a^{*\text{T}}(\boldsymbol{Y} - X_1\beta_1)$ be an arbitrary element from D_{UP}, that is, $X_2^{\text{T}}a = X_2^{\text{T}}a^* = c$ is fulfilled. Next we find

$$\text{var}\left[c^{\text{T}}\boldsymbol{\beta}_2 - a^{*\text{T}}(\boldsymbol{Y}-X_1\beta_1)\right] = \text{var}\left[c^{\text{T}}\boldsymbol{\beta}_2 - a^{*\text{T}}\boldsymbol{Y}\right]$$
$$= \text{var}\left[c^{\text{T}}\boldsymbol{\beta}_2 - a^{\text{T}}\boldsymbol{Y} + a^{\text{T}}\boldsymbol{Y} - a^{*\text{T}}\boldsymbol{Y}\right].$$

Since

$$\text{var}(\boldsymbol{Y}) = E[\text{var}(\boldsymbol{Y})|\boldsymbol{\beta}_2] + \text{var}[E(\boldsymbol{Y}|\boldsymbol{\beta}_2)] = V + X_2BX_2^{\text{T}}$$

and analogously

$$\text{cov}\left(\boldsymbol{Y}, c^{\text{T}}\boldsymbol{\beta}_2\right) = E\left[\text{cov}\left(\boldsymbol{Y}, c^{\text{T}}\boldsymbol{\beta}_2|\boldsymbol{\beta}_2\right)\right] + \text{cov}\left[E(\boldsymbol{Y}|\boldsymbol{\beta}_2), c^{\text{T}}\boldsymbol{\beta}_2\right] = X_2Bc$$

holds, it follows

$$\text{cov}\left[c^{\text{T}}\boldsymbol{\beta}_2 - a^{\text{T}}\boldsymbol{Y}, a^{\text{T}}\boldsymbol{Y} - a^{*\text{T}}\boldsymbol{Y}\right] = (a-a^*)^{\text{T}}X_2Bc - (a-a^*)^{\text{T}}Va.$$

Because of $a^{\mathrm{T}}X_2 = a^{*\mathrm{T}}X_2 = c$, the first summand on the right-hand side is equal to 0, and the second summand becomes with a in (4.47):

$$(a-a^*)^{\mathrm{T}}VV^{-1}X_2\left(X_2^{\mathrm{T}}V^{-1}X_2\right)^{-1}c = (a-a^*)^{\mathrm{T}}X_2\left(X_2^{\mathrm{T}}V^{-1}X_2\right)^{-1}c$$

which is also equal to 0. This implies

$$\mathrm{var}\left[c^{\mathrm{T}}\boldsymbol{\beta}_2 - a^{*\mathrm{T}}\boldsymbol{Y}\right] = \mathrm{var}\left[c^{\mathrm{T}}\boldsymbol{\beta}_2 - a^{\mathrm{T}}\boldsymbol{Y}\right] + (a-a^*)^{\mathrm{T}}\mathrm{var}(\boldsymbol{Y})(a-a^*)$$

and consequently

$$\mathrm{var}\left[c^{\mathrm{T}}\boldsymbol{\beta}_2 - a^{*\mathrm{T}}\boldsymbol{Y}\right] \geq \mathrm{var}\left[c^{\mathrm{T}}\boldsymbol{\beta}_2 - a^{\mathrm{T}}\boldsymbol{Y}\right]$$

which completes the first part of the proof.

The equation (4.48) follows by considering

$$\begin{aligned}\mathrm{var}(\boldsymbol{K}-\boldsymbol{L}) &= \mathrm{var}(\boldsymbol{K}) + \mathrm{var}(\boldsymbol{L}) - 2\,\mathrm{cov}(\boldsymbol{K},\boldsymbol{L}) \\ &= c^{\mathrm{T}}Bc + a^{\mathrm{T}}Va + a^{\mathrm{T}}X_2BX_2^{\mathrm{T}}a - 2c^{\mathrm{T}}BX_2^{\mathrm{T}}a = a^{\mathrm{T}}Va\end{aligned}$$

and replacing a by its representation (4.47).

Practical applications of this method are predictions of values concerning the regressand (predictand) in linear regression or predictions of random effects in mixed models of analysis of variance to determine the breeding values of sires, where $X\beta_1$ is often unknown, cf. Rasch and Herrendörfer (1989).

4.2.2 Estimation of Variance Components

In models of type (4.42), the goal is often to estimate the variance $\mathrm{var}(\boldsymbol{\beta}_2)$ of $\boldsymbol{\beta}_2$ in the case $\mathrm{rk}(X_1) < p_1$, $\mathrm{rk}(X_2) < p_2$. If $B = \mathrm{var}(\boldsymbol{\beta}_2)$ is a diagonal matrix, then the diagonal elements are called variance components, and the factor σ^2 in $\mathrm{var}(\boldsymbol{e}) = \sigma^2 I_N$ is called variance component of the residual (of the error) and is to be estimated too. There are important causes to restrict ourselves to so-called quadratic estimators.

Definition 4.7 Let $\boldsymbol{Y}$ be a random vector satisfying the model equation (4.42) and $\mathrm{var}(\boldsymbol{\beta}_2) = B$ be a diagonal matrix with the diagonal elements $\sigma_j^2\ (j = 1,\dots,p_2)$. Further, let $\sigma^2 = \sigma_0^2$ and $\mathrm{cov}(\boldsymbol{\beta}_2, \boldsymbol{e}) = 0$. The random variable $\boldsymbol{Q} = \boldsymbol{Y}^{\mathrm{T}}A\boldsymbol{Y}$ is said to be a quadratic estimator with respect to a linear combination $W = \sum_{i=0}^{p_2} c_i\sigma_i^2$. It is said to be a quadratic unbiased estimator with respect to W if $E(\boldsymbol{Q}) = W$. Further, $\boldsymbol{Q}$ is said to be an invariant quadratic estimator if

$$\boldsymbol{Q} = \boldsymbol{Y}^{\mathrm{T}}A\boldsymbol{Y} = (\boldsymbol{Y} - X_1\beta_1)^{\mathrm{T}}A(\boldsymbol{Y} - X_1\beta_1) \qquad (4.49)$$

(i.e. if $AX_1 = 0$). Finally, using the notations $c^{\mathrm{T}} = (c_0, c_1, \ldots, c_{p_2})$ and $C = \mathrm{diag}(c)$ for the corresponding diagonal matrix, a quadratic estimator $\boldsymbol{Q}$ is said to be of minimal norm if the expression

$$|||C - X_2^{\mathrm{T}} A X_2||| = |||D|||$$

with the matrix A from $\boldsymbol{Q}$ becomes minimal in an arbitrary matrix norm $|||\ |||$. Usually the spectral norm is used (which is induced by the Euclidian vector norm). Rao (1970, 1971a, 1971b, 1971c) introduced for invariant unbiased estimators of minimal norm the name MINQUE (**mi**nimum **n**orm **q**uadratic **u**nbiased **e**stimator).

There are a lot of papers about such estimators. Estimation methods for special models of analysis of variance can be found in Chapter 7. Following are some hints regarding the literature of general theory.

In many cases with positive scalar W, we would hesitate to accept negative estimators (remembering that an estimator is defined as a mapping into the parameter space).

But estimation principles as MINQUE, the method of analysis of variance described in Chapter 6 as well as the maximum likelihood method or a modified maximum likelihood estimation (REML: restricted maximum likelihood) have for normal distributed $\boldsymbol{Y}$ a positive probability that negative estimators occur; see Verdooren (1980, 1988).

Pukelsheim (1981) discusses in a survey possibilities for guaranteeing non-negative unbiased estimators. Using the MINQUE principle, he states a sufficient condition for the existence of corresponding estimators; see also Verdooren (1988).

Henderson (1953) published a first paper about methods for estimating variance components. Anderson et al. (1984) describe optimal estimations of variance components for arbitrary excess (kurtosis) of the distribution of $\boldsymbol{e}$.

The books of Sarhai and Ojeda (2004, 2005) deliver an inspired overview about the state of the art with respect to the special field of estimating variance components.

4.3 Exercises

4.1 Assume that C is a $(n \times p)$-matrix whose columns form an orthonormal basis of the p-dimensional linear subspace Ω of R^n. Prove that the condition $C^{\mathrm{T}} b = 0_p$ ($b \in R^n$) defines the $(n - p)$-dimensional orthogonal complement of Ω in R^n.

4.2 Prove that the solutions β^* of the Gaussian normal equations $X^{\mathrm{T}}X\beta = X^{\mathrm{T}}Y$ supply a minimum of the squared (Euclidian) norm $f(\beta) = ||Y - X\beta||^2$. Hint: Show that the second derivative of $f(\beta)$ is a positive definite matrix.

4.3 Show that $X = XGX^{\mathrm{T}}X$ is fulfilled, where G is a generalised (inner) inverse of $X^{\mathrm{T}}X$.

4.4 Show that the relation $X^{\mathrm{T}}(I_n - XGX^{\mathrm{T}}) = 0$ is satisfied if G is a generalised (inner) inverse of $X^{\mathrm{T}}X$.

4.5 Show that the matrix A in

$$A\sigma^2 = \begin{pmatrix} \frac{\sigma^2}{n} & \cdots & \frac{\sigma^2}{n} \\ \vdots & & \vdots \\ \frac{\sigma^2}{n} & \cdots & \frac{\sigma^2}{n} \end{pmatrix}$$

is idempotent and has the rank 1.

References

Anderson, R. D., Henderson, H. V., Pukelsheim, F. and Searle, S. R. (1984) Best estimation of variance components from balanced data with arbitrary kurtosis. *Math. Oper. Stat. Ser. Stat.*, **15**, 163–176.

Cochran, W. G. (1934) The distribution of quadratic forms in a normal system, with applications to the analysis of covariance. *Math. Proc. Camb. Philos. Soc.*, **30**, 178–191.

Henderson, C. R. (1953) Estimation of variance and covariance components. *Biometrics*, **9**, 226–252.

Pukelsheim, F. (1981) Linear models and convex geometry aspects of non-negative variance estimation. *Math. Oper. Stat. Ser. Stat.*, **12**, 271–286.

Rao, C. R. (1970) Estimation of heteroscedastic variances in linear models. *J. Am. Stat. Assoc.*, **65**, 445–456.

Rao, C. R. (1971a) Estimation of variance and covariance components in linear models. *J. Am. Stat. Assoc.*, **66**, 872–875.

Rao, C. R. (1971b) Minimum variance quadratic estimation of variance components. *J. Multivar. Anal.*, **1**, 257–275.

Rao, C. R. (1971c) Estimation of variance components – MINQUE theory, *J. Multivar. Anal.*, **1**, 257–275.

Rasch, D. and Herrendörfer, G. (1989) *Handbuch der Populationsgenetik und Züchtungsmethodik*, Deutscher Landwirtschaftsverlag, Berlin.

Sarhai, H. and Ojeda M. M. (2004) *Analysis of Variance for Random Models, Balanced Data*, Birkhäuser, Boston, Basel, Berlin.

Sarhai, H. and Ojeda M. M. (2005) *Analysis of Variance for Random Models, Unbalanced Data*, Birkhäuser, Boston, Basel, Berlin.
Verdooren, L. R. (1980) On estimation of variance components, *Stat. Neerl.*, **34**, 83–106.
Verdooren, L. R. (1988) Exact tests and confidence intervals for ratio of variance components in unbalanced two- and three-stage nested designs, *Commun. Stat. – Theory Methods*, **17**, 1197–1230.
Witting, H. and Nölle, G. (1970) *Angewandte mathematische Statistik*, Teubner, Leipzig, Stuttgart.

5

Analysis of Variance (ANOVA) – Fixed Effects Models (Model I of Analysis of Variance)

5.1 Introduction

An experimenter often has to find out in an experiment whether different values of one variable or of several variables have different results on the experimental material. The variables investigated in an experiment are called factors; their values are called factor levels. If the effects of several factors have to be examined, the conventional method means to vary only one of these factors at once and to keep all other factors constant. To investigate the effect of p factors this way, p experiments have to be conducted. This approach is not only very labour intensive, but it can also be that the results at the levels of factor investigated depend on the constant levels of the remaining factors, which means that interactions between the factors exist. The British statistician R. A. Fisher recommended experimental designs by varying the levels of all factors at the same time. For the statistical analysis of the experimental results of such designs (they are called factorial experiments; see Chapter 12), Fisher developed a statistical procedure, the analysis of variance (ANOVA). The first publication about this topic stemmed from Fisher and Mackenzie (1923), a paper about the analysis of field trials in Fisher's workplace at Rothamsted Experimental Station in Harpenden (UK). A good overview is given in Scheffé (1959).

The ANOVA is based on the decomposition of the sum of squared deviations of the observations from the total mean of the experiment into components. Each of the components is assigned to a specific factor or to the experimental error. Further a corresponding decomposition of the degrees of freedom belonging to sums of squared deviations is done. The ANOVA is mainly used to test statistical hypotheses (model I) or to estimate components of variance that can be assigned to the different factors (model II; see Chapter 6).

The ANOVA can be applied on several problems based on mathematical models called model I, model II and mixed model, respectively. The problem leading to model I is as follows: all factor levels have been particularly selected and involved into the experiment because just these levels are of

Mathematical Statistics, First Edition. Dieter Rasch and Dieter Schott.

practical interest. The objective of the experiment is to find out whether the effects of the different levels (or factor level combinations) differ significantly or randomly from each other. The experimental question can be answered by a statistical test if particular assumptions are fulfilled. The statistical conclusion refers to (finite) factor levels specifically selected. The problem leading to model II is as follows: the levels of the factors are a random sample from a universe of possible levels. The objective of the experiment is to make a conclusion about the universe of all levels of a factor by estimating the proportion of the total variance that could be traced back to the variation of the factors or to test a hypothesis about these proportions of the total variance.

The problems in model I are the estimation of the effects and interaction effects of the several factor levels and testing the significance of these effects. The problems in model II are the estimation of the components of variance of several factors or factor combinations and the hypotheses concerning these components. The estimation of components of variance is discussed in Chapter 6.

In all chapters we also give hints concerning the design of experiments.

Remarks about Program Packages

In the analysis of the examples, we also give calculations without program packages although we assume that for the analysis of this data, the reader usually will use program packages like R, SPSS or SAS. We therefore give a short introduction about IBM SPSS Statistics and concerning sample size determination about the R-package OPDOE. IBM SPSS Statistics is very voluminous and with costs. The reader finds more information via www.ibm.com/marketplace/cloud/statistical-analysis-and-reporting/us/en-us.

With the program package R (free via CRAN: http://www.r-project.org or https://cran.r-project.org/), several analyses as well as experimental designs including sample size determination can be done. First one has to install R and then start. To experimental designs one then comes via the command

```
install.packages("OPDOE")
```

and

```
library("OPDOE")
```

Now one can calculate the sample size for analysis of variance (or for short **Anova**) via `size.anova` and find help by

```
help(size.anova).
```

In SPSS for the one-way ANOVA, we use either

Analyze

    Compare Means

        One-Way ANOVA

or by the path mainly for higher classifications with
Analyze
General Linear Models
Univariate

Definition 5.1 We start with a model

$$Y = X\beta + e, \quad R[X] = \Omega \tag{5.1}$$

where $\boldsymbol{Y}$ is a $N(X\beta, \sigma^2 I_N)$-distributed N-dimensional random variable, $\boldsymbol{e}$ is a $N(0_N, \sigma^2 I_N)$-distributed N-dimensional random variable, β is a $[(a+1) \times 1]$ vector of parameters and X a $[N \times (a+1)]$ matrix of rank $p < a+1 < N$. Then (5.1) is the equation of model I of the ANOVA.

If we abdicate the assumption of normal distribution in the parameter estimation, we receive BLUE instead of UVUE (see Chapter 2). That is the case in the sequel. If in point estimation normal distribution is given, then read UVUE in place of BLUE. In hypothesis testing and confidence estimation, normal distribution in Definition 5.1 is essential and will be assumed in those cases.

We explain this definition by a simple example.

Example 5.1 From a populations $G_1, \ldots, G_a$, random samples $\boldsymbol{Y}_1, \ldots, \boldsymbol{Y}_a$ of dimension (or as we also say of size) $n_1, \ldots, n_a$ have been drawn independently from each other. We write $\boldsymbol{Y}_i = \left(\boldsymbol{y}_{i1}, \ldots, \boldsymbol{y}_{in_i}\right)^T$. The $\boldsymbol{y}_i$ are distributed in the populations G_i as $N(\{\mu_i\}, \sigma^2 I_{n_i})$ with $\{\mu_i\} = (\mu_i, \ldots, \mu_i)^T$. Further we write $\mu_i = \mu + a_i (i = 1, \ldots, k)$. Then we have

$$\boldsymbol{y}_{ij} = \mu + a_i + \boldsymbol{e}_{ij} (i = 1, \ldots, k; j = 1, \ldots, n_i). \tag{5.2}$$

Writing $\beta = (\mu, a_1, \ldots, a_k)^T$ and $\boldsymbol{Y}^T = \left(\boldsymbol{Y}_1^T, \ldots, \boldsymbol{Y}_a^T\right)$, then Y is a $(N \times 1)$ vector by putting $N = \sum_{i=1}^{a} n_i$. Now we can write (5.2) in the form (5.1) if $\boldsymbol{e} = (\boldsymbol{e}_{11}, \ldots, \boldsymbol{e}_{1n_1}, \ldots, \boldsymbol{e}_{a1}, \ldots, \boldsymbol{e}_{an_a})^T$ as well as

$$X^T = \begin{pmatrix} 1 & 1 & \ldots & 1 & 1 & 1 & \ldots & 1 & \ldots & 1 & 1 & \ldots & 1 \\ 1 & 1 & \ldots & 1 & 0 & 0 & \ldots & 0 & \ldots & 0 & 0 & \ldots & 0 \\ 0 & 0 & \ldots & 0 & 1 & 1 & \ldots & 1 & \ldots & 0 & 0 & \ldots & 0 \\ \vdots & \vdots & & \vdots & \vdots & \vdots & & \vdots & & \vdots & \vdots & & \vdots \\ \underbrace{0 \;\; 0 \;\; \ldots \;\; 0}_{n_1} & & & & \underbrace{0 \;\; 0 \;\; \ldots \;\; 0}_{n_2} & & & & \cdots & \underbrace{1 \;\; 1 \;\; \ldots \;\; 1}_{n_a} & & & \end{pmatrix}$$

and $X = \left(1_N, \bigoplus_{i=1}^{a} 1_{n_i}\right)$, respectively.

In Example 4.4, we have shown that in general no unique MSE of β exists because the normal equations have infinitely many solutions. Let $\boldsymbol{\beta}^*$ be any solution of the normal equations

$$X^T X\boldsymbol{\beta}^* = X^T \boldsymbol{Y}.$$

Let $G = (X^T X)^-$ be a generalised inverse of $X^T X$. Then we have

$$\boldsymbol{\beta}^* = GX^T \boldsymbol{Y}. \tag{5.3}$$

If we choose a $[(a+1-p)\times(a+1)]$ matrix B of rank $a+1-p$ so that

$$\mathrm{rk}\begin{pmatrix} X \\ B \end{pmatrix} = a+1$$

and

$$B\beta = 0 \tag{5.4}$$

then by the side condition (5.4) the generalised inverse G of $X^T X$ is uniquely determined and equal to $G = (X^T X + B^T B)^{-1}$. By this also $\boldsymbol{\beta}^*$ is uniquely determined (i.e. β in (5.1) is uniquely defined) and equal to the MSE (MLE):

$$\hat{\boldsymbol{\beta}} = \left(X^T X + B^T B\right)^{-1} X^T \boldsymbol{Y}. \tag{5.5}$$

This leads to

Theorem 5.1 If B in (5.4) is a matrix, whose rank space $R[B]$ is orthogonal to the rank space $R[X]$ of the matrix X in (5.1) and if $\mathrm{rk}(H) = \mathrm{rk}\begin{pmatrix} X \\ B \end{pmatrix} = a+1$ and the side condition (5.4) is fulfilled, then β in (5.1) is estimable by (5.5).

Proof: We minimise $r = \|Y - X\beta\|^2 + \lambda^T B\beta$ with $\lambda^T = \left(\lambda_1, \dots, \lambda_{a+1-p}\right)$ by putting the first derivatives of r with respect to β and λ equal to zero. With the notation $\beta = \beta^*$, we obtain

$$2X^T X\beta^* - 2X^T Y + B^T \lambda = 0,$$

$$B\beta^* = 0.$$

Because r is convex we really obtain a minimum in this way. For each $\theta \in R[X] = \Omega$ is β uniquely defined by $\left(\theta^T, 0^T_{a+1-p}\right) = H\beta$, which means that for each $\theta \in \Omega$ we have $\left(\theta^T, 0^T_{a+1-p}\right) \in R[H]$. Because $H(H^T H)^{-1} H^T$ is the matrix of the orthogonal projection from $R^{N+a+1-p}$ on $R[H]$ (see Example 4.3), we obtain

$$H\left(H^T H\right)^{-1} H^T \begin{pmatrix} \theta \\ 0_{a+1-p} \end{pmatrix} = \begin{pmatrix} \theta \\ 0_{a+1-p} \end{pmatrix}$$

or

$$X\left(H^T H\right)^{-1} X^T \theta = \theta, \quad B\left(H^T H\right)^{-1} X^T \theta = 0_{a+1-p}$$

for all $\theta \in \Omega$. Therefore $R\left[X(H^T H)^{-1} B^T\right] \perp R[X]$ and $B(H^T H)^{-1} X^T = 0$. From the equations above it follows that $X(H^T H)^{-1} X^T$ is idempotent and by this the matrix of the orthogonal projection of R^N into a linear vector space V enclosing Ω.

On the other hand, $V = BX(H^T H)^{-1} X^T \subset \Omega$ so that $V = \Omega$ follows. Multiplying $2X^T X\beta_0 - 2X^T Y + B^T \lambda = 0$ from the left by $B(H^T H)^{-1}$, we immediately obtain (because) $B(H^T H)^{-1} B^T \lambda = 0$.

Now B has full rank and $H^T H$ is positive definite, so that $B(H^T H)^{-1} B^T$ is non-singular and $\lambda = 0$ follows. From the normal equations we therefore obtain

$$X^T \theta = X^T X \beta^* = X^T Y.$$

Multiplying both sides with $X(H^T H)^{-1}$, we see that $(H^T H)^{-1}$ is a generalised inverse of $X^T X$. From (5.3) then follows Equation (5.5) because $H^T H = X^T X + B^T B$.

Example 5.2 In Example 5.1 let $a = 2$ and initially $n = n_1 = n_2$. Then we get

$$X^T = \begin{pmatrix} 1 & \dots & 1 & 1 & \dots & 1 \\ 1 & \dots & 1 & 0 & \dots & 0 \\ 0 & \dots & 0 & 1 & \dots & 1 \end{pmatrix},$$

a matrix with $2n$ columns. Without loss of generality we write in (5.4) $B = (0,1,1)$, and by this (5.4) has the form $\sum a_i = 0$. Writing $N = 2n$, it follows

$$X^T X = \begin{pmatrix} N & n & n \\ n & n & 0 \\ n & 0 & n \end{pmatrix}, \quad B^T B = \begin{pmatrix} 0 & 0 & 0 \\ 0 & 1 & 1 \\ 0 & 1 & 1 \end{pmatrix},$$

and

$$X^T X + B^T B = \begin{pmatrix} 2n & n & n \\ n & n+1 & 1 \\ n & 1 & n+1 \end{pmatrix}$$

is a matrix of rank three. The inverse of this matrix is

$$(X^TX+B^TB)^{-1}=\frac{1}{4n}\begin{pmatrix} n+2 & -n & -n \\ -n & n+2 & n-2 \\ -n & n-2 & n+2 \end{pmatrix}.$$

Using (5.5) and $X^T\boldsymbol{Y}=\begin{pmatrix} \boldsymbol{Y}_{..} \\ \boldsymbol{Y}_{1.} \\ \boldsymbol{Y}_{2.} \end{pmatrix}$, we finally receive

$$\hat{\boldsymbol{\beta}}=\begin{pmatrix} \bar{\boldsymbol{y}}_{..} \\ \bar{\boldsymbol{y}}_{1.} - \bar{\boldsymbol{y}}_{..} \\ \bar{\boldsymbol{y}}_{2.} - \bar{\boldsymbol{y}}_{..} \end{pmatrix}=\begin{pmatrix} \hat{\boldsymbol{\mu}} \\ \hat{\boldsymbol{a}}_1 \\ \hat{\boldsymbol{a}}_2 \end{pmatrix}.$$

In the case $n_1 \neq n_2$, we have with $N = n_1 + n_2$:

$$X^TX=\begin{pmatrix} N & n_1 & n_2 \\ n_1 & n_1 & 0 \\ n_2 & 0 & n_2 \end{pmatrix}.$$

For this case in the literature two methods for choosing B can be found. On the one hand, analogously to the case with $n = n_1 = n_2$, one can choose

$$B_1=(0,\ 1,\ 1)$$

and on the other hand

$$B_2=(0,\ n_1,\ n_2).$$

In the first case again

$$\sum a_i = 0.$$

In contrast with the second case, where it follows

$$\sum n_i a_i = 0.$$

In the second case it is implied that the a_i effects of factor levels have the property that after multiplying with the sample sizes n_i and summing up gives 0. Especially in designs with several factors or if n_i are random (as in animal experiments), such an assumption is not plausible.

In the first case (B_1), we have

$$(X^TX+B_1^TB_1)^{-1}=\frac{1}{4n_1n_2}\begin{pmatrix} n_1n_2+N & n_2-n_1-n_1n_2 & n_1-n_2-n_1n_2 \\ n_2-n_1-n_1n_2 & n_1n_2+N & n_1n_2-N \\ n_1-n_2-n_1n_2 & n_1n_2-N & n_1n_2+N \end{pmatrix},$$

and the estimator of β becomes

$$\hat{\beta}_1 = \begin{pmatrix} \hat{\mu}^{(1)} \\ \hat{a}_1{}^{(1)} \\ \hat{a}_2{}^{(1)} \end{pmatrix} = \begin{pmatrix} \frac{1}{2}(\bar{y}_{1.} + \bar{y}_{2.}) \\ \frac{1}{2}(\bar{y}_{1.} - \bar{y}_{2.}) \\ \frac{1}{2}(\bar{y}_{2.} - \bar{y}_{1.}) \end{pmatrix}.$$

In the second case (B_2), we have

$$\left(X^T X + B_2^T B_2\right)^{-1} = \frac{1}{N^2 n_1 n_2} \begin{pmatrix} n_1 n_2(1+N) & -n_1 n_2 & -n_1 n_2 \\ -n_1 n_2 & n_2\left(n_2^2 + n_1 + n_1 n_2\right) & n_1 n_2(1-N) \\ -n_1 n_2 & n_1 n_2(1-N) & n_1\left(n_1^2 + n_2 + n_1 n_2\right) \end{pmatrix},$$

and the estimator of β is

$$\hat{\beta}_2 = \begin{pmatrix} \hat{\mu}^{(2)} \\ \hat{a}_1{}^{(2)} \\ \hat{a}_2{}^{(2)} \end{pmatrix} = \begin{pmatrix} \bar{y}_{..} \\ \bar{y}_{1.} - \bar{y}_{..} \\ \bar{y}_{2.} - \bar{y}_{..} \end{pmatrix}.$$

The reader may ask which form of B he should use. There is no general answer. While the two forms B_1 and B_2 are arbitrary, many others are possible. In the ambiguity of B, the ambiguity of the generalised inverse $(X^T X)^-$ is reflected. Therefore estimates of a_i are less interesting than those for $\mu + a_i$, which are the same for all possible B.

As shown in Chapter 4, the tests of testable hypotheses of the a_i and the estimates of estimable functions of the a_i do also not depend of the special selected B or $(X^T X)^-$.

Because the tests of testable hypotheses and the estimation of estimable functions of the effects of factor levels play an important role in model I, the ambiguity of $(X^T X)^-$ does not influence the final solution. We therefore solve the normal equations under side conditions, resulting in a simple solution.

We now summarise the definitions of estimable functions and testable hypotheses for model I of the ANOVA together with some important theorems and conclusions.

Following Definition 4.4 a linear function $q^T \beta$ of the parameter vector β in (5.1) is estimable, if it equals at least one linear function $t^T E(\boldsymbol{Y})$ of the expectation vector of the random variable $\boldsymbol{Y}$ in (5.1).

Then it follows from Theorem 4.11 for the model equation (5.1):

a) The linear functions of $E(\boldsymbol{Y})$ are estimable.
b) If $q_j^T \beta$ are estimable functions $(j = 1, \ldots, a)$, then also

$$L = \sum_{j=1}^{a} c_j q_j^T \beta \quad (c_j \text{ real})$$

is an estimable function.

c) $q^T\beta$ is an estimable function if q^T can be written in the form $t^T X$ with X from (5.1) $(t \in R^n)$.

d) The BLUE of an estimable function $q^T\beta$ is

$$\widehat{q^T \boldsymbol{\beta}} = q^T \hat{\boldsymbol{\beta}} = t^T X (X^T X)^- X^T \boldsymbol{Y}$$

with $\hat{\boldsymbol{\beta}}$ from (5.3); it is independent of the choice of $\hat{\boldsymbol{\beta}}$ and by this independent of the choice of $(X^T X)^-$.

e) The covariance between the BLUE $q_i^T\hat{\boldsymbol{\beta}}$ and the BLUE $q_j^T\hat{\boldsymbol{\beta}}$ of two estimable functions of $q_i^T\beta$ and $q_j^T\beta$ is given by

$$\operatorname{cov}\left(q_i^T \hat{\boldsymbol{\beta}},\, q_j^T \hat{\boldsymbol{\beta}}\right) = q_i^T (X^T X)^- q_j \sigma^2. \tag{5.6}$$

As we have seen it is indifferent for the estimation of estimable functions which generalised inverse $(X^T X)^-$ in (5.3) is chosen. Even the variance of the estimator does not depend on the choice of $(X^T X)^-$ because $\operatorname{cov}(\boldsymbol{x},\boldsymbol{x}) = \operatorname{var}(\boldsymbol{x})$.

The concept of an estimable function is closely connected with that of a testable hypothesis.

A hypothesis $H: K^T\beta = a^*$ with β from (5.1) is called testable if with $K = (k_1,\dots,k_q)^T$ and $K^T\beta = \{k_i^T\beta\}$ $(i = 1,\dots,q)$ the $k_i^T\beta$ are for all i estimable functions.

Finally we give some results for generalised inverses in form of lemmas. As used already before, each matrix A^- for which

$$AA^-A = A$$

is called a generalised inverse of the matrix A.

Lemma 5.1 If $(X^T X)^-$ is a generalised inverse of the symmetrical matrix $X^T X$, we get

$$X(X^T X)^- X^T X = X, \quad X^T = X^T X (X^T X)^- X^T.$$

Lemma 5.2 For a system of simultaneous linear equations $X^T X x = X^T y$ (normal equations), all solution vectors x have the form

$$x = (X^T X)^- X^T y.$$

Lemma 5.3 If M is a symmetrical matrix of the form

$$M = \begin{pmatrix} A & B \\ B' & D \end{pmatrix},$$

then with $Q = D - B^T A^- B$ the matrix

$$M^- = \begin{pmatrix} A^- + A^- B Q^- B^T A^- & -A^- B Q^- \\ -Q^- B^T A^- & Q^- \end{pmatrix} = \begin{pmatrix} A^- & 0 \\ 0 & 0 \end{pmatrix} + \begin{pmatrix} -A^- B \\ I \end{pmatrix} Q^- \left(-B^T A^-,\ I \right)$$

is a generalised inverse of M, with the identity matrix I.

5.2 Analysis of Variance with One Factor (Simple- or One-Way Analysis of Variance)

In this section we investigate the situation that in an experiment several 'treatments' or levels of a factor A have to be compared with each other. The corresponding analysis is often called 'simple ANOVA'.

5.2.1 The Model and the Analysis

We start with a model equation of the form (5.2) and call μ the total mean and a_i the effect of the ith level of factor A. In Table 5.1 we find the scheme of the observations of an experiment with a levels $A_1, \ldots, A_a$ of factor A and n_i observations for the ith level A_i of A.

Table 5.1 Observations y_{ij} of an experiment with a levels of a factor.

	Number of the levels of the factor					
	1	2	…	i	…	a
	y_{11}	y_{21}	…	y_{i1}	…	y_{a1}
	y_{12}	y_{22}	…	y_{i2}	…	y_{a2}
	⋮	⋮	⋮	⋮	…	⋮
	y_{1n_1}	y_{2n_2}	…	y_{in_i}		y_{an_a}
n_i	n_1	n_2	…	n_i	…	n_a
$Y_{i.}$	$Y_{1.}$	$Y_{2.}$	…	$Y_{i.}$	…	$Y_{a.}$

If an experiment is designed to draw conclusions about the levels A_i occurring in the experiment, then model I is appropriate and we use the introduced mathematical model I in Definition 5.1 as the basis for the design and analysis. If however A_i are randomly selected from a universe of levels, then model II as described in Chapter 6 is used.

We use Equation (5.2) with the side conditions

$$E(\boldsymbol{e}_{ij}) = 0, \quad \operatorname{cov}(\boldsymbol{e}_{ij}, \boldsymbol{e}_{kl}) = \delta_{ik}\delta_{jl}\sigma^2.$$

For testing hypothesis the $\boldsymbol{e}_{ij}$ and by this also the $\boldsymbol{y}_{ij}$ are assumed to be normally distributed. Then it follows from the examples above.

Theorem 5.2 Solutions $\hat{a}_i$ for the a_i $(i = 1,\ldots,a)$ and $\hat{\mu}$ for μ of the normal equation (5.5) for model equation (5.2) are given by

$${}_1\hat{\boldsymbol{\mu}} = \frac{1}{a}\sum_{i=1}^{a} \bar{\boldsymbol{y}}_{i.}, \tag{5.7}$$

$${}_1\hat{\boldsymbol{a}}_i = \frac{a-1}{a}\bar{\boldsymbol{y}}_{i.} - \frac{1}{a}\sum_{j \neq i} \bar{\boldsymbol{y}}_{j.} \tag{5.8}$$

in the case (5.4) for the matrix $B = (0,1,\ldots,1)$.

In the case (5.4) for the matrix $B = (0, n_1, \ldots, n_a)$, they are given by

$${}_2\hat{\boldsymbol{\mu}} = \bar{\boldsymbol{y}}_{..}, \tag{5.9}$$

$${}_2\hat{\boldsymbol{\alpha}}_i = \bar{\boldsymbol{y}}_{i.} - \bar{\boldsymbol{y}}_{..}. \tag{5.10}$$

Both estimations are identical if $n_i = n$ $(i = 1,\ldots,a)$. The variance σ^2 in both cases is unbiasedly estimated by

$$\boldsymbol{s}^2 = \frac{\sum \boldsymbol{y}_{ij}^2 - \hat{\boldsymbol{\beta}}^T X^T \boldsymbol{Y}}{N-a}.$$

The proof of the first part of that theorem follows from (5.5). For $B = (0.1,\ldots,1)$ is

$$X^T X + B^T B = \begin{pmatrix} N & n_1 & n_2 & \cdots & n_a \\ n_1 & n_1+1 & 1 & \cdots & 1 \\ n_2 & 1 & n_2+1 & \cdots & 1 \\ \vdots & \vdots & \vdots & & \vdots \\ n_a & 1 & 1 & \cdots & n_a+1 \end{pmatrix}.$$

For $B = (0, n_1, \dots, n_a)$,

$$X^T X + B^T B = \begin{pmatrix} N & n_1 & n_2 & \cdots & n_n \\ n_1 & n_1^2 + 1 & n_1 n_2 & \cdots & n_1 n_a \\ n_2 & n_2 n_1 & n_2^2 + n_2 & \cdots & n_2 n_a \\ \vdots & \vdots & \vdots & & \vdots \\ n_a & n_a n_1 & n_a n_2 & \cdots & n_a^2 + n_a \end{pmatrix}.$$

Simply we can obtain (5.7) and (5.8) also by minimising

$$\sum_{i=1}^{a} \sum_{j=1}^{n_i} (y_{ij} - \hat{\mu} - \hat{a}_i)^2$$

under the side condition $\sum_{j=1}^{a} a_i = 0$. The solutions (5.9) and (5.10) can be obtained by minimising

$$\sum_{i=1}^{a} \sum_{j=1}^{n_i} (y_{ij} - \hat{\mu} - \hat{a}_i)^2$$

under the side condition $\sum_{i=1}^{a} n_i a_i = 0$. It follows

$$E\left({}_1\hat{\boldsymbol{\mu}}\right) = \frac{1}{a} \sum_{i=1}^{a} E(\bar{\boldsymbol{y}}_{i.}) = \frac{1}{a} \sum_{i=1}^{a} (\mu + a_i) = \mu$$

because of $\sum_{j=1}^{a} a_i = 0$ and

$$E({}_1\hat{\boldsymbol{\alpha}}_i) = \frac{a-1}{a}(\mu + a_i) - \frac{1}{a} \sum_{j \neq i} \left(\mu + a_j\right) = \frac{a-1}{a} a_i + \frac{1}{a} a_i = a_i$$

because of $\sum a_i = 0 \left(\sum_{j \neq i} a_j = -a_i\right)$.

Analogously the unbiasedness of (5.9) and (5.10) under the corresponding side conditions can be shown.

The second part of the theorems is a special case of Theorem 4.4.

Estimable functions of the model parameters are, for instance, $\mu + a_i (i = 1, \dots, a)$ or $a_i - a_j (i, j = 1, \dots, a; i \neq j)$ with the estimators

$$\widehat{\mu + \boldsymbol{\alpha_i}} = \bar{\boldsymbol{y}}_{i.} = {}_1\hat{\boldsymbol{\mu}} + {}_1\hat{\boldsymbol{\alpha}}_i = {}_2\hat{\boldsymbol{\mu}} + {}_2\hat{\boldsymbol{\alpha}}_i$$

and

$$\widehat{\boldsymbol{\alpha_i} - \boldsymbol{\alpha_j}} = \bar{\boldsymbol{y}}_{i.} - \bar{\boldsymbol{y}}_{j.} = {}_1\hat{\boldsymbol{\alpha}}_i - {}_1\hat{\boldsymbol{\alpha}}_j = {}_2\hat{\boldsymbol{\alpha}}_i - {}_2\hat{\boldsymbol{\alpha}}_j,$$

respectively. They are independent from the special choice of B and of $(X^T X)^-$.

One example of an experiment that can be modelled in a model I is to test the null hypothesis $H_0: a_i = a_j$ for all $i \neq j$ against the alternative that at least two a_i differ from each other. This null hypothesis corresponds with the assumption that the effects of a factor considered for all a levels are equal. The basis of the corresponding tests is the fact that the sum of squared deviations SS of the $\boldsymbol{y}_{ij}$ from the total mean of the experiment $\bar{\boldsymbol{y}}_{..}$ can be broken down to independent components. The following trivial conclusion is formulated as a theorem due to its importance.

Theorem 5.3 Let us draw samples from a populations P_i and let $\boldsymbol{y}_{ij}$ be the jth observations of the sample from the i-ten population and $\bar{y}_{i\cdot}$ the mean of this sample. Let N be the total number of observations and $\bar{y}_{..}$ the total mean of the experiment. The sum of squared deviations of the observations from the total mean of the experiment

$$\boldsymbol{SS}_T = \sum_{i=1}^{a}\sum_{j=1}^{n_i}\left(\boldsymbol{y}_{ij} - \bar{\boldsymbol{y}}_{..}\right)^2 = \boldsymbol{Y}^T\boldsymbol{Y} - N\bar{\boldsymbol{y}}_{..}^2 \quad \text{with} \quad \boldsymbol{Y}^T = \left(\boldsymbol{y}_{11}, \ldots, \boldsymbol{y}_{an_a}\right)$$

can be written in the form

$$\boldsymbol{Y}^T\boldsymbol{Y} - N\bar{\boldsymbol{y}}_{..}^2 = \boldsymbol{Y}^T\left[I_N - X\left(X^TX\right)^{-}X^T\right]\boldsymbol{Y} + \boldsymbol{Y}^TX\left(X^TX\right)^{-}X^T\boldsymbol{Y} - N\bar{\boldsymbol{y}}_{..}^2$$

or as

$$\sum_{i=1}^{a}\sum_{j=1}^{n_i}\left(\boldsymbol{y}_{ij} - \bar{\boldsymbol{y}}_{..}\right)^2 = \sum_{i=1}^{a}\sum_{j=1}^{n_i}\left(\bar{\boldsymbol{y}}_{ij} - \bar{\boldsymbol{y}}_{i\cdot}\right)^2 + \sum_{i=1}^{a}\sum_{j=1}^{n_i}\left(\bar{\boldsymbol{y}}_{i\cdot} - \bar{\boldsymbol{y}}_{..}\right)^2.$$

The left-hand side is called **SS** total or for short $\boldsymbol{SS}_T$; the first component of the right-hand side is called SS within the treatments or levels of factor A (short **SS** within or $\boldsymbol{SS}_w = \boldsymbol{SS}_{res}$) and SS between the treatments or levels of factor A ($\boldsymbol{SS}_b = \boldsymbol{SS}_A$), respectively.

We generally write

$$\boldsymbol{SS}_T = \sum_{ij}\boldsymbol{y}_{ij}^2 - \frac{1}{N}\boldsymbol{Y}_{..}^2,$$

$$\boldsymbol{SS}_{res} = \sum_{ij}\boldsymbol{y}_{ij}^2 - \sum_{i}\frac{\boldsymbol{Y}_{i.}^2}{n_i},$$

$$\boldsymbol{SS}_A = \sum_{i}\frac{\boldsymbol{Y}_{i.}{}^2}{n_i} - \frac{1}{N}\boldsymbol{Y}_{..}^2.$$

Theorem 5.4 Under the assumptions of Definition 5.1,

$$\boldsymbol{F} = \frac{(N-a)\boldsymbol{SS}_A}{(a-1)\boldsymbol{SS}_{res}} \tag{5.11}$$

is distributed as $F(a-1,N-a,\lambda)$ with the non-centrality parameter

$$\lambda = \frac{1}{\sigma^2}\beta^T X^T\left(X\left(X^TX\right)^- X^T - \frac{1}{N}1_{N,N}\right)X\beta$$

and $1_{N,N} = 1_{N,N}^T = 1_n 1_n^T$.

If $H_0: a_1 = \cdots = a_a$, then $\boldsymbol{F}$ because of $\lambda = 0$ is $F(a-1,N-a)$-distributed.

Proof: $\boldsymbol{Y} = \left(\boldsymbol{y}_1,\ldots,\boldsymbol{y}_{1n_1},\ldots,\boldsymbol{y}_{a_1},\ldots,\boldsymbol{y}_{an_a}\right)^T$ is $N(X\beta;\sigma^2 I_N)$-distributed. Because of Theorem 5.3 $\boldsymbol{Y}^T\boldsymbol{Y}$ is the sum of three quadratic forms taking $\boldsymbol{Y}_{..} = \frac{1}{N}\boldsymbol{Y}^T 1_{N,N}\boldsymbol{Y}$ into account, namely,

$$\boldsymbol{Y}^T\boldsymbol{Y} = \boldsymbol{Y}^T A_1\boldsymbol{Y} + \boldsymbol{Y}^T A_2\boldsymbol{Y} + \boldsymbol{Y}^T A_3\boldsymbol{Y}$$

with

$$A_1 = I_N - X\left(X^TX\right)^- X^T,\ \ A_2 = X\left(X^TX\right)^- X^T - \frac{1}{N}1_{N,N},\ \ A_3 = \frac{1}{N}1_{N,N}.$$

From Lemma 5.1 $X(X^TX)^- X^T$ is idempotent of rank a and by this A_1 is idempotent of rank $N-a$. Further A_3 is idempotent of rank 1. Because 1_N^T is the first row of X^T, it follows from Lemma 5.1 $1_N^T X(X^TX)^- X^T = 1_N^T$ and from this the idempotence of A_2. The rank of A_2 is $a-1$. By this, for instance, condition 1 of Theorem 4.6 $(N=n,\ n_1 = N-a,\ n_2 = a-1,\ n_3 = 1)$ is fulfilled. Therefore $\frac{1}{\sigma^2}\boldsymbol{Y}^T A_1\boldsymbol{Y}$ is $CS(N-a,\lambda_1)$-distributed and $\frac{1}{\sigma^2}\boldsymbol{Y}^T A_2\boldsymbol{Y}$ is independent of $\frac{1}{\sigma^2}\boldsymbol{Y}^T A_1\boldsymbol{Y}$ distributed as $CS(a-1,\lambda_2)$ with

$$\lambda_1 = \frac{1}{\sigma^2}\beta^T X^T A_1 X\beta = 0$$

and

$$\lambda_2 = \lambda = \frac{1}{\sigma^2}\beta^T X^T\left(X\left(X^TX\right)^- X^T - \frac{1}{N}1_{N,N}\right)X\beta.$$

This completes the proof.

Following Theorem 5.4 the hypothesis $H_0: a_1 = \cdots = a_a$ can be tested by an F-test. The ratios $\boldsymbol{MS}_A = \boldsymbol{MS}_b = \frac{\boldsymbol{SS}_b}{a-1}$ and $\boldsymbol{MS}_R = \boldsymbol{MS}_{res} = \boldsymbol{MS}_w = \frac{\boldsymbol{SS}_w}{N-a}$ are called mean squares between treatments and within treatments or residual mean squares, respectively. The expectations of these $\boldsymbol{MS}$ are

$$E(\boldsymbol{MS}_A) = \sigma^2 + \frac{1}{a-1}\left[\sum_{i=1}^{a} n_i a_i^2 - \frac{1}{N}\left(\sum_{i=1}^{a} n_i a_i\right)^2\right] = \sigma^2 + \frac{1}{a-1}SS_A$$

and

$$E(\boldsymbol{MS}_{res}) = \sigma^2.$$

Under the reparametrisation condition $\sum_{i=1}^{a} n_i a_i = 0$, we receive

$$E(\boldsymbol{MS}_A) = \sigma^2 + \frac{1}{a-1}\sum_{i=1}^{a} n_i a_i^2.$$

Now the several steps in the simple ANOVA for model I can be summarised as follows.

We assumed that from systematically selected normally distributed populations with expectations $\mu + a_i$ and the same variance σ^2, representing the levels of a factor – also called treatments – independent random samples of size n_i have been drawn. For the N observations $\boldsymbol{y}_{ij}$, we assume model equation (5.2) with its side conditions. From the observations in Table 5.1, the column sums $Y_{i.}$ and the number of observations are initially calculated. The corresponding means

$$\bar{\boldsymbol{y}}_{i.} = \frac{\boldsymbol{Y}_{i.}}{n_i}$$

are UVUE under the assumed normal distribution and for arbitrary distributions with finite second moments BLUE of the $\mu + a_i$.

To test the null hypothesis $a_1 = \cdots = a_a$ that all treatments effects are equal and by this all samples stem from the same population, we need the sums $\sum_{i,j} \boldsymbol{y}_{ij}^2, \sum_i \frac{\boldsymbol{Y}_{i.}^2}{n_i}$ and further $\frac{\boldsymbol{Y}_{..}^2}{N}$.

With these sums, a so-called theoretical ANOVA table can be constructed as shown in Table 5.2. In such an ANOVA table occur so-called sources of

Table 5.2 Theoretical analysis of variance table of the one-way analysis of variance model I $\left(\sum a_i = 0\right)$.

Source of variation	SS	df	MS	E(MS)	F
Main effect A	$\boldsymbol{SS}_A = \sum_i \frac{\boldsymbol{Y}_{i.}^2}{n_i} - \frac{1}{N}\boldsymbol{Y}_{..}^2$	$a-1$	$\boldsymbol{MS}_A = \frac{\boldsymbol{SS}_A}{a-1}$	$\sigma^2 + \frac{1}{a-1}\sum n_i a_i^2$	$\boldsymbol{F}_A = \frac{\boldsymbol{MS}_A}{\boldsymbol{MS}_{res}}$
Residual	$\boldsymbol{SS}_{res} = \sum_{ij} \boldsymbol{y}_{ij}^2 - \sum_i \frac{\boldsymbol{Y}_{i.}^2}{n_i}$	$N-a$	$\boldsymbol{MS}_{res} = \frac{\boldsymbol{SS}_{res}}{N-a}$	σ^2	
Total	$\boldsymbol{SS}_T = \sum_{i,j} \boldsymbol{y}_{ij}^2 - \frac{\boldsymbol{Y}_{..}^2}{N}$	$N-1$			

***MS*, mean squares; *SS*, sum of squares.**

variation (between the treatments or levels of factor A, residual (or within the levels) and total). The ***SS*** is in the second column, the degrees of freedom (*df*) in the third, the ***MS*** in the fourth, the E(***MS***) in the fifth and the *F*-statistic in the sixth.

In a practical ANOVA table with data (computer output), the column *E*(***MS***) does not occur, and no random variables but only their realisations appear.

The following functions of the parameters in β are estimable: $\mu + a_i$ and $\widehat{\mu + a_i}$, $i = 1, \ldots, a$ are BLUEs. Further $\sum_{i=1}^{a} c_i(\mu + a_i)$ is estimable by the BLUE $\sum_{i=1}^{a} c_i \bar{\boldsymbol{y}}_{i.}$. (Under normality assumptions they are UVUEs.)

Further all linear contrasts $\left(\sum_{i=1}^{a} c_i a_i \text{ mit } \sum c_i = 0\right)$ as, for instance, differences $a_i - a_j$ $(i \neq j)$ between the components of a $\left(c_i = 1,\ c_j = -1\right)$ or terms of the form $2a_j - a_s - a_r \left(c_j = 2,\ c_s = -1,\ c_r = -1, j \neq s \neq r\right)$ are estimable. The advantage of estimable functions is their independence of the special choice of $(X^T X)^-$ and that a hypothesis $H_0: K^T \beta = a^*$ with the test statistic given in (4.34) is testable if $K^T \beta$ is estimable.

Because the hypothesis $a_1 = \cdots = a_a$ can be written in the form $K^T \beta = 0$ with $\beta = (\mu, \alpha_1, \ldots, \alpha_a)^T$ and the $[(a-1) \times (a+1)]$ matrix

$$K^T = \begin{pmatrix} 0 & 1 & -1 & 0 & \ldots & 0 \\ 0 & 1 & 0 & -1 & \ldots & 0 \\ \vdots & \vdots & \vdots & \vdots & & \vdots \\ 0 & 1 & 0 & 0 & \ldots & -1 \end{pmatrix} = (0_{a-1}, 1_{a-1}, -I_{a-1}),$$

it is testable. The test statistic as introduced in Theorem 5.4 is along with the given K^T, a special case of the test statistic F in (4.34).

Introducing side conditions can change the conclusions about estimability and the BLUE. For example, under the condition $\sum_{i=1}^{a} n_i a_i = 0$, the parameter μ is estimable; the BLUE is $\bar{\boldsymbol{y}}_{..}$. This also means that the hypothesis $H_0: \mu = 0$ is testable.

Also under the side condition $\sum_{i=1}^{a} a_i = 0$, the parameter μ is estimable; but the BLUE is now $\frac{1}{a}\sum_{i=1}^{a} \bar{\boldsymbol{y}}_{i.}$.

For the ambiguousness of $(X^T X)^-$ and the choice of particular side conditions, we make some general remarks, which are also valid for other classifications in the following sections but will not be repeated:

- Independent of the special choice of $(X^T X)^-$ and by this of the choice of the side conditions are:
  - The *SS*, *MS* and *F*-values in the ANOVA tables of testable hypotheses
  - The estimators of estimable functions

- In practical applications we do not need estimates of non-estimable functions. If, for example, three animal feed for pigs have to be analysed and the model $\boldsymbol{y}_{ij} = \mu + a_i + \boldsymbol{e}_{ij}$ is used, the evaluation of these feed can be done by $\mu + a_1, \mu + a_2$ and $\mu + a_3$ and the parameters a_1, a_2 and a_3 are not needed.
- If a problem is independent of the special choice of $(X^T X)^-$, it is often favourable for the derivation of formulae to do this under special side conditions. Normal equations under side conditions can often be relatively simple.

We demonstrate this by an example.

Example 5.3 In an insemination centre, three sires B_1, B_2, B_3. are available. By help of milk yields $y_{ij} (i = 1,2,3; j = 1,\dots,n_i)$ of n_i daughters of these sires, it shall be examined whether differences in the breeding value of these sires concerning the milk fat exist. We assume that the observations y_{ij} are realisations of $N(\mu + a_i, \sigma^2)$-distributed and independent random variables following model (5.2). Table 5.3 contains the performances y_{ij} of the daughters of the three sires. We can ask the following:

- What is the breeding value of the sires?
- Is the null hypothesis $H_0 : a_1 = a_2 = a_3$ valid?
- What are the estimates of $a_1 - a_2$ and $-8a_1 - 6a_2 + 14a_3$?
- Can we accept the null hypothesis $H_0 :\ a_1 - a_2 = 0, -8a_1 - 6a_2 + 14a_3 = 0$?

All tests should be done with a first kind risk of $\alpha = 0.05$.
It follows from (5.1) and (5.2), respectively,

$$y_{11} = 120 = \mu + a_1 + e_{11},$$
$$y_{12} = 155 = \mu + a_1 + e_{12},$$

Table 5.3 Performances (milk fat in kg) y_{ij} of the daughters of three sires.

	Sire		
	B_1	B_2	B_2
y_{ij}	120	153	130
	155	144	138
	131	147	122
	130		
n_i	4	3	3
$Y_{i.}$	536	444	390
$\bar{y}_{i.}$	134	148	130

$$y_{13} = 131 = \mu + a_1 + e_{13},$$
$$y_{14} = 130 = \mu + a_1 + e_{14},$$
$$y_{21} = 153 = \mu + a_2 + e_{21},$$
$$y_{22} = 144 = \mu + a_2 + e_{22},$$
$$y_{23} = 147 = \mu + a_2 + e_{23},$$
$$y_{31} = 130 = \mu + a_3 + e_{31},$$
$$y_{32} = 138 = \mu + a_3 + e_{32},$$
$$y_{33} = 122 = \mu + a_3 + e_{33},$$

and by this it is in (5.1)

$$Y = (120,155,131,130,153,144,147,130,138,122)^T,$$
$$\beta = (\mu, a_1, a_2, a_3)^T, \quad e = (e_{11}, \dots, e_{33})^T,$$
$$X = \begin{pmatrix} 1 & 1 & 0 & 0 \\ 1 & 1 & 0 & 0 \\ 1 & 1 & 0 & 0 \\ 1 & 1 & 0 & 0 \\ 1 & 0 & 1 & 0 \\ 1 & 0 & 1 & 0 \\ 1 & 0 & 1 & 0 \\ 1 & 0 & 0 & 1 \\ 1 & 0 & 0 & 1 \\ 1 & 0 & 0 & 1 \end{pmatrix} = (1_{10}, 1_4 \oplus 1_3 \oplus 1_3),$$
$$a = 3, n_1 = 4, n_2 = n_3 = 3 \text{ and } N = 10.$$

All hypotheses are testable; $a_1 - a_2$ and $-8a_1 - 6a_2 + 14a_3$ are estimable functions.

It is sufficient to calculate any generalised inverse of X^TX. In this example, in solution l once more, a generalised inverse of X^TX is calculated; solution 2 shows the approach by using the formulae derived in this section. In the examples of the following sections, only the simple formulae of the *SS* are used.

Solution l:

To calculate $(X^TX)^-$ an algorithm exploiting the symmetry of X^TX is used:

- Determine $\mathrm{rk}(X^TX) = r$.
- Select a non-singular $(r \times r)$ submatrix of rank r and invert it.
- Replace each element of the submatrix of X^TX by the element of the inverse and the other elements of X^TX by zeros.

We initially calculate

$$X^T X = \begin{pmatrix} 10 & 4 & 3 & 3 \\ 4 & 4 & 0 & 0 \\ 3 & 0 & 3 & 0 \\ 3 & 0 & 0 & 3 \end{pmatrix}.$$

The sum of the last three rows is equal to the first one. Because the submatrix $\begin{pmatrix} 4 & 0 & 0 \\ 0 & 3 & 0 \\ 0 & 0 & 3 \end{pmatrix}$ has rank 3, we get $\text{rk}(X^T X) = r = 3$. The inverse of $\begin{pmatrix} 4 & 0 & 0 \\ 0 & 3 & 0 \\ 0 & 0 & 3 \end{pmatrix}$ equals $\begin{pmatrix} \frac{1}{4} & 0 & 0 \\ 0 & \frac{1}{3} & 0 \\ 0 & 0 & \frac{1}{3} \end{pmatrix}$, and therefore we obtain

$$(X^T X)^- = \begin{pmatrix} 0 & 0 & 0 & 0 \\ 0 & \frac{1}{4} & 0 & 0 \\ 0 & 0 & \frac{1}{3} & 0 \\ 0 & 0 & 0 & \frac{1}{3} \end{pmatrix}.$$

As a check we can show that $(X^T X)(X^T X)^- X^T X = X^T X$.

To calculate $\hat{\beta}$ first we find

$$\left[X^T Y\right]^T = (Y_{..}, Y_{1.}, Y_{2.}, Y_{3.})^T = (1370,536,444,390)^T,$$

and then we obtain

$$\hat{\beta} = (X^T X)^- X^T Y = \begin{pmatrix} 0 \\ \bar{y}_{1.} \\ \bar{y}_{2.} \\ \bar{y}_{3.} \end{pmatrix} = \begin{pmatrix} 0 \\ 134 \\ 148 \\ 130 \end{pmatrix} = \begin{pmatrix} \hat{\mu} \\ \hat{a}_1 \\ \hat{a}_2 \\ \hat{a}_3 \end{pmatrix}.$$

The breeding value of the sires is estimated by $\bar{y}_{i\cdot}$. The estimable functions $\mu + a_i$ are estimated by 134, 148 and 130, respectively.

To test the null hypothesis $H_0 : a_1 = a_2 = a_3$, we calculate the test statistic (4.34), namely,

$$\boldsymbol{F} = \frac{\left(K^T \hat{\boldsymbol{\beta}} - a\right)^T [K^T (X^T X)^- K]^{-1} \left(K^T \hat{\boldsymbol{\beta}} - a\right)}{\boldsymbol{Y}^T (I_n - X(X^T X)^- X^T) \boldsymbol{Y}} \cdot \frac{n-p}{q},$$

where $H_0: a_1 = a_2 = a_3$ is written in the form $K^T\hat{\beta} = a^*$ with $a^* = 0$ and $K^T = \begin{pmatrix} 0 & 1 & -1 & 0 \\ 0 & 1 & 0 & -1 \end{pmatrix}$ and we have $p = a$, $q = a - 1$.

The realisation F of $\boldsymbol{F}$ is

$$F = \frac{\begin{pmatrix} \widehat{a_1 - a_2} \\ \widehat{a_1 - a_3} \end{pmatrix}^T \begin{pmatrix} \frac{7}{12} & \frac{3}{12} \\ \frac{3}{12} & \frac{7}{12} \end{pmatrix}^{-1} \begin{pmatrix} \widehat{a_1 - a_2} \\ \widehat{a_1 - a_3} \end{pmatrix}}{Y^T(I_{10} - X(X^TX)^- X^T)Y} \cdot \frac{7}{2}.$$

The inverse is $\frac{12}{40}\begin{pmatrix} 7 & -3 \\ -3 & 7 \end{pmatrix}$ and the numerator (because of $\widehat{a_i - a_j} = \bar{y}_{i.} - \bar{y}_{j.}$, $\bar{y}_{1.} - \bar{y}_{2.} = -14, \bar{y}_{1.} - \bar{y}_{.3.} = 4$) finally becomes

$$(-14, 4)\frac{12}{40}\begin{pmatrix} 7 & -3 \\ -3 & 7 \end{pmatrix}\begin{pmatrix} -14 \\ 4 \end{pmatrix} = 546.$$

We further have

$$X\left(X^TX\right)^- X^T = \frac{1}{4}1_{44} \oplus \frac{1}{3}1_{33} \oplus \frac{1}{3}1_{33}.$$

In the denominator it is $Y^T I_{10} Y = \sum y_{ij}^2 = 189068$,

$$Y^T\left[\frac{1}{4}1_{44} \oplus \frac{1}{3}1_{33} \oplus \frac{1}{3}1_{33}\right]Y = \sum_{i=1}^{a} \frac{Y_{i.}^2}{n_i} = 188236,$$

and

$$F = \frac{546}{832} \cdot \frac{7}{2} = 2.297.$$

The quantile of the F-distribution in Table A.5 for $\alpha = 0.05$ with 2 and 7 degrees of freedom is 4.74, and therefore the null hypothesis $H_0: a_1 = a_2 = a_3$ is not rejected. The estimate of $a_1 - a_2$ is, as already mentioned, $\bar{y}_1 - \bar{y}_2 = -14$.

By (5.6) we can calculate $\text{var}\left(\widehat{\boldsymbol{a}_1 - \boldsymbol{a}_2}\right)$. Because $a_1 - a_2$ has the form $\boldsymbol{q}^T\beta$ with $q^T = (0, 1, -1, 0)$, it follows from (5.6):

$$\text{var}\left(\widehat{\boldsymbol{a}_1 - \boldsymbol{a}_2}\right) = (0, 1, -1, 0)\begin{pmatrix} 0 & 0 & 0 & 0 \\ 0 & \frac{1}{4} & 0 & 0 \\ 0 & 0 & \frac{1}{3} & 0 \\ 0 & 0 & 0 & \frac{1}{3} \end{pmatrix}\begin{pmatrix} 0 \\ 1 \\ -1 \\ 0 \end{pmatrix}\sigma^2$$

$$= \left(\frac{1}{4} + \frac{1}{3}\right)\sigma^2 = \frac{7}{12}\sigma^2$$

The function $-8a_1 - 6a_2 + 14a_3$ is a linear contrast and estimable. Following Theorem 4.10 and because of $-8a_1 - 6a_2 + 14a_3 = (0, -8, -6, 14)\beta$, the BLUE of this linear contrast is

$$(0, -8, -6, 14)\hat{\beta} = -8\bar{y}_{1.} - 6\bar{y}_{2.} + 14\bar{y}_{3.} = -140.$$

Taking

$$(0, -8, -6, 14)\left(X^T X\right)^- \begin{pmatrix} 0 \\ 1 \\ -1 \\ 0 \end{pmatrix} = 0$$

into account the two contrasts are orthogonal. From (5.6) we obtain the variance of the estimated contrasts as $\frac{93}{3}\sigma^2 = 31\sigma^2$.

The null hypothesis

$$H_0 : \begin{pmatrix} 0 & 1 & -1 & 0 \\ 0 & -8 & -6 & 14 \end{pmatrix} \beta = 0$$

is tested by the test statistic of Theorem 5.4 with

$$K^T = \begin{pmatrix} 0 & 1 & -1 & 0 \\ 0 & -8 & -6 & 14 \end{pmatrix} \text{ and } G = \left(X^T X\right)^-.$$

It follows

$$K^T G K = \begin{pmatrix} \frac{7}{12} & 0 \\ 0 & \frac{280}{3} \end{pmatrix} \text{ and } \left(K^T G K\right)^{-1} = \begin{pmatrix} \frac{12}{7} & 0 \\ 0 & \frac{3}{280} \end{pmatrix} = \frac{1}{280}\begin{pmatrix} 480 & 0 \\ 0 & 3 \end{pmatrix}.$$

The SS in the numerator of F is then

$$(-14, -140)\frac{1}{280}\begin{pmatrix} 480 & 0 \\ 0 & 3 \end{pmatrix}\begin{pmatrix} -14 \\ -140 \end{pmatrix} = 336 + 210 = 546.$$

The realisation F of $\boldsymbol{F}$ in this case is again

$$F = \frac{546}{832} \cdot \frac{7}{2} = 2.297.$$

However in contrast to the null hypothesis written with non-orthogonal contrasts the sub-hypotheses

$$H_0 : a_1 = a_2, \quad H_0 : -8a_1 - 6a_2 + 14a_3 = 0$$

with the numerators *SS*, 336 and 210, respectively (with one degree of freedom each) can be tested separately so that the test of one hypothesis is independent of the validity of the other. For $H_0: a_1 = a_2$ the test statistic is

$$F = \frac{336}{832} \cdot 7 = 2.827$$

and for the hypothesis $H_0: -8a_1 - 6a_2 + 14a_3 = 0$, the test statistic is

$$F = \frac{210}{832} \cdot 7 = 1.767.$$

The two sub-hypotheses are both accepted.

Solution 2:

This solution is the usual one for practical calculations. Initially the values in Tables 5.3 and 5.4 as well as $Y_{..}^2 = 1876900$ and $\frac{1}{10} Y_{..}^2 = 187690$ are calculated. The $\bar{y}_{i.}$ are estimates of $\mu + a_i$ $(i = 1, 2, 3)$. To test the null hypothesis $H_0: a_1 = a_2 = a_3$, we need an ANOVA table such as Table 5.2 without $E(\boldsymbol{MS})$ (Table 5.5). The values of this table can be obtained from Table 5.4 (e.g. 188 236 − 187 690 = 546). The decomposition of the *SS* between sires in additive components concerning the orthogonal contrasts is shown in Table 5.6.

Table 5.4 Results in the analysis of variance of the material in Table 5.3.

Sire	$Y_{i.}$	$Y_{i.}^2$	$\frac{Y_{i.}^2}{n_i}$	$\sum y_{ij}^2$
B_1	536	287 296	71 824	72 486
B_2	444	197 136	65 712	65 754
B_3	390	152 100	50 700	50 828
Sum	1370		188 236	189 068

Table 5.5 Analysis of variance table for testing the hypothesis $a_1 = a_2 = a_3$ of Example 5.3.

Source of variation	*SS*	*df*	*MS*	*F*
Between sires	546	2	273.00	2.297
Within sires	832	7	118.86	
Total	1378	9		

Table 5.6 Table for testing the hypotheses $a_1 = a_2$ and $-8a_1 - 6a_2 + 14a_3 = 0$.

Source of variation	SS	df	MS	F
$a_1 - a_2$	336	1	336.00	2.827
$-8a_1 - 6a_2 + 14a_3$	210	1	210.00	1.767
Between sires	546	2	273.00	2.297
Within sires	832	7	118.86	
Total	1378	9		

Remarks about Program Packages

With statistical program packages like R, SAS or SPSS, calculations can be done safely and simply. In R we use the command lm().

We demonstrate the analysis of Example 5.3 with IBM SPSS 24 (SPSS for short). Initially the data must be brought into a data matrix. After starting SPSS we use the option 'Data input' and define the variable 'Sire' and 'fat'. By this we define two columns of the data matrix. In the second column, we insert the number of the sire to which the daughter performance belongs, in our case four times 1, three times 2 and three times 3. In the first column, the corresponding 10 daughter performances (fat) are listed. In Figure 5.1 we find sire as factor and the data matrix. We now proceed with

Analyze

Compare Means

– One-Way ANOVA

and define 'sire' as factor and 'fat' as dependent variable. By clicking OK we get the result in Figure 5.2.

5.2.2 Planning the Size of an Experiment

For planning the size of an experiment, precision requirements are needed as in Chapter 3. The following approach is valid for all sections of this chapter.

5.2.2.1 General Description for All Sections of This Chapter

At first we repeat the density function of the non-central F-distribution. It reads

$$f_{n_1,n_2,\lambda}(F) = \sum_{j=0}^{\infty} \frac{e^{-\frac{\lambda}{2}}\Gamma\left(\frac{n_1}{2}+\frac{n_2}{2}+j\right)\lambda^j \cdot n_1^{\frac{n_1}{2}+j} \cdot n_2^{\frac{n_2}{2}}}{j!\cdot 2^j \cdot \Gamma\left(\frac{n_1}{2}+j\right)\cdot\Gamma\left(\frac{n_2}{2}\right)} \cdot \frac{F^{\left(\frac{n_1}{2}+j-1\right)}}{(n_2+n_1F)^{\frac{n_1}{2}+\frac{n_2}{2}+j}} I_{(0,\infty)}.$$

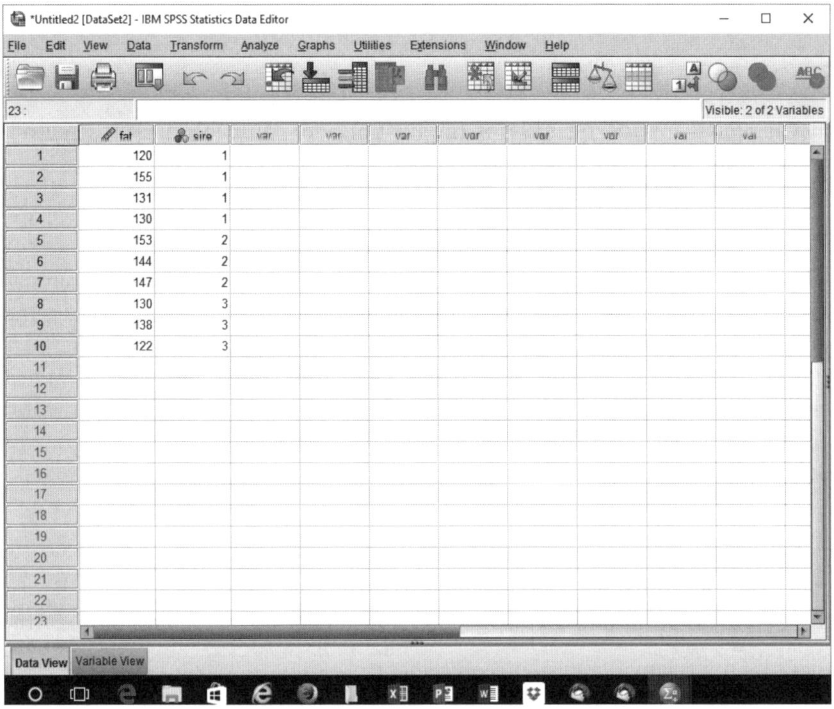

Figure 5.1 Data file of Example 5.3. *Source:* Reproduced with permission of IBM.

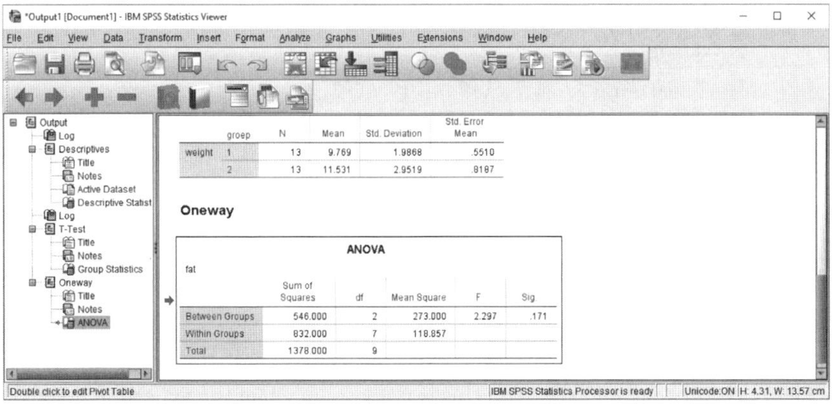

Figure 5.2 ANOVA output for Example 5.3. *Source:* Reproduced with permission of IBM.

Analogously to the relation

$$t(n-1|1-\alpha) = t(n-1,\lambda|\beta)$$

in Chapter 3 for the quantile of the central and non-central F-distribution, respectively, we now use the relation

$$F(f_1,f_2,0|1-\alpha) = F(f_1,f_2,\lambda|\beta), \tag{5.12}$$

where f_1 and f_2 are the degrees of freedom of the numerator and the denominator, respectively, of the test statistic. Further α and β are the two risks and λ is the non-centrality parameter. This equation plays an important role in all other sections of this chapter. Besides f_1, f_2, α and β, the difference δ between the largest and the smallest effect (main effect or in the following sections also interaction effect), to be tested against null, belongs to the precision requirement. We denote the solution λ in (5.12) by

$$\lambda = \lambda(\alpha,\beta,f_1,f_2).$$

Let E_{min}, E_{max} be the minimum and the maximum of q effects $E_1, E_2, \ldots, E_q$ of a fixed factor E or an interaction. Usually we standardise the precision requirement by the relative precision requirement $\tau = \frac{\delta}{\sigma}$.

If $E_{max} - E_{min} \geq \delta$, then for the non-centrality parameter of the F-distribution (for even q) with $\bar{E} = \frac{1}{q}\sum_{i=1}^{q} E_i$ holds

$$\lambda = \sum_{i=1}^{q} (E_i - \bar{E})^2/\sigma^2 \geq \frac{\frac{q}{2}(E_{max} - \bar{E})^2 + \frac{q}{2}(E_{min} - \bar{E})^2}{\sigma^2}$$
$$\geq q(E_{max} - E_{min})^2/(2\sigma^2) \geq q\delta^2/(2\sigma^2).$$

If we omit $\frac{\frac{q}{2}(E_{max} - \bar{E})^2 + \frac{q}{2}(E_{min} - \bar{E})^2}{\sigma^2}$, then it follows

$$\lambda = \sum_{i=1}^{q} (E_i - \bar{E})^2/\sigma^2 \geq q\delta^2/(2\sigma^2). \tag{5.13}$$

The minimal size of the experiment needed depends on λ accordingly to the exact position of all q effects. But this is not known when the experiment starts. We consider two extreme cases, the most favourable (resulting in the smallest minimal size n_{min}) and the least favourable (resulting in the largest minimal size n_{max}). The least favourable case leads to the smallest non-centrality parameter λ_{min} and by this to the so-called maximin size n_{max}. This occurs if the $q-2$

non-extreme effects equal $\frac{E_{\max}+E_{\min}}{2}$. For $\bar{E}=0, \sum_{i=1}^{q}(E_i-\bar{E})^2=qE^2$ this is shown in the following scheme:

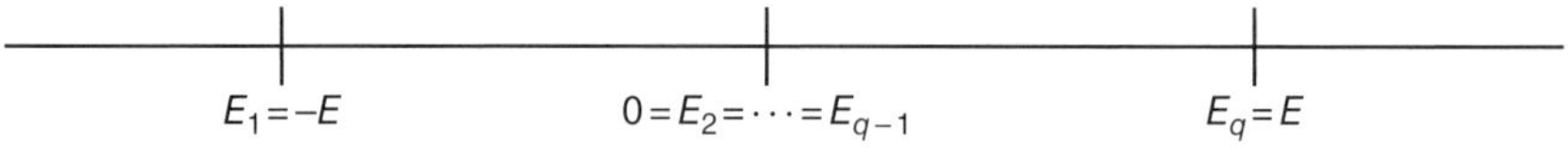

The most favourable case leads to the largest non-centrality parameter $\lambda_{\max}$ and by this to the so-called minimin size $n_{\min}$. For even $q = 2\,m$ this is the case, if m of the E_i equal $E_{\min}$ and the m other E_i equal $E_{\max}$. For odd $q = 2\,m + 1$ again m of the E_i should equal $E_{\min}$ and m other E_i should equal $E_{\max}$, and the remaining effect should be equal to one of the two extremes $E_{\min}$ or $E_{\max}$. For $\bar{E}=0, \sum_{i=1}^{q}(E_i-\bar{E})^2=qE^2$, this is shown in the following scheme for even q:

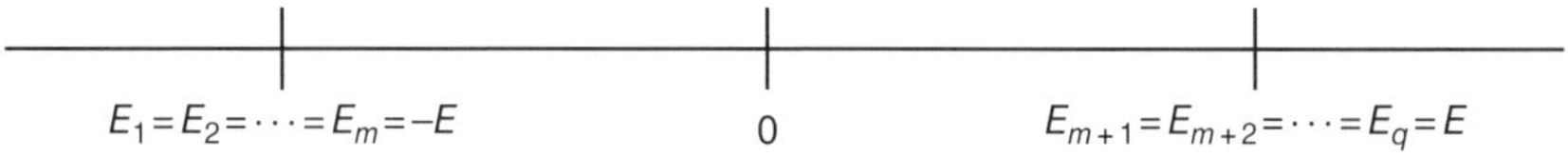

5.2.2.2 The Experimental Size for the One-Way Classification

We now determine the required experimental size for the most favourable as well as for the least favourable case, that is, we are looking for the smallest n (for instance, $n = 2q$) so that for $\lambda_{\max} = \lambda$ and for $\lambda_{\min} = -\lambda$, respectively, (5.13) is fulfilled.

The experimenter must select a size n in the interval $n_{\min} \le n \le n_{\max}$, but if he wants to be on the safe side, he must choose $n = n_{\max}$. The solution of the Equation (5.12) is laborious and done mostly by computer programs. The program OPDOE of R allows the determination of the minimal size for the most favourable and the least favourable case in dependence on α, β, δ and τ and the number a of treatments (levels of factor A) for all cases in this chapter. The corresponding algorithm stems from Lenth (1986) and Rasch et al. (1997). We demonstrate both programs by an example. In any case one can show that the minimal experimental size is smallest if $n_1 = n_2 = \cdots = n_a = n$, which can be reached by planning the experiment. The design function of the R-package OPDOE for the ANOVA is called **size.anova()** and for the one-way ANOVA has the form

```
>size.anova(model="a", a=,alpha=,beta=,delta=,case=).
```

It calculates the minimal size for any of the a levels of factor A for model I in **model** = **"a"** and the number of levels **a=**. Besides the risks, the relative minimal difference $\tau = \delta/\sigma$ **(delta)** and the strategy of optimisation (**case**: "**maximin**" or "**minimin**") must be put in.

We demonstrate all programs by an example.

Example 5.4 Determine n_{min} and n_{max} for $a = 4, \alpha = 0.05, \beta = 0.1$ and $\tau = \delta/\sigma = 2$.
With OPDOE of R we get

```
> size.anova(model="a", a=4, alpha=0.05, beta=0.1,
+delta=2, case="minimin")
n
5
>size.anova(model="a", a=4, alpha=0.05, beta=0.1,
+delta=2, case="maximin")
n
9
```

Now a value of n between 5 and 9 must be used.

5.3 Two-Way Analysis of Variance

The two-way ANOVA is a procedure for experiments to investigate the effects of two factors. Let us investigate a varieties of wheat and b fertilisers in their effect on the yield (per ha). The a varieties as well as the b fertilisers are assumed to be fixed (selected systematically) as always in this chapter with fixed effects. Then factor variety is factor A, and factor fertiliser is factor B. In this and the next chapter, the number of levels of factor X is denoted by the same letter x as factor (a capital letter) but as a small letter. So factor A has a, and factor B has b levels in the experiment. In experiments with two factors, the experimental material is classified in two directions. For this we list the different possibilities:

1) Observations occur in each level of factor A combined with each level of factors B. There are $a \cdot b$ combinations (classes) of factor levels. We say factor A is completely crossed with factor B or we have a complete cross-classification.
   1.1) For each combination (class) of factor levels, there exists one observation ($n_{ij} = 1$ with n_{ij} defined in 1.2).
   1.2) For each combination (class) (i, j) of the level i of factor A with the level j of factor B, we have $n_{ij} \geq 1$ observations, at least one $n_{ij} > 1$. If all $n_{ij} = n$, we have a cross-classification with equal class numbers also called a balanced experimental design.
2) At least one level of factor A occurs together with at least two levels of factors B, and at least one level of factor B occurs together with at least two levels of factors A, but we have no complete cross-classification. Then we say factor A is partially crossed with factor B, or we have an incomplete cross-classification.

3) Each level of factor B occurs together with exactly one level of factor A. This is called a nested classification of factor B within factor A. We also say that factor B is nested within factor A and write $B \prec A$.

The kinds of the two-way classification are as follows:

$n_{ij} = 1$ for all $(i, j) \to$ complete cross-classification with one observation per class.
$n_{ij} \geq 1$ for all $(i, j) \to$ complete cross-classification.
$n_{ij} = n \geq 1$ for all $(i, j) \to$ complete cross-classification with equal class numbers.
$n_{ij_1} \neq 0; n_{ij_2} \neq 0$ for at least one i and at least one $n_{ij} = 0 \to$ incomplete cross-classification.
$n_{i_1j} \neq 0; n_{i_2j} \neq 0$ for at least one j and at least one $n_{ij} = 0 \to$ incomplete cross-classification.
If $n_{kj} \neq 0$, then $n_{ij} = 0$ for $i \neq k$ (at least one $n_{ij} > 1$ and at least two $n_{ij} \neq 0) \to$ nested classification.

5.3.1 Cross-Classification ($A \times B$)

The observations y_{ij} of a complete cross-classification for the ith level A_i of factor A $(i = 1,\dots,a)$ and the jth level B_j of factor B $(j = 1,\dots,b)$ in the case $n_{ij} = 1$ can be written in form of Table 5.7 and in the case of equal class numbers in form of Table 5.8. W.l.o.g. the levels of factor A are the rows and the levels of factor B are the columns of the tables. The special cases of Tables 5.7 and 5.8 are considered at the end of this section. Initially we consider a universal cross-classification where empty classes may occur. Let the random variables $\boldsymbol{y}_{ijk}$ with class (i, j)

Table 5.7 Observations (realisations) y_{ij} of a complete two-way cross-classification with class numbers $n_{ij} = 1$.

		Levels of factor B					
		B_1	B_2	$\dots$	B_j	$\dots$	B_b
Levels of factor A	A_1	y_{11}	y_{12}		y_{1j}		y_{1b}
	A_2	y_{21}	y_{22}		y_{2j}		y_{2b}
	$\vdots$						
	A_i	y_{i1}	y_{i2}		y_{ij}		y_{ib}
	$\vdots$						
	A_a	y_{a1}	y_{a2}		y_{aj}		y_{ab}

Table 5.8 Observations (realisations) y_{ij} of a complete two-way cross-classification with class numbers $n_{ij} = n$.

		Levels of factor B					
		B_1	B_2	…	B_j	…	B_b
Levels of factor A	A_1	y_{111}	y_{121}		y_{1j1}		y_{1b1}
		y_{112}	y_{122}		y_{1j2}		y_{1b2}
		$\vdots$	$\vdots$		$\vdots$		$\vdots$
		y_{11n}	y_{12n}		y_{1jn}		y_{1bn}
	A_2	y_{211}	y_{221}		y_{2j1}		y_{2b1}
		y_{212}	y_{222}		y_{2j2}		y_{2b2}
		$\vdots$	$\vdots$		$\vdots$		$\vdots$
		y_{21n}	y_{22n}		y_{2jn}		y_{2bn}
	$\vdots$						
	A_i	y_{i11}	y_{i21}		y_{ij1}		y_{ib1}
		y_{i12}	y_{i22}		y_{ij2}		y_{ib2}
		$\vdots$	$\vdots$		$\vdots$		$\vdots$
		y_{i1n}	y_{i2n}		y_{ijn}		y_{ibn}
	$\vdots$						
	A_a	y_{a11}	y_{a21}		y_{aj1}		y_{ab1}
		y_{a12}	y_{a22}		y_{aj2}		y_{ab2}
		$\vdots$	$\vdots$		$\vdots$		$\vdots$
		y_{a1n}	y_{a2n}		y_{ajn}		y_{abn}

be a random sample of a population associated with this class. Mean and variance of the population of such a class are called true mean and variance, respectively. The true mean of the class (i, j) is denoted by η_{ij}. Again we consider the case that the levels of factors A and B are chosen systematically (model I).

We call

$$\mu = \bar{\eta}_{..} = \frac{\sum_{i=1}^{a}\sum_{j=1}^{b}\eta_{ij}}{ab}$$

the total mean of the experiment.

Definition 5.2 The difference $a_i = \bar{\eta}_{i.} - \mu$ is called the main effect of the ith level of factor A, and the difference $b_j = \bar{\eta}_{.j} - \mu$ is called the main effect of the jth level of factor B. The difference $a_{i|j} = \eta_{ij} - \bar{\eta}_{.j}$ is called the effect of the ith level of factor A under the condition that factor B occurs in the jth level. Analogously $b_{j|i} = \eta_{ij} - \bar{\eta}_{i.}$ is called the effect of the jth level of factor B under the condition that factor A occurs in the ith level.

The distinction between main effect and 'conditional effect' is important, if the effects of the levels of one factor depend on the number of the level of the other factor. In ANOVA, we then say that an interaction between the two factors exists. We define the effects of these interactions (and use them in place of the conditional results).

Definition 5.3 The interaction $(a, b)_{ij}$ between the ith level of factor A and the jth level of factor B in a two-way cross-classification is the difference between the conditional effect of the level A_i of factor A for a given level B_j of factors B and the main effect of the level A_i of A or, which means the same, the difference between the conditional effect of the level B_j of B for a given level A_i of A and the main effect of the level B_j of B or as formula

$$(a,b)_{ij} = a_{i|j} - a_i = b_{j|i} - b_j = \eta_{ij} - \bar{\eta}_{i.} - \bar{\eta}_{.j} + \mu. \tag{5.14}$$

Under the assumption above the random variable $\boldsymbol{y}_{ij}$ of the cross-classification varies randomly around the class mean in the form

$$\boldsymbol{y}_{ijk} = \eta_{ij} + \boldsymbol{e}_{ijk}$$

We assume that the so-called error variables $\boldsymbol{e}_{ijk}$ are independent of each other $N(0, \sigma^2)$-distributed and write

$$\boldsymbol{y}_{ijk} = \mu + a_i + b_i + (a,b)_{ij} + \boldsymbol{e}_{ijk}, \quad (i = 1,\ldots,a; j = 1,\ldots,b;\ k = 1,\ldots,n_{ij}) \tag{5.15}$$

with $(a,b)_{ij} = 0$ if $n_{ij} = 0$. If in (5.14) all $(a,b)_{ij} = 0$, we call

$$\boldsymbol{y}_{ijk} = \mu + a_i + b_j + \boldsymbol{e}_{ijk}, \quad (i = 1,\ldots,a; j = 1,\ldots,b;\ k = 1,\ldots,n_{ij}) \tag{5.16}$$

a model without interactions.

The models (5.15) and (5.16) are special cases of (5.1). To show this we write

$$\boldsymbol{Y} = \left(\boldsymbol{y}_{111},\ldots,\boldsymbol{y}_{11n_{11}},\ldots,\boldsymbol{y}_{1b1},\ldots,\boldsymbol{y}_{1bn_{1b}},\ldots,\boldsymbol{y}_{abn_{ab}}\right)^T,$$

$$\beta = \left(\mu, a_1,\ldots,a_a,\ b_1,\ldots,b_b,\ (a,b)_{11},\ldots,\ (a,b)_{1b},\ (a,b)_{21},\ldots,\ (a,b)_{2b},\ldots,\ (a,b)_{ab}\right)^T$$

for (5.15) and

$$\beta = (\mu,\ a_1,\ldots,a_a,\ b_1,\ldots,b_b)^T$$

for (5.16). In (5.15) let r of the n_{ij} be equal to 0 and $ab-r=t$ of the n_{ij} be larger than 0.

If (5.15) is written in matrix notation, then β is a $[(t+a+b+1)\times 1]$ vector $[(a+1)(b+1)-r=t+a+b+1]$ and X a $\{N\times[t+a+b+1]\}$ matrix of zeros and ones while $\boldsymbol{e}$ is a $(N\times 1)$ vector of random errors and $N(0, \sigma^2 I_N)$-distributed. Then $\boldsymbol{Y}$ is $N(X\beta, \sigma^2 I_N)$-distributed.

5.3.1.1 Parameter Estimation

Before we generally discuss the estimation of the model parameters, we consider an example.

We demonstrate the choice of the matrix X in (5.1) by

Example 5.5 Let $a=b=n=2$, so that $r=0,\ t=ab=4$ and

$$\boldsymbol{Y}=(\boldsymbol{y}_{111},\boldsymbol{y}_{112},\boldsymbol{y}_{121},\boldsymbol{y}_{122},\boldsymbol{y}_{211},\boldsymbol{y}_{212},\boldsymbol{y}_{221},\boldsymbol{y}_{222})^T,$$
$$\boldsymbol{e}=(\boldsymbol{e}_{111},\boldsymbol{e}_{112},\boldsymbol{e}_{121},\boldsymbol{e}_{122},\boldsymbol{e}_{211},\boldsymbol{e}_{212},\boldsymbol{e}_{221},\boldsymbol{e}_{222})^T,$$
$$\beta=(\mu,a_1,a_2,b_1,b_2,\ (a,b)_{11},\ (a,b)_{12},\ (a,b)_{21},\ (a,b)_{22})^T.$$

Then

$$X=\begin{pmatrix}1&1&0&1&0&1&0&0&0\\1&1&0&1&0&1&0&0&0\\1&1&0&0&1&0&1&0&0\\1&1&0&0&1&0&1&0&0\\1&0&1&1&0&0&0&1&0\\1&0&1&1&0&0&0&1&0\\1&0&1&0&1&0&0&0&1\\1&0&1&0&1&0&0&0&1\end{pmatrix}$$

is a matrix of rank 4. Further $N=abn$ and

$$X^TX=\begin{pmatrix}N&bn&bn&an&an&n&n&n&n\\bn&bn&0&n&n&n&n&0&0\\bn&0&bn&n&n&0&0&n&n\\an&n&n&an&0&n&0&n&0\\an&n&n&0&an&0&n&0&n\\n&n&0&n&0&n&0&0&0\\n&n&0&0&n&0&n&0&0\\n&0&n&n&0&0&0&n&0\\n&0&n&0&n&0&0&0&n\end{pmatrix}=\begin{pmatrix}8&4&4&4&4&2&2&2&2\\4&4&0&2&2&2&2&0&0\\4&0&4&2&2&0&0&2&2\\4&2&2&4&0&2&0&2&0\\4&2&2&0&4&0&2&0&2\\2&2&0&2&0&2&0&0&0\\2&2&0&0&2&0&2&0&0\\2&0&2&2&0&0&0&2&0\\2&0&2&0&2&0&0&0&2\end{pmatrix}.$$

The matrix B in (5.4) following Definitions 5.2 and 5.3 has the form

$$B=\begin{pmatrix}0&N&N&0&0&0&0&0&0\\0&0&0&N&N&0&0&0&0\\0&0&0&0&0&N&N&0&0\\0&0&0&0&0&0&0&N&N\\0&0&0&0&0&N&0&N&0\end{pmatrix}$$

under the side conditions

$$\sum_{i=1}^{a}a_i=\sum_{j=1}^{b}b_i=0,\ \sum_{i=1}^{a}(a,b)_{ij}=0 \text{ for all } j,\ \sum_{j=1}^{b}(a,b)_{ij}=0 \text{ for all } i. \quad (5.17)$$

This leads to

$$B^TB=\begin{pmatrix}0&0&0&0&0&0&0&0&0\\0&N^2&N^2&0&0&0&0&0&0\\0&N^2&N^2&0&0&0&0&0&0\\0&0&0&N^2&N^2&0&0&0&0\\0&0&0&N^2&N^2&0&0&0&0\\0&0&0&0&0&2N^2&N^2&N^2&0\\0&0&0&0&0&N^2&N^2&0&0\\0&0&0&0&0&N^2&0&2N^2&N^2\\0&0&0&0&0&0&0&N^2&N^2\end{pmatrix}$$

with $\text{rk}(B^TB)....(B^TB) = 5$ and further

$$X^TX+B^TB=\begin{pmatrix}N&bn&bn&an&an&n&n&n&n\\bn&N^2+bn&N^2&n&n&n&n&0&0\\bn&N^2&N^2+bn&n&n&0&0&n&n\\an&n&n&N^2+an&N^2&n&0&n&0\\an&n&n&N^2&N^2+an&0&n&0&n\\n&n&0&n&0&2N^2+n&N^2&N^2&0\\n&n&0&0&n&N^2&N^2+n&0&0\\n&0&n&n&0&N^2&0&2N^2+n&N^2\\n&0&n&0&n&0&0&N^2&N^2+n\end{pmatrix}$$

$$= \begin{pmatrix} 8 & 4 & 4 & 4 & 4 & 2 & 2 & 2 & 2 \\ 4 & 68 & 64 & 2 & 2 & 2 & 2 & 0 & 0 \\ 4 & 64 & 68 & 2 & 2 & 0 & 0 & 2 & 2 \\ 4 & 2 & 2 & 68 & 64 & 2 & 0 & 2 & 0 \\ 4 & 2 & 2 & 64 & 68 & 0 & 2 & 0 & 2 \\ 2 & 2 & 0 & 2 & 0 & 130 & 64 & 64 & 0 \\ 2 & 2 & 0 & 0 & 2 & 64 & 66 & 0 & 0 \\ 2 & 0 & 2 & 2 & 0 & 64 & 0 & 130 & 64 \\ 2 & 0 & 2 & 0 & 2 & 0 & 0 & 64 & 66 \end{pmatrix}.$$

Estimators $\hat{\boldsymbol{\beta}}$ for β we obtain under these side conditions as in Section 5.2.1 by calculating $(X^TX + B^TB)^{-1}$ and $\hat{\boldsymbol{\beta}} = (X^TX + B^TB)^{-1} X^T \boldsymbol{Y}$.

The following statements are independent of the special choice under the side conditions.

Theorem 5.5 The matrix X^TX of the model equation (5.15) written in the form

$$\boldsymbol{Y} = X\beta + \boldsymbol{e}$$

with the $[N \times (t + a + b + 1)]$ matrix X has rank $t > 0$, and a solution of the normal equations $X^TX\,\hat{\boldsymbol{\beta}}^* = X^T\boldsymbol{Y}$ is given by

$$\left(\widehat{\boldsymbol{ab}_{ij}}\right) = \bar{\boldsymbol{y}}_{ij.} \text{ for all } i, j \text{ with } n_{ij} > 0$$
$$\hat{\boldsymbol{a}}_i = 0 \quad \text{for all } i, \quad \hat{\boldsymbol{b}}_j = 0 \quad \text{for all } j, \quad \hat{\boldsymbol{\mu}} = 0. \tag{5.18}$$

Proof: We write $X = (x_1, x_2, \ldots, x_{t+a+b+1})$ with the column vectors x_l of X. We easily see that $\sum_{l=2}^{a+1} x_l = \sum_{l=a+2}^{a+b+1} x_l = x_1$. Adding to x_l $(l = a + b + 2, \ldots, a + b + t + 1)$ corresponding to $(a, b)_{ij}$ those corresponding to all $(a, b)_{ij}$ for a given i, then we obtain x_{i+1}. Adding to x_l all those corresponding to $(a, b)_{ij}$ for a given j, we obtain x_{a+1+j}. That means that from the $t + a + b + 1$ rows of X^TX, at least t are linearly independent; because the last t rows and columns of X^TX are a diagonal matrix with t from 0 different elements, we have $\text{rk}(X^TX) = t$. We put $a + b + 1$ values of $\hat{\beta}$ equal to 0, namely, $\mu, a_1, \ldots, a_a, b_1, \ldots, b_b$. The last t equations of the system of normal equations are then the solutions (5.18). When all $(a,b)_{ij} = 0$, that is, when model equation (5.16) has to be used, we obtain

Theorem 5.6 When all $(a,b)_{ij} = 0$, then the matrix X^TX of the model equation (5.16) has, written in the form $\boldsymbol{Y} = X\beta + \boldsymbol{e}$, with the $[N \times (a + b + 1)]$ matrix X the rank $\text{rk}(X^TX) = \text{rk}(X) \le a + b - 1$.

Proof: X^TX is a symmetrical matrix of order $a+b+1$. The sum of the second up to the $(a+1)$th row equals the first row; the $(a+2)$th up to the last row also add up to the first one, so that the rank is at most $a+b+1$.

Before solutions of the normal equations for model equation (5.16) are given, we list some estimable functions and their BLUEs for model (5.15).

5.3.1.1.1 Models with Interactions

We consider the model equation (5.15). Because $E(Y)$ is estimable, then

$$\eta_{ij} = \mu + a_i + b_j + (a,b)_{ij} \quad \text{for all } i,j \quad \text{with } n_{ij} > 0$$

is estimable. The BLUE of η_{ij} is

$$\hat{\boldsymbol{\eta}}_{ij} = \hat{\boldsymbol{\mu}} + \hat{\boldsymbol{a}}_i + \hat{\boldsymbol{b}}_j + \left(\widehat{\boldsymbol{a},\boldsymbol{b}}\right)_{ij}, \tag{5.19}$$

because $\hat{\boldsymbol{\mu}} + \hat{\boldsymbol{a}}_i + \hat{\boldsymbol{b}}_j = 0$ and $\left(\widehat{\boldsymbol{a},\boldsymbol{b}}\right)_{ij} = \bar{\boldsymbol{y}}_{ij\cdot}$. From (5.6) it follows

$$\text{cov}\left(\hat{\boldsymbol{\eta}}_{ij}, \hat{\boldsymbol{\eta}}_{kl}\right) = \frac{\sigma^2}{n_{ij}}\delta_{ik}\delta_{jl}. \tag{5.20}$$

It is now easy to show that differences between a_i and b_j are not estimable. All estimable functions of the components of (5.15) without further side conditions contain interaction effects $(a, b)_{ij}$. It follows the theorem below.

Theorem 5.7 (Searle, 1971) The function

$$L_A = a_i - a_k + \sum_{j=1}^{b} c_{ij}\left(b_j + (a,b)_{ij}\right) - \sum_{j=1}^{b} c_{kj}\left(b_j + (a,b)_{kj}\right) \quad \text{for} \quad i \neq k \tag{5.21}$$

or analogously

$$L_B = b_i - b_k + \sum_{j=1}^{a} d_{ji}\left(a_j + (a,b)_{ji}\right) - \sum_{j=1}^{a} d_{jk}\left(a_j + (a,b)_{jk}\right) \quad \text{for} \quad i \neq k$$

is estimable if $c_{rs} = 0$ for $n_{rs} = 0$ and $d_{rs} = 0$ for $n_{rs} = 0$, respectively, as well as

$$\sum_{j=1}^{b} c_{ij} = \sum_{j=1}^{b} c_{kj} = 1 \quad \left(\text{and } \sum_{j=1}^{a} d_{ji} = \sum_{j=1}^{a} d_{jk} = 1, \text{respectively}\right).$$

The BLUE of an estimable function of the form (5.21) is given by

$$\hat{\boldsymbol{L}}_A = \sum_{j=1}^{b} c_{ij}\bar{\boldsymbol{y}}_{ij.} - \sum_{j=1}^{b} c_{kj}\bar{\boldsymbol{y}}_{kj}, \tag{5.22}$$

and it is

$$\operatorname{var}(\hat{\boldsymbol{L}}_A) = \sigma^2 \sum_{j=1}^{b} \left(\frac{c_{ij}^2}{n_{ij}} + \frac{c_{kj}^2}{n_{kj}} \right). \tag{5.23}$$

Proof: An estimable function is always a linear combination of η_{ij}. Therefore $c_{rs} = 0$, if $n_{rs} = 0$. Now

$$\sum_{j=1}^{b} c_{ij} \eta_{ij.} - \sum_{j=1}^{b} c_{kj} \eta_{kj.}$$

as a linear function of the η_{ij} is estimable. Because of

$$\sum_{j=1}^{b} c_{ij} \eta_{ij} = \sum_{j=1}^{b} c_{ij} \left(\mu + a_i + b_j + (a,b)_{ij} \right) = \mu + a_i + \sum_{j=1}^{b} c_{ij} \left(b_j + (a,b)_{ij} \right)$$

and the analogous relation for the corresponding term in c_{kj}, the estimability of L_A and the validity of (5.22) and (5.23) follow.

If we use model equation (5.16) without interactions and side conditions, then $\eta_{ij} = E\left(y_{ijk}\right) = \mu + a_i + b_j$ is an estimable function; the differences $a_i - a_j$ and $b_i - b_j$ are estimable.

We consider the following example.

Example 5.6 From three test periods of testing pig fattening for male and female offspring of boars, the number of fattening days an animal needed to grow from 40 to 110 kg has been recorded. The values are given in Table 5.9.

Table 5.9 Results of testing pig fattening: fattening days (from 40 to 110 kg for three test periods and two sexes) for the offspring of several boars.

		Sex	
		Male	Female
Test periods	1	91	
		84	99
		86	
	2	94	97
		92	89
		90	
		96	
	3	82	
		86	–

We choose model equation (5.15) as a basis and write it in the form

$$\begin{pmatrix} 91\\84\\86\\99\\94\\92\\90\\96\\97\\89\\82\\86 \end{pmatrix} = \begin{pmatrix} 1&1&0&0&1&0&1&0&0&0&0\\ 1&1&0&0&1&0&1&0&0&0&0\\ 1&1&0&0&1&0&1&0&0&0&0\\ 1&1&0&0&0&1&0&1&0&0&0\\ 1&0&1&0&1&0&0&0&1&0&0\\ 1&0&1&0&1&0&0&0&1&0&0\\ 1&0&1&0&1&0&0&0&1&0&0\\ 1&0&1&0&1&0&0&0&1&0&0\\ 1&0&1&0&0&1&0&0&0&1&0\\ 1&0&1&0&0&1&0&0&0&1&0\\ 1&0&0&1&1&0&0&0&0&0&1\\ 1&0&0&1&1&0&0&0&0&0&1 \end{pmatrix} \cdot \begin{pmatrix} \mu\\a_1\\a_2\\a_3\\b_1\\b_2\\(a,b)_{11}\\(a,b)_{12}\\(a,b)_{21}\\(a,b)_{22}\\(a,b)_{31} \end{pmatrix} + \begin{pmatrix} e_{111}\\e_{112}\\e_{113}\\e_{121}\\e_{211}\\e_{212}\\e_{213}\\e_{214}\\e_{221}\\e_{222}\\e_{311}\\e_{312} \end{pmatrix}.$$

We have $r = 1$, $t = 3\cdot 2 - 1 = 5$ and $N = 12$; X is a (12×11) matrix of rank 5. We obtain

$$X^T X = \begin{pmatrix} 12&4&6&2&9&3&3&1&4&2&2\\ 4&4&0&0&3&1&3&1&0&0&0\\ 6&0&6&0&4&2&0&0&4&2&0\\ 2&0&0&2&2&0&0&0&0&0&2\\ 9&3&4&2&9&0&3&0&4&0&2\\ 3&1&2&0&0&3&0&1&0&2&0\\ 3&3&0&0&3&0&3&0&0&0&0\\ 1&1&0&0&0&1&0&1&0&0&0\\ 4&0&4&0&4&0&0&0&4&0&0\\ 2&0&2&0&0&2&0&0&0&2&0\\ 2&0&0&2&2&0&0&0&0&0&2 \end{pmatrix}$$

and from (5.18)

$$\left(\widehat{a,b}\right)_{11} = \bar{y}_{11.} = 87,\ \left(\widehat{a,b}\right)_{12} = \bar{y}_{12.} = 99,$$

$$\left(\widehat{a,b}\right)_{21} = \bar{y}_{21.} = 93,\ \left(\widehat{a,b}\right)_{22} = \bar{y}_{22.} = 93,$$

$$\left(\widehat{a,b}\right)_{31} = \bar{y}_{31.} = 84.$$

The function $L_1 = b_1 - b_2 + (a,b)_{11} - (a,b)_{12}$ is estimable, because the condition of Theorem 5.7 is fulfilled. The function $L_2 = b_1 - b_2 + (a,b)_{21} - (a,b)_{22}$ is also estimable. We get

$$\hat{L}_1 = \bar{y}_{11.} - \bar{y}_{12.} = -6, \quad \hat{L}_2 = \bar{y}_{21.} - \bar{y}_{22.} = 6.$$

Further $\operatorname{var}(\hat{\boldsymbol{L}}_1) = \frac{4}{3}\sigma^2$ and $\operatorname{var}(\hat{\boldsymbol{L}}_2) = \frac{3}{4}\sigma^2$.

5.3.1.1.2 Models Without Interactions

Model equation (5.16) is simpler than (5.15), but there exists nevertheless no simple solution of the normal equations as for (5.15). The matrix X^TX is

$$X^TX = \begin{pmatrix} N & n_{1.} & \cdots & n_{a.} & n_{.1} & \cdots & n_{.b} \\ n_{1.} & n_{1.} & & & n_{11} & \cdots & n_{1b} \\ \vdots & 0 & \ddots & 0 & \vdots & & \vdots \\ n_{a.} & & & n_{a.} & n_{a1} & \cdots & n_{ab} \\ n_{.1} & n_{11} & \cdots & n_{a1} & n_{.1} & & \\ \vdots & \vdots & & \vdots & 0 & \ddots & 0 \\ n_{.b} & n_{1b} & \cdots & n_{ab} & & & n_{.b} \end{pmatrix}.$$

To obtain a simpler solution, we must rename for $a < b$ the factors w.l.o.g. so that $a \ge b$. Because X^TX following Theorem 5.6 has a rank of at most $a + b - 1$, we can choose two values of β^* arbitrarily. We put $\boldsymbol{\mu}^* = \boldsymbol{\beta}_b^* = 0$ and obtain the reduced system of normal equations

$$\begin{pmatrix} n_{1.} & & 0 & n_{11} & \cdots & n_{1,b-1} \\ & \ddots & & \vdots & & \vdots \\ 0 & & n_{a.} & n_{a1} & \cdots & n_{a,b-1} \\ n_{11} & \cdots & n_{a1} & n_{.1} & & 0 \\ \vdots & & \vdots & 0 & \ddots & \\ n_{1,b-1} & \cdots & n_{a,b-1} & & & n_{.b-1} \end{pmatrix} \begin{pmatrix} \boldsymbol{a}_1^* \\ \vdots \\ \boldsymbol{a}_a^* \\ \boldsymbol{b}_1^* \\ \vdots \\ \boldsymbol{b}_{b-1}^* \end{pmatrix} = \begin{pmatrix} \boldsymbol{Y}_{1.} \\ \vdots \\ \boldsymbol{Y}_{a.} \\ \boldsymbol{Y}_{.1} \\ \vdots \\ \boldsymbol{Y}_{.b-1} \end{pmatrix}.$$

We put

$$D_a = \begin{pmatrix} n_{1.} & & 0 \\ & \ddots & \\ 0 & & n_{a.} \end{pmatrix}, \quad V = \begin{pmatrix} n_{11} & \cdots & n_{1,b-1} \\ \vdots & & \vdots \\ n_{a1} & \cdots & n_{a,b-1} \end{pmatrix}, \quad D_b = \begin{pmatrix} n_{.1} & & 0 \\ & \ddots & \\ 0 & & n_{.b-1} \end{pmatrix}.$$

Now the matrix of coefficients of the reduced system of normal equations can be written as

$$\begin{pmatrix} D_a & V \\ V^T & D_b \end{pmatrix} = R.$$

We put

$$W = -V^T D_a^{-1} V + D_b \tag{5.24}$$

and assume that R has rank $a+b-1$. Then W^{-1} exists. Further, we obtain

$$R^{-1} = \begin{pmatrix} D_a^{-1} + D_a^{-1} VW^{-1}V' D_a^{-1} & -D_a^{-1}VW^{-1} \\ -W^{-1}V^T D_a^{-1} & W^{-1} \end{pmatrix},$$

so that with

$$\boldsymbol{\nu} = \mathcal{Y}_b - V^T D_a^{-1} \mathcal{Y}_a, \quad \boldsymbol{\nu} = (\nu_1,\dots,\nu_{b-1})^T,$$

$$\nu_j = \boldsymbol{Y}_{.j.} - \sum_{i=1}^{a} n_{ij}\bar{y}_{i..}, \quad \bar{\mathcal{Y}}_a = (\bar{y}_{1..},\dots,\bar{y}_{a..})^T,$$

$$\mathcal{Y}_a = (\boldsymbol{Y}_{1..},\dots,\boldsymbol{Y}_{a..})^T, \quad \mathcal{Y}_b = (\boldsymbol{Y}_{.1.},\dots,\boldsymbol{Y}_{.b-1.})^T,$$

the vector

$${}_1\boldsymbol{b}^* = \begin{pmatrix} 0 \\ \bar{\mathcal{Y}}_a - D_a^{-1} VW^{-1}\boldsymbol{\nu} \\ W^{-1}\boldsymbol{\nu} \\ 0 \end{pmatrix} \tag{5.25}$$

is the solution of the system of normal equations and

$$(X^TX)^- = \begin{pmatrix} 0 & 0_a^T & 0_{b-1}^T & 0 \\ 0_a & D_a^{-1} + D_a^{-1}VW^{-1}V^TD_a^{-1} & -D_a^{-1}VW^{-1} & 0_a \\ 0_{b-1} & -W^{-1}V^TD_a^{-1} & W^{-1} & 0_{b-1} \\ 0 & 0_a^T & 0_{b-1}^T & 0 \end{pmatrix} \tag{5.26}$$

is the corresponding generalised inverse.

Definition 5.4 A (incomplete) cross-classification is called connected if $W = \left((a,b)_{ij}\right)$ $(i,j = 1,\dots,b-1)$ in (5.24) is non-singular. If $|W| = 0$, then the cross-classification is disconnected (see also a corresponding definition in Chapter 12).

Example 5.7 We consider a two-way cross-classification with $a = 5$, $b = 4$ and the subclass numbers:

		Levels of B			
		B_1	B_2	B_3	B_4
	A_1	n	n	0	0
	A_2	n	n	0	0
Levels of A	A_3	n	n	0	0
	A_4	0	0	m	m
	A_5	0	0	m	m

Here is $n_{.1} = n_{.2} = 3n,\ n_{.3} = n_{.4} = 2m,\ n_{1.} = n_{2.} = n_{3.} = 2n,\ n_{4.} = n_{5.} = 2m$, and the matrix W is given by

$$W = \begin{pmatrix} \frac{3}{2}n & -\frac{3}{2}n & 0 \\ -\frac{3}{2}n & \frac{3}{2}n & 0 \\ 0 & 0 & m \end{pmatrix}.$$

The first row is (−1) times the second row so that W is singular. The term 'disconnected cross-classification' can be illustrated by this example as follows. From the scheme of the subclass numbers, we see that the levels A_1, A_2, A_3, B_1, B_2 and A_4, A_5, B_3, B_4 form two separate cross-classifications. If we add n further observations in $(A_2\ B_3)$, we obtain $n_{2.} = 3n, n_{.3} = 2m + n$, and W becomes

$$W = \begin{pmatrix} \frac{5}{3}n & -\frac{4}{3}n & -\frac{n}{3} \\ -\frac{4}{3}n & \frac{5}{3n} & -\frac{n}{3} \\ -\frac{n}{3} & -\frac{n}{3} & m + \frac{2}{3}n \end{pmatrix}$$

with $|W| \neq 0$; now the cross-classification is connected.

In SPSS we easily see in a cross-classification of A with a levels and B with b levels directly that there is a disconnected scheme if $df(A) < a - 1$ and/or $df(B) < b - 1$ in the ANOVA table.

For special cases the two-way cross-classification as complete block designs or balanced and partially balanced incomplete block designs is discussed in Chapter 12 where only connected designs are used.

5.3.1.2 Testing Hypotheses

In this section testable hypotheses and tests of such hypotheses are considered. The models (5.15) and (5.16) are handled separately.

5.3.1.2.1 Models without Interactions

We start with model (5.16) and assume a connected cross-classification (W in Definition 5.4 non-singular), that is, $\text{rk}(X^TX) = a + b - 1$. For a testable hypothesis $K^Tb = 0$, we can use the test statistic $\boldsymbol{F}$ in (4.34), that is, the test statistic reads

$$\boldsymbol{F} = \frac{\hat{\boldsymbol{\beta}}^T K[K^T(X^TX)^- K]^{-1} K^T \hat{\boldsymbol{\beta}}}{\boldsymbol{Y}^T[I_N - X(X^TX)^- X^T]\boldsymbol{Y}} \cdot \frac{n-p}{q} \tag{5.27}$$

and is $F(n-p, q, \lambda)$-distributed with non-centrality parameter

$$\lambda = \frac{1}{\sigma^2} b^T K\left[K^T\left(X^TX\right)^- K\right]^{-1} K^T b, \quad p = \text{rk}\left(X^TX\right),\ q = \text{rk}(K).$$

$K^Tb = 0$ leads to $\lambda = 0$. Because $K^Tb = 0$ is assumed to be testable, all rows of K^Tb must be estimable functions. To show how (5.27) is used, we consider an example.

Example 5.8 The hypothesis $H_0 : b_1 = \cdots = b_b$ is to be tested. Initially we investigate whether H_0 is testable. We write H_0 in the form $H_0 : b_j - b_b = 0 (j = 1, \ldots, b-1)$ with

$$K^T = \begin{pmatrix} \overbrace{0\ 0\ \ldots\ 0}^{a} & \overbrace{1 & & & }^{b-1} & -1 \\ & & \ddots & 0 & -1 \\ \vdots\ \vdots\ \ \ \vdots & & 0 & \ddots & \vdots \\ 0\ 0\ \ldots\ 0 & & & 1 & -1 \end{pmatrix} = \left(0_{b-1,a+1}, I_{b-1}, -1_{b-1}\right),$$

so that $K^T(X^TX)^-$ with $(X^TX)^-$ from (5.26) becomes

$$K^T\left(X^TX\right)^- = \left(0_{b-1}, -W^{-1}V^TD_a^{-1}, W^{-1}, 0_{b-1}\right)$$

and $K^T(X^TX)^- K = W^{-1}$. Further with $\hat{\boldsymbol{\beta}}$ from (5.25), we have

$$K^T\hat{\boldsymbol{\beta}} = W^{-1}\boldsymbol{v},$$

and the numerator of $\boldsymbol{F}$ becomes

$$\boldsymbol{v}^T W^{-1}\left(W^{-1}\right)^{-1} W^{-1}\boldsymbol{v} = \boldsymbol{v}^T W^{-1}\boldsymbol{v}.$$

To test the hypothesis $H_0 : a_1 = \cdots = a_a$, we have to use another generalised inverse as in (5.26). We choose ${}_2\hat{\mu} = 0$ and ${}_2\hat{a}_i = 0$ and obtain a reduced system of normal equations; in its matrix the first two rows and columns contain zeros. Let

$$\widetilde{D}_a = \begin{pmatrix} n_{2.} & & 0 \\ & \ddots & \\ 0 & & n_{a.} \end{pmatrix}, \quad \widetilde{D}_b = \begin{pmatrix} n_{.1} & & 0 \\ & \ddots & \\ 0 & & n_{.b} \end{pmatrix}, \quad \widetilde{V} = \begin{pmatrix} n_{21} & \cdots & n_{2b} \\ \vdots & & \vdots \\ n_{a1} & \cdots & n_{ab} \end{pmatrix}$$

and $\widetilde{W} = \widetilde{D}_a - \widetilde{V}\widetilde{D}_b^{-1}\widetilde{V}^T$. The matrix of coefficients

$$\widetilde{R} = \begin{pmatrix} \widetilde{D}_a & \widetilde{V} \\ \widetilde{V}^T & \widetilde{D}_b \end{pmatrix}$$

must have (full) rank $a+b-1$ so that $\widetilde{W}^{-1}$ exist. Then

$$\widetilde{R}^{-1} = \begin{pmatrix} \widetilde{W}^{-1} & -\widetilde{W}^{-1}\widetilde{V}\widetilde{D}_b^{-1} \\ -\widetilde{D}_b\widetilde{V}^T\widetilde{W}^{-1} & \widetilde{D}_b^{-1} + \widetilde{D}_b^{-1}\widetilde{V}^T\widetilde{W}^{-1}\widetilde{V}\widetilde{D}_b^{-1} \end{pmatrix}$$

follows. Putting $\widetilde{\boldsymbol{v}} = (\widetilde{\boldsymbol{v}}_2, \ldots, \widetilde{\boldsymbol{v}}_{\boldsymbol{a}})^T$ with $\widetilde{v}_i = \boldsymbol{Y}_i - \sum_{j=1}^{b} n_{ij}\bar{\boldsymbol{y}}_{.j.}$, we get

$$_2\hat{\boldsymbol{\beta}} = \begin{pmatrix} 0 \\ 0 \\ \bar{\boldsymbol{y}}_b - \widetilde{D}_b^{-1}\widetilde{V}^T\widetilde{W}^{-1}\widetilde{\boldsymbol{v}} \end{pmatrix} \tag{5.25a}$$

analogously to (5.25) with $\bar{\boldsymbol{y}}_b = (\bar{\boldsymbol{y}}_{.1}, \ldots, \bar{\boldsymbol{y}}_{.b})$. In this case

$$(X^TX)^- = \begin{pmatrix} 0_{22} & 0_{2a} & 0_{2b} \\ 0_{a2} & \widetilde{W}^{-1} & \widetilde{W}^{-1}\widetilde{V}\widetilde{D}_b^{-1} \\ 0_{b2} & -\widetilde{D}_b^{-1}\widetilde{V}\,\widetilde{W}^{-1} & \widetilde{D}_b^{-1} + \widetilde{D}_b^{-1}\widetilde{V}\,\widetilde{W}^{-1}\widetilde{D}_b^{-1} \end{pmatrix}. \tag{5.26a}$$

If there is a generalised inverse, then

$$_1\hat{\boldsymbol{\beta}}^T X^T \boldsymbol{Y} = \left(\boldsymbol{Y} - \widetilde{D}_b^{-1} V W^{-1}\boldsymbol{v}\right)^T \mathcal{Y}_a + \left(W^{-1}\boldsymbol{v}\right)^T \mathcal{Y}_b = \sum_{i=1}^{a} \frac{Y_{i..}^2}{n_{i.}} + \boldsymbol{v}^T W^{-1}\boldsymbol{v}$$

and

$$_2\hat{\boldsymbol{\beta}}^T X^T \boldsymbol{Y} = \sum_{j=1}^{b} \frac{\boldsymbol{Y}_{.j.}^2}{n_{.j}} + \widetilde{\boldsymbol{v}}^T \widetilde{W}^{-1}\widetilde{\boldsymbol{v}}.$$

From the special solution $\hat{\boldsymbol{b}}$ of the system of normal equations independent of $\hat{\boldsymbol{\beta}}^* = X^T\boldsymbol{Y}$ and from

$$\sum_{i=1}^{a} \frac{\boldsymbol{Y}_{i..}^2}{n_{i.}} + \boldsymbol{v}^T W^{-1}\boldsymbol{v} = \sum_{j=1}^{b} \frac{\boldsymbol{Y}_{.j.}^2}{n_{.j}} + \boldsymbol{v}^T W^{-1}\boldsymbol{v},$$

it follows that

$$\tilde{\boldsymbol{v}}^T \widetilde{W}^{-1} \tilde{\boldsymbol{v}} = \sum_{i=1}^{a} \frac{\boldsymbol{Y}_{i\cdot\cdot}^2}{n_{i.}} + \tilde{\boldsymbol{v}}^T \widetilde{W}^{-1} \tilde{\boldsymbol{v}} - \sum_{j=1}^{b} \frac{\boldsymbol{Y}_{\cdot j\cdot}^2}{n_{\cdot j}}.$$

Therefore it is sufficient to calculate any generalised inverse and the corresponding solution $\boldsymbol{\beta}^*$.

From the numerator of the F-statistic for the test of $H_0: b_1 = \cdots = b_b$, the numerator of the F-statistic for testing of $H_0: a_1 = \cdots = a_a$ can easily be derived. Because of $\hat{\boldsymbol{\beta}} = (X^T X)^- X^T \boldsymbol{Y}$, it follows that

$$\boldsymbol{Y}^T \left(I_n - X\left(X^T X\right)^- X^T\right) \boldsymbol{Y} = \boldsymbol{Y}^T \boldsymbol{Y} - \hat{\boldsymbol{\beta}}^T X^T \boldsymbol{Y},$$

and the test statistic for $H_0: a_1 = \cdots = a_a$ is

$$\boldsymbol{F} = \frac{\sum_{i=1}^{a} \frac{\boldsymbol{Y}_{i\cdot\cdot}^2}{n_{i.}} - \sum_{j=1}^{b} \frac{\boldsymbol{Y}_{\cdot j\cdot}^2}{n_{\cdot j}} + \tilde{\boldsymbol{v}}^T \widetilde{W}^{-1} \tilde{\boldsymbol{v}}}{\sum_{i,j,k} \boldsymbol{y}_{ijk}^2 - \sum_{i=1}^{a} \frac{\boldsymbol{Y}_{i\cdot\cdot}^2}{n_{i.}} - \tilde{\boldsymbol{v}}^T \widetilde{W}^{-1} \tilde{\boldsymbol{v}}} \cdot \frac{N-a-b+1}{a-1}$$

and for $H_0: b_1 = \cdots = b_b$ correspondingly

$$\boldsymbol{F} = \frac{\tilde{\boldsymbol{v}}^T \widetilde{W}^{-1} \tilde{\boldsymbol{v}}}{\sum_{i,j,k} \boldsymbol{Y}_{ijk}^2 - \sum_{i=1}^{a} \frac{Y_{i\cdot\cdot}^2}{n_{i.}} - \tilde{\boldsymbol{v}}^T \widetilde{W}^{-1} \tilde{\boldsymbol{v}}} \cdot \frac{N-a-b+1}{b-1}.$$

If, as in (5.16), $n_{ij} = n$ (equal subclass numbers), simplifications for the tests of hypotheses about a and b result. We have the possibility further to construct an ANOVA table, in which SS_A, SS_B, $SS_{res} = SS_R$ add to $SS_{total} = SS_T$.

Theorem 5.8 If in model equation (5.16) $n_{ij} = n \geq 1$ for all i and j, then the sum of squared deviations of y_{ijk} from the total mean $\bar{\boldsymbol{y}}_{\ldots}$ of the experiment

$$\boldsymbol{SS}_T = \boldsymbol{Y}^T \boldsymbol{Y} - N\boldsymbol{y}_{\ldots}^2 = \sum_{i=1}^{a} \sum_{j=1}^{b} \sum_{k=1}^{n} \left(\bar{\boldsymbol{y}}_{ijk} - \bar{\boldsymbol{y}}_{\ldots}\right)^2$$

can be written as

$$\boldsymbol{SS}_T = \boldsymbol{SS}_A + \boldsymbol{SS}_B + \boldsymbol{SS}_{res}$$

with

$$\boldsymbol{SS}_A = \frac{1}{bn}\sum_{i=1}^{a}\boldsymbol{Y}_{i..}^2 - \frac{1}{N}\boldsymbol{Y}_{...}^2,\quad \boldsymbol{SS}_B = \frac{1}{an}\sum_{j=1}^{b}\boldsymbol{Y}_{.j.}^2 - \frac{1}{N}\boldsymbol{Y}_{...}^2,$$

$$\boldsymbol{SS}_{res} = \sum_{i=1}^{a}\sum_{j=1}^{b}\sum_{k=1}^{n}\boldsymbol{y}_{ijk}^2 - \frac{1}{bn}\sum_{i=1}^{a}\boldsymbol{Y}_{i..}^2 + \frac{1}{an}\sum_{j=1}^{b}\boldsymbol{Y}_{.j.}^2 + \frac{1}{N}\boldsymbol{Y}_{...}^2.$$

$\boldsymbol{SS}_A, \boldsymbol{SS}_B$ and $\boldsymbol{SS}_{res}$ are independently distributed, and for normally distributed $\boldsymbol{y}_{ijk}$, it is $\frac{1}{\sigma^2}\boldsymbol{SS}_A$ as $CS(a-1, \lambda_a)$, $\frac{1}{\sigma^2}\boldsymbol{SS}_B$ as $CS(b-1, \lambda_b)$ and $\frac{1}{\sigma^2}\boldsymbol{SS}_{res}$ as $CS(N-a-b+1)$ distributed with

$$\lambda_a = \frac{1}{\sigma^2}\sum_{i=1}^{a}(a_i - \bar{a}.)^2,\quad \lambda_b = \frac{1}{\sigma^2}\sum_{j=1}^{b}(b_j - \bar{b}.)^2.$$

These formulae are summarised in Table 5.10.

Example 5.9 Two forage crops (green rye and lucerne) have been investigated concerning their loss of carotene during their storage. For this four storage possibilities (glass jar in a refrigerator, glass jar in a barn, sack in a refrigerator and sack in a barn) are chosen. The loss during storage was defined by the difference between the content of carotene at start and the content of carotene after storing 300 days (in percent of dry mass). The question is whether the kind of storage and/or of forage crop influences the loss during storage. We denote the kind of storage as factor A and the forage crop as factor B and the observations (differences y_{ij}) can be arranged in the form of Table 5.7. Table 5.11 shows these values.

Table 5.10 Analysis of variance table of a two-way cross-classification with single subclass numbers ($n_{ij} = n$).

Source of variation	SS	df	MS	F
Between the levels of A	$SS_A = \frac{1}{bn}\sum_{i=1}^{a} Y_{i..}^2 - \frac{1}{N}Y_{...}^2$	$a-1$	$\frac{SS_A}{a-1} = MS_A$	$\frac{MS_A}{MS_{\text{res}}} = F_A$
Between the levels of B	$SS_B = \frac{1}{an}\sum_{j=1}^{b} Y_{.j.}^2 - \frac{1}{N}Y_{...}^2$	$b-1$	$\frac{SS_B}{b-1} = MS_B$	$\frac{MS_B}{MS_{\text{res}}} = F_B$
Residual	$SS_{res} = SS_T - SS_A - SS_B$	$N-a-b+1$	$\frac{SS_{\text{res}}}{N-a-b+1} = MS_{\text{res}}$	
Total	$SS_T = \sum_{i,j,k} y_{ijk}^2 - \frac{1}{N}Y_{...}^2$	$N-1$		

Table 5.11 Observations (loss during storage in percent of dry mass during storage of 300 days) of the experiment of Example 5.9 and results of first calculations.

		Forage crop				
		Green rye	**Lucerne**	$Y_{i.}$	$Y_{i.}^2$	$\sum_{j=1}^{2} y_{ij}^2$
Kind of storage	Glass in refrigerator	8.39	9.44	17.83	317.9089	159.5057
	Glass in barn	11.58	12.21	23.79	565.9641	283.1805
	Sack in refrigerator	5.42	5.56	10.98	120.5604	60.2900
	Sack in barn	9.53	10.39	19.92	396.8064	198.7730
$Y_{.j}$		34.92	37.60	72.52	1401.2398	
$Y_{.j}^2$		1219.4064	1413.7600	2633.1664		
$\sum_{i=1}^{4} y_{ij}^2$		324.6858	377.0634			701.7402

Because forage crops and kinds of storage have been selected consciously, we use for the y_{ij} model I and (5.16) as the model equation.

The ANOVA assumes that the observations are realisations of random variables that are independent of each other with equal variances and normally distributed. Table 5.11 shows further results of the calculation. Table 5.12 is the ANOVA table following Table 5.10. As the F-tests show, only factor storage has a significant influence on the loss during storage; significant differences could be found only between the kinds of storage, but not between the forage crops $(\alpha = 0.05)$.

How many observations per factor level combination are needed to test the effects of the factors 'kind of storage' with the following precision requirements: $a = 4$, $b = 2$, $\alpha = 0.05$, $\beta = 0.1$ and $\delta/\sigma = 2$?

Table 5.12 Analysis of variance table of Example 5.9.

Source of variation	*SS*	*df*	*MS*	*F*
Between the storages	43.2261	3	14.4087	186.7
Between the forage crops	0.8978	1	0.8978	11.63
Residual	0.2315	3	0.0772	
Total	44.3554	7		

Hints for Programs

With OPDOE of R we put in

```
size.anova(model="axb", hypothesis="a", a=4, b=2,
+alpha=0.05, beta=0.1, delta=2, cases="maximin")
```

and

```
size.anova(model="axb", hypothesis="a", a=4, b=2,
+alpha=0.05, beta=0.1, delta=2, Cases="minimin")
```

We obtain the output

```
n
5
```

and

```
n
3
```

We plan therefore experiments with 3 up to 5 subclass numbers.

5.3.1.2.2 Models with Interactions

We consider now model (5.15) and assume a connected cross-classification. Also in this case a testable hypothesis $K^T b = 0$ can be tested by the statistic (5.27) if the y_{ijk} are $N\left(\mu + a_i + b_j + (a,b)_{ij}, \sigma^2\right)$-distributed. Now β has the form

$$\beta = \left(\mu, a_1 \ldots, a_a, b_1, \ldots, b_b, (a,b)_{11}, \ldots, (a,b)_{ab}\right)^T.$$

Each estimable function is a linear function of

$$E\left(y_{ijk}\right) = \eta_{ij} = \mu + a_i + b_j + (a,b)_{ij}.$$

A testable hypothesis $K^T\beta = 0$ has also the form $K^T\beta = T^T\eta = 0$ with the vector $\eta = (\eta_{11}, \ldots, \eta_{ab})^T$, with the t components η_{ij}, for which $n_{ij} > 0$. By this we obtain (5.27) due to (5.18) and

$$\left(X^T X\right)^- = \begin{pmatrix} 0_{1+a+b,\ 1+a+b} & 0_{1+a+b,\ t} \\ 0_{t,1+a+b} & D \end{pmatrix},$$

with a $(t \times t)$-diagonal matrix D with elements $\frac{1}{n_{ij}}\left(n_{ij} > 0\right)$, that is, because of $K^T (X^T X)^- K = T^T D T$, we get

$$\boldsymbol{F} = \frac{\bar{\boldsymbol{Y}}^T T(T^T DT)^{-1} T^T \bar{\boldsymbol{Y}}}{\bar{\boldsymbol{Y}}^T [I_N - X(X^T X)^- X^T] \bar{\boldsymbol{Y}}} \cdot \frac{N-t}{q} \tag{5.28}$$

with $\bar{\boldsymbol{Y}} = (\bar{\boldsymbol{y}}_{11.}, \dots, \bar{\boldsymbol{y}}_{ab.})^T$. In (5.28) q is the number of (linear independent) rows of K^T or T^T.

Before considering special cases $(n_{ij} = 1,\ n_{ij} = n)$, we look at

Example 5.10 For the values of Table 5.9 of Example 5.5, the hypothesis H_0 : $b_1 - b_2 + (a,b)_{11} - (a,b)_{12} = 0$ for $\alpha = 0.05$ is to be tested. Now H_0 is equivalent to $\eta_{11} - \eta_{12} = 0$, so that $T^T = (1, -1, 0, 0, 0)$. In Example 5.5 we have $(X^T X)^- = O_{5,5} \oplus D$ with

$$D = \begin{pmatrix} \frac{1}{3} & 0 & 0 & 0 & 0 \\ 0 & 1 & 0 & 0 & 0 \\ 0 & 0 & \frac{1}{4} & 0 & 0 \\ 0 & 0 & 0 & \frac{1}{2} & 0 \\ 0 & 0 & 0 & 0 & \frac{1}{2} \end{pmatrix}.$$

Further it is $\bar{\eta} = (87, 99, 93, 93, 84)^T$, $q = 1$, $t = 5$ and $N = 12$. Moreover we find that

$$SS_{res} = Y^T [I_N - X(X^T X)^- X^T] Y = \sum_{i=1}^{a} \sum_{j=1}^{b} \sum_{k=1}^{n_{ij}} y_{ijk}^2 - \hat{b}^T X^T\ Y,$$

and with $\hat{b} = \hat{\beta}$ from (5.5), this is

$$SS_{res} = Y^T [I_N - X(X^T X)^- X^T] Y = \sum_{i=1}^{a} \sum_{j=1}^{b} \sum_{k=1}^{n_{ij}} y_{ijk}^2 - \sum_{i=1}^{a} \sum_{j=1}^{b} \frac{Y_{ij.}^2}{n_{ij}}, \tag{5.29}$$

writing down only summands for $n_{ij} > 0$. In the example it is

$$\sum_{i=1}^{3} \sum_{j=1}^{2} \sum_{k=1}^{n_{ij}} y_{ijk}^2 = 98\ 600, \quad \sum_{i=1}^{3} \sum_{j=1}^{2} \frac{Y_{ij.}^2}{n_{ij}} = 98\ 514$$

and $SS_{res} = 86$. From

$$\bar{Y}^T T = \bar{y}_{11.} - \bar{y}_{12.} = -12, \quad T^T DT = \frac{4}{3}, \quad (T^T DT)^{-1} = \frac{3}{4},$$

we get for F in (5.28)

$$F = \frac{108}{86} \cdot 7 = 8.791.$$

By this the null hypothesis is rejected, because $F(1,7|0.95) = 5.59$.

We consider now some special cases. Initially let $n_{ij} = n$, so that $t = ab, N = abn$ and $N - t = ab(n-1)$. The observations can be written in form of Table 5.8. Because all classes are occupied, we have

$$A_{ik} = a_i - a_k + \frac{1}{b}\left[\sum_{j=1}^{b}(a,b)_{ij} - \sum_{j=1}^{b}(a,b)_{kj}\right] \quad (i,k = 1,\ldots,a; i \neq k)$$

and

$$B_{jl} = b_j - b_l + \frac{1}{a}\left[\sum_{i=1}^{a}(a,b)_{ij} - \sum_{i=1}^{a}(a,b)_{il}\right] \quad (j,l = 1,\ldots,b; j \neq l)$$

as estimable functions. This can easily be shown, because

$$a_i - a_k + \frac{1}{b}\sum_{j=1}^{b}\left((a,b)_{ij} - (a,b)_{kj}\right) = \frac{1}{b}\sum_{j=1}^{b}\left(\eta_{ij} - \eta_{kj}\right).$$

The BLUEs of A_{ik} are

$$\hat{\boldsymbol{A}}_{ik} = \frac{1}{b}\sum_{j=1}^{b}\left(\bar{\boldsymbol{y}}_{ij.} - \bar{\boldsymbol{y}}_{kj.}\right),$$

and BLUEs of B_{jl} are analogously

$$\hat{\boldsymbol{B}}_{jl} = \frac{1}{a}\sum_{i=1}^{a}\left(\bar{\boldsymbol{y}}_{ij.} - \bar{\boldsymbol{y}}_{il.}\right).$$

By this the null hypotheses

$$H_{0A}: a_i + \frac{1}{b}\sum_{j=1}^{b}(a,b)_{ij} = a_a + \frac{1}{b}\sum_{j=1}^{b}(a,b)_{aj}(i = 1,\ldots,a-1),$$

$$H_{0B}: b_j + \frac{1}{a}\sum_{i=1}^{a}(a,b)_{ij} = b_b + \frac{1}{a}\sum_{i=1}^{a}(a,b)_{ib}(j = 1,\ldots,b-1)$$

are testable. W.l.o.g. we consider only H_{0A}. We write H_{0A} in the form

$$H_{0A}: a_i + \frac{1}{b}\sum_{j=1}^{b}(a,b)_{ij} - a_a - \frac{1}{b}\sum_{j=1}^{b}(a,b)_{aj} = 0 \quad (i = 1,\ldots,a-1)$$

or

$$K^T\beta = 0$$

with

$$K^T = \left(0_{a-1}, I_{a-1}, -1_{a-1}, O_{a-1,\ b-1}, \underset{i=1}{\overset{a-1}{\oplus}} \frac{1}{b}1_b^T, -\frac{1}{b}1_{a-1,b}\right).$$

We consider the next example.

Example 5.11 Assume that for the classification in Example 5.5 four observations per class are given. Then $a = 3, b = 2$ and

$$\beta = \left(\mu, a_1, a_2, a_3, b_1, b_2, (a,b)_{11}, (a,b)_{12}, (a,b)_{21}, (a,b)_{22}, (a,b)_{31}, (a,b)_{32}\right)^T.$$

We test the hypothesis

$$H_A : a_1 + \frac{1}{2}\left((a,b)_{11} + (a,b)_{12}\right) = a_2 + \frac{1}{2}\left((a,b)_{21} + (a,b)_{22}\right) = a_3 + \frac{1}{2}\left((a,b)_{31} + (a,b)_{32}\right).$$

K^T in $K^T\beta = 0$ has the form

$$K^T = \begin{pmatrix} 0 & 1 & 0 & -1 & 0 & 0 & \frac{1}{2} & \frac{1}{2} & 0 & 0 & -\frac{1}{2} & -\frac{1}{2} \\ 0 & 0 & 1 & -1 & 0 & 0 & 0 & 0 & \frac{1}{2} & \frac{1}{2} & -\frac{1}{2} & -\frac{1}{2} \end{pmatrix}.$$

If in the general case K^T is given as above, $\boldsymbol{F}$ in (5.27) can be simplified. With $\hat{\boldsymbol{\beta}}$ from (5.18), we have

$$\hat{\boldsymbol{\beta}}^T K = \left(\frac{1}{b}\sum_{j=1}^{b}\left[\bar{y}_{1j.} - \bar{y}_{aj.}\right], \dots, \frac{1}{b}\sum_{j=1}^{b}\left[\bar{y}_{a-1,j.} - \bar{y}_{aj.}\right]\right).$$

Further

$$K^T\left(X^TX\right)^- K = \frac{1}{n}K^{*T}K^* = \frac{1}{bn}M = \frac{1}{bn}\left(I_{a-1} + 1_{a-1,a-1}\right)$$

(and also the multiple M) is a $[(a-1)\times(a-1)]$ matrix of rank $a-1$. K^{*T} is the matrix generated from the ab last columns of K^T. Subtracting in M the $(i+1)$th row from the ith row $(i = 1, \dots, a-2)$ and adding then the first column to the second one, the so generated new second column to the third and so forth, we find that $|M| = a$. By this it is

$$\left|\frac{1}{n}K^{*T}K^*\right| = \frac{a}{b^{a-1}n^{a-1}}.$$

The minors of order $a-2$ belonging to the main diagonal elements of M are $a-1$, and the others are -1, so that

$$[K^T(X^TX)^-K]^{-1} = bn\begin{pmatrix} \frac{a-1}{a} & -\frac{1}{a} & \cdots & -\frac{1}{a} \\ -\frac{1}{a} & \frac{a-1}{a} & \cdots & -\frac{1}{a} \\ \vdots & \vdots & & \vdots \\ -\frac{1}{a} & -\frac{1}{a} & \cdots & \frac{a-1}{a} \end{pmatrix} = bn\left(I_{a-1} - \frac{1}{a}1_{a-1,a-1}\right).$$

By this $\boldsymbol{F}$ in (5.27) becomes

$$\boldsymbol{F}_A = \frac{\left(\frac{1}{bn}\sum_{i=1}^{a}\boldsymbol{Y}_{i..}^2 - \frac{1}{N}\boldsymbol{Y}_{...}^2\right)ab(n-1)}{(a-1)\boldsymbol{SS}_{res}}. \tag{5.30}$$

For H_{0B} we correspondingly receive

$$\boldsymbol{F}_B = \frac{\left(\frac{1}{an}\sum_{j=1}^{b}\boldsymbol{Y}_{.j.}^2 - \frac{1}{N}\boldsymbol{Y}_{...}^2\right)ab(n-1)}{(b-1)\boldsymbol{SS}_{res}}. \tag{5.31}$$

Under the side conditions

$$\sum_{j=1}^{b}(a,b)_{ij} = 0 \text{ for all } i, \quad \sum_{i=1}^{a}(a,b)_{ij} = 0 \text{ for all } j$$

$a_i - a_k$ and $b_i - b_l$ are BLUE with the estimable functions

$$\widehat{\boldsymbol{a}_i - \boldsymbol{a}_k} = \bar{\boldsymbol{y}}_{i..} - \bar{\boldsymbol{y}}_{k..} \quad (i \neq k)$$

and

$$\widehat{\boldsymbol{b}_j - \boldsymbol{b}_l} = \bar{\boldsymbol{y}}_{.j.} - \bar{\boldsymbol{y}}_{.l.} \quad (j \neq l).$$

Then the test statistics (5.30) and (5.31) can be used to test the hypotheses

$$H'_{0A}: a_1 = \cdots = a_a \text{ and } H'_{0B}: b_1 = \cdots = b_b.$$

In the case of equal subclass numbers, we use the side conditions (5.17) and test the hypotheses

$$H^*_{0A}: a_1 = \cdots = a_a(=0),$$
$$H^*_{0B}: b_1 = \cdots = b_b(=0),$$
$$H_{0AB}: (a,b)_{11} = \cdots = (a,b)_{ab}(=0)$$

with the F-statistics (5.30), (5.31) and

$$\boldsymbol{F}_{AB} = \frac{\left(\frac{1}{n}\sum_{i=1}^{a}\sum_{j=1}^{b}\boldsymbol{Y}_{ij.}^{2} - \frac{1}{bn}\sum_{i=1}^{a}\boldsymbol{Y}_{i..}^{2} - \frac{1}{an}\sum_{j=1}^{b}\boldsymbol{Y}_{.j.}^{2} + \frac{1}{N}\boldsymbol{Y}_{...}^{2}\right)ab(n-1)}{(a-1)(b-1)\boldsymbol{SS}_{res}}, \tag{5.32}$$

respectively.

The ANOVA table for this case is Table 5.13. Because of

$$\boldsymbol{SS}_T = \boldsymbol{SS}_A + \boldsymbol{SS}_B + \boldsymbol{SS}_{AB} + \boldsymbol{SS}_{res},$$

the F-statistics (5.30), (5.31) and (5.32) are under the hypotheses H_{0A} , H_{0B} and H_{0AB} central as $F[(a-1),\ ab(n-1)]$, $F[(b-1),\ ab(n-1)]$ and $F[(a-1)(b-1),\ ab(n-1)]$, respectively, distributed. Otherwise they are non-central F-distributed.

Example 5.12 We consider again Example 5.9 and the storages in glass and sack with four observations per subclass as shown in Table 5.14. Table 5.15 shows the calculation and Table 5.16 is the ANOVA table following Table 5.13. Due to the F-test, H_{0A} has to be rejected but not H_{0B} and H_{0AB}.

How many replications in the four subclasses are needed to test the hypothesis

$$H_{0AB}: (a,b)_{11} = \cdots = (a,b)_{22}(=0)$$

with the precision requirements in the following R commands?

The input is

```
size.anova(model="axb", hypothesis="axb", a=2, b=2,
alpha=0.05, beta=0.1, delta=2, cases="minimin")
```

The result is

```
n
4
```

The maximin size is 6.

A further special case is $n_{ij} = n = 1$. We also consider this case under the side conditions (5.17). Then the following theorem can be stated.

Theorem 5.9 (Tukey, 1949).

The random variables $\boldsymbol{y}_{ij}(i = 1,\ldots,a\ ; j = 1,\ldots,b)$ may be represented in the form of Equation (5.15) with $n_{ij} = 1$ for all i, j and (5.17) as well as $(a,b)_{ij} = a_i b_j$ may be fulfilled. The $\boldsymbol{e}_{ij}$ in (5.15) are independent from each other $N(0,\sigma^2)$-distributed for all i, j. Then with the symbolism of Table 5.13, (for $n = 1$) and

Table 5.13 Analysis of variance table of a two-way cross-classification with equal subclass numbers for model I with interactions under the condition (5.17).

Source of variation	*SS*	*df*	*MS*	E(*MS*)	*F*
Between rows (A)	$SS_A = \frac{1}{bn}\sum_i Y_{i..}^2 - \frac{1}{N}Y_{...}^2$	$a-1$	$\frac{SS_A}{a-1}$	$\sigma^2 + \frac{bn}{a-1}\sum_i a_i^2$	$\frac{ab(n-1)SS_A}{(a-1)SS_{\text{res}}}$
Between columns (B)	$SS_B = \frac{1}{an}\sum_j Y_{.j.}^2 - \frac{1}{N}Y_{...}^2$	$b-1$	$\frac{SS_B}{b-1}$	$\sigma^2 + \frac{an}{b-1}\sum_j b_j^2$	$\frac{ab(n-1)SS_B}{(b-1)SS_{\text{res}}}$
Interactions	$SS_{AB} = \frac{1}{n}\sum_{i.j} Y_{ij.}^2 - \frac{1}{bn}\sum_i Y_{i..}^2$	$(a-1)(b-1)$	$\frac{SS_{AB}}{(a-1)(b-1)}$	$\sigma^2 + \frac{n\sum_{i,j}(a,b)_{ij}^2}{(a-1)(b-1)}$	$\frac{ab(n-1)SS_{AB}}{(a-1)(b-1)SS_{\text{res}}}$
	$= \frac{1}{an}\sum_i Y_{.j.}^2 + \frac{Y^2}{N}$				
Within classes (residual)	$SS_{\text{res}} = \sum_{i,j,k} y_{ijk}^2 - \frac{1}{n}\sum_{i,j} Y_{ij.}^2$	$ab(n-1)$	$\frac{SS_{\text{res}}}{ab(n-1)} = s^2\sigma^2$		
Total	$SS_T = \sum_{i,j,k} y_{ijk}^2 - \frac{1}{N}Y_{...}^2$	$N-1$			

Table 5.14 Observations of the carotene storage experiment of Example 5.12.

		Forage crop	
		Green rye	Lucerne
Kind of storage	Glass	8.39	9.44
		7.68	10.12
		9.46	8.79
		8.12	8.89
	Sack	5.42	5.56
		6.21	4.78
		4.98	6.18
		6.04	5.91

Table 5.15 Class sums $Y_{ij.}$ and other results for the observations of Table 5.14.

		Forage crop				
		Green rye	Lucerne	$Y_{i..}$	$Y_{i..}^2$	$\sum_j Y_{ij.}^2$
Kind of storage	Glass	33.65	37.24	70.89	5025.3921	2519.1401
	Sack	22.65	22.43	45.08	2032.2064	1016.1274
$Y_{.j.}$		56.30	59.67	115.97	7057.5985	
$Y_{.j.}^2$		3169.6900	3560.5089	6730.1989		
$\sum_i Y_{ij.}^2$		1645.3450	1889.9225			3535.2675

Table 5.16 Analysis of variance table for the carotene storage experiment of Example 5.12.

Source of variation	*SS*	*df*	*MS*	*F*
Between the kind of storage	41.6347	1	41.6347	101.70
Between the forage crops	0.7098	1	0.7098	1.73
Interactions	0.9073	1	0.9073	2.22
Within classes (residual)	4.9128	12	0.4094	
Total	48.1646	15		

$$SS_N = \left[\sum_{i=1}^{a}\sum_{j=1}^{b}(\bar{y}_{i.} - \bar{\boldsymbol{y}}_{..})\left(\bar{\boldsymbol{y}}_{.j} - \bar{\boldsymbol{y}}_{..}\right)\left(\bar{y}_{ij} - \bar{y}_{i.} - \bar{\boldsymbol{y}}_{.j} + \bar{\boldsymbol{y}}_{..}\right)\right]^2 \frac{ab}{SS_A SS_B} \tag{5.33}$$

the statistic

$$\boldsymbol{F} = \frac{SS_N}{SS_{AB} - SS_N}[(a-1)(b-1)-1] \tag{5.34}$$

is as $F[1,(a-1)(b-1)-1]$-distributed if the null hypothesis $H_{0AB}:\ (a,b)_{ij} = 0$ for all i, j is valid.

Before showing this theorem we prove two lemmas.

Lemma 5.4 Under the assumptions of Theorem 5.9, we have the following:

a) $\hat{\boldsymbol{\mu}} = \bar{\boldsymbol{y}}_{..}$ is independent of $\hat{\boldsymbol{a}}_i = \bar{y}_{i.} - \bar{\boldsymbol{y}}_{..}$, $\hat{\boldsymbol{b}}_i = \boldsymbol{y}_{.j} - \bar{\boldsymbol{y}}_{..}$ and $\left(\widehat{\boldsymbol{a},\boldsymbol{b}}\right)_{ij} = \bar{y}_{ij} - \bar{y}_{i.} - \bar{\boldsymbol{y}}_{.j} + \bar{\boldsymbol{y}}_{..}$ for all i, j.
b) $\hat{\boldsymbol{a}}_i$ and $\hat{\boldsymbol{b}}_j$ are independent for all i, j.
c) $\hat{\boldsymbol{a}}_i$ and $\left(\widehat{\boldsymbol{a},\boldsymbol{b}}\right)_{kl}$ are independent for all i, k, l.
d) $\hat{\boldsymbol{b}}_j$ and $\left(\widehat{\boldsymbol{a},\boldsymbol{b}}\right)_{kl}$ are independent for all j, k, l.
e) $\hat{\boldsymbol{\mu}}$ is $N\left(\mu, \frac{1}{ab}\sigma^2\right)$-distributed, the $\hat{\boldsymbol{a}}_i$ are $N\left(a_i, \frac{a-1}{ab}\sigma^2\right)$-distributed for all i, the $\hat{\boldsymbol{b}}_i$ are $N\left(b_j, \frac{b-1}{ab}\sigma^2\right)$-distributed for all j, the $\left(\widehat{\boldsymbol{a},\boldsymbol{b}}\right)_{ij}$ are $N\left[(a,b)_{ij}, \frac{(a-1)(b-1)}{ab}\sigma^2\right]$- distributed for all i, j and the corresponding $\boldsymbol{SS}$ are χ^2-distributed.

We further have

f) $\operatorname{cov}(\hat{\boldsymbol{a}}_i, \hat{\boldsymbol{a}}_j) = -\frac{1}{ab}\sigma^2$ for $i \neq j$, $\operatorname{cov}\left(\hat{\boldsymbol{b}}_i, \hat{\boldsymbol{b}}_j\right) = -\frac{1}{ab}\sigma^2$ for $i \neq j$,

$$\operatorname{cov}\left(\left(\widehat{\boldsymbol{a},\boldsymbol{b}}\right)_{ij}, \left(\widehat{\boldsymbol{a},\boldsymbol{b}}\right)_{kl}\right) = \frac{\sigma^2}{ab}(a\delta_{ik} - 1)\ (b\delta_{jl} - 1).$$

Proof: By the assumptions $\boldsymbol{y}_{ij}$ are $N\left(\mu + a_i + b_j + (a,b)_{ij}, \sigma^2\right)$-distributed. The estimators are as linear combinations of the $\boldsymbol{y}_{ij}$ also normally distributed. From (5.17) follows $E(\bar{\boldsymbol{y}}_{..}) = \mu$, $E(\hat{\boldsymbol{a}}_i) = a_i$, $E\left(\hat{\boldsymbol{b}}_j\right) = b_j$, $E\left(\widehat{\boldsymbol{a},\boldsymbol{b}}\right)_{ij} = (a,b)_{ij}$ for all i, j. Now we get

$$\operatorname{var}(\bar{\boldsymbol{y}}_{..}) = \operatorname{var}\left[\frac{1}{ab}\sum_{i=1}^{a}\sum_{j=1}^{b}\boldsymbol{y}_{ij}\right] = \frac{1}{a^2b^2}\sum_{i=1}^{a}\sum_{j=1}^{b}\operatorname{var}\left(\boldsymbol{y}_{ij}\right) = \frac{1}{ab}\sigma^2$$

and

$$\operatorname{var}(\hat{\boldsymbol{a}}_i) = \operatorname{var}\left[\frac{a}{ab}\sum_{j=1}^{b}\boldsymbol{y}_{ij} - \frac{1}{ab}\sum_{t=1}^{a}\sum_{j=1}^{b}\boldsymbol{y}_{tj}\right]$$

$$= \frac{1}{a^2b^2}\operatorname{var}\left[(a-1)\sum_{j=1}^{b}\boldsymbol{y}_{ij} - \sum_{t\neq i}^{a}\sum_{j=1}^{b}\boldsymbol{y}_{tj}\right].$$

Because the two terms within the square bracket due to the assumption are independent, we have

$$\operatorname{var}(\hat{\boldsymbol{a}}_i) = \frac{1}{a^2b^2}\left[(a-1)^2 b\sigma^2 + (a-1)b\sigma^2\right] = \frac{a-1}{ab}\sigma^2.$$

Analogously the other relations under (f) follow. By this (e) and (f) are proved. To show the independencies, in (a) to (d) due to (e), we have only to show that the correlations are zero.

For (a) $\operatorname{cov}(\bar{\boldsymbol{y}}_{..}, \hat{\boldsymbol{a}}_i) = \operatorname{cov}(\bar{\boldsymbol{y}}_{..}, \bar{\boldsymbol{y}}_{i.} - \bar{\boldsymbol{y}}_{..}) = \operatorname{cov}(\bar{\boldsymbol{y}}_{..}, \bar{\boldsymbol{y}}_{i.}) - \operatorname{var}(\bar{\boldsymbol{y}}_{i..})$, and because of

$$\operatorname{cov}\left(\frac{1}{ab}\sum_{i=1}^{a}\sum_{j=1}^{b}\boldsymbol{y}_{ij}, \frac{1}{b}\sum_{j=1}^{b}\boldsymbol{y}_{ij}\right) = \frac{1}{ab^2}\operatorname{cov}\left(\sum_{i=1}^{a}\sum_{j=1}^{b}\boldsymbol{y}_{ij}, \sum_{j=1}^{b}\boldsymbol{y}_{ij}\right) = \frac{\sigma^2}{ab}$$

this covariance is zero. The proof of the other statements in (a) to (d) we leave as exercises.

Lemma 5.5 Under the conditions of Theorem 5.9

$$\boldsymbol{u} = \sqrt{\boldsymbol{SS}_N} = \frac{\sum_{i=1}^{a}\sum_{j=1}^{b}(\boldsymbol{a},\boldsymbol{b})_{ij}\boldsymbol{a}_i\boldsymbol{b}_j}{\sqrt{\sum_{i=1}^{a}\boldsymbol{a}_i^2\sum_{j=1}^{b}\boldsymbol{b}_j^2}}$$

is $N(0, \sigma^2)$-distributed, if $(a,b)_{ij} = 0$ for all i, j.

Proof: We consider the $(a+b+1)-$ dimensional random variable $(\boldsymbol{u}, \boldsymbol{a}_1, \ldots, \boldsymbol{a}_a, \boldsymbol{b}_1, \ldots, \boldsymbol{b}_b)$ and show that the conditional distribution of $\boldsymbol{u}$ for given realisations a_i and b_j of $\boldsymbol{a}_i$ and $\boldsymbol{b}_j$ is independent of a_i and b_j and by this equal to the marginal distribution of $\boldsymbol{u}$.

For fixed a_i, b_j the variable $\boldsymbol{u}$ is a linear combination of the normally distributed $(\boldsymbol{a}, \boldsymbol{b})_{ij}$ (from Lemma 5.4), and therefore the conditional distribution of $\boldsymbol{u}$ is a normal distribution. We have $E\left(\left(\widehat{\boldsymbol{a},\boldsymbol{b}}\right)_{ij}\right) = (a,b)_{ij}$, so that under the assumption $(a,b)_{ij} = 0$

$$E(\boldsymbol{u}|a_i,b_j) = 0 \quad (i = 1,\dots,a_i;\ j = 1,\dots,b)$$

is independent of a_i and b_j. From (e) and (f) of Lemma 5.4, $\text{var}(\boldsymbol{u}|a_i,b_j) = \sigma^2$ follows. Because the expectation and the variance of $\boldsymbol{u}$ are independent of the conditions and $\boldsymbol{u}$ is normally distributed, the statement follows.

Proof of Theorem 5.9:

Under the hypothesis H_{0AB} : all $(a,b)_{ij} = 0$ the sum of squares of interactions $\boldsymbol{SS}_{AB}$ is distributed as $CS[(a-1)(b-1)]$. We assume that $(a,b)_{ij} = 0$ for all i, j.

From Lemma 5.5 it follows that $\frac{\boldsymbol{SS}_N}{\sigma^2} = \frac{\boldsymbol{u}^2}{\sigma^2}$ is $CS(1)$-distributed. Because $\frac{\boldsymbol{SS}_{AB}}{\sigma^2} - \frac{\boldsymbol{SS}_N}{\sigma^2}$ is non-negative (Schwarz's inequality), it follows from Theorem 4.6 that this difference is distributed as $CS[(a-1)(b-1)-1]$. From Corollary 4.1 of Theorem 4.6, $\mathbf{SS}_N$ and $\frac{\boldsymbol{SS}_{AB}}{\sigma^2} - \frac{\boldsymbol{SS}_N}{\sigma^2}$ are independent of each other. This completes the proof.

The results of Theorem 5.9 are often in the applications used as follows. With the F-statistic of Theorem 5.9, the hypothesis H_{0AB} : $(a,b)_{ij} = 0$ is tested. If H_{0AB} is rejected, a new experiment to test H_{0A} and H_{0B} with $n > 1$ has to be carried out. If H_{0AB} is accepted, H_{0A} and H_{0B} (often with the same observations) are tested with the test statistic in Table 5.10. Concerning the problems of such an approach, we refer the reader to special literature.

5.3.2 Nested Classification ($A \succ B$)

A nested classification is a classification with super- and subordinated factors, where the levels of a subordinated or nested factor can be considered as further subdivision of the levels of the superordinated factor. Each level of the nested factor occurs in just one level of the superordinated factor. An example is the subdivision of the United States into states (superordinated factor A) and counties (nested factor B). Table 5.17 shows observations of a two-way nested classification.

As for the cross-classification we assume that the random variables $\boldsymbol{y}_{ijk}$ in Table 5.17 vary randomly from the expectations η_{ij}, that is,

$$\boldsymbol{y}_{ijk} = \eta_{ij} + \boldsymbol{e}_{ijk} \quad (i = 1,\dots,a;\ j = 1,\dots,b_i;\ k = 1,\dots,n_{ij}),$$

Table 5.17 Observations y_{ijk} of a two-way nested classification.

Levels of the factor A	A_1		A_2		$\cdots$	A_a	
Levels of the nested factor B	B_{11} ...	B_{1b_1}	B_{21} ...	B_{1b_2}	$\cdots$	B_{a1} ...	B_{ab_a}
Observations	y_{111} ...	$y_{1b_{21}}$	y_{211} ...	$y_{2b_{21}}$	$\cdots$	y_{a11} ...	$y_{ab_{a1}}$
	y_{112} ...	$y_{1b_{22}}$	y_{212} ...	$y_{2b_{22}}$	$\cdots$	y_{a12} ...	$y_{ab_{a2}}$
	$\vdots$	$\vdots$	$\vdots$	$\vdots$		$\vdots$	$\vdots$
	$y_{11n_{11}}$ ...	$y_{1b_{2n_{1b1}}}$	$y_{21n_{21}}$ ...	$y_{2b_2n_{2b_2}}$	$\cdots$	$y_{a1n_{a1}}$ ...	$y_{ab_an_{ab_a}}$

and that $\boldsymbol{e}_{ijk}$ are independent of each other $N(0,\sigma^2)$-distributed. With

$$\mu = \bar{\eta}.. = \frac{\sum_{i=1}^{a}\sum_{j=1}^{b_i}\eta_{ij}n_{ij}}{N},$$

the total mean of the experiment is defined.

In nested classification, interactions cannot be defined.

Analogously to Definition 5.2 we give

Definition 5.5 The difference $a_i = \bar{\eta}_{i.} - \mu$ is called the effect of the ith level of factor A, and the difference $b_{ij} = \eta_{ij} - \eta_i.$ is the effect of the jth level of B within the ith level of A.

By this the model equation for y_{ijk} is

$$\boldsymbol{y}_{ijk} = \mu + a_i + b_{ij} + \boldsymbol{e}_{ijk} \tag{5.35}$$

(interactions do not exist). It is easy to see that (5.35) is a special case of (5.1) if

$$Y = \left(\boldsymbol{y}_{111},\ldots,\boldsymbol{y}_{11n_{11}},\boldsymbol{y}_{121},\ldots,\boldsymbol{y}_{12n_{12}},\ldots,\boldsymbol{y}_{ab_an_{ab_a}}\right)^T,$$
$$\beta = (\mu, a_1,\ldots,a_a,b_{11},\ldots,b_{ab_a})^T,$$
$$\boldsymbol{e} = \left(\boldsymbol{e}_{111},\ldots,\boldsymbol{e}_{11n_{11}},\boldsymbol{e}_{121},\ldots,\boldsymbol{e}_{12n_{12}},\ldots,\boldsymbol{e}_{ab_an_{ab_a}}\right)^T$$

and X is a matrix of zeros and ones so that (5.35) is valid. From assumption it follows that $\boldsymbol{e}$ is $N(0_N,\sigma^2 I_N)$-distributed $\left(N = \sum_{i,j} n_{ij}\right)$. $\boldsymbol{Y}$ and $\boldsymbol{e}$ are $(N \times 1)$ vectors, β *is a* $[(a+1+B.)\times 1]$ vector $\left(B. = \sum_{i=1}^{a} b_i\right)$.

Example 5.13 demonstrates the choice of the matrix X.

Example 5.13 In Table 5.18 observations of a two-way nested classification are given (artificial data). Now we have

$$Y = (14,12,15,18,12,14,6,5,10,7,8,12)^T$$

and

$$\beta = (\mu, a_1, a_2, b_{11}, b_{12}, b_{21}, b_{22}, b_{23})^T.$$

Then

$$X = \begin{pmatrix} 1 & 1 & 0 & 1 & 0 & 0 & 0 & 0 \\ 1 & 1 & 0 & 0 & 1 & 0 & 0 & 0 \\ 1 & 1 & 0 & 0 & 1 & 0 & 0 & 0 \\ 1 & 1 & 0 & 0 & 1 & 0 & 0 & 0 \\ 1 & 0 & 1 & 0 & 0 & 1 & 0 & 0 \\ 1 & 0 & 1 & 0 & 0 & 1 & 0 & 0 \\ 1 & 0 & 1 & 0 & 0 & 0 & 1 & 0 \\ 1 & 0 & 1 & 0 & 0 & 0 & 1 & 0 \\ 1 & 0 & 1 & 0 & 0 & 0 & 1 & 0 \\ 1 & 0 & 1 & 0 & 0 & 0 & 1 & 0 \\ 1 & 0 & 1 & 0 & 0 & 0 & 0 & 1 \\ 1 & 0 & 1 & 0 & 0 & 0 & 0 & 1 \end{pmatrix} = (1_{12}, 1_4 \oplus 1_8, 1_1 \oplus 1_3 \oplus 1_2 \oplus 1_4 \oplus 1_2),$$

Table 5.18 Observations of a two-way nested classification.

	A_1		A_2		
	B_{11}	B_{12}	B_{21}	B_{22}	B_{23}
y_{ijk}	14	12	12	6	8
		15	14	5	12
		18		10	
				7	
n_{ij}	1	3	2	4	2
$N_{i.}$	4		8		

and X^TX becomes

$$X^TX = \begin{pmatrix} 12 & 4 & 8 & 1 & 3 & 2 & 4 & 2 \\ 4 & 4 & 0 & 1 & 3 & 0 & 0 & 0 \\ 8 & 0 & 8 & 0 & 0 & 2 & 4 & 2 \\ 1 & 1 & 0 & 1 & 0 & 0 & 0 & 0 \\ 3 & 3 & 0 & 0 & 3 & 0 & 0 & 0 \\ 2 & 0 & 2 & 0 & 0 & 2 & 0 & 0 \\ 4 & 0 & 4 & 0 & 0 & 0 & 4 & 0 \\ 2 & 0 & 2 & 0 & 0 & 0 & 0 & 2 \end{pmatrix}.$$

The matrix X^TX is of order 8 and, as it can be easily seen, of rank 5, because the second and third rows sum up to the first row, the fourth and fifth rows sum up to the second one and the last three rows sum up to the third one.

This may be generalised. One column of X corresponds to μ; a columns correspond to the levels of A (the $a_i; i = 1,\dots,a$); and $B. = \sum_{i=1}^{a} b_i$ columns correspond to the levels of B within the levels of A. The order of X^TX equals the number of the columns of X, and by this it is $1 + a + B.$. X^TX has with $N = \sum_{ij} n_{ij} N_i$ the form

$$X^TX = \begin{pmatrix} N & N_1 & \cdots & N_a & n_{11} & \cdots & n_{1b_1} & \cdots & n_{a1} & \cdots & n_{ab_a} \\ N_1 & N_1 & \cdots & 0 & n_{11} & \cdots & n_{1b_1} & \cdots & 0 & \cdots & 0 \\ \vdots & \vdots & \cdots & \vdots & \vdots & & \vdots & & \vdots & & \vdots \\ N_a & 0 & \cdots & N_a & 0 & \cdots & 0 & \cdots & n_{a1} & \cdots & n_{ab_a} \\ n_{11} & n_{11} & \cdots & 0 & n_{11} & \cdots & 0 & \cdots & 0 & \cdots & 0 \\ \vdots & \vdots & \cdots & \vdots & \vdots & & \vdots & & \vdots & & \vdots \\ n_{1b_1} & n_{1b_1} & \cdots & 0 & 0 & \cdots & n_{1b_1} & \cdots & 0 & \cdots & 0 \\ n_{a1} & 0 & \cdots & n_{a1} & 0 & \cdots & 0 & \cdots & n_{a1} & \cdots & 0 \\ \vdots & \vdots & \cdots & \vdots & \vdots & & \vdots & & \vdots & & \vdots \\ n_{ab_a} & 0 & \cdots & n_{ab_a} & 0 & \cdots & 0 & \cdots & 0 & \cdots & n_{ab_a} \end{pmatrix}.$$

As we see the first row is the sum of the a following rows, and the ith of these a following rows, that is, the $(i + 1)$th row is the sum of the b_i rows with the row numbers $a + 1 + \sum_{j=1}^{i-1} b_j$ up to $a + 1 + \sum_{j=1}^{i} b_j$. That means there are $a + 1$ linear relations between the rows of X^TX. Therefore $\mathrm{rk}(X^T X)$ of X^TX is smaller or equal to $B_{..}$ But $\text{rk}(X^TX) = B_{..}$, because the B last rows and columns are a non-singular submatrix with the inverse

$$\begin{pmatrix} \frac{1}{n_{11}} & & & & & & \\ & \ddots & & & & & \\ & & \frac{1}{n_{1b_1}} & & & 0 & \\ & & & \ddots & & & \\ & 0 & & & \frac{1}{n_{a1}} & & \\ & & & & & \ddots & \\ & & & & & & \frac{1}{n_{ab_a}} \end{pmatrix},$$

and by this a generalised inverse $(X^TX)^-$ of X^TX is given by a matrix of order $a + 1 + B_{..}$ Their elements apart from the B last ones in the main diagonal are equal to zero. In the main diagonal are $a + 1$ zeros and further the $B_.$ values $\frac{1}{n_{ij}}(i = 1,\dots,a; j = 1,\dots,b_i)$.

Example 5.14 We consider the matrix X^TX of Example 5.13. We derive $(X^TX)^-$ as shown above and obtain

$$(X^TX)^- = \begin{pmatrix} 0 & 0 & 0 & 0 & 0 & 0 & 0 & 0 \\ 0 & 0 & 0 & 0 & 0 & 0 & 0 & 0 \\ 0 & 0 & 0 & 0 & 0 & 0 & 0 & 0 \\ 0 & 0 & 0 & 1 & 0 & 0 & 0 & 0 \\ 0 & 0 & 0 & 0 & \frac{1}{3} & 0 & 0 & 0 \\ 0 & 0 & 0 & 0 & 0 & \frac{1}{2} & 0 & 0 \\ 0 & 0 & 0 & 0 & 0 & 0 & \frac{1}{4} & 0 \\ 0 & 0 & 0 & 0 & 0 & 0 & 0 & \frac{1}{2} \end{pmatrix}.$$

The reader may show as an exercise that $X^TX(X^TX)^-X^TX = X^TX$. As matrix B in (5.4), we may choose, for instance, a $\left[(a+1) \times \left(a + 1 + \sum b_i\right)\right]$-matrix

$$B=\begin{pmatrix} 0 & N & \dots & N & 0 & \dots & 0 & \dots & 0 & \dots & 0 \\ 0 & 0 & \dots & 0 & N_1 & \dots & N_1 & \dots & 0 & \dots & 0 \\ \vdots & \vdots & & \vdots & \vdots & & \vdots & & \vdots & & \vdots \\ 0 & 0 & \dots & 0 & 0 & \dots & 0 & \dots & N_a & \dots & N_a \end{pmatrix}$$

corresponding to the side conditions

$$\sum_{i=1}^{a} a_i = \sum_{j=1}^{b_i} b_{ij} = 0 \quad (\text{for all } i).$$

We see that

$$B^TB=\begin{pmatrix} 0 & 0 & \dots & 0 & 0 & \dots & 0 & \dots & 0 & \dots & 0 \\ 0 & N^2 & \dots & N^2 & 0 & \dots & 0 & \dots & 0 & \dots & 0 \\ \vdots & \vdots & & \vdots & \vdots & & \vdots & & \vdots & & 0 \\ 0 & N^2 & \dots & N^2 & 0 & \dots & 0 & \dots & 0 & \dots & 0 \\ 0 & 0 & \dots & 0 & N_{1.}^2 & \dots & N_{1.}^2 & & & & \\ \vdots & \vdots & & \vdots & \vdots & & \vdots & & 0 & & \\ 0 & 0 & \dots & 0 & N_{1.}^2 & \dots & N_{1.}^2 & & & & \\ \vdots & \vdots & & \vdots & & & & \ddots & & & \\ 0 & 0 & \dots & 0 & & & & & N_{a.}^2 & \dots & N_{a.}^2 \\ \vdots & \vdots & & \vdots & & 0 & & & \vdots & & \vdots \\ 0 & 0 & \dots & 0 & & & & & N_{a.}^2 & \dots & N_{a.}^2 \end{pmatrix}$$

$$= \quad 0 \oplus N^2 1_{aa} \oplus N_1^2 1_{b_1 b_1} \oplus \cdots \oplus N_a^2 1_{b_a b_a}$$

is of rank $a+1$. We choose instead

$$B=\begin{pmatrix} 0 & N_{1.} & \dots & N_{a.} & 0 & \dots & 0 & 0 & \dots & 0 \\ 0 & 0 & \dots & 0 & n_{11} & \dots & n_{1b_1} & 0 & \dots & 0 \\ \vdots & \vdots & & \vdots & \vdots & & \vdots & \vdots & & \vdots \\ 0 & 0 & \dots & 0 & 0 & \dots & 0 & n_{a1} & \dots & n_{ab_a} \end{pmatrix},$$

and this is corresponding to the side conditions

$$\sum_{i=1}^{a} N_{i.} a_i = 0, \quad \sum_{j=1}^{b_i} n_{ij} b_{ij} = 0 \quad (\text{ for all } i). \tag{5.36}$$

Minimising

$$\sum_{i=1}^{a} \sum_{j=1}^{b_i} \sum_{k=1}^{n_{ij}} \left(y_{ijk} - \mu - a_i - b_{ij} \right)^2,$$

under these side conditions without the cumbersome calculation of $(X^T X + B^T B)^{-1}$, we obtain the BLUE (MSE)

$$\hat{\boldsymbol{\mu}} = \bar{\boldsymbol{y}}_{...}, \quad \hat{\boldsymbol{a}}_i = \bar{\boldsymbol{y}}_{i..} - \bar{\boldsymbol{y}}_{...}, \quad \hat{\boldsymbol{b}}_{ij} = \bar{\boldsymbol{y}}_{ij.} - \bar{\boldsymbol{y}}_{i..}. \tag{5.37}$$

Theorem 5.10 In a two-way nested classification, we have

$$\boldsymbol{SS}_T = \sum_{i,j,k} \left(\boldsymbol{y}_{ijk} - \bar{\boldsymbol{y}}_{...} \right)^2$$

$$= \sum_{i,j,k} \left(\bar{\boldsymbol{y}}_{i..} - \bar{\boldsymbol{y}}_{...} \right)^2 + \sum_{i,j,k} \left(\bar{\boldsymbol{y}}_{ij.} - \bar{\boldsymbol{y}}_{i..} \right)^2 + \sum_{i,j,k} \left(\boldsymbol{y}_{ijk} - \bar{\boldsymbol{y}}_{ij.} \right)^2$$

or

$$\boldsymbol{SS}_T = \boldsymbol{SS}_A + \boldsymbol{SS}_{B\ in\ A} + \boldsymbol{SS}_{res}$$

where $\boldsymbol{SS}_A$ is the $\boldsymbol{SS}$ between the A-levels, $\boldsymbol{SS}_{B\ in\ A}$ is the $\boldsymbol{SS}$ between the B-levels within the A-levels and $\boldsymbol{SS}_{res}$ is the $\boldsymbol{SS}$ within the classes (B-levels).

The degrees of freedom of these $\boldsymbol{SS}$ are:

SS	df
$\boldsymbol{SS}_T$	$N-1$
$\boldsymbol{SS}_A$	$a-1$
$\boldsymbol{SS}_{B\ in\ A}$	$B_{.} - a$
$\boldsymbol{SS}_{res}$	$N - B.$

The $\boldsymbol{SS}$ may also be written in the form

$$\boldsymbol{SS}_T = \sum_{i,j,k} \boldsymbol{y}_{ijk}^2 - \frac{\boldsymbol{Y}_{...}^2}{N}, \quad \boldsymbol{SS}_A = \sum_i \frac{\boldsymbol{Y}_{i..}^2}{N_{i.}} - \frac{\boldsymbol{Y}_{...}^2}{N},$$

$$\boldsymbol{SS}_{B\ in\ A} = \sum_{i,j} \frac{\boldsymbol{Y}_{ij.}^2}{n_{ij.}} - \sum_i \frac{\boldsymbol{Y}_{i..}^2}{N_{i.}}, \quad \boldsymbol{SS}_{res} = \sum_{i,j,k} \boldsymbol{y}_{ijk}^2 - \sum_{i,j} \frac{\boldsymbol{Y}_{ij.}^2}{n_{ij}}.$$

The expectations of the $\boldsymbol{MS}$ are given in Table 5.19.

Here and later we assume the side conditions (5.36).

With the results of Chapter 4, we obtain the following theorem

Table 5.19 Analysis of variance table of the two-way nested classification for model I.

Source of variation	*SS*	*df*	*MS*	*E*(*MS*)
Between *A*-levels	$SS_A = \sum_i \frac{Y_{i..}^2}{N_{i.}} - \frac{Y_{...}^2}{N}$	$a-1$	$\frac{SS_A}{a-1}$	$\sigma^2 + \frac{1}{a-1}\sum_i N_{i.}a_i^2$
Between *B*-levels within *A*-levels	$SS_{B\ in\ A} = \sum_{i,j} \frac{Y_{ij.}^2}{n_{ij}} - \sum_i \frac{Y_{i..}^2}{N_{i.}}$	$B.-a$	$\frac{SS_{B\ in\ A}}{B.-a}$	$\sigma^2 + \frac{1}{B.-a}\sum_{i,j} n_{ij}b_{ij}^2$
Within *B*-levels (residual)	$SS_{res} = \sum_{i,j,k} y_{ijk}^2 - \sum_{i,j} \frac{Y_{ij.}^2}{n_{ij}}$	$N-B.$	$\frac{SS_{res}}{N-B.}$	σ^2
Total	$SS_T = \sum_{i,j,k} y_{ijk}^2 - \frac{Y_{...}^2}{N}$	$N-1$	$\frac{SS_T}{N-1}$	

Theorem 5.11 $MS_A + MS_{B\,in\,A}$ and MS_{res} in Table 5.19 are independent of each other distributed as $CS(a-1,\lambda_a)$, $CS(B.-a,\lambda_b)$ and $CQ(N-B.)$, respectively, where

$$\lambda_a = \frac{1}{\sigma^2}E\left(Y^T\right)(B_2-B_3)E(Y),\quad \lambda_b = \frac{1}{\sigma^2}E\left(Y^T\right)(B_1-B_2)E(Y).$$

Here B_1 is the direct sum of $B.$ matrices C_{ij} of order n_{ij}:

$$B_1 = C_{11}\oplus\cdots\oplus C_{ab_a};$$

the elements of C_{ij} are all equal to n_{ij}^{-1}. B_2 is the direct sum of a matrices G_i of order $N_{i.}$:

$$B_2 = G_1\oplus\cdots\oplus G_a;$$

the elements of G_i are all equal to $N_{i.}^{-1}$. B_3 is a matrix of order N; all of its elements are equal to N^{-1}.

Proof: We only have to show that the quadratic forms SS_A, $SS_{B\ in\ A}$ and SS_{res} fulfil the assumptions of Theorem 4.6. We have

$$\sum_{i,j,k} y_{ijk}^2 = Y^TY \quad\text{and}\quad \sum_{i,j}\frac{Y_{ij\cdot}^2}{n_{ij}} = Y^TB_1Y$$

and further

$$\sum_i \frac{Y_{i\cdot\cdot}^2}{N_{i\cdot}} = Y^TB_2Y;$$

B_2 is the direct sum of a matrices of order $N_{i\cdot}$ with the elements $N_{i\cdot}^{-1}$. Finally

$$\frac{Y_{...}^2}{N} = Y^TB_3Y.$$

We have

$$\text{rk}(B_1) = B_., \ \text{rk}(B_2) = a, \ \text{rk}(B_3) = 1.$$

Further B_1, B_2 and B_3 are idempotent (Condition 1 of Theorem 4.6). In

$$\boldsymbol{SS}_A = Y^T(B_2 - B_3)Y, \ \boldsymbol{SS}_{B\,in\,A} = Y^T(B_1 - B_2)Y, \ \boldsymbol{SS}_{res} = Y^T(I_N - B_1)Y,$$

the matrices of the $\boldsymbol{SS}$ have the ranks

$$\text{rk}(B_2 - B_3) = a - 1, \ \text{rk}(B_1 - B_2) = B_. - a, \ \text{rk}(I_N - B_1) = N - B_..$$

Here $I_N - B_1 + B_1 - B_2 + B_2 - B_3 = I_N - B_3$ is the matrix of the quadratic form $\boldsymbol{SS}_T$ of rank $N - 1$. By this two conditions of Theorem 4.6 are fulfilled, and Theorem 5.11 is proven.

Example 5.15 For the data of Example 5.13, we get

$$Y = (14,12,15,18,12,14,6,5,10,7,8,12)^T,$$

$$B_1 = \begin{pmatrix} 1 & 0 & 0 & 0 & 0 & 0 & 0 & 0 & 0 & 0 & 0 & 0 \\ 0 & \frac{1}{3} & \frac{1}{3} & \frac{1}{3} & 0 & 0 & 0 & 0 & 0 & 0 & 0 & 0 \\ 0 & \frac{1}{3} & \frac{1}{3} & \frac{1}{3} & 0 & 0 & 0 & 0 & 0 & 0 & 0 & 0 \\ 0 & \frac{1}{3} & \frac{1}{3} & \frac{1}{3} & 0 & 0 & 0 & 0 & 0 & 0 & 0 & 0 \\ 0 & 0 & 0 & 0 & \frac{1}{2} & \frac{1}{2} & 0 & 0 & 0 & 0 & 0 & 0 \\ 0 & 0 & 0 & 0 & \frac{1}{2} & \frac{1}{2} & 0 & 0 & 0 & 0 & 0 & 0 \\ 0 & 0 & 0 & 0 & 0 & 0 & \frac{1}{4} & \frac{1}{4} & \frac{1}{4} & \frac{1}{4} & 0 & 0 \\ 0 & 0 & 0 & 0 & 0 & 0 & \frac{1}{4} & \frac{1}{4} & \frac{1}{4} & \frac{1}{4} & 0 & 0 \\ 0 & 0 & 0 & 0 & 0 & 0 & \frac{1}{4} & \frac{1}{4} & \frac{1}{4} & \frac{1}{4} & 0 & 0 \\ 0 & 0 & 0 & 0 & 0 & 0 & \frac{1}{4} & \frac{1}{4} & \frac{1}{4} & \frac{1}{4} & 0 & 0 \\ 0 & 0 & 0 & 0 & 0 & 0 & 0 & 0 & 0 & 0 & \frac{1}{2} & \frac{1}{2} \\ 0 & 0 & 0 & 0 & 0 & 0 & 0 & 0 & 0 & 0 & \frac{1}{2} & \frac{1}{2} \end{pmatrix}$$

$$= \left(1 \oplus \frac{1}{3}1_{3,3} \oplus \frac{1}{2}1_{2,2} \oplus \frac{1}{4}1_{4,4} \oplus \frac{1}{2}1_{2,2}\right),$$

where $\oplus$ is again the symbol of a direct sum. We have $\mathrm{rk}(B_1) = 5$ because each summand has rank 1. Further B_2 is the direct sum of the matrix of order 4 with elements $\frac{1}{4}$ and the matrix of order 8 with elements $\frac{1}{8}$, where we have $\mathrm{rk}(B_2) = 2$. B_3 is the matrix of order 12 of rank 1 with elements $\frac{1}{12}$. From this we obtain (matrices as tables)

$\boldsymbol{B}_2 - \boldsymbol{B}_3 =$

<table>
<tr><th></th><th>1</th><th>2</th><th>3</th><th>4</th><th>5</th><th>6</th><th>7</th><th>8</th><th>9</th><th>10</th><th>11</th><th>12</th></tr>
<tr><td>1</td><td colspan="4" rowspan="4">$\frac{1}{6}$</td><td colspan="8" rowspan="4">$-\frac{1}{12}$</td></tr>
<tr><td>2</td></tr>
<tr><td>3</td></tr>
<tr><td>4</td></tr>
<tr><td>5</td><td colspan="4" rowspan="8">$-\frac{1}{12}$</td><td colspan="8" rowspan="8">$\frac{1}{24}$</td></tr>
<tr><td>6</td></tr>
<tr><td>7</td></tr>
<tr><td>8</td></tr>
<tr><td>9</td></tr>
<tr><td>10</td></tr>
<tr><td>11</td></tr>
<tr><td>12</td></tr>
</table>

$\boldsymbol{B}_1 - \boldsymbol{B}_2 =$

<table>
<tr><th></th><th>1</th><th>2</th><th>3</th><th>4</th><th>5</th><th>6</th><th>7</th><th>8</th><th>9</th><th>10</th><th>11</th><th>12</th></tr>
<tr><td>1</td><td>$\frac{3}{4}$</td><td colspan="4">$-\frac{1}{4}$</td><td colspan="7" rowspan="5">0</td></tr>
<tr><td>2</td><td rowspan="4">$-\frac{1}{4}$</td><td colspan="4" rowspan="4">$\frac{1}{12}$</td></tr>
<tr><td>3</td></tr>
<tr><td>4</td></tr>
<tr><td>5</td></tr>
<tr><td>6</td><td colspan="5" rowspan="7">0</td><td colspan="2" rowspan="2">$\frac{3}{8}$</td><td colspan="5" rowspan="2">$-\frac{1}{8}$</td></tr>
<tr><td>7</td></tr>
<tr><td>8</td><td colspan="2" rowspan="5">$-\frac{1}{8}$</td><td colspan="3" rowspan="3">$\frac{1}{8}$</td><td colspan="2" rowspan="3">$-\frac{1}{8}$</td></tr>
<tr><td>9</td></tr>
<tr><td>10</td></tr>
<tr><td>11</td><td colspan="3" rowspan="2">$-\frac{1}{8}$</td><td colspan="2" rowspan="2">$\frac{3}{8}$</td></tr>
<tr><td>12</td></tr>
</table>

$$I_n - B_1 = 0 \oplus \begin{pmatrix} \frac{2}{3} & -\frac{1}{3} & -\frac{1}{3} \\ -\frac{1}{3} & \frac{2}{3} & -\frac{1}{3} \\ -\frac{1}{3} & -\frac{1}{3} & \frac{2}{3} \end{pmatrix} \oplus \begin{pmatrix} \frac{1}{2} & -\frac{1}{2} \\ -\frac{1}{2} & \frac{1}{2} \end{pmatrix}$$

$$\oplus \begin{pmatrix} \frac{3}{4} & -\frac{1}{4} & -\frac{1}{4} & -\frac{1}{4} \\ -\frac{1}{4} & \frac{3}{4} & -\frac{1}{4} & -\frac{1}{4} \\ -\frac{1}{4} & -\frac{1}{4} & \frac{3}{4} & -\frac{1}{4} \\ -\frac{1}{4} & -\frac{1}{4} & -\frac{1}{4} & \frac{3}{4} \end{pmatrix} \oplus \begin{pmatrix} \frac{1}{2} & -\frac{1}{2} \\ -\frac{1}{2} & \frac{1}{2} \end{pmatrix}.$$

The reader may check as an exercise that $B_2 - B_3, B_2 - B_2$ and $I_n - B_1$ are idempotent and that $(B_2 - B_3)(B_1 - B_2) = (B_2 - B_3)(I_n - B_1) = O_{12,12}$.

From Theorem 5.11 it follows that with λ_a and λ_b defined in Theorem 5.11

$$\boldsymbol{F}_A = \frac{\boldsymbol{MS}_A}{\boldsymbol{MS}_{res}} \text{ is distributed as} F(a-1, N-B_., \lambda_a)$$

and

$$\boldsymbol{F}_B = \frac{\boldsymbol{MS}_{B \text{ in } A}}{\boldsymbol{MS}_{res}} \text{ is distributed as} F(B_. - a, N-B_., \lambda_b).$$

If (5.36) is valid, $\boldsymbol{F}_A$ can be used to test the hypothesis $H_{0A} : a_1 = \cdots = a_a$ because under this hypothesis λ_a equals 0 (applying $\sum_{i=1}^{a} N_{i.} a_i = 0$). Analogously $\boldsymbol{F}_B$ can be used to test the hypothesis $H_{0B} : b_{i1} = \cdots = b_{ib}$, for all i, because then (applying $\sum_{j=1}^{bi} n_{ij} b_{ij} = 0$) λ_b equals zero. H_{0B} is also testable, if (5.36) is not true.

Example 5.16 We calculate the analysis of variance table for Example 5.13. We have

$$Y_{11.} = 14, \quad Y_{12.} = 45, \qquad Y_{1..} = 59,$$

$$Y_{21.} = 26, \quad Y_{22.} = 28, \quad Y_{23.} = 20, \quad Y_{2..} = 74,$$

$$Y_{...} = 133.$$

Further it is

$$\sum_{i.j.k} y_{ijk}^2 = 1647, \quad \sum_{i.j} \frac{Y_{ij.}^{\;2}}{n_{ij}} = 1605,$$

$$\sum_i \frac{Y_{i..}^{\;2}}{N_{i.}} = 1545.75; \frac{Y_{...}^{\;2}}{N} = 1474.09.$$

Table 5.20 contains the *SS, df, MS* and the *F*-ratios. The 0.95-quantiles (for $\alpha = 0.05$) are $F(1.7|0.95) = 5.59$ and $F(3.7|0.95) = 4.35$. H_{0A} is rejected, but not the hypothesis H_{0B}.

Hints for Programs

In SPSS a nested classifications can be analysed only if we change the syntax for the DESIGN command. After

Analyze

General Linear Model

Univariate

both factors must be put on 'main effects'. Under 'Model' we choose in 'Sum of Squares' 'Type 1'. Back in the main menu after pressing 'Paste', you can change the syntax for the DESIGN command as shown below:

```
UNIANOVA
y BY a b
/METHOD=SSTYPE(1)
/INTERCEPT=INCLUDE
/CRITERIA=ALPHA(.05)
/DEIGN=a b(a).
```

We now show how for the nested classification the minimal experimental size can be determined. We choose for testing the effects of *A*:

```
>size.anova(model="a>b",hypothesis="a",a=6,b=4,
+alpha=0.05,beta=0.1,delta=1,cases="minimin")
  n
  4
>size.anova(model="a>b",hypothesis="a",a=6,b=4,
+alpha=0.05,beta=0.1,delta=1,cases="maximin")
  n
  9
```

Table 5.20 Analysis of variance table of Example 5.16.

Source of variation	*SS*	*df*	*MS*	*F*
Between *A*	80.66	1	80.66	13.44
Between *B* within *A*	50.25	3	16.75	2.79
Residual	42	7	6.00	
Total	172.91	11		

We have to choose between 4 and 9 observations per level of factor B. For testing the effects of the factor B, we use

```
>size.anova(model="a>b",hypothesis="b",a=6,b=4,
+alpha=0.05,beta=0.2,delta=1,cases="minimin")
n
5
>size.anova(model="a>b",hypothesis="b",a=6,b=4,
+alpha=0.05,beta=0.2,delta=1,cases="maximin")
n
41
```

5.4 Three-Way Classification

The principle underlying the two-way ANOVA (two-way classification) is also useful if more than two factors occur in an experiment. In this section we only give a short overview of the cases with three factors without proving all statements because the principles of proving are similar to those in the case with two factors. Further statements valid for all cases proven in Chapter 4 and at the beginning of this chapter have been proven.

We consider the case with three factors because it often occurs in applications, which can be handled with a justifiable number of pages, and last but not least because besides the cross-classification and the nested classification a mixed classification occurs. At this point some remarks about the numerical analysis of experiments using ANOVA must be made. Certainly a general computer program for arbitrary classifications and numbers of factors following the theory of Chapter 4 with unequal class numbers can be elaborated. However such a program even with modern computers is not easy to apply because the matrices X^TX easily obtain several ten thousands of rows. Therefore we give for some special cases of the three-way ANOVA numerical solutions for which easy-to-use programs can be applied (in SAS, SPSS, R).

Problems with more than three factors are described in Method 3/51/0001 in Rasch et al. (2008).

5.4.1 Complete Cross-Classification ($A \times B \times C$)

We assume that the observations of an experiment are influenced by three factors A, B, C with a, b and c levels $A_1, \ldots, A_a$, $B_1, \ldots, B_b$ and $C_1, \ldots, C_c$, respectively. For each possible combination (A_i, B_j, C_k), let $n \geq 1$ observations y_{ijkl} $(l = 1, \ldots, n)$ be present. Each combination (A_i, B_j, C_k) $(i = 1, \ldots, a; j = 1, \ldots, b;\ k = 1, \ldots, c)$ of factor levels is called a class and characterised by (i, j, k). The expectation in the population associated with the class (i, j, k) is η_{ijk}.

Analogously to Definition 5.2 we define the following:

$\bar{\eta}_{i..} = \frac{\sum_{j,k} \eta_{ijk}}{bc}$ the expectation of the ith level of factor A

$\bar{\eta}_{.j.} = \frac{\sum_{i,k} \eta_{ijk}}{ac}$ the expectation the jth level of factor B

and

$\bar{\eta}_{..k} = \frac{\sum_{i,j} \eta_{ijk}}{ab}$ the expectation of the kth level of factor C

The total expectation is

$$\mu = \bar{\eta}_{...} = \frac{\sum_{i,j,k} \eta_{ijk}}{abc}.$$

The main effects of the factors A, B and C are defined by

$$a_i = \bar{\eta}_{i...} - \mu, \quad b_j = \bar{\eta}_{.j.} - \mu \quad \text{and} \quad c_k = \bar{\eta}_{..k} - \mu.$$

Assuming that the experiment is performed at a particular level C_k of the factors C, we have a two-way classification with the factors A and B, and the conditional interactions between the levels of the factors A and B for fixed k are given by

$$\eta_{ijk} - \bar{\eta}_{i.k} - \bar{\eta}_{.jk} + \bar{\eta}_{..k}. \tag{5.38}$$

The interactions $(a, b)_{ij}$ between the ith A-level and the jth B-level are the means over all C-levels of the terms in (5.38), that is, $(a, b)_{ij}$ is defined as

$$(a,b)_{ij} = \bar{\eta}_{ij.} - \bar{\eta}_{i..} - \bar{\eta}_{.j.} + \mu. \tag{5.39}$$

The interactions between A-levels and C-levels $(a, c)_{ik}$ and between B-levels and C-levels $(b, c)_{jk}$ are defined by

$$(a,c)_{ik} = \bar{\eta}_{i.k} - \bar{\eta}_{i..} - \bar{\eta}_{..k} + \mu \tag{5.40}$$

and

$$(b,c)_{jk} = \bar{\eta}_{.jk} - \bar{\eta}_{.j.} - \bar{\eta}_{..k} + \mu, \tag{5.41}$$

respectively.

The difference between the conditional interactions between the levels of two of the three factors for the given level of the third factor and the (unconditional) interaction of these two factors depends only on the indices of the levels of the factors and not on the fact for which the interaction of two factors is calculated. We call it the second-order interaction $(a, b, c)_{ijk}$ (between the levels of three factors). Without loss of generality we write

$$(a,b,c)_{ijk} = \bar{\eta}_{ijk} - \bar{\eta}_{ij.} - \bar{\eta}_{i.k} - \bar{\eta}_{.jk} + \bar{\eta}_{i..} + \bar{\eta}_{.j.} + \bar{\eta}_{..k} - \mu. \tag{5.42}$$

The interactions defined by (5.39) until (5.41) between the levels of two factors are called first-order interactions. From the definition of the main effect and (5.39) until (5.41), we write for η_{ijk}

$$\eta_{ijk} = \mu + a_i + b_j + c_k + (a,b)_{ij} + (a,c)_{ik} + (b,c)_{jk} + (a,b,c)_{ijk}.$$

Under the definitions above the side conditions for all values of the indices not occurring in the summation at any time are

$$\sum_i a_i = \sum_j b_j = \sum_k c_k = \sum_i (a,b)_{ij} = \sum_j (a,b)_{ij} = \sum_i (a,c)_{ik} = \sum_k (a,c)_{ik}$$
$$= \sum_j (b,c)_{jk} = \sum_k (b,c)_{jk} = \sum_i (a,b,c)_{ijk} = \sum_j (a,b,c)_{ijk} = \sum_k (a,b,c)_{ijk} = 0. \tag{5.43}$$

The n observations $\boldsymbol{y}_{ijkl}$ in each class are assumed to be independent from each other $N(0, \sigma^2)$-distributed. The variable (called error term) $\boldsymbol{e}_{ijkl}$ is the difference between $\boldsymbol{y}_{ijkl}$ and the expectation η_{ijk} of the class, that is, we have

$$\boldsymbol{y}_{ijkl} = \eta_{ijk} + \boldsymbol{e}_{ijkl}$$

or

$$\boldsymbol{y}_{ijkl} = \mu + a_i + b_j + c_k + (a,b)_{ij} + (a,c)_{ik} + (b,c)_{jk} + (a,b,c)_{ijk} + \boldsymbol{e}_{ijkl}. \tag{5.44}$$

By the least squares method, we obtain under (5.43) the following estimators:

$$\bar{\boldsymbol{y}}_{....} = \frac{1}{abcn} \sum_{i,j,k,l} \boldsymbol{y}_{ijkl} \quad \text{for } \mu$$

as well as

$$\hat{\boldsymbol{a}}_i = \bar{\boldsymbol{y}}_{i...} - \bar{\boldsymbol{y}}_{....},$$
$$\hat{\boldsymbol{b}}_j = \bar{\boldsymbol{y}}_{.j..} - \bar{\boldsymbol{y}}_{....},$$
$$\hat{\boldsymbol{c}}_k = \bar{\boldsymbol{y}}_{..k.} - \bar{\boldsymbol{y}}_{....},$$
$$\left(\widehat{\boldsymbol{a},\boldsymbol{b}}\right)_{ij} = \bar{\boldsymbol{y}}_{ij..} - \bar{\boldsymbol{y}}_{i...} - \bar{\boldsymbol{y}}_{.j..} + \bar{\boldsymbol{y}}_{....}$$
$$\left(\widehat{\boldsymbol{a},\boldsymbol{c}}\right)_{ik} = \bar{\boldsymbol{y}}_{i.k.} - \bar{\boldsymbol{y}}_{i...} - \bar{\boldsymbol{y}}_{.k..} + \bar{\boldsymbol{y}}_{....}$$
$$\left(\widehat{\boldsymbol{b},\boldsymbol{c}}\right)_{jk} = \bar{\boldsymbol{y}}_{.jk.} - \bar{\boldsymbol{y}}_{.j..} - \bar{\boldsymbol{y}}_{..k.} + \bar{\boldsymbol{y}}_{....}$$
$$\left(\widehat{\boldsymbol{a},\boldsymbol{b},\boldsymbol{c}}\right)_{ijk} = \bar{\boldsymbol{y}}_{ijk.} - \bar{\boldsymbol{y}}_{ij..} - \bar{\boldsymbol{y}}_{i.k.} - \bar{\boldsymbol{y}}_{.jk.} + \bar{\boldsymbol{y}}_{i...} + \bar{\boldsymbol{y}}_{.j..} + \bar{\boldsymbol{y}}_{..k.} - \bar{\boldsymbol{y}}_{....}$$

We may split $SS\_total = \sum_{i,j,k,l}\left(y_{ijkl} - \bar{y}_{....}\right)^2$ into eight components: three corresponding with the main effects, three with the first-order interactions, one with the second-order interaction and one with the error term or the residual. The corresponding ***SS*** are shown in the ANOVA table (Tables 5.21 and 5.22). In these tables N is again the total number of observations, $N = abcn$.

The following hypotheses can be tested under (5.44) (H_{0x} is one of the hypotheses $H_{0A}, \dots, H_{0ABC}$; $\boldsymbol{SS}_x$ is the corresponding ***SS***):

$$H_{0A} : a_i = 0 \quad (\text{for all } i),$$
$$H_{0B} : b_j = 0 \quad (\text{for all } j),$$
$$H_{0C} : c_k = 0 \quad (\text{for all } k),$$

$$H_{0AB} : (a,b)_{ij} = 0 \quad (\text{for all } i,j),$$
$$H_{0AC} : (a,c)_{ik} = 0 \quad (\text{for all } i,k),$$
$$H_{0BC} : (b,c)_{jk} = 0 \quad (\text{for all } j,k),$$
$$H_{0ABC} : (a,b,c)_{ijk} = 0 \quad (\text{for all } i,j,k,\ if\ n > 1).$$

Under the hypothesis H_{0x}, $\frac{1}{\sigma^2}\boldsymbol{SS}_x$ and $\frac{1}{\sigma^2}\boldsymbol{SS}_{res}$ are independent of each other with the *df* given in the ANOVA table centrally χ^2-distributed. Therefore the test statistics given in the column ***F*** of the ANOVA table are with the corresponding degrees of freedom centrally F-distributed. For $n = 1$ all hypotheses except H_{0ABC} can be tested under the assumption $(a,b,c)_{ijk} = 0$ for all i, j, k because then $\frac{1}{\sigma^2}\boldsymbol{SS}_{ABC} = \frac{1}{\sigma^2}\boldsymbol{SS}_{res}$ and $\frac{1}{\sigma^2}\boldsymbol{SS}_x$ under $H_{0x}(x = A, B, C\ etc.)$ are independent of each other χ^2-distributed. The test statistic $\boldsymbol{F}_x$ is given by

$$\boldsymbol{F}_x = \frac{(a-1)(b-1)(c-1)}{df_x} \cdot \frac{\boldsymbol{SS}_x}{\boldsymbol{SS}_{res}}.$$

The calculation of a three-way ANOVA can be done in such a way as if we have three two-way ANOVA. We demonstrate this by the following example.

Example 5.17 The observations of Example 5.9 can be considered as those of a three-way ANOVA with single class numbers ($n = 1$) if as factors we use the forage crop (A), the kind of storage (B – barn or refrigerator) and the packaging material (C – glass or sack) (Table 5.22). We have $a = b = c = 2$ and $n = 1$. The observations in Table 5.22 can be arranged in three tables of a two-way classification where the new 'observations' are the sums over the third factor of the original observations in the classes defined by the levels of the two factors selected (Tables 5.23, 5.24 and 5.25). Table 5.26 is the ANOVA table of the

Table 5.21 Analysis of variance table of a three-way cross-classification with equal subclass numbers (model i).

Source of variation	SS	df
Between A-levels	$SS_A = \frac{1}{bcn}\sum_i Y_{i...}^2 - \frac{1}{N}Y_{....}^2$	$a-1$
Between B-levels	$SS_B = \frac{1}{acn}\sum_j Y_{.j..}^2 - \frac{1}{N}Y_{....}^2$	$b-1$
Between C-levels	$SS_C = \frac{1}{abn}\sum_k Y_{..k.}^2 - \frac{1}{N}Y_{....}^2$	$c-1$
Interaction $A \times B$	$SS_{AB} = \frac{1}{cn}\sum_{i,j} Y_{ij..}^2 - \frac{1}{bcn}\sum_i Y_{i...}^2 - \frac{1}{acn}\sum_j Y_{.j..}^2 + \frac{Y_{....}^2}{N}$	$(a-1)(b-1)$
Interaction $A \times C$	$SS_{AC} = \frac{1}{bn}\sum_{i,k} Y_{i.k.}^2 - \frac{1}{bcn}\sum_i Y_{i...}^2 - \frac{1}{abn}\sum_k Y_{..k.}^2 + \frac{Y_{....}^2}{N}$	$(a-1)(c-1)$
Interaction $B \times C$	$SS_{BC} = \frac{1}{an}\sum_{j,k} Y_{.jk.}^2 - \frac{1}{acn}\sum_i Y_{.j..}^2 - \frac{1}{abn}\sum_k Y_{..k.}^2 + \frac{Y_{....}^2}{N}$	$(b-1)(c-1)$
Interaction $A \times B \times C$	$SS_{\mathrm{ABC}} = SS_{\mathrm{G}} - SS_{\mathrm{A}} - SS_{\mathrm{B}} - SS_{\mathrm{C}} - SS_{\mathrm{AB}} - SS_{\mathrm{AC}} - SS_{\mathrm{BC}} - SS_{res}$	$(a-1)(b-1)(c-1)$
Within the classes (residual)	$SS_{res} = \sum_{i,j,k,l} y_{ijkl}^2 - \frac{1}{n}\sum_{i,j,k} Y_{ijk.}^2$	$abc(n-1)$
Total	$SS_T = \sum_{i,j,k,l} y_{ijkl}^2 - \frac{Y_{....}^2}{N}$	$(N-1)$

MS	E(MS)	F
$MS_A = \frac{SS_A}{a-1}$	$\sigma^2 + \frac{bcn}{a-1}\sum a_i^2$	$\frac{abc(n-1)}{a-1}\frac{SS_A}{SS_{res}}$
$MS_B = \frac{SS_B}{b-1}$	$\sigma^2 + \frac{acn}{b-1}\sum b_j^2$	$\frac{abc(n-1)}{b-1}\frac{SS_B}{SS_{res}}$
$MS_C = \frac{SS_C}{c-1}$	$\sigma^2 + \frac{abn}{c-1}\sum c_k^2$	$\frac{abc(n-1)}{c-1}\frac{SS_C}{SS_{res}}$
$MS_{AB} = \frac{SS_{AB}}{(a-1)(b-1)}$	$\sigma^2 + \frac{cn}{(a-1)(b-1)}\sum (a,b)_{ij}$	$\frac{abc(n-1)}{(a-1)(b-1)}\frac{SS_{AB}}{SS_{res}}$
$MS_{AC} = \frac{SS_{AC}}{(a-1)(c-1)}$	$\sigma^2 + \frac{bn}{(a-1)(c-1)}\sum (a,c)_{ik}$	$\frac{abc(n-1)}{(a-1)(c-1)}\frac{SS_{AC}}{SS_{res}}$
$MS_{BC} = \frac{SS_{BC}}{(b-1)(c-1)}$	$\sigma^2 + \frac{an}{(b-1)(c-1)}\sum (b,c)_{jk}$	$\frac{abc(n-1)}{(b-1)(c-1)}\frac{SS_{BC}}{SS_{res}}$
$MS_{ABC} = \frac{SS_{ABC}}{(a-1)(b-1)(c-1)}$	$\sigma^2 + \frac{n}{(a-1)(b-1)(c-1)}\sum (a,b,c)_{ijk}$	$\frac{abc(n-1)}{(a-1)(b-1)(c-1)}\frac{SS_{ABC}}{SS_{res}}$
$MS_{\mathrm{res}} = s^2 = \frac{SS_{\mathrm{res}}}{abc(n-1)}$	σ^2	

Table 5.22 Three-way classification of the observations of Table 5.14 with factors kind of storage, packaging material and forage crop.

		Forage crop	
Kind of storage	**Packaging material**	**Green rye**	**Lucerne**
Refrigerator	Glass	8.39	9.44
	Sack	5.42	5.56
Barn	Glass	11.58	12.21
	Sack	9.53	10.39

Table 5.23 Two-way classification of the observations of Table 5.14 with factors kind of storage and forage crop ($Y_{ij.}$).

		Forage crop			
		Green rye	**Lucerne**	$Y_{.j.}$	$Y_{.j.}^2$
Kind of storage	Refrigerator	13.81	15.00	28.81	830.0161
	Barn	21.11	22.60	43.71	1910.5641
$Y_{i..}$		34.92	37.60	72.52	2740.5802
$Y_{i..}^2$		1219.4064	1413.76	2633.1664	

Table 5.24 Two-way classification of the observations of Table 5.14 with factors packaging material and forage crop $(Y_{\cdot jk})$.

		Forage crop			
		Green rye	**Lucerne**	$Y_{..k}$	$Y_{..k}^2$
Packaging material	Glass	19.97	21.65	41.62	1732.2244
	Sack	14.95	15.94	30.90	954.8100
$Y_{.j.}$		34.92	37.60	72.52	2687.0344
$Y_{.j.}^2$		1219.4064	1413.76	2633.1664	

Table 5.25 Two-way classification of the observations of Table 5.14 with factors kind of storage and packaging material $(Y_{i \cdot k})$.

		Packaging material			
		Glass	Sack	$Y_{i..}$	$Y_{i..}^2$
Kind of storage	Refrigerator	17.83	10.98	28.81	830.0161
	Barn	23.79	19.92	43.71	1910.5641
$Y_{..k}$		41.62	30.90	72.52	2740.5802
$Y_{..k}^2$		1732.2244	954.8100	2687.0344	

Table 5.26 Analysis of variance table of Example 5.17.

Source of variation	*SS*	*df*	*MS*	*F*
Between kind of storage	27.7513	1	27.7513	170.78
Between forage crops	0.8978	1	0.8978	5.52
Between packaging material	14.3648	1	14.3648	88.40
Interaction kind of storage × packaging material	1.1100	1	1.11	6.83
Interaction kind of storage × forage crops	0.0112	1	0.0112	<1
Interaction forage crops × packaging material	0.0578	1	0.0578	<1
Residual	0.1625	1	0.1625	
Total	44.3554	7	44.3554	

example. The F-tests are done under the assumption that all second-order interactions vanish using the $\boldsymbol{SS}_{res}$ defined above. Only between the kinds of storage significant differences $(\alpha = 0.05)$ could be found, that is, only the hypothesis H_A is rejected.

Hints for Programs

The size of the experiment is again calculated with the help of OPDOE in R. The syntax is analogous to that of the one- and two-way ANOVA. We demonstrate the calculation for sizes needed for testing the null hypothesis for factor A and the interactions $A \times B$ for a balanced experiment with $a = 3$, $b = 4$ and $c = 3$:

```
> size.anova(model="axbxc",hypothesis="a",a=3,b=4,c=3,
+alpha=0.05,beta=0.1,delta=0.5,cases="minimin")
n
6
```

```
> size.anova(model="axbxc",hypothesis="a",a=3,b=,c=3,
+alpha=0.05,beta=0.1,delta=0.5,cases="maximin")
n
9
> size.anova(model="axbxc",hypothesis="axb",a=3,b=4,
+c=3, alpha=0.05,beta=0.1,delta=1,cases="minimin")
n
3
> size.anova(model="axbxc",hypothesis="axb",a=3,b=4,
+c=3, alpha=0.05,beta=0.1,delta=1,cases="maximin")
n
12
```

5.4.2 Nested Classification ($C \prec B \prec A$)

We speak about a three-way nested classification if factor C is subordinated to factor B (as described in Section 5.3.2) and factor B is subordinated to factor A, that is, if $C \prec B \prec A$. We assume as in Section 5.3.2 that the random variable y_{ijkl} varies randomly with expected value $\eta_{ijk}\left(i = 1,\dots,a; j = 1,\dots,b_i;\ k = 1,\dots,c_{ij}\right)$, that is, we assume

$$\boldsymbol{y}_{ijkl} = \eta_{ijk} + \boldsymbol{e}_{ijkl} \quad \left(l = 1,\dots,n_{ijk}\right),$$

where $\boldsymbol{e}_{ijkl}$ independent from each other are $N(0, \sigma^2)$-distributed. By

$$\mu = \eta_{\dots} = \frac{\sum_{i=1}^{a}\sum_{j=1}^{b_i}\sum_{k=1}^{c_{ij}} \eta_{ijk} n_{ijk}}{N},$$

we define the total mean of the experiment by $N = \sum_{i=1}^{a}\sum_{j=1}^{b_i}\sum_{k=1}^{c_{ij}} n_{ijk}$.

We generalise Definition 5.5 by

Definition 5.6 The difference $a_i = \bar{\eta}_{i\,..} - \mu$ is called the effect of the ith level of A, the difference $b_{ij} = \bar{\eta}_{ij.} - \bar{\eta}_{i\,..}$ is called the effect of the jth level of B within the ith level of A and the difference $c_{ijk} = \eta_{ijk} - \bar{\eta}_{ij.}$ is called the effect of the kth level of C within the jth level of B and the ith level of A.

Then the observations can be modelled by

$$\boldsymbol{y}_{ijkl} = \mu + a_i + b_{ij} + c_{ijk} + \boldsymbol{e}_{ijkl}. \tag{5.45}$$

There exist no interactions. We consider (5.45) with $N_{ij.} = \sum_k n_{ijk}$; $N_{i\,..} = \sum_{jk} n_{ijk}$ under the side conditions

$$\sum_{i=1}^{a} N_{i..} a_i = \sum_{j=1}^{b_i} N_{ij.} b_{ij} = \sum_{k=1}^{c_{ij}} n_{ijk} c_{ijk} = 0. \tag{5.46}$$

Minimising

$$\sum_{i=1}^{a} \sum_{j=1}^{b_i} \sum_{k=1}^{c_{ij}} \sum_{l=1}^{n_{ijk}} \left(y_{ijkl} - \mu - a_i - b_{ij} - c_{ijk}\right)^2, \tag{5.47}$$

under the side conditions (5.46), leads to the BLUE of the parameters as follows:

$$\hat{\boldsymbol{\mu}} = \bar{\boldsymbol{y}}_{....}, \quad \hat{\boldsymbol{a}}_i = \bar{\boldsymbol{y}}_{i...} - \bar{\boldsymbol{y}}_{....}, \quad \hat{\boldsymbol{b}}_{ij} = \bar{\boldsymbol{y}}_{ij..} - \bar{\boldsymbol{y}}_{i...}, \hat{\boldsymbol{c}}_{ijk} = \bar{\boldsymbol{y}}_{ijk.} - \bar{\boldsymbol{y}}_{ij..}.$$

Without proof we give a theorem about the decomposition of the $\boldsymbol{SS}_{total} = \boldsymbol{SS}_T$ where the corresponding non-centrality parameters are calculated analogously to

$$\lambda = \frac{1}{\sigma^2} \beta^T X^T \left(X \left(X^T X\right)^- X^T - \frac{1}{N} 1_{N,N} \right) X\beta$$

in Section 5.1 by multiplying the quadratic form of the ***SS*** with the corresponding expectations.

Theorem 5.12 In a three-way nested classification, we have

$$\boldsymbol{SS}_T = \boldsymbol{SS}_A + \boldsymbol{SS}_{B\ in\ A} + \boldsymbol{SS}_{C\ in\ B\ (and\ A)} + \boldsymbol{SS}_{res}$$

with $N = \sum_{i=1}^{a} \sum_{j=1}^{b} \sum_{k=1}^{c} n_{ijk}$, $N_{ij.} = \sum_k n_{ijk}$; $N_{i..} = \sum_{jk} n_{ijk}$ and

$$\boldsymbol{SS}_T = \sum_{i,j,k,l} \boldsymbol{y}_{ijkl}^2 - \frac{\boldsymbol{Y}_{....}^2}{N}, \quad \boldsymbol{SS}_A = \sum_i \frac{\boldsymbol{Y}_{i...}^2}{N_{i..}} - \frac{\boldsymbol{Y}_{....}^2}{N},$$

$$\boldsymbol{SS}_{B\ in\ A} = \sum_{i,j} \frac{\boldsymbol{Y}_{ij..}^2}{N_{ij.}} - \sum_i \frac{\boldsymbol{Y}_{i...}^2}{N_{i..}}, \quad \boldsymbol{SS}_{C\ in\ B} = \sum_{i,j,k} \frac{\boldsymbol{Y}_{ijk.}^2}{n_{ijk}} - \sum_{i,j} \frac{\boldsymbol{Y}_{ij..}^2}{N_{ij.}},$$

$$\boldsymbol{SS}_{res} = \sum_{i,j,k,l} \boldsymbol{y}_{ijkl}^2 - \sum_{i,j,k} \frac{\boldsymbol{Y}_{ijk.}^2}{n_{ijk}}.$$

$\frac{1}{\sigma^2}\boldsymbol{SS}_A$ up to $\frac{1}{\sigma^2}\boldsymbol{SS}_{C\ in\ B}$ are with $B_{..} = b_{ij}, C_{...} = \sum_{ijk} c_{ijk}$ pairwise independently $CS(a-1,\lambda_a)$, $CS(B.-a,\lambda_b)$, $CS(C..-B.,\lambda_c)$ distributed and $\frac{1}{\sigma^2}\boldsymbol{SS}_{res}$ is $CS(N-C..)$-distributed. The non-centrality parameters λ_a, λ_b and λ_c vanish under the null hypotheses $H_{0A}: a_i = 0\ (i = 1,\dots,a)$, $H_{0B}: b_{ij} = 0$ $(i = 1,\dots,a; j = 1,\dots,b_i)$, $H_{0C}: c_{ijk} = 0\ \left(i = 1,\dots,a; j = 1,\dots,b_i; k = 1,\dots,c_{ij}\right)$, so that the result of Theorem 5.12 for constructing the *F*-statistics can be used. Table 5.27 shows the *SS* and *MS* for calculating the *F*-statistics. If H_{0A} is valid,

$\boldsymbol{F}_A$ is $F(a-1, N-C..)$-distributed. If H_{0B} is valid, then $\boldsymbol{F}_B$ is $F(B-a, N-C..)$-distributed, and if H_{0C} is valid, then $\boldsymbol{F}_C$ is $F(C..-B., N-C..)$-distributed.

Hints for Programs

For the analysis in SPSS analogously to Section 5.3.2, we have to change the syntax in the/DESIGN command. The minimal subclass numbers for the three tests of the main effects with OPDOE in R give the following results:

```
> size.anova(model="a>b>c",hypothesis="a",a=2,b=2,c=3,
+alpha=0.01,beta=0.1,delta=0.5,cases="minimin")
n
21
> size.anova(model="a>b>c",hypothesis="a",a=2,b=2,c=3,
+alpha=0.01,beta=0.1,delta=1,cases="minimin")
n
6
> size.anova(model="a>b>c",hypothesis="b",a=2,b=2,c=3,
+alpha=0.01,beta=0.1,delta=1,cases="minimin")
n
7
> size.anova(model="a>b>c",hypothesis="c",a=2,b=2,c=3,
+alpha=0.01,beta=0.1,delta=1,cases="minimin")
n
10
```

The maximin values are left for the reader as an exercise.

Table 5.27 Analysis of variance table of a three-way nested classification for model i.

Source of variation	*SS*	*df*	*MS*	*E*(*MS*) (under (5.46))	*F*
Between A	$\sum_i \frac{Y_{i...}^2}{N_{i..}} - \frac{Y_{....}^2}{N}$	$a-1$	$\frac{SS}{a-1}$	$\sigma^2 + \frac{1}{a-1}\sum_{i=1}^{n} N_{i..}a_i^2$	$\frac{MS_A}{MS_{res}} = F_A$
Between B in A	$\sum_{i,j} \frac{Y_{ij..}^2}{N_{ij.}} - \sum_i \frac{Y_{....}^2}{N_{i..}}$	$B.-a$	$\frac{SS_{B\,in\,A}}{B.-a}$	$\sigma^2 + \frac{1}{B.-a}\sum_{i,j} N_{ij.}b_{ij}^2$	$\frac{MS_{B\,in\,A}}{MS_{res}} = F_B$
Between C in B and A	$\sum_{i,j,k} \frac{Y_{ijk.}^2}{n_{ijk}} - \sum_{i,j} \frac{Y_{ij..}^2}{N_{ij.}}$	$C..-B.$	$\frac{SS_{C\,in\,B}}{C..-B.}$	$\sigma^2 + \frac{1}{C..-B.}\sum_{i,j,k} n_{ijk}c_{ijk}^2$	$\frac{MS_{C\,in\,B}}{MS_{res}} = F_C$
Residual	$\sum_{i,j,k,l} y_{ijkl}^2 - \sum_{i,j,k} \frac{Y_{ijk.}^2}{n_{ijk}}$	$N-C..$	$\frac{SS_{res}}{N-C..}$	σ^2	
Total	$\sum_{i,j,k,l} y_{ijkl}^2 - \frac{Y_{....}^2}{N}$	$N-1$			

5.4.3 Mixed Classification

In experiments with three or more factors under test besides a cross-classification or a nested classification, we often find a further type of classifications, so-called mixed (partially nested) classifications. In the three-way ANOVA, two mixed classifications occur (Rasch, 1971).

5.4.3.1 Cross-Classification between Two Factors Where One of Them Is Subordinated to a Third Factor $((B \prec A) \times C)$

If in a balanced experiment a factor B is subordinated to a factor A and both are cross-classified with a factor C, then the corresponding model equation is given by

$$\begin{cases} \boldsymbol{y}_{ijkl} = \mu + a_i + b_{ij} + c_k + (a,c)_{ik} + (b,c)_{jk(i)} + \boldsymbol{e}_{ijkl} \\ (i = 1,\dots,a;\ j = 1,\dots,b;\ k = 1,\dots,c;\ l = 1,\dots,n), \end{cases} \tag{5.48}$$

where μ is the general experimental mean, a_i is the effect of the ith level of factor A, b_{ij} is the effect of the jth level of factor B within the ith level of factor A and the c_k is the effect of the kth level of factor C. Further $(a, c)_{ik}$ and $(b, c)_{jk(i)}$ are the corresponding interaction effects and $\boldsymbol{e}_{ijkl}$ are the random error terms.

Model equation (5.48) is considered under the side conditions for all indices not occurring in the summation

$$\sum_{i=1}^{a} a_i = \sum_{j=1}^{b} b_{ij} = \sum_{k=1}^{c} c_k = \sum_{i=1}^{a} (a,c)_{ik} = \sum_{k=1}^{c} (a,c)_{ik} = \sum_{j=1}^{b} (b,c)_{jk} = \sum_{k=1}^{c} (b,c)_{jk(i)} = 0 \tag{5.49}$$

and

$$E(\boldsymbol{e}_{ijkl}) = 0,\quad E(\boldsymbol{e}_{ijkl}\boldsymbol{e}_{i'j'k'l'}) = \delta_{ii'}\delta_{jj'}\delta_{kk'}\delta_{ll'}\sigma^2,\quad \sigma^2 = \mathrm{var}(\boldsymbol{e}_{ijkl}) \tag{5.50}$$

(for all i, j, k, l).

The observations

$$y_{ijkl}\quad (i = 1,\dots,a; j = 1,\dots,b; k = 1,\dots,c; l = 1,\dots,n)$$

are allocated as shown in Table 5.28 (we restrict ourselves to the so-called balanced case where the number of B-levels is equal for all A-levels and the subclass numbers are equal). For the sum of squared deviations of the random variables

$$\boldsymbol{y}_{ijkl}\quad (i = 1,\dots,a; j = 1,\dots,b; k = 1,\dots,c; l = 1,\dots,n)$$

from their arithmetic mean

$$\boldsymbol{SS}_T = \sum_{i,j,k,l} \left(\boldsymbol{y}_{ijkl} - \bar{\boldsymbol{y}}_{....}\right)^2 = \sum_{i,j,k,l} \boldsymbol{y}_{ijkl}^2 - \frac{\boldsymbol{Y}_{....}^2}{N},\quad (N = abcn),$$

Table 5.28 Observations of a mixed three-way classification b with c cross-classified, b in a nested $((B \prec A) \times C)$.

Levels of A	Levels of B	Levels of C			
		C_1	C_2		C_c
A_1	B_{11}	y_{1111}	y_{1121}	$\cdots$	y_{11c1}
		y_{1112}	y_{1122}	$\cdots$	y_{11c2}
		$\vdots$	$\vdots$	$\cdots$	$\vdots$
		y_{111n}	y_{112n}		y_{11cn}
	B_{12}	y_{1211}	y_{1221}	$\cdots$	y_{12c1}
		y_{1212}	y_{1222}	$\cdots$	y_{12c2}
		$\vdots$	$\vdots$	$\cdots$	$\vdots$
		y_{121n}	y_{122n}		y_{12cn}
	$\vdots$	$\vdots$	$\vdots$		$\vdots$
	B_{1b}	y_{1b11}	y_{1b21}	$\cdots$	y_{1bc1}
		y_{1b12}	y_{1b22}	$\cdots$	y_{1bc2}
		$\vdots$	$\vdots$	$\cdots$	$\vdots$
		y_{1b1n}	y_{1b2n}		y_{1bcn}
A_2	B_{21}	y_{2111}	y_{2121}	$\cdots$	y_{21c1}
		y_{2112}	y_{2122}	$\cdots$	y_{21c2}
		$\vdots$	$\vdots$	$\cdots$	$\vdots$
		y_{211n}	y_{212n}		y_{21cn}
	B_{22}	y_{2211}	y_{2221}	$\cdots$	y_{22c1}
		y_{2212}	y_{2222}	$\cdots$	y_{22c2}
		$\vdots$	$\vdots$	$\cdots$	$\vdots$
		y_{221n}	y_{222n}		y_{22cn}
	$\vdots$	$\vdots$	$\vdots$		$\vdots$
	B_{2b}	y_{2b11}	y_{2b21}	$\cdots$	y_{2bc1}
		y_{2b12}	y_{2b22}	$\cdots$	y_{2bc2}
		$\vdots$	$\vdots$	$\cdots$	$\vdots$
		y_{2b1n}	y_{2b2n}		y_{2bcn}
	$\vdots$	$\vdots$	$\vdots$		$\vdots$

(*Continued*)

Table 5.28 (Continued)

Levels of A	Levels of B	Levels of C			
A_a	B_{a1}	y_{a111}	y_{a121}	$\cdots$	y_{a1c1}
		y_{a112}	y_{a122}	$\cdots$	y_{a1c2}
		$\vdots$	$\vdots$	$\cdots$	$\vdots$
		y_{a11n}	y_{a12n}		y_{a1cn}
	B_{a2}	y_{a211}	y_{a221}	$\cdots$	y_{a2c1}
		y_{a212}	y_{a222}	$\cdots$	y_{a2c2}
		$\vdots$	$\vdots$	$\cdots$	$\vdots$
		y_{a21n}	y_{a22n}		y_{a2cn}
	$\vdots$	$\vdots$ $\vdots$			$\vdots$
	B_{ab}	y_{ab11}	y_{ab21}	$\cdots$	y_{abc1}
		y_{ab12}	y_{ab22}	$\cdots$	y_{abc2}
		$\vdots$	$\vdots$	$\cdots$	$\vdots$
		y_{ab1n}	y_{ab2n}		y_{abcn}

we have

$$SS_T = SS_A + SS_{B\ in\ A} + SS_C + SS_{A\times C} + SS_{B\times C\ in\ A} + SS_{res},$$

where

$$SS_A = \frac{1}{bcn}\sum_{i=1}^{a} Y_{i\ldots}^2 - \frac{Y_{\ldots\ldots}^2}{N}$$

are the SS between the levels of A,

$$SS_{B\ in\ A} = \frac{1}{cn}\sum_{i=1}^{a}\sum_{j=1}^{b} Y_{ij\cdot\cdot}^2 - \frac{1}{bcn}\sum_{i=1}^{a} Y_{i\ldots}^2$$

the SS between the levels of B within the levels of A,

$$SS_C = \frac{1}{abn}\sum_{k=1}^{c} Y_{\cdot\cdot k\cdot}^2 - \frac{Y_{\ldots\ldots}^2}{N}$$

the ***SS*** between the levels of C,

$$SS_{A\times C}=\frac{1}{bn}\sum_{i=1}^{a}\sum_{k=1}^{c}Y_{i\cdot k\cdot}^2-\frac{1}{bcn}\sum_{i=1}^{a}Y_{i\cdots}^2-\frac{1}{abn}\sum_{k=1}^{c}Y_{\cdot\cdot k\cdot}^2+\frac{Y_{\cdots\cdot}^2}{N}$$

the ***SS*** for the interactions $A\times C$,

$$SS_{B\times C\ in\ A}=\frac{1}{n}\sum_{i=1}^{a}\sum_{j=1}^{b}\sum_{k=1}^{c}Y_{ijk\cdot}^2-\frac{1}{cn}\sum_{i=1}^{a}\sum_{j=1}^{b}Y_{ij\cdot\cdot}^2$$

$$-\frac{1}{bn}\sum_{i=1}^{a}\sum_{k=1}^{c}Y_{i\cdot k\cdot}^2+\frac{1}{bcn}\sum_{i=1}^{a}Y_{i\cdots}^2$$

the ***SS*** for the interactions $B\times C$ within the levels of A and

$$SS_{res}=\sum_{i,j,k,l}y_{ijk}^2-\frac{1}{n}\sum_{i=1}^{a}\sum_{j=1}^{b}\sum_{k=1}^{c}Y_{ijk\cdot}^2$$

the ***SS*** within the classes. The $N-1$ degrees of freedom of SS_T corresponding with the components of SS_T can be split into six components. These components are shown in Table 5.29; the third column of Table 5.29 contains the ***MS*** gained from the ***SS*** by division with the degrees of freedom.

If hypotheses have to be tested about the constants in model equation (5.47), we additionally have to assume that e_{ijkl} are normally distributed. The hypotheses can then again be tested with of F-tests. The choice of the correct test statistic for a particular hypothesis can easily be found heuristically. For this the expectation $E(\boldsymbol{MS})$ of the ***MS*** must be known. The $E(\boldsymbol{MS})$ can be found in the last column of Table 5.29. Representatively for the derivation of an $E(\boldsymbol{MS})$ we show the approach for $E(\boldsymbol{MS_A})$. We have

$$E(\boldsymbol{MS}_A)=\frac{1}{a-1}E(\mathbf{SQ}_A)=\frac{1}{a-1}\left[E\left(\frac{1}{bcn}\sum_{i=1}^{a}Y_{i\cdots}^2\right)-E\left(\frac{Y_{\cdots\cdot}^2}{N}\right)\right].$$

Now we replace the $\boldsymbol{y}_{ijkl}$ by the right side of the model equation (5.48) and obtain

$$\boldsymbol{Y}_{i\cdots}=bcn\mu+bcna_i+cn\sum_{j=1}^{b}b_{ij}+bn\sum_{k=1}^{c}c_k+bn\sum_{k=1}^{c}(a,c)_{ik}$$

$$+n\sum_{j=1}^{b}\sum_{k=1}^{c}(b,c)_{ijk}+\sum_{j=1}^{b}\sum_{k=1}^{c}\sum_{l=1}^{n}\boldsymbol{e}_{jikl}$$

and using (5.49)

$$\boldsymbol{Y}_{i\cdots}=bcn\mu+bcna_i+\sum_{j=1}^{b}\sum_{k=1}^{c}\sum_{l=1}^{n}\boldsymbol{e}_{ijkl}.$$

Table 5.29 Analysis of variance table for a balanced three-way mixed classification $(B<A)xC$ for model I $(n>1)$.

Source of variation	SS
Between the levels of A	$SS_A = \frac{1}{bcn}\sum_{i=1}^{a} Y_{i\cdots}^2 - \frac{1}{N}Y_{\cdots\cdot}^2$
Between the levels of B within the levels of A	$SS_{B\text{ in}1} = \frac{1}{cn}\sum_{i=1}^{a}\sum_{j=1}^{b} Y_{ij\cdot\cdot}^2 - \frac{1}{bcn}\sum_{i=1}^{a} Y_{i\cdots}^2$
Between the levels of C	$SS_c = \frac{1}{abn}\sum_{k=1}^{a} Y_{\cdot\cdot k\cdot}^2 - \frac{1}{N}Y_{\cdots\cdot}^2$
Interaction $A\times C$	$SS_{A\times C} = \frac{1}{bn}\sum_{i=1}^{a}\sum_{k=1}^{a} Y_{i\cdot k\cdot}^2 - \frac{1}{bcn}\sum_{i=1}^{a} Y_{i\cdots}^2 - \frac{1}{abn}\sum_{k=1}^{e} Y_{\cdot\cdot k\cdot}^2 + \frac{1}{N}Y_{\cdots\cdot}^2$
Interaction $B\times C$ within the levels of A	$SS_{B\times C\text{ in }A} = \frac{1}{n}\sum_{i=1}^{a}\sum_{j=1}^{b}\sum_{k=1}^{c} Y_{ijk\cdot}^2 - \frac{1}{cn}\sum_{i=1}^{a}\sum_{j=1}^{b} Y_{ij\cdot\cdot}^2 - \frac{1}{bn}\sum_{i=1}^{a}\sum_{k=1}^{e} Y_{i\cdot k\cdot}^2 - \frac{1}{bcn}\sum_{i=1}^{a} Y_{i\cdots}^2$
Residual	$SS_{res} = \sum_{i=1}^{a}\sum_{j=1}^{b}\sum_{k=1}^{c}\sum_{l=1}^{n} y_{ijkl}^2 - \frac{1}{n}\sum_{i=1}^{a}\sum_{j=1}^{b}\sum_{k=1}^{c} Y_{ijk\cdot}^2$
Total	$SS_T = \sum_{i=1}^{a}\sum_{j=1}^{b}\sum_{k=1}^{c}\sum_{l=1}^{n} y_{ijkl}^2 - \frac{1}{N}Y_{\cdots\cdot}^2$

df	MS	E(MS) under (5.49)
$a-1$	$MS_A = \frac{SS_A}{a-1}$	$\sigma^2 + \frac{bcn}{a-1}\sum_{i=1}^{a} a_i^2$
$a(b-1)$	$MS_{B\text{ in}A} = \frac{SS_{B\text{ in}A}}{a(b-1)}$	$\sigma^2 + \frac{cn}{a(b-1)}\sum_{i=1}^{a}\sum_{j=1}^{b} b_{ij}^2$
$c-1$	$MS_C = \frac{SS_c}{c-1}$	$\sigma^2 + \frac{abn}{c-1}\sum_{k=1}^{c} c_k^2$
$(a-1)(c-1)$	$MS_{A\times C} = \frac{SS_{A\times C}}{(a-1)(c-1)}$	$\sigma^2 + \frac{bn}{(a-1)(c-1)}\sum_{i=1}^{a}\sum_{k=1}^{c} (a,c)_{ik}^2$
$a(b-1)(c-1)$	$MS_{B\times C\text{ in}A} = \frac{SS_{B\times C\text{ in}A}}{a(b-1)(c-1)}$	$\sigma^2 + \frac{n}{a(b-1)(c-1)}\sum_{i=1}^{a}\sum_{j=1}^{b}\sum_{k=1}^{c} (b,c)_{jk(i)}^2$
$N-abc$	$MS_{\text{res}} = \frac{SS_{res}}{abc(n-1)}$	σ^2
$N-1$		

Analogously we receive under (5.49)

$$\boldsymbol{Y}_{....} = abcn\mu + \sum_{i,j,k,l} \boldsymbol{e}_{ijkl}.$$

Now we obtain for $E\left(\boldsymbol{Y}_{i...}^2\right)$ the equation

$$E\left(\boldsymbol{Y}_{i...}^2\right) = b^2c^2n^2\mu^2 + b^2c^2n^2a_i^2 + 2b^2c^2n^2\mu a_i + bcn\sigma^2$$

and for $E\left(\boldsymbol{Y}_{....}^2\right)$ the equation

$$E\left(\boldsymbol{Y}_{....}^2\right) = N^2\mu^2 + N\sigma^2.$$

With these two equations we get

$$E(\boldsymbol{MS}_A) = \frac{bcn}{a-1}\sum_{i=1}^{a} a_i^2 + \sigma^2.$$

The hypothesis $H_{0A}: a_i = 0$ can be tested by the help of the statistic $\boldsymbol{F}_A = \frac{\boldsymbol{MS}_{\text{A}}}{\boldsymbol{MS}_{\text{res}}}$, which under H_{0A} with $a-1$ and $N-abc$ degrees of freedom is F-distributed. If the null hypothesis is correct, numerator and denominator of $\boldsymbol{F}_A$ (from Table 5.29) have the same expectation. In general there is a ratio of two $\boldsymbol{MS}$ of a particular null hypothesis with the corresponding degrees of freedom centrally F-distributed, if the numerator and the denominator in case that the hypothesis is valid have the same expectation. This equality is however not sufficient if unequal subclass numbers occur; for instance, it is not sufficient if the $\boldsymbol{MS}$ are not independent from each other. In this case we obtain in the way shown above only a test statistic that is approximately F-distributed. We will in the following not differentiate between exact and approximately F-distributed test statistics. From Table 5.29 we see that in our model, the hypothesis over all effects ($a_i, b_{ij}, \ldots, (a, b, c)_{ijk}$) can be tested by using the ratios of the corresponding $\boldsymbol{MS}$ and $\boldsymbol{MS}_{res}$ as test statistic.

As an example we consider again testing pig fattening for male and female (factor C) offspring of sows (factor B) nested in boars (factor A). The observed character is the number of fattening days an animal needed to grow up from 40 to 110 kg (compare Example 5.6).

We again give an example for the calculation of the experimental size using the symbolism of OPDOE in R as in the other sections:

```
>size.anova(model="(axb)>c", hypothesis="a",a=6, b=5,
+c=4, alpha=0.05, beta=0.1, delta=0.5, case="minimin")
n
3
```

5.4.3.2 Cross-Classification of Two Factors in Which a Third Factor Is Nested $(C \prec (A \times B))$

If two cross-classified factors ($A \times B$) are super-ordered to a third factor (C), we have another mixed classification. The model equation for the random observations in a balanced design is given by

$$\boldsymbol{y}_{ijkl} = \mu + a_i + b_j + c_{ijk} + (a,b)_{ij} + \boldsymbol{e}_{ijkl},\ (i = 1,\dots, a;\ j = 1,\dots,b;\ k = 1,\dots,c;\ l = 1,\dots,n). \tag{5.51}$$

This is again the situation of model I, where the error terms $\boldsymbol{e}_{ijkl}$ may fulfil condition (5.50).

Analogously to (5.48) we assume that for all values of the indices not occurring in the summation, we have the side conditions

$$\sum_{i=1}^{a} a_i = \sum_{j=1}^{b} b_j = \sum_{k=1}^{c} c_{ijk} = \sum_{i=1}^{a} (a,b)_{ij} = \sum_{j=1}^{b} (a,b)_{ij} = 0. \tag{5.52}$$

The total sum of squared deviations can be split into components

$$\boldsymbol{SS}_T = \boldsymbol{SS}_A + \boldsymbol{SS}_B + \boldsymbol{SS}_{C\,in AB} + \boldsymbol{SS}_{A\times B} + \boldsymbol{SS}_{res}$$

with

$$\boldsymbol{SS}_A = \sum_{i=1}^{a} \frac{\boldsymbol{Y}_{i\cdots}^2}{bcn} - \frac{\boldsymbol{Y}_{\cdots\cdot}^2}{N},$$

the SS between the A-levels,

$$\boldsymbol{SS}_B = \sum_{j=1}^{a} \frac{\boldsymbol{Y}_{\cdot j\cdot\cdot}^2}{acn} - \frac{\boldsymbol{Y}_{\cdots\cdot}^2}{N},$$

the SS between the B-levels,

$$\boldsymbol{SS}_{C \text{ in } AB} = \sum_{i=1}^{a}\sum_{j=1}^{b}\sum_{k=1}^{c} \frac{\boldsymbol{Y}_{ijk\cdot}^2}{n} - \sum_{i=1}^{a}\sum_{j=1}^{b} \frac{\boldsymbol{Y}_{ij\cdot\cdot}^2}{cn},$$

the SS between the C-Levels within the $A \times B$ combinations,

$$\boldsymbol{SS}_{A\times B} = \sum_{i=1}^{a}\sum_{j=1}^{b} \frac{\boldsymbol{Y}_{ij\cdot\cdot}^2}{cn} - \sum_{i=1}^{a} \frac{\boldsymbol{Y}_{i\cdots}^2}{bcn} - \sum_{j=1}^{b} \frac{\boldsymbol{Y}_{\cdot j\cdot\cdot}^2}{acn} + \frac{\boldsymbol{Y}_{\cdots\cdot}^2}{N},$$

the *SS* for the interactions between factor *A* and factor *B* and

$$SS_{res} = \sum_{i=1}^{a}\sum_{j=1}^{b}\sum_{k=1}^{c}\sum_{l=1}^{n} y_{ijkl}^2 - \sum_{i=1}^{a}\sum_{j=1}^{b}\sum_{k=1}^{c} \frac{Y_{ijk\cdot}^2}{n}.$$

The expectations of the *MS* in this model are shown in Table 5.31, and the hypotheses

$$H_{0A}: \ a_i = 0, \ \ H_{0B}: \ b_j = 0, \ \ H_{0C}: \ c_{ijk} = 0, \ \ H_{0AB}: \ (a,b)_{ij} = 0$$

can be tested by using the corresponding *F*-statistic as the ratios of MS_A, MS_B, MS_C and $MS_{A\times B}$, respectively (as numerator) and MS_{res} (as denominator).

As an example consider the mast performance of offspring of beef cattle (factor *C*) of different genotypes (factor *A*) in several years (factor *B*). If each sire occurs just once, then the structure of Table 5.30 is given.

Table 5.30 Observations of a balanced three-way mixed classification, *a* with *b* cross-classified and *c* nested in the $a \times b$-combinations.

	Levels of *B*			
Levels of *A*	**B_1 Levels of *C***	**B_2 Levels of *C***	…	**B_b Levels of *C***
	C_{111} C_{112} … C_{11c}	C_{121} C_{122} … C_{12c}	…	C_{1b1} C_{1b2} … C_{1bc}
A_1	y_{1111} y_{1121} ⋯ y_{11c1}	y_{1211} y_{1221} ⋯ y_{12c1}	…	y_{1b11} y_{1b21} ⋯ y_{1bc1}
	y_{1112} y_{1122} ⋯ y_{11c2}	y_{1212} y_{1222} ⋯ y_{12c2}		y_{1b12} y_{1b22} ⋯ y_{1bc2}
	⋮ ⋮ ⋮	⋮ ⋮ ⋮		⋮ ⋮ ⋮
	y_{111n} y_{112n} ⋯ y_{11cn}	y_{121n} y_{122n} ⋯ y_{12cn}		y_{1b1n} y_{1b2n} ⋯ y_{1bcn}
	C_{211} C_{212} … C_{21c}	C_{221} C_{222} … C_{22c}	…	C_{2b1} C_{2b2} … C_{2bc}
A_2	y_{2111} y_{2121} ⋯ y_{21c1}	y_{2211} y_{2221} ⋯ y_{22c1}	…	y_{2b11} y_{2b21} ⋯ y_{2bc1}
	y_{2112} y_{2122} ⋯ y_{21c2}	y_{2212} y_{2222} ⋯ y_{22c2}		y_{2b12} y_{2b22} ⋯ y_{2bc2}
	⋮ ⋮ ⋮	⋮ ⋮ ⋮		⋮ ⋮ ⋮
	y_{211n} y_{212n} ⋯ y_{12cn}	y_{221n} y_{222n} ⋯ y_{22cn}		y_{2b1n} y_{2b2n} ⋯ y_{2bcn}
	⋮ ⋮ ⋮	⋮ ⋮ ⋮		⋮ ⋮ ⋮
	C_{a11} C_{a12} … C_{a1c}	C_{a21} C_{a22} … C_{a2c}	…	C_{ab1} C_{ab2} … C_{abc}
A_a	y_{a111} y_{a121} ⋯ y_{a1c1}	y_{a211} y_{a221} ⋯ y_{a2c1}	…	y_{ab11} y_{ab21} ⋯ y_{abc1}
	y_{a112} y_{a122} ⋯ y_{a1c2}	y_{a212} y_{a222} ⋯ y_{a2c2}		y_{ab12} y_{ab22} ⋯ y_{abc2}
	⋮ ⋮ ⋮	⋮ ⋮ ⋮		⋮ ⋮ ⋮
	y_{a11n} y_{a12n} ⋯ y_{a1cn}	y_{a21n} y_{a22n} ⋯ y_{a2cn}		y_{ab1n} y_{ab2n} ⋯ y_{abcn}

Table 5.31 Analysis of variance table and expectations of the ***MS*** for model I of a balanced three-way analysis of variance A with B cross-classified, C in the $A \times B$-combinations nested.

Source of variation	SS
Between A-levels	$SS_A = \sum_{i=1}^{a} \frac{Y_{i...}^2}{bcn} - \frac{Y_{....}^2}{N}$
Between B-levels	$SS_B = \sum_{j=1}^{a} \frac{Y_{.j..}^2}{acn} - \frac{Y_{....}^2}{N}$
Between C-levels in $A \times B$ combinations	$SS_{C \text{ in } AB} = \sum_{i=1}^{a} \sum_{j=1}^{b} \sum_{k=1}^{c} \frac{Y_{ijk.}^2}{n} - \sum_{i=1}^{a} \sum_{j=1}^{b} \frac{Y_{ij..}^2}{cn}$
Interaction $A \times B$	$SS_{A \times B} = \sum_{i=1}^{a} \sum_{j=1}^{b} \frac{Y_{ij..}^2}{cn} - \sum_{i=1}^{a} \frac{Y_{i...}^2}{bcn} - \sum_{j=1}^{b} \frac{Y_{.j..}^2}{acn} + \frac{Y_{....}^2}{N}$
Residual	$SS_{\text{res}} = \sum_{i=1}^{a} \sum_{j=1}^{b} \sum_{k=1}^{c} \sum_{l=1}^{n} y_{ijkl}^2 - \sum_{i=1}^{a} \sum_{j=1}^{b} \sum_{k=1}^{c} \frac{Y_{ijk.}^2}{n}$
Total	$SS_T = \sum_{i=1}^{a} \sum_{j=1}^{b} \sum_{k=1}^{c} \sum_{l=1}^{n} y_{ijkl}^2 - \frac{Y_{....}^2}{N}$

DF	*MS*	E(*MS*) under (5.52)
$a-1$	$MS_A = \frac{SS_A}{a-1}$	$\sigma^2 + \frac{bcn}{a-1} \sum_{i=1}^{a} a_i^2$
$b-1$	$MS_B = \frac{SS_B}{b-1}$	$\sigma^2 + \frac{acn}{b-1} \sum_{j=1}^{b} b_j^2$
$ab(c-1)$	$MS_{C \text{in} AB} = \frac{SS_{C \text{in} AB}}{ab(c-1)}$	$\sigma^2 + \frac{n}{ab(c-1)} \sum_{i=1}^{a} \sum_{j=1}^{b} \sum_{k=1}^{c} c_{ijk}^2$
$(a-1)(b-1)$	$MS_{A \times B} = \frac{SS_{A \times B}}{(a-1)(b-1)}$	$\sigma^2 + \frac{cn}{(a-1)(b-1)} \sum_{i=1}^{a} \sum_{j=1}^{b} (a,b)_{ij}^2$
$N-abc$	$MS_{res} = \frac{SS_{res}}{N-abc}$	σ^2
$N-1$		

Hints for Programs

Again we determine the minimal experimental size by OPDOE of R using the procedure as in sections above:

```
> size.anova(model="(axb)>c", hypothesis="b",a=6, b=5,
+c=4, alpha=0.05, beta=0.1, delta=0.5, case="minimin")
n
3
> size.anova(model="(axb)>c", hypothesis="b",a=6, b=5,
+c=4,+ alpha=0.05, beta=0.1, delta=0.5, case="maximin")
n
6.
```

5.5 Exercises

5.1 Prove (a) to (d) of Lemma 5.4.

5.2 Analyse the data of Table 5.14 by SPSS or R, that is, compute the analysis of variance table and all F-values.

5.3 Analyse the data of Table 5.18 by SPSS or R, that is, compute the analysis of variance table and all F-values.

5.4 Prove that $X(X^TX)^-X^TX = X$.

5.5 Show that in Example 5.15 the differences $B_2 - B_3, B_1 - B_2$ and $I_n - B_1$ are idempotent and that $(B_2 - B_3)(B_1 - B_2) = (B_2 - B_3)(I_n - B_1) = 0$.

5.6 Install and load in R the program package OPDOE.

5.7 Compute with OPDOE of R for $\alpha = 0.025$, $\beta = 0.1$ and $\delta/\sigma = 1$ maximin and minimin of the one-way analysis of variance for $a = 6$.

5.8 Compute with OPDOE of R for $\alpha = 0.05$, $\beta = 0.1$ and $\delta/\sigma = 1$ maximin and minimin of the two-way cross-classification for testing factor A for $a = 6$ and $b = 4$.

5.9 Compute with OPDOE of R for $\alpha = 0.05$, $\beta = 0.1$ and $\delta/\sigma = 1$ maximin and minimin of the two-way nested classification for testing the factors A and B for $a = 6$ and $b = 4$.

5.10 Compute with OPDOE of R for $\alpha = 0.05$, $\beta = 0.1$ and $\delta/\sigma = 1$ maximin and minimin of the two-way cross-classification for testing the interactions $A \times B$ for $a = 6$ and $b = 4$.

5.11 Compute with OPDOE of R for $\alpha = 0.05$, $\beta = 0.1$ and $\delta/\sigma = 0.5$ maximin and minimin of the three-way cross-classification for testing factor A for $a = 6$, $b = 5$ and $c = 4$.

References

Fisher, R. A. and Mackenzie, W. A. (1923) Studies in crop variation. II. The manorial response of different potato varieties. *J. Agric. Sci.*, **13**, 311–320.

Lenth, R. V. (1986) Computing non-central beta probabilities. *Appl. Stat.*, **36**, 241–243.

Rasch, D. (1971) Mixed classification the three-way analysis of variance. *Biom. Z.*, **13**, 1–20.

Rasch, D., Wang, M. and Herrendörfer, G. (1997) Determination of the size of an experiment for the *F*-test in the analysis of variance. Model I, *Advances in Statistical Software* 6. *The 9th Conference on the Scientific Use of Statistical Software*, Heidelberg, March 3–6, 1997.

Rasch, D., Herrendörfer, G., Bock, J., Victor, N. and Guiard, V. (Eds.) (2008) *Verfahrensbibliothek Versuchsplanung und - auswertung*, 2. verbesserte Auflage in einem Band mit CD, R. Oldenbourg Verlag, München, Wien.

Scheffé, H. (1959) *The Analysis of Variance*, John Wiley & Sons, Inc., New York, Hoboken.

Searle, S. R. (1971, 2012) *Linear Models*, John Wiley & Sons, Inc., New York.

Tukey, J. W. (1949) One degree of freedom for non-additivity. *Biometrics*, **5**, 232–242.

6

Analysis of Variance: Estimation of Variance Components (Model II of the Analysis of Variance)

6.1 Introduction: Linear Models with Random Effects

In this chapter models of the analysis of variance (ANOVA) where all factors are random are considered; we call the model in this case model II. Our aim in such models is not only as in Chapter 5 the testing of particular hypotheses but also the methods of estimating the components of variance. For the latter we first of all consider the best elaborated case of the one-way analysis of variance. We again use the notation of Section 5.1 and consider formally the same models as in Chapter 5. The difference between Chapters 5 and 6 is that the effects of model II are random. We assume that, for instance, for a factor A, say, exactly a levels are randomly selected from a universe P_A of (infinite) levels of the factor A so that $\boldsymbol{\alpha}_1, \dots, \boldsymbol{\alpha}_a$; the effects of these levels are random variables.

The terms main effect and interaction effect are defined analogously as in Chapter 5, but these effects are now random variables and not parameters that could be estimated.

Models, in which some effects are fixed and other are random, are discussed in Chapter 7. In Chapter 6 some terms defined in Chapter 5 are used, without defining them once more.

Definition 6.1 Let $\boldsymbol{Y} = (\boldsymbol{y}_1, \dots, \boldsymbol{y}_N)^T$ be an N-dimensional random vector and $\boldsymbol{\beta} = (\mu, \boldsymbol{\beta}_1, \dots, \boldsymbol{\beta}_k)^T$ a vector, of elements that except for μ are random variables. Further X as in (5.1) is a $N \times (k+1)$ matrix of rank $p < k+1$. The vector $\boldsymbol{e}$ is also an N-dimensional random vector of error terms. Then we call

$$\boldsymbol{Y} = X\boldsymbol{\beta} + \boldsymbol{e} \tag{6.1}$$

a model II of the ANOVA if

$$\operatorname{var}(\boldsymbol{e}) = \sigma^2 I_N, \quad \operatorname{cov}(\boldsymbol{\beta}, \boldsymbol{e}) = O_{k+1,N} \text{ and } E(\boldsymbol{e}) = 0_N, \quad E(\boldsymbol{\beta}) = \begin{pmatrix} \mu \\ 0_k \end{pmatrix}.$$

Mathematical Statistics, First Edition. Dieter Rasch and Dieter Schott.

We write (6.1) in the form

$$Y = \mu 1_N + Z\gamma + e \tag{6.2}$$

where Z is the columns two up to $(k + 1)$- of X and γ contains the second up to the $(k + 1)$-th element of β. Then we have $E(Y) = \mu 1_N$.

If in β of (6.1) effects of r factors and factor combinations occur, we may write

$$\gamma^{\mathrm{T}} = \left(\gamma_{A1}^{\mathrm{T}}, \gamma_{A2}^{\mathrm{T}}, \ldots, \gamma_{Ar}^{\mathrm{T}}\right) \text{ and } Z = \left(Z_{A1}, Z_{A2}, \ldots, Z_{Ar}\right).$$

In a two-way cross-classification with interactions and the factors A and B, we have, for instance, $r = 3$, $A = A_1$, $B = A_2$ and $AB = A_3$. In general, we have

$$Y = \mu 1_N + \sum_{i=1}^{r} Z_{A_i}\gamma_{A_i} + e. \tag{6.3}$$

Definition 6.2 Equation (6.3) under the side conditions of Definition 6.1 and the additional assumption that all elements of γ_{A_i} are uncorrelated and have the same variance σ_i^2 so that $\operatorname{cov}\left(\gamma_{A_i}\gamma_{A_j}\right) = O_{a_i a_j}$ for all $i, j (i \neq j)$ $\operatorname{var}\left(\gamma_{A_i}\right) = \sigma_i^2 I_{a_i}$ if a_i is the number of levels of the factors, A_i is called a special model II of the analysis of variance; σ_i^2 and σ^2 are called components of variance or variance components.

From Definition 6.2 we get

$$\operatorname{var}(Y) = \sum_{i=1}^{r} Z_{A_i} Z_{A_i}^{\mathrm{T}} \sigma_i^2 + \sigma^2 I_N. \tag{6.4}$$

Theorem 6.1 If Y is an N-dimensional random variable so that (6.3) is a model II of the ANOVA following Definition 6.2 for the quadratic form $Y^T A Y$ with an $N \times N$-matrix A, we have

$$E\left(Y^{\mathrm{T}} A Y\right) = \mu^2 1_N^{\mathrm{T}} A 1_N + \sum_{i=1}^{r} \sigma_i^2 \operatorname{tr}\left(A Z_{A_i} Z_{A_i}^{\mathrm{T}}\right) + \sigma^2 \operatorname{tr}(A). \tag{6.5}$$

Proof: We see that

$$E\left(Y^{\mathrm{T}} A Y\right) = \operatorname{tr}[A \ \operatorname{var}(Y)] + E\left(Y^{\mathrm{T}}\right) A E(Y)$$

and because $E(Y) = \mu 1_N$ and (6.4) now follows (6.5).

Theorem 6.1 allows us to calculate the expectations of the mean squares of an ANOVA based on model II, which is of importance to one of the methods for estimating the components σ_i^2 and σ^2.

Henderson (1953), Rao (1970, 1971a), Hartley and Rao (1967), Harville (1977), Drygas (1980) and Searle et al. (1992) developed methods for the estimation of variance components. A part of these methods can also be used for mixed models (Chapter 7). Henderson's ANOVA method works as follows: at first ANOVA for the corresponding model I is calculated including the ANOVA table excluding the calculation of the $E(\boldsymbol{MS})$. Then using (6.5) the $E(\boldsymbol{MS})$ is calculated for model II. These are functions of the variance components σ_i^2. The $E(\boldsymbol{MS})$ then equalised with the observed MS and the resulting equations are solved for σ_i^2. The solutions are used as estimates $\hat{\sigma}_i^2 = s_i^2$ of σ_i^2.

In this way differences occur between the MS, and these can be negative; consequently, negative estimates of the variance components can result as a consequence of this method. That means that the method gives no estimators (or estimates) as defined in Definition 2.1. If the value of a variance component is small (near 0), negative estimates may often occur. Negative estimates may either mean that the estimated component is very small or they may be a signal of an inappropriately chosen model, for instance, if effects are nonadditive. The interpretation of negative estimates is discussed by Verdooren (1982).

In the following sections the method of Henderson is applied for several classifications. The estimator of a component is reached by replacing all observed values in the corresponding equation of the estimate by the corresponding random variables.

Simultaneously with the estimation tests of hypotheses about the variance, components are described.

Besides the method of Henderson, three further methods are mainly in use. For normally distributed $\boldsymbol{Y}$ we can use the maximum likelihood method or a special version of it, the restricted maximum likelihood method (REML). Further we have the MINQUE method (Figure 6.1), minimising a matrix norm. We propose always to use REML.

Each of these four methods can be performed by SPSS via (see next page)

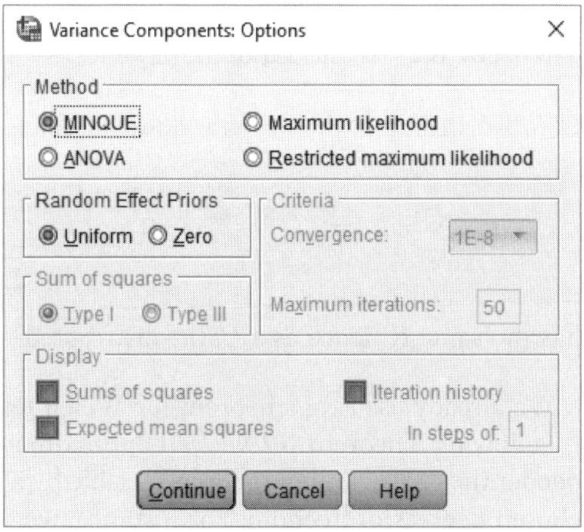

Figure 6.1 Methods of variance component estimation available in SPSS. *Source:* Reproduced with permission of IBM.

Analyze
General Linear Model
Variance Components
Options

Before we discuss special cases, some statements about the approximate distribution of linear combinations of χ^2-distributed random variables must be made. The random variables $\boldsymbol{u}_1, \ldots, \boldsymbol{u}_k$ may be independent of each other, and $\frac{n_i \boldsymbol{u}_i}{\sigma_i^2}$ may be $CS(n_i)$-distributed. The variance of the linear combination

$$\boldsymbol{z} = \sum_{i=1}^{k} c_i \boldsymbol{u}_i \left(c_i \text{ so that } \sum_{i=1}^{k} c_i \sigma_i^2 > 0 \right)$$

is then

$$\operatorname{var}(\boldsymbol{z}) = 2 \sum_{i=1}^{k} c_i^2 \frac{\sigma_i^4}{n_i}.$$

We divide $\boldsymbol{z}$ by the weighted variance $\sigma_W^2 = \sum_{i=1}^{k} c_i \sigma_i^2, \sigma_W^2 > 0$, and we will approximate the distribution of $\frac{n\boldsymbol{z}}{\sigma_W^2}$ for certain n by a χ^2-distribution, which has the same variance as $\frac{n\boldsymbol{z}}{\sigma_W^2}$. This we achieve by putting (following Satterthwaite, 1946)

$$n = \frac{\sigma_W^4}{\sum_{i=1}^{k} c_i^2 \frac{\sigma_i^4}{n_i}}.$$

by Theorem 6.2 below.

Theorem 6.2 If the random variables $\frac{n_i \boldsymbol{u}_i}{\sigma_i^2}$ are independent of each other $CS(n_i)$-distributed, then the random variable $\frac{n\boldsymbol{z}}{\sigma_W^2}$ with

$$\boldsymbol{z} = \sum_{i=1}^{k} c_i \boldsymbol{u}_i,\ \sigma_W^2 = \sum_{i=1}^{k} c_i \sigma_i^2 > 0 \text{ and } n = \frac{\sigma_W^4}{\sum_{i=1}^{k} c_i^2 \frac{\sigma_i^4}{n_i}}$$

has the same variance as a $CS(n)$-distributed variable.

We already used this theorem for Welch test in Chapter 3.

Following Theorem 6.2 we can approximate a linear combination of independently χ^2-distributed random variable by a χ^2-distribution with appropriate chosen degrees of freedom. For instance, we see in Theorem 6.2 that

$$\left[\frac{nz}{CS\left(n,1-\frac{\alpha}{2}\right)},\frac{nz}{CS\left(n,\frac{\alpha}{2}\right)}\right] \tag{6.6}$$

an approximate confidence interval with coefficient $1-\alpha$ for σ_W^2 if σ_W^2, z and n are chosen as in Theorem 6.2. Welch (1956) showed that for $n>0$ a better approximate confidence interval as (6.6) can be found as

$$\left[\frac{nz}{A_{1-\frac{\alpha}{2}}},\frac{nz}{A_{\frac{\alpha}{2}}}\right] \tag{6.7}$$

where

$$A_\gamma = CS(n,\gamma)-\frac{2}{3}\left(2z_{1-\alpha}^2+1\right)\left[z\frac{\sum_{i=1}^{k}\frac{c_i^3 z_i^3}{n_i^2}}{\left(\sum_{i=1}^{k}\frac{c_i^2 z_i^2}{n_i}\right)^2}-1\right].$$

For some cases Graybill and Wang (1980) found a further improvement.

6.2 One-Way Classification

We consider Equation (6.3) for the case $r=1$ and put $\gamma_{A1}=(\boldsymbol{\alpha}_1,\ \dots,\boldsymbol{\alpha}_\alpha)^T$ and $\sigma_i^2=\sigma_a^2$. Then (6.3) can be written in the form

$$\boldsymbol{y}_{ij}=\mu+\boldsymbol{\alpha}_i+\boldsymbol{e}_{ij}\quad (i=1,\dots,a;j=1,\dots,n_i). \tag{6.8}$$

The side conditions of Definition 6.1 are $\operatorname{var}(\boldsymbol{e}_{ij})=\sigma^2$, $\operatorname{var}(\boldsymbol{a}_i)=\sigma_a^2$ and that $\boldsymbol{a}_i$ are independent of each other and $\boldsymbol{e}_{ij}$ are independent of each other and of $\boldsymbol{\alpha}_i$.

From Example 5.1 with X in (6.8) and (6.4) it follows that

$$V=\operatorname{var}(\boldsymbol{Y})=\bigoplus_{i=1}^{a}\left(1_{n_i,n_i}\sigma_a^2+I_{n_i}\sigma^2\right). \tag{6.9}$$

For the case $a=3$, $n_1=n_2=n_3=2$, the direct sum in (6.9) has the form

$$V=\left(1_{2,2}\sigma_a^2+I_2\sigma^2\right)\oplus\left(1_{2,2}\sigma_a^2+I_2\sigma^2\right)\oplus\left(1_{2,2}\sigma_a^2+I_2\sigma^2\right)$$

$$=\begin{bmatrix} \sigma_a^2+\sigma^2 & \sigma_a^2 & 0 & 0 & 0 & 0 \\ \sigma_a^2 & \sigma_a^2+\sigma^2 & 0 & 0 & 0 & 0 \\ 0 & 0 & \sigma_a^2+\sigma^2 & \sigma_a^2 & 0 & 0 \\ 0 & 0 & \sigma_a^2 & \sigma_a^2+\sigma^2 & 0 & 0 \\ 0 & 0 & 0 & 0 & \sigma_a^2+\sigma^2 & \sigma_a^2 \\ 0 & 0 & 0 & 0 & \sigma_a^2 & \sigma_a^2+\sigma^2 \end{bmatrix}$$

Lemma 6.1 If the square matrix V of order n has the form

$$V_n(a,b) = bI_n + a1_{nn},$$

then its determinant is if $b \neq -na$ and $b \neq 0$

$$|V_n(a,b)| = b^{n-1}(b+na) \tag{6.10}$$

and further its inverse is

$$V_n^{-1}(a,b) = V_n\left(\frac{1}{b}, -\frac{a}{b(+na)}\right) = -\frac{a}{b(+na)}1_{n,n} + \frac{1}{b}I_n.$$

Proof (Rasch and Herrendörfer, 1986): We subtract the last column of $V_n(a, b)$ from all the other columns and add the $n - 1$ first rows of the matrix generated in this way to the last row. Then the determinant of the resulting matrix equals (6.10).

If the inverse of $V_n(a, b)$ has the form $dI_n + c1_{n,n}$, then $V_n(a, b)(dI_n + c1_{n,n}) = I_n$, and this leads to $d = \frac{1}{b}; c = -\frac{a}{b(b+an)}$.

Lemma 6.2 The eigenvalues of

$$V = \bigoplus_{i=1}^{a} \left(1_{n_i,n_i}\sigma_a^2 + I_{n_i}\sigma^2\right)$$

are with $N = \sum_{i=1}^{a} n_i$.

$$\lambda_k = \begin{cases} n_k\sigma_a^2 + \sigma^2 & (k = 1,\ldots,a), \\ \sigma^2 & (k = a+1,\ldots,N). \end{cases}$$

The orthogonal eigenvectors are

$$\tau_k = \begin{cases} 1_{n_k} & (k = 1,\ldots,a), \\ s_k & (k = a+1,\ldots,N) \end{cases}$$

with $(s_k) = S_N = \oplus_{i=1}^{a} S_i$ where S_i is the matrix

$$S_i = \begin{pmatrix} 1 & 1 & 1 & \cdots & 1 \\ -1 & 1 & 1 & \cdots & 1 \\ & -2 & 1 & \cdots & 1 \\ & & -3 & \cdots & 1 \\ & & & & \vdots \\ & 0 & & & -(n_i-1) \end{pmatrix}$$

Proof: We have

$$|V-\lambda I_N| = \prod_{i=1}^{a} |V_i - \lambda I_{n_i}| = \prod_{i=1}^{a} |1_{n_i,n_i}\sigma_a^2 + I_{n_i}(\sigma^2-\lambda)|$$

and due to Lemma 6.1

$$|V-\lambda I_N| = \prod_{i=1}^{a}\left\{(\sigma^2-\lambda)^{n_i-1}(n_i\sigma_a^2+\sigma^2-\lambda)\right\} = (\sigma^2-\lambda)^{N-a}\prod_{i=1}^{a}(n_i\sigma_a^2+\sigma^2-\lambda).$$

This term has the $(N-a)$-fold multiple zero $\lambda=\sigma^2$ and a zeros $\lambda_i = n_i\sigma_a^2+\sigma^2$, and this proves the first part.

Orthogonal eigenvectors must fulfil the conditions $V\mathrm{r}_k = \lambda_k \mathrm{r}_k$ and $r_k^T \mathrm{r}_{k'} = 0$ $(k \neq k')$. We put $R=(r_1, \ldots, r_N) = (T_N, S_N)$, where T_N is a $(N\times a)$-matrix and S_N a $N\times(N-a)$-matrix.

With $T_N = \oplus_{i=1_{ni}}^{a}$ we get

$$VT_N = T_N \bigoplus_{i=1}^{a} \lambda_i.$$

Further the columns of T_N are orthogonal, and by this the columns of T_N are the eigenvectors of the first k eigenvalues. For the $N-a$ eigenvalues $\lambda_k=\sigma^2(k=a+1, \ldots, N)$, we have

$$Vr_k = \sigma^2 r_k\,(k=a+1,\ldots,N)$$

or

$$\left(V-\sigma^2 I_N\right)r_k = 0 \;\;(k=a+1,\ldots,N)$$

or

$$\sigma_a^2 T_N T_N^T r_k = 0 \;\;(k=a+1,\ldots,N).$$

With $S_N = \oplus_{i=1}^{a} S_i = \left(r_{a+1},\ldots,r_N\right)$ the last condition is fulfilled if $1_{n_i-1,n_i}S_i = 0$. From the orthogonality property, it follows that

$$S_i^T S_i = \begin{pmatrix} 2 & 0 & \cdots & 0 \\ 0 & 6 & \cdots & 0 \\ \vdots & \vdots & & \vdots \\ 0 & 0 & \cdots & n_i(n_i-1) \end{pmatrix}.$$

With

$$S_i = \begin{pmatrix} 1 & 1 & 1 & \cdots & 1 \\ -1 & 1 & 1 & \cdots & 1 \\ & -2 & 1 & \cdots & 1 \\ & & -3 & \cdots & 1 \\ & & & & \vdots \\ & 0 & & & -(n_i-1) \end{pmatrix}$$

all conditions are fulfilled. Further the columns of T_N and S_N are orthogonal.

6.2.1 Estimation of Variance Components

For the one-way classification, several methods of estimation are described and compared with each other. The ANOVA method is the simplest one and stems from the originator of the ANOVA R. A. FISHER. In HENDERONS fundamental paper from 1953, it was mentioned as method I.

6.2.1.1 Analysis of Variance Method

In Table 5.2 we find the ***SS***, *df* and ***MS*** of a one-way analysis of variance. These terms are independent of the model; they are the same for model I and model II. But $E(\boldsymbol{MS})$ for model II differs from those of model I. Further we have to respect that for model II; $\boldsymbol{y}_{ij}$ in (6.8) within the classes are not independent. We have, namely,

$$\operatorname{cov}\left(\boldsymbol{y}_{ij},\boldsymbol{y}_{ik}\right)=E\left[\left(\boldsymbol{y}_{ij}-\mu\right)\left(\boldsymbol{y}_{ik}-\mu\right)\right]=E\left[\left(\boldsymbol{a}_i+\boldsymbol{e}_{ij}\right)\left(\boldsymbol{a}_i+\boldsymbol{e}_{ik}\right)\right]$$

and from the side conditions of model II it follows:

$$\operatorname{cov}\left(\boldsymbol{y}_{ij},\boldsymbol{y}_{ik}\right)=E\left(\boldsymbol{a}_i^2\right)=\sigma_a^2.$$

We call $\operatorname{cov}(\boldsymbol{y}_{ij},\boldsymbol{y}_{ik})$ the covariance within classes.

Definition 6.3 The correlation coefficient between two random variables $\boldsymbol{y}_{ij}$ and $\boldsymbol{y}_{ik}$ in the same class *i* of an experiment for which model II of the ANOVA as in (6.8) can be used is called within-class correlation coefficient and is given by

$$\rho_I=\frac{\sigma_a^2}{\sigma_a^2+\sigma^2}.$$

The within-class correlation coefficient ρ_I is independent of the special class *i*.

We now derive $E(\boldsymbol{MS})$ for model II. $E(\boldsymbol{MS}_I)=E(\boldsymbol{MS}_{res})$ is as in model I equal to σ^2. For $E(\boldsymbol{MS}_A)$ follows from model (6.8)

$$E(\boldsymbol{MS}_A)=\frac{1}{a-1}\left(E\left[\sum_i\frac{\boldsymbol{Y}_{i.}^2}{n_i}\right]-E\left[\frac{\boldsymbol{Y}_{..}^2}{N}\right]\right).$$

At first, we calculate $E\left(\boldsymbol{Y}_{i.}^2\right)$. The model assumptions supply

$$E\left(\boldsymbol{Y}_{i\cdot}^2\right)=E\left[\left(n_i\mu+n_i\boldsymbol{a}_i+\sum_j\boldsymbol{e}_{ij}\right)^2\right]=n_i^2\mu^2+n_i^2\sigma_a^2+n_i\sigma^2$$

and by this

$$E\left(\sum_i\frac{\boldsymbol{Y}_{i\cdot}^2}{n_i}\right)=N\mu^2+N\sigma_a^2+a\sigma^2.$$

For $E\left(\frac{\boldsymbol{Y}_{..}^2}{N}\right)$ we obtain due to $\boldsymbol{Y}_{..} = N\mu + \sum_{i=1}^{a} n_i \boldsymbol{a}_i + \sum_{i,j} \boldsymbol{e}_{ij}$

$$E\left(\frac{\boldsymbol{Y}_{..}^2}{N}\right) = N\mu^2 + \frac{1}{N}\sum n_i^2 \sigma_a^2 + \sigma^2.$$

And therefore we obtain

$$E(\boldsymbol{MS}_A) = \frac{1}{a-1}\left[\sigma_a^2\left(N - \frac{\sum n_i^2}{N}\right)\right] + \sigma^2. \tag{6.11}$$

If in (6.8) all $n_i = n$, then because $\sum n_i^2 = n^2 a$ and $N = a \cdot n$

$$E(\boldsymbol{MS}_A) = \sigma^2 + n\sigma_a^2.$$

Then an unbiased estimator of σ_a^2 simply can be gained from $\boldsymbol{MS}_A$ and $\boldsymbol{MS}_{res}$ by

$$\boldsymbol{s}_a^2 = \frac{1}{n}(\boldsymbol{MS}_A - \boldsymbol{MS}_{res})$$

or

$$\boldsymbol{s}_a^2 = \frac{1}{n}\left[\frac{1}{a-1}\left(\sum_j \frac{\boldsymbol{Y}_{i.}^2}{n} - \frac{\boldsymbol{Y}_{..}^2}{n}\right) - \frac{1}{N-a}\left(\sum_{i,j} \boldsymbol{y}_{ij}^2 - \sum_i \frac{\boldsymbol{Y}_{i.}^2}{n}\right)\right].$$

In general $\boldsymbol{s}_a^2$ is given by

$$\boldsymbol{s}_a^2 = \frac{a-1}{N - \frac{\sum n_i^2}{N}}(\boldsymbol{MS}_A - \boldsymbol{MS}_{res}). \tag{6.12}$$

This corresponding estimates are negative if $MS_{res} > MS_A$.

As already mentioned this approach to put the calculated *MS* equal to the $E(\boldsymbol{MS})$ is called the ANOVA method and can be used for any higher or nested classification. The corresponding estimators gained by transition to random variables; these unbiased estimators can give negative estimates. Later we will not use unbiased estimators, which sometimes give non-negative estimates of the variance components, see, for example, REML.

If we are interested in an estimation in the sense of Definition 2.1 (mapping into R^+) and use $\mathrm{Max}(0, \boldsymbol{s}_a^2)$ as estimator, the unbiasedness is lost, but the *MQD* becomes smaller as for $\boldsymbol{s}_a^2$. The matrix A of the quadratic form $Y^T A Y = \sum_{i=1}^{a} \frac{Y_{i.}^2}{n_i} - \frac{Y_{..}^2}{N}$ is

$$A = \oplus_i \frac{1}{n_i} 1_{n_i,n_i} - \frac{1}{N} 1_{N,N}.$$

From (6.9) we obtain

$$\text{tr}[A\ \text{var}(\boldsymbol{Y})] = \sum_{i=1}^{a}\sum_{j=1}^{n_i}\left[\sigma_a^2 - \frac{n_i}{N}\sigma_a^2 + \frac{1}{n_i}\sigma^2 - \frac{1}{N}\sigma^2\right]$$

$$= \sigma_a^2\left(N - \sum_{i=1}^{a}\frac{n_i^2}{N}\right) + \sigma^2(a-1).$$

Further $E[\boldsymbol{Y}^{\mathrm{T}}]AE[\boldsymbol{Y}] = 0$ because $E[\boldsymbol{Y}] = \mu 1_N$. By this we obtain again (6.11) from (6.5). The matrices A and var($\boldsymbol{Y}$) are confusing for higher classifications. For the case of equal subclass numbers, simple rules for the calculation of the $E(\boldsymbol{MS})$ exist, which will be described in Chapter 7 for the general case of the mixed model as well as for specialisations for model II. The two methods presented below only for the case of unequal subclass numbers are really needed.

6.2.1.2 Estimators in Case of Normally Distributed *Y*

We assume now that the vector $\boldsymbol{Y}$ of $\boldsymbol{y}_{ij}$ in (6.8) are $N(\mu 1_N, V)$-distributed with V from (6.9). Further we assume $n_i = n(i = 1, \ldots, a)$, that is, $N = an$. From (6.10) and Lemma 6.1, it follows that

$$|V| = (\sigma^2)^{a(n-1)}(\sigma^2 + n\sigma_a^2)^a$$

and

$$V^{-1} = \oplus_i \frac{1}{\sigma^2}\left(I_n - \frac{\sigma_a^2}{\sigma^2 + n\sigma_a^2}1_{n,n}\right)$$

with a summands in the direct sum. The density function of $\boldsymbol{Y}$ is

$$f(Y|\mu,\sigma^2,\sigma_a^2) = \frac{1}{(2\pi)^{\frac{N}{2}}|V|^{\frac{1}{2}}}e^{\left[\frac{1}{2}(Y-\mu 1_N)^T V^{-1}(Y-\mu 1_N)\right]}$$

$$= \frac{e^{\frac{1}{2\sigma^2}(Y-\mu 1_N)^T(Y-\mu 1_N) + \frac{\sigma_a^2}{2\sigma^2(\sigma^2+n\sigma_a^2)}(Y-\mu 1_N)^T \bigoplus_{i=1}^{a} 1_{n,n}(Y-\mu 1_N)}}{(2\pi)^{\frac{N}{2}}(\sigma^2)^{\frac{a}{2}(n-1)}(\sigma^2+n\sigma_a^2)^{\frac{a}{2}}}.$$

Because

$$(Y-\mu 1_N)^T(Y-\mu 1_N) = \sum_{i,j}\left(y_{ij} - \bar{y}_{i.} + \bar{y}_{.j} - \mu\right)^2$$

$$= \sum_{i,j}(y_{ij} - \bar{y}_{i\cdot})^2 + n\sum_{i=1}^{a}(\bar{y}_{i\cdot} - \mu)^2$$

and

$$(Y-\mu 1_N)^T \oplus 1_{n,n}(Y-\mu 1_N) = n^2\sum_{i,j}\left(y_{ij} - \bar{y}_{i.} + \bar{y}_{.j} - \mu\right)^2,$$

$$= n^2\sum\nolimits_{i=1}^{a}(\bar{y}_{i.} - \bar{y}_{..})^2 + an^2(\bar{y}_{..} - \mu)^2$$

this density becomes

$$f\left(Y|\mu,\sigma^2,\sigma_a^2\right) = \frac{\exp\left[-\frac{1}{2}\left(\frac{SS_{res}}{\sigma^2} + \frac{SS_A}{\sigma^2 + n\sigma_a^2} + \frac{an(\bar{y}_{..} - \mu)^2}{\sigma^2 + n\sigma_a^2}\right)\right]}{(2\pi)^{\frac{N}{2}}(\sigma^2)^{\frac{a}{2}(n-1)}\left(\sigma^2 + n\sigma_a^2\right)^{\frac{a}{2}}} = L$$

with SS_{res} and SS_A from Theorem 5.3.

The maximum likelihood estimates $\tilde{\sigma}^2$, $\tilde{\sigma}_a^2$ and $\tilde{\mu}$ are obtained, by zeroing the derivations of ln L with respect to the three unknown parameters and obtain

$$0 = \frac{-an}{\tilde{\sigma}^2 + n\tilde{\sigma}_a^2}(\bar{y}_{..} - \tilde{\mu})$$

$$0 = -\frac{a(n-1)}{2\tilde{\sigma}^2} - \frac{a}{2\left(\tilde{\sigma}^2 + n\tilde{\sigma}_a^2\right)} + \frac{SS_{res}}{2\tilde{\sigma}^4} + \frac{SS_A}{2\left(\tilde{\sigma}^2 + n\tilde{\sigma}_a^2\right)^2}$$

$$0 = -\frac{na}{2\left(\tilde{\sigma}^2 + n\tilde{\sigma}_a^2\right)} + \frac{nSS_A}{2\left(\tilde{\sigma}^2 + n\tilde{\sigma}_a^2\right)^2}.$$

From the first equation (after transition to random variables), it follows for the estimators

$$\tilde{\mu} = \bar{y}_{..}$$

and from the two other equations

$$a\left(\tilde{\boldsymbol{\sigma}}^2 + n\tilde{\boldsymbol{\sigma}}_a^2\right) = \boldsymbol{SS}_A$$

or

$$\tilde{\boldsymbol{\sigma}}^2 = \frac{\boldsymbol{SS}_{res}}{a(n-1)} = \boldsymbol{s}^2 = \boldsymbol{MS}_{res} \tag{6.13}$$

and

$$\tilde{\boldsymbol{\sigma}}_a^2 = \frac{1}{n}\left[\frac{\boldsymbol{SS}_{\mathrm{A}}}{a} - \boldsymbol{MS}_{res}\right] = \frac{1}{n}\left[\left(1 - \frac{1}{a}\right)\boldsymbol{MS}_{\mathrm{A}} - \boldsymbol{MS}_{res}\right]. \tag{6.14}$$

Because the matrix of the second derivations is negative definite, we reach maxima.

As it is easy to see, $\tilde{\mu}$ and s^2 are for μ and σ^2 unbiased. But $\tilde{\sigma}_a^2$ has following (6.11) the expectation

$$E\left(\tilde{\sigma}_a^2\right) = \frac{1}{n}\left[\left(1 - \frac{1}{a}\right)\left(\sigma^2 + n\sigma_a^2\right) - \sigma^2\right] = \sigma_a^2 - \frac{1}{an}\left(\sigma^2 + n\sigma_a^2\right).$$

Because $\tilde{\sigma}_a^2$ for $\left(1 - \frac{1}{a}\right)\boldsymbol{MS}_A < \boldsymbol{MS}_{res}$ is negative, $\left(\tilde{\sigma}^2, \tilde{\sigma}_a^2\right)$ is in general no *MLS* concerning $\left(\sigma^2, \sigma_a^2\right)$ because following Chapter 2 the maximum must be taken with respect to Ω, that is, for all $\theta \in R^1 \times (R^+)^2$.

Herbach (1959) could show that besides this maximum $\breve{\mu} = \bar{y}_{..}$ leads to

$$\breve{\sigma}_a^2 = \begin{cases} \frac{1}{n}\left[\left(1-\frac{1}{a}\right)\boldsymbol{MS}_A - \boldsymbol{MS}_{\text{res}}\right], & \text{if } \left(1-\frac{1}{a}\right)\boldsymbol{MS}_{\text{A}} \geq \boldsymbol{MS}_{\text{res}} \\ 0 & \text{otherwise} \end{cases} \tag{6.15}$$

and

$$\breve{\sigma}^2 = \begin{cases} s^2, & \text{if } \left(1-\frac{1}{a}\right)\boldsymbol{MS}_{\text{A}} \geq \boldsymbol{MS}_{\text{res}} \\ 0 & \text{otherwise} \end{cases}. \tag{6.16}$$

Both estimators are biased.

Using the given notation of SS_{res} after Theorem 5.3 and SS_A, the exponents in the exponential function of $f\left(Y|\mu,\sigma^2,\sigma_a^2\right)$ are equal to

$$M = -\frac{1}{2\sigma^2}\left[\sum_i\sum_j y_{ij}^2 - \frac{1}{n}\sum_{i=1}^{a} Y_{i\cdot}^2\right] - \frac{1}{2\left(\sigma^2+n\sigma_a^2\right)}\left[\frac{1}{n}\sum_{i=1}^{a} Y_{i\cdot}^2 - \frac{1}{an}Y_{..}^2\right] - \frac{an(\bar{y}_{..}-\mu)^2}{2\left(\sigma^2+n\sigma_a^2\right)}$$

$$= \eta_1 M_1(\boldsymbol{Y}) + \eta_2 M_2(\boldsymbol{Y}) + \eta_3 M_3(\boldsymbol{Y}) + A(\eta)$$

where $A(\eta)$ only depends on θ.

This is the canonical form of a three parametric exponential family of full rank with

$$\eta_1 = -\frac{1}{2\sigma^2},\ \eta_2 = \frac{n}{2\left(\sigma^2+n\sigma_a^2\right)},\ \eta_3 = \frac{n}{2\left(\sigma^2+n\sigma_a^2\right)}$$

and

$$M_1(\boldsymbol{Y}) = \sum_{i=1}\sum_{j=1}\mathbf{y}_{ij}^2,\ M_2(\boldsymbol{Y}) = \sum_{i=1}\bar{y}_{i.}^2,\ M_3(\boldsymbol{Y}) = \boldsymbol{y}_{..}.$$

By this is $(M_1(Y), M_2(Y), M_3(Y))$ following the conclusion of the Chapters 1 and 2 an UVUE of (η_1, η_2, η_3).

6.2.1.3 REML Estimation

The method REML can be found in Searle et al. (1992). We describe this estimation generally in Chapter 7 for mixed models. The method means that the likelihood function of TY is maximised, where T is a $(N - a - 1) \times N$-matrix, whose rows are $N - a - 1$ linear independent rows of $I_N - X(X^TX)^- X^T$.

The (natural) logarithm of the likelihood function of TY is

$$\ln L - \frac{1}{2}(N-a-1)\ln(2\pi) - \frac{1}{2}(N-a-1)\ln\sigma^2 - \frac{1}{2}\ln\left(\left|\frac{\sigma_a^2}{\sigma^2}TVT^T\right|\right)$$
$$-\frac{1}{2\sigma^2 Y^T T^T \frac{\sigma_a^2}{\sigma^2}TVT^T TY}\frac{\sigma_a^2}{\sigma^2}TVT^T.$$

Now we differentiate this function with respect to σ^2 and $\frac{\sigma_a^2}{\sigma^2}$ and zeroing this derivation. The arising equation we solve iteratively and gain the estimates.

Because the matrix of second derivatives is negative definite, we find maxima.

This method is increasingly in use in the applications; even for not normally distributed variables, the REML method is equivalent to an iterative MINQUE; it is discussed in the next section.

6.2.1.4 Matrix Norm Minimising Quadratic Estimation

We look now for quadratic estimators for σ_a^2 and σ^2 that are unbiased and invariant against translation of the vector $\boldsymbol{Y}$ and have minimal variance for the case that $\sigma_a^2 = \lambda\sigma^2$ with known $\lambda > 0$. By this the estimators are in the sense of Definition 2.3 *LVES* in the class the translation invariant quadratic estimators.

We start with the general model (6.8) with the covariance matrix var($\boldsymbol{Y}$) = V in (6.9) and put

$$\frac{\sigma_a^2}{\sigma^2} = \lambda, \quad \lambda \in R^+.$$

Theorem 6.3 For model (6.8) under the corresponding side conditions

$$\boldsymbol{S}_a^2 = \frac{1}{(N-1)K - L^2}\left\{\left[N-1-2\lambda+\lambda^2K\right]\boldsymbol{Q}_1 - (L-\lambda K)\boldsymbol{Q}_2\right\}, \tag{6.17}$$

$$\boldsymbol{S}^2 = \frac{1}{(N-1)K - L^2}\left[K\boldsymbol{Q}_2 - (L-\lambda K)\boldsymbol{Q}_1\right] \tag{6.18}$$

is at $\lambda \in R^+$ an LVUE concerning $\begin{pmatrix}\sigma_a^2\\ \sigma^2\end{pmatrix}$ in class $\mathcal{K}$ of all estimators of the quadratic form $\boldsymbol{Q} = \boldsymbol{Y}^T A\boldsymbol{Y}$, having finite second moments, and are invariant against transformations of the form $\boldsymbol{X} = \boldsymbol{Y} + a$ with a constant $(n \times 1)$-vector a. Here the symbols $L, K, \boldsymbol{Q}_1, \boldsymbol{Q}_2$ in (6.17) and (6.18) are defined as follows:

Initially let

$$\tilde{\tilde{y}}_{..} = \left(\sum_{i=1}^{a} \frac{n_i}{n_i\lambda + 1}\right)^{-1} \sum_{i=1}^{a} \frac{n_i}{n_i\lambda + 1}\bar{y}_{i.}$$

and

$$k_t \sum_{i=1}^{a} \left(\frac{n_i}{n_i\lambda + 1}\right)^t \quad (t = 1,2,3).$$

Then we have

$$L = k_1 - \frac{k_2}{k_1},\ K = K_2 - 2\frac{k_3}{k_1} + \frac{k_2^2}{k_1^2},$$

$$\boldsymbol{Q}_1 = \sum_{i=1}^{a} \frac{n_i^2}{(n_i\lambda + 1)^2}\left(\bar{y}_{i\cdot} - \tilde{\tilde{y}}_{..}\right)^2 \tag{6.19}$$

and

$$\boldsymbol{Q}_2 = \boldsymbol{Q}_1 + \boldsymbol{SS}_I \tag{6.20}$$

with $\boldsymbol{SS}_I$ from Section 5.2.

The proof of this theorem is from Rao (1971b) and is not repeated here.

6.2.1.5 Comparison of Several Estimators

Which of the estimators offered should be applied in practice? Methods leading to negative estimates for positive defined quantities are not estimators because they do not map into the parameter space and are often not accepted. In practice the estimation of σ_a^2 is often done following Herbach's approach with a truncated estimation analogous to (6.15), but contrary to (6.16), $s^2 = \boldsymbol{MS}_I$ is always used. We lose by this the unbiasedness of the estimator of σ_a^2.

For the special case equal subclass numbers $n_i = n(i = 1, \dots, a)$, we have

Theorem 6.4 The estimators of the ANOVA method

$$\boldsymbol{s}^2 = \boldsymbol{MS}_I \tag{6.21}$$

and $\boldsymbol{s}_a^2$ following (6.12) and the LVUE (6.17) and (6.18) for (σ^2, σ_a^2) are for $n_i = n$ identical. In this case the LVUE do not depend on $\lambda = \frac{\sigma_a^2}{\sigma^2}$ and because of this are also UVUE in class $\mathcal{K}$.

Proof: Initially from $n_i = n$, (6.12) becomes

$$\boldsymbol{s}_a^2 = \frac{1}{n}(\boldsymbol{MS}_A - \boldsymbol{MS}_I). \tag{6.22}$$

The constants in (6.17) and (6.18) simplify for $n_i = n$ as follows:

$$k_1 = \frac{an}{n\lambda+1},\ k_2 = \frac{an^2}{(n\lambda+1)^2},\ k_3 = \frac{an^3}{(n\lambda+1)^3}.$$

By this we obtain

$$L = \frac{(a-1)n}{n\lambda+1},\ K = \frac{(a-1)n^2}{(n\lambda+1)^2}$$

and further

$$L - \lambda K = \frac{n(a-1)}{(n\lambda+1)^2},\quad (N-1)K - L^2 = \frac{(N-a)(a-1)n^2}{(n\lambda+1)^2}.$$

Finally

$$N - 1 - 2\lambda L + \lambda^2 K = N - a + \frac{a-1}{(n\lambda+1)^2}.$$

Because in our special case $\tilde{\tilde{\bar{y}}}_{..} = \bar{y}_{..}$ (6.19) and (6.20) simplify to

$$\boldsymbol{Q}_1 = \frac{n}{(n\lambda+1)^2}\boldsymbol{SS}_A,\quad \boldsymbol{Q}_2 = \frac{1}{(n\lambda+1)^2}\boldsymbol{SS}_A + \boldsymbol{SS}_{\mathrm{I}}.$$

By this $\boldsymbol{S}_a^2$ in (6.17) becomes

$$\begin{aligned}\boldsymbol{S}_a^2 &= \frac{(n\lambda+1)^2}{(N-a)(a-1)n^2}\left\{\left[N-a+\frac{a-1}{(n\lambda+1)^2}\right]\frac{n}{(n\lambda+1)^2}\boldsymbol{SS}_A\right.\\ &\quad\left.-\frac{n(a-1)}{(n\lambda+1)^2}\left[\frac{1}{(n\lambda+1)^2}\boldsymbol{SS}_{\mathrm{A}} + \boldsymbol{SS}_{\mathrm{I}}\right]\right\}\\ &= \frac{1}{n}(\boldsymbol{MS}_{\mathrm{A}} - \boldsymbol{MS}_{\mathrm{I}})\end{aligned}$$

and this is independent of λ and identical with $\boldsymbol{s}_a^2$ in (6.12). Analogously follows from (6.18) the relation

$$\boldsymbol{S}^2 = \boldsymbol{MS}_{\mathrm{I}} = \boldsymbol{s}^2.$$

By this we propose to proceed in the case of equal subclass numbers ($n_i = n$) by estimating analogue to (6.15) σ_a^2 by

$$\boldsymbol{S}_a^{*2} = \begin{cases} \boldsymbol{s}_a^2 = \frac{1}{n}(\boldsymbol{MS}_A - \boldsymbol{MS}_{\mathrm{I}}), & \text{if } \boldsymbol{MS}_A > \boldsymbol{MS}_{\mathrm{I}} \\ 0 & \text{otherwise} \end{cases}$$

and σ^2 by $\boldsymbol{MS}_i$ via (6.21). These estimators are biased but have small *MSD*.

But how to act in the case of unequal subclass numbers? How good are the MINQUE-estimators if we use a wrong λ-value? Often we have no idea how to choose λ. What is the consequence of 'unbalancedness' (the inequality of n_i) to the UVUE-property? Empirical results are given in Ahrens (1983). MINQUE can of course be used iteratively or adaptively by starting with some a priori values for the variance components and choose the new estimates as a priori information for the next step. Such an 'iterative MINQUE' converges often to the REML estimates given in Section 6.2.1.3. For this, see Searle et al. (1992). Rasch and Mašata (2006) compared the four methods above and some more by simulation with unbalanced data. They found nearly no differences; the total variance was best estimated by REML and MINQUE.

6.2.2 Tests of Hypotheses and Confidence Intervals

To construct confidence intervals for σ_a^2 and σ^2 and to test hypotheses about these variance components, we need as in Section 6.2.1.2 a further side condition in model equation (6.6) about the distribution of $\boldsymbol{y}_{ij}$. We assume again that $\boldsymbol{y}_{ij}$ are $N\left(\mu, \sigma_a^2 + \sigma^2\right)$-distributed. Then for the distribution of $\boldsymbol{MS}_B$ and $\boldsymbol{MS}_{res}$, use the following theorem for the special case of equal subclass numbers.

Theorem 6.5 The random vector $\boldsymbol{Y}$ following the model equation (6.8) for $n_1 = \cdots = n_a = n$ for its components $\boldsymbol{y}_{ij}$ may be $N(\mu e_N, V)$-distributed. Here, $V = \text{var}(\boldsymbol{Y})$ is given by (6.9). Then the quadratic forms $\frac{\boldsymbol{SS}_1}{\sigma^2} = \boldsymbol{u}_1$ and $\frac{\boldsymbol{SS}_A}{\sigma^2 + n\sigma_a^2} = \boldsymbol{u}_2$ are independent of each other and are $CS[a\ (n - 1)]$- and $CS[a - 1]$-distributed, respectively.

Proof: We write

$$\boldsymbol{u}_1 = \boldsymbol{Y}^{\mathrm{T}} A_1 \boldsymbol{Y} \quad \text{with} \quad A_1 = \frac{1}{\sigma^2}\left[I_N - \frac{1}{n}\bigoplus_{i=1}^{a} 1_{n,n}\right]$$

and

$$\boldsymbol{u}_2 = \boldsymbol{Y}^{\mathrm{T}} A_2 \boldsymbol{Y} \quad \text{with} \quad A_2 = \frac{1}{\sigma^2 + \sigma_a^2}\left[\frac{1}{n}\bigoplus_{i=1}^{a} 1_{n,n} - \frac{1}{N} 1_{N,N}\right].$$

Now, from (6.9) with $n_i = n$,

$$\begin{aligned} A_1 V &= \frac{1}{\sigma^2}\left\{\bigoplus_{i=1}^{a}\left[\sigma^2 I_n + \sigma_a^2 1_{n,n}\right] - \frac{\sigma^2}{n}\bigoplus_{i=1}^{a} 1_{n,n} - \sigma_a^2 \bigoplus_{i=1}^{a} 1_{n,n}\right\} \\ &= I_N - \frac{1}{n}\bigoplus_{i=1}^{a} 1_{n,n} \end{aligned} \tag{6.23}$$

and this is an idempotent matrix using

$$1_{nm} 1_{mr} = m 1_{nr}, \tag{6.24}$$

Further

$$A_2 = \frac{1}{\sigma^2 + \sigma_a{}^2}\left[\bigoplus_{i=1}^{a} 1_{n,n}\sigma_a{}^2 - \frac{n}{N}1_{N,N}\sigma_a{}^2 + \bigoplus_{i=1}^{a} 1_{n,n}\sigma^2 - \frac{n}{N}1_{N,N}\sigma^2 = \bigoplus_{i=1}^{a} 1_{n,n} - \frac{n}{N}1_{N,N}\right] \tag{6.25}$$

and this is idempotent.

We now only have to show that $A_1 V A_2 = 0$, but this follows from

$$\left(\sigma^2 + n\sigma_a^2\right)A_1 V A_2 = \left(I_N - \frac{1}{n}\bigoplus_{i=1}^{a} 1_{n,n}\right)\left(\frac{1}{n}\bigoplus_{i=1}^{a} 1_{n,n} - \frac{1}{N}1_{N,N}\right) = 0.$$

Because rk $(A_1) = N - a = a(n-1)$ and rk $(A_2) = a - 1$, the proof of Theorem 6.5 is completed.

From Theorem 6.5, it follows that

Corollary 6.1 Under the assumptions of Theorem 6.5 is

$$\boldsymbol{F} = \frac{\boldsymbol{SS}_{\mathrm{A}}}{\boldsymbol{SS}_{\mathrm{I}}}\frac{a(n-1)\sigma^2}{(a-1)\left(\sigma^2 + n\sigma_a^2\right)} \tag{6.26}$$

and under the null hypothesis $H_0 : \sigma_a^2 = 0$, this becomes

$$\boldsymbol{F} = \frac{\boldsymbol{SS}_{\mathrm{A}}}{\boldsymbol{SS}_{\mathrm{I}}}\frac{a(n-1)}{a-1}, \tag{6.27}$$

and this is $F[a-1, a(n-1)]$-distributed.

Corollary 6.1 allows us to use $\boldsymbol{F}$ in (6.27) to test the null hypothesis $H_0 : \sigma_a^2 = 0$. The test statistic (6.27) is identical with that in (5.11) and under the corresponding null hypothesis both test statistics have the same distribution. If the null hypothesis is wrong then, $\boldsymbol{F}$ in (5.11) is in the case $\sigma_a^2 > 0$ the $\frac{\sigma^2 + n\sigma_a^2}{\sigma^2}$-fold of a centrally F-distributed random variable. By this we can construct confidence intervals for the variance components. Because $\boldsymbol{u}_1$ is $CS[a(n-1)]$-distributed,

$$\left[\frac{\boldsymbol{SS}_{\mathrm{I}}}{\chi^2\left[a(n-1)|\,1 - \frac{\alpha}{2}\right]}, \frac{\boldsymbol{SS}_{\mathrm{I}}}{\chi^2\left[a(n-1)\,|\frac{\alpha}{2}\right]}\right] \tag{6.28}$$

is a $(1-\alpha)$-confidence interval for σ^2 if $n = n_1 = \cdots = n_a$. From Corollary 6.1 it follows that

$$\left[\frac{\boldsymbol{MS}_{\mathrm{A}} - \boldsymbol{MS}_{\mathrm{I}}F_{1-\frac{\alpha}{2}}}{\boldsymbol{MS}_{\mathrm{A}} + (n-1)\boldsymbol{MS}_{\mathrm{I}}F_{1-\frac{\alpha}{2}}}, \frac{\boldsymbol{MS}_{\mathrm{A}} - \boldsymbol{MS}_{\mathrm{I}}F_{\frac{\alpha}{2}}}{\boldsymbol{MS}_{\mathrm{A}} + (n-1)\boldsymbol{MS}_{\mathrm{I}}F_{\frac{\alpha}{2}}}\right] \tag{6.29}$$

with $F_\varepsilon = F[a+1, a(n-1) \mid \varepsilon]$ is a $(1-\alpha)$-confidence interval for $\frac{\sigma_a^2}{\sigma^2+\sigma_a^2}$. An approximate confidence interval for σ_a^2 in the case of unequal subclass numbers is obtained (Seely and Lee, 1994).

6.2.3 Variances and Properties of the Estimators of the Variance Components

As we have seen, estimators from the ANOVA method are unbiased concerning the two variance components. From (6.11) and (6.12) we get

$$E\left(\boldsymbol{s}_a^2\right) = \sigma_a^2,$$

and

$$E\left(\boldsymbol{s}^2\right) = \sigma^2.$$

Now we need the variance of the estimators $\boldsymbol{s}_a^2$ and $\boldsymbol{s}^2$. By the analysis of variance method, all estimators of the variance components are linear combinations of the $\boldsymbol{MS}$. From Theorem 6.5 it follows that $\boldsymbol{MS}_A$ and $\boldsymbol{MS}_{res}$ are stochastically independent if all subclass numbers are equal. In this case we have $\mathrm{cov}(\boldsymbol{MS}_{res}, \boldsymbol{MS}_A) = 0$:

$$\left.\begin{aligned} \mathrm{var}(\boldsymbol{s}^2) &= \mathrm{var}(\boldsymbol{MS}_{\mathrm{res}}) \\ \mathrm{var}\left(\boldsymbol{s}_a^2\right) &= \frac{1}{n^2}\left[\mathrm{var}(\boldsymbol{MS}_A) + \mathrm{var}(\boldsymbol{MS}_{\mathrm{res}})\right] \end{aligned}\right\}. \tag{6.30}$$

In the case where $\boldsymbol{Y}$ is $N(\mu 1_n, \mathrm{V})$-distributed, it follows from Theorem 6.5 that

$$\mathrm{var}\left(\frac{\boldsymbol{SS}_{res}}{\sigma^2}\right) = 2a(n-1) = \mathrm{var}\left[\frac{a(n-1)}{\sigma^2}\boldsymbol{MS}_{res}\right].$$

This immediately leads to

$$\mathrm{var}\left(\boldsymbol{s}^2\right) = \mathrm{var}(\boldsymbol{MS}_{res}) = \frac{2\sigma^4}{a(n-1)}. \tag{6.31}$$

Analogously

$$\mathrm{var}\left(\frac{\boldsymbol{SS}_{\mathrm{A}}}{\sigma^2 + n\sigma_a^2}\right) = 2(a-1) = \mathrm{var}\left[\frac{a-1}{\sigma^2+n\sigma_a^2}\boldsymbol{MS}_{\mathrm{A}}\right]$$

and

$$\mathrm{var}(\boldsymbol{MS}_A) = 2\frac{\left(\sigma^2+n\sigma_a^2\right)^2}{a-1}. \tag{6.32}$$

From (6.31), (6.32) and (6.30), we obtain, if $\boldsymbol{Y}$ is $N(\mu 1_N, V)$-distributed,

$$\operatorname{var}\left(\boldsymbol{s}_a^2\right) = \frac{2}{n^2}\left[\frac{\left(\sigma^2 + n\sigma_a^2\right)^2}{a-1} + \frac{\sigma^4}{a(n-1)}\right]. \tag{6.33}$$

We summarise this in

Theorem 6.6 Under the conditions of Theorem 6.5, the variances of $\boldsymbol{s}_a^2 = \frac{1}{n}(\boldsymbol{MS}_A - \boldsymbol{MS}_I)$ and $\boldsymbol{s}^2 = \boldsymbol{MS}_I$ are given by (6.33) and (6.31), respectively. Further

$$\operatorname{cov}\left(\boldsymbol{s}^2, \boldsymbol{s}_a^2\right) = \frac{-2\sigma^4}{na(n-1)}. \tag{6.34}$$

The relation for the covariance follows because

$$\operatorname{cov}\left(\boldsymbol{s}^2, \boldsymbol{s}_a^2\right) = \operatorname{cov}\left[\boldsymbol{MS}_I, \frac{1}{n}(\boldsymbol{MS}_A - \boldsymbol{MS}_I)\right] = -\frac{1}{n}\operatorname{var}(\boldsymbol{MS}_I)$$

and from (6.31).

Estimators for the variances and covariances in (6.31), (6.33) and (6.34) can be obtained, by replacing the quantities σ^2 and σ_a^2 occurring in these formulae by their estimators $\hat{\boldsymbol{\sigma}}^2 = \boldsymbol{s}^2$ and $\hat{\boldsymbol{\sigma}}_a^2 = \boldsymbol{s}_a^2$. These estimators of the variances and covariance components are biased. It can easily be seen that

$$\widehat{\operatorname{var}\left(\boldsymbol{s}^2\right)} = \frac{2\boldsymbol{s}^4}{a(n-1)+2}, \tag{6.35}$$

$$\widehat{\operatorname{var}\left(\boldsymbol{s_a}^2\right)} = \frac{2}{n^2}\left[\frac{\boldsymbol{s}^2 + \boldsymbol{s}_a^2}{a+1} - \frac{\boldsymbol{s}^2}{a(n-1)+2}\right] \tag{6.36}$$

and

$$\widehat{\operatorname{cov}\left(\boldsymbol{s}^2, \boldsymbol{s}_a^2\right)} = \frac{-2\boldsymbol{s}^4}{n[a(n-1)+2]} \tag{6.37}$$

are unbiased concerning $\operatorname{var}(\boldsymbol{s}^2)$, $\operatorname{var}\left(\boldsymbol{s}_a^2\right)$ and $\operatorname{cov}\left(\boldsymbol{s}^2, \boldsymbol{s}_a^2\right)$ because, if $\boldsymbol{z} = \frac{f\boldsymbol{MS}_K}{E(\boldsymbol{MS}_K)}$ is $CS(f)$-distributed then $\operatorname{var}(\boldsymbol{z}) = 2f$ and by this

$$\operatorname{var}(\boldsymbol{MS}_K) = \frac{2}{f}[E(\boldsymbol{MS}_K)]^2.$$

Further

$$E\left(\boldsymbol{MS}_K^2\right) - [E(\boldsymbol{MS}_K)]^2 = \frac{2}{f}[E(\boldsymbol{MS}_K)]^2$$

(in more detail see the proof of Theorem 6.10).

In the case of unequal subclass numbers, Theorem 6.5 cannot be applied. But Formula (6.31) was derived independently of Theorem 6.5 and is therefore valid for unequal subclass numbers if we replace $a(n-1)$ by $N-a$.

Deriving the formulae for $\mathrm{var}(\boldsymbol{s}_a^2)$ and $\mathrm{cov}(\boldsymbol{s}^2, \boldsymbol{s}_a^2)$ for unequal n_i is cumbersome. The derivation can be found in Hammersley (1949) and by another method in Hartley (1967). Townsend (1968, appendix IV) gives a derivation for the case $\mu = 0$. For the proof of the following theorems, we therefore refer to these references.

Theorem 6.7 The random vector $\boldsymbol{Y}$ with the components in model equation (6.8) is assumed to be $N(\mu 1_n, V)$-distributed; $V = \mathrm{var}(\boldsymbol{Y})$ is given by (6.9). Then for $\boldsymbol{s}_a^2$ in (6.12), we receive

$$\mathrm{var}(\boldsymbol{s}_a^2) = \frac{2\left[N^2 - \sum n_i^2 + \left(\sum n_i^2\right)^2 - 2N\sum n_i^3\right]}{\left(N^2 - \sum n_i^2\right)^2}\sigma_a^4 + \frac{4N}{N^2 - \sum n_i^2}\sigma_a^2\sigma^2 + \frac{2N^2(N-1)(a-1)}{\left(N^2 - \sum n_i^2\right)^2(N-a)}\sigma^4. \tag{6.38}$$

Further

$$\mathrm{var}(\boldsymbol{s}^2) = \frac{2\sigma^4}{N-a}, \tag{6.39}$$

$$\mathrm{cov}(\boldsymbol{s}^2, \boldsymbol{s}_a^2) = \frac{-2(a-1)N}{(N-a)\left(N^2 - \sum n_i^2\right)}\sigma^4. \tag{6.40}$$

For $n_i = n$ we obtain the Formulae (6.31) to (6.33). If $\mu = 0$, we get

$$\mathrm{var}(\boldsymbol{s}_a^{*2}) = \frac{2}{N^2}\left[\sigma_a^4 \sum n_i^2 + 2\sigma_a^2\sigma^2 N + \sigma^4 \frac{aN}{N-a}\right], \tag{6.41}$$

where

$$\boldsymbol{s}_a^{*2} = \frac{a}{N}\left(\sum_{i=1}^{a} \frac{\boldsymbol{Y}_{i.}^2}{n_i} - \frac{1}{N-a}\boldsymbol{SS}_{\mathrm{I}}\right) \tag{6.42}$$

is the ML-estimator of σ_a^2 if $\mu = 0_N$.

Example 6.1 Table 6.1 shows milk fat performances (in kg) y_{ij} of the daughters of ten sires randomly selected from a corresponding population. The portion of the fathers in the variance of this trait in the population shall be estimated as well as the variances of this estimator and the estimator of the residual variance and the covariance between the two estimators. Table 6.2 is the ANOVA table, and Table 6.3 contains the estimates.

Table 6.1 Milk fat performances y_{ij} of daughters of 10 sires.

	Sire (bull)									
	B_1	B_2	B_3	B_4	B_5	B_6	B_7	B_8	B_9	B_{10}
	120	152	130	149	110	157	119	150	144	159
	155	144	138	107	142	107	158	135	112	105
	131	147	123	143	124	146	140	150	123	103
	130	103	135	133	109	133	108	125	121	105
	140	131	138	139	154	104	138	104	132	144
	140	102	152	102	135	119	154	150	144	129
	142	102	159	103	118	107	156	140	132	119
	146	150	128	110	116	138	145	103	129	100
	130	159	137	103	150	147	150	132	103	115
	152	132	144	138	148	152	124	128	140	146
	115	102	154		138	124	100	122	106	108
	146	160			115	142		154	152	119
n_i	12	12	11	10	12	12	11	12	12	12
$y_{i\cdot}$	1647	1584	1538	1227	1559	1576	1492	1593	1538	1452
$\bar{y}_{i\cdot}$	137.25	132.00	139.82	122.70	129.92	131.33	135.64	132.75	128.17	121.00

Table 6.2 Analysis of variance table (SPSS output) for the data of Table 6.1 of Example 6.1.

Tests of between subjects effects						
Dependent variable: milk						
Source		**Type III sum of squares**	***df***	**Mean square**	***F***	**Sig.**
Sire	Hypothesis	3609.106	9	401.012	1.272	.261
	Error	33426.032	106	315.340		

Table 6.3 Results the variance component estimation using four methods.

Method	s_a^2	s^2	$\widehat{\text{var}(s^2)}$	$\widehat{\text{var}(s_a^2)}$	$\widehat{\text{cov}(s^2, s_a^2)}$
Analysis of variance	7.388	315.34			
MINQUE	8.171	315.35			
ML	3.248	316.03	199.45	1883.95	−161.99
REML	6.802	315.90	271.26	1882.59	−162.06

According to the ANOVA method is $s_a^2 = 7.388$. In the ANOVA table, we find $s^2 = 315.34$. The inner-class correlation coefficient ρ_I is estimated as

$$r_I = \frac{7.388}{322.728} = 0.023.$$

The ANOVA table is calculated by SPSS via

Analyze

General Linear Model

Variance Components

At first we receive the data and via OPTIONS the possible methods of estimation as shown in Figure 6.2.

By using the button 'model' putting the sum of squares to 1, we get this result.

In the SPSS output in Table 6.2, Sig leads to the rejection of the null hypothesis, in cases where the value is smaller or equal to the first kind risk α chosen.

We now will estimate the variance components with SPSS using all available methods in this program given in Figure 6.1.

Again we put *SS* to type I. In the window arising after this, we select the corresponding method (Figure 6.1). We obtain the results of Table 6.3.

As we see, the results, except for the variance of factor A in *ML*, differ unessentially from each other.

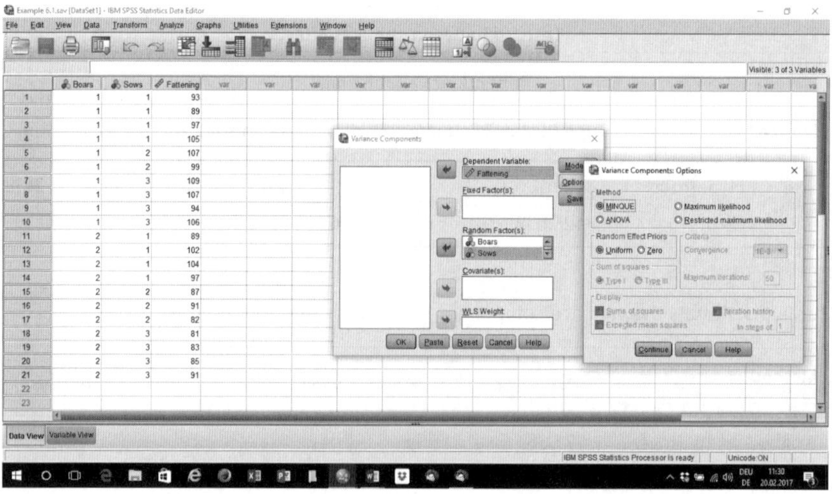

Figure 6.2 The data of Example 6.1 and the possible methods of estimation in SPSS.

6.3 Estimators of Variance Components in the Two-Way and Three-Way Classification

In this chapter, we consider only the ANOVA method. In case of unequal subclass numbers, there are methods already shown in Section 6.2, which can be calculated with SPSS. But as in Section 6.1.2 for the one-way ANOVA, we also cannot say here that one of these methods is uniformly better than the ANOVA method; however in practice the REML method is increasingly used. Readers interested in this method are referred to Searle et al. (1992) and Ahrens (1983).

For the following we need

Definition 6.4 Let Y be a random variable with a distribution independent of the parameter (vector) θ.

$\hat{\theta}$ is an unbiased estimator of θ, being a quadratic function of $\boldsymbol{Y}$. If $\hat{\theta}$ has minimal variance amongst all unbiased estimators quadratic in $\boldsymbol{Y}$ with finite second moments, then $\hat{\theta}$ is called best quadratic unbiased estimator (BQUE) of θ.

6.3.1 General Description for Equal and Unequal Subclass Numbers

Definition 6.5 For a special model II of the ANOVA in Definition 6.2 and for correspondingly structured other models, we speak about a balanced case; if for each factor the subclass numbers in the levels are equal and in nested classifications the number of nested factors is equal for each level of the superior factor.

Balanced cases are, for instance, the cross-classification with equal subclass numbers and nested classification with equal number of levels of the inferior factor and equal subclass number.

In the one-way classification in (6.3) is $r = 1$ and for $Z_{A_1} = Z = \oplus_{i=1}^{a} e_n$, we have

$$e_N^T Z = n e_a^T, \quad Z e_a = e_N. \tag{6.43}$$

The general approach of the ANOVA method in the balanced case as already said is to look for the ANOVA table (except the column for $E(\boldsymbol{MS})$) for the corresponding model I in Chapter 5. Now the $E(\boldsymbol{MS})$ for model II are calculated, and the MS are formally equated to the $E(\boldsymbol{MS})$. The solutions of the then arising simultaneous equations are the estimates of the variance components. The estimators are given by transition to the corresponding random variables. The factors of the variance components in the $E(\boldsymbol{MS})$ can be found by using the rules of Chapter 7. We denote by $\boldsymbol{q} = (\boldsymbol{MS}_1, \ldots, \boldsymbol{MS}_r)^{\mathrm{T}}$ the vector of the $\boldsymbol{MS}$ in an

ANOVA table and with $\left(\sigma_1^2,\ldots,\sigma_r^2\right)^T$ the vector of the variance components and with K the non-singular matrix of the factors k_{ij} so that

$$\left(\boldsymbol{s}_1^2,\ldots,\boldsymbol{s}_r^2\right)^{\mathrm{T}} = K^{-1}\boldsymbol{q}. \tag{6.44}$$

The random solutions of $q = K\left(\sigma_1^2,\ldots,\sigma_r^2\right)^T$ are used as an estimator $\left(\boldsymbol{s}_1^2,\ldots,\boldsymbol{s}_r^2\right)^T$ of $\left(\sigma_1^2,\ldots,\sigma_r^2\right)^T$, and we get $\left(\boldsymbol{s}_1^2,\ldots,\boldsymbol{s}_r^2\right)^T = K^{-1}\boldsymbol{q}$.

Without proof we give the following theorem:

Theorem 6.8 (Graybill)
In an ANOVA for a special linear model of the form (6.3), we have in any balanced case:

1) The estimator (6.44) is in the case that γ_{A_i} in (6.3) have finite third and fourth moments and are equal for all elements of γ_{A_i} (and for each i) a BQUE.
2) The estimator (6.44) for normally distributed random variables $\boldsymbol{Y}$ is the best (unbiased) estimator.

The proof of this theorem can be found in Graybill (1954).

The unbiasedness follows immediately from

$$E\left[\left(\boldsymbol{s}_1^2,\ldots,\boldsymbol{s}_r^2\right)^T\right] = K^{-1}E(\boldsymbol{q}) = K^{-1}K\left(\sigma_1^2,\ldots,\sigma_r^2\right)^T = \left(\sigma_1^2,\ldots,\sigma_r^2\right)^T.$$

The covariance matrix of the estimator (6.44) is

$$\operatorname{var}\left[\left(\boldsymbol{s}_1^2,\ldots,\boldsymbol{s}_r^2\right)^T\right] = K^{-1}\operatorname{var}(\boldsymbol{q})K^{-1}.$$

Theorem 6.9 Let (6.3) be a special model of the ANOVA following Definition 6.2 and $\boldsymbol{Y}$ in (6.3) is N-dimensional normally distributed. Then in the balanced case for $\boldsymbol{SS}_i$ of the corresponding ANOVA (see Chapter 5) with the degrees of freedom $v_i(i = 1, \ldots, r+1$, $\boldsymbol{SS}_{r+1} = \boldsymbol{SS}_{\text{res}})$

$$\frac{\boldsymbol{SS}_i}{E(\boldsymbol{MS}_i)} = \boldsymbol{Y}^{\mathrm{T}}A_i\boldsymbol{Y}$$

with the positive definite matrices A_i of rank v_i that are independent of each other $CS(v_i)$-distributed.

The proof of this theorem can be obtained with the help of Theorem 4.6 showing that A_iV is idempotent and $A_iVA_j = 0$ for $i \neq j$ and $\mu 1_n^{\mathrm{T}}A_i1_n\mu = 0$.

Theorem 6.10 If $\bar{y}$ in (6.3) in the balanced case is N-dimensional normally distributed, then for $\left(s_1^2,\ldots,s_r^2\right)^T$ in (6.44)

$$\operatorname{var}\left[\left(s_1^2,\ldots,s_r^2\right)^T\right] = K^{-1}D\left(K^T\right)^{-1}$$

with the diagonal matrix D, having elements equal to $\frac{2}{\nu_i}[E(\boldsymbol{DQ_i})]^2$. The ν_i are the degrees of freedom of $\boldsymbol{MS}_i$ for $i = 1, \ldots, r+1$. Further

$$\widehat{\operatorname{var}\left[\left(s_1^2,\ldots,s_r^2\right)^T\right]} = K^{-1}\hat{\boldsymbol{D}}\left(K^T\right)^{-1}$$

is with the diagonal matrix $\hat{\boldsymbol{D}}$ with elements $\frac{2}{\nu_i+2}\boldsymbol{MS}_i^2$ an unbiased estimator for $\operatorname{var}\left[\left(s_1^2,\ldots,s_r^2\right)^T\right]$.

Proof: From Theorem 6.9 follows that in the balanced case, $\frac{\nu_i \boldsymbol{MS}_i}{E(\boldsymbol{MS}_i)}$ $(i = 1,\ldots,r+1)$ are independent of each other $CS(\nu_i)$-distributed. Therefore from $\operatorname{var}(\chi^2) = 2n$ for each $CS(n)$-distributed random variable χ^2

$$\operatorname{var}\left[\frac{\nu_i \boldsymbol{MS}_i}{E(\boldsymbol{MS}_i)}\right] = \frac{\nu_i^2}{[E(\boldsymbol{MS}_i)]^2}\operatorname{var}(\boldsymbol{MS}_i) = 2\nu_i$$

and from this follows because of $\operatorname{cov}(\boldsymbol{MS}_i, \boldsymbol{MS}_j) = 0$ for all $i \neq j$ the stated form of D.

Because

$$\operatorname{var}(\boldsymbol{MS}_i) = E\left(\boldsymbol{MS}_i^2\right) - [E(\boldsymbol{MS}_i)]^2 = \frac{2}{\nu_i}[E(\boldsymbol{MS}_i)]^2,$$

we have

$$E\left(\boldsymbol{MS}_i^2\right) = \left[E(\boldsymbol{MS})_i\right]^2\frac{2+\nu_i}{\nu_i},$$

and $\frac{2}{\nu_i+2}\boldsymbol{MS}_i^2$ is unbiased concerning $\frac{2}{\nu_i}[E(\boldsymbol{MS}_i)]^2$, and we get $E\left(\hat{\boldsymbol{D}}\right) = D$.

We consider now the unbalanced case, that is, such models for which (6.43) is not valid. We restrict as already said on the ANOVA method because it is simple to calculate and no uniformly better method exists – but see Ahrens (1983).

The analogy is as follows. The $\boldsymbol{SS}_i$ in the balanced case can be written as linear combinations of squares of the components of $\boldsymbol{Y}$ and of partial sums of this components. We denote now these elements in the $\boldsymbol{SS}_i$ written as linear combinations by $\boldsymbol{s}_{A_i}$, where the A_i are the factors or factor combinations in (6.3) ($\boldsymbol{S}_{A_0} = \boldsymbol{S}_\mu$

is assigned to μ). Analogously to the $\boldsymbol{s}_{A_i}$ for the balanced case, the corresponding $\boldsymbol{S}_{A_i}$ for the unbalanced case are calculated as follows:

$$\begin{aligned} \boldsymbol{S}_\mu &= \frac{Y_{\dots}^2}{N}, \quad S_{\text{res}} = \boldsymbol{Y}^{\mathrm{T}}\boldsymbol{Y} = \boldsymbol{S}_{A_{r+1}} \\ \boldsymbol{S}_{A_i} &= \sum_{j=1}^{a_i} \frac{\boldsymbol{Y}_{\cdot}^2(A_{ij})}{N_{\cdot}(A_{ij})} \quad (i = 1,\dots,r). \end{aligned} \tag{6.45}$$

In (6.45) $Y_{\cdot}(A_{ij})$ is the sum of components of $\boldsymbol{Y}$ in the j-th level of the factors (or factor combination) A_i, $N_{\cdot}(A_{ij})$ is the number of summands in $\boldsymbol{Y}_{\cdot}(A_{ij})$ and a_i are the number of levels of A_i.

$\boldsymbol{S}_{A_i}$ are transformed to quasi-*SS* with the help of the linear combinations derived for the balanced case. Putting these quasi-*SS* or the corresponding quasi-*MS* equal to their expectations leads to simultaneous equations. The solutions are the estimates the variance components from the ANOVA method for the unbalanced case. The denotation quasi-*SS* was chosen, because these quadratic forms are not always positive definite and by this not a sum of squared deviations. For the estimation of the variance component, that is, however, irrelevant.

For the derivation of the simultaneous equations, we need the expectations of the quasi-***SS*** and by this the $\boldsymbol{S}_{A_i}$. Denoting by $k\left(\sigma_j^2, \boldsymbol{S}_{A_i}\right)$ the coefficients of σ_j^2 in expectation of $\boldsymbol{S}_{A_i}(i, j = 1,\dots,r)$, we can calculate these coefficients following Hartley (1967) (see also Hartley and Rao, 1967). We put

$$\boldsymbol{S}_{A_i} = \sum_{j=1}^{a_i} \frac{1}{N_{\cdot}(A_{ij})} \boldsymbol{Y}^2(A_{ij}) = \boldsymbol{Y}^T B_i \boldsymbol{Y} = \boldsymbol{S}_{A_i}(\boldsymbol{Y})$$

and use $Z_{A_i} = [z_1(A_i),\dots,z_{a_i}(A_i)]$ with the column vectors $z_j(A_i)(j = 1, \dots, a_i)$. Then we have

$$k_{ij} = k\left(\sigma_j^2, S_{A_i}\right) = \sum_{j=1}^{a_i} S_{A_i}\left[z_j(A_i)\right]. \tag{6.46}$$

For the derivation of (6.46), we refer to Hartley (1967). The coefficients of σ^2 are equal to a_i, and we have further

$$E(\boldsymbol{S}_{res}) = E\left(\boldsymbol{Y}^{\mathrm{T}}\boldsymbol{Y}\right) = N\left(\mu^2 + 1_{r+1}^{\mathrm{T}}\left(\sigma_1^2,\dots,\sigma_r^2\right)^{\mathrm{T}}\right).$$

If in the balanced case for the calculation of the *SS* the formulae

$$\begin{aligned} \boldsymbol{SS}_i &= \sum_{j=1}^{r} c_{ij}\boldsymbol{S}_{A_j} + c_{r+1,i}\boldsymbol{S}_{\text{res}} \\ \boldsymbol{SS}_{\text{res}} &= \sum_{j=1}^{r} c_{j,r+1}\boldsymbol{S}_{A_j} + c_{r+1,r+1}\boldsymbol{S}_{\text{res}} \end{aligned} \tag{6.47}$$

are valid the quasi-SS ($\boldsymbol{QSS}_i$), also in the non-balanced case are calculated by (6.47). Let C be the positive definite matrix of the coefficients of σ_j^2 in the expectations of $\boldsymbol{QSS}_i$ (i row index, j column index), a^* be the vector of the coefficients of σ^2, Σ the vector of the variance components $\sigma_i^2 (i = 1,\ldots,r)$ and $\boldsymbol{S}$ be the vector of $\boldsymbol{QSS}_i$, so we gain the simultaneous equations

$$E\begin{bmatrix} \boldsymbol{S} \\ \boldsymbol{SS}_{\text{res}} \end{bmatrix} = \begin{pmatrix} C & a^* \\ 0_r^{\text{T}} & N-p \end{pmatrix} \cdot \begin{pmatrix} \Sigma \\ \sigma^2 \end{pmatrix} \tag{6.48}$$

where p is the number the subclasses with at least one observation. The matrix of coefficients comes from (6.47), (6.46) and the corresponding formulae for the $\boldsymbol{SS}$ in the balanced case. From (6.48) we get the estimation equations by the ANOVA method in the form

$$\begin{pmatrix} \boldsymbol{S} \\ \boldsymbol{SS}_{\text{res}} \end{pmatrix} = \begin{pmatrix} C & a^* \\ 0_r^{\text{T}} & N-p \end{pmatrix} \cdot \begin{pmatrix} \hat{\Sigma} \\ \boldsymbol{s}^2 \end{pmatrix} \tag{6.49}$$

where $\hat{\Sigma}^T = \left(\boldsymbol{s}_1^2,\ldots,\boldsymbol{s}_r^2\right)$. From (6.49) we obtain

$$\boldsymbol{s}^2 = \frac{1}{N-p}\boldsymbol{SS}_{\text{res}} \tag{6.50}$$

and

$$\hat{\boldsymbol{\Sigma}} = C^{-1}\left(\boldsymbol{S} - \boldsymbol{s}^2 a^*\right). \tag{6.51}$$

Formulae for the variances (and estimators of the variances) of $\boldsymbol{s}^2$ and $\boldsymbol{s}_i^2$ can be found in Searle (1971).

6.3.2 Two-Way Cross-Classification

In the two-way cross-classification, our model following Definition 6.2 is

$$\boldsymbol{y}_{ijk} = \mu + \boldsymbol{a}_i + \boldsymbol{b}_j + (\boldsymbol{a},\boldsymbol{b})_{ij} + \boldsymbol{e}_{ijk} \quad \left(i = 1,\ldots,a; j = 1,\ldots,b; k = 1,\ldots,n_{ij}\right) \tag{6.52}$$

with side conditions that $\boldsymbol{a}_i$, $\boldsymbol{b}_j$, $(\boldsymbol{a}, \boldsymbol{b})_{ij}$ and $\boldsymbol{e}_{ijk}$ are uncorrelated and

$$E(\boldsymbol{a}_i) = E(\boldsymbol{b}_j) = E\left((\boldsymbol{a},\boldsymbol{b})_{ij}\right) = E(\boldsymbol{a}_i\boldsymbol{b}_j) = E\left(\boldsymbol{a}_i(\boldsymbol{a},\boldsymbol{b})_{ij}\right) = E\left(\boldsymbol{b}_j(\boldsymbol{a},\boldsymbol{b})_{ij}\right) = 0$$

$$E\left(\boldsymbol{e}_{ijk}\right) = E\left(\boldsymbol{a}_i\boldsymbol{e}_{ijk}\right) = E\left(\boldsymbol{b}_j\boldsymbol{e}_{ijk}\right) = E\left((\boldsymbol{a},\boldsymbol{b})_{ij}\boldsymbol{e}_{ijk}\right) = 0 \text{ for all } i,j,k$$

$$\text{var}(\boldsymbol{a}_i) = \sigma_a^2 \text{ for all } i, \quad \text{var}\left(\boldsymbol{b}_j\right) = \sigma_b^2 \text{ for all } j$$

$$\text{var}\left((\boldsymbol{a},\boldsymbol{b})_{ij}\right) = \sigma_{ab}^2 \text{ for all } i,j, \quad \text{var}\left(\boldsymbol{e}_{ijk}\right) = \sigma^2 \text{ for all } i,j,k.$$

For testing and constructing confidence intervals, we additionally assume that $\boldsymbol{y}_{ijk}$ is normally distributed.

A special case of Theorem 6.9 is

Theorem 6.11 In a balanced two-way cross-classification ($n_{ij} = n$ for all i, j), model II and normally distributed $\boldsymbol{y}_{ijk}$ the sum of squares in Table 5.13 are stochastically independent, and we have

$$\frac{\boldsymbol{SS}_A}{bn\sigma_a^2 + n\sigma_{ab}^2 + \sigma^2} \text{ is } CS(a-1)$$

$$\frac{\boldsymbol{SS}_B}{an\sigma_b^2 + n\sigma_{ab}^2 + \sigma^2} \text{ is } CS(b-1)$$

$$\frac{\boldsymbol{SQ}_{AB}}{n\sigma_{ab}^2 + \sigma^2} \text{ is } CS[(a-1)(b-1)]\text{-distributed.}$$

Theorem 6.11 allows us to test the hypotheses

$$H_{A0}: \sigma_a^2 = 0, \quad H_{B0}: \sigma_b^2 = 0, \quad H_{AB0}: \sigma_{ab}^2 = 0.$$

Theorem 6.12 With the assumptions of Theorem 6.11, the test statistic

$$\boldsymbol{F}_A = \frac{\boldsymbol{SS}_A}{\boldsymbol{SS}_{AB}}(b-1)$$

is the $\frac{bn\sigma_a^2 + n\sigma_{ab}^2 + \sigma^2}{n\sigma_{ab}^2 + \sigma^2}$ fold of a random variable distributed as $F[a-1, (a-1)(b-1)]$. If H_{A0} is true, $\boldsymbol{F}_A$ is $F[a-1, (a-1)(b-1)]$-distributed.

The statistic

$$\boldsymbol{F}_B = \frac{\boldsymbol{SS}_B}{\boldsymbol{SS}_{AB}}(a-1)$$

is the $\frac{an\sigma_b^2 + n\sigma_{ab}^2 + \sigma^2}{n\sigma_{ab}^2 + \sigma^2}$-fold of a random variable distributed as $F[b-1, (a-1)(b-1)]$. If H_{B0} is true, $\boldsymbol{F}_B$ is $F[b-1, (a-1)(b-1)]$-distributed.

The statistic

$$\boldsymbol{F}_{AB} = \frac{\boldsymbol{SS}_{AB}}{\boldsymbol{SS}_{\text{Rest}}} \cdot \frac{ab(n-1)}{(a-1)(b-1)}$$

is the $\frac{n\sigma_{ab}^2 + \sigma^2}{\sigma^2}$ fold of a random variable distributed as $F[(a-1)(b-1), ab(n-1)]$. If H_{AB0} is true, $\boldsymbol{F}_{AB}$ is $F[(a-1)(b-1), ab(n-1)]$-distributed.

The proof follows from Theorem 6.11. The hypotheses H_{A0}, H_{B0} and H_{AB0} are tested by the statistics $\boldsymbol{F}_A$, $\boldsymbol{F}_B$ and $\boldsymbol{F}_{AB}$, respectively. If the observed F-values are larger than the $(1 - a)$-quantiles of the central F-distribution with the corresponding degrees of freedom, we may conjecture that the corresponding variance component is positive and not zero.

To derive Theorem 6.11 from Theorem 6.9, we need (for the balanced case)

$$\left.\begin{aligned} E(\boldsymbol{MS}_A) &= bn\sigma_a^2 + n\sigma_{ab}^2 + \sigma^2 \\ E(\boldsymbol{MS}_B) &= an\sigma_b^2 + n\sigma_{ab}^2 + \sigma^2 \\ E(\boldsymbol{MS}_{AB}) &= \qquad n\sigma_{ab}^2 + \sigma^2 \\ E(\boldsymbol{MS}_{\text{Rest}}) &= \qquad\qquad \sigma^2 \end{aligned}\right\} \tag{6.53}$$

(see Exercise 6.4).

Table 6.4 is the ANOVA table of the balanced case. With (6.53) the ANOVA method provides the variance components of the balanced case

$$\left.\begin{aligned} \boldsymbol{s}^2 &= \boldsymbol{MS}_{\text{res}}, & \boldsymbol{s}_{ab}^2 &= \frac{1}{n}(\boldsymbol{MS}_{AB} - \boldsymbol{MS}_{\text{res}}) \\ \boldsymbol{s}_b^2 &= \frac{1}{an}(\boldsymbol{MS}_B - \boldsymbol{MS}_{AB}) & \boldsymbol{s}_a^2 &= \frac{1}{bn}(\boldsymbol{MS}_A - \boldsymbol{MS}_{AB}) \end{aligned}\right\} \tag{6.54}$$

Formula (6.54) is a special case of (6.44), because (6.53) generates K in (6.44) as

$$K = \begin{pmatrix} bn & 0 & n & 1 \\ 0 & an & n & 1 \\ 0 & 0 & n & 1 \\ 0 & 0 & 0 & 1 \end{pmatrix}.$$

We get $|K| = abn^3$ and

$$K^{-1} = \frac{1}{abn}\begin{pmatrix} a & 0 & -a & 0 \\ 0 & b & -b & 0 \\ 0 & 0 & ab & -ab \\ 0 & 0 & 0 & abn \end{pmatrix}.$$

From Theorem 6.10 the variances of the estimators $\boldsymbol{s}_a^2, \boldsymbol{s}_b^2, \boldsymbol{s}_{ab}^2$ and $\boldsymbol{s}^2$ are obtained as follows. At first we calculate the diagonal matrix D from (6.53) or from Table 6.4:

$$d_{11} = \frac{2}{a-1}\left(bn\sigma_a^2 + n\sigma_{ab}^2 + \sigma^2\right)^2,\ d_{22} = \frac{2}{b-1}\left(an\sigma_b^2 + n\sigma_{ab}^2 + \sigma^2\right)^2,$$

$$d_{33} = \frac{2}{(a-1)(b-1)}\left(n\sigma_{ab}^2 + \sigma^2\right)^2,\ d_{44} = \frac{2}{ab(n-1)}\sigma^4$$

Table 6.4 Supplement of Table 5.13 for model II.

Source of variation	E(MS)	F
Between levels of A	$\sigma^2 + n\sigma_{ab}^2 + bn\sigma_a^2$	$(b-1)\dfrac{\boldsymbol{SS}_A}{\boldsymbol{SS}_{AB}}$
Between levels of B	$\sigma^2 + n\sigma_{ab}^2 + an\sigma_b^2$	$(a-1)\dfrac{\boldsymbol{SS}_B}{\boldsymbol{SS}_{AB}}$
Interactions	$\sigma^2 + n\sigma_{ab}^2$	$\dfrac{ab(n-1)}{(a-1)(b-1)}\dfrac{\boldsymbol{SS}_{AB}}{\boldsymbol{SS}_{res}}$
Residual	σ^2	

From this we obtain the covariance matrix V of the vector $\left(s_a^2, s_b^2, s_{ab}^2, s^2\right)$

$$\begin{bmatrix} \frac{d_{11}+d_{33}}{b^2n^2} & \frac{d_{33}}{abn^2} & \frac{-d_{33}}{bn^2} & 0 \\ \frac{d_{33}}{abn^2} & \frac{d_{22}+d_{33}}{a^2n^2} & \frac{-d_{33}}{an^2} & 0 \\ \frac{-d_{33}}{bn^2} & \frac{-d_{33}}{an^2} & \frac{d_{33}+d_{44}}{n^2} & \frac{-d_{44}}{n} \\ 0 & 0 & \frac{-d_{44}}{n} & d_{44} \end{bmatrix}.$$

For instance, $\operatorname{var}(s^2) = \frac{2}{ab(n-1)}\sigma^4$ and

$$\operatorname{cov}\left(s_a^2, s_b^2\right) = \frac{2}{a(a-1)b(b-1)n^2}\left(n^2\sigma_{ab}^4 + 2n\sigma_{ab}^4\sigma^2 + \sigma^4\right).$$

Estimators of the elements of the covariance matrix V are the elements of the matrix $\hat{V}$, which is gained from V by replacing the d_{ii} by $\hat{d}_{ii}$, where

$$\hat{d}_{11} = \frac{2}{a+1}MS_A^2,\ \hat{d}_{22} = \frac{2}{b+1}MS_B^2$$
$$\hat{d}_{33} = \frac{2}{(a-1)(b-1)+2}MS_{AB}^2,\ \hat{d}_{44} = \frac{2}{ab(n-1)+2}MS_{\text{res}}.$$

For instance, we have

$$\widehat{\operatorname{var}\left(s^2\right)} = \frac{2}{ab(n-1)+2}MS_{\text{res}}^2$$

and

$$\widehat{\operatorname{cov}\left(s_a^2, s_b^2\right)} = \frac{2}{[(a-1)(b-1)+2]abn^2}MS_{AB}^2.$$

The unbalanced case: In p classes at least one observation may be present ($0 < p \le ab$). If $p = ab$, we assume that not all n_{ij} are equal. The quasi-***SS*** (with the aid of 6.37) are analogue to the ***SS*** in Table 5.13

$$\left.\begin{aligned} QSS_A &= S_A - S_\mu \\ QSS_B &= S_B - S_\mu \\ QSS_{AB} &= S_{AB} - S_A - S_B + S_\mu \\ QSS_{\text{res}} &= S_{\text{res}} - S_{AB} \end{aligned}\right\} \tag{6.55}$$

with

$$\left.\begin{aligned} &S_\mu = \frac{1}{N}Y_{\ldots}^2,\ S_{\text{res}} = \sum_{i=1}^{a}\sum_{j=1}^{b}\sum_{k=1}^{n_{ij}} y_{ijk}^2 \\ &S_A = \sum_{i=1}^{a}\frac{Y_{i..}^2}{N_{i.}},\ S_B = \sum_{j=1}^{b}\frac{Y_{.j.}^2}{N_{.j}},\ S_{AB} = \sum_{i=1}^{a}\sum_{j=1}^{b}{}^{*}\frac{Y_{ij.}^2}{n_{ij}} \end{aligned}\right\}. \tag{6.56}$$

Here Σ^* means that only summands with a non-zero denominator have been taken. Equation (6.56) is a special case of (6.45). The expectations of $\boldsymbol{S}_\mu$, $\boldsymbol{S}_A$, $\boldsymbol{S}_B$, $\boldsymbol{S}_{AB}$ and $E(\boldsymbol{QSS}_{\text{res}}) = (N - p)\sigma^2$ can be obtained by utilising the model equation (6.52) in the Formula (6.56) or from Formula (6.46). In the present case we get

$$E(\boldsymbol{S}_A) = \sum_{i=1}^{a}\left\{N_{i.}\mu^2 + E\left[N_{i.}\boldsymbol{a}_i^2 + \frac{\sum_{j=1}^{b} n_{ij}^2\boldsymbol{b}_j^2}{N_{i.}} + \frac{\sum_{j=1}^{b} n_{ij}^2(\boldsymbol{a},\boldsymbol{b})_{ij}^2}{N_{i.}} + \frac{\boldsymbol{E}_{i..}^2}{N_{i.}}\right]\right\}$$

$$= N\mu^2 + N\sigma_a^2 + \sum_{i=1}^{a}\frac{\sum_{j=1}^{b} n_{ij}^2}{N_{i.}}\sigma_b^2 + \sum_{i=1}^{a}\frac{\sum_{j=1}^{b} n_{ij}^2}{N_{i.}}\sigma_{ab}^2 + a\sigma^2$$

$$\left(\boldsymbol{E}_{i..} = \sum_{j,k}\boldsymbol{e}_{ijk}\right)$$

$$E(\boldsymbol{S}_B) = N\mu^2 + N\sigma_b^2 + \sum_{j=1}^{b}\frac{\sum_{i=1}^{a} n_{ij}^2}{N_{.j}}\sigma_a^2 + \sum_{j=1}^{b}\frac{\sum_{i=1}^{a} n_{ij}^2}{N_{.j}}\sigma_{ab}^2 + b\sigma^2$$

$$E(\boldsymbol{S}_{AB}) = N\left(\mu^2 + \sigma_a^2 + \sigma_b^2 + \sigma_{ab}^2\right) + p\sigma^2$$

$$E(\boldsymbol{S}_\mu) = N\mu^2 + \frac{\sum_{i=1}^{a} N_{i.}^2}{N}\sigma_a^2 + \frac{\sum_{j=1}^{b} N_{.j}^2}{N}\sigma_b^2 + \frac{\sum_{i=1}^{a}\sum_{j=1}^{a} n_{ij}^2}{N}\sigma_{ab}^2 + \sigma^2$$

$$E(\boldsymbol{S}_{\text{res}}) = N\left(\mu^2 + \sigma_a^2 + \sigma_b^2 + \sigma_{ab}^2 + \sigma^2\right)$$

and by this

$$E(\boldsymbol{QSS}_A) = \sigma^2\left[\frac{N - \sum_{i=1}^{a} N_{i.}^2}{N}\right] + \sigma_b^2\left[\sum_{i=1}^{a}\frac{\sum_{j=1}^{b} n_{ij}^2}{N_{i.}} - \frac{\sum_{j=1}^{b} N_{.j}^2}{N}\right]$$

$$+ \sigma_{ab}^2\sum_{i=1}^{a}\sum_{j=1}^{b} n_{ij}^2\left(\frac{1}{N_{i.}} - \frac{1}{N}\right) - (a-1)\sigma^2$$

$$E(\boldsymbol{QSS}_B) = \sigma_a^2\left[\sum_{i=1}^{a}\frac{\sum_{i=1}^{a} n_{ij}^2}{N_{.j}} - \frac{\sum_{i=1}^{a} N_{i.}^2}{N}\right] + \sigma_b^2\left[N - \frac{\sum_{j=1}^{b} N_{.j}^2}{N}\right]$$

$$+ \sigma_{ab}^2\sum_{j=1}^{b}\sum_{i=1}^{a} n_{ij}^2\left(\frac{1}{N_{.j}} - \frac{1}{N}\right) + (b-1)\sigma^2 \quad .$$

$$E(\boldsymbol{QSS}_{AB}) = \sigma_a^2\left[\frac{\sum_{i=1}^{a} N_{i.}^2}{\mathrm{N}} - \sum_{j=1}^{b}\frac{\sum_{i=1}^{a} {n_{ij}}^2}{N_{.j}}\right] + \sigma_b^2\left[\frac{\sum_{j=1}^{b} N_{.j}^2}{N} - \sum_{i=1}^{a}\frac{\sum_{j=1}^{b} n_{ij}^2}{N_{i.}}\right]$$

$$+ \sigma_{ab}^2\left[N - \sum_{i=1}^{a}\frac{\sum_{j=1}^{b} n_{ij}^2}{N_{i.}} - \sum_{j=1}^{b}\frac{\sum_{i=1}^{a} n_{ij}^2}{N_{.j}} + \frac{1}{N}\sum_{i=1}^{a}\sum_{j=1}^{b} n_{ij}^2\right]$$

$$+ \sigma^2(p - a - b + 1)$$

$$E(\boldsymbol{QSS}_{\text{res}}) = (N - p)\sigma^2$$

If all classes are occupied, we have $p = ab$.

We obtained estimators $\boldsymbol{s}_a^2, \boldsymbol{s}_b^2, \boldsymbol{s}_{ab}^2$ and $\boldsymbol{s}^2$ with the ANOVA method by replacing the $E(\boldsymbol{QSS})$ by the $\boldsymbol{QSS}$ and the variance components by their estimators.

6.3.3 Two-Way Nested Classification

The two-way nested classification is a special case of the incomplete two-way cross-classification; it is maximal disconnected. The formulae for the estimators of the variance components become very simple. We use the notation of Section 5.3.2, but now a_i and b_j in (5.33) are random variables. The model equation (5.33) then becomes

$$\boldsymbol{y}_{ijk} = \mu + \boldsymbol{a}_i + \boldsymbol{b}_{ij} + \boldsymbol{e}_{ijk}, \left(i = 1,\ldots,a; j = 1,\ldots,b_i; k = 1,\ldots,n_{ij}\right) \tag{6.57}$$

with the side conditions of uncorrelated $\boldsymbol{a}_i$, $\boldsymbol{b}_{ij}$ and $\boldsymbol{e}_{ijk}$ and

$$0 = E(\boldsymbol{a}_i) = E\left(\boldsymbol{b}_{ij}\right) = \operatorname{cov}\left(\boldsymbol{a}_i, \boldsymbol{b}_{ij}\right) = \operatorname{cov}\left(\boldsymbol{a}_i, \boldsymbol{e}_{ijk}\right) = \operatorname{cov}\left(\boldsymbol{b}_i, \boldsymbol{e}_{ijk}\right)$$

for all i, j, k.

The quasi-SS of the sections so far become real *SS*, because Theorem 5.10 is valid and independent of the special model. In Table 6.5 we find the $E(\boldsymbol{MS})$. In this table occur positive coefficients λ_i defined by

$$\left.\begin{aligned} \lambda_1 &= \frac{1}{B. - a}\left(N - \sum_{i=1}^{a}\frac{\sum_{j=1}^{b_i} n_{ij}^2}{N_{i\cdot}}\right), \\ \lambda_2 &= \frac{1}{a-1}\sum_{i=1}^{a}\sum_{j=1}^{b} n_{ij}^2\left(\frac{1}{N_{i\cdot}} - \frac{1}{N}\right), \\ \lambda_3 &= \frac{1}{a-1}\left(N - \frac{1}{N}\sum_{i=1}^{a} N_{i\cdot}^2\right) \end{aligned}\right\}. \tag{6.58}$$

Table 6.5 Column the $E(MS)$ of the two-way nested classification for model II (the other part of the analysis of variance table is given in Table 5.19).

Source of variation	$E(MS)$
Between A-levels	$\sigma^2 + \lambda_2\sigma_b^2 + \lambda_3\sigma_a^2$
Between B-levels within A-levels	$\sigma^2 + \lambda_1\sigma_b^2$
Within B-levels	σ^2

We gain the coefficients in (6.58) either by deriving the $E(\boldsymbol{MS})$ with the help of the model equation (6.57) or as special cases of the coefficients in the $E(\boldsymbol{QSS})$ of the last sections. From the analysis of variance method, we obtain the estimators of the variance components by

$$\left.\begin{aligned} \boldsymbol{s}^2 &= \boldsymbol{MS}_{\text{res}} \\ \boldsymbol{s}_b &= \frac{1}{\lambda_1}(\boldsymbol{MS}_{B \text{ in } A} - \boldsymbol{MS}_{\text{res}}) \\ \boldsymbol{s}_a^2 &= \frac{1}{\lambda_3}\left(\boldsymbol{MS}_A - \frac{\lambda_2}{\lambda_1}\boldsymbol{MS}_{B \text{ in } A} - \left(1 - \frac{\lambda_2}{\lambda_1}\right)\boldsymbol{MS}_{\text{res}}\right) \end{aligned}\right\}. \tag{6.59}$$

With

$$\lambda_1' = (B. - a)\lambda_1,\ \ \lambda_2' = (a-1)\lambda_2,\ \ \lambda_3' = (a-1)\lambda_3,$$

$$\lambda_4 = N + \frac{1}{N}\sum_{i=1}^{a} N_{i.}^2,\ \ \lambda_5 = \lambda_1'^2\left[(\lambda_4 - N)\lambda_4 - \frac{2}{N}\sum_{i=1}^{a} N_{i.}^3\right],$$

$$\lambda_6 = (\lambda_2' - \lambda_1' + N)\left[N\lambda_2'^2 + \left(\lambda_2'^2 - \lambda_1' + N\right)\lambda_2'^2\right] + (\lambda_1' + \lambda_2')\sum_{i=1}^{a}\frac{\left(\sum_{j=1}^{b} n_{ij}^2\right)^2}{N_{i.}^2}$$

$$-2\left(\lambda_1^2 + \lambda_2^2\right)\left[\lambda_2'\sum_{i=1}^{a}\frac{\sum_{j=1}^{b} n_{ij}^2}{N} + 2\frac{\lambda_1'\lambda_2'}{N}\sum_{i=1}^{a}\sum_{j=1}^{b} n_{ij}^3 + \lambda_1'\sum_{i=1}^{a}\frac{\left(\sum_{j=1}^{b} n_{ij}^2\right)^2}{NN_{i.}}\right],$$

$$\lambda_7 = \frac{1}{N - B.}\left[\lambda_1'^2(N-1)(a-1) - (\lambda_1' + \lambda_2')^2(a-1)(B. - a) + \lambda_1'^2(N-1)(B. - a)\right],$$

$$\lambda_8 = \lambda_1'^2\left[\frac{1}{N}\sum_{i=1}^{a}\sum_{j=1}^{b} n_{ij}^2\lambda_4 - \frac{2}{N}\sum_{i=1}^{a} N_{i.}\sum_{j=1}^{b} n_{ij}^2\right]$$

and

$$\lambda_9 = \lambda_1'^2\lambda_3', \lambda_{10} = \lambda_1'\lambda_2'\left(\lambda_1' + \lambda_2'\right),$$

the following formulae for the variances of the variance components result under the assumption that $\boldsymbol{y}_{ijk}$ are normally distributed

$$\left.\begin{aligned} \operatorname{var}(\boldsymbol{s}^2) &= \frac{2}{N - B_{i.}}\sigma^4, \\ \operatorname{var}\left(\boldsymbol{s}_a^2\right) &= \frac{2}{\lambda_1'^2\lambda_3'^2}\left(\lambda_5\sigma_a^4 + \lambda_6\sigma_b^4 + \lambda_7\sigma^4 + 2\lambda_8\sigma_a^2\sigma_b^2 + 2\lambda_9\sigma_a^2\sigma^2 + 2\lambda_{10}\sigma_b^2\sigma^2\right), \\ \operatorname{var}\left(\boldsymbol{s}_a^2\right) &= \frac{2}{\lambda_1'^2\lambda_1'^2}\left[\sum_{i=1}^{a}\frac{\left(\sum_{j=1}^{b} n_{ij}^2\right)^2}{N_{i.}^2} + \sum_{i=1}^{a}\sum_{j=1}^{b} n_{ij}^2 - 2\sum_{i=1}^{a}\frac{\sum_{j=1}^{b} n_{ij}^3}{N_{i.}}\right]\sigma_b^2 \\ &\quad + 4\lambda_1'\sigma_b^2\sigma^2 + \frac{2(B. - a)(N - a)\sigma^4}{N - B} \end{aligned}\right\} \tag{6.60}$$

and the covariances

$$\left.\begin{aligned}
\operatorname{cov}\left(s_a^2, s^2\right) &= \left[\frac{\lambda_2'(B.-a)}{\lambda_1'} - (a-1)\right]\frac{\operatorname{var}\left(s^2\right)}{\lambda_3'},\\
\operatorname{cov}\left(s_b^2, s^2\right) &= \frac{(B.-a)\operatorname{var}(s^2)}{\lambda_1'}\\
\operatorname{cov}\left(s_a^2, s_b^2\right) &= \frac{2}{\lambda_1'\lambda_3'}\left\{\sum_{i=1}^{a}\left[\sum_{j=1}^{b}\frac{n_{ij}^3}{N_{i.}^2} - \frac{\left(\sum_{j=1}^{b} n_{ij}^2\right)^2}{N_{i.}^2} + \frac{\left(\sum_{j=1}^{b} n_{ij}^2\right)^2}{N_{i.}N} - \frac{1}{N}\sum_{j=1}^{b} n_{ij}^3\right]\sigma_b^4\right.\\
&\qquad \left. + \frac{2(a-1)(B.-a)}{N-B.}\sigma^4 - \lambda_1'\lambda_2'\operatorname{var}\left(s_b^2\right)\right\}.
\end{aligned}\right\}. \tag{6.61}$$

6.3.4 Three-Way Cross-Classification with Equal Subclass Numbers

We start with the model equation

$$\begin{gathered}
\boldsymbol{y}_{ijkl} = \mu + \boldsymbol{a}_i + \boldsymbol{b}_j + \boldsymbol{c}_k + (\boldsymbol{a},\boldsymbol{b})_{ij} + (\boldsymbol{b},\boldsymbol{c})_{jk} + (\boldsymbol{a},\boldsymbol{b},\boldsymbol{c})_{ijk} + \boldsymbol{e}_{ijkl}\\
(i = 1,\dots,a; j = 1,\dots,b; k = 1,\dots,c; l = 1,\dots,n)
\end{gathered} \tag{6.62}$$

with the side conditions that the expectations of all random variable of the right hand side of (6.62) are equal to zero and all covariances between different random variables of the right side of (6.62) vanish. Further we assume for tests that $\boldsymbol{y}_{ijkl}$ are normally distributed. Table 6.6 is the ANOVA table for this case.

Table 6.6 The column *E*(*MS*) as supplement for model II to the analysis of variance Table 5.21.

Source of variation	*E*(*MS*)
Between *A*-levels	$\sigma^2 + n\sigma_{abc}^2 + cn\sigma_{ab}^2 + bn\sigma_{ac}^2 + bcn\sigma_a^2$
Between *B*-levels	$\sigma^2 + n\sigma_{abc}^2 + cn\sigma_{ab}^2 + an\sigma_{bc}^2 + acn\sigma_b^2$
Between *C*-levels	$\sigma^2 + n\sigma_{abc}^2 + an\sigma_{bc}^2 + bn\sigma_{ac}^2 + abn\sigma_c^2$
Interaction $A \times B$	$\sigma^2 + n\sigma_{abc}^2 + cn\sigma_{ab}^2$
Interaction $A \times C$	$\sigma^2 + n\sigma_{abc}^2 + bn\sigma_{ac}^2$
Interaction $B \times C$	$\sigma^2 + n\sigma_{abc}^2 + an\sigma_{bc}^2$
Interaction $A \times B \times C$	$\sigma^2 + n\sigma_{abc}^2$
Within the subclasses (residual)	σ^2

Following the ANOVA method, we obtain the estimators for the variance components

$$\begin{aligned}
\boldsymbol{MS}_A &= \boldsymbol{s}^2 + n\boldsymbol{s}^2_{abc} + cn\boldsymbol{s}^2_{ab} + bn\boldsymbol{s}^2_{ac} + bcn\boldsymbol{s}^2_a \\
\boldsymbol{MS}_B &= \boldsymbol{s}^2 + n\boldsymbol{s}^2_{abc} + cn\boldsymbol{s}^2_{ab} + an\boldsymbol{s}^2_{bc} + acn\boldsymbol{s}^2_b \\
\boldsymbol{MS}_C &= \boldsymbol{s}^2 + n\boldsymbol{s}^2_{abc} + an\boldsymbol{s}^2_{bc} + bn\boldsymbol{s}^2_{ac} + abn\boldsymbol{s}^2_c \\
\boldsymbol{MS}_{AB} &= \boldsymbol{s}^2 + n\boldsymbol{s}^2_{abc} + cn\boldsymbol{s}^2_{ab} \\
\boldsymbol{MS}_{AC} &= \boldsymbol{s}^2 + n\boldsymbol{s}^2_{abc} + bn\boldsymbol{s}^2_{ac} \\
\boldsymbol{MS}_{BC} &= \boldsymbol{s}^2 + n\boldsymbol{s}^2_{abc} + an\boldsymbol{s}^2_{bc} \\
\boldsymbol{MS}_{ABC} &= \boldsymbol{s}^2 + n\boldsymbol{s}^2_{abc} \\
\boldsymbol{MS}_{\text{rest}} &= \boldsymbol{s}^2.
\end{aligned}$$

Under the assumption of a normal distribution of $\boldsymbol{y}_{ijkl}$, it follows from Theorem 6.9 that

Theorem 6.13 If for the $\boldsymbol{y}_{ijkl}$ model equation (6.62) including its side conditions about expectations and covariances of the components of $\boldsymbol{y}_{ijkl}$ is valid and $\boldsymbol{y}_{ijkl}$ are multivariate normally distributed with the marginal distributions

$$N\left(\mu, \sigma_a^2 + \sigma_b^2 + \sigma_c^2 + \sigma_{ab}^2 + \sigma_{ac}^2 + \sigma_{bc}^2 + \sigma_{abc}^2 + \sigma^2\right),$$

then $\dfrac{\boldsymbol{SS}_X}{E(\boldsymbol{MS}_X)}$ are $CS(df_X)$-distributed (X=A, B, C, AB, AC, BC, ABC) with $\boldsymbol{SS}_X$, $E(\boldsymbol{MS}_X)$ and df_X from Table 5.21.

From Theorem 6.13 it follows that the F-values of the first column of Table 6.7 have the distribution given in the third column. By this we can test the hypotheses $H_{AB}: \sigma_{ab}^2 = 0, H_{AC}: \sigma_{ac}^2 = 0$, $H_{BC}: \sigma_{bc}^2 = 0, H_{ABC}: \sigma_{abc}^2 = 0$ with an F-test.

For testing the hypothesis $H_A: \sigma_a^2 = 0, H_B: \sigma_b^2 = 0, H_C: \sigma_c^2 = 0$, we need

Lemma 6.3 (Satterthwaite, 1946)

If $\boldsymbol{z}_1, \dots, \boldsymbol{z}_k$ are independent of each other as $\dfrac{CS(n_i)E(\boldsymbol{z}_i)}{n_i}$-distributed, ($i = 1, \dots, k$), so for real a_i

$$\boldsymbol{z} = \sum_{i=1}^{k} a_i \boldsymbol{z}_i$$

is with

$$n' = \frac{\left(\sum_{i=1}^{k} a_i z_i\right)^2}{\sum_{i=1}^{k} \frac{a_i^2}{n_i} z_i^2} \tag{6.63}$$

Table 6.7 Test statistics for testing hypotheses and distributions of these test statistics.

Test statistic	H_0	Distribution of the test statistic	Distribution of the test statistic under H_0
$F_{AB}=\frac{MS_{AB}}{MS_{ABC}}$	$\sigma^2_{ab}=0$	$\frac{cn\sigma^2_{ab}+n\sigma^2_{abc}+\sigma^2}{n\sigma^2_{abc}+\sigma^2}\cdot F[(a-1)(b-1),(a-1)(b-1)(c-1)]$	$F[(a-1)(b-1),$ $(a-1)(b-1)(c-1)]$
$F_{AC}=\frac{MS_{AC}}{MS_{ABC}}$	$\sigma^2_{ac}=0$	$\frac{bn\sigma^2_{ac}+n\sigma^2_{abc}+\sigma^2}{n\sigma^2_{abc}+\sigma^2}\cdot F[(a-1)(c-1),(a-1)(b-1)(c-1)]$	$F[(a-1)(c-1),$ $(a-1)(b-1)(c-1)]$
$F_{BC}=\frac{MS_{BC}}{MS_{ABC}}$	$\sigma^2_{bc}=0$	$\frac{an\sigma^2_{bc}+n\sigma^2_{abc}+\sigma^2}{n\sigma^2_{abc}+\sigma^2}\cdot F[(b-1)(c-1),(a-1)(b-1)(c-1)]$	$F[(b-1)(c-1),$ $(a-1)(b-1)(c-1)]$
$F_{ABC}=\frac{MS_{ABC}}{MS_{\text{res}}}$	$\sigma^2_{abc}=0$	$\frac{n\sigma^2_{abc}+\sigma^2}{\sigma^2}\cdot F[(a-1)(b-1),(c-1),N-abc]$	$F[(a-1)(b-1)$ $(c-1),N-abc]$

approximately $\frac{CS(n')E(\boldsymbol{z})}{n'}$-distributed, if $E(\boldsymbol{z})>0$.

This means that each realisation z of $\boldsymbol{z}$ is a realisation of an approximately $CS(n')$-distributed random variable. The approximation is relatively good for positive a_i (see also the remarks after Theorem 6.2).

We further need the following corollary to this lemma:

Corollary 6.2 If $\boldsymbol{MS}_i$ are independent of each other and if $\boldsymbol{z}_i=\frac{\boldsymbol{MS}_i n_i}{E(\boldsymbol{MS}_i)}$ are $CS(n_i)$-distributed (i = 1, …, k), so is

$$\boldsymbol{F}=\frac{\sum_{i=r}^{s}\boldsymbol{MS}_i}{\sum_{i=u}^{v}\boldsymbol{MS}_i}$$

under the null hypothesis $H_0:\sigma_x^2=0$ approximately $F(n',m')$-distributed with

$$n'=\frac{\left(\sum_{i=r}^{s}MS_i\right)^2}{\sum_{i=r}^{s}\frac{MS_i^2}{n_i}},\quad m'=\frac{\left(\sum_{i=u}^{v}MS_i\right)^2}{\sum_{i=u}^{v}\frac{MS_i^2}{n_i}},$$

if

$$E\left[\sum_{i=r}^{s}\boldsymbol{MS}_i\right]=c\sigma_x^2+E\left[\sum_{i=\text{u}}^{v}\boldsymbol{MS}_i\right]$$

and the second summand of the right side is positive.

Gaylor and Hopper (1969) could show by simulation experiments that the difference $\boldsymbol{MS_D}$ between $\boldsymbol{MS}_{\text{I}}$ and $\boldsymbol{MS}_{\text{II}}$, which are independent of each other with degrees of freedom n_I and n_{II}, respectively, multiplied with $\frac{n_D}{E(\boldsymbol{MS}_D)}$, that is, $\boldsymbol{MS}_D = (\boldsymbol{MS}_I - \boldsymbol{MS}_{II})\frac{n_D}{E(\boldsymbol{MS}_D)}$ are exact (or approximately) $CS(n_{II})$-distributed if n_D are the degrees of freedom of $\boldsymbol{MS}_D$ and $\frac{\boldsymbol{MS}_{\text{I}} n_{\text{I}}}{E(\boldsymbol{MS}_{\text{I}})}$ is exact (or approximately) $CS(n_{II})$-distributed and $\frac{\boldsymbol{MS}_{\text{II}} n_{\text{II}}}{E(\boldsymbol{MS}_{\text{II}})}$ is exact (or approximately) $CS(n_{II})$-distributed with

$$n_D = \frac{(MS_{\text{I}} - MS_{\text{II}})^2}{\frac{MS_{\text{I}}^2}{n_{\text{I}}} + \frac{MS_{\text{II}}^2}{n_{\text{II}}}}.$$

The approximation is sufficient as long as

$$\frac{\boldsymbol{MS}_{\text{I}}}{\boldsymbol{MS}_{\text{II}}} > F(n_{\text{II}}, n_{\text{I}}, 0{,}975)F(n_{\text{I}}, n_{\text{II}}, 0.50).$$

We use this corollary, to construct test statistics for the null hypotheses $H_{A0}: \sigma_a^2 = 0, H_{B0}: \sigma_b^2 = 0$ and $H_{C0}: \sigma_c^2 = 0$, which are approximately F-distributed.

From Table 6.6 we find

$$E(\boldsymbol{MS}_A) = bcn\sigma_a^2 + E(\boldsymbol{MS}_{AB} + \boldsymbol{MS}_{AC} - \boldsymbol{MS}_{ABC})$$
$$E(\boldsymbol{MS}_B) = acn\sigma_b^2 + E(\boldsymbol{MS}_{AB} + \boldsymbol{MS}_{BC} - \boldsymbol{MS}_{ABC})$$
$$E(\boldsymbol{MS}_C) = abn\sigma_c^2 + E(\boldsymbol{MS}_{AC} + \boldsymbol{MS}_{BC} - \boldsymbol{MS}_{ABC})$$

so that

$$\boldsymbol{F}_A = \frac{\boldsymbol{MS}_A}{\boldsymbol{MS}_{AB} + \boldsymbol{MS}_{AC} - \boldsymbol{MS}_{ABC}}$$

is under H_{A0} approximately $F(a_1, a_2)$-distributed,

$$\boldsymbol{F}_B = \frac{\boldsymbol{MS}_B}{\boldsymbol{MS}_{AB} + \boldsymbol{MS}_{BC} - \boldsymbol{MS}_{ABC}}$$

is under H_{B0} approximately $F(b_1, b_2)$-distributed, and

$$\boldsymbol{F}_C = \frac{\boldsymbol{MS}_C}{\boldsymbol{MS}_{BC} + \boldsymbol{MS}_{AC} - \boldsymbol{MS}_{ABC}}$$

is under H_{C0} approximately $F(c_1, c_2)$-distributed. From (6.63) we get

$$a_1 = a - 1, \quad b_1 = b - 1, \quad c_1 = c - 1$$

$$a_2 = \frac{(MS_{AB} + MS_{AC} - MS_{ABC})^2}{\frac{MS_{AB}^2}{(a-1)(b-1)} + \frac{MS_{AC}^2}{(a-1)(c-1)} + \frac{MS_{ABC}^2}{(a-1)(b-1)(c-1)}}.$$

Analogue formulae are valid for b_2 and c_2.

As Davenport and Webster (1973) show, it is sometimes better to use in place of $\boldsymbol{F}_A$, $\boldsymbol{F}_B$ and $\boldsymbol{F}_C$, respectively, the test statistics

$$\boldsymbol{F}_A^* = \frac{\boldsymbol{MS}_A + \boldsymbol{MS}_{ABC}}{\boldsymbol{MS}_{AB} + \boldsymbol{MS}_{AC}}, \quad \boldsymbol{F}_B^* = \frac{\boldsymbol{MS}_B + \boldsymbol{MS}_{ABC}}{\boldsymbol{MS}_{AB} + \boldsymbol{MS}_{BC}}$$

and

$$\boldsymbol{F}_C^* = \frac{\boldsymbol{MS}_C + \boldsymbol{MS}_{ABC}}{\boldsymbol{MS}_{AC} + \boldsymbol{MS}_{BC}},$$

respectively. Here again the Satterthwaite approximation is used; for instance, F_A^* is approximately $F(a_1^*, a_2^*)$-distributed with

$$a_1^* = \frac{(MS_A + MS_{ABC})^2}{\dfrac{MS_A^2}{(a-1)} + \dfrac{MS_{ABC}^2}{(a-1)(b-1)(c-1)}}$$

and

$$a_2^* = \frac{MS_{AB}^2 + MS_{AC}^2}{\dfrac{MS_{AB}^2}{(a-1)(b-1)} + \dfrac{MS_{AC}^2}{(a-1)(c-1)}}.$$

For the case of unequal subclass numbers, we use model equation (6.62) but now l runs $l = 1, \ldots, n_{ijk}$. Analogously to the two-way cross-classification, we construct quasi-SS (corresponds with the SS of Table 6.6) as, for instance,

$$\boldsymbol{QSS}_A = \sum_{i=1}^{a} \frac{\boldsymbol{Y}_{i\ldots}^2}{N_{i..}} - \frac{1}{N}\boldsymbol{Y}_{\ldots.}^2$$

$$\boldsymbol{QSS}_{AB} = \sum_{i=1}^{a}\sum_{j=1}^{b}{}^{*} \frac{\boldsymbol{Y}_{ij..}^2}{N_{ij.}} - \sum_{i=1}^{a} \frac{\boldsymbol{Y}_{i\ldots}^2}{N_{i..}} - \sum_{j=1}^{b} \frac{\boldsymbol{Y}_{.j..}^2}{N_{.j.}} + \frac{1}{N}\boldsymbol{Y}_{\ldots}^2$$

where Σ^* means summing up only over subclasses with $N_{ij.} > 0$. So we obtain the ANOVA table (Table 6.8).

In Table 6.8 is

$$\lambda_{a,b} = \sum_{i=1}^{a} \frac{\sum_{j=1}^{b} N_{ij.}^2}{N_{i..}}, \quad \lambda_{a,c} = \sum_{i=1}^{a} \frac{\sum_{k=1}^{c} N_{i.k}^2}{N_{i..}}$$

$$\lambda_{b,a} = \sum_{j=1}^{b} \frac{\sum_{i=1}^{a} N_{ij.}^2}{N_{.j.}}, \quad \lambda_{b,c} = \sum_{j=1}^{b} \frac{\sum_{k=1}^{c} N_{.jk}^2}{N_{.j.}}$$

$$\lambda_{a,bc} = \sum_{i=1}^{a} \frac{\sum_{j=1}^{b}\sum_{k=1}^{c} n_{ijk}^2}{N_{i..}}, \quad \lambda_{b,ac} = \sum_{j=1}^{b} \frac{\sum_{i=1}^{a}\sum_{k=1}^{c} n_{ijk}^2}{N_{.j.}}$$

Table 6.8 Analysis of variance table of a three-way cross-classification for model II.

				Coefficients the variance components in *E*(*QMS*)	
Source of variation	**Quasi-*SS***	**Quasi-*df***	**Quasi-*MS***	σ_a^2	σ_b^2
Between A-levels	QSS_A	$a-1$	QMS_A	$\frac{N-k_a}{a-1}$	$\frac{\lambda_{a,b}-k_b}{a-1}$
Between B-levels	QSS_B	$b-1$	QMS_B	$\frac{\lambda_{b,a}-k_a}{b-1}$	$\frac{N-k_a}{a-1}$
Between C-levels	QSS_C	$c-1$	QMS_C	$\frac{\lambda_{c,a}-k_a}{c-1}$	$\frac{\lambda_{c,b}-k_c}{c-1}$
Interaction $A \times B$	QSS_{AB}	$p_{ab}-a-b+1$[a]	QSM_{AB}	$\frac{k_a-\lambda_{b,a}}{p_{ab}-a-b+1}$	$\frac{k_b-\lambda_{a,b}}{p_{ab}-a-b+1}$
Interaction $A \times C$	QSS_{AC}	$p_{ac}-a-c+1$[a]	QMS_{AC}	$\frac{k_a-\lambda_{c,a}}{p_{ac}-a-c+1}$	$\frac{\lambda_{ac,b}-\lambda_{a,b}+k_b-\lambda_{c,a}}{p_{ac}-a-c+1}$
Interaction $B \times C$	QSS_{BC}	$p_{bc}-b-c+1$[a]	QMS_{BC}	$\frac{\lambda_{bc,a}-\lambda_{b,a}-\lambda_{c,a}+k_a}{p_{bc}-b-c+1}$	$\frac{k_c-\lambda_{b,c}}{p_{bc}-b-c+1}$
Interaction $A \times B \times C$	QSS_{ABC}	$p-p_{ab}-p_{ac}-p_{bc}+a+b+c-1$[a]	QMS_{ABC}	c_A	c_B
Residual	QSS_{res}	$N-p$	QMS_{res}	0	0

(Continued)

Table 6.8 (Continued)

σ_c^2	σ_{ab}^2	σ_{ac}^2	σ_{bc}^2	σ_{abc}^2	σ^2
$\frac{\lambda_{a,c}-k_c}{a-1}$	$\frac{\lambda_{a,b}-k_{ab}}{a-1}$.	$\frac{\lambda_{a,c}-k_{ac}}{a-1}$	$\frac{\lambda_{a,bc}-k_{bc}}{a-1}$	$\frac{\lambda_{a,bc}-k_{abc}}{a-1}$	1
$\frac{\lambda_{b,c}-k_c}{b-1}$	$\frac{\lambda_{b,a}-k_{ab}}{b-1}$	$\frac{\lambda_{b,ac}-k_{ac}}{b-1}$.	$\frac{\lambda_{b,c}-k_{bc}}{b-1}$	$\frac{\lambda_{b,ac}-k_{abc}}{b-1}$	1
$\frac{N-k_c}{c-1}$	$\frac{\lambda_{c,ab}-k_{ab}}{c-1}$	$\frac{\lambda_{c,a}-k_{ac}}{c-1}$.	$\frac{\lambda_{b,c}-k_{bc}}{c-1}$	$\frac{\lambda_{c,ab}-k_{abc}}{c-1}$	1
$\frac{\lambda_{ab,c}-\lambda_{a,c}-\lambda_{b,c}+k_{abc}}{p_{ab}-a-b+1}$	$\frac{N-\lambda_{a,b}-\lambda_{b,a}+k_{ab}}{p_{ab}-a-b+1}$	$\frac{\lambda_{ab,c}-\lambda_{a,c}-\lambda_{b,ac}+k_{ac}}{p_{ab}-a-b+1}$	$\frac{\lambda_{ab,c}-\lambda_{a,bc}-\lambda_{b,c}+k_{bc}}{p_{ab}-a-b+1}$	$\frac{\lambda_{ab,c}-\lambda_{a,bc}-\lambda_{b,ac}+k_{abc}}{p_{ab}-a-b+1}$	1
$\frac{k_c-\lambda_{a,c}}{p_{ac}-a-c+1}$	$\frac{\lambda_{ac,b}-\lambda_{a,b}-\lambda_{c,ab}+k_{ab}}{p_{ac}-a-c+1}$	$\frac{N-\lambda_{a,c}-\lambda_{c,a}+k_{ac}}{p_{ac}-a-c+1}$	$\frac{\lambda_{ac,b}-\lambda_{a,bc}-\lambda_{c,b}+k_{bc}}{p_{ac}-a-c+1}$	$\frac{\lambda_{ac,b}-\lambda_{a,bc}-\lambda_{c,ab}+k_{abc}}{p_{ac}-a-c+1}$	1
$\frac{k_c-\lambda_{b,c}}{p_{bc}-b-c+1}$	$\frac{\lambda_{bc,a}-\lambda_{b,a}-\lambda_{c,ab}+k_{ab}}{p_{bc}-b-c+1}$	$\frac{\lambda_{bc,a}-\lambda_{b,ac}-\lambda_{c,a}+k_{ac}}{p_{bc}-b-c+1}$	$\frac{N-\lambda_{b,c}-\lambda_{c,a}+k_{bc}}{p_{bc}-b-c+1}$	$\frac{\lambda_{bc,a}-\lambda_{b,ac}-\lambda_{c,ab}+\lambda_{abc}}{p_{bc}-b-c+1}$	1
c_C	c_{AB}	c_{AC}	c_{BC}	c_{ABC}	1
0	0	0	0	0	1

[a] p = number of subclasses with at least one observation, p_{ab} = number of $N_{ij.} > 0$, p_{bc} = number of $N_{.jk} > 0$, p_{ac} = number of $N_{i\,.\,k} > 0$.

$$\lambda_{c,ab} = \sum_{k=1}^{c} \frac{\sum_{i=1}^{a}\sum_{j=1}^{b} n_{ijk}^2}{N_{..k}}$$

$$\lambda_{ab,c} = \sum_{i=1}^{a}\sum_{j=1}^{b} \frac{\sum_{k=1}^{c} n_{ijk}^2}{N_{ij.}}, \quad \lambda_{ac,b} = \sum_{i=1}^{a}\sum_{k=1}^{c} \frac{\sum_{j=1}^{b} n_{ijk}^2}{N_{i.k}}$$

$$\lambda_{ab,c} = \sum_{i=1}^{a}\sum_{j=1}^{b} \frac{\sum_{k=1}^{c} n_{ijk}^2}{N_{ij\cdot}}, \lambda_{ac,b} = \sum_{i=1}^{a}\sum_{k=1}^{c} \frac{\sum_{j=1}^{b} n_{ijk}^2}{N_{i\cdot k}}, \lambda_{bc,a} = \sum_{j=1}^{b}\sum_{k=1}^{c} \frac{\sum_{i=1}^{a} n_{ijk}^2}{N_{\cdot jk}},$$

$$k_a = \frac{1}{N}\sum_{i=1}^{a} N_{i..}^2, \quad k_b = \frac{1}{N}\sum_{j=1}^{b} N_{.j.}^2, \quad k_c = \frac{1}{N}\sum_{k=1}^{c} N_{..k}^2$$

$$k_{ab} = \frac{1}{N}\sum_{i=1}^{a}\sum_{j=1}^{b} N_{ij.}^2, \quad k_{ac} = \frac{1}{N}\sum_{i=1}^{a}\sum_{k=1}^{c} N_{i.k}^2, \quad k_{bc} = \frac{1}{N}\sum_{j=1}^{b}\sum_{k=1}^{c} N_{.jk}^2$$

$$vc_A = \lambda_{b,a} + \lambda_{c,a} - \lambda_{bc,a} - k_a,$$
$$vc_B = \lambda_{a,b} + \lambda_{c,b} - \lambda_{ac,b} - k_b,$$
$$vc_C = \lambda_{a,c} + \lambda_{b,c} + \lambda_{ab,c} - k_c,$$
$$vc_{AB} = \lambda_{a,b} + \lambda_{b,a} + \lambda_{c,ab} - \lambda_{ac,b} - \lambda_{bc,a} - k_{ab},$$
$$vc_{AC} = \lambda_{a,c} + \lambda_{b,ac} + \lambda_{c,a} - \lambda_{ab,c} - \lambda_{bc,a} - k_{ac},$$
$$vc_{BC} = \lambda_{a,bc} + \lambda_{b,c} + \lambda_{c,b} - \lambda_{ab,c} - \lambda_{ac,b} - k_{bc},$$
$$vc_{ABC} = N + \lambda_{a,bc} + \lambda_{b,ac} + \lambda_{c,ab} - \lambda_{ab,c} - \lambda_{ac,b} - \lambda_{bc,a} - k_{abc},$$

where

$$v = p - p_{ab} - p_{ac} - p_{bc} + a + b + c - 1.$$

From the equations, gained from the coefficients of the $E(\boldsymbol{QMS})$ when replacing the $E(\boldsymbol{QMS})$ by the $\boldsymbol{QMS}$ and σ_x^2 by $\boldsymbol{s}_x^2$, we receive the estimator of σ_x^2.

6.3.5 Three-Way Nested Classification

For the three-way nested classification $C \prec B \prec A$, the following model equation is assumed:

$$\boldsymbol{y}_{ijkl} = \mu + \boldsymbol{a}_i + \boldsymbol{b}_{ij} + \boldsymbol{c}_{ijk} + \boldsymbol{e}_{ijkl}$$
$$\left(i = 1,\ldots,a; j = 1,\ldots,b_i; k = 1,\ldots,c_{ij}; l = 1,\ldots,n_{ijk}\right). \tag{6.64}$$

The side conditions are that all random variables of the right-hand side of (6.64) have expectation 0 and are pairwise uncorrelated and $\text{var}(\boldsymbol{a}_i) = \sigma_a^2$ for all i, $\text{var}(\boldsymbol{b}_{ij}) = \sigma_b^2$ for all i,j, $\text{var}(\boldsymbol{c}_{ijk}) = \sigma_c^2$ for all i, j, k and $\text{var}(\boldsymbol{e}_{ijkl}) = \sigma^2$ for all i, j, k, l.

Because Theorem 5.12 is independent of the model, we find the ***SS***, *df* and ***MS*** of the three-way nested ANOVA in Table 5.27. For the calculation the $E(\boldsymbol{MS})$ we need

$$D = \sum_i N_{i..}^2, \quad E_i = \sum_j N_{ij.}^2, \quad E = \sum_i E_i,$$

$$F_{ij} = \sum_k n_{ijk}^2, \quad F_i = \sum_j F_{ij}, \quad F = \sum_i F_i,$$

$$\lambda_1 = \sum_{i,j} \frac{F_{ij}}{N_{ij.}}, \quad \lambda_2 = \sum_i \frac{F_i}{N_{i..}} \quad \lambda_3 = \sum_i \frac{E_i}{N_{i..}}.$$

The $E(\boldsymbol{MS})$ can be found in Table 6.9. By the ANOVA method, we gain the following estimators for the variance components:

$$\boldsymbol{s}^2 = \boldsymbol{MS}_{\text{res}}$$

$$\boldsymbol{s}_c^2 = \frac{C_{..} - B_.}{N - \lambda_1}\left(\boldsymbol{MS}_{C \text{ in } B} - \boldsymbol{MS}_{\text{res}}\right)$$

$$\boldsymbol{s}_b^2 = \frac{B_. - a}{N - \lambda_3}\left(\boldsymbol{MS}_{B \text{ in } A} - \boldsymbol{MS}_{\text{res}} - \frac{\lambda_1 - \lambda_2}{B_. - a}\boldsymbol{s}_c^2\right)$$

$$\boldsymbol{s}_a^2 = \frac{a-1}{n - \frac{D}{N}}\left(\boldsymbol{MS}_A - \boldsymbol{MS}_{\text{res}} - \frac{\lambda_2 - \frac{F}{N}}{a-1}\boldsymbol{s}_c^2 - \frac{\lambda_3 - \frac{F}{N}}{a-1}\boldsymbol{s}_b^2\right).$$

The variances of these variance components can be found in Searle (1971) and will not repeat here for space considerations.

Table 6.9 Expectations of the *MS* of a three-way nested classification for model II.

Source of variation	E(MS)
Between the *A*-levels	$\sigma^2 + \sigma_c^2 \frac{\lambda_2 - \frac{F}{N}}{a-1} + \sigma_b^2 \frac{\lambda_3 - \frac{E}{N}}{a-1} + \sigma_a^2 \frac{N - \frac{D}{N}}{a-1}$
Between the *B*-levels within the *A*-levels	$\sigma^2 + \sigma_c^2 \frac{\lambda_1 - \lambda_2}{B_. - a} + \sigma_b^2 \frac{N - \lambda_3}{B_. - a}$
Between the *C*-levels within the *B*- and *A*-levels	$\sigma^2 + \sigma_c^2 \frac{N - \lambda_1}{C_{..} - B_.}$
Residual	σ^2

6.3.6 Three-Way Mixed Classification

We consider the mixed classifications in Sections 5.4.3.1 and 5.4.3.2 for model II of the ANOVA. The model equation for the type $(B \prec A) \times C$ is

$$\begin{aligned} \boldsymbol{y}_{ijkl} &= \mu + \boldsymbol{a}_i + \boldsymbol{b}_{ij} + \boldsymbol{c}_k + (\boldsymbol{a},\boldsymbol{b})_{ik} + (\boldsymbol{b},\boldsymbol{c})_{i\cdot jk} + \boldsymbol{e}_{ijkl} \\ &(i = 1,\ldots,a; j = 1,\ldots,b; k = 1,\ldots,c; l = 1,\ldots,n). \end{aligned} \tag{6.65}$$

The model equation for the type $C \prec AB$ is

$$\boldsymbol{y}_{ijkl} = \mu + \boldsymbol{a}_i + \boldsymbol{b}_j + \boldsymbol{c}_{ijk} + (\boldsymbol{a},\boldsymbol{b})_{ij} + \boldsymbol{e}_{ijkl} \quad (i = 1,\ldots,a; j = 1,\ldots,b; k = 1,\ldots,c; l = 1,\ldots,n). \tag{6.66}$$

Again we assume that the random components of the right hand side of (6.65) and of (6.66) that have expectation zero are pairwise uncorrelated and have for all indices the same variances

$$\mathrm{var}(\boldsymbol{a}_i) = \sigma_a^2,\ \mathrm{var}(\boldsymbol{b}_{ij}) = \sigma_{b\,in\,a}^2 \left[\mathrm{var}(\boldsymbol{b}_j) = \sigma_b^2\right],$$

$$\mathrm{var}(\boldsymbol{c}_k) = \sigma_c^2 \left[\mathrm{var}(\boldsymbol{c}_{ijk}) = \sigma_c^2\right],\ \mathrm{var}\left((\boldsymbol{a},\boldsymbol{b})_{ik}\right) = \sigma_{ac}^2 \left[\mathrm{var}\left((\boldsymbol{b},\boldsymbol{c})_{i,jk}\right) = \sigma_{bc\,in\,a}^2\right]$$

$$\mathrm{var}(\boldsymbol{e}_{ijkl}) = \sigma^2 \left[\mathrm{var}\left((\boldsymbol{a},\boldsymbol{b})_{ij}\right) = \sigma_{ab}^2\right].$$

The decomposition of the ***SS*** and *df* can be given in Sections 5.4.3.1 and 5.4.3.2. To estimate the variance components by the ANOVA method, we need $E(\boldsymbol{DQ})$. Following Rasch (1971) we obtained the type $(B \prec A) \times C$

$$\left.\begin{aligned} E(\boldsymbol{MS}_A) &= \sigma^2 + n\sigma_{bc\text{ in }a}^2 + bn\sigma_{ac}^2 + cn\sigma_{b\text{ in }a}^2 + bcn\sigma_a^2 \\ E(\boldsymbol{MS}_{B\text{ in }A}) &= \sigma^2 + n\sigma_{bc\text{ in }a}^2 + cn\sigma_{b\text{ in }a}^2 \\ E(\boldsymbol{MS}_C) &= \sigma^2 + bn\sigma_{ac}^2 + abn\sigma_c^2 + n\sigma_{bc\text{ in }a}^2 \\ E(\boldsymbol{MS}_{AC}) &= \sigma^2 + bn\sigma_{ac}^2 + n\sigma_{bc\text{ in }a}^2 \\ E(\boldsymbol{MS}_{BC\text{ in }A}) &= \sigma^2 + n\sigma_{bc\text{ in }a}^2 \\ E(\boldsymbol{MS}_{\text{res}}) &= \sigma^2 \end{aligned}\right\} \tag{6.67}$$

and for the type $C \prec AB$

$$\left.\begin{aligned} E(\boldsymbol{QMS}_A) &= \sigma^2 + cn\sigma_{ab}^2 + n\sigma_{c\text{ in }ab}^2 + bcn\sigma_a^2 \\ E(\boldsymbol{QMS}_B) &= \sigma^2 + cn\sigma_{ab}^2 + n\sigma_{a\text{ in }ab}^2 + acn\sigma_b^2 \\ E(\boldsymbol{QMS}_{C\text{ in }AB}) &= \sigma^2 + n\sigma_{c\text{ in }ab}^2 \\ E(\boldsymbol{QMS}_{AB}) &= \sigma^2 + cn\sigma_{ab}^2 - n\sigma_{c\text{ in }ab}^2 \\ E(\boldsymbol{QMS}_{\text{res}}) &= \sigma^2 \end{aligned}\right\}. \tag{6.68}$$

By the ANOVA method, we obtain the estimators of the variance components, by replacing in (6.67) and (6.68), σ_x^2 by $\boldsymbol{s}_x^2$ and $E(\boldsymbol{MS}_x)$ by $\boldsymbol{MS}_x$ and solve the equations for $\boldsymbol{s}_x^2$.

6.4 Planning Experiments

Systematic descriptions of designing the experiments for the one-way ANOVA and definitions of several optimality criteria gives Herrendörfer (1976) on which the results of this section are based. We start with model equation (6.8) with its side conditions. Further, all random effects in (6.8) may be normally distributed. As estimators of σ_a^2 and σ^2, we choose (6.12) and $\boldsymbol{MS}_{res}$, respectively. We use the following notations: $\Sigma^{\mathrm{T}} = \left(\sigma_a^2, \sigma^2\right)$ and $\hat{\Sigma}^{\mathrm{T}} = \left(\boldsymbol{s}_a^2, \boldsymbol{s}^2\right)$ with $\boldsymbol{s}_a^2$ from (6.12) and $\boldsymbol{s}^2 = \boldsymbol{MS}_{res}$ from Table 5.2.

Definition 6.6 The vector $V_N = (a, n_1, \dots, n_a)_N$ is called concrete experimental design for the estimation of Σ, if $2 \le a \le N-1, n_i \ge 1, \sum_{i=1}^{a} n_i = N$ where a and n_i are integers. ${}_0V_N = (a, N_1, \dots, n_a)_N$ is called discrete experimental design for the estimation of Σ, if $2 \le a \le N-1, n_i \ge 1, \sum_{i=1}^{a} n_i = N$ where a and N are integers, but the n_i may be real. With $\{V_N\}$ and $\{{}_0V_N\}$ we denote the set of possible concrete or discrete experimental design, respectively, for fixed N.

We see that $\{V_N\} \subset \{{}_0V_N\}$.

Definition 6.7 An experimental design ${}_0V_N^* \in \{{}_0V_N\}\left(V_N^* \in \{V_N\}\right)$ is called discrete (concrete) A-optimal experimental design for given N, if for this experimental design

$$\operatorname{var}\left(\boldsymbol{s}_a^2\right) + \operatorname{var}\left(\boldsymbol{s}^2\right) = \frac{1}{w^2}\left[\operatorname{var}(\boldsymbol{MS}_{\mathrm{A}}) + \operatorname{var}(\boldsymbol{MS}_{res})\right] + \operatorname{var}(\boldsymbol{MS}_{res})$$

with

$$w = \frac{1}{a-1}\left(N - \frac{1}{N}\sum_{i=1}^{a} n_i^2\right)$$

in the set $\{{}_0V_N\}(\{V_N\})$ is minimal.

Theorem 6.14 (Herrendörfer)
The discrete A-optimal experimental design in $\{{}_0V_N\}$ for estimating Σ must be found amongst the designs with equal subclass numbers ($n_i = n$).

Proof: The formulae (6.38) and (6.39) are initially defined only for natural a and $n_i (i = 1, \dots, a)$. For a discrete experimental design, we allow real $n_i \ge 1$. For fixed N and a the w is maximum for $n_i = \frac{N}{a} = \bar{n}$. Because $n_i = \bar{n}$ is minimising

$\frac{1}{w^2}\text{var}(\boldsymbol{MS}_{\text{A}})$ (Hammersley, 1949) and var($\boldsymbol{MS}_{\text{res}}$) are independent of the decomposition of N into n_i, the theorem is proved because this is true for all pairs (a, N).

Thus for the determination of a discrete A-optimal experimental design, the term

$$A(N,a) = \frac{1}{\bar{n}^2}\left[\frac{2\left(\bar{n}\sigma_a^2+\sigma^2\right)^2}{a-1}+\frac{2\sigma^4}{a(\bar{n}-1)}\right]+\frac{2\sigma^4}{a(\bar{n}-1)} \tag{6.69}$$

must be minimised. Putting for $\rho_I = \rho$, for ρ_I in Definition 6.3, (6.69) becomes

$$A(N,a) = \left\{\frac{2}{a+1}\left(\rho+\frac{1-\rho}{N}\right)^2+\frac{4}{N-a}(1-\rho)^2\right.$$

$$\left.+2\left(\frac{1-\rho}{N}\right)\left[2\rho-\frac{1-\rho}{N}(N-1)\right]\right\}\left(\sigma_a^2+\sigma^2\right)^2.$$

From Definition 6.3 and $\sigma^2 > 0$, it follows always $0 < \rho < 1$.

Looking at the second partial derivation of $A(N, a)$ with respect to a, we see that $A(N, a)$ for $1 \le a \le N$ is convex from below and for $0 < \varrho < 1$ therefore $A(N, a)$ has exactly one relative minimum. Putting $\frac{\partial A(N,a)}{\partial a}$ equal to zero gives the two solutions

$$a_1 = 1+\frac{(N-1)\left(\rho+\frac{1-\rho}{N}\right)}{\rho+\frac{1-\rho}{N}-\sqrt{2}(1-\rho)} \tag{6.70}$$

and

$$a_2 = 1+\frac{(N-1)\left(\rho+\frac{1-\rho}{N}\right)}{\rho+\frac{1-\rho}{N}+\sqrt{2}(1-\rho)}. \tag{6.71}$$

But a_1 is not in the interval $0 \le a \le N-1$, and with this only the solution a_2 in (6.71) is acceptable. If a_2 is an integer and $2 \le a_2 \le N$, the A-optimal design is given by

$$a = \frac{N\left[N\rho+\left(1+\sqrt{2}\right)(1-\rho)\rho\right]}{N\rho+\left(1+N\sqrt{2}\right)(1-\rho)} \tag{6.72}$$

and $\bar{n} = -\frac{N}{a}$.

If $a' \le a_2 \le a''$ with $a'' = a' + 1$ (a', a'' integer) and only a' or a'' is in the interval [2,*N*], then the integer in this interval is *a* of the *A*-optimal discrete design. If both numbers a' and a'' are in [2,*N*], we calculate for both $A(N, a)$ and choose that one as optimal, for which $A(N, a)$ is minimum.

We find the concrete *A*-optimal experimental design by systematical search in the neighbourhood of the discrete *A*-optimal experimental design. By this systematical search, we also vary *a* and of course unequal n_i can occur. Theorems about optimal experimental designs to minimise the variance of a variance component (so called *C*-optimal designs) and the cost optimal choice of *N* can be found in Herrendörfer (1976). There and in Rasch et al. (2008), tables of optimal designs and experimental sizes are given.

6.5 Exercises

6.1 For testing performances of boars, offspring of boars under unique feeding fattening and slaughter performances are measured. From the results of such testing, two boars b_1 and b_2 have been randomly selected. For each boar, the offspring of several sows have been observed. As well as from b_1 and also from b_2, three observations *y* (number of the fattening days from 40 kg up to 110 kg) are available. The variance components for boars and sows (within boars) and within sows must be estimated.

Table 6.10 shows the observations $y_{ijk\cdot}$. This case is $a = 2$, $b_1 = 3$, $b_2 = 3$. The *E*(***MS***) are given in Table 6.6.

Table 6.10 Data of Example 6.1.

Number the fattening days $y_{ijk\cdot}$	Boars		b_1			b_2	
	Sows	s_1	s_2	s_3	s_4	s_5	s_6
Offspring	y_{ijk}	93	107	109	89	87	81
		89	99	107	102	91	83
		97		94	104	82	85
		105		106	97		91
	n_{ij}	4	2	4	4	3	4
			10			11	

6.2 Determine the *A*-optimal experimental design by (6.71) for $N = 200$ and $\rho = 0.5$.

6.3 Add in Example 6.1 for boar 5 the missing value by the corresponding mean (2 decimal places) and add for boar 2 the mean twice. Estimate the variance components for the new data set D.

6.4 Derive

$$\left.\begin{aligned} E(\boldsymbol{MS}_A) &= bn\sigma_a^2 + n\sigma_{ab}^2 + \sigma^2 \\ E(\boldsymbol{MS}_B) &= an\sigma_b^2 + n\sigma_{ab}^2 + \sigma^2 \\ E(\boldsymbol{MS}_{AB}) &= \qquad\quad n\sigma_{ab}^2 + \sigma^2 \\ E(\boldsymbol{MS}_{res}) &= \qquad\qquad\qquad \sigma^2 \end{aligned}\right\}$$

using the rule of Chapter 7.

References

Ahrens, H. (1983) MINQUE und ANOVA, Schätzer für Varianzkomponenten – Effizienz und Vergleich der Unbalanciertheit. *Probl. Angew. Stat. Forsch.–zentr. Tierprod. Dummerstorf*, Heft 8.

Davenport, J. M. and Webter, J. T. (1973) A comparison of some approximate *F*-tests. *Technometrics*, **15**, 779–789.

Drygas, H. (1980) Hsu's theorem in variance components models, in *Mathematical Statistics, Banach Centre Publications*, WN – Polish Scientific Publishers, Warsaw, **6**, pp. 95–107.

Gaylor, D. W. and Hopper, F. N. (1969) Estimating the degree of freedom for linear combinations of mean squares by Satterthwaite's formula. *Technometrics*, **11**, 691–706.

Graybill, F. A. (1954) On quadratic estimates of variance components. *Ann. Math. Stat.*, **25**, 367–372.

Graybill, F. A. and Wang, C. M. (1980) Confidence intervals on nonnegative linear combinations of variances. *J. Am. Stat. Assoc.*, **75**, 869–873.

Hammersley, J. M. (1949) The unbiased estimate and standard error of the interclass variance. *Metron*, **15**, 189–204.

Hartley, H. O. (1967) Expectations, variances and covariances of ANOVA mean squares by "synthesis". *Biometrics*, **23**, 105–114.

Hartley, H. O. and Rao, J. N. K. (1967) Maximum likelihood estimation for the mixed analysis of variance model. *Biometrika*, **54**, 92–108.

Harville, D. A. (1977) Maximum-likelihood approaches to variance component estimation and to related problems. *J. Am. Stat. Assoc.*, **72**, 320–340.

Henderson, C. R. (1953) Estimation of variance and covariance components. *Biometrics*, **9**, 226–252.

Herbach, L. H. (1959) Properties of type II analysis of variance tests. *Ann. Math. Stat.*, **30**, 939–959.

Herrendörfer, G. (1976) *Versuchsplanung im linearen Modell – Schätzung von Varianzkomponenten und eingeschränkte Messwerterfassung bei der Schätzung und dem Vergleich von Mittelwerten und Regressionskoeffizienten*, Habilitation, Martin-Luther-Universität Halle-Wittenberg.

Rao, C. R. (1970) Estimation of heteroscedastic variances in linear models. *J. Am. Stat. Assoc.*, **65**, 445–456.

Rao, C. R. (1971a) Estimation of variance and covariance components in linear models. *J. Am. Stat. Assoc.*, **66**, 872–875.

Rao, C. R. (1971b) Minimum variance quadratic estimation of variance components. *J. Multivar. Anal.*, **1**, 257–275.

Rasch, D. (1971) Gemischte Klassifikation der dreifachen Varianzanalyse. *Biom. Z.*, **13**, 1–20.

Rasch, D. and Herrendörfer, G. (1986) *Experimental Design-Sample Size Determination and Block Designs*, Reidel, Dordrecht.

Rasch, D. and Mašata, O. (2006) Methods of variance component estimation. *Czech. J. Anim. Sci.*, **51**, 227–235.

Rasch, D., Herrendörfer, G., Bock, J., Victor, N. and Guiard, V. (Eds.) (2008) Verfahrensbibliothek Versuchsplanung und -auswertung, *2.* verbesserte Auflage in einem Band mit CD, R. Oldenbourg Verlag München, Wien.

Satterthwaite, F. E. (1946) An approximate distribution of estimates of variance components. *Biom. Bull.*, **2**, 110–114.

Searle, S. R. (1971, 2012) *Linear Models*, John Wiley & Sons, Inc., New York.

Searle, S. R., Casella, E. and McCullock, C. E. (1992) *Variance Components*, John Wiley & Sons, Inc., New York.

Seely, J. F. and Lee, Y. (1994) A note on Satterthwaite confidence interval for a variance. *Commun. Stat.*, **23**, 859–869.

Townsend, E. C. (1968) *Unbiased estimators of variance components in simple unbalanced designs*, PhD thesis, Cornell University, Ithaca, NY.

Verdooren, L. R. (1982) How large is the probability for the estimate of variance components to be negative? *Biom. J.*, **24**, 339–360.

Welch, B. L. (1956) On linear combinations of several variances. *J. Am. Stat. Assoc.*, **51**, 1144–1157.

7

Analysis of Variance – Models with Finite Level Populations and Mixed Models

In the present chapter we consider models with factor levels from a finite population of factor levels, or as we call them in short 'finite level populations'. Covering at least in the case of equal subclass numbers, model I and model II of the analysis of variance (ANOVA) as special cases. Even the mixed models, also introduced in this chapter, are limiting cases of models with finite level populations.

In mixed models as well, problems of variance component estimation and also of estimating and testing fixed effects occur. In Section 7.3, some special methods are presented, which are demonstrated for some special cases in Section 7.4.

7.1 Introduction: Models with Finite Level Populations

Models with finite level populations are of interest because we meet practical situations where the selection of factor levels covers a finite number of levels but not all levels in a population with a finite number of levels and further because other models are special or limiting cases of such models.

Definition 7.1 Let the elements $\gamma_{A_k,j}(j = 1,\dots,a_k)$ of the vectors γ_{A_k} in model equation (6.3) be a_k random variables. The realisations of those a_k random variables are a_k effects, sampled (without replacement) from a population of $N(A_k)$ effects. Then we call the model equation (6.3) (under the side conditions that the effects in the populations sum up to zero) an ANOVA model with finite level populations. If (6.43) holds, we speak about a balanced case of the model with finite level populations.

This means we assume that the a_k effects in an experiment are selected randomly from a level population with $N(A_k) \geq a_k$ effects and that each level can be selected only once. If $N(A_k) = a_k$, all levels are selected and the factor A_k is a fixed factor in a model I. If $N(A_k) \to \infty$, then the factor A_k is a random factor of model II.

Mathematical Statistics, First Edition. Dieter Rasch and Dieter Schott.

In the balanced case we present in Section 7.2 simple rules for the derivation of ***SS***, *df*, ***MS*** and *E*(***MS***).

For simplicity we call an n-dimensional random variable with identically distributed (but not independent) components a type 2 random sample as it is usual in the theory of sampling from finite populations.

In the theory of sampling from finite populations, the variance in the population is defined as the sum of squared deviations from the expectation divided by the population size N. But in the ANOVA, we use for simplification of further formulae quasi-variance by dividing the sum of squared deviations from the expectation by $N - 1$. We denote the quasi-variances by σ^2, σ_a^2 and so on but the real variances by $\sigma^{*2}, \sigma_a^{*2}$. The conversion of varainces into quasi-variances and vice versa is demonstrated by the example below.

Example 7.1 Let the cross-classified factors A and B be nested in the factor C. Then the variance component of the interaction $A \times B$ is

$$\sigma^2_{ab\,\text{in}\,c} = \frac{N(A) \cdot N(B)}{(N(A)-1)(N(B)-1)} \sigma^{*2}_{ab\,\text{in}\,c} \tag{7.1}$$

where

$$\sigma^2_{ab\,\text{in}\,c} = \frac{\sum_{i,j} (ab)^2_{ij \cdot k}}{(N(A)-1)(N(B)-1)} \text{ for all } k. \tag{7.2}$$

Example 7.2 We consider a model with finite level populations and two factors and a factor level combination, where $A_1 = A, A_2 = B, A_3 = A \times B, a_1 = a$, $a_2 = b$ and $a_3 = ab$ and R stands for the residual. The model equation for the balanced case is

$$\boldsymbol{y}_{ijk} = \mu + \boldsymbol{a}_i + \boldsymbol{b}_j + (\boldsymbol{ab})_{ij} + \boldsymbol{e}_{ijk} \quad (i = 1,\ldots,a; j = 1,\ldots,b; k = 1,\ldots,n). \tag{7.3}$$

The side conditions are

$$n < N(R), \ \sum_{i=1}^{N(A)} a_i = \sum_{j=1}^{N(B)} b_i = \sum_{i=1}^{N(A)} (ab)_{ij} = \sum_{j=1}^{N(B)} (ab)_{ij} = \sum_{k=1}^{N(R)} e_{ijk} = 0,$$

$$\frac{1}{N(A)-1} \sum_{i=1}^{N(A)} a_i^2 = \sigma_a^2, \quad \frac{1}{N(B)-1} \sum_{j=1}^{N(B)} b_j^2 = \sigma_b^2,$$

$$\frac{1}{[N(A)-1][N(B)-1]} \sum_{i=1}^{N(A)} \sum_{j=1}^{N(B)} (ab)^2_{ij} = \sigma^2_{ab}, \quad \frac{1}{N(R)-1} \sum_{k=1}^{N(R)} e^2_{ijk} = \sigma^2$$

for all i and j not at the bounds of the sigma signs.

We can derive the following expectations after inserting the right-hand side of the model equation for $\boldsymbol{y}$ in the $E(\boldsymbol{MS})$.

With the notations of Table 5.13, this results in

$$\left.\begin{aligned}
E(\boldsymbol{MS}_A) &= \left(1-\frac{n}{N(R)}\right)\sigma^2 + n\left(1-\frac{b}{N(B)}\right)\sigma_{ab}^2 + nb\sigma_a^2,\\
E(\boldsymbol{MS}_B) &= \left(1-\frac{n}{N(R)}\right)\sigma^2 + n\left(1-\frac{a}{N(A)}\right)\sigma_{ab}^2 + na\sigma_b^2,\\
E(\boldsymbol{MS}_{AB}) &= \left(1-\frac{n}{N(R)}\right)\sigma^2 + n\sigma_{ab}^2.\\
E(\boldsymbol{MS}_{res}) &= \sigma^2.
\end{aligned}\right\} \tag{7.4}$$

If $N(R)\to\infty$, $N(B)\to\infty$ and $N(A)\to\infty$, we obtain the $E(\boldsymbol{MS})$ of model II in (6.53). For $N(R)\to\infty$, $a = N(A)$ and $b = N(B)$, we obtain the $E(\boldsymbol{MS})$ of model I in Table 5.13. For $N(R)\to\infty$, $N(B)\to\infty$ and $a = N(A)$, we get the model of Example 7.2, a mixed model (A fixed, B random), and we receive

$$\left.\begin{aligned}
E(\boldsymbol{MS}_A) &= \sigma^2 + n\sigma_{ab}^2 + nb\sigma_a^2,\\
E(\boldsymbol{MS}_B) &= \sigma^2 + na\sigma_b^2,\\
E(\boldsymbol{MS}_{AB}) &= \sigma^2 + n\sigma_{ab}^2,\\
E(\boldsymbol{MS}_{res}) &= \sigma^2.
\end{aligned}\right\} \tag{7.5}$$

In (7.5) we put $\sigma_a^2 = \frac{1}{a-1}\sum_{i=1}^{a} a_i^2$ but this is no variance.

Example 7.2 shows the potential of models with finite level populations. In the balanced case simple rules for the calculation of the $E(\boldsymbol{MS})$ exist.

7.2 Rules for the Derivation of *SS*, *df*, *MS* and *E*(*MS*) in Balanced ANOVA Models

In Chapters 5 and 6, we could see that the derivation of $E(\boldsymbol{MS})$ even in simple cases is elaborate. Now we give rules by which formulae for $\boldsymbol{SS}$, df, $\boldsymbol{MS}$ and $E(\boldsymbol{MS})$ for a balanced case can be easily derived.

Let us consider t factors A_k; $k = 1, \ldots, t$ in an ANOVA with the size of the level populations $N(a_k)$ and the number of selected levels a_k. (If there are few factors, we rename $A_1 = A$, $A_2 = B$, $A_3 = C$.) If a factor A_{k_1} is subordinated to a factor A_{k_2}, we write $A_{k_1} \prec A_{k_2}$. The indices of the effects in the model equations are split into two groups. The indices in any suffix of subordinated (nested) factors are given first; then the indices of the superordinate factors or factor combinations follow in a bracket such as $e_{k(i,j)}$ or $e_{l(i,j,k)}$. In the ANOVA table for each factor (including residual), a row exists. Further, there are rows for factor combinations (interactions). If a factor X is not subordinated to any factor, we write $X\prec$.

Rule 1 Interactions between two factors or factor combinations are obtained formally by symbolic multiplication of the factors (or factor combinations) both left of the $\prec$ sign and right of the $\prec$ sign. If the same letter occurs, the right of the $\prec$ sign more than once, it will be noted only once ($X \cdot X = X$). An interaction is not defined if the same letter occurs at both sides of the $\prec$ sign.

Rule 2 The degrees of freedom in a row are obtained, reducing the number of occurring levels of the A_k left of the $\prec$ sign by 1 and multiplying with analogously reduced numbers of other factors left of the $\prec$ sign as well as with the number of the selected levels of factors right of the $\prec$ sign.

Rule 3 The ***SS*** in a row are obtained by performing a product of the $\boldsymbol{S}_{A_k}$ in (6.45) analogue to the product of the degrees of freedom, which means that an A_k left of the $\prec$ sign gives a factor $\boldsymbol{S}_{A_k} - \boldsymbol{e}$, in the product, but the right of the $\prec$ sign gives A_k a factor $\boldsymbol{S}_{A_k}$ in the product. The error term $\boldsymbol{e}$ is the identity element of this multiplication ($\boldsymbol{S}_{A_k}\boldsymbol{e} = \boldsymbol{e}\boldsymbol{S}_{A_k} = \boldsymbol{S}_{A_k}$) and defined by $\boldsymbol{e} = \frac{1}{N}\boldsymbol{Y}^2_{\ldots}$. Further as a result of that symbolic multiplication, S_{A_i}, S_{A_j} is to read as $S_{A_iA_j}$ and $S_{R,A_1,\ldots,A_t}$ as S_R.

Rule 4 The $E(\boldsymbol{MS})$ are calculated as follows: define a table with rows defined by the components (except μ) of the right-hand side of the model equation. The columns correspond with the indices in the suffix. If in a cell of the table the index defining the column does not occur in the effect defining the row, we fill the cell with the number of selected levels of the factor defining the column. If the index defining the column occurs in the bracket of the row effect, we put a 1 into the corresponding cell of the table; otherwise we put there

$$1 - \frac{\text{number of selected levels of the column}}{\text{number of levels of the column}}.$$

Now each $E(\boldsymbol{MS})$ is written as a linear combination of σ^2 and all the variance components whose suffixes contain the upper bounds of those indices occurring in the effect corresponding to the ***MS*** in front of the bracket. The coefficients of the linear combination are generated in that row of the table corresponding with the variance component by multiplication of the contents of those cells in that row defined by a column suffix not in the bracket of the effect defining the ***MS***. Finally, we convert as shown in Example 7.1.

Example 7.3 Given a two-way cross-classification with finite level populations as in Example 7.2. At first, the model equation (7.3) is assumed. The ***SS***, ***df*** and ***MS*** of Table 5.13 as well as the E(**MS**) are to state according to the rules of this section. The ANOVA table has to contain the rows for A, B, AB and residual. At first, we have to put the indices of superordinate factors in brackets. Because only the error terms are subordinated to all factors, (7.3) becomes

$$\boldsymbol{y}_{ijk} = \mu + \boldsymbol{a}_i + \boldsymbol{b}_j + (\boldsymbol{ab})_{ij} + \boldsymbol{e}_{k(i,j)}.$$

Only one interaction exists, $(A\prec)(B\prec) = AB\prec$. The degrees of freedom following rule 2 (*res* and R means residual)

$$\begin{aligned}
A\prec &: a-1,\\
B\prec &: b-1,\\
AB\prec &: (a-1)(b-1),\\
R\prec AB &: (n-1)ab.
\end{aligned}$$

The ***SS*** following rule 3 are

$$A\prec\ :\ \boldsymbol{S}_A - \frac{\boldsymbol{Y}_{\cdots}^2}{N} = \sum_{i=1}^{a}\frac{\boldsymbol{Y}_{i\cdot\cdot}^2}{bn} - \frac{\boldsymbol{Y}_{\cdots}^2}{N},$$

$$B\prec\ :\ \boldsymbol{S}_B - \frac{\boldsymbol{Y}_{\cdots}^2}{N} = \sum_{j=1}^{b}\frac{\boldsymbol{Y}_{\cdot j\cdot}^2}{an} - \frac{\boldsymbol{Y}_{\cdots}^2}{N},$$

$$AB\prec : (\boldsymbol{S}_A - \boldsymbol{e})(\boldsymbol{S}_B - \boldsymbol{e}) = \boldsymbol{S}_{AB} - \boldsymbol{S}_A - \boldsymbol{S}_B + \boldsymbol{e} = \sum_{i=1}^{a}\sum_{j=1}^{b}\frac{\boldsymbol{Y}_{ij\cdot}^2}{n} - \sum_{i=1}^{a}\frac{\boldsymbol{Y}_{i\cdot\cdot}^2}{bn} - \sum_{j=1}^{b}\frac{\boldsymbol{Y}_{\cdot j\cdot}}{an} + \frac{\boldsymbol{Y}_{\cdots}^2}{N},$$

$$R\prec AB : (\boldsymbol{S}_R - \boldsymbol{e})\boldsymbol{S}_A\boldsymbol{S}_B = \boldsymbol{S}_{R,AB} - \boldsymbol{S}_{AB} = \sum_{i=1}^{a}\sum_{j=1}^{b}\sum_{k=1}^{n}\boldsymbol{y}_{ijk}^2 - \sum_{i=1}^{a}\sum_{j=1}^{b}\frac{\boldsymbol{Y}_{ij\cdot}}{n}.$$

Then $\boldsymbol{SS}_T$ is the sum of all $\boldsymbol{S}$:

$$\boldsymbol{SS}_T = \sum_{i=1}^{a}\sum_{j=1}^{b}\sum_{k=1}^{n}\boldsymbol{y}_{ijk}^2 - \frac{\boldsymbol{Y}_{\cdots}^2}{N}$$

To determine the $E(\boldsymbol{MS})$ following rule 4, we first construct the table defined there:

	i	*j*	*k*
a_i	$1-\frac{a}{N(A)}$	b	n
b_j	a	$1-\frac{b}{N(B)}$	n
$(ab)_{ij}$	$1-\frac{a}{N(A)}$	$1-\frac{b}{N(B)}$	n
$e_{k(i,j)}$	1	1	$1-\frac{n}{N(res)}$

The column subscript i of the first column does not occur in β_j, so that in the first cell in the second row, a is placed. Because j is not in a_i, b is placed in the second cell of the first row. k does not occur in a_i, b_j and $(ab)_{ij}$; therefore the first three cells in the third column contain n. The indices i and j are for $e_{k(i,j)}$ in the bracket; therefore 1 stands in the first two cells of the last row. In all other cells of the first column, we put $1-\frac{a}{N(A)}$, in the still free cells of the second column, we put $1-\frac{b}{N(B)}$ and in the free cell of the last column, we put $1-\frac{n}{N(res)}$.

Now, following rule 4,

$$E(\boldsymbol{MS}_A) = c_1\sigma_a^2 + c_2\sigma_{ab}^2 + c_3\sigma^2$$

with

$$c_1 = bn,\quad c_2 = n\left(1-\frac{b}{N(B)}\right),\quad c_3 = 1-\frac{n}{N(res)}$$

and

$$E(\boldsymbol{MS}_B) = c_4\sigma_b^2 + c_5\sigma_{ab}^2 + c_6\sigma^2$$

with

$$c_4 = an,\quad c_5 = n\left(1-\frac{a}{N(A)}\right),\quad c_6 = 1-\frac{n}{N(res)}$$

and

$$E(\boldsymbol{MS}_{AB}) = c_7\sigma_{ab}^2 + c_8\sigma^2$$

with

$$c_7 = n,\quad c_8 = 1-\frac{n}{N(res)}$$

and finally

$$E(\boldsymbol{MS}_{res}) = \sigma^2.$$

Example 7.4 We determine *df*, ***SS*** and *E*(***MS***) for the ANOVA of type $C \prec AB$ in Section 5.4.3.2. We write model equation (5.48) as

$$\boldsymbol{y}_{ijkl} = \mu + \boldsymbol{a}_i + \boldsymbol{b}_j + \boldsymbol{c}_{k(i,j)} + (\boldsymbol{ab})_{ij} + \boldsymbol{e}_{l(i,j,k)}$$
$$(i = 1,\dots,a; j = 1,\dots,b; k = 1,\dots,c; l = 1,\dots,n).$$

In the ANOVA table, we have rows for $A \prec, B \prec, C \prec AB, AB \prec$ and *res*.

The degrees of freedom are (rule 2)

$$A \prec \quad : \; a-1,$$

$$B \prec \quad : \; b-1,$$

$$C \prec AB \; : \; (c-1)ab,$$

$$AB \prec \quad : \; (a-1)(b-1),$$

$$R \prec CAB \; : \; (n-1)abc.$$

The ***SS*** (rule 3) are

$$A \prec \quad : \quad \boldsymbol{SS}_A = \boldsymbol{S}_A - \boldsymbol{e} = \sum_{i=1}^{a} \frac{\boldsymbol{Y}_{i\cdots}^2}{bcn} - \frac{1}{N}\boldsymbol{Y}_{\cdots\cdot}^2,$$

$$B \prec \quad : \quad \boldsymbol{SS}_B = \boldsymbol{S}_B - \boldsymbol{e} = \sum_{j=1}^{b} \frac{\boldsymbol{Y}_{\cdot j\cdot\cdot}^2}{acn} - \frac{1}{N}\boldsymbol{Y}_{\cdots\cdot}^2,$$

$$C \prec AB \; : \quad \boldsymbol{SS}_C \text{ in } AB = (\boldsymbol{S}_C - \boldsymbol{e})\boldsymbol{S}_A\boldsymbol{S}_B = \boldsymbol{S}_{CAB} - \boldsymbol{S}_{AB}$$

$$= \sum_{i=1}^{a}\sum_{j=1}^{b}\sum_{k=1}^{c} \frac{\boldsymbol{Y}_{ijk\cdot}^2}{n} - \sum_{i=1}^{a}\sum_{j=1}^{b} \frac{\boldsymbol{Y}_{ij\cdot\cdot}}{cn},$$

$$R \prec \quad : \quad \boldsymbol{SS}_R = (\boldsymbol{S}_R - \boldsymbol{e})\boldsymbol{S}_A\boldsymbol{S}_B\boldsymbol{S}_C = \boldsymbol{S}_R - \boldsymbol{S}_{ABC}$$

$$= \sum_{i=1}^{a}\sum_{j=1}^{b}\sum_{k=1}^{c}\sum_{l=1}^{n} \boldsymbol{y}_{ijkl}^2 - \frac{1}{n}\sum_{i=1}^{a}\sum_{j=1}^{b}\sum_{k=1}^{c} \boldsymbol{Y}_{ijk\cdot}^2$$

Following rule 4 we construct the table:

	i	*j*	*k*	*l*
a_i	$1-\frac{a}{N(A)}$	b	c	n
b_j	a	$1-\frac{b}{N(B)}$	c	n
$c_{k(i,j)}$	1	1	$1-\frac{c}{N(C)}$	n
$(ab)_{ij}$	$1-\frac{a}{N(A)}$	$1-\frac{b}{N(B)}$	c	n
$e_{l(i,j,k)}$	1	1	1	$1-\frac{n}{N(res)}$

Then we get

$$E(\boldsymbol{MS}_A) = bcn\sigma_a^2 + n\left(1 - \frac{c}{N(C)}\right)\sigma_{c\,\mathrm{in}\,ab}^2 + cn\left(1 - \frac{b}{N(B)}\right)\sigma_{ab}^2 + \left(1 - \frac{n}{N(res)}\right)\sigma^2,$$

$$E(\boldsymbol{MS}_B) = acn\sigma_b^2 + n\left(1 - \frac{c}{N(C)}\right)\sigma_{c\,\mathrm{in}\,ab}^2 + cn\left(1 - \frac{a}{N(A)}\right)\sigma_{ab}^2 + \left(1 - \frac{n}{N(res)}\right)\sigma^2,$$

$$E(\boldsymbol{MS}_{C\,\mathrm{in}\,AB}) = n\sigma_{c\,\mathrm{in}\,ab}^2 + \left(1 - \frac{n}{N(res)}\right)\sigma^2,$$

$$E(\boldsymbol{MS}_{AB}) = n\left(1 - \frac{c}{N(C)}\right)\sigma_{c\,\mathrm{in}\,ab}^2 + cn\sigma_{ab}^2 + \left(1 - \frac{n}{N(res)}\right)\sigma^2,$$

$$E(\boldsymbol{MS}_{res}) = \sigma^2.$$

7.3 Variance Component Estimators in Mixed Models

Mixed models in the ANOVA are such models where in Equation (6.3) at least one but not all γ_{A_i} are random variables. More general mixed models are defined so that models I and II are special cases. We use this general rule, but nevertheless find it reasonable to consider model I and model II in separate chapters as done in Chapters 5 and 6 and to use the methods developed for mixed models only if neither model I nor model II can be used.

Definition 7.2 Let $Y = (y_1, \ldots, y_N)^T$ be an N-dimensional random vector depending on the effects $\gamma_{A_i}, \ldots, \gamma_{A_s}, \boldsymbol{\gamma}_{A_{s+1}}, \ldots, \boldsymbol{\gamma}_{A_r}$ of r factors or factor combinations A_i ($i = 1, \ldots, r$) with a_i levels as

$$Y = \mu 1_N + \sum_{i=1}^{s} Z_{A_i}\gamma_{A_i} + \sum_{i=s+1}^{r} Z_{A_i}\boldsymbol{\gamma}_{A_i} + \boldsymbol{e}. \tag{7.6}$$

Equation (7.6) under the side conditions

$$\mathrm{var}(\boldsymbol{e}) = \sigma^2 E_N,\ E(\boldsymbol{e}) = 0_N,\ \mathrm{cov}\left(\boldsymbol{\gamma}_{A_i}, \boldsymbol{e}\right) = \mathrm{O}_{a_i,N}\ (i = s+1, \ldots, r),$$

$$\mathrm{cov}\left(\boldsymbol{\gamma}_{A_i}, \boldsymbol{\gamma}_{A_j}\right) = \mathrm{O}_{a_i,a_j}(i,j = s+1, \ldots, r; i \neq j),\ E\left(\boldsymbol{\gamma}_{A_i}\right) = 0_{a_i}$$

is called a mixed model of the ANOVA.

Inserting in Definition 7.2,

$$\beta_1 = \left(\mu, \gamma_{A_1}^T, \ldots, \gamma_{A_s}^T\right)^T,\ \boldsymbol{\beta}_2 = \left(\boldsymbol{\gamma}_{A_{s+1}}^T, \ldots, \boldsymbol{\gamma}_{A_r}^T\right)^T,$$

$$X_1 = (1_N, Z_{A_i}, \ldots, Z_{A_s}) \text{ and } X_2 = (Z_{A_{s+1}}, \ldots, Z_{A_r}),$$

then (7.6) becomes

$$\boldsymbol{Y} = X_1\beta_1 + X_2\boldsymbol{\beta}_2 + \boldsymbol{e} \tag{7.7}$$

analogue to (5.1) and (6.1). For $X_1 = X, \beta_1 = \beta(X_2 = 0)$ (7.7) becomes equation (5.1), and for $\beta_1 = \mu, X_1 = 1_N, X_2 = Z, \beta_2 = \boldsymbol{\gamma}$ (7.7) becomes Equation (6.2). Hereafter we are interested in the so-called real mixed models where β_1 except μ contains at least one further component and X_2 and β_2 are not zero. New with the real mixed models is the fact that fixed effects are estimated or tested and variance components are estimated.

Do not confuse mixed models with mixed classifications.

7.3.1 An Example for the Balanced Case

Example 7.5 (Mixed model in the two-way cross-classification with equal subclass numbers)

Two cross-classified factors A (fixed) and B and their interactions AB in Equation (7.6) lead us to

$$A_1 = A,\ A_2 = B,\ A_3 = AB.$$

Then $s = 1$ and $r = 3$ and we put $a_1 = a, a_2 = b,$ and consequently $a_3 = ab$. Because we have equal subclass numbers with $n > 1$, it follows $N = abn$. Equations (7.6) and (7.7) become

$$\boldsymbol{y}_{ijk} = \mu + a_i + \boldsymbol{b}_j + (\boldsymbol{ab})_{ij} + \boldsymbol{e}_{ijk}. \tag{7.8}$$

From the side conditions of Definition 7.2, we get side conditions for (7.8). Let additionally (case I)

$$\operatorname{var}(\boldsymbol{b}_j) = \sigma_b^2 \text{ for all } j, \quad \operatorname{cov}(\boldsymbol{b}_j,\ \boldsymbol{b}_k) = 0, \quad \text{for all } j,k \text{ with } j \neq k,$$

$$\operatorname{var}\left((\boldsymbol{ab})_{ij}\right) = \sigma_{ab}^2 \text{ for all } i, j, \quad \sum_{i=1}^{a} a_i = 0,$$

$$\operatorname{cov}\left(\boldsymbol{b}_j, (\boldsymbol{ab})_{ij'}\right) = \operatorname{cov}\left(\boldsymbol{b}_{j'}, \boldsymbol{e}_{ijk}\right) = \operatorname{cov}\left((\boldsymbol{ab})_{i'j'}, \boldsymbol{e}_{ijk}\right) = 0$$

and

$$\operatorname{cov}\left((\boldsymbol{ab})_{ij}, (\boldsymbol{ab})_{ij'}\right) = 0 \quad (j \neq j').$$

The columns ***SS***, *df* and ***MS*** in the corresponding ANOVA table are model independent and given in Table 5.13. The expectations of the ***MS*** for model (7.8) can be found in Table 7.1.

Table 7.1 Expectations of the *MS* in Table 5.13 for a mixed model (levels of *A* fixed) for two side conditions.

Source of variation	$E(MS)$, if $\text{cov}\left((ab)_{ij},(ab)_{i,j'}\right)=0$ (Case I)	$E(MS)$, if $\sum_{i=1}^{a}(ab)_{ij}=0$ for all j (Case II)
Between the levels of *A*	$\frac{bn}{a-1}\sum_{i=1}^{a}(a_i-\bar{a})^2+n\sigma_{ab}^2+\sigma^2$	$\frac{bn}{a-1}\sum_{i=1}^{a}a_i^2+\frac{na}{a-1}\sigma_{ab}^{*2}+\sigma^2$
Between the levels of *B*	$an\sigma_b^2+n\sigma_{ab}^2+\sigma^2$	$an\sigma_b^{*2}+\sigma^2$
Interactions $A\times B$	$n\sigma_{ab}^2+\sigma^2$	$\frac{na}{a-1}\sigma_{ab}^{*2}+\sigma^2$
Residual	σ^2	σ^2

If the additional side conditions (case II) are

$$\sum_{i=1}^{a}(\boldsymbol{ab})_{ij}=0 \quad \text{for all } j \tag{7.9}$$

the term $\bar{a}=\frac{1}{a}\sum_{i=1}^{a}a_i$ vanishes and $(\boldsymbol{ab})_{ij}$ and $(\boldsymbol{ab})_{i'j}(i\neq i';j=1,\dots,b)$ are correlated; the covariance is $\text{cov}\left((\boldsymbol{ab})_{ij},(\boldsymbol{ab})_{i'j}\right)=\sigma_{ab}$ for all j and $i\neq i'$. Because $\text{var}\left(\sum_{i=1}^{a}(\boldsymbol{ab})_{ij}\right)=\text{var}(0)=0$ $\quad 0=\text{var}\left(\sum_{i=1}^{a}(\boldsymbol{ab})_{ij}\right)=\sum_{i=1}^{a}\sigma_{ab}^{*2}+\sum_{i=1}^{a}\sum_{\substack{i'=1\\ i\neq i'}}^{a}\sigma_{ab}=a\sigma_{ab}^{*2}+a(a-1)\sigma_{ab}$, the relations and $\sigma_{ab}=-\frac{1}{a-1}\sigma_{ab}^{*2}$ follow.

The conditions (7.9) lead to the $E(\boldsymbol{MS})$ in the last column in Table 7.1. Searle (1971) clearly recorded the relations between the two cases. He showed that σ_b^2 in the two cases changed in meaning. To show this, we write down Equation (7.8) separately for both cases: (7.8) as it stands for case I and (7.8) with effects complemented by $''$ for case II (side conditions (7.9))

$$\boldsymbol{y}_{ijk}=\mu''+a_i''+\boldsymbol{b}_j''+(\boldsymbol{ab})_{ij}''+\boldsymbol{e}_{ijk}.$$

(7.8) can be written as

$$\boldsymbol{y}_{ijk}=\mu+a_i+\boldsymbol{b}_j+(\boldsymbol{ab})_{.j}+(\boldsymbol{ab})_{ij}-(\boldsymbol{ab})_{.j}+\boldsymbol{e}_{ijk}$$

with $(\boldsymbol{ab})_{.j}=a^{-1}\sum_{i=1}^{a}(\boldsymbol{ab})_{ij}$. Then we obtain $\mu''=\mu+\bar{a}, a_i''=a_i-\bar{a}, \boldsymbol{b}_j''=\boldsymbol{b}_j-\bar{\boldsymbol{b}}$ and $(\boldsymbol{ab})_{ij}''=(\boldsymbol{ab})_{ij}-(\boldsymbol{ab})_{.j}$. Then we have

$$\sigma_b^{*2}=\sigma_{b'}^{*2}=\sigma_b^2=\frac{1}{a}\sigma_{ab}^2,\quad \sigma_{ab}^{*2}=\sigma_{a'b'}^2=\frac{a-1}{a}\sigma_{ab}^2$$

and

$$\sigma_{ab} = \operatorname{cov}\left(\left(\boldsymbol{ab}\right)_{ij}, \left(\boldsymbol{ab}\right)_{i'j'}\right) = -\frac{1}{a-1}\sigma^2_{a'b'} = -\frac{1}{a-1}\sigma^{*2}_{ab}.$$

The variance components $\sigma_b^2, \sigma_{ab}^2$ and σ^2 can be estimated due to balancedness. For case I from column 2 in Table 7.1 by the ANOVA method,

$$\left.\begin{aligned} s^2 &= \boldsymbol{MS}_{res}, \\ s^2_{ab} &= \frac{1}{n}(\boldsymbol{MS}_{AB} - \boldsymbol{MS}_{res}), \\ s^2_b &= \frac{1}{an}(\boldsymbol{MS}_B - \boldsymbol{MS}_{AB}). \end{aligned}\right\} \tag{7.10}$$

For case II from the last column in Table 7.1,

$$\left.\begin{aligned} s^2 &= \boldsymbol{MS}_{res}, \\ s^{*2}_{ab} &= \frac{a-1}{na}(\boldsymbol{MS}_{AB} - \boldsymbol{MS}_{res}), \\ s^{*2}_b &= \frac{1}{an}(\boldsymbol{MS}_B - \boldsymbol{M}_{res}). \end{aligned}\right\} \tag{7.10a}$$

Analolously to this example, cross-classified, nested or mixed-classified balanced designs of the mixed model can be treated; the general statements of Section 6.3 are still valid.

7.3.2 The Unbalanced Case

In unbalanced cases we use Hendersons method III (in Henderson, 1953), starting with the model equation (7.7). A quadratic form in $\boldsymbol{Y}$ must be found so that its expectation independent of β_1 contains only the variance components we are looking for, if the covariances between the random effects of each factor are zero.

Theorem 7.1 For $\boldsymbol{Y}$ let the mixed model in Definition 7.2 have the form (7.7). With $X = (X_1, X_2)$, the expectation of the quadratic form

$$\boldsymbol{Y}^T\left[X\left(X^TX\right)^- X^T - X_1\left(X_1^TX_1\right)^- X_1^T\right]\boldsymbol{Y} = \boldsymbol{Y}^T(U - V)\boldsymbol{Y}$$

depends only on the unknown σ^2 and on var($\boldsymbol{\beta}_2$), but not on β_1 even if $\boldsymbol{\beta}_1$ is random.

Proof: We rewrite (6.5) for $\boldsymbol{Y}$ in (7.7) as

$$E\left(\boldsymbol{Y}^TA\boldsymbol{Y}\right) = \operatorname{tr}\left[X^TAXE\left(\boldsymbol{\beta}\boldsymbol{\beta}^T\right)\right] + \sigma^2\operatorname{tr}(A). \tag{7.11}$$

with $\boldsymbol{\beta} = \begin{pmatrix} \beta_1 \\ \boldsymbol{\beta}_2 \end{pmatrix}$ $(E(\boldsymbol{\beta}_1) = \beta_1)$. Due to the idempotence of $X(X^T X)^- X^T$ and $X^T X(X^T X)^- X^T X = X^T X$, we get

$$E\left(\boldsymbol{Y}^T U \boldsymbol{Y}\right) = E\left[\boldsymbol{Y}^T X\left(X^T X\right)^- X^T \boldsymbol{Y}\right] = \operatorname{tr}\left[\left(X^T X\right) E\left(\boldsymbol{\beta}\boldsymbol{\beta}^T\right)\right] + \sigma^2 \operatorname{rk}(X)$$

and

$$\begin{aligned} E\left(\boldsymbol{Y}^T V \boldsymbol{Y}\right) &= E\left[\boldsymbol{Y}^T \left(X_1^T \left(X_1^T X_1\right)^- X_1^T\right) \boldsymbol{Y}\right] \\ &= \operatorname{tr}\left[X^T X_1 \left(X_1^T X_1\right)^- X_1^T X E\left(\boldsymbol{\beta}\boldsymbol{\beta}^T\right)\right] + \sigma^2 \operatorname{rk}(X_1). \end{aligned}$$

We write

$$X^T X = \begin{pmatrix} X_1^T X_1 & X_1^T X_2 \\ X_2^T X_1 & X_2^T X_2 \end{pmatrix}$$

with $(X_1,\ X_2) = X$ and obtain

$$E\left(\boldsymbol{Y}^T U \boldsymbol{Y}\right) = \operatorname{tr}\left[\begin{pmatrix} X_1^T X_1 & X_1^T X_2 \\ X_2^T X_1 & X_2^T X_2 \end{pmatrix} E\left(\boldsymbol{\beta}\boldsymbol{\beta}^T\right)\right] + \sigma^2 \operatorname{rk}(X) \tag{7.12}$$

or because $X^T X(X^T X)^- X^T = X^T$,

$$\begin{aligned} E\left(\boldsymbol{Y}^T V \boldsymbol{Y}\right) &= \operatorname{tr}\left[\begin{pmatrix} X_1^T X_1 \\ X_2' X_1 \end{pmatrix} \left(X_1^T X_1\right)^- \left(X_1^T X_1,\ X_1^T X_2\right) E\left(\boldsymbol{\beta}\boldsymbol{\beta}^T\right)\right] + \sigma^2 \operatorname{rk}(X_1) \\ &= \operatorname{tr}\left[\begin{pmatrix} X_1^T X_1 & X_1^T X_2 \\ X_2^T X_1 & X_2^T X_1 \left(X_1^T X_1\right)^- X_1^T X_2 \end{pmatrix} E\left(\boldsymbol{\beta}\boldsymbol{\beta}^T\right)\right] + \sigma^2 \operatorname{rk}(X_1). \end{aligned}$$

With $\boldsymbol{\beta} = \begin{pmatrix} \beta_1 \\ \boldsymbol{\beta}_2 \end{pmatrix}$ we obtain

$$\begin{aligned} &E\left\{\boldsymbol{Y}^T \left[X\left(X^T X\right)^- X^T - X_1 \left(X_1^T X_1\right)^- X_1^T\right] \boldsymbol{Y}\right\} \\ &= \operatorname{tr}\left\{X_2' \left[I_N - X_1 \left(X_1^T X_1\right)^- X_1^T\right] X_2 E\left(\boldsymbol{\beta}_2 \boldsymbol{\beta}_2^T\right)\right\} + \sigma^2 [\operatorname{rk}(X) - \operatorname{rk}(X_1)], \end{aligned} \tag{7.13}$$

and this completes the proof.

Hendersons method III uses Theorem 7.1.

The partitioning of $\boldsymbol{\beta}$ into two vector components $\beta_1, \boldsymbol{\beta}_2$ (and X in X_1, X_2) is independent of β_1 containing only fixed effects or not. Theorem 7.1 is valid for all partitioning of $\boldsymbol{\beta}$, as long as $\operatorname{rk}(X) - \operatorname{rk}(X_1) > 0$.

We now build for a mixed model with $r - s$ random components all quadratic forms of type $\boldsymbol{Y}^T(U - V)\boldsymbol{Y}$, in which $\boldsymbol{\beta}_2$ has one, two, …, $r - s$ random groups of elements and by this X_2 one, two, …, $r - s$ groups of columns. Together with $E(\boldsymbol{MS}_{res}) = \sigma^2$, the expectations of these quadratic forms result in $r - s + 1$ equations with the variance components being obtained, as long as $E\left(\boldsymbol{\beta}_2\boldsymbol{\beta}_2^T\right)$ is a diagonal matrix.

In these equations we replace $E[\boldsymbol{Y}^T(U - V)\boldsymbol{Y}]$ by $\boldsymbol{Y}^T(U - V)\boldsymbol{Y}$ and the variance components $\sigma_i^2(\sigma^2)$ with their estimators $\boldsymbol{s}_i^2(\boldsymbol{s}^2)$ and receive equations with the estimators of the variance components as unknown quantities. The estimates can become negative, but the estimators are unbiased and independent of the fixed effects. Note that due to the unbiasedness, we can get negative estimates that of course are nonsense. If we replace the negative estimates by zero, the unbiasedness property is no longer true.

In mixed models, it may happen that variance components of the random effects as well as the fixed effects have to be estimated. If the distribution of $\boldsymbol{Y}$ is normal, we can use the maximum likelihood method. The likelihood function is differentiated concerning the fixed effects and the variance components. The derivatives are set at zero, and we get simultaneous equations that can be solved by iteration. The formulae and proposals for numerical solutions of the simultaneous equations are given in Hartley and Rao (1967). The numerical solution is elaborated (see the remarks about REML in Section 6.2.1.2).

7.4 Tests for Fixed Effects and Variance Components

We assume now that all random entries in (7.6) are normally distributed and that the side conditions (7.9) are defined so that all fixed effects are estimable. W.l.o.g. we restrict ourselves to $i = 1$ and $i = s + 1$

$$H_{0F}: \gamma_{A1} = 0_{a1}, \text{ against } H_{AF}: \gamma_{A1} \neq 0_{a1} \tag{7.14}$$

and

$$H_{0V}: \sigma_{s+1}^2 = 0 \text{ against } H_{AV}: \sigma_{s+1}^2 \neq 0.$$

By

$$\boldsymbol{SS}_i = \boldsymbol{Y}^T T_i \boldsymbol{Y} \quad (i = 1, \ldots, r+1) \tag{7.15}$$

we denote the $\boldsymbol{SS}$ of factor A_i ($\boldsymbol{SS}_{r+1} = \boldsymbol{SS}_{res}$) where the T_i are idempotent matrices of rank $\text{rk}(T_i) = f_i$. In the special cases of Section 7.5, $\boldsymbol{SS}_i$ and f_i are given in the ANOVA tables. The magnitudes

$$\boldsymbol{MS}_i = \frac{1}{f_i}\boldsymbol{SS}_i \quad (i = 1, \ldots, r+1) \tag{7.16}$$

are the corresponding $\boldsymbol{MS}$, $\boldsymbol{MS}_{r+1} = \boldsymbol{MS}_{res}$, $f_{r+1} = f_{res}$.

To test H_{0F} we construct a test statistic

$$F_1 = \frac{Y^T T_1 Y}{f_1 \sum_{i=s+1}^{r+1} c_i MS_i} = \frac{MS_1}{\sum_{i=s+1}^{r+1} c_i MS_i} \tag{7.17}$$

so that if H_{0F} is true for suitable c_i,

$$E(MS_1 | H_{0F}) = E\left[\sum_{i=s+1}^{r+1} c_i MS_i\right].$$

A corresponding construction is used if the test statistic for H_{0V} is given by

$$F_{s+1} = \frac{MS_{s+1}}{\sum_{j=s+2}^{r+1} k_j MS_j} \tag{7.18}$$

where we have to choose the k_j so that

$$E(MS_{s+1} | H_{0V}) = E\left[\sum_{j=s+2}^{r+1} k_j MS_j\right].$$

The degrees of freedom of the test statistics (7.17) and (7.18) in all cases, in which only one of c_i ($i = s + 1, \ldots, r + 1$) or k_j ($j = s + 2, \ldots, r + 1$) differs from zero and is equal to 1, are given by (f_1, f_i) or (f_{s+1}, f_j). In all other cases we approximate the degrees of freedom by $\left(f_1, f_F^*\right)$ or $\left(f_{s+1}, f_V^*\right)$ using the corollary of Lemma 6.3. For testing H_{0F} and H_{0V}, Seifert (1980, 1981) used another approach leading to mixed models to exact α-tests and in balanced cases to simple formulae. The principle of Seifert was to use test statistics that are ratios of two independent quadratic forms $Y^T B_1 Y$ and $Y^T B_2 Y$ where $Y^T B_2 Y$ is centrally χ^2-distributed with g_2 degrees of freedom and $Y^T B_1 Y$ – if H_{0F} (H_{0V}) is true – is centrally χ^2-distributed with g_1 degrees of freedom. Then

$$F = \frac{Y^T B_1 Y g_2}{Y^T B_2 Y g_1} \tag{7.19}$$

– if H_{0F} (H_{0V}) is true – is $F(g_1, g_2)$ centrally F-distributed. This procedure for model I and model II may also be used ($s = 0, r = 0$).

7.5 Variance Component Estimation and Tests of Hypotheses in Special Mixed Models

Below we discuss simple cases (mainly balanced designs) of the two- and three-way analyses with mixed models. If we discuss statistical tests of fixed effects or variance components, we assume that all random variables in (7.6) are normally distributed and that side conditions are fulfilled so that all fixed effects are estimable.

7.5.1 Two-Way Cross-Classification

W.l.o.g. we assume that in a mixed model of the two-way cross-classification, factor A is fixed, and factor B is random as in model equation (7.8). The variance component estimation already was handled in Example 7.5.

The ***SS***, ***MS*** and *df* are given in Table 5.10; there we have single subclass numbers and $R = AB$. We have $s = 1,\ r = 2,\ a_1 = a,\ a_2 = b,\ a_3 = n$, and the model equation is (7.8) for $(ab)_{ij} = 0$·

The null hypotheses that can be tested are

H_{01}: 'All a_i are zero'.
H_{02}: '$\sigma_b^2 = 0$'.
H_{03}: '$\sigma_{ab}^2 = 0$'.

If H_{01} is true, then $E(\boldsymbol{MS}_A)$ equals $E(\boldsymbol{MS}_{AB})$, and we test H_{01} using

$$\boldsymbol{F}_A = \frac{\boldsymbol{MS}_A}{\boldsymbol{MS}_{AB}},$$

which has under H_{01} an F-distribution with $(a - 1)$ and $(a - 1)(b - 1)$ degrees of freedom.

To find the minimum size of the experiment that will satisfy given precision requirements, we must remember that only the degrees of freedom of the corresponding F-statistic influence the power of the test and by this the size of the experiment. To test the hypothesis H_{01} that the fixed factor has no influence on the observations, we have $(a - 1)$ and $(a - 1)(b - 1)$ degrees of freedom of numerator and denominator, respectively. Thus the subclass number n does not influence the size needed and therefore should be chosen as small as possible. If we know that there are no interactions, we choose $n = 1$, but if interactions may occur we choose $n = 2$. Because the number a of levels of the factor under test is fixed, we can only choose b, the size of the sample of B-levels to fulfil precision requirements.

Here is an example.

Example 7.6 We want to test the null hypothesis that six wheat varieties do not differ in their yields.

For the precision requirements $\alpha = 0.05$, $\beta = 0.2$, $\sigma = 1$, $\delta = 1.6$, we receive a maximin number of levels of factor B as $b = 12$.

For the experiment we randomly selected 12 farms. The varieties are the levels of a fixed factor A and the twelve farms are levels of a random factor B. Both are cross-classified. The yield in *dt*/*ha* was measured. The results are shown in Table 7.2.

Table 7.2 Yields of 6 varieties tested on 12 farms.

	A: varieties					
B: farms	1	2	3	4	5	6
1	32	48	25	33	48	29
2	28	52	25	38	27	27
3	30	47	34	44	38	31
4	44	55	28	39	21	31
5	43	53	26	38	30	33
6	48	57	33	37	36	26
7	42	64	40	53	38	27
8	42	64	42	41	29	33
9	39	64	47	47	23	32
10	44	59	34	54	33	31
11	40	58	27	50	36	30
12	42	57	32	46	36	35

We can perform the two-way ANOVA with SPSS by using the procedure:

Analyze
General Linear Model
Univariate

We input 'variety' as fixed factors, 'farm' as random factor and 'yield' as the dependent variable as shown in Figure 7.1. With the model key, we use a model without interactions. The results of the SPSS calculations are shown in Table 7.3. Note that because we have single-cell observation, the residual is equal to the interaction AB.

We received $\boldsymbol{F}_A = \frac{MS_A}{MS_{AB}} = 1139.247/33.144 = 34.372$, and therefore we found significant differences between the varieties. This follows directly from Table 7.3 because sig. < 0.05. For $\boldsymbol{F}_B$ we receive $\boldsymbol{F}_B = 2.007$,and this means that the variance component for the farms is with $\alpha = 0.05$, not significantly different from the error variance.

To estimate the variance component for the farms, we use in SPSS:

Analyze
General Linear Models
Variance Components

Again we use a model without interactions and select the ANOVA method. The result shows Table 7.4.

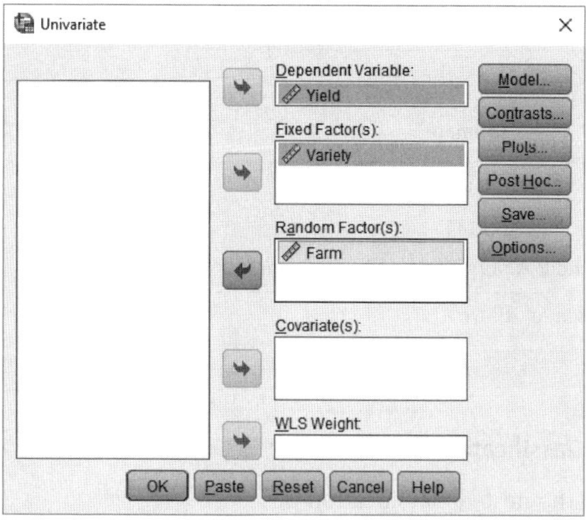

Figure 7.1 The factors of Example 7.6. *Source:* Reproduced with permission of IBM.

Table 7.3 ANOVA table for Example 7.6.

Tests of between-subjects effects

Dependent variable:yield

Source		Type III sum of squares	df	Mean square	F	Sig.
Intercept	Hypothesis	110842.014	1	110842.014	1666.070	.000
	Error	731.819	11	66.529[a]		
Variety	Hypothesis	5696.236	5	1139.247	34.372	.000
	Error	1822.931	55	33.144[b]		
Farm	Hypothesis	731.819	11	66.529	2.007	.045
	Error	1822.931	55	33.144[b]		

[a]MS(farm)
[b]MS(error)
Source: Reproduced with permission of IBM.

Table 7.4 Variance component estimates of Example 7.6.

Variance estimates	
Component	**Estimate**
Var(farm)	5.564
Var(error)	33.144
Dependent variable: yield method: ANOVA (type I sum of squares)	

Source: Reproduced with permission of IBM.

7.5.2 Two-Way Nested Classification $B \prec A$

Let the model equation of the nested classification be

$$\boldsymbol{y}_{ijk} = \mu + \boldsymbol{a}_i + \boldsymbol{b}_{j(i)} + \boldsymbol{e}_{k(i.j)} \quad (i = 1,\ldots,a; j = 1,\ldots,b; k = 1,\ldots,n) \tag{7.20}$$

with the side conditions that the levels of A,B and the residuals stem from finite level populations and for all i and i, j, we assume

$$\left.\begin{aligned} 0 = E(\boldsymbol{a}_i) &= \frac{1}{N(A)}\sum_{i=1}^{N(A)} a_i \\ = E\left(\boldsymbol{b}_{j(i)}\right) &= \frac{1}{N(B)}\sum_{j=1}^{N(B)} b_{j(i)} = \frac{1}{N(res)}\sum_{k=1}^{N(res)} e_{k(i,j)} \\ \frac{1}{N(A)-1}\sum_{i=1}^{N(A)} a_i^2 &= \sigma_a^2, \\ \frac{1}{N(B)-1}\sum_{j=1}^{N(B)} b_{j(i)}^2 &= \sigma_{b \text{ in } a}^2, \\ \frac{1}{N(res)-1}\sum_{k=1}^{N(res)} e_{k(i,j)}^2 &= \sigma^2. \end{aligned}\right\} \tag{7.21}$$

Further, all covariances between the components in (7.14) (e . g . between $\boldsymbol{a}_i$ and $\boldsymbol{b}_{j(i)}$) shall be zero. We use the rules in Section 7.2 to generate the ANOVA table. This ANOVA table has three rows: levels of A, levels of B in A, and residual. The ANOVA table is Table 7.5. By rule 2 the degrees of freedom have been found and by rule 3, the ***SS***.

Table 7.5 ANOVA table of a two-way balanced nested classification for a model with finite level populations.

Source of variation	SS	df	MS	E(MS)
Between the levels of A	$SS_A = \sum_{i=1}^{a} \frac{Y_{i..}^2}{bn} - \frac{Y_{...}^2}{abn}$	$a-1$	$\frac{SS_A}{a-1} = MS_A$	$bn\sigma_a^2 + n\left(1-\frac{b}{N(B)}\right)\sigma_{bina}^2 + \left(1-\frac{n}{N(res)}\right)\sigma^2$
Between the levels of B within A	$SS_{BinA} = \sum_{i=1}^{a}\sum_{j=1}^{b} \frac{Y_{ij.}^2}{n} - \sum_{i=1}^{a} \frac{Y_{i..}^2}{bn}$	$a(b-1)$	$\frac{SS_{BinA}}{a(b-1)} = MS_{BinA}$	$n\sigma_{bina}^2 + \left(1-\frac{n}{N(res)}\right)\sigma^2$
Residual	$SS_{res} = \sum_{i=1}^{a}\sum_{j=1}^{b}\sum_{k=1}^{n} y_{ijk}^2 - \sum_{i=1}^{a}\sum_{j=1}^{b} \frac{Y_{ij.}^2}{n}$	$ab(n-1)$	$\frac{SS_{res}}{ab(n-1)} = MS_{res}$	$\left(1-\frac{n}{N(res)}\right)\sigma^2$

We use rule 4 to determine the $E(\boldsymbol{MS})$ using the table below:

	i	j	k
a_i	$1-\frac{a}{N(A)}$	b	n
$b_{j(i)}$	1	$1-\frac{b}{N(B)}$	n
$e_{k(i,j)}$	1	1	$1-\frac{n}{N(res)}$

(7.22)

Then $E(\boldsymbol{MS})$ are

$$\left.\begin{aligned} E(\boldsymbol{MS}_A) &= bn\sigma_a^2 + n\left(1-\frac{b}{N(B)}\right)\sigma_{bina}^2 + \left(1-\frac{n}{N(res)}\right)\sigma^2, \\ E(\boldsymbol{MS}_{BinA}) &= n\sigma_{bina}^2 + \left(1-\frac{n}{N(res)}\right)\sigma^2, \\ E(\boldsymbol{MS}_{res}) &= \left(1-\frac{n}{N(res)}\right)\sigma^2. \end{aligned}\right\} \tag{7.23}$$

If $N(res)$ tends to infinity (∞), then we receive

$$\left.\begin{aligned} E(\boldsymbol{MS}_A) &= bn\sigma_a^2 + n\left(1-\frac{b}{N(B)}\right)\sigma_{b\,\text{in}\,a}^2 + \sigma^2,\\ E(\boldsymbol{MS}_B) &= n\sigma_{b\,\text{in}\,a}^2 + \sigma^2,\\ E(\boldsymbol{MS}_{res}) &= \sigma^2. \end{aligned}\right\} \tag{7.24}$$

Putting in (7.24) $N(B) = b$, we obtain the $E(\boldsymbol{MS})$ of model I in Table 5.19 for the balanced case ($n_i = n,\ b_i = b$). For $N(B)\rightarrow\infty$ we obtain from (7.24) the $E(\boldsymbol{MS})$ of model II in Table 6.5 for the balanced case (with a corresponding definition of σ_a^2 and $\sigma_{b\text{ in }a}^2$). Nevertheless, here we are interested in mixed models. In the nested classification, there exist two mixed models.

7.5.2.1 Levels of *A* Random

Let the levels of A be randomly selected from the level population and the levels of B fixed, the model equation is then

$$\boldsymbol{y}_{ijk} = \mu + \boldsymbol{a}_i + b_{j(i)} + \boldsymbol{e}_{k(i,j)} \quad (i = 1,\ldots,a; j = 1,\ldots,b; k = 1,\ldots,n) \tag{7.25}$$

with corresponding side conditions.

(Expectations of all random variables are zero, $\text{var}(\boldsymbol{a}_i) = \sigma_a^2$ for all i; $\text{var}(\boldsymbol{e}_{k(i,j)}) = \sigma^2$ for all i, j, k; all covariances between different random variables on the right-hand side of (7.25) are zero, $\sum_{j=1}^{b} b_{j(i)} = 0$.)

The $E(\boldsymbol{MS})$ we get from (7.24) for $N(B) = b$ is given in column 2 of Table 7.6.

The estimators of the variance components are

$$\boldsymbol{s}^2 = \boldsymbol{MS}_{res},\ \boldsymbol{s}_a^2 = \frac{1}{bn}(\boldsymbol{MS}_A - \boldsymbol{MS}_{res}). \tag{7.26}$$

Table 7.6 $E(\boldsymbol{MS})$ of mixed models of the two-way nested classification.

Source of variation	*A* random, *B* fixed	*A* fixed, *B* random
Between the levels of A	$bn\sigma_a^2 + \sigma^2$	$\frac{bn}{a-1}\sum_{i=1}^{a} a_i^2 + n\sigma_{b\,\text{in}\,a}^2 + \sigma^2$
Between the levels of B within the levels of A	$\frac{n}{a(b-1)}\sum_{i=1}^{a}\sum_{j=1}^{b} b_{j(i)}^2 + \sigma^2$	$n\sigma_{b\,\text{in}\,a}^2 + \sigma^2$
Residual	σ^2	σ^2

7.5.2.2 Levels of *B* Random

If the levels of *A* are fixed and those of *B* random, the model equation is

$$\boldsymbol{y}_{ijk} = \mu + a_i + \boldsymbol{b}_{j(i)} + \boldsymbol{e}_{k(i,j)}, \quad (i = 1,\dots,a; j = 1,\dots,b; k = 1,\dots,n) \tag{7.27}$$

with the side conditions

$$\text{var}\left(\boldsymbol{b}_{j(i)}\right) = \sigma^2_{b\,\text{in}\,a} \quad \text{for all } i,j,$$

$$\text{var}\left(\boldsymbol{e}_{k(i,j)}\right) = \sigma^2 \quad \text{for all } i,j,k,$$

$$\sum_{i=1}^{a} a_i = 0,$$

and the expectations of the random variables on the right-hand side of (7.27), and all covariances between the random variables on the right-hand side of (7.27) are equal to zero.

The $E(\boldsymbol{MS})$ for this case follows from (7.24) and can be found in the last column of Table 7.6. The estimators of σ^2 and $\sigma^2_{b\,\text{in}\,a}$ are

$$\boldsymbol{s}^2 = \boldsymbol{MS}_{res}, \quad \boldsymbol{s}^2_{b\,\text{in}\,a} = \frac{1}{n}\left(\boldsymbol{MS}_{B\,\text{in}\,A} - \boldsymbol{MS}_{res}\right). \tag{7.28}$$

The null hypothesis that the effects of all the levels of factor *A* are equal is tested using the test statistic:

$$\boldsymbol{F}_A = \frac{\boldsymbol{MS}_A}{\boldsymbol{MS}_{B\,\text{in}\,A}},$$

which under H_0 has an *F*-distribution with $(a - 1)$ and $a(b - 1)$ degrees of freedom.

For the mixed model we use conditions analogue to case II of Section 7.3.1 but corresponding to the remarks below. In Example 7.1 in place of $\frac{a}{a-1}\sigma^{*2}_{ab}$, we write the quasi-variance component σ^2_{ab}. Then we get

$$E(\boldsymbol{MS}_A) = \frac{bn}{a-1}\sum_{i=1}^{a} a_i^2 + n\sigma^2_{ab} + \sigma^2,$$

$$E(\boldsymbol{MS}_B) = an\sigma^2_b + \sigma^2, \quad E(\boldsymbol{MS}_{AB}) = n\sigma^2_{ab} + \sigma^2.$$

To find the minimum size of the experiment, which will satisfy given precision requirements, we have to find the minimum number *b* of levels of factor *B*. Because in nested models we have no interactions, we can fix $n = 1$. Consider the following example.

Example 7.7 It shall be tested whether an amino acid supplementation in the rearing rations of young boars (7 months old) causes a significant increase in sperm production (total number of spermatozoa per ejaculate in *Mrd.*) of the boars. There were $a = 2$ feeding groups (with and without supplementation) formed; in each group $b_1 = b_2 = b$ randomly selected boars from an animal population have been investigated. From each boar spermatozoa from c ejaculates have been counted. This is a two-way nested classification with factor feeding (fixed) and boar (random). In the SPSS procedure it reads:

Analyze
General Linear Model
Univariate

We click the model button, and then under 'build terms' we enter both factors as 'main effects'; under 'sum of squares' we choose 'type I' After clicking 'continue', click on 'paste', and in the syntax window change '/design feeding boar' to '/design feeding boar(feeding)' (signifying boar nested in feeding). The new syntax is

```
DATASET ACTIVATE DataSet1.
UNIANOVA Yield BY Feeding Boar
  /RANDOM=Boar
  /METHOD=SSTYPE(1)
  /INTERCEPT=EXCLUDE
  /CRITERIA=ALPHA(0.05)
  /DESIGN=Feeding Boar(Feeding).
```

Then start the programme with 'Run'.

7.5.3 Three-Way Cross-Classification

To calculate the variance components, we must calculate as in Example 7.4 the quasi-variance components; we call them variance components for short.

The reader corresponding to the procedure in the sections above can derive the side conditions of the models. We only give the $E(\boldsymbol{MS})$ for a model with finite level populations and for both types of balanced mixed models. In the unbalanced case the method Henderson III with two random factors does not lead to a unique solution; this case is not included in this text. We recommend in those situations to use the REML method in Section 6.2.1.2.

The model for finite level populations is

$$\left.\begin{array}{l}\boldsymbol{y}_{ijkl} = \mu + \boldsymbol{a}_i + \boldsymbol{b}_j + \boldsymbol{c}_k + (\boldsymbol{ab})_{ij} + (\boldsymbol{ac})_{ik} + (\boldsymbol{bc})_{jk} + (\boldsymbol{abc})_{ijk} + \boldsymbol{e}_{ijkl} \\ (i = 1,\dots,a; j = 1,\dots,b; k = 1,\dots,c; l = 1,\dots,n),\end{array}\right\} \quad (7.29)$$

the sums of overall effects of the level populations are assumed to be zero.

The ***SS***, *df* and ***MS*** of the three-way cross-classification are given in Table 5.21. To derive the *E*(***MS***) for model equation (7.29), we first use the table below:

	i	*j*	*k*	*l*
a_i	$1-\frac{a}{N(A)}$	b	c	n
b_j	a	$1-\frac{b}{N(B)}$	c	n
c_k	a	b	$1-\frac{c}{N(C)}$	n
$(ab)_{ij}$	$1-\frac{a}{N(A)}$	$1-\frac{b}{N(B)}$	c	n
$(ac)_{ik}$	$1-\frac{a}{N(A)}$	b	$1-\frac{c}{N(C)}$	n
$(bc)_{jk}$	a	$1-\frac{b}{N(B)}$	$1-\frac{c}{N(C)}$	n
$(abc)_{ijk}$	$1-\frac{a}{N(A)}$	$1-\frac{b}{N(B)}$	$1-\frac{c}{N(C)}$	n
$e_{l(i,j,k)}$	1	1	1	$1-\frac{n}{N(res)}$

If $N(res)\to\infty$, we obtain the *E*(***MS***) of the second column in Table 7.7. In the three-way cross-classification exist two types of mixed models. In the first type the levels of one factor (we choose w.l.o.g. factor *C*) are randomly selected. In the second type the levels of two factors (we choose w.l.o.g. factors *B* and *C*) are randomly selected.

The model equation of the first type (*A*, *B* fixed, *C* random) is

$$\begin{aligned}&\boldsymbol{y}_{ijkl} = \mu + a_i + b_j + \boldsymbol{c}_k + (ab)_{ij} + (\boldsymbol{ac})_{ik} + (\boldsymbol{bc})_{jk} + (\boldsymbol{abc})_{ijk} + \boldsymbol{e}_{l(i,j,k)}\\&(i = 1,\ldots,a; j = 1,\ldots,b; k = 1,\ldots,c; l = 1,\ldots,n).\end{aligned} \tag{7.30}$$

For $N(A) = a$, $N(B) = b$ and $N(C)\to\infty$, we receive the *E*(***MS***) for the model equation (7.30) in the third column of Table 7.7. From this the estimators of the variance components become

$$\left.\begin{aligned}&\boldsymbol{s}^2 = \boldsymbol{MS}_{res}, && \boldsymbol{s}_{abc}^2 = \frac{1}{n}(\boldsymbol{MS}_{ABC} - \boldsymbol{MS}_{res}),\\&\boldsymbol{s}_{bc}^2 = \frac{1}{an}(\boldsymbol{MS}_{BC} - \boldsymbol{MS}_{res}), && \boldsymbol{s}_{ac}^2 = \frac{1}{bn}(\boldsymbol{MS}_{AC} - \boldsymbol{MS}_{res}),\\&\boldsymbol{s}_c^2 = \frac{1}{abn}(\boldsymbol{MS}_C - \boldsymbol{MS}_{res}).\end{aligned}\right\} \tag{7.31}$$

Table 7.7 $E(MS)$ of a three-way cross-classification for a model with finite level populations and two mixed models.

Source of variation	Model with finite level populations $[N(R)\to\infty]$
Levels of A	$bcn\sigma_a^2 + cn\left(1 - \dfrac{b}{N(B)}\right)\sigma_{ab}^2 + bn\left(1 - \dfrac{c}{N(C)}\right)\sigma_{ac}^2$
	$+ n\left(1 - \dfrac{b}{N(B)}\right)\left(1 - \dfrac{c}{N(C)}\right)\sigma_{abc}^2 + \sigma^2$
Levels of B	$acn\sigma_b^2 + cn\left(1 - \dfrac{a}{N(A)}\right)\sigma_{ab}^2 + an\left(1 - \dfrac{c}{N(C)}\right)\sigma_{bc}^2$
	$+ n\left(1 - \dfrac{a}{N(A)}\right)\left(1 - \dfrac{c}{N(C)}\right)\sigma_{abc}^2 + \sigma^2$
Levels of C	$abn\sigma_c^2 + bn\left(1 - \dfrac{a}{N(A)}\right)\sigma_{ac}^2 + an\left(1 - \dfrac{b}{N(B)}\right)\sigma_{bc}^2$
	$+ n\left(1 - \dfrac{a}{N(A)}\right)\left(1 - \dfrac{b}{N(B)}\right)\sigma_{abc}^2 + \sigma^2$
Interaction $A \times B$	$cn\sigma_{ab}^2 + n\left(1 - \dfrac{c}{N(C)}\right)\sigma_{abc}^2 + \sigma^2$
Interaction $A \times C$	$bn\sigma_{ac}^2 + n\left(1 - \dfrac{b}{N(B)}\right)\sigma_{abc}^2 + \sigma^2$
Interaction $B \times C$	$an\sigma_{bc}^2 + n\left(1 - \dfrac{a}{N(A)}\right)\sigma_{abc}^2 + \sigma^2$
Interaction $A \times B \times C$	$n\sigma_{abc}^2 + \sigma^2$
Residual	σ^2

A, B fixed, C random; model equation (7.30)	A fixed, B, C random; model equation (7.32)
$\dfrac{bcn}{a-1}\displaystyle\sum_{i=1}^{a} a_i^2 + bn\sigma_{ac}^2 + \sigma^2$	$\dfrac{bcn}{a-1}\displaystyle\sum_{i=1}^{a} a_i^2 + cn\sigma_{ab}^2 + bn\sigma_{ac}^2 + n\sigma_{abc}^2 + \sigma^2$
$\dfrac{acn}{b-1}\displaystyle\sum_{j=1}^{b} b_j^2 + an\sigma_{bc}^2 + \sigma^2$	$acn\sigma_b^2 + an\sigma_{bc}^2 + \sigma^2$
$abn\sigma_c^2 + \sigma^2$	$abn\sigma_c^2 + an\sigma_{bc}^2 + \sigma^2$
$\dfrac{cn}{(a-1)(b-1)}\displaystyle\sum_{i=1}^{a}\sum_{j=1}^{b} (ab)_{ij}^2 + n\sigma_{abc}^2 + \sigma^2$	$cn\sigma_{ab}^2 + n\sigma_{abc}^2 + \sigma^2$
$bn\sigma_{ac}^2 + \sigma^2$	$bn\sigma_{ac}^2 + n\sigma_{abc}^2 + \sigma^2$
$an\sigma_{bc}^2 + \sigma^2$	$an\sigma_{bc}^2 + \sigma^2$
$n\sigma_{abc}^2 + \sigma^2$	$n\sigma_{abc}^2 + \sigma^2$
σ^2	σ^2

The model equation of the second type (*A* fixed, *B*, *C* random) is

$$\begin{aligned}&\boldsymbol{y}_{ijkl}=\mu+a_i+\boldsymbol{b}_j+\boldsymbol{c}_k+(\boldsymbol{ab})_{ij}+(\boldsymbol{ac})_{ik}+(\boldsymbol{bc})_{jk}+(\boldsymbol{abc})_{ijk}+\boldsymbol{e}_{l(i,j,k)}\\&(i=1,\dots,a;j=1,\dots,b;k=1,\dots,c;l=1,\dots,n).\end{aligned}\tag{7.32}$$

If we put in the $E(\boldsymbol{MS})$ of the model with finite level populations $N(A)=a$ and let $N(B)$ and $N(C)$ tend to ∞, then we obtain the $E(\boldsymbol{MS})$ for the model equation (7.32), in the last column of Table 7.7. From this we obtain the estimators of the variance components:

$$\left.\begin{aligned}&s^2=\boldsymbol{MS}_{res} && s^2_{abc}=\frac{1}{n}(\boldsymbol{MS}_{ABC}-\boldsymbol{MS}_{res}),\\&s^2_{bc}=\frac{1}{an}(\boldsymbol{MS}_{BC}-\boldsymbol{MS}_{res}),\\&s^2_{ac}=\frac{1}{bn}(\boldsymbol{MS}_{AC}-\boldsymbol{MS}_{ABC}), && s^2_{ab}=\frac{1}{cn}(\boldsymbol{MS}_{AB}-\boldsymbol{MS}_{ABC}),\\&s^2_{a}=\frac{1}{abn}(\boldsymbol{MS}_{C}-\boldsymbol{MS}_{BC}), && s^2_{b}=\frac{1}{acn}(\boldsymbol{MS}_{B}-\boldsymbol{MS}_{BC}).\end{aligned}\right\}\tag{7.33}$$

7.5.4 Three-Way Nested Classification

In the three-way nested classification, there are six mixed models. To save space here, side conditions are not given. They are analogue to those in Section 7.5.2. (The sums over all fixed effects of a factor are zero; covariances between random model components are zero.) At first the model for finite level populations is discussed, and then the $E(\boldsymbol{MS})$ are derived by the rules in Section 7.2. Then the six balanced mixed models are treated; their $E(\boldsymbol{MS})$ can be derived by the reader from those of the model with finite level populations. They are summarised in Table 7.8 and 7.9. We further give the estimators. The ***SS***, *df* and ***MS*** are given in Table 5.27. The model with finite level populations has the model equation

$$\begin{aligned}&\boldsymbol{y}_{ijkl}=\mu+\boldsymbol{a}_i+\boldsymbol{b}_{j(i)}+\boldsymbol{c}_{k(i,j)}+\boldsymbol{e}_{l(i,j,k)}\\&(i=1,\dots,a;j=1,\dots,b;k=1,\dots,c;l=1,\dots,n).\end{aligned}\tag{7.34}$$

Following rule 4 in Section 7.2, we obtain the following scheme:

	i	*j*	*k*	*l*
a_i	$1-\frac{a}{N(A)}$	b	c	n
$b_{j(i)}$	1	$1-\frac{b}{N(B)}$	c	n
$c_{k(i,j)}$	1	1	$1-\frac{c}{N(C)}$	n
$e_{l(i,j,k)}$	1	1	1	$1-\frac{n}{N(res)}$

and from this the $E(\boldsymbol{MS})$ in Tables 7.8 and 7.9. The six mixed models are as follows:

I) Levels of A randomly selected, levels of B, C fixed
II) Levels of B randomly selected, levels of A, C fixed
III) Levels of C randomly selected, levels of A, B fixed
IV) Levels of B and C randomly selected, levels of A fixed
V) Levels of A and C randomly selected, levels of B fixed
VI) Levels of A and B randomly selected, levels of C fixed

The estimators are
$s^2 = MS_{\text{Rest}}$ for all six cases and further:
For case I,

$$s_a^2 = \frac{1}{bcn}(MS_A - MS_{res}). \tag{7.35}$$

For case II,

$$s_{b\,\text{in}\,a}^2 = \frac{1}{cn}(MS_{B\,\text{in}\,A} - MS_{res}). \tag{7.36}$$

For case III,

$$s_{c\,\text{in}\,b}^2 = \frac{1}{n}(MS_{C\,\text{in}\,B} - MS_{res}). \tag{7.37}$$

For case IV,

$$\left.\begin{aligned} s_{c\,\text{in}\,b}^2 &= \frac{1}{n}(MS_{C\,\text{in}\,B} - MS_{res}), \\ s_{b\,\text{in}\,a}^2 &= \frac{1}{cn}(MS_{B\,\text{in}\,A} - MS_{C\,\text{in}\,B}). \end{aligned}\right\} \tag{7.38}$$

For case V,

$$\left.\begin{aligned} s_{c\,\text{in}\,b}^2 &= \frac{1}{n}(MS_{C\,\text{in}\,B} - MS_{res}), \\ s_a^2 &= \frac{1}{bcn}(MS_A - MS_{C\,\text{in}\,B}). \end{aligned}\right\} \tag{7.39}$$

For case VI,

$$\left.\begin{aligned} s_{b\,\text{in}\,a}^2 &= \frac{1}{cn}(MS_{B\,\text{in}\,A} - MS_{res}), \\ s_a^2 &= \frac{1}{bcn}(MS_A - MS_{B\,\text{in}\,A}). \end{aligned}\right\} \tag{7.40}$$

Table 7.8 *E*(*MS*) for a balanced three-way nested classification – models with finite level populations and mixed models with one random facto.

Source of variation	Model with finite level populations (see Example 7.1)	*A* random, *B*, *C* fixed (I)	*B* random, *A*, *C* fixed (II)	*C* random, *A*, *B*, fixed (III)
Between A	$bcn\sigma_a^2 + cn\left(1-\frac{b}{N(B)}\right)\sigma_{b \text{ in } a}^2 + n\left(1-\frac{c}{N(C)}\right)\sigma_{c \text{ in } b}^2 + \sigma^2$	$bcn\sigma_a^2 + \sigma^2$	$\frac{bcn}{a-1}\sum_{i=1}^{a} a_i^2 + cn\sigma_{b \text{ in } a}^2 + \sigma^2$	$\frac{bcn}{a-1}\sum_{i=1}^{a} a_i^2 + n\sigma_{c \text{ in } b}^2 + \sigma^2$
Between B in A	$cn\sigma_{b \text{ in } a}^2 + n\left(1-\frac{c}{N(C)}\right)\sigma_{c \text{ in } b}^2 + \sigma^2$	$\frac{cn}{a(b-1)}\sum_{i,j} b_{j(i)}^2 + \sigma^2$	$cn\sigma_{b \text{ in } a}^2 + \sigma^2$	$\frac{cn}{a(b-1)}\sum_{i,j} b_{j(i)}^2 + n\sigma_{c \text{ in } b}^2 + \sigma^2$
Between C in B *and* A	$n\sigma_{c \text{ in } a}^2 + \sigma^2$	$\frac{n}{ab(c-1)}\sum_{i,j,k} c_{k(i,j)}^2 + \sigma^2$	$\frac{n}{ab(c-1)}\sum_{i,j,k} c_{k(i,j)}^2 + \sigma^2$	$n\sigma_{c \text{ in } b}^2 + \sigma^2$
Residual	σ^2	σ^2	σ^2	σ^2

Table 7.9 *E*(*MS*) for a balanced three-way nested classification – models with one fixed factor.

Source of variation	*A* fixed, *B*, *C* random (IV)	*B* fixed, *A*, *C* random (V)	*C* fixed, *A*, *B* random (VI)
Between A	$\frac{bcn}{a-1}\sum_{i=1}^{a} a_i^2 + cn\sigma_{b \text{ in } a}^2 + n\sigma_{c \text{ in } b}^2 + \sigma^2$	$bcn\sigma_a^2 + n\sigma_{c \text{ in } b}^2 + \sigma^2$	$bcn\sigma_a^2 + cn\sigma_{b \text{ in } a}^2 + \sigma^2$
Between B in A	$cn\sigma_{b \text{ in } a}^2 + n\sigma_{c \text{ in } b}^2 + \sigma^2$	$\frac{cn}{a(b-1)}\sum_{i,j} b_{j(i)}^2 + n\sigma_{c \text{ in } b}^2 + \sigma^2$	$cn\sigma_{b \text{ in } a}^2 + \sigma^2$
Between C in B and A	$n\sigma_{c \text{ in } b}^2 + \sigma^2$	$n\sigma_{c \text{ in } b}^2 + \sigma^2$	$\frac{n}{ab(c-1)}\sum_{i,j,k} c_{k(i,j)}^2 + \sigma^2$
Residual	σ^2	σ^2	σ^2

7.5.5 Three-Way Mixed Classification

In Chapters 5 and 6, mixed classifications with three factors have been considered. Mixed models for the two types of mixed three-way classification are now discussed.

7.5.5.1 The Type $(B \prec A) \times C$

For the balanced case of the mixed classification (model equation (5.45) discussed in Section 5.4.3.1), first the $E(\boldsymbol{MS})$ for the model with finite level populations are derived. The $E(\boldsymbol{MS})$ are given in Table 7.10.

For the six mixed models, the $E(\boldsymbol{MS})$ can be found in Tables 7.10 and 7.11. The estimators for the variance components besides $\boldsymbol{s}^2 = \boldsymbol{MS}_{res}$ are given below.

With rule 4 in Section 7.2, we receive

	i	j	k	l
a_i	$1-\frac{a}{N(A)}$	b	c	n
$b_{j(i)}$	1	$1-\frac{b}{N(B)}$	c	n
c_k	a	b	$1-\frac{c}{N(C)}$	n
$(ac)_{ik}$	$1-\frac{a}{N(A)}$	b	$1-\frac{c}{N(C)}$	n
$(bc)_{jk(i)}$	1	$1-\frac{b}{N(B)}$	$1-\frac{c}{N(C)}$	n
$e_{l(i,j,k)}$	1	1	1	$1-\frac{n}{N(res)}$

For the six models the estimators of the variance components are the following:

– A fixed, B, C random

$$\left.\begin{aligned}
s^2_{bc\,\mathrm{in}\,a} &= \frac{1}{n}\left(\boldsymbol{MS}_{B\times C\,\mathrm{in}\,A} - \boldsymbol{MS}_{res}\right),\\
s^2_a &= \frac{1}{bn}\left(\boldsymbol{MS}_{A\times C} - \boldsymbol{MS}_{B\times C\,\mathrm{in}\,A}\right),\\
s^2_c &= \frac{1}{abn}\left(\boldsymbol{MS}_{C} - \boldsymbol{MS}_{B\times C\,\mathrm{in}\,A}\right),\\
s^2_{b\,\mathrm{in}\,a} &= \frac{1}{cn}\left(\boldsymbol{MS}_{B\,\mathrm{in}\,A} - \boldsymbol{MS}_{B\times C\,\mathrm{in}\,A}\right);
\end{aligned}\right\} \tag{7.41}$$

Table 7.10 *E*(*MS*) of the mixed classification $(B \prec A) \times C$ – model with finite level populations ($N(R) \to \infty$) (see Example 7.1).

Source of variation	E(MS)
Between A	$bn\sigma_a^2 + cn\left(1-\frac{b}{N(B)}\right)\sigma_{b \text{ in } a}^2 + bn\left(1-\frac{c}{N(C)}\right)\sigma_{ac}^2 + n\left(1-\frac{b}{N(B)}\right)\left(1-\frac{c}{N(C)}\right)\sigma_{bc \text{ in } a}^2 + \sigma^2$
Between B in A	$cn\sigma_{b \text{ in } a}^2 + n\left(1-\frac{c}{N(C)}\right)\sigma_{bc \text{ in } a}^2 + \sigma^2$
Between C	$abn\sigma_c^2 + bn\left(1-\frac{a}{N(A)}\right)\sigma_{ac}^2 + n\left(1-\frac{b}{N(B)}\right)\sigma_{bc \text{ in } a}^2 + \sigma^2$
Interaction $A \times C$	$bn\sigma_{ac}^2 + n\left(1-\frac{b}{N(B)}\right)\sigma_{bc \text{ in } a}^2 + \sigma^2$
Interaction $B \times C$ in A	$n\sigma_{bc \text{ in } a}^2 + \sigma^2$
Residual	σ^2

– B fixed, A, C random

$$\left.\begin{aligned} \boldsymbol{s}_{bc\,\text{in}\,a}^2 &= \frac{1}{n}(\boldsymbol{MS}_{B\times C\,\text{in}\,A} - \boldsymbol{MS}_{res}),\\ \boldsymbol{s}_{ac}^2 &= \frac{1}{bn}(\boldsymbol{MS}_{A\times C} - \boldsymbol{MS}_{res}),\\ \boldsymbol{s}_{c}^2 &= \frac{1}{abn}(\boldsymbol{MS}_{C} - \boldsymbol{MS}_{A\times C}),\\ \boldsymbol{s}_{a}^2 &= \frac{1}{bcn}(\boldsymbol{MS}_{A} - \boldsymbol{MS}_{A\times C}); \end{aligned}\right\} \tag{7.42}$$

– C fixed, A, B random

$$\left.\begin{aligned} \boldsymbol{s}_{bc \text{ in } a}^2 &= \frac{1}{n}(\boldsymbol{MS}_{B\times C\,\text{in}\,A} - \boldsymbol{MS}_{res}),\\ \boldsymbol{s}_{ac}^2 &= \frac{1}{bn}(\boldsymbol{MS}_{A\times C} - \boldsymbol{MS}_{B\times C\,\text{in}\,A}),\\ \boldsymbol{s}_{b \text{ in } a}^2 &= \frac{1}{cn}(\boldsymbol{MS}_{B\,\text{in}\,A} - \boldsymbol{MS}_{res}),\\ \boldsymbol{s}_{a}^2 &= \frac{1}{bcn}(\boldsymbol{MS}_{A} - \boldsymbol{MS}_{B \text{ in } A}); \end{aligned}\right\} \tag{7.43}$$

– A random B, C fixed

Table 7.11 *E(MS)* of the mixed classification $(B \prec A) \times C$ – models with one fixed factor.

Source of variation	*A* fixed, *B*, *C* random	*B* fixed, *A*, *C* random	*C* fixed, *A*, *B* random
Between A	$\frac{bcn}{a-1}\sum_i a_i^2 + cn\sigma_{b\,\text{in}\,a}^2 + bn\sigma_{ac}^2 + n\sigma_{bc\ \text{in}\ a}^2 + \sigma^2$	$bcn\sigma_a^2 + bn\sigma_{ac}^2 + \sigma^2$	$bcn\sigma_a^2 + cn\sigma_{b\,\text{in}\,a}^2 + \sigma^2$
Between B in A	$cn\sigma_{b\ \text{in}\ a}^2 + n\sigma_{bc\ \text{in}\ a}^2 + \sigma^2$	$\frac{cn}{a(b-1)}\sum_{i,j} b_{j(i)}^2 + n\sigma_{bc\,\text{in}\ a}^2 + \sigma^2$	$cn\sigma_{b\ \text{in}\ a}^2 + \sigma^2$
Between C	$abn\sigma_c^2 + n\sigma_{bc\,\text{in}\,a}^2 + \sigma^2$	$abn\sigma_c^2 + bn\sigma_{ac}^2 + \sigma^2$	$\frac{abn}{c-1}\sum_k c_k^2 + bn\sigma_{ac}^2 + n\sigma_{bc\ \text{in}\ a}^2 + \sigma^2$
Interaction $A \times C$	$bn\sigma_{ac}^2 + n\sigma_{bc\,\text{in}\,a}^2 + \sigma^2$	$bn\sigma_{ac}^2 + \sigma^2$	$bn\sigma_{ac}^2 + n\sigma_{bc\,\text{in}\,a}^2 + \sigma^2$
Interaction $B \times C$ in A	$n\sigma_{bc\,\text{in}\,a}^2 + \sigma^2$	$n\sigma_{bc\,\text{in}\ a}^2 + \sigma^2$	$n\sigma_{bc\,\text{in}\,a}^2 + \sigma^2$
Residual	σ^2	σ^2	σ^2

$$\left.\begin{aligned} s_{ac}^2 &= \frac{1}{bn}(MS_{A\times C} - MS_{res}),\\ s_a^2 &= \frac{1}{bcn}(MS_A - MS_{res}); \end{aligned}\right\} \tag{7.44}$$

– B random A, C fixed

$$\left.\begin{aligned} s_{bc \text{ in } a}^2 &= \frac{1}{n}(MS_{B\times C \text{in} A} - MS_{res}),\\ s_b^2 &= \frac{1}{cn}(MS_{B \text{ in } A} - MS_{res}); \end{aligned}\right\} \tag{7.45}$$

– C random A, B fixed

$$\left.\begin{aligned} s_{bc \text{ in } a}^2 &= \frac{1}{n}(MS_{B\times C \text{in } A} - MS_{res}),\\ s_{ac}^2 &= \frac{1}{bn}(MS_{A\times C} - MS_{res}),\\ s_c^2 &= \frac{1}{abn}(MS_C - MS_{res}). \end{aligned}\right\} \tag{7.46}$$

The $E(\boldsymbol{MS})$ of the six models are given in Table 7.11 and in Table 7.12.

7.5.5.2 The Type $C \prec AB$

For the type $C \prec AB$ of the mixed classification the $E(\boldsymbol{MS})$ for the model with finite level populations have been derived in Section 7.2, Example 7.4. The following mixed models exist: The $E(\boldsymbol{MS})$ for these four cases can be found in Table 7.13. The estimators of the variance components are given below:

Fall I: Cfixed, A or B (W.l.o.g. we choose A) fixed,

Fall II: Cfixed, A and B random,

Fall III: Crandom, A and B fixed,

Fall IV: Crandom, A or B (W.l.o.g. we choose A) random.

Case I:

$$\left.\begin{aligned} s_{ab}^2 &= \frac{1}{cn}(MS_{A\times B} - MS_{res}),\\ s_b^2 &= \frac{1}{acn}(MS_B - MS_{res}); \end{aligned}\right\} \tag{7.47}$$

Table 7.12 *E(MS)* of the mixed classification $(B \prec A) \times C$ – models with one random factor.

Source of variation	*A* random, *B*, *C* fixed	*B* random, *A*, *C* fixed	*C* random, *A*, *B* fixed
Between A	$bcn\sigma_a^2 + \sigma^2$	$\frac{bcn}{a-1}\sum_i a_i^2 + cn\sigma_{b\ in\ a}^2 + \sigma^2$	$\frac{bcn}{a-1}\sum_i a_i^2 + bn\sigma_{ac}^2 + \sigma^2$
Between B in A	$\frac{cn}{a(b-1)}\sum_{i,j} b_{j(k)}^2 + \sigma^2$	$cn\sigma_{b\text{in}a}^2 + \sigma^2$	$\frac{cn}{a(b-1)}\sum_{i,j} b_{j(i)}^2 + n\sigma_{bc\text{in}\ a}^2 + \sigma^2$
Between C	$\frac{abn}{c-1}\sum_k c_k^2 + bn\sigma_{ac}^2 + \sigma^2$	$\frac{abn}{c-1}\sum_k c_k^2 + n\sigma_{bc\ \text{in}\ a}^2 + \sigma^2$	$abn\sigma_c^2 + \sigma^2$
Interaction $A \times C$	$bn\sigma_{ac}^2 + \sigma^2$	$\frac{bn}{(a-1)\ (c-1)}\sum_{i,k} (ab)_{ik}^2 + n\sigma_{bc\ \text{in}\ a}^2 + \sigma^2$	$bn\sigma_{ac}^2 + \sigma^2$
Interaction $B \times C$ in A	$\frac{n}{a(b-1)(c-1)}\sum_{i,j,k} (bc)_{jk(i)}^2 + \sigma^2$	$n\sigma_{bc\text{in}\ a}^2 + \sigma^2$	$n\sigma_{bc\ \text{in}\ a}^2 + \sigma^2$
Residual	σ^2	σ^2	σ^2

Table 7.13 *E*(*MS*) for mixed models of the mixed three-way classification of type $C \prec AB$.

Source of variation	*B* random, *A*, *C* fixed	*A*, *B* random, *C* fixed	*C* random, *A*, *B* fixed	*A*, *C* random, *B* fixed
Between A	$\frac{bcn}{a-1}\sum_i a_i^2 + cn\sigma_{ab}^2 + \sigma^2$	$bcn\sigma_a^2 + cn\sigma_{ab}^2 + \sigma^2$	$\frac{bcn}{a-1}\sum_i a_i^2 + n\sigma_{c \text{ in } ab}^2 + \sigma^2$	$bcn\sigma_a^2 + n\sigma_{c \text{ in } ab}^2 + \sigma^2$
Between B	$acn\sigma_b^2 + \sigma^2$	$acn\sigma_b^2 + cn\sigma_{ab}^2 + \sigma^2$	$\frac{acn}{b-1}\sum_j b_j^2 + n\sigma_{c \text{ in } ab}^2 + \sigma^2$	$\frac{acn}{b-1}\sum_j b_j^2 + n\sigma_{c \text{ in } ab}^2 + cn\sigma_{ab}^2 + \sigma^2$
Between C in $A \times B$	$\frac{n}{ab(c-1)}\sum_{i,j,k} c_{k(i,j)}^2 + \sigma^2$	$\frac{n}{ab(c-1)}\sum_{i,j,k} c_{k(i,j)}^2 + \sigma^2$	$n\sigma_{c \text{ in } ab}^2 + \sigma^2$	$n\sigma_{c \text{ in } ab}^2 + \sigma^2$
Interaction $A \times B$	$cn\sigma_{ab}^2 + \sigma^2$	$cn\sigma_{ab}^2 + \sigma^2$	$n\sigma_{c \text{ in } ab}^2 + \frac{cn\sum_{i,j}(ab)_{ij}^2}{(a-1)(b-1)} + \sigma^2$	$n\sigma_{c \text{ in } ab}^2 + cn\sigma_{ab}^2 + \sigma^2$
Residual	σ^2	σ^2	σ^2	σ^2

Case II:

$$\left.\begin{aligned} s_{ab}^2 &= \frac{1}{cn}(MS_{A \times B} - MS_{res}), \\ s_b^2 &= \frac{1}{acn}(MS_B - MS_{A \times B}), \\ s_a^2 &= \frac{1}{bcn}(MS_A - MS_{A \times B}); \end{aligned}\right\} \tag{7.48}$$

Case III:

$$s_{c \text{ in } ab}^2 = \frac{1}{n}(DQ_{MSC \text{ in} A \times C} - MS_{res}); \tag{7.49}$$

Case IV:

$$\left.\begin{aligned} s_{c \text{ in } ab}^2 &= \frac{1}{n}(MS_{C \text{ in } A \times C} - MS_{res}), \\ s_{ab}^2 &= \frac{1}{cn}(MS_{A \times B} - MS_{C \text{ in } A \times B}), \\ s_a^2 &= \frac{1}{bcn}(MS_A - MS_{C \text{ in } A \times C}). \end{aligned}\right\} \tag{7.50}$$

7.6 Exercises

7.1 Use the data set D in Exercise 6.3. The six boars must randomly be split into two groups. The groups are now understood as two locations as levels of a fixed factor. What model do we now have? Call the generated file D1.

7.2 For file D1 test the null hypothesis that there are no differences between the locations.

7.3 Estimate from D1 the variance component of the factor 'boar'.

References

Hartley, H. O. and Rao, J. N. K. (1967) Maximum likelihood estimation for the mixed analysis of variance model. *Biometrika*, **54**, 92–108.

Henderson, C. R. (1953) Estimation of variance and covariance components. *Biometrics*, **9**, 226–252.

Searle, S. R. (1971, 2012) *Linear Models*, John Wiley & Sons, Inc., New York.

Seifert, B. (1980) *Prüfung linearer Hypothesen über die festen Effekte in balancierten gemischten Modellen der Varianzanalyse.* Diss. Sektion Mathematik, Humboldt Universität Berlin.

Seifert, B. (1981) Explicit formulae of exact tests in mixed balanced ANOVA-models. *Biom. J.*, **23**, 535–550.

8

Regression Analysis – Linear Models with Non-random Regressors (Model I of Regression Analysis) and with Random Regressors (Model II of Regression Analysis)

8.1 Introduction

In this chapter we describe relations between two or more magnitudes with statistical methods.

Dependencies between magnitutes can be found in several laws of nature. There is a dependency of the height h of a physical body falling under the influence of gravity (in a vacuum) and the case time t in the form $h = \alpha t^2$, and the relationship provided by this formula is a special function, a so-called functional relationship. Similar equations can be given for the relationship between pressure and temperature or between brightness and distance from a light source. The relationship is strict, that is, for each value of t, there is a unique h-value, or in other words, with appropriate accuracy from the same t-value, there always results a unique h-value. One could calculate α by the formula above by setting t and measuring h, if there is no measurement error. The h-values for various t-values lie on a curve (parabola) when t is plot on the abscissa and h on the ordinate of a coordinate system. In this example, you could give h as well and measure the time. In functional relationships, therefore, it doesn't matter which variable is given and which is measured, if no other aspects (accuracy, effort in the measurement), which have nothing to do with the context itself, lead to the preference of one of these variables.

There are events in nature and variables, between which there is no functional relationship but they are well dependent on each other. For instance, let's consider height at withers and age or height at withers and chest girth of cattle. Although there is obviously no formula by which you can calculate the chest girth or the age of cattle from the height at withers, nevertheless there is obviously a connection between both. You can see this in some animals when both measurements are present and a point represents the value pair of each animal in a coordinate system. All these points are not, as in the case of a functional dependency, on a curve; it is rather a point cloud or as we say a scatter diagram.

Mathematical Statistics, First Edition. Dieter Rasch and Dieter Schott.

In such a cloud, a clear trend is frequently recognizable, which suggests the existence of a relationship. Such relationships, which are not strictly functional, are called stochastic, and their investigation is the main subject of the regression analysis.

Even if a functional relationship between two features exists, it may happen that the graphic representation of the measured value pairs is a point cloud; this is the case if the characteristic values cannot be observed without greater measurement errors.

The cloud itself is only a clue to the nature of the relationship between two variables and suggests their existence. It is required, however, to discern the relationship precisely through a formula-based representation. In all cases, the estimation target is a function of the independent variable called the regression function. In regression analysis, it is also of interest to characterize the variation of the dependent variable around the regression function, which can be described by a probability distribution. One must distinguish two important special cases that should be characterized for the case of two variables x, y – the generalization to more than two variables is left to the reader. In the first case, x is a non-random variable. Most commonly, regression analysis estimates the conditional expectation $E(\boldsymbol{y}|x_i) = f(x_i)$ of the regressand variable given the value of the regressor variable – that is, the average value of the dependent variable when the independent variable is fixed. The relationship is modelled by

$$\boldsymbol{y}_i = \boldsymbol{y}(x_i) = f(x_i) + \boldsymbol{e}_i \tag{8.1}$$

or

$$\boldsymbol{y} = f(x) + \boldsymbol{e}.$$

$\boldsymbol{e}_i$ are random variables with $E(\boldsymbol{e}_i) = 0$, $\text{var}(\boldsymbol{e}_i) = \sigma^2$ and $\text{cov}(\boldsymbol{e}_i, \boldsymbol{e}_j) = 0$ for $i \neq j$. Often the distribution of $\boldsymbol{e}_i$ is assumed to be normal $N(0, \sigma^2)$.

This we call a model I of regression analysis. As an example, you could call the relationship between the height of withers and the age of the cattle. Of course, you can also write the functional relationship in which only the measured values of a variable (as y) are strongly influenced by measurement errors, in the form of Equation (8.1) and treat analysis with the model I regression. The functional relationship is between and through $y = g(x)$.

In this chapter all occurring functions are assumed to be differentiable for all their arguments.

In the second case both $\boldsymbol{x}$ and $\boldsymbol{y}$ are random variables distributed by a two-dimensional distribution with density function $g(x, y)$, marginal expectation μ_x, μ_y, marginal variances σ_x^2, σ_y^2 and covariance σ_{xy}. Regression of $\boldsymbol{x}$ on $\boldsymbol{y}$ or of $\boldsymbol{y}$ on $\boldsymbol{x}$ means the conditional expectations $E(\boldsymbol{x}|y)$ and $E(\boldsymbol{y}|x)$, respectively. If $g(x, y)$ is the density function of a two-dimensional normal distribution, then the conditional expectations $E(\boldsymbol{x}|y)$ and $E(\boldsymbol{y}|x)$, respectively, are linear functions of y and x, respectively,

$$E(\boldsymbol{x}|y) = \alpha + \beta y,\ E(\boldsymbol{y}|x) = \alpha^* + \beta^* x.$$

The random variables $\boldsymbol{x}$ or $\boldsymbol{y}$ deviate by $\boldsymbol{e}$ or $\boldsymbol{e}^*$, respectively, from $E(\boldsymbol{x}|y)$ or $E(\boldsymbol{y}|x)$; therefore the stochastic dependency between $\boldsymbol{x}$ and $\boldsymbol{y}$ is either of the form

$$\boldsymbol{x} = E(\boldsymbol{x}|y) + \boldsymbol{e} = \alpha + \beta y + \boldsymbol{e} \tag{8.2}$$

or of the form

$$\boldsymbol{y} = E(\boldsymbol{y}|x) + \boldsymbol{e}^* = \alpha^* + \beta^* x + \boldsymbol{e}^*. \tag{8.3}$$

The Equations (8.2) and (8.3) are not transferable into each other, which means that

$$\boldsymbol{y} = \frac{x - \alpha - \boldsymbol{e}}{\beta} = \frac{x}{\beta} - \frac{\alpha}{\beta} - \frac{\boldsymbol{e}}{\beta} \text{ and } \boldsymbol{y} = \alpha^* + \beta^* x + \boldsymbol{e}^*$$

differ from each other. This is easy to see, if we look at the meaning of α, β, α^* and β^*.

An equation of the form $\boldsymbol{x} = E(\boldsymbol{x}|\boldsymbol{y}) + \boldsymbol{e}$ or $\boldsymbol{y} = E(\boldsymbol{y}|\boldsymbol{x}) + \boldsymbol{e}^*$ is called a model II of regression analysis if the conditioning variables are written as random. An example is the relationship between wither height and chest girth mentioned above.

The difference between both models becomes clear by looking at the examples above. In the dependency wither height–age, the age can be understood as a non-random variable (chosen in advance by the investigator). The wither height is considered dependent on age and not the age dependent on the wither height. For the dependency wither height–chest girth we model by two random variables. Therefore two equation analogues (8.2) and (8.3) are possible.

The function $y = f(x)$ in (8.1) and the functions $E(\boldsymbol{x}|\boldsymbol{y})$ and $E(\boldsymbol{y}|\boldsymbol{x})$ are called regression functions. The argument variable is called the regressor or the influencing variable (in program packages often the misleading expression 'independent variable' is used, but in model II both variables are dependent on each other). The variables $\boldsymbol{y}$ in (8.1), $\boldsymbol{x}$ in (8.2) and $\boldsymbol{y}$ in (8.3) are called regressand (or dependent variable). In this chapter we assume that regression functions are a special case of the theory of linear models in Chapter 4.

Definition 8.1 Let X be a $[n \times (k+1)]$ matrix of rank $k+1 < n$ and $\Omega = R[X]$ the rank space of X. Further, let $\beta \in \Omega$ be a vector of parameters β_j $(j = 0, \dots, k)$ and $\boldsymbol{Y} = \boldsymbol{Y}_n$ an n-dimensional random variable. If the relations $E(\boldsymbol{e}) = 0_n$ and $\text{var}(\boldsymbol{e}) = \sigma^2 I_n$ for the error term $\boldsymbol{e}$ are valid, then

$$\boldsymbol{Y} = X\beta + \boldsymbol{e}\ (Y \in R^n, \beta \in \Omega = R[X]) \tag{8.4}$$

with $X = (1_n, X^*)$ is called model I of the linear regression with k regressors in standard form.

Equation (8.4) is, as shown in Example 4.3, a special case of Equation (4.1). As shown in Chapter 4, $\text{var}(\boldsymbol{e}) = V\sigma^2$ with positive definite matrix V can be reduced to (8.4). At first we consider (8.4) and later use also $\text{var}(\boldsymbol{e}) = V\sigma^2$.

8.2 Parameter Estimation

8.2.1 Least Squares Method

For model I of regression we can prove the following.

Theorem 8.1 The BLUE of β and an unbiased estimator of σ^2 are given by

$$\boldsymbol{b} = \hat{\boldsymbol{\beta}} = \left(X^T X\right)^{-1} X^T \boldsymbol{Y} \tag{8.5}$$

and

$$\begin{aligned} \boldsymbol{s}^2 &= \frac{1}{n-k-1} \| \boldsymbol{Y} - X(X^T X)^{-1} X^T \boldsymbol{Y} \|^2 \\ &= \frac{1}{n-k-1} \boldsymbol{Y}^T (I_n - X(X^T X)^{-1} X^T) \boldsymbol{Y}. \end{aligned} \tag{8.6}$$

The proof follows from Example 4.3.

Theorem 8.2 If Y in (8.4) is $N(X\beta, \sigma^2 I_n)$-distributed, then the MLE of β is given by (8.5) and the MLE of σ^2 is given by

$$\tilde{\boldsymbol{\sigma}}^2 = \frac{1}{n} \left\| \boldsymbol{Y}^T - X\left(X^T X\right)^{-1} X^T \boldsymbol{Y} \right\|^2. \tag{8.7}$$

$\boldsymbol{b}$ in (8.5) and $\boldsymbol{s}^2$ in (8.6) are sufficient with respect to β and σ^2. $\boldsymbol{b}$ is $N[\beta, \sigma^2(X^T X)^{-1}]$-distributed and $\frac{n-k-1}{\sigma^2}\boldsymbol{s}^2$ is independent of $\boldsymbol{b}$ $CS(n-k-1)$-distributed.

Proof: $\boldsymbol{b}$ in (8.5) and $\tilde{\boldsymbol{\sigma}}^2$ in (8.7) are MLE of β, and σ^2 follows from Example 4.3 together with (4.13) and (4.14). With $\mu = X\beta$, $\Sigma = \sigma^2 I_n$ and $A = (X^T X)^{-1} X^T$, it follows that $\boldsymbol{b}$ with

$$E(\boldsymbol{b}) = A\mu = (X^T X)^{-1} X^T X\beta = \beta$$

and

$$\operatorname{var}(\boldsymbol{b}) = A\,\Sigma A' = (X^T X)^{-1} X^T \sigma^2 I_n X (X^T X)^{-1} = \sigma^2 (X^T X)^{-1}$$

is $(k+1)$-dimensional normally distributed.

To show that $\frac{(n-k-1)\boldsymbol{s}^2}{\sigma^2}$ is $CS(n-k-1)$-distributed, we have to show that $I_n - X(X^T X)^{-1} X^T = K$ is idempotent of rank $n-k-1$ and $\lambda = \frac{1}{\sigma^2}\beta^T X^T K X \beta = 0$. The idempotence of K is obvious. Because with X also $X^T X$ and $X(X^T X)^{-1} X^T$ are of rank $k+1$, due to the idempotence of $X(X^T X)^{-1} X^T = B$, there exists an orthonormal matrix T, so that $T^T B T$ is a diagonal matrix with $k+1$ values 1 and $n-k-1$ zeros. Therefore $\operatorname{rk}(K) = n-k-1$. Finally

$$\sigma^2\lambda = \beta^T X^T (I_n - X(X^T X)^{-1} X^T) X\beta = 0.$$

By this $\boldsymbol{b}$ and $\boldsymbol{Y}^T K \boldsymbol{Y}$ are independent because $(X^T X)^{-1} X^T K = O_{k+1,n}$.

We only have to show the sufficiency of $\boldsymbol{b}$ and $\boldsymbol{s}^2$. This is done, if the likelihood function $L(Y, \beta, \sigma^2)$ can be written in the form (1.3). From our assumption it follows that

$$L(Y,\beta,\sigma^2) = \frac{1}{\left(\sigma\sqrt{2\pi}\right)^n} \exp\left[-\frac{1}{2\sigma^2}\|Y - X\beta\|^2\right].$$

From (8.6) the identity

$$\begin{aligned}\|Y - X\beta\|^2 &= \|Y - Xb\|^2 + (b-\beta)^T \left(X^T X\right)(b-\beta)\\ &= (n-k-1)s^2 + f(\beta,b),\end{aligned}$$

follows for a certain $f(\beta, b)$ such that

$$L(Y,\beta,\sigma^2) = \frac{1}{\left(\sigma\sqrt{2\pi}\right)^n} \exp\left[-\frac{1}{2\sigma^2}\left[(n-k-1)s^2 + f(\beta,b)\right]\right]$$

has the form (1.3) and the theorem is proven.

Example 8.1 If the number of regressors in (8.4) is $k = 1$, we have a linear regression with one regressor or a so-called simple linear regression. With $k = 1$ we get

$$X^T = \begin{pmatrix} 1 & 1 & \dots & 1 \\ x_1 & x_2 & \dots & x_n \end{pmatrix} \quad \text{and} \quad \beta^T = (\beta_0, \beta_1)$$

and (8.4) becomes

$$\boldsymbol{y}_i = \beta_0 + \beta_1 x_i + \boldsymbol{e}_i \ (i = 1, \dots, n). \tag{8.8}$$

$\mathrm{rk}(X) = k + 1 = 2$ means that at least two of the x_i must be different. We look for the estimators of the coefficients β_0 and β_1. By the least squares method, an empirical regression line $\hat{y} = b_0 + b_1 x$ in the (x, y)-coordinate system as an estimate of the 'true' regression line $y = \beta_0 + \beta_1 x$ has to be found in such a way that if it is put into the scatter diagram (x_i, y_i) that $S = \sum_{i=1}^n (y_i - \beta_0 - \beta_1 x_i)^2$ is minimised.

The values of β_0 and β_1, minimising S are denoted by b_0 and b_1. We receive the following equations by putting the partial derivations of S with respect to β_0 and β_1 equal to zero and replacing all y by the random variables $\boldsymbol{y}$

$$\boldsymbol{b}_1 = \frac{\sum_{i=1}^{n} x_i \boldsymbol{y}_i - \frac{\sum_{i=1}^{n} x_i \sum_{i=1}^{n} \boldsymbol{y}_i}{n}}{\sum_{i=1}^{n} x_i^2 - \frac{\left(\sum_{i=1}^{n} x_i\right)^2}{n}} = \frac{n\sum_{i=1}^{n} x_i \boldsymbol{y}_i - \sum_{i=1}^{n} x_i \sum_{i=1}^{n} \boldsymbol{y}_i}{n\sum_{i=1}^{n} x_i^2 - \left(\sum_{i=1}^{n} x_i\right)^2}, \tag{8.9}$$

$$\boldsymbol{b}_0 = \bar{\boldsymbol{y}} - \boldsymbol{b}_1 \bar{x} \text{ with } \bar{\boldsymbol{y}} = \frac{\sum \boldsymbol{y}_i}{n} \text{ and } \bar{x} = \frac{\sum x_i}{n}. \tag{8.10}$$

Because S is convex, we really get a minimum.

We call estimators obtained in this way least squares estimators or in short LS estimators. As in the analysis of variance (ANOVA), we use for the sums of squared deviations for short:

$$SS_x = \sum x^2 - \frac{\left(\sum x\right)^2}{n}$$

and

$$\boldsymbol{SS}_y = \sum \boldsymbol{y}^2 - \frac{\left(\sum \boldsymbol{y}\right)^2}{n}.$$

And analogously for the sum of products

$$\boldsymbol{SP}_{xy} = \sum x\boldsymbol{y} - \frac{\sum x \sum \boldsymbol{y}}{n} = \sum (x - \bar{x})(\boldsymbol{y} - \bar{\boldsymbol{y}}),$$

the symbol $\boldsymbol{SP}_{xy}$ ($SP_{xx} = SS$) is used. Then $\boldsymbol{b}$ can be written as

$$\boldsymbol{b}_1 = \frac{\boldsymbol{SP}_{xy}}{\boldsymbol{SS}_x}.$$

Equations (8.9) and (8.10) are special cases of (8.5). We get (all summation from $i = 1$ to $i = n$)

$$X^T X = \begin{pmatrix} n & \sum x_i \\ \sum x_i & \sum x_i^2 \end{pmatrix}, \quad X^T \boldsymbol{Y} = \begin{pmatrix} \sum \boldsymbol{y}_i \\ \sum x_i \boldsymbol{y}_i \end{pmatrix}.$$

Now $(X^T X)^{-1} = \frac{1}{|X^T X|} \begin{pmatrix} \sum x_i^2 & -\sum x_i \\ -\sum x_i & n \end{pmatrix}$ with the determinant $|X^T X| = n\sum x_i^2 - \left(\sum x_i\right)^2$, and this leads to (8.9) and (8.10).

An estimator $\boldsymbol{s}^2$ of σ^2 is by (8.6) equal to

$$\boldsymbol{s}^2 = \frac{\sum_{i=1}^{n} (\boldsymbol{y}_i - \boldsymbol{b}_0 - \boldsymbol{b}_1 x_i)^2}{n-2} = \frac{\boldsymbol{SS}_y - \frac{\boldsymbol{SP}_{xy}^2}{\boldsymbol{SS}_x}}{n-2}. \tag{8.11}$$

(8.11) follows from (8.6) because

$$\boldsymbol{Y}^T\left(I_n - X\left(X^TX\right)^{-1}X^T\right)\boldsymbol{Y} = \sum y_i^2 - \boldsymbol{Y}^TX\left(X^TX\right)^{-1}X^T\boldsymbol{Y},$$

if we insert $X^T\boldsymbol{Y} = (\boldsymbol{Y}^TX)^T$ and $(X^TX)^{-1}$ as given above.

Putting the variables x and $\boldsymbol{y}$ in a coordinate system, the values of the variable x at the abscissa and the realisations y_i of the random variables $\boldsymbol{y}$ (or their expectations $\hat{y}_i$) at the ordinate, we obtain by

$$\hat{y}_i = b_0 + b_1x_i \quad (i = 1,\ldots,n) \tag{8.12}$$

a straight line with slope b_1 and intercept b_0. This straight line is called an (estimated) regression line. It is connecting the estimated expected values $\hat{y}_i$ for the values of x_i. If we put the observed values (x_i, y_i) as points in the coordinate system, we receive a scatter diagram. Amongst all lines, which could be put into this scatter diagram, the regression line is that one for which the sum of the squares of all distances parallel to the ordinate between the points and the straight is minimum. The value, respectively, of b_1 and β_1 shows us by how many units y is changing in mean if x is increasing by one unit. The distribution of $\boldsymbol{b}_0$ and $\boldsymbol{b}_1$ is given by the corollary of Theorem 8.1.

Corollary 8.1 The estimators $\boldsymbol{b}_0$ given by (8.10) and $\boldsymbol{b}_1$ given by (8.9) are under model equation (8.8) and its side conditions with expectations

$$E(\boldsymbol{b}_0) = \beta_0,\ E(\boldsymbol{b}_1) = \beta_1, \tag{8.13}$$

the variances

$$\sigma_0^2 = \text{var}(\boldsymbol{b}_0) = \frac{\sigma^2\sum x_j^2}{n\sum\left(x_j - \bar{x}\right)^2},\quad \sigma_1^2 = \text{var}(\boldsymbol{b}_1) = \frac{\sigma^2}{\sum\left(x_j - \bar{x}\right)^2} \tag{8.14}$$

and the covariance

$$\text{cov}\ (\boldsymbol{b}_0, \boldsymbol{b}_1) = -\frac{\sigma^2\sum x_j}{n\sum\left(x_j - \bar{x}\right)^2} \tag{8.15}$$

distributed, with $\boldsymbol{Y}$ are also $\boldsymbol{b}_0$ and $\boldsymbol{b}_1$ normally distributed.

Of course we can use other loss functions than the quadratic one. At the place of the sum of squared deviations, we could, for instance, use the sum of the p-th powers of the module of the deviations

$$S^* = \sum_{i=1}^{n}\left|y_i - \beta_0^* - \beta_1^*x_i\right|^p$$

and minimise them (L_p-norm). Historically this happened before using the LS estimators already by Bošković, an astronomer in Ragusa (Italy). Between 1750 and 1753 for astronomical calculations, he used a method for fitting functions by minimising the absolute sum of residuals, that is, a L_1-norm. Carl Friedrich Gauss made notice of Boškovićs work on 'orbital determination of luminaries' (see Eisenhart, 1961). A modern description of the parameter estimation using the L_1 loss function, iteration methods and asymptotic properties of the estimates can be found in Bloomfield and Steiger (1983) and for other p-values in Gonin and Money (1989).

If k regressors $x_1, \dots, x_k$ are given with the values $x_{i1}, \dots, x_{ik}$ $(i = 1, \dots, n)$, we call (8.4) for $k > 1$ the model equation of the multiple linear regression. Then the components $\boldsymbol{b}_i = \hat{\boldsymbol{\beta}}_i$ of $\hat{\boldsymbol{\beta}}$ in (8.5) are the estimators with realisations minimising

$$S = \sum_{i=1}^{n} \left(y_i - \beta_0 - \sum_{j=1}^{k} \beta_j x_{ij} \right)^2 .$$

The right-hand side of the equation

$$\hat{\boldsymbol{y}}_i = \widehat{\boldsymbol{E}(\boldsymbol{y}_i)} = \boldsymbol{b}_0 + \boldsymbol{b}_1 x_{i1} + \dots + \boldsymbol{b}_k x_{ik} \tag{8.16}$$

is the estimator of the expectation of the $\boldsymbol{y}_i$. An equation in the realisations

$$\hat{y}_i = \widehat{E(\boldsymbol{y}_i)} = b_0 + b_1 x_{i1} + \dots + b_k x_{ik}$$

defines a hyperplane, called the estimated regression plane. The $\boldsymbol{b}_i$ and b_i are called regression coefficients, respectively. For the estimator of σ^2, we write

$$\boldsymbol{s}^2 = \frac{\sum_{i=1}^{n} (\boldsymbol{y}_i - \hat{\boldsymbol{y}}_i)^2}{n-k-1}. \tag{8.17}$$

By (8.4) we also can describe non-linear dependencies between $\boldsymbol{Y}$ and one regressor $x = x_1$ (but also more regressors) if this non-linearity is of a special form.

We restrict ourselves to one regressor; the transmission to more regressors is simple and is left to the reader. In generalisation of (8.8), let

$$\boldsymbol{y}_i = f(x_i) + \boldsymbol{e}_i. \tag{8.18}$$

In (8.8) we had $f(x) = \beta_0 + \beta_1 x$.

Definition 8.2 Let $k+1$ linear independent functions $g_i(x)$ $(x \in B \subset R^1;$ $g_0(x) \equiv 1)$ be given and amongst them at least one is non-linear in x. If the non-linear regression function with $k + 1$ (unknown) parameters α_i can be written as

$$f(x) = f(x, \alpha_0, \dots, \alpha_k) = \sum_{i=1}^{k} \alpha_i g_i(x), \tag{8.19}$$

and the (known) functions $g_i(x)$ are independent of the parameters α_i, we call the in α_i linear function $f(x)$ a quasilinear regression function.

The non-linearity of a quasilinear regression function refers this to the regressor only and not to the parameters.

Regression analysis with quasilinear regression functions can be easily led back to the multiple linear regression analysis. Setting in (8.19)

$$g_i(x) = x_i, \tag{8.20}$$

we receive

$$f(x,\alpha_0,\ldots,\alpha_k) = \sum_{i=0}^{k} \alpha_i x_i \quad \text{with} \quad x_0 = 1,$$

so that (8.18) becomes

$$\boldsymbol{y}_i = \sum_{j=0}^{k} \alpha_j x_{ij} + \boldsymbol{e}_i \quad (i = 1,\ldots,n; x_{0j} = 1).$$

In addition, this model equation is apart from symbolism identical with (8.4). By this quasilinear regression function can be handled as a multiple linear regression function.

Nevertheless a practical important special case will be considered in some detail, because we can find simplifications in computation. This special case is the polynomial regression function.

Definition 8.3 If the $g_i(x)$ in (8.19) are polynomials of degree i in x, that is, if $f(x, \alpha_0, \ldots, \alpha_k)$ can be written as

$$f(x,\alpha_0,\ldots,\alpha_k) = \sum_{j=0}^{k} \alpha_j P_j(x) = \sum_{j=0}^{k} \beta_j x^j = P(x,\beta_0,\ldots,\beta_k), \tag{8.21}$$

then we call $f(x, \alpha_0, \ldots, \alpha_k)$ and $P(x, \beta_0, \ldots, \beta_k)$ polynomial regression functions, respectively.

With Definition 8.2 we can write the model equation of the polynomial regression as follows:

$$\boldsymbol{y}_i = \sum_{j-0}^{k} \beta_j x_i^j + \boldsymbol{e}_i \quad (i = 1,\ldots,n). \tag{8.22}$$

If the n values x_i of the regressor x are prespecified equidistantly, that is, we have

$$x_i = a + ih \; (i = 1,\ldots,n; h = \text{const}), \tag{8.23}$$

computation becomes easy in (8.21), if we replace the $P_j(x)$ and use the orthogonal polynomials in $i - \bar{i}$, because the values of these polynomials are tabulated (Fisher and Yates, 1974).

We demonstrate this at first for a regression function quadratic in x; afterwards the procedure is explained for regression functions of arbitrary degree in x.

Example 8.2 Orthogonal polynomials for a polynomial regression of second degree
If the regression function in (8.21) is a polynomial of the second degree in x, we get from (8.22)

$$\boldsymbol{y}_i = \beta_0 + \beta_1 x_i + \beta_2 x_i^2 + \boldsymbol{e}_i \quad (i = 1,\dots,n), \tag{8.24}$$

and this shall be written in the form

$$\boldsymbol{y}_i = \alpha_0 + \alpha_1 P_1(i-\bar{i}) + \alpha_2 P_2(i-\bar{i}) + \boldsymbol{e}_i. \tag{8.25}$$

P_1 and P_2 shall be orthogonal polynomials, that is, the relations

$$\sum_{i=1}^{n} P_1(i-\bar{i}) P_2(i-\bar{i}) = 0$$

and $\sum_{i=1}^{n} P_j(i-\bar{i}) = 0,\ \ j = 1,2$ shall be valid.

Because $\bar{i} = \frac{1}{n}\sum_{i=1}^{n} i = \frac{n(n+1)}{2n} = \frac{n+1}{2}$, we have

$$P_1(i-\bar{i}) = P_1\left(i - \frac{n+1}{2}\right) = c_0 + c_1\left(i - \frac{n+1}{2}\right) \quad (c_1 \neq 0) \tag{8.26}$$

and

$$P_2(i-\bar{i}) = P_2\left(i - \frac{n+1}{2}\right) = d_0 + d_1\left(i - \frac{n+1}{2}\right) + d_2\left(i - \frac{n+1}{2}\right)^2 (d_2 \neq 0). \tag{8.27}$$

The values c_0, c_1, d_0, d_1 and d_2 have to be chosen in such a way that the conditions

$$\sum_{i=1}^{n} P_1 P_2 = 0$$

and

$$\sum_{i=1}^{n} P_1 = \sum_{i=1}^{n} P_2 = 0$$

are fulfilled using short notation.

From (8.23) and (8.24), we get

$$\boldsymbol{y}_i = \beta_0 + \beta_1(a + ih) + \beta_2(a + ih)^2 + \boldsymbol{e}_i. \tag{8.28}$$

Further it follows from (8.25) to (8.27):

$$\boldsymbol{y}_i = \alpha_0 + \alpha_1\left[c_0 + c_1\left(i - \frac{n+1}{2}\right)\right] + \alpha_2\left[d_0 + d_1\left(i - \frac{n+1}{2}\right) + d_2\left(i - \frac{n+1}{2}\right)^2\right] + \boldsymbol{e}_i. \tag{8.29}$$

Extending (8.28) and (8.29) and arranging the result according to powers of i, by comparing the coefficients of these powers, leads to

$$\begin{aligned}&\beta_0 + a\beta_1 + a^2\beta_2 = \alpha_0 + \alpha_1\left(c_0 - c_1\frac{n+1}{2}\right) + \alpha_2\left(d_0 - d_1\frac{n+1}{2} + d_2\frac{(n+1)^2}{4}\right),\\ &h\beta_1 + 2ah\beta_2 = \alpha_1 c_1 + \alpha_2(d_1 - (n+1)d_2), h^2\beta_2 = d_2\alpha_2.\end{aligned} \tag{8.30}$$

Estimating the α_i by the least squares method and replacing the parameters in (8.30) by their estimates (Theorem 4.3) model equation (8.24) can be replaced by (8.25). Equation (8.30) simplifies when choosing c_0, c_1, d_0, d_1, d_2, as mentioned above. Due to
$\sum P_1 = \sum P_2 = \sum P_1P_2 = 0$, $c_1 \neq 0$, $d_2 \neq 0$, we obtain

$$\sum P_1 = \sum c_0 + c_1 \sum\left(i - \frac{n+1}{2}\right) = nc_0,$$

that is, it follows $c_0 = 0$. Further it is

$$\sum P_2 = nd_0 + d_2\left(\sum_{i=1}^{n} i^2 - \frac{n(n+1)^2}{4}\right) = nd_0 + d_2\frac{n(n-1)(n+1)}{12},$$

because the sum of the squares of 1 to n equals $\frac{n(n+1)(2n+1)}{6}$. Because of $d_2 \neq 0$ from $\sum P_2 = 0$ follows $d_0 = -d_2\frac{(n-1)(n+1)}{12}$.

From $\sum P_1P_2 = 0$ follows $d_1 = 0$.

Hence, orthogonal polynomials P_1 and P_2 have the form

$$P_1\left(i - \frac{n+1}{2}\right) = c_1\left(i - \frac{n+1}{2}\right) \tag{8.31}$$

and

$$P_2\left(i - \frac{n+1}{2}\right) = d_2\left[\left(i - \frac{n+1}{2}\right)^2 - \frac{(n-1)(n+1)}{12}\right]. \tag{8.32}$$

We should choose c_1 and d_2 so that each polynomial has integer coefficients. Fisher and Yates (1974) give tables of $P_1, P_2, \sum P_1^2$ and $\sum P_2^2$.

We now consider the general case of the quasilinear polynomial regression and assume w.l.o.g. that (8.21) is already written as

$$P(x, \beta_0, \ldots, \beta_k) = \sum_{j=0}^{k} \beta_j x_i^j.$$

If we put $x_{ij} = x_i^j$, we have shown that

$$\boldsymbol{y}_i = \beta_0 + \sum_{j=1}^{k} \beta_j x_i^j + \boldsymbol{e}_i \quad (i = 1, \ldots, n) \tag{8.33}$$

is of the form (8.4). It only must be shown that

$$X = \begin{pmatrix} 1 & x_1 & x_1^2 & \ldots & x_1^k \\ 1 & x_2 & x_2^2 & \ldots & x_2^k \\ \vdots & \vdots & \vdots & & \vdots \\ 1 & x_n & x_n^2 & \ldots & x_n^k \end{pmatrix}$$

is of rank $k + 1$. This certainly is the case, if at least $k + 1$ of the x_i are different from each other and $k + 1 < n$. Let these conditions be fulfilled (assumption of the polynomial regression). Now Theorem 8.1 can also be applied on the quasi-linear regression. For $X^T X$ and $X^T \boldsymbol{Y}$, we obtain

$$X^T X = \begin{pmatrix} n & \sum x_i & \sum x_i^2 & \ldots & \sum x_i^k \\ \sum x_i & \sum x_i^2 & \sum x_i^3 & \ldots & \sum x_i^{k+1} \\ \sum x_i^2 & \sum x_i^3 & \sum x_i^4 & \ldots & \sum x_i^{k+2} \\ \vdots & \vdots & \vdots & & \vdots \\ \sum x_i^k & \sum x_i^{k+1} & \sum x_i^{k+2} & \ldots & \sum x_i^{2k} \end{pmatrix} \text{ and } X^T \boldsymbol{Y} = \begin{pmatrix} \sum \boldsymbol{y}_i \\ \sum x_i \boldsymbol{y}_i \\ \sum x_i^2 \boldsymbol{y}_i \\ \vdots \\ \sum x_i^k \boldsymbol{y}_i \end{pmatrix}.$$

For equidistant x_i as in Example 8.2 for $k = 2$, the use of orthogonal polynomials has numerical advantages.

With modern computer programs, of course, the work does not need such transformations. Nevertheless, we consider this special case.

Theorem 8.3 If in model (8.33), the x_i are equidistant, that is, it can be written in the form (8.23), and if $P_j\left(i - \frac{n+1}{2}\right)$ are polynomials of degree j in $i - \frac{n+1}{2}$ $(j = 0, \ldots, k; i = 1, \ldots, n)$ so that

$$\sum_{j=0}^{k} \beta_j x_i^j = \sum_{j=0}^{k} \alpha_j P_j\left(i - \frac{n+1}{2}\right), \quad P_0\left(i - \frac{n+1}{2}\right) \equiv 1, \tag{8.34}$$

then an LS estimator of the vector $\alpha = (\alpha_0, \dots, \alpha_k)^T$ is given by

$$\hat{\boldsymbol{a}} = \left(\bar{\boldsymbol{y}}, \frac{\sum \boldsymbol{y}_i P_1\left(i - \frac{n+1}{2}\right)}{\sum P_1^2\left(i - \frac{n+1}{2}\right)}, \dots, \frac{\sum \boldsymbol{y}_i P_k\left(i - \frac{n+1}{2}\right)}{\sum P_k^2\left(i - \frac{n+1}{2}\right)} \right)^T. \tag{8.35}$$

The LS estimator of $\beta = (\beta_0, \dots, \beta_k)^T$ is

$$\boldsymbol{b} = U^{-1} W \boldsymbol{a} \tag{8.36}$$

with U and W obtained from (8.34) (by comparison of coefficients) so that

$$U\beta = W\alpha. \tag{8.37}$$

Proof: With

$$X = \begin{pmatrix} 1 & P_1\left(1 - \frac{n+1}{2}\right) & \cdots & P_k\left(1 - \frac{n+1}{2}\right) \\ 1 & P_1\left(2 - \frac{n+1}{2}\right) & \cdots & P_k\left(2 - \frac{n+1}{2}\right) \\ \vdots & \vdots & & \vdots \\ 1 & P_1\left(n - \frac{n+1}{2}\right) & \cdots & P_k\left(n - \frac{n+1}{2}\right) \end{pmatrix},$$

the right-hand side of (8.34) with $P_0 \equiv 1$ presentable as $X\alpha$ and by this (8.33) is presentable as $\boldsymbol{Y} = X\alpha + \boldsymbol{e}$, and this is an equation of the form (8.4). Further (8.35) is a special case of (8.5), because

$$X^T X = \begin{pmatrix} n & \sum P_{1i} & \dots & \sum P_{ki} \\ \sum P_{1i} & \sum P_{1i}^2 & \dots & \sum P_{1i} P_{ki} \\ \vdots & & & \vdots \\ \sum P_{ki} & \sum P_{ki} P_{1i} & \dots & \sum P_{ki}^2 \end{pmatrix} \quad \left(P_{ji} = P_j\left(i - \frac{n+1}{2}\right)\right)$$

and

$$(X^T X)^{-1} = \begin{pmatrix} \frac{1}{n} & & & 0 \\ & \frac{1}{\sum P_{1i}^2} & & \\ & & \ddots & \\ 0 & & & \frac{1}{\sum P_{ki}^2} \end{pmatrix}$$

is a diagonal matrix, and

$$X^T \boldsymbol{Y} = \begin{pmatrix} \sum \boldsymbol{y}_i \\ \sum \boldsymbol{y}_i P_{1i} \\ \vdots \\ \sum \boldsymbol{y}_i P_{ki} \end{pmatrix}.$$

Equation (8.36) follows from the Gauss–Markov theorem (Theorem 4.3). $X^T X$ becomes by suitable choice of coefficients in

$$P_j\left(i - \frac{n+1}{2}\right) = k_{0j} + k_{1j}\left(i - \frac{n+1}{2}\right) + k_{2j}\left(i - \frac{n+1}{2}\right)^2 + \cdots + k_{jj}\left(i - \frac{n+1}{2}\right)^j$$

a diagonal matrix. The values of the polynomials are uniquely fixed by i, j and n, and they are tabulated.

Example 8.3 In a carotene storage experiment, it should be investigated whether the change of the carotene content in grass depends upon the method of storage. For this the grass was stored in a sack and in a glass. During the time of storage in days, samples were taken and their carotene content was ascertained. Table 8.1 shows the results for both kinds of storage.

The relationship between carotene content and time of storage is modelled by Equation (8.8); the side conditions and the additional assumptions may be fulfilled. This relationship must be modelled by model I, because the time of storage is not a random variable; its values are chosen by the experimenter.

Hints for SPSS

Contrary to the ANOVA in SPSS, no distinction is made between models with fixed and random factors. Therefore the correlation coefficient defined in Section 8.5 is always calculated, also for model I where it is fully meaningless and must be dropped.

Table 8.1 Carotene content (in mg/100 g dry matter) *y* of grass in dependency of the time of storage *x* (in days) for two kinds of storage as SPSS input (columns 4 until 7 are explained later).

Time	Sack	Glass	Pre_1	Res_1	Pre_2	Res_2
1	31.2500	31.2500	31.16110	31.16110	31.16110	0.08890
60	28.7100	30.4700	27.94238	27.94238	27.94238	0.76762
124	23.6700	20.3400	24.45089	24.45089	24.45089	−0.78089
223	18.1300	11.8400	19.04999	19.04999	19.04999	−0.91999
303	15.5300	9.4500	14.68563	14.68563	14.68563	0.84437

We choose in SPSS (after data input):

Analyze
Regression
Linear

and consider both storage in a sack and in a glass. The data matrix of Table 8.1 estimates of $\beta_{i0}, \beta_{i1}, E\left(\boldsymbol{y}_{ij}\right)$, $\text{var}\left(\hat{\boldsymbol{\beta}}_{i0}\right)$, $\text{var}\left(\hat{\boldsymbol{\beta}}_{i1}\right)$ and $\text{cov}\left(\hat{\boldsymbol{\beta}}_{i0}, \hat{\boldsymbol{\beta}}_{i1}\right)$ for $i = 1$ should be given. The case $i = 2$ is left to the reader.

By (8.9) and (8.10) the estimates b_{10} and b_{11} as well b_{20} and b_{21} can be calculated by SPSS. The estimates $\hat{y}_{1j}$ of $E\left(\boldsymbol{y}_{1j}\right)$ and $\hat{y}_{2j}$ of $E\left(\boldsymbol{y}_{2j}\right)$ are given in Table 8.1 as PRE_1 and PRE_2, respectively. RES_1 and RES_2 are the differences $\hat{y}_{1j} - y_{1j}$ and $\hat{y}_{2j} - y_{2j}$, respectively.

To obtain these values we must in SPSS (Figure 8.1) go to the button 'Save' and activate there 'Predicted values' and 'Residuals'. Do this for each kind of storage. The results appear in the data matrix. To get the covariance matrix under statistics, we choose 'Covariance matrix'.

The regression coefficients, standard deviations of the estimates and the covariance cov ($\boldsymbol{b}_{10}$, $\boldsymbol{b}_{11}$)between both coefficients are shown in Figure 8.2.

The ANOVA table, in SPSS also named ANOVA table, is explained in the next section. We obtain

$$b_{11} = -0.055,$$

$$b_{10} = 31.216.$$

Further we find in the column right of that of the coefficients in Figure 8.2 (Std error) the coefficients in $\sigma_{10} = \sqrt{\text{var}(\boldsymbol{b}_{10})} = 0.706\sigma$ and $\sigma_{11} = \sqrt{\text{var}(\boldsymbol{b}_{11})} = 0.004\sigma$.

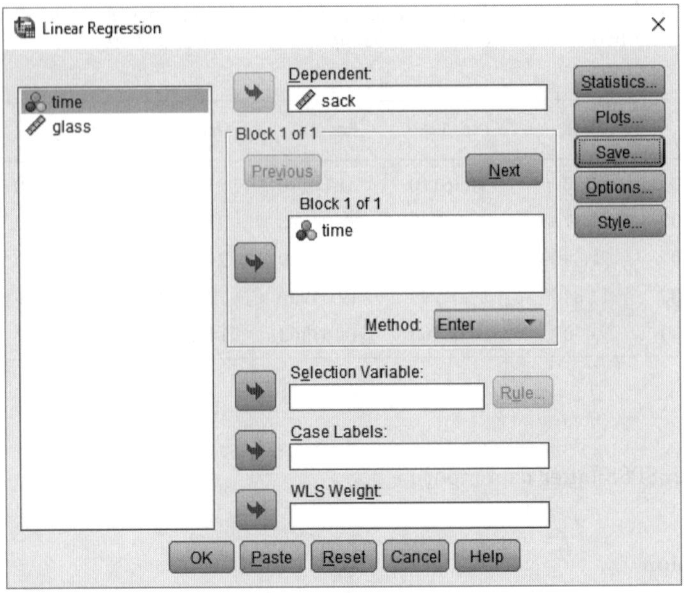

Figure 8.1 Introduction to regression analysis in SPSS. *Source:* Reproduced with permission of IBM.

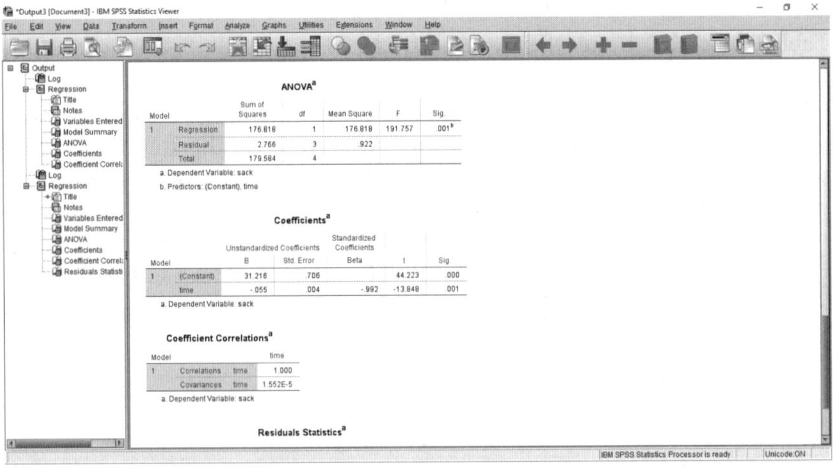

ANOVA[a]

Model		Sum of Squares	df	Mean Square	F	Sig.
1	Regression	176.818	1	176.818	191.757	.001[b]
	Residual	2.766	3	.922		
	Total	179.584	4			

a. Dependent Variable: sack
b. Predictors: (Constant), time

Coefficients[a]

Model		Unstandardized Coefficients B	Std. Error	Standardized Coefficients Beta	t	Sig.
1	(Constant)	31.216	.706		44.223	.000
	time	-.055	.004	-.992	-13.848	.001

a. Dependent Variable: sack

Coefficient Correlations[a]

Model			time
1	Correlations	time	1.000
	Covariances	time	1.552E-5

a. Dependent Variable: sack

Residuals Statistics[a]

Figure 8.2 SPSS output for storage in sack of Example 8.6. *Source:* Reproduced with permission of IBM.

Because in the ANOVA table in the row residual $s^2 = \frac{2.7666}{3} = 0.922$ is found, the estimates of σ^2_{10}, σ^2_{11} and $\sigma^{(1)}_{12}$ can easily be calculated by multiplying the results above by s in place of σ.

For $i = 2$ we obtain analogously

$$b_{21} = -0.081,\ b_{20} = 32.185.$$

The equations of the estimated regression lines are

$$\begin{aligned} i = 1: \hat{y}_{1j} &= 31.216 - 0.055x_j, \\ i = 2: \hat{y}_{2j} &= 32.185 - 0.081x_j \end{aligned} \quad \left(1 \le x_j \le 303\right).$$

For the estimated regression function, we should always give the region of the values of the regressor because any extrapolation of the regression curve outside this region is dangerous. Both estimated regression lines are shown in Figure 8.3.

If we are not sure, whether the regression is linear or not, we use another branch of SPSS, namely,

Analyze
Regression
Curve Estimation

to calculate polynomials or

Analyze
Regression
Nonlinear

for intrinsically non-linear regression as described in Chapter 9.

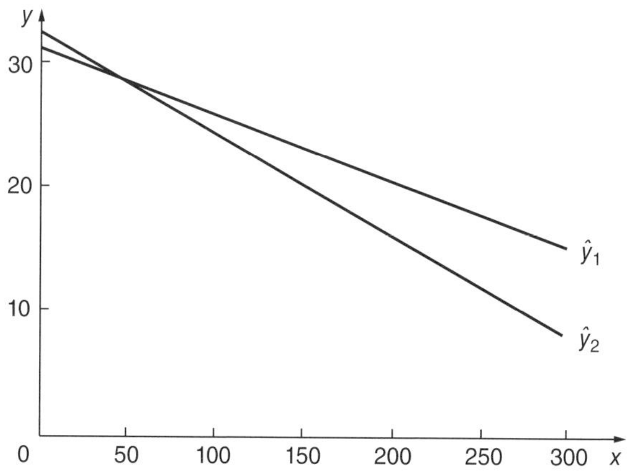

Figure 8.3 Estimated regression lines of Example 8.3.

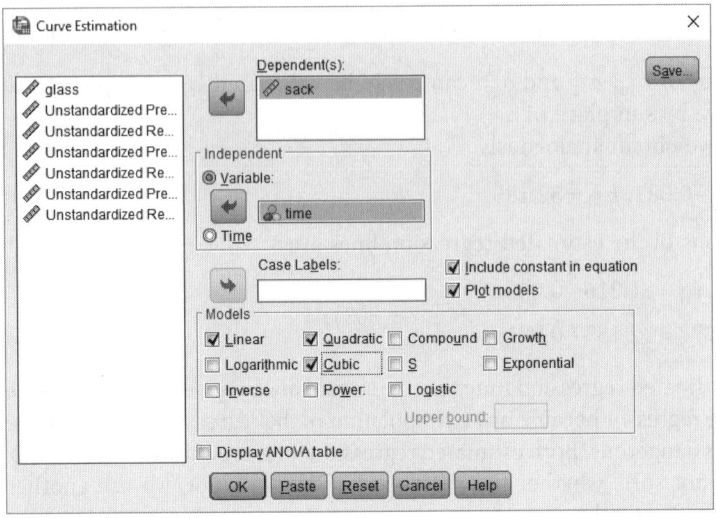

Figure 8.4 SPSS–window for curvilinear regression. *Source:* Reproduced with permission of IBM.

We demonstrate the first way for linear, quadratic and cubic polynomials for the trait 'sack'. At first we come to the sheet in Figure 8.4.

The graphical output is shown in Figure 8.5.

Here we can see how dangerous an extrapolation outside the interval [1, 303] can be. The cubic regression function goes up after 303 days; this certainly is impossible. Between the three lines within the interval [1, 303] no large differences can be found, the coefficients of the quadratic and cubic terms are not significant; the numerical output may be done by the reader.

8.2.2 Optimal Experimental Design

In this section, the optimal choice of X in model equation (8.4) for estimating β is described. We assume that the size n of the experiment is already given and β or $X\beta$ must be estimated by its LS estimator. Rasch and Herrendörfer (1982) discuss the problem that X, n and the estimator of β have to be chosen simultaneously.

Let $X = (x_1, \ldots, x_n)^T$ and B the domain of the R^{k+1}, in which the row vectors x_i^T of X are located. B is called experimental region. $\{L_n\}$ is that set of the X, for which $x_i \in B$. We now call X a design matrix. Contrary to discrete and continuous experimental designs, introduced now, we understand by X the design matrix of a concrete (existing) designs. In the sequel we call X briefly a design.

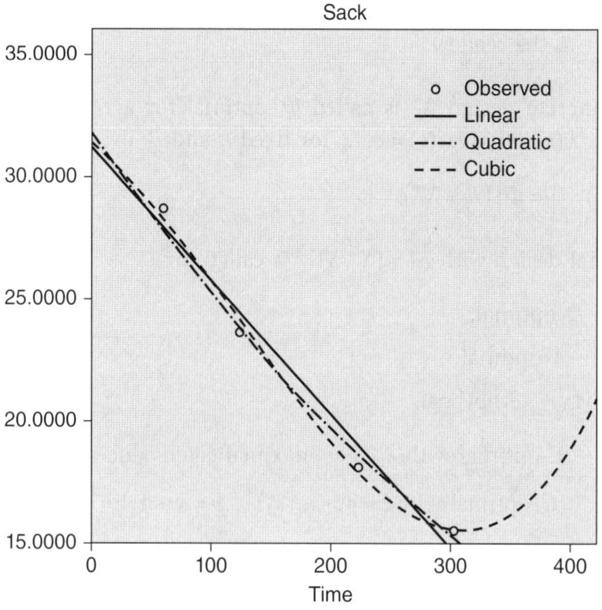

Figure 8.5 The regression curves for the linear, quadratic and cubic regression.
Source: Reproduced with permission of IBM.

In the theory of optimal experimental designs as in Kiefer (1959), Fedorov (1971) and Melas (2008), the following definitions are important.

Definition 8.4 Each set of pairs

$$\xi_m = \begin{Bmatrix} x_1 & x_2 & \cdots & x_m \\ p_1 & p_2 & \cdots & p_m \end{Bmatrix} \tag{8.38}$$

with $x_i \in B$, $0 < p_i \le 1$ $(i = 1,\ldots,m)$, $x_i \neq x_j$ for $i \neq j$ $(i,j = 1,\ldots,m)$ and $\sum_{i=1}^{m} p_i = 1$ is called a discrete m-point design, p_i are called weights and $x_1, \ldots, x_m$ is called support of ξ_m.

Definition 8.5 Each probability measure ξ on the measurable space $(B, \mathcal{B})$ is called a continuous design.

Y, as a discrete design, is a special case of a continuous design for a discrete probability measure. A concrete design has the form of a discrete design with $p_i = \frac{k_i}{n}, \sum k_i = n$ and k_i integer.

The problem is to construct a discrete or continuous design in such a way that the covariance matrix of $\hat{\boldsymbol{\beta}}$ meets some optimality criteria. The optimality criteria in this section concern a functional Φ mapping $(X^TX)^{-1}$ in the R^1. We define the optimality for concrete designs; the definitions for discrete and continuous designs are left to the reader.

Definition 8.6 A concrete design X^* is called Φ–optimal for a regression model $\boldsymbol{Y} = X\beta + \boldsymbol{e}$ with $E(\boldsymbol{e}) = 0_n$, $\text{var}(\boldsymbol{e}) = \sigma^2 I_n$, for fixed n and B, if

$$\min_{X \in \{L_n\}} \Phi\left[\left(X^T X\right)^{-1}\right] = \Phi\left[\left(X^{*T} X^*\right)^{-1}\right]. \tag{8.39}$$

Especially a Φ-optimal design with $M = (X^T X)^{-1}$ is called for

- $\Phi(M) = |M|$ D-optimal,
- $\Phi(M) = \text{tr}\,(M)$ A-optimal,
- $\Phi(M) = \underset{x \in B}{\text{Max}}\, x^T M x$ G-optimal,
- $\Phi(M) = \lambda_{\max}(M)$ E-optimal with $\lambda_{\max}$ as maximal eigenvalue of M,
- $\Phi(M) = c^T M c$ C-optimal with $c = \left(c_1, \ldots, c_p\right)^T$, $p = k + 1$.

The C-optimality is of importance if the variance of a linear contrast $c^T\beta$ of the parameter vector must be minimised. If we wish to make an extrapolation, the results of an experiment from the experimental region B in a region B^* (prediction), we replace in the G-optimality B by B^*.

From a theorem of Kiefer (1959), we know that discrete or continuous (but not always concrete!) designs are exactly D-optimal, if they are G-optimal. From the same theorem it follows that for special B (e.g. in R^2) the support of a discrete D- (G-) optimal design contains only points where $\text{var}(\hat{y})$ is maximal, that is, for which

$$\max_{x \in B} x^T \left(X^T X\right)^{-1} x = \frac{k+1}{n}.$$

We restrict ourselves to the G- or D-optimality for the simple linear regression. Jung (1973) gives a systematic investigation of the construction of concrete optimal designs. Some of his results for a special case of model equation (8.4) are given in the sequel. Concerning proofs see his paper.

At first we consider the case of Examples 8.1 for $k = 1$ $(p = 2)$, with

$$\left(X^T X\right)^{-1} = \frac{1}{|X^T X|} \begin{pmatrix} \sum_{i=1}^n x_i^2 & -\sum_{i=1}^n x_i \\ -\sum_{i=1}^n x_i & n \end{pmatrix} \tag{8.40}$$

and the experimental region $B = \langle a, b \rangle$. Then the design with $m = 2$, the support $\{a, \; b\}$ and the weights $p_1 = p_2 = 1/2$ is a discrete D-optimal design. For integer n this is of course also a concrete D- (and G-) optimal design, where half of the

y-values lie at the boundary of the interval. This fact is a special case of the following theorems.

Theorem 8.4 A concrete design with the matrix $X = (x_1, \dots, x_n)^T$ with $x_i^T = (1, x_i)$, $B = \{x_i | x_i \in [a, b]\}$ and $n \geq 2$ is then and only then G-optimal, if

a) For integer n, $\frac{n}{2}$ of the x_i are equal to a and b, respectively.

b) For odd n, $\frac{n-1}{2}$ x_i values are equal to a and $\frac{n-1}{2}$ x_i values are equal to b and one x_i equals $\frac{a+b}{2}$.

It can be shown that for odd n concrete D- and G-optimal designs are not identical.

Theorem 8.5 Under the assumptions of Theorem 8.4, X is then and only then D-optimal, if

a) For integer n, $\frac{n}{2}$ of the x_i are equal to a and b, respectively.

b) For odd n, $\frac{n-1}{2}$ of the x_i are a and b, respectively, and the remaining x_i is either a or b.

For the case $n = 5$ is for $a = -1$ and $b = 1$,

$$X_G^T = \begin{pmatrix} 1 & 1 & 1 & 1 & 1 \\ -1 & -1 & 0 & 1 & 1 \end{pmatrix}$$

a G-optimal design and

$$X_D^T = \begin{pmatrix} 1 & 1 & 1 & 1 & 1 \\ -1 & -1 & 1 & 1 & 1 \end{pmatrix}$$

a D-optimal design in $\langle -1, 1 \rangle$. We have $|X_G^T X_G| = 20$ and $|X_D^T X_D| = 24$.

8.3 Testing Hypotheses

The parameter vector $\beta = (\beta_0, \dots, \beta_k)^T$ lies in a $(k+1)$-dimensional vector space Ω. If $q < k+1$ of the β_j $(j = 0, \dots, k)$ equal zero (or some other fixed number) has the consequence that β lies in a $(k+1-q)$-dimensional subspace ω of Ω. In Theorem 4.7 it was shown that the components of β always can be renumbered in such a way that the first q components are the restricted ones. We then say that the conditions are given in canonical form (Definition 4.2). We restrict ourselves to the case $\beta_0 = \cdots = \beta_{q-1} = 0$, but remember that β_0 must not further be the constant of the regression equation.

The hypothesis H_0 that these conditions hold, that is,

$$\beta \in \omega \text{ or } \beta_0 = \cdots = \beta_{q-1} = 0 \tag{8.41}$$

shall be tested against the alternative $\beta \in \Omega \backslash \omega$.

Theorem 8.6 If Y in (8.4) is $N(X\beta, \sigma^2 I_n)$ distributed, the null hypothesis H_0, that (8.41) holds, against the alternative hypothesis that H_0 does not hold can be tested with the test statistic

$$\boldsymbol{F} = \frac{n-k-1}{q} \cdot \frac{\boldsymbol{Y}^T \left[X(X^T X)^{-1} X^T - X_1 \left(X_1^T X_1 \right)^{-1} X_1^T \right] \boldsymbol{Y}}{\boldsymbol{Y}^T \left(I_n - X(X^T X)^{-1} X^T \right) \boldsymbol{Y}}. \tag{8.42}$$

If H_0 holds, $\boldsymbol{F}$ is in (8.42) central F – distributed with q and $n - k - 1$ degrees of freedom. X_1 is the $[n \times (k + 1 - q)]$–matrix with the last $k + 1 - q$ columns of X.

Proof: The statement of this theorem follows from Example 4.3.

This result can be summarised by an ANOVA table (Table 8.2) putting $\hat{\boldsymbol{\beta}} = (X^T X)^{-1} X^T \boldsymbol{Y}$ and $\hat{\boldsymbol{\gamma}} = X^T \left(X_1^T X_1 \right)^{-1} X^T \boldsymbol{Y}$. This table is a special case of Table 4.1.

If $q = 1$, then $\boldsymbol{F} = \boldsymbol{t}^2$ and if the null hypothesis holds F is the square of a central t-distributed random variable with $n - k - 1$ degrees of freedom. In this case (8.42) becomes very simple.

Corollary 8.2 If $\boldsymbol{Y}$ in (8.4) is $N(X\beta, \sigma^2 I_n)$-distributed, the null hypothesis $H_0 : \beta_j = 0$ against $H_A : \beta_j \neq 0$ $(j = 0, \ldots, k)$ can be tested by the test statistic

$$\boldsymbol{t}_j = \frac{\boldsymbol{b}_j}{\boldsymbol{s}\sqrt{c_{jj}}}. \tag{8.43}$$

In (8.43) is the $\boldsymbol{b}_j = \hat{\boldsymbol{\beta}}_j$ the $(j + 1)$ – th component of the estimated parameter vector, $\boldsymbol{s}$ the square root of $\boldsymbol{s}^2$ in (8.6) and c_{jj} the $(j + 1)$ – th main diagonal

Table 8.2 Analysis of variance table for testing the hypothesis $H_0 : \beta_0 = \beta_1 = \cdots = \beta_{q-1} = 0$.

Source of variation	SS	df	MS	Test statistic
Total	$\boldsymbol{Y}^T\boldsymbol{Y}$	n		
H_0: $\beta_0 = \cdots = \beta_{q-1} = 0$	$\boldsymbol{Y}^T X\hat{\boldsymbol{\beta}} - \boldsymbol{Y}^T X_1 \hat{\boldsymbol{\gamma}} = \boldsymbol{Z}$	q	$\frac{\boldsymbol{Z}}{q}$	$\boldsymbol{F} = \frac{n-k-1}{q} \cdot \frac{\boldsymbol{Z}}{\boldsymbol{N}}$
Residual	$\boldsymbol{Y}^T\boldsymbol{Y} - \boldsymbol{Y}^T X\hat{\boldsymbol{\beta}} = \boldsymbol{N}$	$n - k - 1$	$\frac{\boldsymbol{N}}{n-k-1}$	
Regression	$\boldsymbol{Y}^T X_1 \hat{\boldsymbol{\gamma}}$	$k + 1 - q$		

element of $C = (X^T X)^{-1}$; $\boldsymbol{t}_j$ if $H_0 : \beta_j = 0$ holds is central t-distributed with $n - k - 1$ degrees of freedom.

Proof: We assume that the null hypothesis is in canonical form $H_0 : \beta_0 = 0$. If x_0 is the first column of X and X_1 the matrix of the k remaining columns of X, then $X = (x_0, X_1)$ and

$$X^T X = \begin{pmatrix} x_0^T x_0 & x_0^T X_1 \\ X_1^T x_0 & X_1^T X_1 \end{pmatrix}.$$

We decompose the symmetric inverse C in submatrices of the same type and obtain

$$C = \begin{pmatrix} C_{11} & C_{12} \\ C_{21} & C_{22} \end{pmatrix}.$$

(C_{11} is a scalar). Then we have

$$\left(X_1^T X_1\right)^{-1} = C_{22} - C_{21} C_{11}^{-1} C_{12},$$

and $\boldsymbol{Z} = \boldsymbol{Y}^T \left[X(X^T X)^{-1} X^T - X_1 \left(X_1^T X_1\right)^{-1} X_1^T\right] \boldsymbol{Y}$ in (8.42) becomes

$$\boldsymbol{Y}^T \left(x_0 C_{11} x_0^T + X_1 C_{21} x_0^T + x_0 C_{12} X_1^T 1 + X_1 C_{21} C_{11}^{-1} C_{12} X_1^T 1\right) \boldsymbol{Y}.$$

It follows now from (8.5)

$$\boldsymbol{b}_0 = \left(C_{11} \quad C_{12}\right) \begin{pmatrix} x_0^T \\ X_1^T \end{pmatrix} \boldsymbol{Y}$$

or

$$\boldsymbol{b}_0^2 = \boldsymbol{Y}^T \left(x_0 C_{11} C_{11} x_0^T + x_0 C_{11} C_{12} X_1^T + X_1 C_{21} C_{11} x_0^T + X_1 C_{21} C_{12} X_1^T\right) \boldsymbol{Y}.$$

Using $\boldsymbol{b}_0^2 = \left(\boldsymbol{b}_0^2\right)^T$ and $C_{12}^T = C_{21}$ shows that $\boldsymbol{Z}$ can be written as $C_{11}^{-1} \boldsymbol{b}_0^2$. C_{11} contains one element c_{00} only so that $c_{00}^{-1} = \frac{1}{c_{00}}$. Therefore (8.42) becomes

$$\boldsymbol{F} = \frac{\boldsymbol{b}_0^2}{c_{00} \boldsymbol{s}^2}$$

or going back to the original hypothesis

$$\boldsymbol{F} = \frac{\boldsymbol{b}_j^2}{c_{jj} \boldsymbol{s}^2},$$

and this completes the proof.

It is easy to see that under the hypothesis $\beta_j = \beta_j^*$ the test statistic

$$\boldsymbol{t} = \frac{\boldsymbol{b}_j - \beta_j^*}{\boldsymbol{s}\sqrt{c_{jj}}} \tag{8.44}$$

is with $n - k - 1$ degrees of freedom central t-distributed.

To test hypotheses of the form $H_0 : \beta = \beta^*$, for which ω contains only one point and has dimension 0 we need the following theorem.

Theorem 8.7 If Y in (8.4) is $N(X\beta, \sigma^2 I_n)$-distributed, the hypothesis $H_0 : \beta = \beta^*$ can be tested against the alternative hypothesis $\beta \neq \beta^*$ due to $X(X^TX)^{-1}X^T = X(X^TX)^{-1}X^TX(X^TX)^{-1}X^T$ and (8.5) with the test statistic

$$\boldsymbol{F} = \frac{n-k-1}{k+1}\frac{(\boldsymbol{Y}-X\beta^*)^T X(X^TX)^{-1}X^T(\boldsymbol{Y}-X\beta^*)}{\boldsymbol{Y}^T(I_n - X(X^TX)^{-1}X^T)\boldsymbol{Y}} = \frac{1}{\boldsymbol{s}^2(k+1)}(\boldsymbol{b}-\beta^*)^T(X^TX)(\boldsymbol{b}-\beta^*). \tag{8.45}$$

$\boldsymbol{F}$ in (8.45) is non-central F-distributed ($F(k + 1, n - k - 1, \lambda)$) with non-centrality parameter

$$\lambda = \frac{1}{\sigma^2}(\beta-\beta^*)^T(X^TX)(\beta-\beta^*).$$

Proof: Because for $\theta = \theta^*$

$$\max_{\theta\in\Omega} L(\theta,\sigma^2|Y) = \frac{n^{n/2}\, e^{-n/2}}{(2\pi)^{n/2}\|Y-\theta^*\|^n}$$

holds, Q in (4.18) becomes

$$\boldsymbol{Q} = \left[\frac{\|\boldsymbol{Y}-A\boldsymbol{Y}\|^2}{\|\boldsymbol{Y}-X\beta^*\|^2}\right]^{n/2}.$$

The orthogonal projection A of R^n on Ω is idempotent and therefore is $\theta^* = A\theta^*$ and $\|\boldsymbol{Y} - \theta^*\|^2 - \|\boldsymbol{Y} - A\boldsymbol{Y}\|^2 = (\boldsymbol{Y} - \theta^*)^TA(\boldsymbol{Y} - \theta^*)$.

The test statistic F in (4.19) has via Example 4.3 the form (8.45).

Example 8.4 We consider the simple linear regression of Examples 8.1 and use its symbols. We assume that the $\boldsymbol{e}_i$ in (8.8) are independent from each other and $N(0, \sigma^2)$-distributed. If σ^2 is known, the hypothesis $H_0 : \beta_0 = \beta_0^*$ can be tested with the test statistic

$$\boldsymbol{z}_0 = \frac{\boldsymbol{b}_0 - \beta_0^*}{\sigma_0} = \frac{\boldsymbol{b}_0 - \beta_0^*}{\sigma}\sqrt{\frac{n\sum(x_i - \bar{x})^2}{\sum x_i^2}} \tag{8.46}$$

and the hypothesis $H_0: \beta_1 = \beta_1^*$ with the test statistic

$$\boldsymbol{z}_1 = \frac{\boldsymbol{b}_1 - \beta_1^*}{\sigma_1} = \frac{\boldsymbol{b}_1 - \beta_1^*}{\sigma}\sqrt{\sum (x_i - \bar{x})^2}.$$

$\boldsymbol{z}_0$ and $\boldsymbol{z}_1$ due to Corollary 8.1 are $N(0, 1)$-distributed if the corresponding null hypothesis holds. If σ^2 is not known, it follows from (8.44) that if the hypothesis $\beta_0 = \beta_0^*$ holds, then

$$\boldsymbol{t} = \frac{\boldsymbol{b}_0 - \beta_0^*}{\boldsymbol{s}}\sqrt{\frac{n\sum (x_i - \bar{x})^2}{\sum x_i^2}} = \frac{\boldsymbol{b}_0 - \beta_0^*}{\boldsymbol{s}_0} \tag{8.47}$$

is central $t(n-2)$-distributed because in Example 8.1 it was shown that in the simple linear regression

$$C = \left(X^T X\right)^{-1} = \frac{1}{n\sum (x_i - \bar{x})^2}\begin{pmatrix} \sum x_i^2 & -\sum x_i \\ -\sum x_i & n \end{pmatrix},$$

that is, $c_{00} = \sum \frac{x_i^2}{n\sum (x_i - \bar{x})^2}$. Because $c_{11} = \frac{1}{\sum (x_i - \bar{x})^2}$, it follows from (8.44) that if $H_0: \beta_1 = \beta_1^*$ holds, the test statistic

$$\boldsymbol{t} = \frac{\boldsymbol{b}_1 - \beta_1^*}{\boldsymbol{s}}\sqrt{\sum (x_i - \bar{x})^2} \tag{8.48}$$

is $t(n-2)$-distributed.

The null hypothesis $H_0: \beta_0 = \beta_0^*$ (or $H_0: \beta_1 = \beta_1^*$) is rejected with a first kind risk α and the alternative hypothesis $\beta_0 > \beta_0^*$ (or $\beta_1 > \beta_1^*$) accepted, if for t in (8.47) (or in (8.48)) $t > t(n-2 \mid 1-\alpha)$. We accept the alternative hypothesis $\beta_0 < \beta_0^* \left(\beta_1 < \beta_1^*\right)$ if for t in (8.47) (or in (8.48)), $t < t(n-2 \mid \alpha)$. For a two-sided alternative hypothesis $H_A:\ \beta_0 \neq \beta_0^*$ (or $\beta_1 \neq \beta_1^*$), the null hypothesis is rejected with a first kind risk α, if with t in (8.47) (or in (8.48)) $|t| > t\left(n-2 \,\middle|\, 1-\frac{\alpha}{2}\right)$.

The hypothesis $\beta_1 = 0$ means that the random variable $\boldsymbol{y}$ is independent of the regressor.

To test the null hypothesis $\beta = \beta^*$, that is, $\beta_0 = \beta_0^*$, $\beta_1 = \beta_1^*$, we use the test statistic (8.45) of Theorem 8.5, and because

$$X^T X = \begin{pmatrix} n & \sum x \\ \sum x & \sum x^2 \end{pmatrix},$$

we receive the test statistic

$$F = \frac{n\left(b_0 - \beta_0^*\right)^2 + 2\sum x_i\left(b_0 - \beta_0^*\right)\left(b_1 - \beta_1^*\right) + \sum x_i^2\left(b_1 - \beta_1^*\right)^2}{2s^2}. \tag{8.49}$$

F is if the null hypothesis holds central $F(2,\ n-2)$-distributed. The null hypothesis $H_0: \beta_0 = \beta_0^*,\ \beta_1 = \beta_1^*$ is rejected with the first kind risk α if $F > F(2, n-2 \mid 1-\alpha)$, here $F(2,\ n-2 \mid 1-\alpha)$ is the $(1-\alpha)$–quantile of the F-distribution with 2 and $n-2$ degrees of freedom.

Usually the steps in calculating the F-test statistic are presented in an ANOVA table, as already discussed in Chapter 5. We decompose SS-total, that is, $SS_T = \sum_{i=1}^n \left(y_i - \beta_0^* - \beta_1^* x_i\right)^2$, that is, the sum of squared deviations of the observed values of the corresponding values of the regression function if the null hypothesis holds

$$E(y_i)^* = \beta_0^* + \beta_1^* x_i,$$

into two components. The first component contains that part of SS_T, originated by the deviations of the estimated regression line $\hat{y} = b_0 + b_1 x$ from the regression line given by the null hypothesis. This first component is called SS-regression($SS_{\text{Regr.}}$). The other component contains that part of SS_T, originated by the deviations of the observed values y_i of the values $\hat{y}_i$ from the estimated regression function; this component is called SS – residual (SS_{res}). Analogously the degrees of freedom are decomposed. The SS after transition to random variables are

$$\boldsymbol{SS}_T = \sum_{i=1}^n \left(\boldsymbol{y}_i - \beta_0^* - \beta_1^* x_i\right)^2,$$

$$\boldsymbol{SS}_{\text{Regr.}} = \sum_{i=1}^n \left(\boldsymbol{b}_0 + \boldsymbol{b}_1 x_i - \beta_0^* - \beta_1^* x_i\right)^2 = \sigma^2 \boldsymbol{Q}_1\left(\beta_0^*, \beta_1^*\right),$$

$$\boldsymbol{SS}_{\text{res}} = \sum_{i=1}^n \left(\boldsymbol{y}_i - \boldsymbol{b}_0 - \boldsymbol{b}_1 x_i\right)^2.$$

Because $\boldsymbol{SS}_{\text{Regr.}}$ is the numerator of $\boldsymbol{F}$ in (8.49) and $\boldsymbol{SS}_{res} = (n-2)\boldsymbol{s}^2$, the relation $\boldsymbol{SS}_T = \boldsymbol{SS}_{\text{Regr.}} + \boldsymbol{SS}_{res}$ follows. The ANOVA table is Table 8.3.

Often we are interested to test if two regression equations

$$\boldsymbol{y}_{1i} = \beta_{10} + \beta_{11} x_{1i} + \boldsymbol{e}_{1i},\quad \boldsymbol{y}_{2i} = \beta_{20} + \beta_{21} x_{2i} + \boldsymbol{e}_{2i},$$

with parameters from two groups of n_1 and n_2 observed pairs $(\boldsymbol{y}_{1i}, x_{1i})$ and $(\boldsymbol{y}_{2i}, x_{2i})$, respectively, have the same slope. The hypothesis $H_0: \beta_{11} = \beta_{21}$ has to be tested. For the model equations for $\boldsymbol{y}_{1i}$ and $\boldsymbol{y}_{2i}$, the side conditions and the additional assumptions of the model equation (8.8) may be fulfilled. From (8.9) estimates b_{i1} for $\beta_{i1} (i = 1,\ 2)$ are

Table 8.3 Analysis of variance table for testing the hypothesis $H_0: \beta_0 = \beta_0^*,\ \beta_1 = \beta_1^*$.

Source of variation	SS	df	MS	F
Total	SS_T	n		
Regression	$SS_{Regr.}$	2	$\frac{SS_{Regr.}}{2}$	$\frac{SS_{Regr.}}{2s^2}$
Residual	SS_{res}	$n-2$	s^2	

$$b_{i1} = \frac{n_i \sum_{j=1}^{n_i} x_{ij} y_{ij} - \sum_{j=1}^{n_j} x_{ij} \sum_{j=1}^{n_i} y_{ij}}{n_i \sum_{j=1}^{n_i} x_{ij}^2 - \left(\sum_{j=1}^{n_i} x_{ij}\right)^2} \quad (i = 1,\ 2)$$

and from (8.7) estimates b_{i0} for β_{i0} are given by

$$b_{i0} = \bar{y}_i - b_{i1}\bar{x}_i \quad (i = 1,\ 2)$$

with

$$\bar{y}_i = \frac{\sum_{j=1}^{n_i} y_{ij}}{n_i} \text{ and } \bar{x}_i = \frac{\sum_{j=1}^{n_i} x_{ij}}{n_i} \quad (i = 1,2).$$

We have shown that the $\boldsymbol{b}_{i1}$ are

$$N\left(\beta_{i1}, \frac{\sigma_i^2}{\sum_{j=1}^{n_i}\left(x_{ij} - \bar{x}_i\right)^2}\right)$$

distributed, if the $\boldsymbol{Y}_i$ are $N\left(X_i\beta_i, \sigma_i^2 I_{n_i}\right)$-distributed ($i = 1, 2$). Under the assumption that the two samples $(\boldsymbol{y}_{1i},\ x_{1i})$ and $(\boldsymbol{y}_{2i},\ x_{2i})$ are independent from each other, it follows that $\boldsymbol{b}_{11}$ and $\boldsymbol{b}_{21}$ are also independent from each other. We assume here the independency of both samples. Further we assume that these samples stem from populations with equal variances, that is, we have $\sigma_1^2 = \sigma_2^2 = \sigma^2$. Then the difference $\boldsymbol{b}_{11} - \boldsymbol{b}_{21}$ with expectation $\beta_{11} - \beta_{21}$ is normally distributed and

$$\boldsymbol{t} = \frac{\boldsymbol{b}_{11} - \boldsymbol{b}_{21} - (\beta_{11} - \beta_{21})}{\boldsymbol{s}_d}$$

is $t(n_1 + n_2 - 4)$-distributed, with $\boldsymbol{s}_d$ as the square root of

$$\boldsymbol{s}_d^2 = \frac{\sum_{j=1}^{n_1}\left(\boldsymbol{y}_{1j} - \boldsymbol{b}_{10} - \boldsymbol{b}_{11}x_{1j}\right)^2 + \sum_{j=1}^{n_2}\left(\boldsymbol{y}_{2j} - \boldsymbol{b}_{20} - \boldsymbol{b}_{21}x_{2j}\right)^2}{n_1 + n_2 - 4}\left[\frac{1}{\sum_{j=1}^{n_1}\left(x_{1j} - \bar{x}_1\right)^2} + \frac{1}{\sum_{j=1}^{n_2}\left(x_{2j} - \bar{x}_2\right)^2}\right].$$

If the null hypothesis $\beta_{11} = \beta_{21}$ holds,

$$\boldsymbol{t} = \frac{\boldsymbol{b}_{11} - \boldsymbol{b}_{21}}{\boldsymbol{s}_d}$$

is $t(n_1 + n_2 - 4)$-distributed, and $\boldsymbol{t} = \frac{\boldsymbol{b}_{11} - \boldsymbol{b}_{21}}{\boldsymbol{s}_d}$ can be used as a test statistic for the corresponding t–test of this null hypothesis against the alternative hypothesis $\beta_{11} \neq \beta_{21}$ (or one-sided alternatives).

But also in this case, we recommend not to trust in the equality of both variances, but to use the approximate test with the test statistic

$$\boldsymbol{t}^* = \frac{\boldsymbol{b}_{11} - \boldsymbol{b}_{21}}{\boldsymbol{s}_d^*} \tag{8.50}$$

with

$$\boldsymbol{s}_d^{*2} = \frac{\sum_{i=1}^{n_1} (\boldsymbol{y}_{1j} - \boldsymbol{b}_{10} - \boldsymbol{b}_{11} x_{1j})^2}{(n_1 - 2) \sum_{j=1}^{n_1} (x_{1j} - \bar{x}_1)^2} + \frac{\sum_{i=1}^{n_1} (\boldsymbol{y}_{2j} - \boldsymbol{b}_{20} - \boldsymbol{b}_{21} x_{2j})^2}{(n_2 - 2) \sum_{j=1}^{n_1} (x_{2j} - \bar{x}_2)^2} = \boldsymbol{s}_1^{*2} + \boldsymbol{s}_2^{*2}$$

and reject H_0, if $|\ t^*|$ exceeds the corresponding quantile of the central t-distribution with f degrees of freedom with

$$f = \frac{\left(s_1^{*2} + s_2^{*2}\right)^2}{\frac{s_1^{*4}}{(n_1 - 2)} + \frac{s_2^{*4}}{(n_2 - 2)}}$$

We give a simple example for $n = 5$.

Example 8.5 For the data of Example 8.3, we will test each of the hypotheses

$$\begin{aligned}
&H_0 : \beta_{10} = 30 && \text{against} \quad H_A : \beta_{10} \neq 30, \\
&H_0 : \beta_{11} = 0 && \text{against} \quad H_A : \beta_{11} < 0, \\
&H_0 : \beta_1 = \begin{pmatrix} \beta_{10} \\ \beta_{11} \end{pmatrix} = \begin{pmatrix} 30 \\ 0 \end{pmatrix} && \text{against} \quad H_A : \beta_1 \neq \begin{pmatrix} 30 \\ 0 \end{pmatrix}, \\
&H_0 : \beta_{11} = \beta_{21} && \text{against} \quad H_A : \beta_{11} \neq \beta_{21}
\end{aligned}$$

with a first kind risk $\alpha = 0.05$. The one-sided alternative $H_A:\ \beta_{11} < 0$ stems from the fact that the carotene content cannot increase during storage.

Table 8.4 is the ANOVA table for this example. $\boldsymbol{MS}_{res} = 0.922\,03$ is the estimate s_1^2 of σ^2.

The test statistic for the hypothesis $\beta_{10} = 30$ becomes via (8.47)

$$t = \frac{31.215 - 30}{0.7059} = 1.72 < t(3|0.975).$$

Table 8.4 Analysis of variance table for testing the hypothesis $H_0:\beta_{10}=30,\; b_{11}=0$ in Example 8.6 for $i=1$.

Source of variation	*SS*	*df*	*MS*	*F*
Total	393.5733	5		
Regression	390.8072	2	195.50	211.9
Residual	2.7661	3	0.92203	

For a first kind risk 0.05, the hypothesis $\beta_{10}=30$ is not rejected.

From (8.48) we obtain the test statistic for the hypothesis $\beta_{11}=0$

$$t=-\frac{0.05455}{0.00394}=-13.85<t(3\mid 0.05),$$

and this hypothesis is rejected with a first kind risk 0.05. This result was already given in the SPSS output of Figure 8.2. In the fifth column we find the value of the test statistic, and the sig-value in column six is below 0.05, and this means rejection.

The hypothesis, as we see from the F–test statistic in Table 8.4, is also rejected.

Finally we test the hypothesis that both (theoretical) regression lines have the same slope, that is, the hypothesis $\beta_{11}=\beta_{21}$. For this we use the test statistic given in (8.50) and obtain with $f=6.24$

$$t=\frac{-0.05455+0.08098}{\sqrt{0.004^2+0.011^2}}=2.17>t(6.24\mid 0.975),$$

and the null hypothesis of parallelism of both regression lines (i.e. that the loss of carotene is the same for both kinds of storage), is rejected by this approximate test.

The test of the hypothesis that some components of β in (8.4) equal zero is often used to find out whether some of the regressors, that is, some columns of X in (8.4) can be dropped.

This method can be applied to test the degree of a polynomial.

From (8.34) it follows that β_k equals zero if $\alpha_k=0$. By this the hypothesis $H_0:\beta_k=0$ is identical with $H_0:\alpha_k=0$ and can be tested from the corollary of Theorem 8.6.

Corollary 8.3 (of Theorem 8.6): Let

$$Y=X\alpha+\boldsymbol{e}$$

be a quasilinear polynomial regression model of degree k, where X has the form as shown in the proof of Theorem 8.3 and α depends on β in (8.33) by (8.34). Let Y be $N(X\alpha,\sigma^2 I_n)$-distributed. The hypothesis $H_0:\alpha_k=\beta_k=0$ can be tested with the test statistic

$$F = \frac{\boldsymbol{a}_k^2 \sum_{i=1}^n P_{ki}^2 (n-k-1)}{\sum_{i=1}^n \boldsymbol{y}_i^2 - \sum_{j=0}^k \boldsymbol{a}_j^2 \sum_{i=1}^n P_{ji}^2} = \frac{(n-k-1)\dfrac{\left(\sum_{i=1}^n \boldsymbol{y}_i P_{ki}\right)^2}{\sum_{i=1}^n P_{ki}^2}}{\sum_{i=1}^n \boldsymbol{y}_i^2 - \sum_{j=1}^k \dfrac{\left(\sum_{i=1}^n \boldsymbol{y}_i P_{ji}\right)^2}{\sum_{i=1}^n P_{ji}^2}} \tag{8.51}$$

if the $\boldsymbol{a}_i$ are the components of the MLE $\hat{\boldsymbol{a}}$ in (8.35).

Proof: X has the form given in the proof of Theorem 8.3; $X^T X$ is a diagonal matrix. Because $\boldsymbol{Y}^T X(X^T X)^{-1} X^T \boldsymbol{Y} = \boldsymbol{Y}^T X(X^T X)^{-1} X^T X(X^T X)^{-1} X^T \boldsymbol{Y}$ and from (8.35) Equation (8.42) with $q = 1$ becomes

$$\boldsymbol{F} = (n-k-1)\frac{\boldsymbol{a}^T X^T X \boldsymbol{a} - \boldsymbol{c}^T X_1^T X_1 \boldsymbol{c}}{\boldsymbol{Y}^T \boldsymbol{Y} - \boldsymbol{a}^T (X^T X)\boldsymbol{a}}. \tag{8.52}$$

Here X_1 is the matrix, which arises, if the last column in X is dropped and $\boldsymbol{c}$ is the LS estimator if the null hypothesis $\alpha_k = 0$ holds. Further

$$\boldsymbol{a}^T X^T X \boldsymbol{a} = \sum_{j=0}^{k} \boldsymbol{a}_j^2 \sum_{i=1}^{n} P_{ji}^2. \tag{8.53}$$

and

$$\boldsymbol{c}^T X_1^T X_1 \boldsymbol{c} = \sum_{j=0}^{k-1} \boldsymbol{a}_j^2 \sum_{i=1}^{n} P_{ji}^2, \tag{8.54}$$

and this completes the proof.

8.4 Confidence Regions

When we know the distributions of the estimators of the parameters, confidence regions for these parameters can be constructed. In this section we will construct confidence regions (intervals) for the components β_i of β, the variance σ^2, the expectations $E(\boldsymbol{y}_i)$ and the vector $\beta \in \Omega$. Again we make the assumption that $\boldsymbol{Y}$ is $N(X\beta, \sigma^2 I_n)$-distributed and the model equation (8.4) holds. From (8.44) it follows that

$$P\left\{ t\left(n-k-1 \,\middle|\, \frac{\alpha}{2}\right) \le \frac{\boldsymbol{b}_j - \beta_j}{\boldsymbol{s}\sqrt{c_{jj}}} \le t\left(n-k-1 \,\middle|\, 1-\frac{\alpha}{2}\right) \right\} = 1-\alpha \tag{8.55}$$

and due to the symmetry of the t-distribution is

$$\left[\boldsymbol{b}_j - t\left(n-k-1 \,\middle|\, 1-\frac{\alpha}{2}\right) \boldsymbol{s}\sqrt{c_{jj}},\ \boldsymbol{b}_j + t\left(n-k-1 \,\middle|\, 1-\frac{\alpha}{2}\right) \boldsymbol{s}\sqrt{c_{jj}}\right] \tag{8.56}$$

a confidence interval of the component β_j with a confidence coefficient $1-\alpha$. In (8.56) is c_{jj} the j-th main diagonal element of $(X^T X)^{-1}$ and $\boldsymbol{s}=\sqrt{\boldsymbol{s}^2}$ the square root of the residual variance in (8.6). Due to the assumptions $\frac{\boldsymbol{s}^2(n-k-1)}{\sigma^2}$ is $CS(n-k-1)$-distributed. If χ^2 is a $CS(n-k-1)$-distributed random variable and $\chi^2(n-k-1 \mid \alpha_1)$ and $\chi^2(n-k-1 \mid 1-\alpha_2)$ are chosen in such a way that with $\alpha_1+\alpha_2=\alpha$,

$$P\left(\chi^2 < \chi^2(n-k-1|\alpha_1)\right) = \alpha_1$$

and

$$P\left(\chi^2 > \chi^2(n-k-1|1-\alpha_2)\right) = \alpha_2,$$

then

$$P\left\{\chi^2(n-k-1|\alpha_1) \le \frac{\boldsymbol{s}^2(n-k-1)}{\sigma^2} \le \chi^2(n-k-1|1-\alpha_2)\right\} = 1-\alpha,$$

and a confidence interval for σ^2 with a confidence coefficient $1-\alpha$ is given by

$$\left[\frac{\boldsymbol{s}^2(n-k-1)}{\chi^2(n-k-1|1-\alpha_2)}, \frac{\boldsymbol{s}^2(n-k-1)}{\chi^2(n-k-1|\alpha_1)}\right]. \tag{8.57}$$

If we choose a vector $x=(x_0, \ldots, x_k)^T$ of the values of the regressor so that

$$\min_i x_{ij} \le x_j \le \max_i x_{ij}$$

holds for $j=0, \ldots, k$, then by the Gauss–Markov theorem (Theorem 4.3) an estimator $\hat{\boldsymbol{y}}$ of $y=x^T\beta$ in the regression function is given by

$$\boldsymbol{y} = x^T \boldsymbol{b}$$

with $\boldsymbol{b}$ in (8.5).

Now $\boldsymbol{b}$ is $N(\beta, \sigma^2(X^TX)^{-1})$-distributed (independent of $\boldsymbol{s}^2$) so that $x^T\boldsymbol{b}$ is $N[x^T\beta, x^T(X^TX)^{-1}x\sigma^2]$-distributed. From this it follows that

$$\boldsymbol{z} = \frac{x^T(\boldsymbol{b}-\beta)}{\sigma\sqrt{x^T(X^TX)^{-1}x}}$$

is $N(0,1)$-distributed, and because $\frac{\boldsymbol{s}^2(n-k-1)}{\sigma^2}$ is independent of $\boldsymbol{z}$ as $CS(n-k-1)$-distributed the test statistic

$$\boldsymbol{t} = \frac{x^T(\boldsymbol{b}-\beta)}{\boldsymbol{s}\sqrt{x^T(X^TX)^{-1}x}}$$

is $t(n-k-1)$-distributed. From this it follows that

$$\left[\hat{y}-t\left(n-k-1\middle|1-\frac{\alpha}{2}\right)s\sqrt{x^T(X'X)^{-1}x},\hat{y}+t\left(n-k-1\middle|1-\frac{\alpha}{2}\right)s\sqrt{x^T(X'X)^{-1}x}\right] \tag{8.58}$$

is a confidence interval for $y = x^T\beta$ with a confidence coefficient $1-\alpha$. The confidence intervals (8.56) give for each j an interval that covers β_j with probability $1-\alpha$. From these confidence intervals, no conclusions can be drawn, in which region the parameter vector β rests with a given probability.

A region in Ω, covering β with the probability $1-\alpha$, is called a simultaneous confidence region for $\beta_0, \ldots, \beta_k$. With the test statistic $\boldsymbol{F}$ in (8.45) for the test of $\beta=\beta^*$, we construct this a simultaneous confidence region. From (8.45) it follows

$$P\left\{\frac{1}{\boldsymbol{s}^2(k+1)}(\boldsymbol{b}-\beta)^T X^T X(\boldsymbol{b}-\beta) \le F(k+1,n-k-1|1-\alpha)\right\} = 1-\alpha,$$

so that the interior and the boundary of the ellipsoid

$$(\boldsymbol{b}-\beta)^T X^T X(\boldsymbol{b}-\beta) = (k+1)\boldsymbol{s}^2 F(k+1,n-k-1|1-\alpha)$$

is our confidence region.

Example 8.6 For Example 8.3 ($i=1$) confidence regions for β_0,β_1,σ^2, $\hat{y}=\beta_0+\beta_1 x$ and $\beta^T=(\beta_0,\beta_1)$ each with a confidence coefficient 0.95 shall be found. By SPSS after activating 'confidence interval' after pressing the button 'Statistics', we receive Table 8.5 with the confidence intervals for β_{10} and β_{11}.

From (8.57) [0.26;12.82] is a confidence interval for σ^2 with $\alpha_1=\alpha_2=\frac{\alpha}{2}$. But this is because of the skewness of the χ^2–distribution, not the decomposition of α into two components, leading to the smallest expected width of the confidence interval.

To calculate a 95%-confidence interval for $E(\hat{y})$ in (8.58), we need for some

Table 8.5 SPSS output with confidence intervals (analogue to Figure 8.2).

	Unstandardized coefficients				95.0% confidence interval for B	
Model	B	Std. error	*t*	Sig.	Lower bound	Upper bound
1 (Constant)	31.216	0.706	44.223	0.000	28.969	33.462
time	−0.055	0.004	−13.848	0.001	−0.067	−0.042

Source: Reproduced with permission of IBM.

$x_0 \in B$ the values of

Table 8.6 95%-confidence bounds for $E(\hat{\boldsymbol{y}}_i)$ in Example 8.6.

x_j	$\hat{y}_j$	K_j	Confidence bounds	
			Lower	Upper
1	31.16	0.73184	28.92	33.40
60	27.94	0.56012	26.23	29.65
124	24.45	0.45340	23.06	25.84
223	19.05	0.55668	17.35	20.75
303	14.69	0.79701	12.25	17.12

$$K_0 = \sqrt{\frac{\sum x_j^2 - 2x_0 \sum x_j + n x_0^2}{n\sum (x_j - \bar{x})^2}} = \sqrt{x^T (X^T X)^{-1} x},$$

given in Table 8.6 together with the confidence bounds for $E(\hat{\boldsymbol{y}}_i)$. Figure 8.6 shows the estimated regression line for $i = 1$ and the confidence belt, obtained by mapping the confidence bounds for $E(\hat{\boldsymbol{y}})$ in Table 8.5 and splicing them.

A confidence region for $\beta = \begin{pmatrix} \beta_0 \\ \beta_1 \end{pmatrix}$ is an ellipse, given by

$$n(b_0 - \beta_0)^2 + 2\Sigma x_i (b_0 - \beta_0)(b_1 - \beta_1) + \Sigma x_i^2 (b_1 - \beta_1)^2 = 2s^2 F(2, n-2|1-\alpha).$$

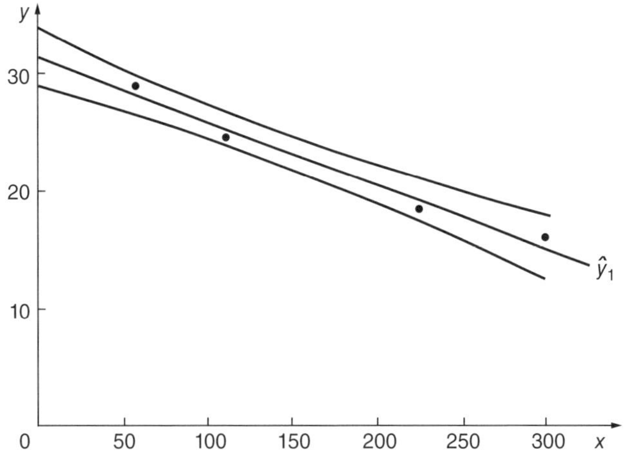

Figure 8.6 Estimated regression line and confidence belt of Example 8.6.

Using the data of Example 8.3 gives

$$5(31.25-\beta_0)^2-1422(31.215-\beta_0)(0.05455+\beta_1)+160515(0.05455+\beta_1)^2=1.84406\cdot 9.522$$

8.5 Models with Random Regressors

For random regressors we only consider the linear case.

8.5.1 Analysis

Definition 8.7 If $\boldsymbol{x}^T=(\boldsymbol{x}_1,\ldots\boldsymbol{x}_{k+1})$ is a $(k+1)$-dimensional normally distributed random vector and is $\boldsymbol{X}=(\boldsymbol{x}_{ij})$ $(i=1,\ldots,k+1;j=1,\ldots,n)$ a random sample of n such vectors, distributed as $\boldsymbol{x}$ then equation

$$\boldsymbol{y}_j=\boldsymbol{x}_{k+1,j}=\sum_{i=0}^{k}\beta_i\boldsymbol{x}_{ij}+\boldsymbol{e}_j,\quad x_{0j}\equiv 1, \tag{8.59}$$

with the additional assumption that $\boldsymbol{e}_j$ are independent of each other, $N(0,\sigma^2)$-distributed and are independent of the $\boldsymbol{x}_{ij}$, is called a model II of the (multiple) linear regression. Definition 8.4 can be generalised by neglecting the assumption that $\boldsymbol{Y}$ is normally distributed. Nevertheless, for tests and confidence estimation, the assumption is necessary. Correlation coefficients are always defined, as long as (8.59) holds, and the distribution has finite second moments.

To estimate the parameter of (8.59), we use the same formulae as for model I.

An estimator for $\rho_{x_iy}=\sigma_{x_iy}/\sigma_{x_i}\sigma_y$ by (5.33), we obtain by replacing $\sigma_{x_iy},\sigma^2_{x_i}$ and σ^2_y by the unbiased estimators $\boldsymbol{s}_{x_iy},\boldsymbol{s}^2_{x_i}$ and $\boldsymbol{s}^2_y$ of the covariances and variances, respectively. Then we get the (not unbiased) estimator

$$\boldsymbol{r}_{x_iy}=\frac{\boldsymbol{s}_{x_iy}}{\boldsymbol{s}_{x_i}\boldsymbol{s}_y}=\frac{\boldsymbol{SP}_{x_iy}}{\sqrt{\boldsymbol{SS}_{x_i}\boldsymbol{SS}_y}} \tag{8.60}$$

of the correlation coefficient.

At first the special case $k=2$ is considered. The random variable $(\boldsymbol{x}_1,\boldsymbol{x}_2,\boldsymbol{x}_3)$ shall be three-dimensional normally distributed; it can be shown that the three conditional two-dimensional distributions $f_k(x_i,x_j|x_k)$ $(i\neq j\neq k;\ i,j,k=1,2,3)$ are two-dimensional normal distributions with correlation coefficients

$$\rho_{ij.k}=\frac{\rho_{ij}-\rho_{ik}\rho_{jk}}{\sqrt{(1-\rho^2_{ik})(1-\rho^2_{jk})}}\quad (i\neq j\neq k;i,j,k=1,2,3) \tag{8.61}$$

Here ρ_{ij},ρ_{ik} and ρ_{jk} are the correlation coefficients of the three two-dimensional (normal) marginal distributions of $(\boldsymbol{x}_i,\boldsymbol{x}_j,\boldsymbol{x}_k)$. It can easily be shown that these marginal distributions are two-dimensional normal distributions (Exercise 8.1).

The correlation coefficient (8.61) of the conditional two-dimensional normal distribution of $(\boldsymbol{x}_i, \boldsymbol{x}_j)$ for x_k given is called partial correlation coefficient between $\boldsymbol{x}_i$ and $\boldsymbol{x}_j$ after the cut-off of $\boldsymbol{x}_k$.

The name partial correlation coefficient stems from applications and is of general use even if the name conditional correlation coefficient seems to be better.

It follows from (8.61) that the value of $\boldsymbol{x}_k$ has no influence on the correlation coefficient of the conditional distribution of $(\boldsymbol{x}_i, \boldsymbol{x}_j)$ and therefore $\rho_{ij\cdot k}$ is independent of x_k. We say that $\rho_{ij\cdot k}$ is a measure of the relationship between $\boldsymbol{x}_i$ and $\boldsymbol{x}_j$ after the cut-off of the influence of x_k or after the elimination of x_k. This interpretation of $\rho_{ij\cdot k}$ can be illustrated as follows. Starting with the marginal distributions of $(\boldsymbol{x}_i, \boldsymbol{x}_k)$ and $(\boldsymbol{x}_j, \boldsymbol{x}_k)$ because these marginal distributions are normal distributions as conditional random expectations (in dependency on $\boldsymbol{x}_k$) of these marginal distributions, we receive

$$E(\boldsymbol{x}_i|\boldsymbol{x}_k) = \mu_i + \beta_{ik}(\boldsymbol{x}_k - \mu_k) \tag{8.62}$$

and

$$E(\boldsymbol{x}_j|\boldsymbol{x}_k) = \mu_j + \beta_{jk}(\boldsymbol{x}_k - \mu_k), \tag{8.63}$$

where $\mu_l = E(\boldsymbol{x}_l)$ is the expectation of the one-dimensional marginal distribution of $\boldsymbol{x}_l$ and $\beta_{i\,k}$ and $\beta_{j\,k}$ are the regression coefficients of the marginal distributions. Calculating the differences,

$$\boldsymbol{d}_i = \boldsymbol{d}_{i\cdot k} = \boldsymbol{x}_i - \mu_i - \beta_{ik}(\boldsymbol{x}_k - \mu_k)$$

and

$$\boldsymbol{d}_j = \boldsymbol{d}_{j\cdot k} = \boldsymbol{x}_j - \mu_j - \beta_{j\,k}(\boldsymbol{x}_k - \mu_k),$$

leads to a normally distributed two-dimensional random variable $(\boldsymbol{d}_{i\cdot k}, \boldsymbol{d}_{j\cdot k})$. It is to be shown that the correlation coefficient ρ_{d_i,d_j} is given by (8.61).

We have

$$\rho_{d_i,d_j} = \frac{\operatorname{cov}(\boldsymbol{d}_i, \boldsymbol{d}_j)}{\sqrt{\operatorname{var}(\boldsymbol{d}_i)\operatorname{var}(\boldsymbol{d}_j)}}. \tag{8.64}$$

Because $\operatorname{cov}(\boldsymbol{d}_i, \boldsymbol{d}_j) = E(\boldsymbol{d}_i \cdot \boldsymbol{d}_j) - E(\boldsymbol{d}_i)E(\boldsymbol{d}_j)$, we obtain

$$\operatorname{cov}(\boldsymbol{d}_i, \boldsymbol{d}_j) = E(\boldsymbol{x}_i, \boldsymbol{x}_j) - \mu_i\mu_j - \beta_{ik}\sigma_{jk} - \beta_{jk}\sigma_{ik} + \beta_{ik}\beta_{jk}\sigma_k^2$$

and

$$\operatorname{cov}(\boldsymbol{d}_i, \boldsymbol{d}_j) = \sigma_{ij} - \frac{\sigma_{ik}\sigma_{jk}}{\sigma_k^2}.$$

Further

$$\sigma_{i\cdot k}^2 = \operatorname{var}(\boldsymbol{d}_i) = \sigma_i^2 + \rho_{ik}^2\sigma_i^2 - 2\rho_{ik}^2\sigma_i^2 = \sigma_i^2\left(1-\rho_{ik}^2\right)$$

and analogous

$$\sigma_{j\cdot k}^2 = \operatorname{var}(\boldsymbol{d}_j) = \sigma_j^2\left(1-\rho_{ik}^2\right),$$

so that $\rho_{d_i,d_i} = \rho_{ji\cdot k}$.

Between the regression coefficients of $\boldsymbol{x}$ on $\boldsymbol{y}$ and the correlation coefficients of the normally distributed random variables $(\boldsymbol{x}, \boldsymbol{y})$ one has

$$\beta_{xy} = \rho\frac{\sigma_x}{\sigma_y} \tag{8.65}$$

In the three-dimensional case it can be shown that the relation

$$\beta_j^{(i)} = \rho_{ij\cdot k}\frac{\sigma_{i\cdot k}}{\sigma_{j\cdot k}} \quad (i \neq j \neq k; i,j,k = 1,2,3)$$

holds, where the multiple (partial) regression coefficients $\beta_j^{(i)}$ are the coefficients in case $k = 2$. The $\beta_j^{(i)}$ can be interpreted as regression coefficients between $\boldsymbol{d}_{i\cdot k}$ and $\boldsymbol{d}_{j\cdot k}$ and are therefore often called partial regression coefficients. The $\beta_j^{(i)}$ show, by how many units x_i changes, if x_j increases by one unit, while all other regressors remain unchanged. For the four-dimensional normally distributed random variable $(\boldsymbol{x}_1, \boldsymbol{x}_2, \boldsymbol{x}_3, \boldsymbol{x}_4)$, we can define a partial correlation coefficient between two components for fixed values of the both residual components.

We call the expression

$$\rho_{ij\cdot lk} = \rho_{ij\cdot kl} = \frac{\rho_{ij\cdot k} - \rho_{il\cdot k}\rho_{jl\cdot k}}{\sqrt{\left(1-\rho_{i\cdot lk}^2\right)\left(1-\rho_{j\cdot lk}^2\right)}} \quad (i \neq j \neq k \neq l; i,j,k,l = 1,2,3,4), \tag{8.66}$$

defined for the four-dimensional normally distributed random variable $(\boldsymbol{x}_1, \boldsymbol{x}_2, \boldsymbol{x}_3, \boldsymbol{x}_4)$ a partial correlation coefficient (of second order) between $\boldsymbol{x}_i$ and $\boldsymbol{x}_j$ after the cut-off of $\boldsymbol{x}_k$ and $\boldsymbol{x}_l$.

Analogous partial correlation coefficients of higher order can be defined.

We obtain estimators $\boldsymbol{r}_{ij\cdot k}$ and $\boldsymbol{r}_{ij\cdot kl}$ for partial correlation coefficients by replacing the simple correlation coefficients in (8.61) and (8.66) by their estimators. For instance, we get

$$\boldsymbol{r}_{ij\cdot k} = \frac{\boldsymbol{r}_{ij} - \boldsymbol{r}_{ik}\boldsymbol{r}_{jk}}{\sqrt{\left(1-\boldsymbol{r}_{ik}^2\right)\left(1-\boldsymbol{r}_{jk}^2\right)}}. \tag{8.67}$$

Without proof we give the following theorem.

Theorem 8.8 If $(\boldsymbol{x}_1, \ldots, \boldsymbol{x}_k)$ is k-dimensional normally distributed and for some partial correlation coefficients of s-th order $(s = k - 2)$ the hypothesis $\rho_{ij\cdot u_1\ldots u_s} = 0$ ($u_1, \ldots, u_s$ are $s = k - 2$ different numbers from $1, \ldots, k$, different from i and j) is true, then

$$\boldsymbol{t} = \frac{\boldsymbol{r}_{ij\cdot u_1\ldots u_s}\sqrt{n-k}}{\sqrt{1-\boldsymbol{r}^2_{ij\cdot u_1\ldots u_s}}} \tag{8.68}$$

is $t(n-k)$-distributed, if n values of the k-dimensional variables are observed. Especially for $k = 3$ $(s = 1)$ under $H_0: \rho_{ij\cdot k} = 0$

$$\boldsymbol{t} = \frac{\boldsymbol{r}_{ij\cdot k}\sqrt{n-3}}{\sqrt{1-\boldsymbol{r}^2_{ij\cdot k}}}$$

is $t(n-3)$-distributed and for $k = 4$ under $H_0: \rho_{\{ij\}\cdot kl} = 0$

$$\boldsymbol{t} = \frac{\boldsymbol{r}_{ij\cdot kl}\sqrt{n-4}}{\sqrt{1-\boldsymbol{r}^2_{ij\cdot kl}}}$$

is $t(n-4)$-distributed.

By Theorem 8.8 for $k = 2$, the hypothesis $\rho = 0$ can be tested with the test statistic (8.68). For a two-sided alternative $(\rho \neq 0)$, the null hypothesis is rejected, if $|t| > t\left(n-2\middle|1-\frac{\alpha}{2}\right)$.

To test the hypothesis $H_0: \rho = \rho^* \neq 0$, we replace $\boldsymbol{r}$ by the Fisher transform

$$\boldsymbol{z} = \frac{1}{2}\ln\frac{1+r}{1-r} \tag{8.69}$$

that is approximately normally distributed with expectation

$$E(\boldsymbol{z}) \approx \frac{1}{2}\ln\frac{1+\rho}{1-\rho} + \frac{\rho}{2(n-1)}$$

and variance $\operatorname{var}(\boldsymbol{z}) \approx \frac{1}{n-3}$. If the hypothesis $\rho=\rho^*$ is valid,

$$\boldsymbol{u} = \left[\boldsymbol{z} - \ln\frac{1+\rho^*}{1-\rho^*} - \frac{\rho^*}{2(n-1)}\right]\sqrt{n-3}$$

is approximately $N(0,\ 1)$-distributed. For large n in place of $\boldsymbol{u}$ also

$$\boldsymbol{u}^* = \left[\boldsymbol{z} - \ln\frac{1+\rho^*}{1-\rho^*}\right]\sqrt{n-3}$$

can be used. An approximate $(1-\alpha)$. 100% confidence interval for ρ is

$$\left[\tanh\left(\boldsymbol{z}-\frac{z_{1-\frac{\alpha}{2}}}{\sqrt{n-3}}\right),\ \tanh\left(\boldsymbol{z}+\frac{z_{1-\frac{\alpha}{2}}}{\sqrt{n-3}}\right)\right],$$

with the $\left(1-\frac{\alpha}{2}\right)$- quantile $z_{1-\frac{\alpha}{2}}$ of the standard normal distribution $\left[P\left(\mathbf{z}>z_{1-\frac{\alpha}{2}}\right)=\frac{\alpha}{2}\right]$.

A sequential test of the hypothesis $\rho = 0$ was already given in Chapter 3 using $\boldsymbol{z}=\frac{1}{2}\ln\frac{1+r}{1-r}$ in place of $\boldsymbol{z}=\ln\frac{1+r}{1-r}$.

To interpret the value of ρ (and also of r), we again consider the regression function $f(\boldsymbol{x}) = E(\boldsymbol{y}|\boldsymbol{x}) = \alpha_0 + \alpha_1\boldsymbol{x}$. ρ^2 can now be explained as a measure of the proportion of the variance of $\boldsymbol{y}$, explainable by the regression on $\boldsymbol{x}$. The conditional variance of $\boldsymbol{y}$ is

$$\text{var}(\boldsymbol{y}|x) = \sigma_y^2\left(1-\rho^2\right)$$

and

$$\frac{\text{var}(\boldsymbol{y}|x)}{\sigma_y^2} = 1-\rho^2$$

is the proportion of the variance of $\boldsymbol{y}$, not explainable by the regression on x, and by this the statement above follows. We call $\rho^2 = B$ measure of determination.

To construct confidence intervals for β_0 and β_1 or to test hypotheses about these parameters seems to be difficult, but the methods for model I can also be applied for model II. We demonstrate this as example of the confidence interval for β_0. The argumentation for confidence intervals for other parameters and for the statistical tests is analogue.

The probability statement

$$P\left[\boldsymbol{b}_0 - t\left(n-2\middle|1-\frac{\alpha}{2}\right)\boldsymbol{s}_0 \le \beta_0 \le \boldsymbol{b}_0 + t\left(n-2\middle|1-\frac{\alpha}{2}\right)\boldsymbol{s}_0\right] = 1-\alpha,$$

leading to the confidence interval (8.56) for $j = 0$ is true, if for fixed values $x_1, \ldots, x_n$ samples of y-values are selected repeatedly. Using the frequency interpretation, β_0 is covered in about $(1-\alpha)\cdot 100\%$ of the cases by the interval (8.56). This statement is valid for each arbitrary n-tuple $x_{i1}, \ldots, x_{in}$, also for an n-tuple $x_{i1}, \ldots, x_{in}$, randomly selected from the distribution because (8.56) is independent of $x_1, \ldots, x_n$, if the conditional distribution of the $\boldsymbol{y}_j$ is normal. But this is the case, because $(\boldsymbol{y}, \boldsymbol{x}_1, \ldots, \boldsymbol{x}_k)$ was assumed to be normally distributed. By this the construction of confidence intervals and testing of hypotheses can be done by the methods and formulae given above. But the expected width of the confidence intervals and the power function of the tests differ for both models.

That $\left[\boldsymbol{b}_i - t\left(n-2\middle|1-\frac{\alpha}{2}\right)\boldsymbol{s}_i, \boldsymbol{b}_i + t\left(n-2\middle|1-\frac{\alpha}{2}\right)\boldsymbol{s}_i\right]$ is really a confidence interval with a confidence coefficient $1 - \alpha$ also for model II can of course be proven exactly, using a theorem of Bartlett (1933) by which

$$\boldsymbol{t}_i = \frac{\boldsymbol{s}_x\sqrt{n-2}}{\sqrt{\boldsymbol{s}_y^2 - \boldsymbol{b}_i^2\,\boldsymbol{s}_x^2,}}(\boldsymbol{b}_i - \beta_i)$$

is $t(n-2)$-distributed.

8.5.2 Experimental Designs

The experimental design for model II of the regression analysis differs fundamentally from that of model I. Because $\boldsymbol{x}$ in model II is a random variable, the problem of the optimal choice of x does not occur. Experimental design in model II means only the optimal choice of n in dependency of given precision requirements. A systematic description about that is given in Rasch et al. (2008). We repeat this in the following.

At first we restrict in (8.59) on $k = 1$ and consider the more general model of the regression within of $a \geq 1$ groups with the same slope β_1:

$$\boldsymbol{y}_{hj} = \beta_{h0} + \beta_1 x_{hj} + \boldsymbol{e}_{hj} \quad \left(h = 1,\ldots,a; j = 1,\ldots,n_h \geq 2\right). \tag{8.70}$$

We estimate β_1 for $a > 1$ not by (8.9), but by

$$\boldsymbol{b}_{I1} = \frac{\sum_{h=1}^{a}\mathbf{SP}_{x,y}^{(h)}}{\sum_{h=1}^{a}\mathbf{SS}^{(h)}} = \frac{\mathbf{SP}_{Ixy}}{\mathbf{SS}_{Ix}}, \tag{8.71}$$

with $\mathbf{SP}_{xy}^{(h)}$ and $\mathbf{SS}_x^{(h)}$ for each of the a groups as defined in Example 8.1.

If we look for a minimal $n = \sum_{h=1}^{a} n_h$ so that $V(\boldsymbol{b}_{11}) \leq C$, we find in Bock (1998)

$$n - a - 2 = \left\lceil \frac{\sigma^2}{C\,\sigma_k^2} \right\rceil.$$

If in (8.59) for $k = 1$ for the expectation $E(\boldsymbol{y}|x) = \beta_0 + \beta_1 x$ a $(1 - \alpha)$–confidence interval is to be given so that the expectation of the square of the half of the width of the interval (8.58) (for $k = 1$) does not exceed the value d^2, then

$$n - 3 = \left\lceil \frac{\sigma^2}{d^2}\left[1 - \frac{2}{n\,\sigma_x^2}\mathrm{Max}\left[(x_0-\mu_0)^2, (x_1-\mu_1)^2\right] t^2\left(n-2\middle|1-\frac{\alpha}{2}\right)\right]\right\rceil \tag{8.72}$$

must be chosen.

The theorem of Bock (1998) is important and given without proof.

Theorem 8.9 The minimal sample size n for the test of the hypothesis $H_0 : \beta_1 = \beta_{10}$ with the t-test statistic (8.48) should be determined so that for a given first kind risk α and a second kind risk β not larger than β^* as well as $|\beta_1 - \beta_{10}| \le d$ is given by

$$n \approx \frac{4\left(z_P + z_{1-\beta^*}\right)^2}{\left(\ln\left(1 + \frac{d\sigma_x}{\sqrt{d^2\,\sigma_x^2 - \sigma^2}}\right) - \ln\left(1 - \frac{d\sigma_x}{\sqrt{d^2\,\sigma_x^2 + \sigma^2}}\right)\right)^2} \tag{8.73}$$

Here is $P = 1 - \alpha$ for one-sided and $P = 1 - \alpha/2$ for two-sided alternatives.

Concerning the optimal choice of the sample size for comparing two or more slopes (test for parallelism), we refer to Rasch et al. (2008).

8.6 Mixed Models

If the conditional expectation of the component $\boldsymbol{y}$ of an r-dimensional random variable $(\boldsymbol{y}, \boldsymbol{x}_{k-r+2}, \ldots, \boldsymbol{x}_k)$ is a function of $k - r$ further (non-random) regressors, in place of (8.59)

$$\boldsymbol{y}_j = \sum_{i=0}^{k-r+1} \beta_i x_{ij} + \sum_{i=k-r+2}^{k} \beta_i \boldsymbol{x}_{ij} + \boldsymbol{e}_j,\ \ x_{0j} \equiv 1 \tag{8.74}$$

must be used.

Definition 8.8 Model equation (8.74) under the assumption that $\boldsymbol{e}_j$ are independent of each other and of the $\boldsymbol{x}_{ij}$ and $N(0, \sigma^2)$-distributed, and the vectors $\boldsymbol{x}_j = (\boldsymbol{y}_j, \boldsymbol{x}_{k-r+2,j}, \ldots, \boldsymbol{x}_{k,j})$ are independent of each other and $N(\mu, \Sigma)$-distributed with the vector of marginal expectations

$$\mu^* = \left(\sum_{i=0}^{k-r+1} \beta_i x_{ij}, \mu_{k-r+2}, \ldots, \mu_k\right)^T.$$

It is called mixed model of the linear regression ($|\Sigma| \neq 0$).

Estimators and tests can formally be used as for model II. The problem of the experimental design consists in the optimal choice of the matrix of the $x_{ij} (i = 0, \ldots, k - r + 1; j = 1, \ldots, n)$ and the optimal experimental size n. Results can be found in Bartko (1981).

8.7 Concluding Remarks about Models of Regression Analysis

Because the estimators for β_0 and β_1 for model I and model II equal each other and tests and confidence intervals are constructed by the same formulae, the reader may think that no distinction between both models has to be made. Indeed in many instructions for the statistical analysis of observed material and in nearly all program packages, a delicate distinction is missed. But the equality of the numerical treatment for both models does not justify to neglect both models in a mathematical treatment. Further, between both models there are differences that must also be considered in a pure numerical analysis. We describe this shortly for $k = 1$.

1) In model I only one regression function is reasonable:

$$E(\boldsymbol{y}) = \alpha_0 + \alpha_1 x.$$

But for model II two regression functions are possible:

$$E(\boldsymbol{y}|\boldsymbol{x}) = \alpha_0 + \alpha_1\, \boldsymbol{x} \text{ and } E(\boldsymbol{x}|\boldsymbol{y}) = \beta_0 + \beta_1 \boldsymbol{y}.$$

For model II the question, which regression function, should be used arises. If the parameters of the regression function are estimated to predict the values of one variable from observed values of the other one, we recommend using that variable as regressand, which should be predicted. That is, because the corresponding regression line by the least squares method is estimated, the deviations parallel to the axis of the regressand are squared and the sums are minimized.

But if the two-dimensional normal distribution is truncated in such a way that only for one variable the region of the universe is restricted (in breeding by selection concerning one variable), that variable could not be used as regressand. We illustrate this by an example.

Example 8.7 We consider a fictive finite universe, as shown in Figure 8.7 with linear regression functions $f(x)$ and $g(y)$. If we truncate with respect to x (regressor), the samples stem from that part of the population where $x > 3$. For simplification, we assume that the sample is the total remaining population. Then the regression function is identical with the regression function for the total universe $\left(\alpha_0 = 0, \alpha_1 = \frac{1}{2}\right)$ and given by the function $E(\boldsymbol{y}|\boldsymbol{x}) = \frac{1}{2}\boldsymbol{x}$. Truncation with respect to y, leads to different regression lines. Truncation for $\boldsymbol{y} > 3$ $(\boldsymbol{y} < -3)$ is shown in Figure 8.7 and leads to the regression functions

$$f_1(\boldsymbol{x}) = E(\boldsymbol{y}|\boldsymbol{x}, y > 3) = 3.25 + 0.25\boldsymbol{x}$$

$$f_2(\boldsymbol{x}) = E(\boldsymbol{y}|\boldsymbol{x}, y < -3) = -3.25 + 0.25\boldsymbol{x},$$

respectively, with wrong estimates $\alpha_0 = 0$ and $\alpha_1 = 0.5$.

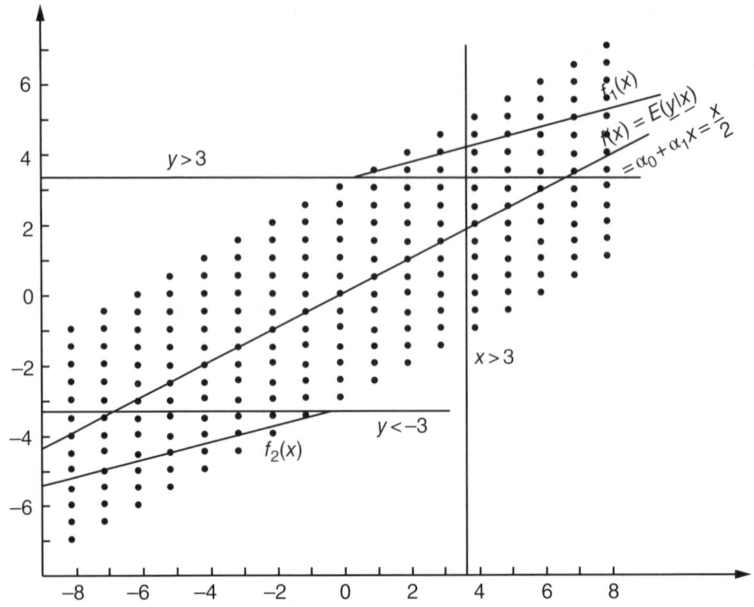

Figure 8.7 Fictive population with truncation shown in Example 8.7.

The example shows that truncation with respect to the regressand results in inacceptable estimates, while truncation with respect to the regressors causes no problem.

2) While for model I the estimators $\boldsymbol{b}_0$ and $\boldsymbol{b}_1$ are normally distributed, this is not the case for model II.
3) The confidence intervals for both parameters are calculated by the same formulae, but the expected width of these intervals differs for both models.
4) The hypotheses for both models are tested by the same test statistic, but nevertheless the tests are different because they have different power functions, leading to different sample sizes.
5) In case of model II, the regression analysis can be completed by calculating the correlation coefficient. For model I it may also be calculated, but cannot be interpreted as a statistical estimate of any population parameter and should be avoided. That the calculation of a sample correlation coefficient (as done by program packages) for model I is unreasonable follows from the fact that its value can be manipulated by a suitable choice of the x_i.
6) In the experimental design of model I, although the optimal choice the matrix X is important for model II, only the optimal choice of the sample size is important.

We considered only the most important models of the linear regression. Models with errors in the regressor variables are not included in this book. Models with random regression coefficients $\boldsymbol{\beta}$ as occurring in the population mathematics if each individual has its own regression coefficient are discussed in Swamy (1971) and Johansen (1984).

8.8 Exercises

8.1 Derive Equations (8.9) and (8.10) using the partial derivations of S given in the text before these equations.

8.2 Prove Corollary 8.1.

8.3 Estimate the parameters in the quasilinear regression model:

$$\boldsymbol{y}_i = \beta_0 + \beta_1 \cos(2x) + \beta_2 \ln(6x) + \boldsymbol{e}_i \ (i = 1, \dots, n).$$

8.4 Calculate in Example 8.3 all estimates for the storage in glass with SPSS.

8.5 Determine for Example 8.3 the G- and D-optimal design in the experimental region and calculate the determinants $\left|X_G^T X_G\right|$ and $\left|X_D^T X_D\right|$.

References

Bartko, M. (1981) *Versuchsplanung für Schätzungen im gemischten Modell der linearen Regression, Dissertation*, Universität Rostock.

Bartlett, M. S. (1933) On the theory of statistical regression. *Proc. Royal Soc. Edinburgh*, **53**, 260–283.

Bloomfield, P. and Steiger, W. L. (1983) *Least Absolute Deviations, Theory, Applications and Algorithms*, Birkhäuser, Boston and Basel.

Bock, J. (1998) *Bestimmung des Stichprobenumfanges für biologische Experimente und kontrollierte klinische Studien*, R. Oldenbourg Verlag, München, Wien.

Eisenhart, C. (1961) Bošković and the combination of observations, in Whyte, L. L. (Ed.), *Roger Joseph Bošković*, Fordham University Press, New York.

Fedorov, V. V. (1971) *Teorija optimalnych eksperimentov*, Nauka, Moskva.

Fisher, R. A. and Yates, F. (1949) *Statistical Tables for Biological, Agricultural and Medical Research*, 1st edition, Oliver and Boyd, Edinburgh, 6th edition 1963, 1974, Longman.

Gonin, R. and Money, A. H. (1989) Nonlinear L_p-norm Estimation, *M. Dekker, New York.*

Johansen, S. (1984) *Functional Relations, Random Coefficients and Nonlinear Regression with Applications in Kinetic Data*, Springer, New York.

Jung, W. (1973) *Optimale Versuchsplanung im Hinblick auf die Konstruktion konkreter optimaler Versuchspläne*. Dissertation, Bergakademie Freiberg.
Kiefer, J. (1959) Optimum experimental designs. *J. R. Stat. Soc. Ser. B*, **21**, 272–319.
Melas, V. B. (2008) *Functional Approach to Optimal Experimental Design*, Springer, New York.
Rasch, D. and Herrendörfer, G. (1986) *Experimental Designs – Sample Size Determination and Block Designs*, Reidel Publ. Co. Dortrecht, Boston, Lancaster, Tokyo.
Rasch, D., Herrendörfer, G., Bock, J., Victor, N. and Guiard, V. (Eds.) (2008) *Verfahrensbibliothek Versuchsplanung und - auswertung*, 2. verbesserte Auflage in einem Band mit CD, R. Oldenbourg Verlag München, Wien.
Swamy, P. S. (1971) *Statistical Inference in Random Coefficient Regression Models*, Springer, New York.

9

Regression Analysis – Intrinsically Non-linear Model I

In this chapter estimates are given for parameters in such regression functions that are non-linear in $x \in B \subset R$ and are not presentable in the form (8.19). We restrict ourselves to cases of real-valued regressands x; generalisations for vectoriale x are simple.

Definition 9.1 Regression functions $f(x, \theta)$ in a regressor $x \in B \subset R$ and with the parameter vector

$$\theta = \left(\theta_1,\dots,\theta_p\right)^T, \quad \theta \in \Omega \subset R^p,$$

which are non-linear in x and in at least one of the θ_i and cannot be made linear or quasilinear by any continuous transformation of the non-linearity parameter are called intrinsically non-linear regression functions. Correspondingly we also say intrinsically non-linear regression. More precisely, supposing $f(x, \theta)$ is concerning θ differentiable and

$$\frac{\partial f(x,\theta)}{\partial \theta}$$

is the first derivative of $f(x,\theta)$ with respect to θ, the regression function is said to be partially non-linear, if

$$\frac{\partial f(x,\theta)}{\partial \theta} = C(\theta)g(x,\varphi), \quad \varphi = (\theta_{i_1},\dots,\theta_{i_r})^T \tag{9.1}$$

and $0 < r < p$ where $C(\theta)$ is a $(p \times p)$–matrix not depending on x and φ, chosen in such a way that r is minimal ($r = 0$ means a quasilinear regression). If $r = p$, then $f(x, \theta)$ is called completely non-linear.

$\theta_{i_j}(j = 1,\dots,r)$ are called non-linearity parameters, and the other components of θ are called linearity parameter.

Mathematical Statistics, First Edition. Dieter Rasch and Dieter Schott.

We illustrate the definition by some examples.

Example 9.1 We consider

$$f(x,\theta) = \frac{\theta_1 x}{(1-x)[\theta_2 + (1-\theta_2)x]}.$$

That is, we have $p = 2, \theta = (\theta_1, \theta_2)^T$ and obtain

$$\frac{\partial f(x,\theta)}{\partial \theta} = \begin{pmatrix} \dfrac{x}{(1-x)[\theta_2 + (1-\theta_2)x]} \\ \dfrac{-\theta_1 x}{[\theta_2 + (1-\theta_2)x]^2} \end{pmatrix} = \begin{pmatrix} 1 & 0 \\ 0 & \theta_1 \end{pmatrix} \begin{pmatrix} \dfrac{x}{(1-x)[\theta_2 + (1-\theta_2)x]} \\ \dfrac{-x}{[\theta_2 + (1-\theta_2))]^2} \end{pmatrix}.$$

Here θ_1 is a linearity parameter and $\varphi = \theta_2$ a non-linearity parameter. Further $r = 1$.

Example 9.2 We consider

$$f(x,\theta) = \theta_1\left(x + \mathrm{e}^{-\theta_3 x}\right) - \theta_2 x \mathrm{e}^{-\theta_3 x},$$

that is, we have $\theta^T = (\theta_1, \theta_2, \theta_3)$ and $p = 3$ and obtain

$$\frac{\partial f(x,\theta)}{\partial \theta} = \begin{pmatrix} x + \mathrm{e}^{-\theta_3 x} \\ -x\mathrm{e}^{-\theta_3 x} \\ (-x\theta_1 + x^2\theta_2)\mathrm{e}^{-\theta_3 x} \end{pmatrix} = \begin{pmatrix} 1 & 0 & 0 \\ 0 & 1 & 0 \\ 0 & \theta_1 & \theta_2 \end{pmatrix} \begin{pmatrix} x + \mathrm{e}^{-\theta_3 x} \\ -x\mathrm{e}^{-\theta_3 x} \\ x^2\mathrm{e}^{-\theta_3 x} \end{pmatrix};$$

$\varphi = \theta_3$ is a non-linearity parameter, and θ_1 and θ_2 are linearity parameters (i.e. $r = 1$).

For linear models a general theory of estimating and tests was possible, and by the Gauss–Markov theorem, we could verify optimal properties of the least squares (LS) estimator $\hat{\boldsymbol{\theta}}$. A corresponding theory for the general (non-linear) case does not exist. For quasilinear regression the theory of linear models, as shown in Section 8.2, can be applied. For intrinsically non-linear problems the situation can be characterised as follows:

- The existing theoretical results are not useful for solutions of practical problems; many results about distributions of the estimates are asymptotic: for special cases simulation results exist.
- The practical approach leads to numerical problems for iterative solutions; concerning properties of distributions of the estimators, scarcely anything is known. The application of the methods of the intrinsically non-linear regression is more a problem of numerical mathematics than of mathematical statistics.

- A compromise could be to go over to quasilinear approximations for the non-linear problem and conduct the parameter estimation for the approximative model. But by this we lose the interpretability of the parameters, desired in many applications by practitioners.

We start with the model equation

$$\boldsymbol{Y} = \eta + \boldsymbol{e} \tag{9.2}$$

with the side conditions $E(\boldsymbol{e}) = 0_n$ $(E(\boldsymbol{Y}) = \eta)$ and $\text{var}(\boldsymbol{e}) = \sigma^2 I_n$ $(\sigma^2 > 0)$. $\boldsymbol{Y}$, η and $\boldsymbol{e}$ are vectors with components $\boldsymbol{y}_i$, η_i and $\boldsymbol{e}_i$ $(i = 1, \ldots, n)$, respectively, where η_i are intrinsically non-linear functions

$$\eta_i = f(x_i, \theta), \quad \theta \in \Omega \subset R^p \quad (i = 1, \ldots, n) \tag{9.3}$$

in the regressor values $x_i \in B \subset R$. We use the abbreviations

$$\left.\begin{aligned}
&\eta(\theta) = (f(x_1,\theta),\ldots,f(x_n,\theta))^T,\\
&f_j(x,\theta) = \frac{\partial f(x,\theta)}{\partial\theta_j},\quad f_{jk}(x,\theta) = \frac{\partial^2 f(x,\theta)}{\partial\theta_j \partial\theta_k},\\
&F_i = F_i(\theta) = \left(f_1(x_i,\theta),\ldots,f_p(x_i,\theta)\right),\\
&F = F(\theta) = (F_1(\theta),\ldots,F_n(\theta))^{\mathrm{T}} = \left(f_j(x_i,\theta)\right),\\
&K_i = K_i(\theta) = \left(k_{jk}(x_i,\theta)\right) = \left(\frac{\partial^2 f(x_i,\theta)}{\partial\theta_j \partial\theta_k}\right),
\end{aligned}\right\} \tag{9.4}$$

and assume always that $n > p$. Further

$$R(\theta) = \|\boldsymbol{Y} - \eta(\theta)\|^2 = \sum_{i=1}^{n} [y_i - f(x_i,\theta)]^2. \tag{9.5}$$

The first question is whether different values of θ always lead to different parameters of the distribution of $\boldsymbol{Y}$ – in other words, whether the parameter is identifiable. Identifiability is a necessary assumption for estimability of θ. But the identifiability condition is often very drastic in the intrinsically non-linear case, so that we will not discuss this further. Instead we choose a pragmatic approach that also can be used if identifiability is established.

Definition 9.2 The random variable $\hat{\boldsymbol{\theta}}$ is called LS estimator of θ, if its realisation $\hat{\theta}$ is a unique solution of

$$R(\hat{\theta}) = \min_{\theta \in \Omega} R(\theta). \tag{9.6}$$

In (9.6) $R(\theta)$ is given by (9.5). Further let

$$\hat{\boldsymbol{\eta}} = f\left(x, \hat{\boldsymbol{\theta}}\right)$$

be the LS estimator of $\eta = f(x,\theta)$. In place of (9.6) we also write

$$\hat{\theta} = \arg \min_{\theta \in \Omega} R(\theta).$$

We always assume for all discussions that for (9.6) a unique solution exists. The estimators of the parameters by the LS method for the intrinsically non-linear regression function are in general not unbiased. The exact distribution is usually unknown, and therefore for confidence estimations and tests, we have to use asymptotic distributions.

The possibility to approximate function (9.3) by replacing $f(x, \theta)$ by a quasi-linear function was discussed in the literature (see Box and Lucas, 1959; Box and Draper, 1963; Karson et al., 1969; Ermakoff, 1970; Bunke, 1973; Petersen, 1973). For instance, a continuous differentiable function $f(x, \theta)$ could be developed in a Taylor series stopping after a designated number of terms and choosing the design (the x_i) in such a way that the discrepancy between $f(x, \theta)$ and the approximate quasilinear function is a minimum. The approximate function can then be estimated by the methods of Chapter 8.

We now assume that the experimenter is interested in the parameters of a special intrinsically non-linear function obtained from a subject-specific differential equation. We must therefore use direct methods even if we know only few about the statistical properties of the estimators.

For the case that $f(x, \theta)$ concerning θ is a continuously differentiable function, we obtain by zeroing the first partial derivations of (9.5) with respect to the components of θ

$$R_j(\hat{\theta})\eta_j(\hat{\theta})\left[Y - \eta(\hat{\theta})\right] = 0 \tag{9.7}$$

with

$$R_j(\theta) = \frac{\partial R(\theta)}{\partial \theta_j} \quad \text{and} \quad \eta_j(\theta) = \left[f_j(x_1,\theta),\ldots,f_j(x_n,\theta)\right]^T$$

and

$$f_j(x,\theta) = \frac{\partial f(x,\theta)}{\partial \theta_j}.$$

9.1 Estimating by the Least Squares Method

At first we give numerical methods for an approximate solution of (9.7) or (9.6). The existence of a unique solution is from now on assumed. In Section 9.1.2 we give methods without an exact knowledge of the function to be minimised and without using the first derivations. In Section 9.1.3 we present methods applied directly to the differential equation and not to its integral $f(x, \theta)$.

9.1.1 Gauss–Newton Method

We assume that $f(x, \theta)$ is twice continuously differentiable concerning $\theta \in \Omega$ and that for a given x exactly one local minimum concerning θ exists. We develop $f(x, \theta)$ around $\theta_0 \in \Omega$ in a Taylor series stopping after the first-order terms. If $f_j(x, \theta_0)$ is the value of $f_j(x, \theta)$ at $\theta = \theta_0$, then

$$f(x,\theta) \approx f(x,\theta_0) + (\theta-\theta_0)\frac{\partial f(x,\theta)}{\partial\theta}|_{\theta=\theta_0} = \tilde{f}^0(x,\theta). \tag{9.8}$$

We approximate (9.2) written in its realisations starting with $l = 0$ by

$$Y = \tilde{\eta}^{(l)} + e^{(l)},\ \ \tilde{\eta}^{(l)} = \left[\tilde{f}^{(l)}(x_1,\theta),\ldots,\tilde{f}^{(l)}(x_n,\theta)\right]^T. \tag{9.9}$$

Equation (9.9) is linear in $\theta - \theta_l = \Delta\theta_l$ $(l = 0)$. The Gauss–Newton method means that $\Delta\theta_0$ in (9.9) is estimated by the LS method, and from a Taylor series expansion and with the estimate $\Delta\hat{\theta}_0$, we build up the vector $\theta_1 = \theta_0 + \nu_0\Delta\hat{\theta}_0$. Around θ_1 once more a Taylor series analogously to (9.8) is constructed, and model (9.9) is now used with $l = 1$.

Now $\Delta\theta_1 = \theta - \theta_1$ has to be estimated by the LS method. If θ_0 is near enough to the solution $\hat{\theta}$ of (9.7) (in the algorithm of Hartley below, it is exactly explained what 'near enough' means), the sequence $\theta_0, \theta_1, \ldots$ converges against $\hat{\theta}$. If we however start with a bad initial vector θ_0, a cutting of a Taylor series expansion after the first terms leads to large differences between f and $\tilde{f}$, and the method converges not to the global minimum but to a relative minimum (see Figure 9.4). If $\theta_l \in \Omega$ is the vector in the l-th step of a Taylor series and $\tilde{f}^{(l)}(x,\theta)$ is determined analogously to (9.8), then in the l-th step the simultaneous equations become

$$F_l^{\mathrm{T}} F_l \Delta\hat{\theta}_l = F_l^{\mathrm{T}}\left(Y - \tilde{\eta}^{(l)}\right) \tag{9.10}$$

with $F_l = \left(f_{ij}^{(l)}\right) = \left(\tilde{f}_j^{(l)}(x_i,\vartheta)\right)$. We assume that x_i are chosen in such a way that $F_l^T F_l$ is non-singular and by this that (9.10) has a unique solution. The iteration method calculating vectors for a Taylor series by

$$\theta_{l+1} = \theta_l + \nu_l\Delta\hat{\theta}_l \tag{9.11}$$

can be (convergence assumed) continued as long as for all j

$$\left|\theta_{j,l-1} - \theta_{jl}\right| < \delta_j\ \ \left(\theta_l = (\theta_{1l},\ldots,\theta_{pl})^T\right).$$

But because the objective of the iteration is the solution of (9.6), it makes more sense to continue the iteration with θ_{l+1} and θ_l for $\hat{\theta}$ in (9.6) as long as

$$|R(\theta_l) - R(\theta_{l+1})| < \varepsilon$$

is reached for the first time.

In the original form of the Gauss–Newton iteration, $v_l = 1$ ($l = 0, 1, \ldots$) was used. But then the convergence of the procedure is not sure or can be very slow.

Some proposals are known to modify the Gauss–Newton method, for example, by Levenberg (1944) and Hartley (1961), and the latter will be described now.

The method of Hartley offers a quicker convergence and has the advantage that assumptions for a reliable convergence can be formulated.

Let the following assumptions be fulfilled:

V1: $f(x, \theta)$ has for all x continuous first and second derivatives concerning θ.
V2: For all $\theta_0 \in \Omega_0 \subset \Omega$ (Ω_0 restricted, convex) with $F = [\eta_1(\theta), \ldots, \eta_p(\theta)]$, the matrix $F^T F$ is positive definite.
V3: There exists a $\theta_0 \in \Omega_0 \subset \Omega$ so that

$$R(\theta_0) < \inf_{\theta \in \Omega \setminus \Omega_0} R(\theta).$$

Hartley's modification of the Gauss–Newton method means to choose v_l in (9.11) so that $R(\theta_l + v_l \Delta\theta_l)$ for a given θ_l as a function of v_l for $0 \le v_l \le 1$ is a minimum.

Hartley proved the following theorems, and we give them without proof.

Theorem 9.1 (Existence theorem)
If V1 to V3 are fulfilled, then a subsequence $\{\theta_u\}$ of the sequence $\{\theta_l\}$ of vectors in (9.11) always exists with v_l that minimises $R(\theta_l + v_l \Delta\theta_l)$ for given θ_l and $0 \le v_l \le 1$ that converges against a solution of

$$R(\theta^*) = \min_{\theta \in \Omega_0} R(\theta).$$

For restricted and convex Ω_0, by this theorem, Hartley's method converges against a solution of (9.6).

Theorem 9.2 (Uniqueness theorem)
If the assumptions of Theorem 9.1 are valid and with the notations of this section the quadratic form $a^T R a$ with $R = \left(R_{ij}(\theta)\right)$ and $R_{ij}(\theta) = \dfrac{\partial^2 R(\theta)}{\partial\theta_i \partial\theta_j}$ is positive definite in Ω_0, then there is only one stationary point of $R(\theta)$.

A problem of Hartley's method is the suitable choice of a point θ_0 in a restricted convex set Ω_0. The numerical determination of v_{l+1} is often elaborate. Approximately v_{l+1} can be found by quadratic interpolation with the values $v_{l+1}^* = 0$, $v_{l+1}^{**} = \frac{1}{2}$ and $v_{l+1}^{***} = 1$ from

$$v_{l+1} = \frac{1}{2} + \frac{1}{4} \frac{R(\theta_l) - R(\theta_l + \Delta\theta_l)}{R(\theta_l + \Delta\theta_l) - 2R\left(2\theta_l + \frac{1}{2}\Delta\theta_l\right) + R(\theta_l)}. \tag{9.12}$$

Further modifications of the Gauss–Newton method are given by Marquardt (1963, 1970) and Nash (1979).

Program Hint

With SPSS an LS estimator can be obtained as follows.

A data file with (x, y)-values, as for instance that for the growth of hemp plants given by Barath et al. (1996), is needed and shown in Figure 9.1. The value 20 for age was added for later calculations and does not influence the parameter estimation, due to the missing height value.

We choose at first

Regression
 Nonlinear

and get the window in Figure 9.2.

In this window first the parameters with their initial values must be put in, and then the regression function must be programmed. We choose at first the logistic regression from Section 9.6.3. The programmed function and the initial values of the parameters are given in Figure 9.3.

After many iterations an unsatisfactory result with an error MS of 506.9 as shown in Figure 9.4 came out.

With a bad choice of the initial values, we get a relative (but no absolute) minimum of $R(\theta)$.

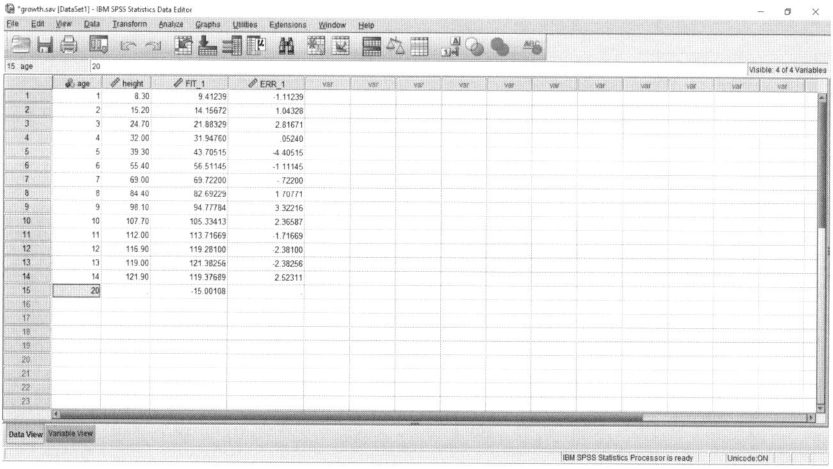

	age	height	FIT_1	ERR_1
1	1	8.30	9.41239	-1.11239
2	2	15.20	14.15672	1.04328
3	3	24.70	21.88329	2.81671
4	4	32.00	31.94760	.05240
5	5	39.30	43.70515	-4.40515
6	6	55.40	56.51145	-1.11145
7	7	69.00	69.72200	-.72200
8	8	84.40	82.69229	1.70771
9	9	98.10	94.77784	3.32216
10	10	107.70	105.33413	2.36587
11	11	112.00	113.71669	-1.71669
12	12	116.90	119.28100	-2.38100
13	13	119.00	121.38256	-2.38256
14	14	121.90	119.37689	2.52311
15	20	.	-15.00108	.

Figure 9.1 Data file of hemp data with additional calculated values (explained later). *Source:* Reproduced with permission of IBM.

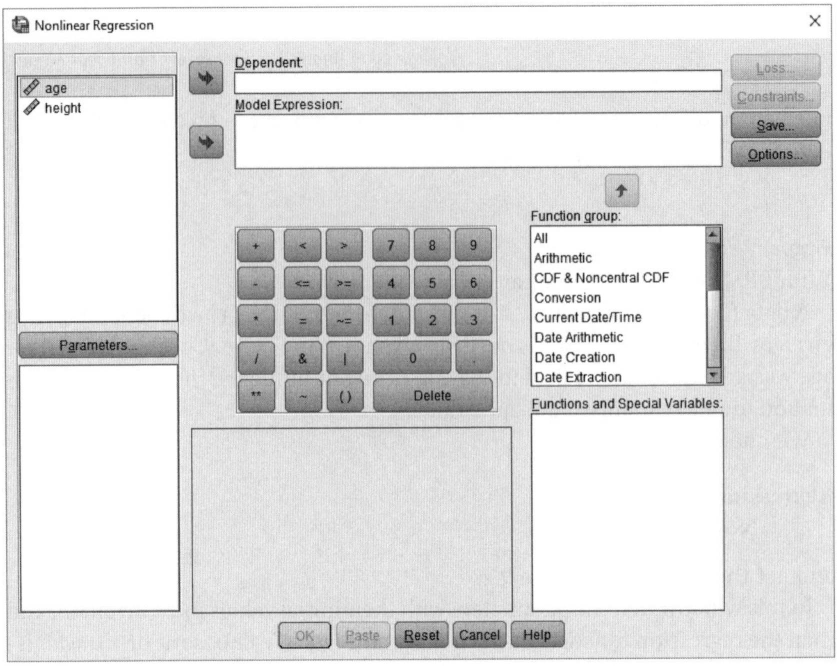

Figure 9.2 Start of non-linear regression in SPSS. *Source:* Reproduced with permission of IBM.

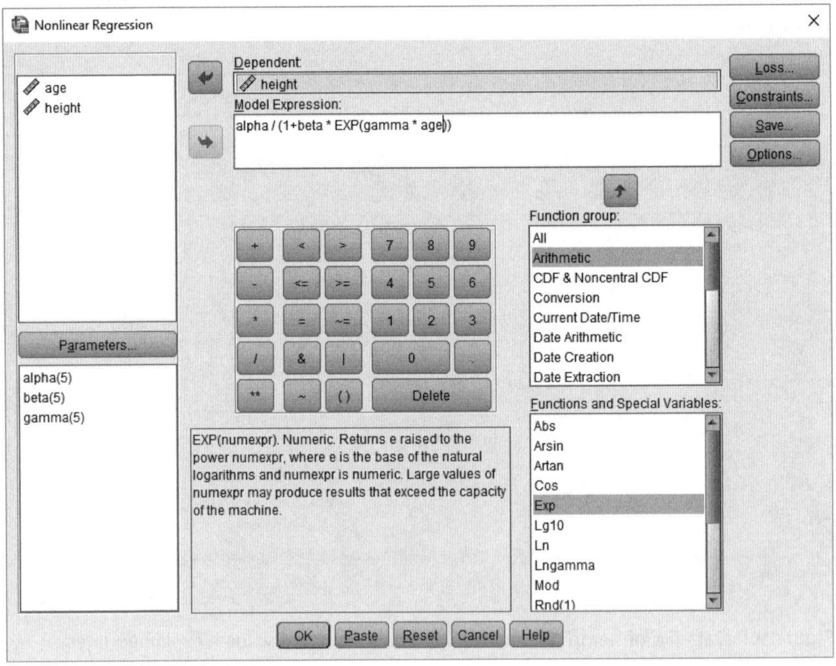

Figure 9.3 Programmed function and the initial values. *Source:* Reproduced with permission of IBM.

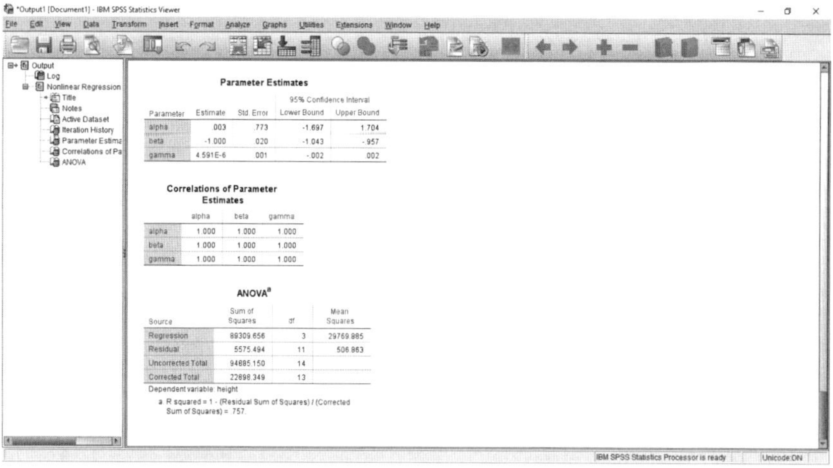

Parameter Estimates

Parameter	Estimate	Std. Error	95% Confidence Interval Lower Bound	Upper Bound
alpha	.003	.773	-1.697	1.704
beta	-1.000	.020	-1.043	-.957
gamma	4.591E-6	.001	-.002	.002

Correlations of Parameter Estimates

	alpha	beta	gamma
alpha	1.000	1.000	1.000
beta	1.000	1.000	1.000
gamma	1.000	1.000	1.000

ANOVA[a]

Source	Sum of Squares	df	Mean Squares
Regression	89309.656	3	29769.885
Residual	5575.494	11	506.863
Uncorrected Total	94885.150	14	
Corrected Total	22898.349	13	

Dependent variable: height

a. R squared = 1 - (Residual Sum of Squares) / (Corrected Sum of Squares) = .757.

Figure 9.4 Result of the calculation with the initial parameter values from Figure 9.3. *Source:* Reproduced with permission of IBM.

We now choose other initial values by help of the data. For $x = \infty$ the logistic function reaches for negative γ the value α; we replace it by the rounded maximum 122. If the growth starts ($x = 0$), the value of the function is about $\frac{122}{1+\beta}$, and we replace this by the rounded smallest value 8 and receive $\beta = 14.25$. Finally we choose $\gamma = -0.1$ and change the initial values correspondingly.

Now we gain in Figure 9.5 a global minimum with a residual MS error of 3.71, and the estimated logistic regression function is $\frac{126.22}{1 + 19.681e^{-0.46x}}$.

Another possibility is to try some function already programmed in SPSS. For this we use the commands

Analyze
  Regression
    Curve Estimation

We receive Figure 9.6 where we already have chosen the cubic regression (logistic does not mean the function used before), and with *save* we gave the command to calculate the predicted values FIT and the error terms RES, which are already shown in Figure 9.1.

In Figure 9.7 we find the graph of the fitted cubic regression with extrapolation to age 20.

The MS error is the sum of squares of the residuals divided by $df = 10 = 14 - 4$, because four parameters have been estimated in the cubic regression. We

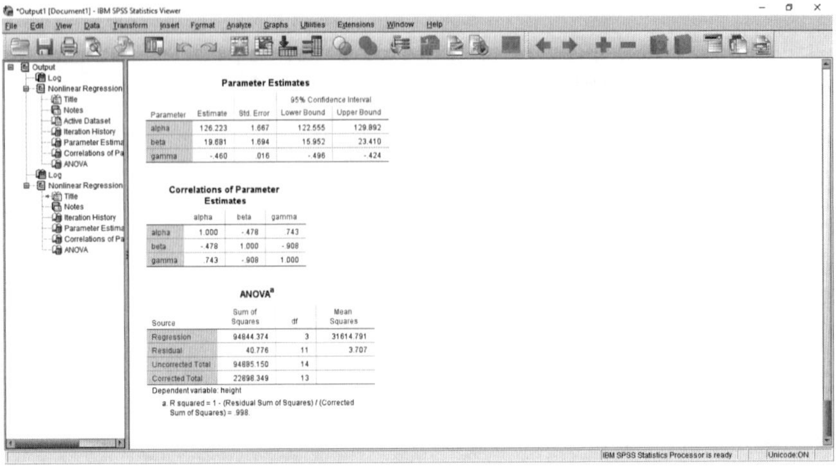

Figure 9.5 Result of the calculation with improved initial parameter values. *Source:* Reproduced with permission of IBM.

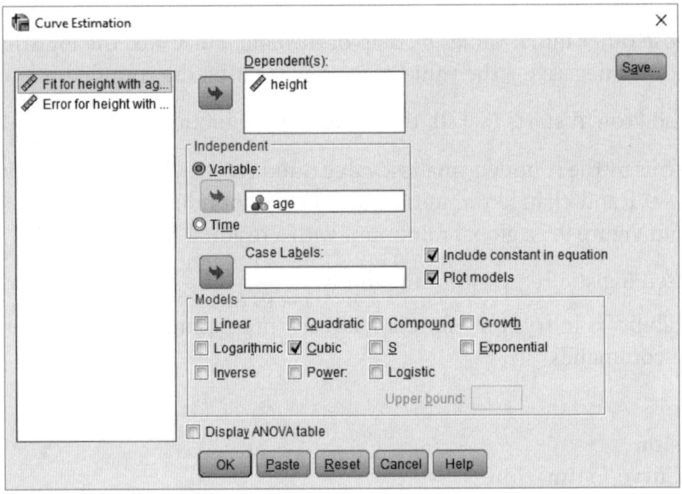

Figure 9.6 Start of curve estimation in the regression branch of SPSS. *Source:* Reproduced with permission of IBM.

receive MS error = 7.16. This is larger than the corresponding value 3.71 in Figure 9.6, and it seems that the logistic regression makes a better fit to the data. But even if the MS error for the cubic regression would be smaller than that of the logistic regression, one may have doubts about using the latter. From the graph we find that extrapolation (to an age of 20) gives a terrible result for the cubic function that can be seen in the graph as well as in the predicted value

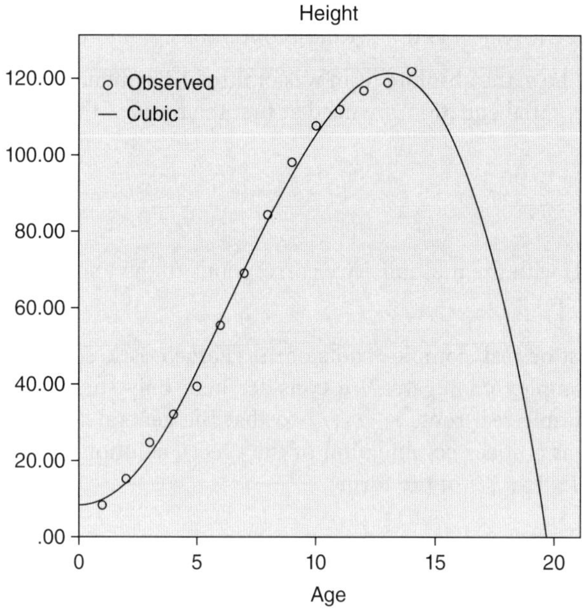

Figure 9.7 Graph of the fitted cubic regression with extrapolation to age 20.
Source: Reproduced with permission of IBM.

−15.0 in Figure 9.1. The predicted height for the logistic regression at age 20 is 129.3, a reasonable result. The reason is that the logistic regression is an integral of a differential equation for growth processes.

9.1.2 Internal Regression

The principle of internal regression starts with Hotelling (1927). Later Hartley (1948) developed for the simple intrinsically non-linear regression with equidistant x_i a method for the case that $f(x)$ is integral of a linear differential equation of first order. The observed values y_i are not fit to the regression function but approximatively to the generating differential equation by approximating the differential quotient. This method has later been extended to intrinsically non-linear regression functions, which are integral of a linear homogeneous differential equation of higher order with constant coefficients, and to non-linear differential equations (Scharf, 1970).

We restrict ourselves in the following outline to the generalised method of internal regression for homogeneous linear differential equations of order k of the form

$$f^{(k)} + \sum_{l=1}^{k} b_l f^{(l-1)} = 0 \quad (k > 0, \text{integer}) \tag{9.13}$$

with $f^{(l)} = \frac{d^l f(x)}{dx^l}$, unknown real b_l and a k-times continuously differentiable function $f(x, \theta) = f^{(0)}$. (W.l.o.g. the absolute term was omitted.) To obtain a general solution of this differential equation, we need at first the roots of the characteristic equation

$$r^k + \sum_{l=1}^{k} b_l r^{l-1} = 0. \tag{9.14}$$

Each real root r^* of (9.14) with the multiplicity ν corresponds to the ν solutions

$$x^l e^{r^* x} \quad (l = 0, \dots, \nu - 1) \tag{9.15}$$

of (9.13). Because b_l must be real, complex roots of the characteristic equation can only be pairwise complex conjugate. We consider here only such cases where (9.14) has only simple real roots $r_1, \dots, r_t$, so that the general solution of (9.13) can be written as a linear combination of the special solutions (9.15) with real coefficients $c_i (i = 1, \dots, t)$ of the form

$$f(x,\theta) = f(x) = \sum_{l=1}^{t} c_l e^{r_l x} \quad \left(\theta = (c_1, \dots, c_t, r_1, \dots, r_t)^T\right). \tag{9.16}$$

We now have to estimate the coefficients b_l in (9.13) in place of the parameters of $f(x, \theta) = f(x)$ in (9.16) and use the stochastic model

$$f^{(t)}(x_i) + \sum_{l=1}^{t} b_l f^{(l-1)}(x_i) = \tilde{\boldsymbol{e}}_i \quad (i = 1, \dots, n), \quad n > t. \tag{9.17}$$

We assume that the vector $\tilde{\boldsymbol{e}}_i$ is $N(0_n, \tilde{\sigma}^2 I_n)$-distributed and we wish to determine the LS estimators $\hat{\boldsymbol{b}}_l$ of the b_l in (9.17). Of course we now consider a model different from (9.2), and we now assume additive error terms in the differential equation and not, as in (9.2), for the integral. The applicability of the internal regression depends on the tenability of model (9.17) (at least approximately). $\hat{b}_l$ are calculated so that

$$\sum_{i=1}^{n} \left[f^{(t)}(x_i) + \sum_{l=1}^{t} \hat{b}_l f^{(l-1)}(x_i) \right]^2 = \min_{-\infty < b_l < \infty} \sum_{i=1}^{n} \left[f^{(t)}(x_i) + \sum_{l=1}^{t} b_l f^{(l-1)}(x_i) \right]^2 \tag{9.18}$$

holds.

To replace the differential quotient by a difference quotient, we need the following notations assuming $x_1 < x_2 < \cdots$:

$$\begin{aligned} \Delta_i^0 &= y_i, \\ \Delta_i^s &= \frac{\Delta_{i+1}^{s-1} - \Delta_{i-1}^{s-1}}{x_{i+1} - x_{i-1}} \\ &\left(i = \left[\frac{s-2}{2}\right] + 1, \dots, n - \left[\frac{s+2}{2}\right]; \quad s = 1, \dots, k \right). \end{aligned} \tag{9.19}$$

y_i are the observed values or in case of multiple measurements the means of the observed values at x_i. From (9.18) we obtain by derivation with respect to b_i and zeroing these derivatives and by

$$f^{(u)}(x_i) \approx \Delta_l^u \tag{9.20}$$

the approximate equations for $\hat{b}_l$:

$$\sum_{i=t+1}^{n-t} \left(\Delta_i^t + \sum_{l=1}^{t} \hat{b}_l \Delta_i^{l-1} \right) \Delta_i^{h-1} = 0 \quad (h = 1,\ldots,t). \tag{9.21}$$

These linear simultaneous equations in $\hat{b}_l$ can easily be solved. From (9.14) and with $\hat{b}_l$, we obtain estimates $\hat{r}_1,\ldots,\hat{r}_t$ for r and correspondingly by (9.16) the general solution

$$\widehat{f(x_i)} = \sum_{l=1}^{t} c_l e^{\hat{r}_i x_i}.$$

With $z_{li} = e^{\hat{r}_i x_i}$ the c_l are estimated as regression coefficients of a multiple linear regression problem with the model equation

$$y_i = \sum_{l=1}^{t} c_l z_{li} + e_i^*. \tag{9.22}$$

Transition to random variables gives

$$\hat{\boldsymbol{\theta}} = (\hat{\boldsymbol{c}}_1,\ldots,\hat{\boldsymbol{c}}_t,\hat{\boldsymbol{r}}_1,\ldots,\hat{\boldsymbol{r}}_t)^T$$

as estimator of θ by internal regression.

9.1.3 Determining Initial Values for Iteration Methods

The convergence of iteration methods to minimise non-linear functions and to solve non-linear simultaneous equations depends extremely on the choice of initial values. If the parameters of a function can be interpreted from a practical application or if the parameters can roughly be determined from a graphical representation, then a heuristic choice of an initial value θ_0 of θ can be reasonable. But if we only have the (y_i, x_i) values in our computer, we can try to find initial values by some specific methods. Unfortunately some of those methods like Verhagen's trapezial method (Verhagen, 1960) (Section 9.6) are applicable only for special functions. A more general method is the 'internal regression' (Section 9.1.2) that is recommended to determine initial values.

The residual variance σ^2 is either estimated by $s^2 = \frac{1}{n-p} R\left(\hat{\boldsymbol{\theta}}\right)$ or by $\tilde{\sigma}^2 = \frac{1}{n} R\left(\hat{\boldsymbol{\theta}}\right)$. The motivation is given in Section 9.4.2.

9.2 Geometrical Properties

The problem of minimising $R(\theta)$ in (9.5) will now be discussed geometrically. For a fixed support $(x_1, \ldots, x_n)$ in (9.4), the function $\eta(\theta)$ defines an expectation surface in R^n.

9.2.1 Expectation Surface and Tangent Plane

First we give the following definition.

Definition 9.3 The set

$$\mathrm{ES} = \{Y^* | \exists\theta : Y^* = \eta(\theta)\} \tag{9.23}$$

is called the expectation surface of the regression function $\eta(\theta)$ in R^n.

If $\sigma^2 > 0$, an observed value Y with probability 1 does not lie in ES.

From Definition 9.2 we may conclude that the LS estimate $\hat{\theta}$ is just that value in Ω that has a minimum distance between Y and θ or in other words $f(x,\hat{\theta}) = \hat{f}$ is the orthogonal projection of Y on ES.

The distance is the length of the vector orthogonal to the tangent plane of the expectation surface at the point $\eta(\theta)$ and has its (not necessarily unique) minimum at $\eta(\hat{\theta})$.

Example 9.3 Let $\theta \in \Omega = R^1$, that is, $p = 1$, $x \in R^1$, and let us consider the function

$$f(x,\theta) = \frac{10}{1 + 2e^{\theta x}}. \tag{9.24}$$

Further let $n = 2, x_1 = 1, x_2 = 2$. For

$$Y = \begin{pmatrix} y_1 \\ y_2 \end{pmatrix} \text{ and } \eta(\theta) = \begin{pmatrix} \dfrac{10}{1+2e^{\theta}} \\ \dfrac{10}{1+2e^{2\theta}} \end{pmatrix},$$

we consider the four cases: $(y_{11}; y_{21}) = (4; 8)$; $(y_{12}; y_{22}) = (7; 3)$; $(y_{13}; y_{23}) = (1.25; 2.5)$ and $(y_{14}; y_{24}) = (1.25; 1.5)$.

For each case (i = 1, 2, 3, 4), $\mathrm{R}(\theta | y_{1i}, y_{2i})$ was calculated as a function of θ (Table 9.1), and the graphs of these four functions are shown in Figure 9.8.

The coordinates of the expectation surface in R^2, which because $p = 1$ is an expectation curve (and the tangent plane is a tangent), are shown in Table 9.2. From Definition 9.3 it follows that the expectation surface does not depend on the observations. The expectation curve, observed points Y_i (i = 1, 2, 3, 4) and two tangents are shown in Figure 9.9.

Table 9.1 Values of the functionals $R(\theta)$ for four pairs of values of Example 9.3 (values > 30 are not included).

θ	$R(\theta \mid y_{11}, y_{21})$	$R(\theta \mid y_{12}, y_{22})$	$R(\theta \mid y_{13}, y_{23})$	$R(\theta \mid y_{14}, y_{24})$
−1.5	9.69			
−1.4	8.1			
−1.3	6.61			
−1.2	5.24			
−1.1	4.05	27.9		
−1.0	3.12	25.3		
−0.9	2.53	22.6		
−0.8	2.37	20.0		
−0.7	2.73	17.6		
−0.6	3.68	15.5	26.4	
−0.5	5.28	13.8	21.3	28.8
−0.4	7.54	12.6	16.8	23.3
−0.3	10.45	11.9	12.9	18.4
−0.2	13.94	11.9	9.60	14.1
−0.1	17.91	12.5	7.00	10.6
0	22.22	13.6	5.03	7.70
0.1	26.75	15.1	3.64	5.45
0.2		17.0	2.74	3.76
0.3		19.2	2.23	2.54
0.4		21.5	2.03	1.70
0.5		23.9	2.06	1.16
0.6		26.4	2.23	0.85
0.7		28.7	2.51	0.71
0.8			2.85	0.68
0.9			3.21	0.74
1			3.57	0.84
1.2				1.31

The θ scale at the expectation curve can be found in the first column of Table 9.2. For in θ non-linear regression functions $f(y, \theta)$ the expectation surface is curved, and the curvature depends on θ and in an environment $U(\theta_0)$ of θ_0 it can be defined as the degree of the deviation of the expectation surface from the

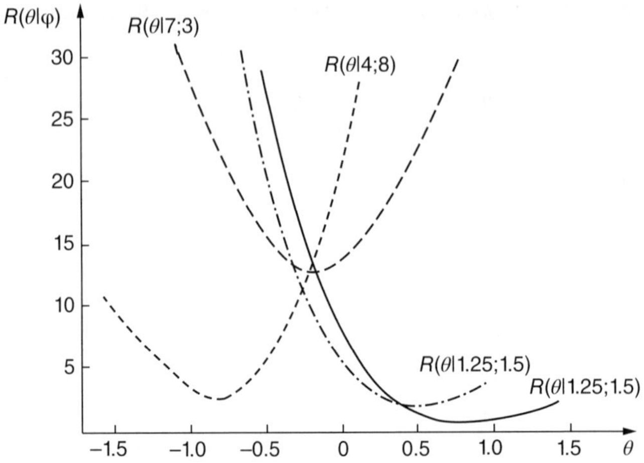

Figure 9.8 Graphs of $R(\theta,Y)$ from Table 9.1.

Table 9.2 Coordinates of the expectation curve of Example 9.6.

θ	$\frac{10}{1+2e^{\theta}}$	$\frac{10}{1+2e^{2\theta}}$
−1.9	7.70	9.57
−1.7	7.32	9.37
−1.5	6.91	9.09
−1.3	6.47	8.71
−1.1	6.00	8.18
−0.9	5.52	7.52
−0.82	5.32	7.20
−0.7	5.02	6.70
−0.5	4.52	5.76
−0.3	4.03	4.77
−0.24	3.89	4.47
−0.1	3.56	3.79
0	3.33	3.33
0.2	2.90	2.51
0.4	2.51	1.83
0.44	2.44	1.72
0.6	2.15	1.31
0.78	1.86	0.95
1.0	1.55	0.63
2.0	0.63	0.09

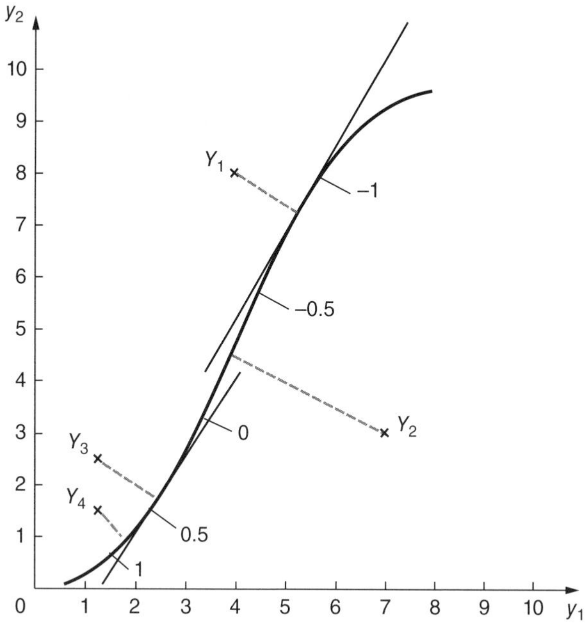

Figure 9.9 Expectation vectors of the function $\frac{10}{1+2e^{\theta x}}$ for x = 1 and x = 2 and four observations.

tangent plane at θ_0. The coordinate system on the expectation surface is not uniform, that is, as $\theta_3 \neq \theta_4$ from

$$\theta_4 - \theta_3 = \theta_2 - \theta_1,\quad \theta_i \in \Omega \quad (i = 1,2,3,4)$$

not necessarily follows

$$\|\eta(\theta_4) - \eta(\theta_3)\| = \|\eta(\theta_2) - \eta(\theta_1)\| .$$

In this chapter, we assume that $R(\theta)$ has for all Y a unique minimum.

The expectation surface is independent of the parametrisation of the function defined below.

Definition 9.4 A regression function $f(x, \theta)$ is reparametrised, if a one-to-one transformation $g: \Omega \to \Omega^*$ is applied on θ and $f(x, \theta)$ is written as

$$f(x,\theta) = \psi[x, g(\theta)] = \psi(x, \theta^*) \tag{9.25}$$

with $\theta^* = g(\theta)$ $[\theta = g^{-1}(\theta^*)]$.

Example 9.4 The function

$$f(x,\theta) = \frac{\alpha}{1+\beta e^{\gamma x}} \quad \text{with} \quad \theta = (\alpha,\beta,\gamma)^T, \beta > 0 \tag{9.26}$$

is called logistic function. We assume $\alpha\beta\gamma \neq 0$. The first and second derivatives by x are

$$\frac{df(x,\theta)}{dx} = -\frac{\alpha\beta\gamma e^{\gamma x}}{(1+\beta e^{\gamma x})^2}$$

and

$$\frac{d^2 f(x,\theta)}{dx^2} = -\alpha\beta\gamma \frac{(1+\beta e^{\gamma x})^2 \gamma e^{\gamma x} - 2e^{\gamma x}(1+\beta e^{\gamma x})\beta\gamma e^{\gamma x}}{(1+\beta e^{\gamma x})^4} = -\alpha\beta\gamma^2 e^{\gamma x} \frac{1-\beta e^{\gamma x}}{(1+\beta e^{\gamma x})^3}.$$

Let the parameter vector lie in a subspace Ω_0 of Ω where $f(x, \theta)$ has an inflection point (x_ω, η_ω). In this point the numerator of the second derivative has to be zero so that

$$1-\beta e^{\gamma x_\omega} = 0; \quad \beta = e^{-\gamma x_\omega} \text{ and } x_\omega = -\frac{1}{\gamma}\ln\beta$$

and

$$f_\omega = \frac{\alpha}{2} = \eta_\omega.$$

At x_ω the second derivative $\frac{d^2 f(x,\theta)}{dx^2}$ changes its sign and we really get an inflection point.

Because $\beta = e^{-\gamma x_\omega}$, we get

$$f(x,\theta) = \frac{\alpha}{1+\beta e^{\gamma x}} = \alpha\frac{e^{\gamma x_\omega}}{e^{\gamma x_\omega} + e^{\gamma x}} = \frac{\alpha}{2}\left\{1 + \frac{e^{-\gamma x}e^{\gamma x_\omega} - 1}{e^{-\gamma x}e^{\gamma x_\omega} + 1}\right\}$$
$$= \frac{\alpha}{2}\left\{1 + \tanh\left[-\frac{\gamma}{2}(x - x_\omega)\right]\right\} = a\{1 + \tanh[b(x-c)]\} = \psi(x,\vartheta^*) \tag{9.27}$$

with

$$\theta^{*T} = (a,b,c), \quad a = \frac{\alpha}{2}, \quad b = -\frac{\gamma}{2}, \quad c = -\frac{1}{\gamma}\ln\beta.$$

Both versions of function can also be written as a three-parametric hyperbolic tangent function. The parameter c is the x-coordinate of the inflection point and $\alpha/2$ the y-coordinate of the inflection point.

Both versions of $f(x, \theta)$ in (9.27) have, of course, the same expectation curve.

The function of Example 9.3 written as hyperbolic tangent function is

$$f(x,\theta) = 5\left\{1+ \tanh\left[-\frac{\theta}{2}x - \frac{\ln 2}{2}\right]\right\},$$

and the vector of the coordinates of the expectation curve is

$$\eta(\theta) = \left\{5\left(1+ \tanh\left[-\frac{1}{2}(\ln 2+\theta)\right]\right), 5\left(1+ \tanh\left[-\frac{1}{2}(\ln 2+2\theta)\right]\right)\right\}^T.$$

If as the result of parametrisation a new parameter depends non-linear on the original parameter, the curvature changes as is shown below.

Example 9.5 The function

$$f(x,\theta) = (\alpha+\beta e^{\gamma x})^{\delta} \quad \left(\theta = (\alpha,\beta,\gamma,\delta)^T, \alpha\beta<0, \gamma>0, \delta>0\right) \tag{9.28}$$

was used in Richards (1959) to model the growth of plants under the restrictions in (9.28) of the parameter space. The function (9.28) is called Richard's function. We rewrite this function also as

$$\psi(x,\theta^*) = A\left\{1+B\ \exp[(D-x)]\frac{C}{A}(B+1)^{1+1/B}\right\}^{-1/B}.$$

The connection between θ and $\theta^* = (A,B,C,D)^T$ is given by the relationship $(\theta = g^{-1}(\theta^*))$:

$$\alpha = A^{-B},\ \ \beta = BA^{-B}\exp\left[\frac{C}{A}D(B+1)^{1+1/B}\right],$$

$$\gamma = -\frac{C}{A}(B+1)^{1+1/B},\ \ \delta = -\frac{1}{B}$$

or $(\theta^* = g(\theta))$

$$A = \alpha^{\delta},\ \ B = -\frac{1}{\delta},\ \ C = -\gamma\alpha^{\delta}\left(1-\frac{1}{\delta}\right)^{\delta-1},\ \ D = -\frac{1}{\gamma}\ln\left(\frac{-\beta\delta}{\alpha}\right).$$

The parameters A, C and D can be interpreted as follows:

A: Final value $\left(A = \lim_{x\to\infty}\psi(x,\theta^*)\right)$.
D: x-coordinate of the inflection point of the curve of $\psi(x,\theta^*)$.
C: y-coordinate of $\frac{d\psi(x,\theta^*)}{dx}$ at $x = D$ (maximal growth).

This can be used to determine initial values for the iterative calculation of LS estimates.

For the case $p = 1$ Hougaard (1982) showed that a parameter transformation $\tilde{g}$ exists, leading to a parameter $\alpha = \tilde{g}(\theta)$, so that the asymptotic variance of its estimator is nearly independent of θ. Asymptotic skewness and asymptotic bias of $\hat{\boldsymbol{\alpha}}$ are zero, and the likelihood function of $\hat{\boldsymbol{\alpha}}$ is approximately that of a normal distribution, if the error terms $\boldsymbol{e}$ in (9.2) are normally distributed (this parametrisation has further parameter-dependent curvature; see Section 9.2.2).

This transformation is given by

$$\frac{\partial g(\theta)}{\partial \theta} = c\sqrt{F^{\mathrm{T}}F} \tag{9.29}$$

with an arbitrary constant c.

Generalisations of this result for $p > 1$ are given in Holland (1973) and Hougaard (1984).

Theorem 9.3 Let $\eta = \eta(\theta)$ in (9.2) in Ω three times continuously (concerning θ) differentiable and Ω be connected. Then a covariance-stabilising transformation $g = g(\theta_1, \ldots, \theta_p)$ is a solution of

$$\frac{\partial g}{\partial \theta_i \partial \theta_j} = \left(\frac{\partial g}{\partial \theta_1}, \ldots, \frac{\partial g}{\partial \theta_p}\right)\left(F^T F\right)^{-1}\left(\frac{\partial g}{\partial \theta_1}, \ldots, \frac{\partial g}{\partial \theta_p}\right)^T k_{ij}$$

with k_{ij} and F defined in (9.4).

Hougaard (1984) could show that for functions of type $f(x_i, \theta) = \theta_1 + \theta_2 h(x_i, \theta_3)$ such a parameter transformation exists.

9.2.2 Curvature Measures

As shown in Example 9.4, the same function can by reparametrisation be written in several forms.

Definition 9.5 Given two continuous differentiable functions $f(x, \theta)$ and $h(x, \delta)$, $\theta \in \Omega, \partial \in \Delta$ and let $g(\theta)$ be a one-to-one mapping of Ω on Δ. For all $x \in R$ let $f(x, g(\theta)) = h(x, \delta)$. Then we call $h(x, \delta)$ a reparametrisation of $f(x, \theta)$ (and vice versa).

Are there different properties of the estimators in a non-linear regressions model if different reparametrisation are used? Does a reparametrisation exist leading to a smaller bias as for the original parametrisation? In general there are questions about the influence of reparametrisation on the curvature. To answer such questions, we first need a definition of curvature.

The second derivative of the regression function (concerning the parameter vector) often defines curvature measures. Such a locally measure (depending on the parameter) must suitably be globalised, for instance, by a supremum (see Beale, 1960; Bates and Watts, 1988).

Morton (1987) proposed a statistically motivated curvature measures based on higher moments of a symmetrical error distribution. We follow Morton's approach.

In model equation (9.2) we assume now that the error terms $\boldsymbol{e}_i$ are identically and independently distributed (i.i.d.) with expectation 0 and positive finite variance σ^2. Further we assume that this distribution is symmetric. We write the LS estimator $\hat{\boldsymbol{\theta}} = \left(\hat{\boldsymbol{\theta}}_1,\ldots,\hat{\boldsymbol{\theta}}_p\right)^T$ in dependency of the error terms as

$$\hat{\boldsymbol{\theta}} = \hat{\boldsymbol{\theta}}(e) = \left(\hat{\boldsymbol{\theta}}_1(e),\ldots,\hat{\boldsymbol{\theta}}_p(e)\right)^T.$$

By

$$\boldsymbol{u}_j = \frac{1}{2}\left\{\hat{\boldsymbol{\theta}}_j(\boldsymbol{e}) - \hat{\boldsymbol{\theta}}_j(-\boldsymbol{e})\right\}, \tag{9.30}$$

$$\boldsymbol{v}_j = \frac{1}{2}\left\{\hat{\boldsymbol{\theta}}_j(\boldsymbol{e}) + \hat{\boldsymbol{\theta}}_j(-\boldsymbol{e})\right\} - \theta_j, \tag{9.31}$$

we receive

$$\hat{\boldsymbol{\theta}}_j = \theta_j + \boldsymbol{u}_j + \boldsymbol{v}_j\cdot$$

From these assumptions, it follows $E\left[\left\{\hat{\boldsymbol{\theta}}_j(\hat{\boldsymbol{e}})\right\}\right] = E\left[\left\{\hat{\boldsymbol{\theta}}_j(-\hat{\boldsymbol{e}})\right\}\right]$ and from this $E(\boldsymbol{u}_j) = 0$. Then the bias of the jth component of the LS estimator is

$$b_j = E\left(\hat{\boldsymbol{\theta}}_j - \theta\right) = E\left(\boldsymbol{v}_j\right)\cdot$$

Now curvature measures for the components of θ following Morton (1987) can be defined.

Definition 9.6 A measure for the curvature (non-linearity) of the jth component θ_j ($j = 1, \ldots, p$) of θ in (9.3) with the symbols in (9.30) and (9.31) is given by

$$N_j = \frac{\operatorname{var}(\boldsymbol{v}_i)}{\operatorname{var}\left(\hat{\boldsymbol{\theta}}_j\right)} = \frac{\operatorname{var}(\boldsymbol{v}_i)}{\operatorname{var}(\boldsymbol{u}_i) + \operatorname{var}(\boldsymbol{v}_i)}.$$

We define by linear regression of $\boldsymbol{v}_j$ on all $\binom{p}{2}$ products $\boldsymbol{u}_k\boldsymbol{u}_l$ for each j a $(p \times p)$–matrix C_j so that $\operatorname{cov}(\boldsymbol{u}_k\boldsymbol{u}_l, \boldsymbol{v}_{2j}) = 0$ for each pair (k, l) if

$$\boldsymbol{v}_{2j} = \boldsymbol{v}_j - \boldsymbol{v}_{1j} \text{ and } \boldsymbol{v}_{1j} = \frac{1}{2}\boldsymbol{u}^{\mathrm{T}} C_j \boldsymbol{u}$$

with $\boldsymbol{u} = (\boldsymbol{u}_1, \ldots, \boldsymbol{u}_p)^T$. The special choice of C_j has the advantage that the two components $\boldsymbol{v}_1$ and $\boldsymbol{v}_2$ of $\boldsymbol{v}$ are uncorrelated and

$$\operatorname{var}(\boldsymbol{v}_j) = \operatorname{var}(\boldsymbol{v}_{1j}) + \operatorname{var}(\boldsymbol{v}_{2j})$$

follows.

In the following we decompose the curvature measure in Definition 9.6 into two parts. The first part becomes small by a suitable reparametrisation (theoretically 0).

Definition 9.7 We call

$$N_{1j} = \frac{\operatorname{var}(\boldsymbol{v}_{1j})}{\operatorname{var}(\hat{\boldsymbol{\theta}}_j)} \tag{9.32}$$

reparametrisation-dependent curvature (non-linearity) of the component θ_j of θ and

$$N_{2j} = \frac{\operatorname{var}(\boldsymbol{v}_{2j})}{\operatorname{var}(\hat{\boldsymbol{\theta}}_j)} \tag{9.33}$$

intrinsically curvature (non-linearity) of the component θ_j of θ.

It is easy to see that $N_j = N_{1j} + N_{2j}$.

Morton (1987) made a proposal for finding a suitable reparametrisation. With the matrix F in (9.4), we write

$$\frac{1}{n} F^T(\theta) F(\theta) = I_n(\theta) = (m_{ij}) \text{ and } I_n^{-1}(\theta) = (m^{ij}).$$

Let L be defined by $L^T I_n(\theta) L = I_p$. With the symbols in (9.4) we define

$$t_{uvj} = \sum_{i=1}^{n} k_{uv}(x_i,\theta) f_i(x_i,\theta), \quad t_{uvjl} = \sum_{i=1}^{n} k_{uv}(x_i,\theta) k_{jl}(x_i,\theta),$$

$$D_j = (d_{uvj}) \text{ with } d_{uvj} = \sum_{l=1}^{n} m^{jl} t_{uvl}$$

and

$$a_j = \sqrt{N_{1j}} \frac{\operatorname{cov}(\boldsymbol{u}_j^2, \boldsymbol{v}_j)}{\sqrt{\operatorname{var}(\boldsymbol{u}_j^2) \operatorname{var}(\boldsymbol{v}_j)}}.$$

Morton showed that the first-order approximation below pertains to

$$N_{1j} \approx \frac{\sigma^2}{2nm^{jj}} \operatorname{tr}\left\{ \left(L^T D_j L\right)^2 \right\},$$

$$N_{2j} \approx \frac{\sigma^2}{nm^{jj}} \sum_{u,v,k,l} m^{ju} m^{jv} m^{kl} \left\{ t_{kluv} + \sum_s t_{kus} d_{lvs} \right\},$$

$$b_j \approx -\frac{\sigma^2}{2n} \operatorname{tr}\left\{ L^T D_j L \right\}.$$

Further

$$\operatorname{var}\left(\boldsymbol{u}_j\right) \approx \frac{\sigma^2}{n} m^{jj},$$

$$\operatorname{var}\left(\boldsymbol{v}_{1j}\right) \approx \frac{\sigma^4}{2n^2} \operatorname{tr}\left\{ \left(L^T D_j L\right)^2 \right\},$$

$$\operatorname{var}\left(\boldsymbol{v}_{2j}\right) \approx \frac{\sigma^2}{n^2} \sum_{u,v,k,l} m^{ju} m^{jv} m^{kl} \left\{ t_{kluv} + \sum_s t_{kus} d_{lvs} \right\}.$$

Definition 9.8 $\theta^* = \left(\vartheta_1^*, \dots, \vartheta_p^*\right)^T$ is called with

$$\theta_j^* = \frac{a_j}{\sqrt{2\left\{ \operatorname{var}\left(\boldsymbol{u}_j\right) - \operatorname{var}\left(\boldsymbol{v}_{1j}\right) - \operatorname{var}\left(\boldsymbol{v}_{2j}\right) \right\}}} \quad (j = 1, \dots, p)$$

N_{1j} – optimal reparametrisation of $f(\theta)$ in (9.3) for all N_{ij}.

9.3 Asymptotic Properties and the Bias of LS Estimators

The properties of LS estimators differ strongly between intrinsically non-linear and linear (including quasilinear) regression. In the intrinsically non-linear regression, we know nearly nothing about the distribution of $\hat{\boldsymbol{\theta}} = \hat{\boldsymbol{\theta}}_n$, $\boldsymbol{s}^2$ and $\tilde{\boldsymbol{\sigma}}^2$.

The magnitude

$$\frac{1}{\sigma^2} \boldsymbol{s}^2 (n-p) = \frac{1}{\sigma^2} \tilde{\boldsymbol{\sigma}}^2 \cdot n$$

is not chi-squared-distributed, even if the error terms are normally distributed. Also the bias of

$$v_n(\theta) = E\left[\hat{\boldsymbol{\theta}}_n - \theta\right] \tag{9.34}$$

is only approximatively known. Nevertheless, in the next section we propose confidence estimators and tests, which hold pregiven risks approximately, and the larger the sample size, the better. We show this in this section by presenting the important results of Jennrich (1969) and Johansen (1984) and concerning the bias of the estimators' results from Box (1971).

We assume that the parameter space $\Omega \subset R^p$ is compact and $f(x, \theta)$ is twice continuously differentiable concerning θ.

At first we introduce Jennrich's tail products that simplify the presentation below.

For finite n the set of measurement points $(x_1, \ldots, x_n)$ (the support of a discrete design) can formally be considered as a discrete probability measure with a distribution function $F_n(x)$ (even if here is no random variable x). If n tends to infinity (∞), then $F_n(x)$ tends against a limiting distribution function $F(x)$. Then for some restricted continuous functions s and t with $s, t: R \otimes \Omega \to R$ and $(\theta, \theta) \in \Omega \otimes \Omega$ we define

$$\int_x s(x,\theta)t(x,\theta^*)\mathrm{d}F(x) = (s(\theta), t(\theta^*)). \tag{9.35}$$

Definition 9.9 We say the sequence $\{g_i\}$ $(i = 1, 2, \ldots)$ of functions $g_i : R \otimes \Omega \to R$ has a tail product (g, g) as in (9.35), if

$$\frac{1}{n}\sum_{i=1}^{n} g_i(\theta)g_i(\theta^*), \quad \theta,\theta^* \in \Omega,$$

tends for $n \to \infty$ uniformly in $(\theta, \theta^*) \in \Omega \times \Omega$ against (g, g). If $\{g_i\}$ and $\{h_i\}$ are two sequences $g_i : R \otimes \Omega \to R$, $h_i ; R \otimes \Omega \to R$, we say that these sequences have a tailed cross product (g, h), if

$$\frac{1}{n}\sum_{i=1}^{n} g_i(\theta)\, h_i(\theta^*), \quad \theta,\theta^* \in \Omega,$$

converges for all $(\theta, \theta^*) \in \Omega \otimes \Omega$ uniformly against (g, h).

From the continuity of all g_i and h_i and the uniform convergence, the continuity of (g, g) and (g, h) follows.

To understand better the following theorems, we need an extended definition of an almost sure convergence for the case of a sequence of random functions depending on a parameter θ.

In the non-stochastic case, uniform convergence (for all $\theta \in \Omega$) is defined by the demand that for a function sequence $\{f_i(\theta)\}$ the quantity

$$\sup_{\theta \in \Omega} |f_i(\theta) - f(\theta)|$$

for $i \to \infty$ tends against zero.

For random functions we extend this by

Definition 9.10 If $f(\theta)$ and $f_i(\theta)$ $(i = 1, 2, \ldots)$ are random functions for $\theta \in \Omega \subset R^p$ and if $(\{Y\}, \Omega, P)$ is the common probability space of the arguments of f and all f_i, then f_i converge uniformly almost sure in Ω against f, if all

$$\sup_{\theta \in \Omega} |f_i(\theta) - f(\theta)| \quad \text{for} \quad i = 1,2,\ldots$$

are random variables except for a set of P-measure 0 for all elements $Y \in \{Y\}$ (i.e. for all $Y \in \{Y\} \backslash N$ with P-null set N) and for all $\varepsilon > 0$ there exists a $n_0(Y, \varepsilon)$ so that for $i \geq n_0(Y, \varepsilon)$

$$\sup_{\theta \in \Omega} |f_i(\theta, Y) - f(\theta, Y)| \leq \varepsilon.$$

The proof of Theorem 9.4 is based on

Lemma 9.1 (Borel–Cantelli)
If y and $y_1, y_2, \ldots$ are random variables with a common probability space $(\{Y\}, \Omega, P)$ and if for all $\varepsilon > 0$

$$\sum_i P\{|y_i - y|\} > \varepsilon\} < \infty,$$

the sequence $\{y_i\}$ converges almost sure against y.

The proof can, for instance, be found in Bauer (2002, p. 73) or in Feller (1961). The proof of the lemma below can be found in Jennrich (1969, p. 637).

Lemma 9.2 Let $R = R(Y, \theta)$ be a real-valued function on $R^n \times \Omega$ and Ω a compact subset of R^p and let $R(Y, \theta)$ for all $\theta \in \Omega$ be a (continuous in θ) measurable function of Y for all $Y \in \{Y\}$, then a measurable mapping $\hat{\theta}$ of $\{Y\}$ in Ω exists so that for all $Y \in \{Y\}$

$$R\left[Y, \hat{\theta}(Y)\right] = \inf_{\theta \in \Omega} R(Y, \theta).$$

From this lemma it follows that LS estimators really are random variables.

Theorem 9.4 Let $g_i : \Omega \to R$ be continuous mappings of the parameter space Ω in R; the sequence $\{g_i\}$ may have a tail product. If $\{u_i\}$ is a sequence of independent $N(0, \sigma^2)$-distributed random variables, then

$$z_n = \frac{1}{n} \sum_{i=1}^{n} u_i g_i(\theta) \quad (n = 1,2,\ldots)$$

converges almost certain uniformly in Ω against 0.

We write the LS estimator introduced in Definition 9.2 now as $\hat{\boldsymbol{\theta}} = \hat{\boldsymbol{\theta}}_n$ (it is under normality assumption a MLS), and with $R(\theta)$ in (9.5) we write

$$\tilde{\sigma}_n^2 = \frac{1}{n} R\left(\hat{\boldsymbol{\theta}}_n\right).$$

Theorem 9.5 (Jennrich)
If with the notations of (9.2) to (9.5) $\boldsymbol{e}_i$ are pairwise independent and identical with $E(\boldsymbol{e}_i) = 0$ and $\text{var}(\boldsymbol{e}_i) = \sigma^2$ normally distributed and Ω is compact and if

$$(f_i) = (f_i(\theta)) = (f(x_i, \theta))$$

has a tail product and

$$S(\theta^*) = \left|(f(\theta^*), f(\theta))\right|^2, \quad (\theta, \theta^*) \in \Omega \otimes \Omega$$

has a unique minimum at $\theta^* = \theta$, then $\hat{\boldsymbol{\theta}}_n$ converges uniformly in θ almost certain against θ and $\tilde{\boldsymbol{\sigma}}_n^2$ converges uniformly in θ almost certain against σ^2.

For testing hypotheses and for confidence estimations, Theorem 9.6 is important.

Theorem 9.6 (Jennrich)
Let the assumption of Theorem 9.5 be fulfilled and $f(Y, \theta)$ be twice continuously differentiable concerning θ. Let the function sequences $\{f(\boldsymbol{y}_i, \theta)\}$, $\{f_j(\boldsymbol{y}_i, \theta)\}$ $(j = 1, \ldots, p)$ and $\{k_{jl}(\boldsymbol{y}_i, \theta)\}$ $(j, l = 1, \ldots, p)$ in (9.4) have tail products and tailed cross products, respectively. Let

$$I(\theta) = \lim_{n \to \infty} \frac{1}{n} \sum_{i=1}^{n} F_i^T(\theta) F_i(\theta) \tag{9.36}$$

be non-singular.
Then for each θ from the inward of Ω

$$\sqrt{n}\left(\hat{\boldsymbol{\theta}}_n - \theta\right) \tag{9.37}$$

is asymptotically $N(0_p, \sigma^2 I^{-1}(\theta))$-distributed.

The proof is given in Jennrich (1969, p. 639).

We formulate the message of Theorem 9.6 so that $\hat{\boldsymbol{\theta}}_n$ is asymptotically $N(\theta, \sum)$-distributed with $\sum = \lim_{n \to \infty} n \cdot \text{var}_A(\theta)$ and

$$\text{var}_A(\theta) = \sigma^2 \left[F^T(\theta) F(\theta)\right]^{-1} = \sigma^2 \left[\sum_{i=1}^{n} F_i^T(\theta) F_i(\theta)\right]^{-1}. \tag{9.38}$$

We call $\text{var}_A(\theta)$ the asymptotic covariance matrix of $\hat{\boldsymbol{\theta}}_n$ and

$$\widehat{\text{var}_A\left(\hat{\theta}_n\right)} = \frac{1}{n-p} R\left(\hat{\theta}_n\right) \left[F^T\left(\hat{\theta}_n\right) F\left(\hat{\theta}_n\right)\right]^{-1} = s_n^2 \left[F^T\left(\hat{\theta}_n\right) F\left(\hat{\theta}_n\right)\right]^{-1} \tag{9.39}$$

the estimated asymptotic covariance matrix of $\hat{\boldsymbol{\theta}}_n$. Moreover,

$$\boldsymbol{s}_n^2 = \frac{1}{n-p} R\left(\hat{\boldsymbol{\theta}}_n\right) \tag{9.40}$$

is an estimator of σ^2 and asymptotic equivalent with $\tilde{\boldsymbol{\sigma}}_n^2$.

Simulation results (see Rasch and Schimke, 1983) show that $\boldsymbol{s}_n^2$ has a smaller bias than $\tilde{\boldsymbol{\sigma}}_n^2$. In the paper of Malinvaud (1970) it was shown (independent of Jennrich) that $\hat{\boldsymbol{\theta}}_n$ is a consistent estimator concerning θ.

Generalisations of the results of Jennrich (mainly with general error distributions) are given in Wu (1981) and in Ivanov and Zwanzig (1983).

Next we will discuss the bias $v_n(\theta)$ of $\hat{\boldsymbol{\theta}}_n$.

Theorem 9.7 (Box, 1971)
With the assumptions of Theorem 9.6 and if

$$\Delta = \hat{\theta}_n - \theta$$

approximately (first order) ($e = Y - \eta(\theta)$ with $\eta(\theta)$ from (9.4)) with certain matrices $A_{p,n}, B_{n,n}^{(1)}, \ldots, B_{n,n}^{(p)}$ has the form

$$\Delta = A_{p,n}e + \left(e^T B_{n,n}^{(1)} e, \ldots, e^T B_{n,n}^{(p)} e\right)^T, \tag{9.41}$$

then we get with the notations in (9.4) approximately

$$v_n(\theta) = \frac{1}{2\sigma^2} \mathrm{var}_A(\theta) \sum_{i=1}^{n} F_i^T(\theta) \mathrm{tr}\left\{ \left[F^T(\theta) F(\theta)\right]^{-1} K_i(\theta) \right\}. \tag{9.42}$$

The proof can be found in Box (1971), where it is further shown how the matrices $A_{p,n}, B_{n,n}^{(1)}, \ldots, B_{n,n}^{(p)}$ suitably can be chosen.

Close relations exist between (9.42) and the curvature measures (see Morton, 1987).

9.4 Confidence Estimations and Tests

Confidence estimations and tests for the parameters of intrinsically non-linear regression functions or even for regression functions cannot so easily be constructed as in the linear case. The reason is that the estimators of θ and of functions of θ cannot be explicitly written down in closed form and their distribution is unknown.

9.4.1 Introduction

Special intrinsically non-linear regression functions in the applications are not written in the form $f(x, \theta)$ as in Definition 9.1 and the theoretical first part of this chapter. If there are only few (two, three or four) parameters as in Section 9.5, we often write $\theta_1 = \alpha\,; \theta_2 = \beta\,; \theta_3 = \gamma\,; \theta_4 = \delta$.

Therefore the first kind risk from now on in this chapter is denoted by α^* and consequently we have $(1 - \alpha^*)$ – confidence intervals. Analogous the second kind risk is β^*.

Regarding properties of $(1 - \alpha^*)$-confidence intervals and α^*-tests, nearly nothing is known, we are glad if we can construct them in some cases. We restrict ourselves at first on the construction of confidence estimations $K(Y)$ concerning θ and define a test of $H_0: \theta = \theta_0$ as

$$k(\boldsymbol{Y}) = \begin{cases} 1, & \text{if } \theta_0 \in K(\boldsymbol{Y}) \\ 0, & \text{otherwise.} \end{cases}$$

Concerning confidence estimators for $\eta(\theta)$, we refer to Maritz (1962).

Williams (1962) developed a method for construction of confidence intervals for the parameter γ in non-linear functions of the type

$$f(x,\theta) = \alpha + \beta g(x,\gamma), \theta = (\alpha,\beta,\gamma)^T, \tag{9.43}$$

with a real-valued function $g(x, \gamma)$, which is twice continuously differentiable for γ. Halperin (1963) generalised this method in such a way that confidence intervals for all components of θ can be constructed.

We consider the vector

$$[f(x_1,\theta),\ldots,f(x_n,\theta)]^T = B\lambda \tag{9.44}$$

with

$$\theta = (\lambda_1,\ldots,\lambda_{p-r},\varphi_1,\ldots,\varphi_r)^T \in \Omega = \Lambda \otimes \Gamma,$$

$$\varphi = (\varphi_1,\ldots,\varphi_r)^T \in \Lambda,\ \lambda = (\lambda_1,\ldots,\lambda_{p-r})^T \in \Gamma$$

and $p < n$. The $p < n(n \times (p - r))$-matrix B contains the elements $b_j(x_i, \varphi)$. The $b_j(x_i, \varphi)$ do not depend on λ and concerning φ are twice continuously differentiable. The matrix B for $\varphi \neq 0_r$ has the rank r.

We start with the model

$$\boldsymbol{Y} = B\lambda + \boldsymbol{e},\ \ \boldsymbol{e} \sim N(0_n, \sigma^2 I_n). \tag{9.45}$$

with $\beta^T(\lambda^T, 0_r^T) = (\theta_r^T, 0_r^T)$ and a $(n \times r)$-matrix D, so that(B, D) is of rank p and (9.45) can be written as

$$\boldsymbol{Y} = (B,D)\beta + \boldsymbol{e}. \tag{9.46}$$

By Theorem 8.1 we obtain LS estimates of θ_l and 0_r from

$$\hat{\boldsymbol{\theta}}_l = (B^T B)^{-1} B^T \boldsymbol{Y} - (B^T B)^{-1}(B^T B)(U^T U)^{-1} U^T \boldsymbol{Y}$$

and

$$\hat{\mathbf{0}}_r = (U^T U)^{-1} U^T \boldsymbol{Y},$$

respectively, where

$$U^T = D^T\left(I_n - B\left(B^T B\right)^{-1} B^T\right),$$

as the solutions

$$\hat{\boldsymbol{\beta}} = \begin{pmatrix} B^T B & B^T D \\ D^T B & D^T D \end{pmatrix}^{-1} \cdot \begin{pmatrix} B^T \\ D^T \end{pmatrix} \boldsymbol{Y}$$

of the simultaneous normal equations. These estimates depend on φ.

It follows from Theorem 8.2 that $\hat{\theta}_I$ and $\hat{\mathbf{0}}_r$ are BLUE (because we assumed normal distribution even LVUE) concerning θ_I and 0_r if λ is known. From Theorem 8.6 it follows that

$$\boldsymbol{F}_1 = \frac{n-p}{p} \frac{\left(\hat{\boldsymbol{\beta}} - \beta\right)^T (B,D)^T (B,D) \left(\hat{\boldsymbol{\beta}} - \beta\right)}{\boldsymbol{Y}^T \boldsymbol{Y} - \hat{\boldsymbol{\beta}}^T (B,D)^T (B,D) \hat{\boldsymbol{\beta}}} \tag{9.47}$$

is $F(p, n-p)$-distributed and

$$\boldsymbol{F}_2 = \frac{n-p}{r} \frac{\hat{\mathbf{0}}_r^T U^T Y \hat{\mathbf{0}}_r}{\boldsymbol{Y}^T \boldsymbol{Y} - \hat{\boldsymbol{\beta}}^T (B,D)^T (B,D) \hat{\boldsymbol{\beta}}} \tag{9.48}$$

is $F(r, n-p)$-distributed.

With $\boldsymbol{F}_1$ confidence regions concerning θ and with $\boldsymbol{F}_2$ concerning φ can be constructed, due to

Theorem 9.8 The set of all $\theta \in \Omega$ of model (9.45) with

$$F_1 \le F(p, n-p | 1-\alpha^* |) \tag{9.49}$$

defines a $(1-\alpha^*)$ confidence region concerning θ, and the set of all $\varphi \in \Gamma$ with

$$F_2 \le F(r, n-p | 1-\alpha^* |) \tag{9.50}$$

defines a $(1-\alpha^*)$-confidence region concerning φ if D is independent of λ.

Williams (1962) and Halperin (1963) proposed to choose D in such a way that F_2 disappears, if $\varphi = \hat{\varphi}$, that is, φ equals its LS estimate. From (9.7) we see that $\hat{\varphi}$ is solution of

$$\left.\begin{aligned} &\lambda^T \frac{\partial B^T}{\partial \varphi_j}(Y - B\lambda) = 0 \quad (j = 1, \ldots, r), \\ &e_l(B^T Y - B^T B\lambda) = 0. \end{aligned}\right\} \tag{9.51}$$

With the additional assumption that in each column of B exactly one component of φ occurs so that

$$\frac{\partial b_k\left(x_j, \varphi\right)}{\partial \varphi_j}$$

differs from zero for exactly one $k^* = k(j)$, it follows from (9.51)

$$\frac{\partial B^T}{\partial \varphi_j}(Y - B\lambda) = 0, \ 1_l\left(B^T Y - B^T B\lambda\right) = 0,$$

and we choose for D

$$d = \left(d_{ij}\right) = \left(\sum_{k=1}^{p-r} \frac{\partial b_k\left(x_j, \varphi\right)}{\partial \varphi_j}\right). \tag{9.52}$$

In (9.52) each sum has exactly one summand different from zero. The calculation of confidence regions as described above is laborious as shown in

Example 9.6 (Williams, 1962). Let

$$f(x_i, \theta) = \alpha + \beta \mathrm{e}^{\gamma x_1}.$$

With

$$g(x, \gamma) = \mathrm{e}^{\gamma x}, \ B^T = \begin{pmatrix} 1 & 1 & \dots & 1 \\ \mathrm{e}^{\gamma x_1} & \mathrm{e}^{\gamma x_2} & \dots & \mathrm{e}^{\gamma x_n} \end{pmatrix} \text{ and } \beta^{\mathrm{T}} = (\alpha, \beta, 0),$$

the model has the form (9.46) $(p = 3, r = 1, n > 3)$. Because

$$d_{il} = d_i = \frac{\partial}{\partial \gamma} 1 + \frac{\partial}{\partial \gamma} \mathrm{e}^{\gamma x_i} = x_i \mathrm{e}^{\gamma x_i},$$

we get

$$D^T = (x_i \mathrm{e}^{\gamma x_i}, \dots, x_n \mathrm{e}^{\gamma x_n}),$$

$$B^T B = \begin{pmatrix} n & \sum \mathrm{e}^{\gamma x_i} \\ \sum \mathrm{e}^{\gamma x_i} & \sum \mathrm{e}^{2\gamma x_i} \end{pmatrix}, \ B^T D = \left(\sum x_i \mathrm{e}^{\gamma x_i} \ \sum x_i \mathrm{e}^{2\gamma x_i}\right),$$

$$D^T D = \sum x_i^2 \mathrm{e}^{2\gamma x_i},$$

$$\left(B^T B\right)^{-1} = \frac{1}{n \sum \mathrm{e}^{2\gamma x_i} - (\sum \mathrm{e}^{\gamma x_i})^2} \begin{pmatrix} \sum \mathrm{e}^{2\gamma x_i} & -\sum \mathrm{e}^{\gamma x_i} \\ -\sum \mathrm{e}^{\gamma x_i} & n \end{pmatrix}.$$

The elements u_{lk} of U are

$$u_{lk} = x_l \mathrm{e}^{\gamma x_i} - \frac{\sum \mathrm{e}^{2\gamma x_i} - (x_l + x_k) \sum \mathrm{e}^{\gamma x_i'} + x_l x_k}{n \sum \mathrm{e}^{2\gamma x_i} - (\sum \mathrm{e}^{\gamma x_i})^2}.$$

The corresponding quantities are now inserted in F_2 and the values $\gamma_l(1 - \alpha^*)$ as lower and $\gamma_u(1 - \alpha^*)$ as upper bound of a realised $(1 - \alpha^*)$-confidence interval, respectively, can now be calculated iteratively.

9.4.2 Tests and Confidence Estimations Based on the Asymptotic Covariance Matrix

For practical applications the solutions given above are strongly restricted and awkward. A way out could be to use the asymptotic covariance matrix (9.38) or the estimated asymptotic covariance matrix (9.39) and in analogy to the linear case to construct simple tests and confidence estimations with the quantiles of the central t-distribution or with the normal distribution. This was done by Bliss and James (1968) for hyperbolic models.

It is not clear whether these tests really are α^*-tests and the confidence intervals are $(1-\alpha^*)$-confidence intervals and what the power of those tests is. Because there is no theoretical solution, such questions can only be answered by simulation experiments. We here demonstrate the method and in Section 9.4.3 the verification by simulation experiments.

In Section 9.6 we present results of simulation experiments for special functions. Heuristically tests and confidence estimations based on the asymptotic covariance matrix $\text{var}_A(\theta)$ of the LS estimator $\hat{\boldsymbol{\theta}}$ can be introduced as follows:

In

$$\text{var}_A(\theta) = \left(\sigma^2 v_{jk}\right)\ (j,k=1,\ldots,p),$$

let us replace θ by its LS estimator $\hat{\boldsymbol{\theta}}$ and estimate σ^2 by

$$\boldsymbol{s}^2 = \frac{R\left(\hat{\boldsymbol{\theta}}\right)}{n-p}$$

with $R(\theta)$ in (9.5). This gives us the estimated asymptotic covariance matrix in (9.39) now as

$$\text{var}_A\left(\hat{\boldsymbol{\theta}}\right) = \left(\boldsymbol{s}^2\hat{\boldsymbol{v}}_{jk}\right). \tag{9.53}$$

To test for an arbitrary j ($j=1,\ \ldots,p$) the null hypothesis $H_{0j}:\theta_j=\theta_{j0}$ against H_{Aj}: $\theta_j \neq \theta_{j0}$ analogously to the linear case, we propose to use the test statistic

$$\boldsymbol{t}_j = \frac{\hat{\boldsymbol{\theta}}_j - \theta_{j0}}{\boldsymbol{s}\sqrt{\hat{\boldsymbol{v}}_{jj}}} \tag{9.54}$$

to define a test with a nominal risk of first kind α_{nom},

$$k_j(Y) = \begin{cases} 1, & \text{if} \quad |\,t_j\,| > t\left(n-p\,|\,1-\dfrac{\alpha_{nom}}{2}\right) \\ 0, & \text{otherwise} \end{cases} \tag{9.55}$$

A confidence interval concerning the component θ_j of θ is analogously defined as

$$\left[\hat{\boldsymbol{\theta}}_j - \boldsymbol{s}\sqrt{\hat{\boldsymbol{v}}_{jj}}\,t\left(n-p\,|\,1-\frac{\alpha_{nom}}{2}\right);\hat{\boldsymbol{\theta}}_j + \boldsymbol{s}\sqrt{\hat{\boldsymbol{v}}_{jj}}\,t\left(n-p\,|\,1-\frac{\alpha_{nom}}{2}\right)\right]. \tag{9.56}$$

Schmidt (1979) proposed to use in place of (9.54) a z-test statistic

$$z_j = \frac{\hat{\theta}_j - \theta_{j0}}{\sigma\sqrt{v_{jj}}}. \tag{9.57}$$

But the corresponding test is often non-recommendable if $n < 20$ but just these cases are often of interest.

9.4.3 Simulation Experiments to Check Asymptotic Tests and Confidence Estimations

In mathematics, if we cannot obtain results in an analytic way, we are in the same situation as scientists in empirical sciences. The most important means of knowledge acquisition in empirical sciences is the experiment (a trial). To get from experiments statements with pregiven precision, an experiment has to be planned. Experiments in statistics are often based on simulated samples; we could speak about empirical mathematics. How important such an approach has become in the meantime can be seen from the fact that in 2016 the Eighth International Workshop on Simulation was held in Vienna, the series of workshops started in May 1994 in *St. Petersburg* (Russia) (see also Chapter 1).

Most information below is based on research project by more than 20 statisticians during the years 1980–1990. A summary of the robust results is given in Rasch and Guiard (2004).

The number of samples (simulations), also called runs, has – in the same way as in real experiments – to be derived in dependency on precision requirements demanded in advance.

If it is the aim of a simulation experiment to determine the risk of the first kind α^* of tests or the confidence level $1 - \alpha^*$ of a confidence estimation based on asymptotic distributions, we fix a nominal value α_{nom}, called nominal α^* in the t-quantile of the tests or the confidence estimator. By simulating the situation of the null hypothesis repeatedly, we count the relative frequency of rejecting the null hypothesis (wrongly of course) and call it the actual risk of the first kind α_{act}. We remember the Definition 3.8 of 'robustness'.

If a probability α_{act} shall be estimated by a confidence interval so that the half expected width of that interval is not above 0.005, then we need about $N = 10\,000$ runs, if $\alpha_{act} = 0.05$. Simulation experiments in Section 9.6 are therefore based on 10 000 runs. In the parameter space Ω, we define a subspace of practically relevant parameter values $\theta^{(r)}$ $(r = 1, \ldots, R)$, and for each r a simulation experiment with $N = 10\,000$ runs was made. (Components of θ with no influence on the method have been fixed at some arbitrary value.) If a method is $100(1 - \varepsilon)\%$ robust (in the sense of Definition 3.8) for the R extreme point, then we argue that this is also the case in the practically relevant part of Ω.

Let θ^* be an arbitrary of these vectors $\theta^{(r)}$ with $\theta^* = \left(\theta_1^*, \ldots, \theta_p^*\right)^T$. Due to the connections between confidence estimations and tests, we restrict ourselves to tests in the following (Chapter 3).

The hypothesis $H_{0j} : \theta_j = \theta_j^* = \theta_{j0}$ has to be tested against $H_{Aj} : \theta_j^* \neq \theta_{j0}$ with the test statistic (9.54). For each of the 10 000 runs, we use the same sample size $n \geq p + 1$ for each test and add to the function $f(x_i, \theta^*)(i = 1, \ldots, n)$ at n in $[x_l, x_o]$ fixed support points x_i pseudorandom numbers e_i from a distribution with expectation 0 and variance σ^2. Then for each i

$$y_i = f(x_i, \theta^*) + e_i \; (i = 1, \ldots, n; x_i \in [x_l, x_u])$$

is a simulated observation. We calculate then from the n simulated observation the LS estimate $\hat{\theta}$ and the estimate s^2 of σ^2 and the test statistic (9.54). We obtain 10 000 estimations $\hat{\theta}$ and s^2 calculate the empirical means, variances, covariances, skewness and kurtosis of the components of $\hat{\theta}$ and of s^2. Then we count how often for a test statistic t_j from (9.54)

$$t_j < -t\left(n-1 \mid 1 - \frac{\alpha_{nom}}{2}\right); -t\left(n-1 \mid 1 - \frac{\alpha_{nom}}{2}\right) \leq t_j \leq t\left(n-1 \mid 1 - \frac{\alpha_{nom}}{2}\right)$$

and

$$t_j > t\left(n-1 \mid 1 - \frac{\alpha_{nom}}{2}\right) \quad (j = 1, \ldots, 10\,000)$$

occurred (the null hypothesis was always correct), divided by 10 000, giving an estimate of α_{act}. Further 10 000 runs to test $H_0 : \theta_j = \theta_j^* = \Delta_l$ with three Δ_l values have been performed to get information about the power. The most simulation experiment used besides normally distributed $\boldsymbol{e}_i$ also error terms $\boldsymbol{e}_i$ with the following pairs of skewness γ_1 and kurtosis γ_2

γ_1	0	1	0	1.5	0	2
γ_2	1.5	1.5	3.75	3.75	7	7

to investigate the robustness of statistical methods against non-normality. For the generation of pseudorandom numbers with these moments, we used the following distribution system.

Definition 9.11 A distribution belongs to the Fleishman system (Fleishman, 1978) H if its first four moments exist and if it is the distribution of the transform

$$\boldsymbol{y} = a + b\boldsymbol{x} + c\boldsymbol{x}^2 + d\boldsymbol{x}^3$$

where $\boldsymbol{x}$ is a standard normal random variable (with mean 0 and variance 1).

By a proper choice of the coefficient a, b, c and d, the random variable $\boldsymbol{y}$ will have any quadruple of first four moments $(\mu, \sigma^2, \gamma_1, \gamma_2)$. For instance, any normal distribution (i.e. any element of G) with mean μ and variance σ^2 can be represented as a member of the Fleishman system by choosing $a = \mu$, $b = \sigma$ and $c = d = 0$. This shows that we really have $H \supset G$ as demanded in Definition 3.8.

Nowadays we have convenient computer packages for simulation. A package described in Yanagida (2017) demonstrates simulation tools and the package Yanagida (2016) allows the generation of any member of the Fleishman system H. More information about statistics and simulation can be found in Rasch and Melas (2017).

Results of simulation experiments for several regression functions are given in Section 9.6.

9.5 Optimal Experimental Design

We reflect on the definition of experimental design problems in Section 1.5 but do not consider cost. If we use a quadratic loss function based on $R(\theta)$ in (9.5) and as the statistical approach point estimators concerning θ with the LS estimator $\hat{\theta}$, in Definition 9.2, the choice of a suitable risk function is the next step. A good overview about the choice of risk functions is given in Melas (2008).

A functional of the covariance matrix of $\hat{\boldsymbol{\theta}}$ cannot be used because it is unknown. We may choose a risk function based on the asymptotic covariance matrix $\text{var}_A(\theta)$ in (9.39) or on the approximate covariance matrix of Clarke (1980) or on an asymptotic covariance matrix derived from asymptotic expansions of higher order (see Pazman, 1985).

We use the first possibility for which already many results can be found. First we consider the optimal choice of the support points for a given number n of measurements and give later hints to the minimal choice of n in such a way that the value of the risk function is just below a given bound. A disadvantage of the experimental design in intrinsically non-linear regression is the fact that the optimal design depends on at least one value of the unknown parameter vector. For practical purposes we proceed as follows. We use a priori knowledge θ_0 about θ defining a region $U(\theta_0)$ where the parameters θ is conjectured. Then we determine the optimal design at that value $\theta \in U(\theta_0)$, leading to the maximal risk of the optimal designs in $U(\theta_0)$. The size of the experiment n at this place gives an upper bound for the risk in $U(\theta_0)$, because the position of the support points often only slightly depends on θ (see Rasch, 1993). This must be checked for each special function separately as done in Section 9.6.

Definition 9.12 A scheme

$$V_{n,m} = \begin{pmatrix} x_1, \dots, x_m \\ n_1, \dots, n_m \end{pmatrix}, x_i \in [x_l, x_u], n_i \text{ integer}, \sum_{i=1}^{m} n_i = n$$

is called a concrete m-point design or a m-point design for short with the support $S_m = (x_1, \dots, x_m)$ and the allocation $N_m = (n_1, n_2, \dots, n_m)$. The interval $[x_l, x_o]$ is called the experimental region.

If a special regression function

$$f(x,\theta), x \in [x_l, x_u], \theta \in \Omega \subset R^p$$

is given, then $V_{n,m}$ is element of all possible concrete designs:

$$\mathcal{V}_n = \left\{ V_n = V_{n,m} : p \le m \le n, \quad \text{card}(S_m) = m, \quad \sum_{j=1}^{m} n_j = n, \quad n_j \ge 0 \right\}.$$

If $Z : \mathcal{V}_n \to R^+$ is a mapping $Z_0(V_n) = Z[\text{var}_A(\theta | V_n)]$ with $\theta_0 \in \Omega, V_n \in \mathcal{V}_n$, $Z : R^{p \times p} \to R^1$ and the asymptotic covariance matrix (9.38) can now be written in dependency on θ_0 and $V_{n,m}$

$$\text{var}_A(\theta) = \text{var}_A(\theta | V_n),$$

then $V^*_{n,m}$ is called a locally Z-optimal m-point design at $\theta = \theta_0$, if

$$Z_0\left(V^*_{n,m}\right) = \inf_{V_{n,m} \in \mathcal{V}_n} Z_0(V_{n,m}). \tag{9.58}$$

If $\mathcal{V}_{n,m}$ is the set of concrete m-point designs, then $V^*_{n,m}$ is called concrete locally Z-optimal m-point design, if

$$Z_0\left(V^*_{n,m}\right) = \inf_{V_{n,m} \in \mathcal{V}_n} \{Z_0(V_{n,m})\}. \tag{9.59}$$

The mapping $V_n \to \text{var}_A(\theta_0 | V_n)$ is symmetric concerning S_m. Therefore we focus on supports with $x_1 < x_2 < \cdots < x_m$. In place of minimising the $(\rho \times \rho)$-matrix $Z[\text{var}_A(\theta, V_{n,m})] = M$, we can maximise its inverse, the so-called asymptotic information matrix M^{-1}.

Especially $V^*_{n,m}$ for $r = 1, \ldots, p + 2$ with the functionals Z_r and the $(p \times p)$-Matrix $M = (m_{ij})$ is called

$Z_r(M) = m_{rr} \quad (r = 1, \ldots, p)$	locally C_{θ_r}-optimal,
$Z_{p+1}(M) = \lvert M \rvert$	locally D-optimal,
$Z_{p+2}(M) = \text{Sp}(M)$	locally A-optimal,

and in general for $r = 1, \ldots, p + 2$ then Z_r - optimal.

For some regression functions and optimality criteria, analytical solutions in closed form of the problems could be found. Otherwise search methods must be applied.

The first analytical solution can be found in Box and Lucas (1959) as well as in

Theorem 9.9 (Box and Lucas, 1959).
For the regression model

$$f(x,\theta) = \alpha + \beta e^{\gamma x} \tag{9.60}$$

with $n = 3$, $\theta = (\alpha, \beta, \gamma)^T$ and $x \in [x_l, x_u]$, the locally D-optimal concrete design V_3 depends only on the component γ_0 of $\theta_0 = (\alpha_0, \beta_0, \gamma_0)^T$ and has the form

$$V_3 = \begin{pmatrix} x_u & x_2 & x_0 \\ 1 & 1 & 1 \end{pmatrix}$$

with

$$x_2 = \frac{1}{\gamma_0} + \frac{x_u e^{\gamma_0 x_u} - x_o e^{\gamma_0 x_o}}{e^{\gamma_0 x_u} - e^{\gamma_0 x_0}}. \tag{9.61}$$

Atkinson and Hunter (1968) gave sufficient and for $n = kp$ sufficient and necessary conditions for the function $f(x, \theta)$ that the support of a locally D-optimal design of size n *is* $p = \dim(\Omega)$. These conditions are difficult to verify for $p > 2$.

Theorem 9.10 (Rasch, 1990).
The support of a concrete locally D-optimal p-point design of size n is independent of n; the n_i of this design are as equal as possible (i.e. if $n = ap$, then $n_i = a$; otherwise n_i differ maximal by 1).

Proof: If $m=p$, the asymptotic covariance matrix in (9.38) is (after dropping the design independent factor σ^2) equal to $\frac{1}{\sigma^2}\text{var}_A(\theta) = \left[F^T(\theta)F(\theta)\right]^{-1} = G^T(\theta)\ \tilde{N}G(\theta)$ with $G(\theta) = \{f_j(x_i, \theta)\}, i = 1, \ldots, p$ and $\tilde{N} = \text{diag}(n_1,\ldots,n_p)$. Now minimising $|[F^T(\theta)F(\theta)]^{-1}|$ means maximising $\left|G^T(\theta)\tilde{N}G(\theta)\right|$. For quadratic matrices A and B of the same order, $|AB| = |A||B|$ and $|A| = |A^T|$ always hold; therefore we obtain $\left|G^T(\theta)\tilde{N}G(\theta)\right| = \left|G^T\right|^2\left|\tilde{N}\right|$. This completes the proof, because $|G^T|$ can be maximised independently of $\tilde{N}$ and $\left|\tilde{N}\right| = \prod_{i=1}^{p} n_i$ is a maximum if n_i are equal or as equal as possible.

In Rasch (1990) further theorems concerning the D-optimality can be found.

Theorem 9.11 Let $f(x, \theta)$ be an intrinsically non-linear regression function, $x \in R, \theta \in \Omega \subset R^p$, with the non-linearity parameter $\varphi = (\theta_{i_1},\ldots,\theta_{i_r})^T, 0 < r < p$ in Definition 9.1 and let F be non-singular. Then the concrete D-optimal design of size $n \geq p$ only depends on $\varphi = (\theta_{i_1},\ldots,\theta_{i_r})^T, 0 < r < p$, and not on the linearity parameters.

Proof: In Definition 9.1,

$$\frac{\partial f(x,\theta)}{\partial \theta} = C(\theta)g(x,\varphi)$$

with $g^T(x, \varphi) = (g_1(x, \varphi), \ldots, g_p(x, \varphi))$. If we put $G = (g_j(x_i, \varphi))$, then

$$|F^T F| = |C(\theta)G^T GC^T(\theta)| = |C(\theta)||C^T(\theta)||G^T G| = |C(\theta)|^2|G^T G|.$$

$|F^T F|$ is maximal, if $|G^T G|$ is maximal, and G only depends on $\varphi = (\theta_{i_1}, \ldots, \theta_{i_r})^T, 0 < r < p$.

If $n > 2p$, the D-optimal concrete designs are approximately G-optimal in the sense that the value of the G-criterion for the concrete D-optimal p-point design even for $n \neq tp$ (t integer) does nearly not differ from that of the concrete G-optimal design. For the functions in (9.6), we found optimal designs by search methods, and for $n > p + 2$ we often found p-point designs. Searching D-optimal designs in the class of p-point designs, then $\text{var}_A(\theta)$ in (9.39) becomes

$$\text{var}_A(\theta) = \left[B^T \text{diag}(n_1, \ldots, n_p)B\right]^{-1} \sigma^2$$

because

$$V_n = \begin{pmatrix} x_1, \ldots, x_p \\ n_1, \ldots, n_p \end{pmatrix}, \quad F^T F = \left(\sum_{i=1}^{n} f_j(x_i, \theta) f_k(x_i, \theta) \right) \quad (j, k = 1, \ldots, p)$$

and

$$\sum_{i=1}^{n} f_j(x_i, \theta) f_k(x_i, \theta) = \sum_{i=1}^{p} n_l f_j(x_l, \theta) f_k(x_l, \theta)$$

with

$$B = (f_j(x_i, \theta)) \quad (i, j = 1, \ldots, p).$$

Theorem 9.12 The minimal experimental size n_{min} so that for a given $K > 0$

$$|\text{var}_A(\theta)| \leq K|B|^2$$

can be determined as follows. Find the smallest positive integer z with

$$z \geq \sqrt{\frac{1}{\sqrt[p]{K}}}$$

In the case of equality above $n_{min} = pz$. Otherwise calculate the largest integer r, so that

$$\frac{1}{z^{p-r}(z-1)^r} \leq K.$$

Then

$$n_{\min} = pz - 1.$$

Proof: From the proof of Theorem 9.10 we know that

$$|\mathrm{var}_A(\theta)| = \frac{\sigma^2}{|B|^2 \prod_{i=1}^{p} n_i},$$

for D-optimal p-point designs and the final proof is left to the reader.

9.6 Special Regression Functions

In this section we discuss some regression functions important in the applications in biosciences as well as in engineering. Each of the special functions is discussed by a unique approach. We determine the asymptotic covariance matrix, determine the experimental size N_0 for testing parameters and determine parts of the parameter space, for which the actual risk of the first kind of a tests is between 0.04 and 0.06 if the nominal risk $\alpha_{nom} = 0.05$.

9.6.1 Exponential Regression

The exponential regression is discussed extensively as a kind of pattern; the other functions follow the same scheme, but their treatment is shorter.

Model (9.2) is called the model of the exponential regression, if $f_E(x, \theta)$ is given by (9.60). The derivation of $f_E(x, \theta)$ concerning θ ($\theta = (\alpha, \beta, \gamma)^T$) is

$$\frac{\partial f_E(x,\theta)}{\partial \theta} = \begin{pmatrix} 1 \\ \mathrm{e}^{\gamma x} \\ \beta x \mathrm{e}^{\gamma x} \end{pmatrix} = \begin{pmatrix} 1 & 0 & 0 \\ 0 & 1 & 0 \\ 0 & 0 & \beta \end{pmatrix} \begin{pmatrix} 1 \\ \mathrm{e}^{\gamma x} \\ x \mathrm{e}^{\gamma x} \end{pmatrix}, \tag{9.62}$$

so that γ by Definition 9.1 is a non-linearity parameter.

9.6.1.1 Point Estimator

For $R(\theta)$ in (9.5) we get

$$R(\theta) = \sum_{i=1}^{n} (y_i - \alpha - \beta \mathrm{e}^{\gamma x_i})^2$$

and determine $\hat{\theta}$ so that

$$R(\hat{\theta}) = \min_{\theta \in \Omega} R(\theta).$$

Because

$$\left.\begin{aligned} A &= \sum_{i=1}^{n} \mathrm{e}^{\gamma x_i}, \quad B = \sum_{i=1}^{n} x_i \mathrm{e}^{\gamma x_i}, \quad C = \sum_{i=1}^{n} \mathrm{e}^{2\gamma x_i}, \\ D &= \sum_{i=1}^{n} x_i \mathrm{e}^{2\gamma x_i}, \quad E = \sum_{i=1}^{n} x_i^2 \mathrm{e}^{2\gamma x_i}, \end{aligned}\right\}, \tag{9.63}$$

it follows

$$|F^T F| = \beta^2 \{ n(CE - D^2) + 2ABD - B^2 C - A^2 E \} = \beta^2 \Delta,$$

which means that $\beta \neq 0$. For fixing Ω_0 we either choose $\beta > 0$ or $\beta < 0$ depending on the practical problem. For growth processes, it follows because $\gamma < 0$ immediately $\beta < 0$; then in this case we choose

$$\Omega_0 \subset R^- \times R^- \times R^1 \subset \Omega \subset R^3.$$

The region Ω_0 must be chosen so that the assumptions V2 and V3 in Section 9.1.1 are fulfilled.

The inverse of $F^T F$ has the form

$$(F^T F)^{-1} = \frac{1}{\Delta} \begin{pmatrix} CE - D^2 & BD - AE & \frac{1}{\beta}(AD - BC) \\ BD - AE & nE - B^2 & \frac{1}{\beta}(AD - nD) \\ \frac{1}{\beta}(AD - BC) & \frac{1}{\beta}(AD - nD) & \frac{1}{\beta^2}(nC - A^2) \end{pmatrix}. \tag{9.64}$$

Next we describe a method of Verhagen (1960) that can be used to find initial values for calculating iteratively the LS estimators by the Gauss–Newton method. We start with the integrals

$$J(x_i) = \int_0^{x_i} (\alpha + \beta \mathrm{e}^{\gamma \tau}) d\tau = \alpha x_i + \frac{\eta_i - \alpha}{\gamma} - \frac{\beta}{\gamma}$$

with $\eta_i = \alpha + \beta \mathrm{e}^{\gamma x_i} (i = 1,\ldots,n)$ and approximate them by

$$T_i = T(x_i) = \frac{1}{2} \sum_{j=2}^{i} (y_{j-1} + y_j)(x_j - x_{j-1}) \ (i = 2,\ldots,n).$$

Now we put

$$\eta_i \sim \gamma T_i - \alpha \gamma x_i + \alpha + \beta \ (i = 2,\ldots,n)$$

and estimate the parameters of the approximate linear model

$$\boldsymbol{y}_i = \gamma T_i - \alpha \gamma x_i + \alpha + \beta + \boldsymbol{e}_i^* \ (i = 2,\ldots,n)$$

with the methods of Chapter 8. The LS estimators are

$$c_v = \hat{\gamma}_v = \frac{SP_{Ty}SS_x - SP_{Tx}SP_{xy}}{SS_T SS_x - SP_{Tx}^2}$$

$$a_v = \hat{\alpha}_v = \frac{c_v SP_{Tx} - SP_{xy}}{c_v SS_x}$$

and

$$b_v = \hat{\beta}_v = \bar{y} - c_v \bar{T} - a_v(c_v \bar{x} - 1)$$

with

$$SP_{uv} = \sum_{i=2}^{n} u_i v_i - \frac{1}{n-1}\left(\sum_{i=2}^{n} u_i\right)\left(\sum_{i=2}^{n} v_i\right), \quad SS_u = SP_{uu}$$

and the arithmetic means $\bar{y}, \bar{T}$ and $\bar{x}$ of the $n - 1$ values y_i, T_i and x_i for $i = 2, \ldots,$ n, respectively.

9.6.1.2 Confidence Estimations and Tests

Let the assumptions of Section 9.3 be given. The asymptotic covariance matrix

$$\text{var}_A(\theta) = \sigma^2 \left(F^T F\right)^{-1}$$

with $(F^T F)^{-1}$ in (9.64) and the abbreviations (9.63) can be used for the construction of confidence intervals for α, β and γ and tests of hypotheses about α, β and γ may be used, respectively.

Following Section 9.5 we test

$$H_{0\alpha}: \alpha = \alpha_0 \;\; \text{against} \;\; H_{A\alpha}: \alpha \neq \alpha_0$$

with the test statistic

$$t_\alpha = \frac{(\boldsymbol{\alpha} - \alpha_0)\sqrt{\hat{\Delta}}}{s\sqrt{\hat{C}\hat{E} - \hat{D}^2}}. \tag{9.65}$$

Further

$$H_{0\beta}: \beta = \beta_0 \;\; \text{against} \;\; H_{A\beta}: \beta \neq \beta_0$$

is tested with

$$t_\beta = \frac{(\boldsymbol{\beta} - \beta_0)\sqrt{\hat{\Delta}}}{s\sqrt{n\hat{E} - \hat{B}^2}} \tag{9.66}$$

and

$$H_{0\gamma}: \gamma = \gamma_0 \text{ against } H_{A\gamma}: \gamma \neq \gamma_0$$

with

$$\boldsymbol{t}_\gamma = \frac{(\boldsymbol{c} - \gamma_0)\sqrt{\hat{\Delta}}}{\boldsymbol{s}\sqrt{n\hat{\boldsymbol{C}} - \hat{\boldsymbol{A}}^2}}. \tag{9.67}$$

$\hat{\boldsymbol{A}}, \ldots, \hat{\boldsymbol{E}}$ in the formulae of the test statistics are gained from $A, \ldots, E$ in (9.63) by replacing there the parameter γ by its estimator $\hat{\boldsymbol{\gamma}} = \boldsymbol{c}$. Further

$$\Delta = n(CE - D^2) + 2ABD - A^2E - B^2C,$$

and here also we obtain $\hat{\boldsymbol{\Delta}}$ from Δ by replacing γ by $\boldsymbol{c}$. Finally $\boldsymbol{s}$ is the square root of

$$\boldsymbol{s}^2 = \frac{1}{n-3}\sum_{i=1}^{n}(\boldsymbol{y}_i - \boldsymbol{a} - \boldsymbol{b}e^{\boldsymbol{c}x_i})^2,$$

the estimator of σ^2.

Tests have the form

$$k_l(Y) = \begin{cases} 1, & \text{if } |t_l| > t(n-3|1-\alpha_{nom}/2);\ l = \alpha, \beta, \gamma \\ 0, & \text{otherwise} \end{cases}$$

with the $(1-\alpha_{nom}/2)$-quantile of the central t-distribution with $n-3$ degrees of freedom. Here α^*_{nom} is the nominal risk of the first kind of the tests. Confidence intervals with a nominal confidence coefficient $1-\alpha_{nom}$ are defined as follows putting $t(n-3|1-\alpha_{nom}/2) = T(n, \alpha_{nom})$

Parameter α:

$$\left[\boldsymbol{a} - \boldsymbol{s}\frac{\sqrt{\hat{\boldsymbol{C}}\hat{\boldsymbol{E}} - \hat{\boldsymbol{D}}^2}}{\sqrt{\hat{\boldsymbol{\Delta}}}}T(n, \alpha_{nom}), \boldsymbol{a} + \boldsymbol{s}\frac{\sqrt{\hat{\boldsymbol{C}}\hat{\boldsymbol{E}} - \hat{\boldsymbol{D}}^2}}{\sqrt{\hat{\boldsymbol{\Delta}}}}T(n, \alpha_{nom})\right]$$

Parameter β:

$$\left[\boldsymbol{b} - \boldsymbol{s}\frac{\sqrt{n\hat{\boldsymbol{E}} - \hat{\boldsymbol{B}}^2}}{\sqrt{\hat{\boldsymbol{\Delta}}}}T(n, \alpha_{nom}), \boldsymbol{b} + \boldsymbol{s}\frac{\sqrt{n\hat{\boldsymbol{E}} - \hat{\boldsymbol{B}}^2}}{\sqrt{\hat{\boldsymbol{\Delta}}}}T(n, \alpha_{nom})\right]$$

Parameter γ:

$$\left[\boldsymbol{c} - \frac{\boldsymbol{s}}{\boldsymbol{b}}\frac{\sqrt{n\hat{\boldsymbol{C}} - \hat{\boldsymbol{A}}^2}}{\sqrt{\hat{\boldsymbol{\Delta}}}}T(n, \alpha_{nom}), \boldsymbol{c} + \frac{\boldsymbol{s}}{\boldsymbol{b}}\sqrt{n\hat{\boldsymbol{C}} - \hat{\boldsymbol{A}}^2}\hat{\boldsymbol{\Delta}}T(n, \alpha_{nom})\right]$$

Example 9.7 Let us consider a numerical example. Table 9.3 shows the growth of leaf surfaces of oil palms observed in Indonesia.

Program Hint

Most calculations and graphs in Section 9.6 have been done by our own special program *Growth*, which can be found in the program package CADEMO (see http://www.swmath.org/software/1144). The program determines initial values for the iteration to calculate the LS estimates from the data. In Figure 9.10 the LS estimates with the asymptotic confidence intervals and the estimated residual variance are given. In Figure 9.11 the curve of the estimated regression function together with the scatter plot of the observation is given.

Table 9.3 Leaf surface (y_i) in m^2 of oil palms on a trial area in dependency of age x_i in years.

x_i	1	2	3	4	5	6	7	8	9	10	11	12
y_i	2.02	3.62	5.71	7.13	8.33	8.29	9.81	11.3	12.18	12.67	12.62	13.01

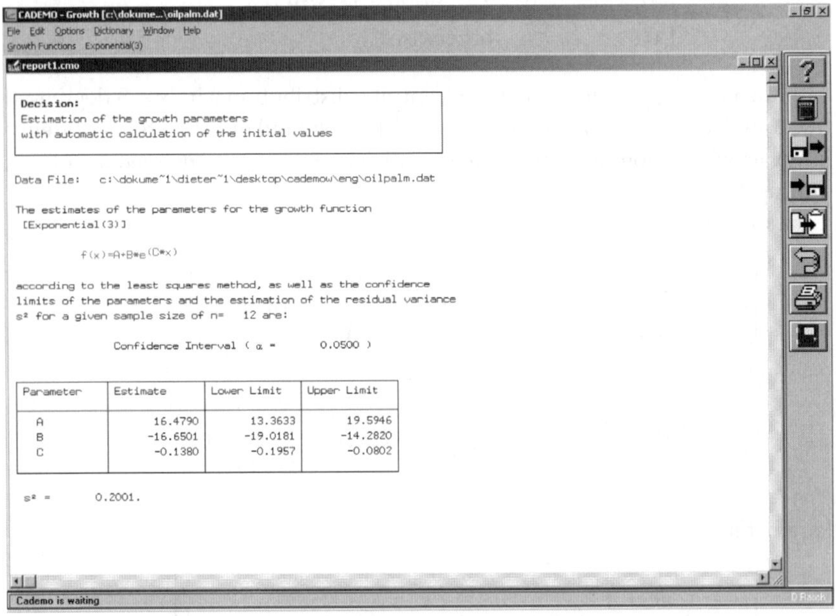

Figure 9.10 LS estimates, asymptotic confidence intervals and estimated residual variance of the exponential regression with data of Example 9.7.

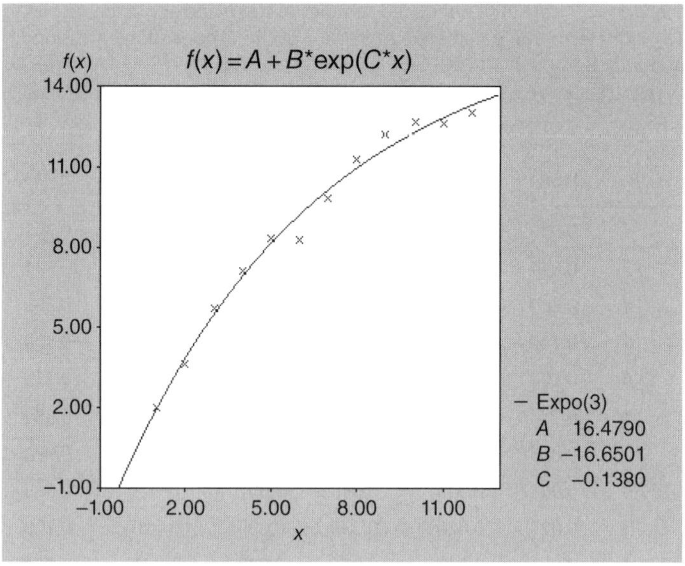

Figure 9.11 Curve of the estimated exponential regression function together with the scatter plot of the observations of Example 9.7.

9.6.1.3 Results of Simulation Experiments

For the exponential regression, we perform simulation experiments described in Section 9.4.3. The questions are as follows:

- Is the bias of $\boldsymbol{a}, \boldsymbol{b}$ and $\boldsymbol{c}$ important?
- Differ the asymptotic variances from the empirical ones?
- Is the denominator $n - 3$ of the estimator of σ^2 appropriate?

The results in Rasch et al. (2008) and Rasch and Schimke (1983) for equidistant $x_i \in [0, 65], i = 1, \ldots, n$ and $n = 4, 6, 14$ as well as 12 (β, γ)-combinations are summarised below. W.l.o.g. we choose $\alpha = 0$ and further $\sigma^2 = 1$. The number of runs was 5000. In each α, β and γ have been estimated, and from the 5000 estimates, the empirical means $\bar{a}, \bar{b}$ and $\bar{c}$ and the empirical variances s_a^2, s_b^2 and s_c^2 and covariances have been calculated.

Table 9.4 shows the empirical bias $v_{E,n}$ representing $\bar{a} - \alpha, \bar{b} - \beta$ and $\bar{c} - \gamma$ for $n =$ 4, 6 and 14 in comparison with the by (9.42) calculated approximative bias $v_n(\theta)$.

To calculate $v_n(\theta)$ by (9.42), we use the notations of (9.4) and the vector $F_i(\theta) = (1, e^{\gamma x_i}, \beta x_i e^{\gamma x_i})^T$ and the inverse $(F^T F)^{-1}$ from (9.64). For $K_i(\theta)$ we get

$$K_i(\theta) = \begin{pmatrix} 0 & 0 & 0 \\ 0 & 0 & x_i e^{\gamma x_i} \\ 0 & x_i e^{\gamma x_i} & \beta x_i^2 e^{\gamma x_i} \end{pmatrix}.$$

Table 9.4 Empirical bias $v_{E,n}$ from 5000 simulated samples of size n and approximate bias v_n from (9.42); the LS estimates of the parameters α, ß and γ of the exponential regression for $n = 4, 6$ and 10 and $\sigma^2 = 1$.

			α		ß			γ
$-\beta$	$-10^2\gamma$	n	$v_{E,n}$	v_n	$-v_{E,n}$	$-v_n$	$-10^3 v_{E,n}$	$-10^3 v_n$
30	3	4	0.520	0.523	0.252	0.526	0.419	0.251
		6	0.644	0.614	0.625	0.622	0.131	0.134
		14	0.263	0.238	0.287	0.248	0.003	0.055
	5	4	0.147	0.137	0.135	0.139	0.441	0.470
		6	0.125	0.102	0.142	0.107	0.096	0.215
		14	0.128	0.057	0.166	0.066	−0.170	0.084
	7	4	0.055	0.070	0.059	0.071	1.139	0.990
		6	0.052	0.048	0.048	0.052	0.338	0.363
		14	0.027	0.026	0.035	0.035	0.120	0.129
	9	4	0.035	0.047	0.035	0.048	2.821	2.184
		6	0.002	0.031	0.019	0.033	0.685	0.610
		14	0.012	0.016	−0.025	0.025	0.320	0.189
50	3	4	0.310	0.314	0.323	0.316	0.117	0.091
		6	0.279	0.249	0.307	0.253	0.048	0.048
		14	0.210	0.143	0.190	0.149	0.058	0.020
	5	4	0.070	0.082	0.041	0.083	0.236	0.169
		6	0.077	0.061	0.090	0.064	0.050	0.077
		14	0.025	0.034	0.042	0.040	0.035	0.030
	7	4	0.030	0.042	0.048	0.042	0.358	0.356
		6	0.045	0.029	0.033	0.031	0.011	0.131
		14	0.010	0.016	0.032	0.021	0.122	0.047
	9	4	0.023	0.028	0.039	0.029	0.888	0.786
		6	0.020	0.018	0.023	0.020	0.182	0.219
		14	0.001	0.010	0.021	0.015	0.077	0.068
70	3	4	0.301	0.224	0.297	0.225	0.015	0.045
		6	0.124	0.178	0.125	0.181	0.054	0.025
		14	0.021	0.102	0.173	0.106	0.106	0.010
	5	4	0.075	0.059	0.081	0.059	0.079	0.086
		6	0.054	0.044	0.154	0.046	0.000	0.040
		14	0.022	0.025	0.061	0.028	0.044	0.015

Table 9.4 (Continued)

			α		ß			γ
$-\beta$	$-10^2\gamma$	n	$v_{E,n}$	v_n	$-v_{E,n}$	$-v_n$	$-10^3 v_{E,n}$	$-10^3 v_n$
	7	4	0.027	0.029	0.054	0.030	0.194	0.182
		6	0.020	0.021	0.014	0.022	0.043	0.067
		14	0.038	0.011	0.027	0.015	0.008	0.024
	9	4	0.020	0.020	0.018	0.021	0.462	0.401
		6	0.026	0.013	0.020	0.014	0.071	0.112
		14	0.029	0.007	0.038	0.011	0.060	0.035

Adding to (9.63) the abbreviations

$$G = \sum_{i=1}^{n} x_i^2 e^{\gamma x_i},\ H = \sum_{i=1}^{n} x_i^3 e^{2\gamma x_i}$$

because of $\sigma^2 = 1$ and (9.64) we receive

$$\operatorname{tr}\left\{ \left(F^{\mathrm{T}}F\right)^{-1} K_i(\theta) \right\} = \frac{1}{\Delta\beta}\left(2(AB - nD)x_i e^{\gamma x_i} + \left(nC - A^2\right)x_i^2 e^{\gamma x_i}\right)$$

and finally

$$v_n(\theta) \approx \frac{1}{2\Delta\beta}\left(F^T F\right)^{-1} \begin{pmatrix} 2B(AB - nD) + G(nC - A^2) \\ 2D(AB - nD) + E(nC - A^2) \\ 2\beta E(AB - nD) + H\beta(nC - A^2) \end{pmatrix}. \tag{9.68}$$

We see in Tables 9.5 and 9.6 that the empirical variances do not differ strongly from the main diagonal elements of the asymptotic covariance matrix even for $n = 4$.

The choice of the denominator $n - 3$ in estimate s^2 of σ^2 is analogous to the linear case. There $n - 3$ (or in general $n - p$) is the number degrees of freedom of the χ^2-distribution of the nominator of s^2. If we compare expectation, variance, skewness and kurtosis of a χ^2-distribution with $n - 3$ degrees of freedom with the corresponding empirical values from the simulation experiment, we see that even for the smallest possible $n = 4$, a good accordance is found. This means that $n - 3$ is a good choice for the denominator in the estimator of σ^2.

Table 9.7 shows the relative frequencies of confidence estimations and tests with $\alpha_{nom} = 0.05$ and $\alpha_{nom} = 0.1$ for a special parameter configuration from 10 000 runs, respectively. As we can see already with $n = 4$, a sufficient alignment is found between α_{nom} and α_{act}. Therefore the tests in Section 9.6.1.2 can be used as approximative α_{nom}-tests and the confidence intervals as approximative

Table 9.5 Empirical variances s_a^2 and s_b^2 with the asymptotic variances $\text{var}_A(a)$ and $\text{var}_A(b)$ of the estimates of α and $\beta(\sigma^2 = 1)$ for $n = 4$ and $n = 6$.

$-10^2\gamma$	n	$10^5 s_a^2$	$10^5\text{var}_A(a)$	$10^5 s_b^2$	$10^5\text{var}_A(b)$
3	4	878224	800768	853658	780404
	6	680837	613678	611028	547540
5	4	197339	187157	266298	260565
	6	130694	129982	178729	182512
7	4	105017	98990	197639	191016
	6	64415	63300	145588	144533
9	4	71968	73079	170001	170312
	6	44366	44152	137567	135566

Table 9.6 Empirical variances (upper value) and asymptotic variances (lower value) of the estimate of γ for $n = 4$ multiplied by 10^9, $\sigma^2 = 1$.

$-10^2\gamma$	$\beta = -70$	$\beta = -50$	$\beta = -30$
3	7751	15535	42825
	7512	14723	40897
5	10475	20416	57989
	10199	19989	55251
7	19087	38258	117038
	18873	36992	102754
9	39407	80225	281944
	39407	77238	214550

$(1 - \alpha_{nom})$ confidence intervals. The power function of the tests was evaluated in Rasch and Schimke (1983) as well as the behaviour of the tests for non-equidistant supports. Summarising it can be stated that the methods based on the asymptotic covariance matrix are satisfactory already for $n = 4$ and about 90% robust against non-normality in the Fleishman system.

9.6.1.4 Experimental Designs

To find locally D-optimal designs, we can use Theorem 9.9. During extensive searches of optimal designs, not only in the class of three-point designs optimal

Table 9.7 Relative frequencies of 10 000 simulated samples for the (incorrect) rejection (left hand n_l, right hand $n_{u)}$ and for the (correct) acception (n_M) of H_0 for the exponential regression with $\alpha = 0$, $\beta = -50$, $\gamma = -0.05$, $n = 10(-1)4$ and $\alpha_{nom} = 0.05$ and $\alpha_{nom} = 0.1$.

	$\alpha_{nom} = 0.05$			$\alpha_{nom} = 0.1$		
n	n_u	n_o	n_M	n_u	n_o	n_M
$H_{0\alpha}: \alpha = 0$						
10	2.71	2.03	95.26	5.36	4.01	90.63
9	3.17	2.07	94.76	6.23	4.44	89.33
8	2.51	2.03	95.46	5.06	4.28	90.66
7	2.59	2.03	95.38	5.24	4.54	90.22
6	2.98	2.04	94.98	5.52	4.26	90.22
5	2.80	2.19	95.01	5.57	4.22	90.21
4	2.66	2.41	94.93	5.12	4.96	89.92
$H_{0\beta}: \beta = -50$						
10	2.44	2.31	95.25	4.97	4.48	90.55
9	2.43	2.44	95.13	5.11	4.85	90.04
8	2.46	2.21	95.33	5.01	4.38	90.61
7	2.74	2.01	95.25	5.26	4.46	90.28
6	2.63	2.48	94.89	5.32	4.92	89.76
5	2.37	2.49	95.14	4.87	5.03	90.10
4	2.59	2.27	95.14	5.34	4.80	89.86
$H_{0\gamma}: \gamma = -0.05$						
10	2.50	2.26	95.24	4.99	4.48	90.53
9	2.76	2.52	94.72	5.72	4.90	89.38
8	2.85	2.08	95.07	5.39	4.40	90.21
7	2.79	1.82	95.39	5.26	4.33	90.41
6	2.68	2.39	94.93	5.17	4.59	90.24
5	2.63	2.43	94.94	4.80	4.97	90.23
4	2.56	2.35	95.09	5.42	4.72	89.86

Table 9.8 Optimal experimental designs in the experimental region [1,12] and $n = 12$.

Criterion	$(\beta, \gamma) = (-17, -0.14)$	$(\beta, \gamma) = (-19, -0.2)$	$(\beta, \gamma) = (-14, -0.08)$
D	$\begin{pmatrix} 1 & 5.14 & 12 \\ 4 & 4 & 4 \end{pmatrix}$	$\begin{pmatrix} 1 & 4.63 & 12 \\ 4 & 4 & 4 \end{pmatrix}$	$\begin{pmatrix} 1 & 5.7 & 12 \\ 4 & 4 & 4 \end{pmatrix}$
C_α	$\begin{pmatrix} 1 & 5.08 & 12 \\ 2 & 5 & 5 \end{pmatrix}$	$\begin{pmatrix} 1 & 4.61 & 12 \\ 2 & 4 & 6 \end{pmatrix}$	$\begin{pmatrix} 1 & 5.69 & 12 \\ 2 & 6 & 4 \end{pmatrix}$
C_β	$\begin{pmatrix} 3.87 & 11.59 & 12 \\ 6 & 2 & 4 \end{pmatrix}$	$\begin{pmatrix} 1.48 & 11.29 & 12 \\ 6 & 1 & 5 \end{pmatrix}$	$\begin{pmatrix} 1 & 5.44 & 12 \\ 2 & 6 & 4 \end{pmatrix}$
C_γ	$\begin{pmatrix} 1 & 5.02 & 12 \\ 3 & 6 & 3 \end{pmatrix}$	$\begin{pmatrix} 1 & 4.48 & 12 \\ 3 & 6 & 3 \end{pmatrix}$	$\begin{pmatrix} 1 & 5.63 & 12 \\ 3 & 6 & 3 \end{pmatrix}$

designs have been found, which are three-point designs as derived by Box for $n = 3$ in Theorem 9.9. By search methods concerning the locally C_α-, C_γ- and A-optimality, we found that the optimal designs always have been three-point designs in $[x_l, x_u]$ with $x_1 = x_l$ and $x_3 = x_u$. For the C_β-optimality often one of the bounds of the experimental region did not belong to the support of the locally C_β-optimal design, but they always have been three-point designs. In Table 9.8 we report some results of our searches using the parameters and experimental regions of Example 9.7.

We can now compare the criterion values of the D-optimal design $\begin{pmatrix} 1 & 5.14 & 12 \\ 4 & 4 & 4 \end{pmatrix}$ (it was $0.00015381\,\sigma^6$) with the design used in the experiment $\begin{pmatrix} 1 & 2 & 3 & 4 & 5 & 6 & 7 & 8 & 9 & 10 & 11 & 12 \\ 1 & 1 & 1 & 1 & 1 & 1 & 1 & 1 & 1 & 1 & 1 & 1 \end{pmatrix}$ that had a criterion value $0.0004782\,\sigma^6$, and this is 3.11 times larger than that of the optimal design. The criterion of the C_γ-optimal design is $0.00154\,\sigma^2$ and that for the design used in the experiment is $0.00305\,\sigma^2$.

It can generally be stated for all models and optimality criteria that an equidistant design in the experimental region with one observation at each support point is far from being optimal.

9.6.2 The Bertalanffy Function

The regression function $f_B(x)$ of the model

$$y_i = (\alpha + \beta e^{\gamma x_i})^3 + e_i = f_B(x_i) + e_i, \quad i = 1, \ldots, n, \quad n > 3, \tag{9.69}$$

is called Bertalanffy function and was used by Bertalanffy (1929) to describe the growth of body weight of animals. This function has two inflection points if α and β have different signs and are located at $x_{I1} = \frac{1}{\gamma}\ln\left(-\frac{\alpha}{\beta}\right)$ and $x_{I2} = \frac{1}{\gamma}\ln\left(-\frac{\alpha}{3\beta}\right)$, respectively, with $f_B(x_{I1}) = 0$ and $f_B(x_{I2}) = \left(\frac{2}{3}\alpha\right)^3$.

With $\theta = (\theta_1, \theta_2, \theta_3)^T = (\alpha, \beta, \gamma)^T$ and with the notation of Definition 9.1, we obtain

$$\frac{\partial f_B(x,\theta)}{\partial \theta} = \begin{pmatrix} 3(\alpha+\beta e^{\gamma x})^2 \\ 3(\alpha+\beta e^{\gamma x})^2 e^{\gamma x} \\ 3(\alpha+\beta e^{\gamma x})^2 e^{\gamma x}\beta x \end{pmatrix},$$

and by this all components of θ are non-linearity parameters. Analogous to (9.63) we use abbreviations like

$$z_i = (\alpha+\beta e^{\gamma x_i})^4,$$

$$A = \sum_{i=1}^n z_i,\quad B = \sum_{i=1}^n z_i e^{\gamma x_i},\quad C = \sum_{i=1}^n x_i z_i e^{\gamma x_i},$$

$$D = \sum_{i=1}^n z_i e^{2\gamma x_i},\quad E = \sum_{i=1}^n z_i x_i e^{2\gamma x_i},\quad G = \sum_{i=1}^n z_i x_i^2 e^{2\gamma x_i}.$$

Then

$$F^T F = 9\begin{pmatrix} A & B & \beta C \\ B & ED & \beta E \\ \beta C & \beta E & \beta^2 G \end{pmatrix}$$

and

$$F^T F = 9^3\beta^2\left[ADG + 2BCE - C^2D - E^2A - B^2G\right] = 9^3\beta^2\Delta.$$

The asymptotic covariance matrix is therefore

$$\operatorname{var}_A(\theta) = \sigma^2\left(F^T F\right)^{-1}$$

$$= \frac{\sigma^2}{9\Delta}\begin{pmatrix} DG - E^2 & EC - BG & \frac{1}{\beta}(BE - CD) \\ EC - BG & AG - C^2 & \frac{1}{\beta}(BE - AE) \\ \frac{1}{\beta}(BE - CD) & \frac{1}{\beta}(BC - AE) & \frac{1}{\beta^2}(AD - B^2) \end{pmatrix}. \tag{9.70}$$

To determine the initial values, it is recommended to transform the y_i- values of (x_i, y_i) $(i = 1, \ldots, n)$ to

$$v_i = \sqrt[3]{y_i}$$

and estimate from (x_i, v_i) the parameters α, β, γ of an exponential regression in Section 9.6.1. These estimates a^*, b^*, c^* are used as initial values for the iterative determination of the LS estimates a,b,c of the Bertalanffy function.

Concerning the hypothesis testing, we receive from (9.70) for $n > 3$ the test statistics

for $H_{0\alpha}: \alpha = \alpha_0$ against $H_{A\alpha}: \alpha \neq \alpha_0$

$$\boldsymbol{t}_\alpha = \frac{(\boldsymbol{a} - a_0)3\sqrt{\hat{\Delta}}}{s\sqrt{\hat{\boldsymbol{D}}\hat{\boldsymbol{G}} - \hat{\boldsymbol{E}}}},$$

for $H_{0\beta}: \beta = \beta_0$ against $H_{A\beta}: \beta \neq \beta_0$

$$\boldsymbol{t}_\beta = \frac{(\boldsymbol{b} - \beta_0)3\sqrt{\hat{\Delta}}}{s\sqrt{\hat{\boldsymbol{A}}\hat{\boldsymbol{G}} - \hat{\boldsymbol{C}}^2}},$$

for $H_{0\gamma}: \gamma = \gamma_0$ against $H_{0\gamma}: \gamma \neq \gamma_0$

$$\boldsymbol{t}_\gamma = \frac{(\boldsymbol{c} - \gamma_0)3\boldsymbol{b}\sqrt{\hat{\Delta}}}{s\sqrt{\hat{\boldsymbol{A}}\hat{\boldsymbol{D}} - \hat{\boldsymbol{B}}^2}},$$

and the confidence intervals in Section 9.4. The symbols $\hat{\boldsymbol{A}}, \ldots, \hat{\Delta}$ are defined as in Section 9.6.1 and $\boldsymbol{s}$ is the square root of

$$\boldsymbol{s}^2 = \frac{1}{n-3}\sum_{i=1}^{n}\left[y_i - \left(\boldsymbol{a} + \boldsymbol{b}\mathrm{e}^{cx_i}\right)^3\right]^2 \quad (n > 3).$$

Example 9.7 – Continued

We now use the oil palm data to estimate the parameters of the Bertalanffy function. The results (of the CADEMO package) are shown in Figure 9.12.

The estimated regression curve is shown in Figure 9.13.

Schlettwein (1987) did the simulation experiments described in Section 9.4.3 with normally distributed $\boldsymbol{e}_i$ and for several parameter combinations and n-values. Some results are shown in Table 9.9.

These and the other results of Schlettwein allow the conclusion that for normally distributed $\boldsymbol{e}_i$ in model (9.69), the asymptotic tests and confidence for all $n \geq 4$ are appropriate.

We now give in Table 9.10 optimal designs analogously to those in Section 9.6.3.

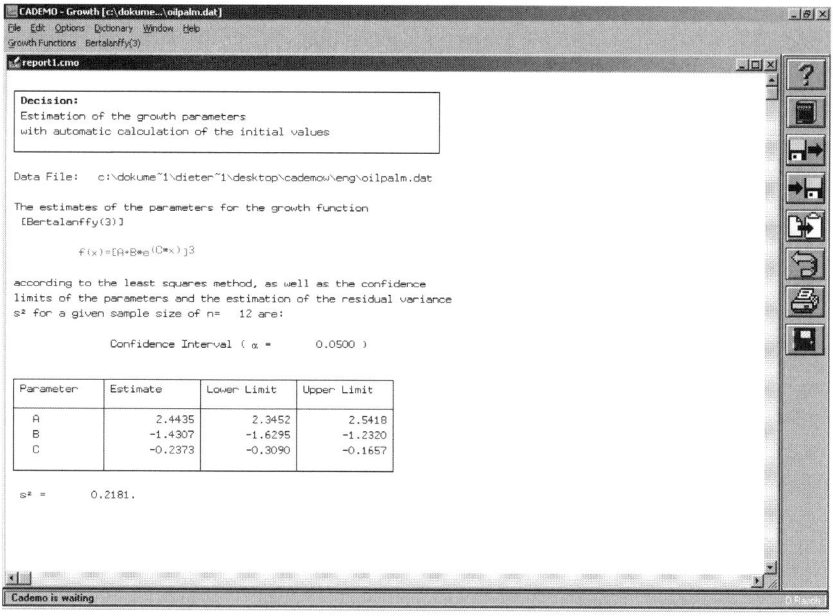

CADEMO - Growth [c:\dokume...\oilpalm.dat]

report1.cmo

Decision:
Estimation of the growth parameters
with automatic calculation of the initial values

Data File: c:\dokume~1\dieter~1\desktop\cademow\eng\oilpalm.dat

The estimates of the parameters for the growth function
[Bertalanffy(3)]

f(x)=[A+B*e^(C*x)]^3

according to the least squares method, as well as the confidence
limits of the parameters and the estimation of the residual variance
s² for a given sample size of n= 12 are:

Confidence Interval (α = 0.0500)

Parameter	Estimate	Lower Limit	Upper Limit
A	2.4435	2.3452	2.5418
B	-1.4307	-1.6295	-1.2320
C	-0.2373	-0.3090	-0.1657

s² = 0.2181.

Cademo is waiting

Figure 9.12 LS estimates, asymptotic confidence intervals and the estimated residual variance of the Bertalanffy function with data of Example 9.7.

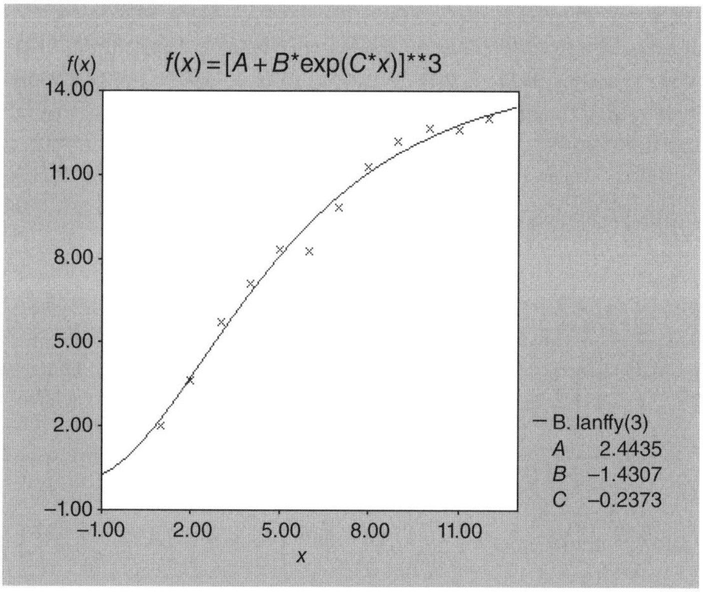

Figure 9.13 Curve of the estimated Bertalanffy regression function together with the scatter plot of the observations of Example 9.7.

Table 9.9 Relative frequencies of 10 000 simulated samples for the (incorrect) rejection (left hand n_l) right hand n_u and for the (correct) acception (n_M) of H_0 for the Bertalanffy function with several parameter values and for $\alpha_{nom} = 0.01$, 0.05 and 0.1 and n = 4.

	α_{nom} = 0.01			α_{nom} = 0.05			α_{nom} = 0.10		
H_0	n_u	n_o	n_M	n_u	n_o	n_M	n_u	n_o	n_M
$\alpha = 5$	0.46	0.48	99.06	2.32	2.40	95.28	4.73	5.07	90.20
$\beta = -2$	0.43	0.44	99.13	2.48	2.26	95.26	5.10	4.81	90.09
$\gamma = -0.05$	0.54	0.42	99.04	2.55	2.39	95.06	5.06	4.51	90.43
$\alpha = 5$	0.51	0.58	98.91	2.53	2.59	94.88	5.08	4.96	89.96
$\beta = -2$	0.61	0.57	98.82	2.65	2.34	95.01	4.98	4.75	90.27
$\gamma = -0.06$	0.49	0.44	99.07	2.64	2.26	95.10	5.16	5.78	90.06
$\alpha = 5$	0.53	0.59	98.88	2.61	2.60	94.79	5.02	5.28	89.70
$\beta = -3$	0.49	0.69	98.82	2.46	2.88	94.66	4.97	5.40	89.63
$\gamma = -0.05$	0.57	0.66	98.77	2.59	2.64	94.77	5.18	5.17	89.65
$\alpha = 5$	0.44	0.57	98.99	2.33	2.53	95.14	4.62	5.51	89.87
$\beta = -3$	0.47	0.59	98.94	2.32	2.60	95.08	4.81	5.28	89.91
$\gamma = -0.06$	0.52	0.58	89.90	2.47	2.40	95.13	5.24	4.72	90.04
$\alpha = 6$	0.51	0.51	98.98	2.52	2.75	94.73	4.88	5.30	89.82
$\beta = -4$	0.50	0.53	98.97	2.38	2.75	94.87	4.47	5.34	90.19
$\gamma = -0.07$	0.49	0.52	98.99	2.65	2.38	94.97	5.16	4.85	89.99
$\alpha = 6$	0.47	0.53	99.00	2.32	2.37	95.31	4.73	4.62	90.65
$\beta = -2$	0.54	0.50	98.96	2.54	2.20	95.26	5.04	4.82	90.14
$\gamma = -0.06$	0.57	0.57	98.86	2.37	2.33	95.30	4.98	4.78	90.24

Table 9.10 Optimal experimental designs in the experimental region [1,12] and n = 12.

Criterion	$(\alpha, \beta, \gamma) =$ (2.44, −1.43, −0.24)	$(\alpha, \beta, \gamma) =$ (2.35, −1.63, −0.31)	$(\alpha, \beta, \gamma) =$ (2.54, −1.23, −0.17)
D	$\begin{pmatrix} 1 & 5.39 & 12 \\ 4 & 4 & 4 \end{pmatrix}$	$\begin{pmatrix} 1.19 & 5.05 & 12 \\ 4 & 4 & 4 \end{pmatrix}$	$\begin{pmatrix} 1 & 5.88 & 12 \\ 4 & 4 & 4 \end{pmatrix}$
C_α	$\begin{pmatrix} 1 & 5.55 & 12 \\ 2 & 4 & 6 \end{pmatrix}$	$\begin{pmatrix} 1 & 5.09 & 8.07 \\ 2 & 3 & 7 \end{pmatrix}$	$\begin{pmatrix} 1 & 5.72 & 12 \\ 2 & 5 & 5 \end{pmatrix}$
C_β	$\begin{pmatrix} 1 & 10.27 & 12 \\ 9 & 1 & 2 \end{pmatrix}$	$\begin{pmatrix} 1 & 6.97 & 8.07 \\ 10 & 1 & 1 \end{pmatrix}$	$\begin{pmatrix} 1.94 & 2.69 & 12 \\ 4 & 4 & 4 \end{pmatrix}$
C_γ	$\begin{pmatrix} 1 & 5.54 & 12 \\ 4 & 5 & 3 \end{pmatrix}$	$\begin{pmatrix} 1 & 5.23 & 12 \\ 4 & 5 & 3 \end{pmatrix}$	$\begin{pmatrix} 1 & 5.92 & 12 \\ 4 & 5 & 3 \end{pmatrix}$

9.6.3 The Logistic (Three-Parametric Hyperbolic Tangent) Function

The function $f_L(x, \theta)$ of the model

$$\begin{aligned} \boldsymbol{y}_i &= \frac{\alpha}{1+\beta e^{\gamma x_i}} + e_i = f_L(x_i, \theta) + \boldsymbol{e}_i, \\ i &= 1,\ldots,n, \quad n > 3, \quad \alpha \neq 0, \quad \beta > 0, \quad \gamma \neq 0, \end{aligned} \tag{9.71}$$

is called logistic function. It has an inflection point at

$$x_I = -\frac{1}{\gamma}\ln\beta$$

with $f_L(x_I, \theta) = \alpha/2$.

The function in (9.71) can be written as three-parametric hyperbolic tangent function with the parameters

$$\alpha_T = \frac{\alpha}{2}, \beta_T = -\frac{1}{\gamma}\ln\beta \text{ and } \gamma_T = -\frac{\gamma}{2}$$

(see Example 9.4). With $\theta_1 = \alpha_T, \theta_2 = \beta_T$ and $\theta_3 = \gamma_T$

$$\begin{aligned} \boldsymbol{y}_i &= \alpha_T\{1 + \tanh[\gamma_T(x_i - \beta_T)]\} + \boldsymbol{e}_i, \\ i &= 1,\ldots,n, \quad n \geq 3, \quad \alpha_T \neq 0, \quad \beta_T \neq 0, \quad \gamma_T \neq 0, \end{aligned} \tag{9.72}$$

is the regression model of a three-parametric hyperbolic tangent function.

From Section 9.2 follows that as the consequence of a reparametrisation other non-linearity properties can be created. It therefore seems reasonable to find a reparametrisation with a small curvature measure. We first treat model (9.71) and receive

$$\frac{\partial f_L(x,\theta)}{\partial \theta} = \begin{pmatrix} \dfrac{1}{1+\beta e^{\gamma x}} \\ \dfrac{-\alpha e^{\gamma x}}{(1+\beta e^{\gamma x})^2} \\ \dfrac{-\alpha\beta x e^{\gamma x}}{(1+\beta e^{\gamma x})^2} \end{pmatrix}$$

and the model can be written with

$$C(\theta) = \begin{pmatrix} 1 & 0 & 0 \\ 0 & -\alpha & 0 \\ 0 & 0 & -\alpha \end{pmatrix}$$

in the form (9.1). We see that β and γ are non-linearity parameters. Analogously this can also be stated for the function in (9.72).

The information matrix for model (9.71) is

$$F^T F = \begin{pmatrix} A & -\alpha B & -\alpha\beta C \\ -\alpha B & \alpha^2 D & \alpha^2\beta E \\ -\alpha\beta C & \alpha^2\beta E & \alpha^2\beta^2 G \end{pmatrix}$$

with

$$Z_i = (1 + \beta e^{\gamma x_i})^{-1}$$

and

$$A = \sum_{i=1}^{n} Z_i^2, \quad B = \sum_{i=1}^{n} Z_i^3 e^{\gamma x_i}, \quad C = \sum_{i=1}^{n} Z_i^3 x_i e^{\gamma x_i},$$

$$D = \sum_{i=1}^{n} Z_i^4 e^{2\gamma x_i}, \quad E = \sum_{i=1}^{n} Z_i^4 x_i e^{2\gamma x_i}, \quad G = \sum_{i=1}^{n} Z_i^4 x_i^2 e^{2\gamma x_i}.$$

Then

$$\left|F^T F\right| = \alpha^4\beta^2 \left[ADG + 2BCE - C^2 D - AE^2 - B^2 G\right] = \alpha^4\beta^2 \Delta$$

follows and the asymptotic covariance matrix is

$$\mathrm{var}_A = \left(F^T F\right)^{-1} \sigma^2$$

$$= \frac{\sigma^2}{\Delta} \begin{pmatrix} DG - E^2 & -\frac{1}{\alpha}(EC - BG) & -\frac{1}{\alpha\beta}(BE - CD) \\ -\frac{1}{\alpha}(EC - BG) & \frac{1}{\alpha^2}(AG - C^2) & \frac{1}{\alpha^2\beta}(BC - AE) \\ -\frac{1}{\alpha\beta}(BE - CD) & \frac{1}{\alpha^2\beta}(BC - AE) & \frac{1}{\alpha^2\beta^2}(AD - B^2) \end{pmatrix}.$$

Initial values for the iterative calculation of the LS estimates are found by internal regression (see Section 9.1.2). The differential equation with integral $f_L(x, \theta)$ is given by

$$\frac{\partial f_L(x,\theta)}{\partial \theta} = -\gamma f_L(x,\theta)\left(1 - \frac{1}{\alpha} f_L(x,\theta)\right).$$

Minimising

$$S_1 = \sum_{i=1}^{n-1} \left(c_1 y_i + c_2 y_i^2 + y_i^*\right)^2, c_1 \neq 0, c_2 \neq 0 \text{ with}$$

$$y_i^* = \frac{y_{i+1} - y_i}{x_{i+1} - x_i} (i = 1,\ldots, n-1)$$

by the LS method results in $\hat{c}_1$ and $\hat{c}_2$. From these values, initial values $\hat{a}$ and $\hat{c}$ for the estimator of α and γ are given by

$$\hat{c} = c_1, \quad \hat{a} = -\frac{c_1}{c_2}.$$

The initial value $\hat{\beta}$ for the estimator of β is that value $b = \hat{b}$, minimising

$$S_2 = \sum_{i=1}^{n-1} \left(y_i^* + b \frac{\hat{c}}{\hat{\alpha}} y_i^2 e^{\hat{c} x_i} \right)^2.$$

The initial values a_T, b_T and c_T for the hyperbolic tangent function can be gained from those of the logistic function using the parameter transformation in front of (9.72).

The information matrix $F^T F$ of the model (9.72) has with the abbreviations

$$u_i = \tanh[\gamma(x_i - \beta)],$$

$$A_T = \sum_{i=1}^{n} u_i^2, \quad B_T = \sum_{i=1}^{n} (1 - u_i^2) u_i, \quad C_T = \sum_{i=1}^{n} (x_i - \beta)(1 - u_i^2) u_i,$$

$$D_T = \sum_{i=1}^{n} (1 - u_i^2)^2, \quad E_T = \sum_{i=1}^{n} (x_i - \beta)(1 - u_i^2)^2,$$

$$G_T = \sum_{i=1}^{n} (x_i - \beta)^2 (1 - u_i^2)^2$$

the form

$$F^T F = \begin{pmatrix} A_T & -\alpha\gamma B_T & \alpha C_T \\ -\alpha\gamma B_T & \alpha^2\gamma^2 D_T & -\alpha^2\gamma E_T \\ \alpha C_T & -\alpha^2\gamma E_T & \alpha^2 G_T \end{pmatrix}$$

with

$$|F^T F| = \alpha^4\gamma^2 \left[A_T D_T G_T + 2 B_T C_T E_T - C_T^2 D_T - A_T^2 E_T - B_T^2 G_T \right] = \alpha^4 \gamma^2 \Delta_T.$$

The asymptotic covariance matrix of the estimator $\hat{\theta}_T^T$ of $\theta_T^T = (\alpha_T, \beta_T, \gamma_T)$ is given by

$$\operatorname{var}_A(\theta_T) = \frac{\sigma^2}{\Delta_T} \begin{pmatrix} D_T G_T - E_T^2 & -\frac{1}{\alpha\gamma}(E_T C_T - B_T G_T) & \frac{1}{\alpha}(B_T E_T - D_T C_T) \\ -\frac{1}{\alpha\gamma}(E_T C_T - B_T G_T) & \frac{1}{\alpha^2\gamma^2}(A_1 G_1 - C_T^2) & \frac{1}{\alpha^2\gamma}(B_T C_T - A_T E_T) \\ \frac{1}{\alpha}(B_T E_T - C_T D_T) & \frac{1}{\alpha^2\gamma}(B_T C_T - A_T E_T) & \frac{1}{\alpha^2}(A_T D_T - B_T^2) \end{pmatrix}$$

Table 9.11 Optimal experimental designs in the experimental region [1,14] and n = 14.

Criterion	$(\alpha,\beta,\gamma) = (126, 20, -0.46)$	$(\alpha,\beta,\gamma) = (123, 16, -0.5)$	$(\alpha,\beta,\gamma) = (130, 23, -0.42)$
D	$\begin{pmatrix} 3.93 & 8.29 & 14 \\ 5 & 5 & 4 \end{pmatrix}$	$\begin{pmatrix} 3.30 & 7.39 & 14 \\ 5 & 5 & 4 \end{pmatrix}$	$\begin{pmatrix} 4.42 & 9.00 & 14 \\ 5 & 5 & 4 \end{pmatrix}$
C_α	$\begin{pmatrix} 1.15 & 8{,}07 & 14 \\ 2 & 3 & 9 \end{pmatrix}$	$\begin{pmatrix} 1.73 & 7.34 & 14 \\ 2 & 2 & 10 \end{pmatrix}$	$\begin{pmatrix} 2.81 & 9.20 & 14 \\ 3 & 3 & 8 \end{pmatrix}$
C_β	$\begin{pmatrix} 2.70 & 8.95 & 14 \\ 11 & 2 & 1 \end{pmatrix}$	$\begin{pmatrix} 2.16 & 2.37 & 14 \\ 11 & 1 & 2 \end{pmatrix}$	$\begin{pmatrix} 3.11 & 3.46 & 9.75 \\ 11 & 1 & 2 \end{pmatrix}$
C_γ	$\begin{pmatrix} 2.38 & 8.31 & 14 \\ 8 & 4 & 2 \end{pmatrix}$	$\begin{pmatrix} 1.81 & 7.37 & 14 \\ 8 & 4 & 2 \end{pmatrix}$	$\begin{pmatrix} 2.80 & 9.12 & 14 \\ 8 & 4 & 2 \end{pmatrix}$

For $n > 3$ test statistics and confidence intervals can be written down correspondingly to the sections above.

Further

$$s_T^2 = \frac{1}{n-3}\sum_{i=1}^{n}\left(y_i - a_T - a_T \tan\text{h}\left[c_T(x_i - b_T)\right]\right)^2$$

is the residual variance. In Example 9.3 the curve fitting of a logistic function was demonstrated by SPSS.

In simulation experiments as described in Section 9.4.3 for 15 (α,β,γ)-combinations (with inflection points at 10, 30 and 50 respectively), x_i-values in $[0, 65]$, normal-distributed $\boldsymbol{e}_i$ and $\alpha_{nom} = 0.05$ and 0.1 have been performed. For all parameter combinations, the result was that tests and confidence estimations based on the asymptotic covariance matrix can be recommended not only for $n > 3$ as well as for normally distributed $\boldsymbol{e}_i$ but also for $\boldsymbol{e}_i$ following some Fleishman distributions.

All concrete optimal designs have been three-point designs. In Table 9.11 we give optimal designs for the estimates of the parameters and for the confidence bounds in Figure 9.5 (hemp growth of Example 9.3).

9.6.4 The Gompertz Function

The regression function $f_G(x,\theta)$ of the model

$$\boldsymbol{y}_i = \alpha e^{\beta e^{\gamma x_i}} + \boldsymbol{e}_i = f_G(x_i,\theta) + \boldsymbol{e}_i \quad (i = 1,\ldots,n, n > 3, \alpha \neq 0, \gamma \neq 0, \beta < 0) \qquad (9.73)$$

is called Gompertz function. In Gompertz (1825) it was used to describe the population growth. The function has an inflection point at

$$x_I = -\frac{\ln(-\beta)}{\gamma} \quad \text{with } f_G(x_I) = \frac{\alpha}{e}.$$

The vector

$$\frac{\partial f_G(x,\theta)}{\partial\theta} = \begin{pmatrix} \frac{1}{\alpha} f_G(x) \\ f_G(x) e^{\gamma x} \\ f_G(x)\beta x e^{\gamma x} \end{pmatrix}; \theta^T = (\alpha,\beta,\gamma)$$

can be written with

$$C(\theta) = \begin{pmatrix} 1 & 0 & 0 \\ 0 & 1/\alpha & 0 \\ 0 & 0 & 1/\alpha \end{pmatrix}$$

in the form (9.1), where β and γ are non-linearity parameters. We again use abbreviations

$$A = \sum_{i=1}^{n} e^{2\beta e^{\gamma x_i}},\quad B = \sum_{i=1}^{n} e^{\gamma x_i} e^{2\beta e^{\gamma x_i}},\quad C = \sum_{i=1}^{n} x_i e^{\gamma x_i} e^{2\beta e^{\gamma x_i}},$$

$$D = \sum_{i=1}^{n} e^{2\gamma x_i} e^{2\beta e^{\gamma x_i}},\quad E = \sum_{i=1}^{n} x_i e^{2\gamma x_i} e^{2\beta e^{\gamma x_i}},\quad G = \sum_{i=1}^{n} x_i^2 e^{2\gamma x_i} e^{2\beta e^{\gamma x_i}},$$

so that

$$F^T F = \begin{pmatrix} A & \alpha B & \alpha\beta C \\ \alpha B & \alpha^2 D & \alpha^2\beta E \\ \alpha\beta C & \alpha^2\beta E & \alpha^2\beta^2 G \end{pmatrix}$$

and

$$|F^T F| = \alpha^4\beta^2\Delta = \alpha^4\beta^2\left[ADG + 2BCE - C^2 D - AE^2 - B^2 G\right] \neq 0.$$

The asymptotic covariance matrix is therefore

$$\text{var}_A(\theta) = \sigma^2\left(F^T F\right)^{-1} = \frac{\sigma^2}{\Delta}\begin{pmatrix} DG - E^2 & \frac{1}{a}(EC - BG) & \frac{1}{\alpha\beta}(BE - CD) \\ \frac{1}{\alpha}(EC - BG) & \frac{1}{a^2}\left(AG - C^2\right) & \frac{1}{\alpha^2\beta}(BC - CD) \\ \frac{1}{\alpha\beta}(BE - CD) & \frac{1}{\alpha^2\beta}(BC - AE) & \frac{1}{\alpha^2\beta^2}\left(AD - B^2\right) \end{pmatrix}.$$

Initial values for the iterative calculation of parameter estimates can be found by reducing the problems to that of the exponential regression using $z_i = \ln y_i$ ($y_i > 0$).

Because

$$\ln f_G(x,\theta) = \ln\alpha + \beta e^{\gamma x} = \alpha_E + \beta e^{\gamma x} \quad \text{with} \quad \alpha_E = \ln\alpha$$

Table 9.12 Relative frequencies of 10 000 simulated samples for the (incorrect) rejection (left hand n_l, right hand n_u) and for the (correct) acception (n_M) of H_0 for the Gompertz function with several parameter values and for $\alpha_{nom} = 0.01$, 0.05 and 0.1 and $n = 4$.

	α_{nom} = **0.01**			α_{nom} = **0.05**			α_{nom} = **0.10**		
H_0	n_l	n_u	n_M	n_l	n_u	n_M	n_l	n_u	n_M
$\alpha = 33.33$	0.56	0.48	98.96	2.58	2.03	95.39	5.27	4.17	90.56
$\beta = -6.05$	0.58	0.46	98.96	3.33	2.44	94.23	6.11	4.75	89.14
$\gamma = -0.06$	0.45	0.52	99.03	2.26	2.73	95.01	4.76	5.65	89.59
$\alpha = 33.33$	0.47	0.41	99.12	2.61	2.07	95.32	5.01	4.22	90.77
$\beta = -11.023$	0.46	0.53	99.01	2.46	2.16	95.38	4.84	4.54	90.62
$\gamma = -0.08$	0.50	0.45	99.05	2.17	2.36	95.47	4.47	4.77	90.76
$\alpha = 33.33$	0.49	0.50	99.01	2.65	2.53	94.82	5.07	5.11	89.82
$\beta = -20.09$	0.46	0.45	99.09	2.41	2.80	94.79	4.77	5.28	89.95
$\beta = -0.10$	0.44	0.43	99.13	2.64	2.46	94.90	5.24	4.85	89.91
$\alpha = 100$	0.61	0.51	98.88	2.77	2.49	94.74	5.35	4.53	90.12
$\beta = -36.6$	0.49	0.47	99.04	2.68	2.37	94.95	5.21	5.01	89.78
$\gamma = -0.06$	0.50	0.56	98.94	2.48	2.55	94.97	4.77	5.20	90.03
$\alpha = 25$	0.52	0.39	99.09	2.67	1.96	95.37	5.89	3.75	90.36
$\beta = -6.05$	0.65	0.43	98.92	2.93	2.43	94.64	6.13	4.73	89.14
$\beta = -0.06$	0.44	0.60	98.96	2.46	2.69	94.85	4.70	5.29	90.01
$\alpha = 25$	0.59	0.50	98.91	2.59	2.12	95.29	5.26	4.39	90.35
$\beta = -11.023$	0.47	0.50	99.03	2.68	2.19	95.13	5.20	4.61	90.19
$\gamma = -0.08$	0.51	0.47	99.02	2.19	2.43	95.38	4.35	4.86	90.79
$\alpha = 25$	0.55	0.44	99.01	2.69	1.98	95.33	5.28	4.24	90.48
$\beta = -20.09$	0.43	0.50	99.07	2.16	2.24	95.60	4.61	4.39	91.00
$\gamma = -0.10$	0.58	0.47	98.95	2.27	2.13	95.60	4.48	4.52	91.00

from the initial values (or LS estimates), a_E, b_E, c_E of the exponential regression for the (z_i, x_i) initial values $a' = e^{a_E}$, $b' = b_E$ and $c' = c_E$ for the Gompertz function can be obtained.

Tests and confidence regions can analogously be constructed as in the sections above. Numerous simulation experiments have been performed to show how good those tests are even for small sample sizes. In Table 9.12 we give the results for $n = 4$. In general it can be said that tests and confidence regions approximately hold the nominal risks and can be always recommended.

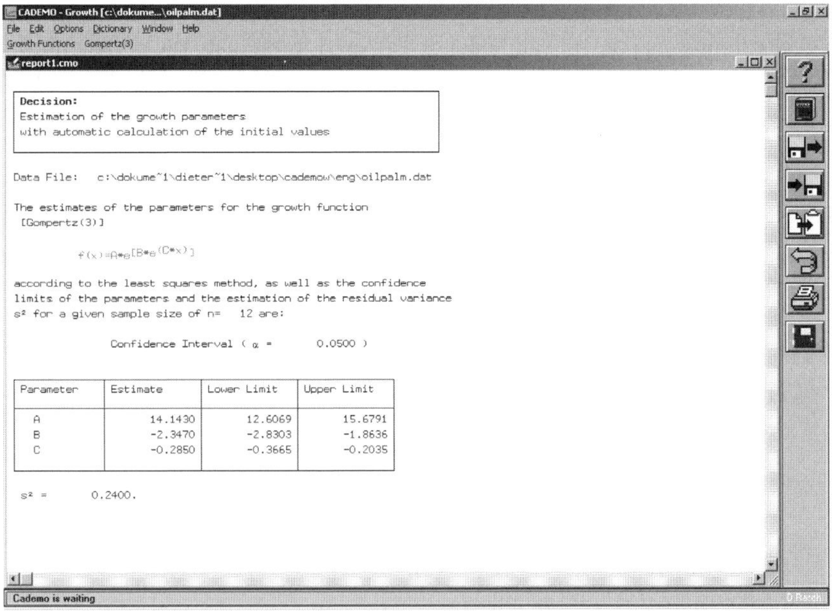

Figure 9.14 LS estimates, asymptotic confidence intervals and the estimated residual variance of the Gompertz function with data of Example 9.7.

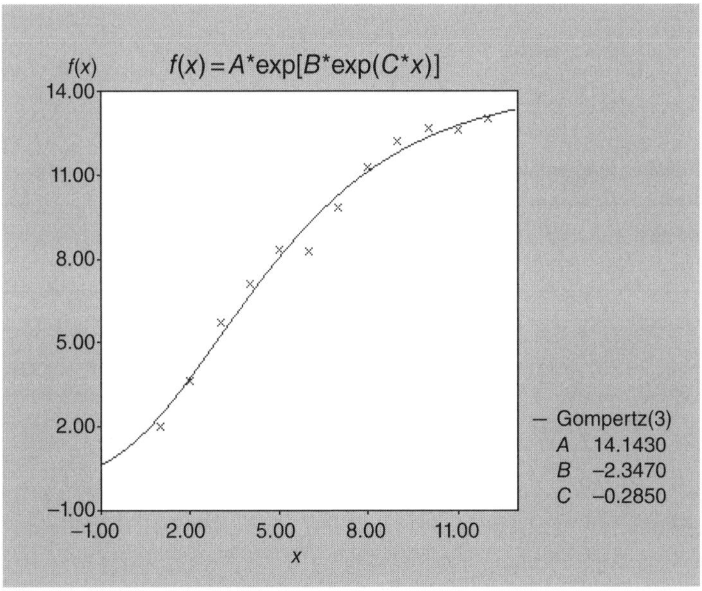

Figure 9.15 Curve of the estimated Gompertz regression function together with the scatter plot of the observations of Example 9.7.

Example 9.7 – Continued

We can now use the oil palm data of Table 9.3 to estimate the parameters of the Gompertz function by the program *Growth* (see Section 9.6.1.2). The results are shown in Figure 9.14. The estimated regression curve is shown in Figure 9.15.

All concrete optimal designs have been three-point designs. In Table 9.13 we give optimal designs for the estimates of the parameters and for the confidence bounds in Figure 9.14 (oil palm growth).

9.6.5 The Hyperbolic Tangent Function with Four Parameters

We consider the regression model

$$\boldsymbol{y}_i = \alpha + \beta \tan\mathrm{h}(\gamma + \delta x_i) + \boldsymbol{e}_i = f_T(x_i, \theta) + \boldsymbol{e}_i \quad (i = 1, \dots, n,\ n > 4,\ \beta > 0,\ \delta > 0). \tag{9.74}$$

$f_T(x, \theta)$ has an inflection point at $x_I = -\frac{\gamma}{\delta}$ with $f(x_I, \theta) = \alpha$ and two asymptotes at $y = \alpha + \beta$ and $y = \alpha - \beta$, respectively.

Because

$$\frac{\partial f_T(x,\theta)}{\partial \theta} = \begin{pmatrix} 1 \\ \tanh(\gamma + \delta x) \\ \beta\left[1 - \tanh^2(\gamma + \delta x)\right] \\ \beta x\left[1 - \tanh^2(\gamma + \delta x)\right] \end{pmatrix},$$

Table 9.13 Optimal experimental designs in the experimental region [1,12] and $n = 12$.

Criterion	$(\alpha, \beta, \gamma) =$ (14.14, −2.35, −0.285)	$(\alpha, \beta, \gamma) =$ (12.61, −2.83, −0.37)	$(\alpha, \beta, \gamma) =$ (15.68, −1.86, −0.2)
D	$\begin{pmatrix} 1 & 5.56 & 12 \\ 4 & 4 & 4 \end{pmatrix}$	$\begin{pmatrix} 1.29 & 5.14 & 12 \\ 4 & 4 & 4 \end{pmatrix}$	$\begin{pmatrix} 1 & 6.09 & 12 \\ 4 & 4 & 4 \end{pmatrix}$
C_α	$\begin{pmatrix} 1 & 5.65 & 12 \\ 2 & 4 & 6 \end{pmatrix}$	$\begin{pmatrix} 1 & 4.90 & 12 \\ 2 & 2 & 8 \end{pmatrix}$	$\begin{pmatrix} 1 & 6.22 & 12 \\ 2 & 5 & 5 \end{pmatrix}$
C_β	$\begin{pmatrix} 1 & 7.94 & 9.09 \\ 10 & 1 & 1 \end{pmatrix}$	$\begin{pmatrix} 1 & 5.47 & 5.95 \\ 10 & 1 & 1 \end{pmatrix}$	$\begin{pmatrix} 1.51 & 11.21 & 12 \\ 9 & 1 & 1 \end{pmatrix}$
C_γ	$\begin{pmatrix} 1 & 5.86 & 12 \\ 4 & 5 & 3 \end{pmatrix}$	$\begin{pmatrix} 1 & 5.39 & 12 \\ 4 & 5 & 3 \end{pmatrix}$	$\begin{pmatrix} 1 & 6.26 & 12 \\ 4 & 5 & 3 \end{pmatrix}$

$C(\theta)$ in (9.1) can be chosen so that γ and δ are non-linearity parameters.

With $v_i = \tan h(\gamma + \delta x_i)$ and

$$A = \sum_{i=1}^{n} v_i,\ B = \sum_{i=1}^{n} v_i^2,\ C = \sum_{i=1}^{n} x_i,\ D = \sum_{i=1}^{n} x_i v_i^2,\ E = \sum_{i=1}^{n} v_i^3,$$

$$G = \sum_{i=1}^{n} v_i^4,\ H = \sum_{i=1}^{n} x_i v_i^4,\ I = \sum_{i=1}^{n} x_i v_i,\ K = \sum_{i=1}^{n} x_i^2,$$

$$L = \sum_{i=1}^{n} x_i^2 v_i^2,\ M = \sum_{i=1}^{n} x_i v_i^3,\ N = \sum_{i=1}^{n} x_i^2 v_i^4,$$

$F^T F$ becomes

$$F^T F = \begin{pmatrix} n & A & \beta(n-B) & \beta(c-D) \\ A & B & \beta(A-E) & \beta(I-M) \\ \beta(n-B) & \beta(A-E) & \beta^2(n-2B+G) & \beta^2(C-B-D+H) \\ \beta(C-D) & \beta(I-M) & \beta^2(C-B-D+H) & \beta^2(K-2L+N) \end{pmatrix}$$

$$= \begin{pmatrix} n & A & \beta P & \beta Q \\ A & B & \beta R & \beta S \\ \beta P & \beta R & \beta^2 T & \beta^2 U \\ \beta Q & \beta S & \beta^2 U & \beta^2 W \end{pmatrix},$$

and the asymptotic covariance matrix is

$$\mathrm{var}_A(\theta) = \sigma^2 \left(F^T F\right)^{-1} = \left(\sigma_{\xi\eta}\right),\quad \xi,\ \eta = a, b, c, d.$$

To reach numerical stability in simulation experiments, it is favourable to invert $F^T F$ analytically and then to input the x_i-values. The formulae for the elements of $\mathrm{var}_A(\theta)$ are given in Gretzebach (1986). We give below the main diagonal elements $\sigma_{\xi\xi} = \sigma_\xi (\xi = a,\ b,\ c,\ d)$, that is, the asymptotic variances of the LS estimators ***a*, *b*, *c*** and ***d***, writing

$$\Delta = \frac{1}{\beta^4}\left|F^T F\right| :$$

$$\sigma_a^2 = \frac{\sigma^2}{\Delta}\left[BTW + 2RSU - R^2W - S^2T - U^2B\right],$$

$$\sigma_b^2 = \frac{\sigma^2}{\Delta}\left[nTW + 2QPU - RQ^2 - P^2W - nU^2\right],$$

$$\sigma_c^2 = \frac{\sigma^2}{\Delta\beta^2}\left[nBW + 2AQS - BQ^2 - A^2W - nS^2\right],$$

$$\sigma_d^2 = \frac{\sigma^2}{\Delta\beta^1}\left[nBT + 2ARP - BP^2 - A^2T - nS^2\right].$$

The initial values should be found by internal regression, leading to the LS estimates a, b, c and d for α, β, γ and δ. Test statistics and confidence estimations are obtained analogously to the sections above.

Example 9.7 – Continued

Now we can use the oil palm data to estimate the parameters of the hyperbolic tangent function with four parameters. The results are shown in Figure 9.16, and the estimated regression curve is shown in Figure 9.17.

The simulation experiments described in Section 9.4.3 to check the tests and confidence estimations based on the asymptotic covariance matrix for small n are performed for $\alpha = \beta = 50$ and

- $\delta = 0.15$ with $\gamma = -2.25, -4.5$ and -6.75
- $\delta = 0.1$ with $\gamma = -1.5, -3$ and -4.5
- $\delta = 0.05$ with $\gamma = -0.75, -1.5$ and -2.25

normally distributed $\boldsymbol{e}_i$ and n equidistant x_i-values in the interval $[0, 65]$; $n = 5$ (1)15.

It was found that the actual risk α_{act} of tests and confidence estimations differed by maximal 20% from the nominal risk $\alpha_{nom} = 0.05$ if at least $n = 10$ measurements were present. For $\alpha_{act} = 0.1$ this was already the case from $n =$ 9, but for $\alpha_{nom} = 0.01$ at least 25 measurements have been needed.

For these results let us conjecture that tests and confidence estimations based on the asymptotic covariance matrix for four-parametric functions (not as for three-parametric functions) can already be recommended from $n = p + 1$ on. This conjecture is supported in the two following sections.

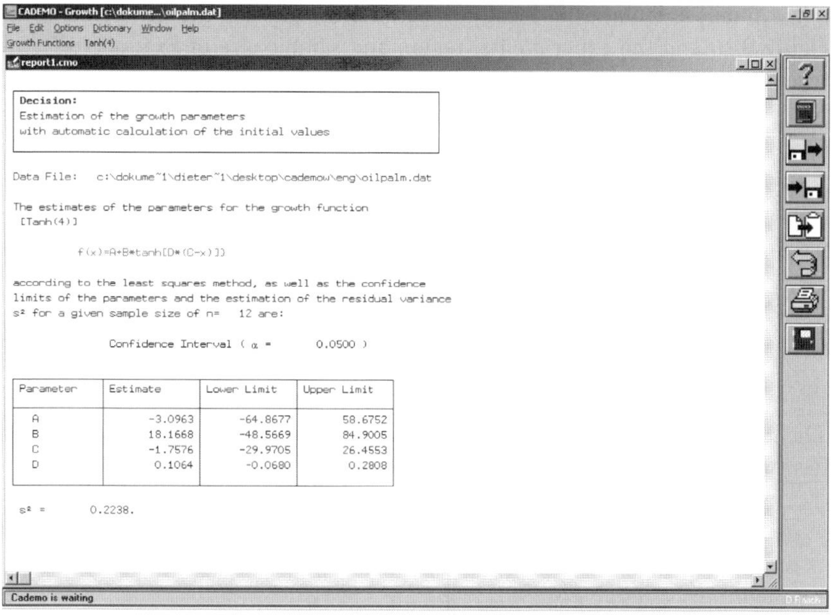

Parameter	Estimate	Lower Limit	Upper Limit
A	-3.0963	-64.8677	58.6752
B	18.1668	-48.5669	84.9005
C	-1.7576	-29.9705	26.4553
D	0.1064	-0.0680	0.2808

Figure 9.16 LS estimates, asymptotic confidence intervals and the estimated residual variance of the hyperbolic tangent function with data of Example 9.7.

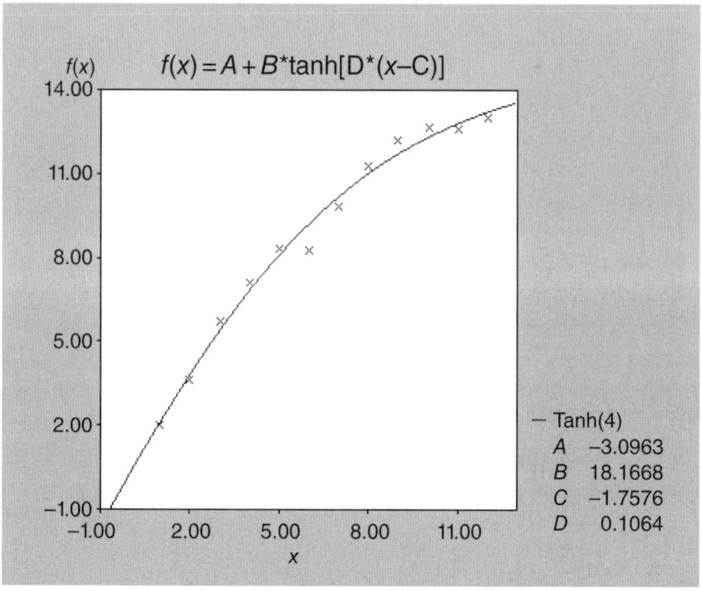

Figure 9.17 Curve of the estimated hyperbolic tangent regression function together with the scatter plot of the observations of Example 9.7.

We finally calculate the locally *D*-optimal designs for the LS estimates in Figure 9.16 and the experimental region and sample size of Example 9.7 and receive

$$\begin{pmatrix} 1 & 3.39 & 8.36 & 14 \\ 4 & 4 & 3 & 3 \end{pmatrix}.$$

9.6.6 The Arc Tangent Function with Four Parameters

We consider the regression model

$$\begin{gathered} \boldsymbol{y}_i = \alpha + \beta \arctan[\gamma(x_i - \delta)] + \boldsymbol{e}_i = f_A(x_i, \theta) + \boldsymbol{e}_i \\ (i = 1, \dots, n,\ n > 4,\ \beta \neq 0, \gamma > 0,\ \delta \neq 0). \end{gathered} \tag{9.75}$$

The function $f_A(x, \theta)$ has an inflection point at $x_I = \delta$ where $f_A(x_I, \theta) = \alpha$. $f_A(x, \theta)$ has two asymptotes at $\alpha + \beta\pi/2$ and $\alpha - \beta\pi/2$. We receive

$$\frac{\partial f_A(x, \theta)}{\partial \theta} = \begin{pmatrix} 1 \\ \arctan[\gamma(x - \delta)] \\ \dfrac{\beta(x - \delta)}{1 + \gamma^2 (x - \delta)^2} \\ \dfrac{-\beta\gamma}{1 + \gamma^2 (x - \delta)^2} \end{pmatrix}.$$

Writing this in the form (9.1) shows that γ and δ are non-linearity parameters.

We put

$$u_i = x_i - \delta,\quad v_i = 1 + \gamma^2 (x_i - \delta),\quad w_i = \arctan[\gamma(x_i - \delta)]$$

and get

$$F^T F = \begin{pmatrix} n & A & \beta C & -\beta\gamma D \\ A & B & \beta E & -\beta\gamma G \\ \beta C & \beta E & \beta^2 H & -\beta^2 \gamma J \\ -\beta\gamma D & -\beta\gamma G & -\beta^2 \gamma J & \beta^2 \gamma^2 K \end{pmatrix}$$

with

$$A = \sum_{i=1}^n w_i,\quad B = \sum_{i=1}^n w_i^2,\quad C = \sum_{i=1}^n \frac{u_i}{v_i},$$

$$D = \sum_{i=1}^n \frac{1}{v_i},\quad E = \sum_{i=1}^n \frac{u_i w_i}{v_i},\quad G = \sum_{i=1}^n \frac{w_i}{v_i},$$

$$H = \sum_{i=1}^n \frac{u_i^2}{v_i^2},\quad J = \sum_{i=1}^n \frac{u_i}{v_i^2},\quad K = \sum_{i=1}^n \frac{1}{v_i^2}.$$

The asymptotic covariance matrix

$$\mathrm{var}_A(\theta) = \sigma^2\left(F^T F\right)^{-1}$$

is estimated by

$$V = s^2\left(\hat{F}^T \hat{F}\right)^{-1} = s^2\left(k_{ij}\right).$$

Further

$$s^2 = \frac{1}{n-4}\sum_{i=1}^{n}\{y_i - a - b\ \arctan\left[c(x_i - d\right\}^2.$$

Initial values should be found by internal regression, leading to the LS estimates a, b, c and d for α, β, γ and δ. Test statistics and confidence estimations are obtained analogous to the sections above.

Example 9.7 – Continued

Now we use the oil palm data to estimate the parameters of the arcustangens function with four parameters. The results are shown in Figure 9.18, and the estimated regression curve is shown in Figure 9.19.

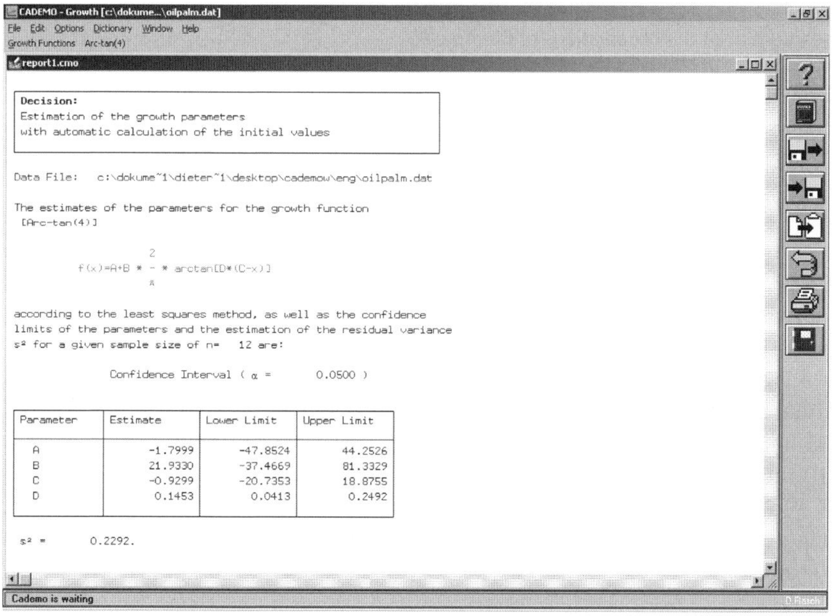

Figure 9.18 LS estimates, asymptotic confidence intervals and the estimated residual variance of the arcustangens function with data of Example 9.7.

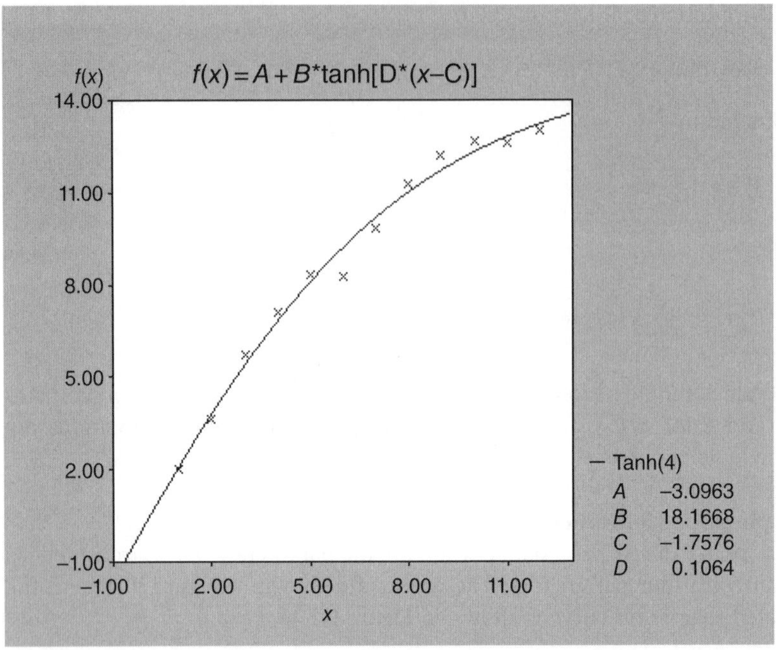

Figure 9.19 17 Curve of the estimated Arcustangens regression function together with the scatter plot of the observations of Example 9.7.

The simulation experiments described in Section 9.4.3 to check the tests and confidence estimations based on the asymptotic covariance matrix for small n are performed for $\alpha = \beta = 50$ and $\alpha = 40$, $\beta = 20$ with

$$\gamma = 0.05, 0.1, 0.2 \text{ and } \delta = -50, -30, -10,$$

normally distributed $\boldsymbol{e}_i$ and n equidistant x_i-values in the interval $[0, 65]$ using $n = 4(1)20$.

It was found that the actual risk α_{act} of tests and confidence estimations differed by maximal 20% from the nominal risk $\alpha_{nom} = 0.05$ if at least n = 11 measurements were present. For $\alpha_{act} = 0.1$ this was already from n = 10 the case.

We finally calculate the locally D-optimal designs for the LS estimates in Figure 9.18 and the experimental region and sample size of Example 9.7 and receive

$$\begin{pmatrix} 1 & 2.97 & 7.68 & 14 \\ 4 & 4 & 3 & 3 \end{pmatrix}.$$

9.6.7 The Richards Function

The function $f_R(x, \theta)$ in the regression model

$$\boldsymbol{y}_i = \alpha\left[1 + e^{\frac{\gamma}{\alpha}(\beta - x_i)(\delta+1)^{1+1/\delta}}\right]^{-1/\delta} + \boldsymbol{e}_i = f_R(x_i) + \boldsymbol{e}_i$$
$$(i = 1,\dots,n, n > 4, \alpha \neq 0,\ \gamma \neq 0,\ \delta < 0) \tag{9.76}$$

in Richards (1959) was used to model the growth of plants; the parametrisation in (9.76) stems from Schönfelder (1987) and was introduced because the iterative calculation of the LS estimates have been relatively easy and the suitability of the asymptotic covariance matrix for tests and confidence estimations was given. Further some parameters can be interpreted: α is the value of the asymptote and β the x-coordinate of the inflection point.

The parameters β, γ, δ are non-linearity parameters. Writing $f_R(x, \theta)$ in its original form $f_R(x,\theta) = \left(\alpha^* + \beta^* e^{\gamma * x}\right)^{\delta^*}$, in Richards (1959), then all parameters are non-linearity parameters.

There are enormous numerical problems with this function, especially for gaining initial values and for the iterative calculations of the LS estimates. We recommend the interested reader to read the PhD thesis in Schönfelder (1987) where FORTRAN programs are given.

Tests and confidence estimations have been checked by Schönfelder as described in Section 9.4.3 for equidistant x_i and for the x_i of a locally D-optimal designs in [1; 65] and the parameter combinations (α, β, γ, δ):

$$(35;\ 27;\ 1;\ 0.7), (20;\ 27;\ 1;\ 0.7), (35;\ 15;\ 1;\ 0.7), (35;\ 27;\ 5;\ 0.7), (35;\ 27;\ 1 - 0.5),$$
$$(50;\ 27;\ 1;\ 0.7), (35;\ 45;\ 1;\ 0.7), (35;\ 27;\ 3;\ 0.7), (35;\ 27;\ 1;\ 10)$$

for normally distributed $\boldsymbol{e}_i$.

The tests and confidence estimations based on the asymptotic covariance can be used if for $\alpha_{nom} = 0.05$ there is $n > 14$. For the locally D-optimal design already with $n > 8$ satisfying results have been found.

9.6.8 Summarising the Results of Sections 9.6.1–9.6.7

First we summarise the results of the simulation experiments.

The simulation experiments described in Section 9.4.3 to check the tests and confidence estimations based on the asymptotic covariance matrix give the inducement to conjecture that for three-parametric regression functions, $n = 4$ observations are sufficient for the approximation with the asymptotic covariance matrix.

For four-parametric regression functions, the minimal number sufficient for the approximation with the asymptotic covariance matrix depends strongly on the function and on the allocation of the support points. Seldom $n < 10$ is sufficient.

Next we summarise results concerning locally optimal designs.

For three-parametric regression functions, we conjecture that locally optimal designs are always three-point designs. For four-parametric regression functions, we conjecture that the D-optimal designs are three-point designs.

Paulo and Rasch (2002) investigated the sensitivity of D-optimal designs if parameters differ from values used in the locally optimal designs and found that the support is relatively non-sensitive.

9.6.9 Problems of Model Choice

As we can see from Example 9.7, it is not easy to select a proper regression function for given (x_i, y_i) values. Numerical criteria have been proposed by several authors. Given a class $F = \{f_1(x, \theta), \ldots, f_r(x, \theta)\}$ of functions from which one has to be chosen as the 'best', which of the criteria should be used? Rasch and van Wijk (1994) considered in a simulation experiment the $r = 8$ in Sections 9.6.1–9.6.7 handled functions and five criteria. In the following $\hat{f}_j(x_i, \theta)$ are the values of the LS method to get a (x_i, y_i) fitted function with the estimated parameters. The regression function is $f_j(x, \theta) \in F$ and p_j is the number of the estimated parameters.

The criteria are $(n > p, j = 1,\ldots,r)$:

K1: s_j^2 (residual variance by fitting $f_j(x, \theta) \in F$), with

$$s_j^2 = \frac{1}{n-p_j}\sum_{i=1}^{n}\left[y_i - \hat{f}_j(x_i,\theta)\right]^2.$$

K2: $C_{pj} = \dfrac{(n-p_j)\, s_j^2}{n\,\hat{\sigma}_j^2} + 2p_j - n$ (C_p criterion in Mallows (1973))

$\hat{\sigma}_j^2$ is an estimate of σ^2 different from s_j^2. This could be the MS_{res} in a simple analysis of variance, if several measurements at support points are available.

K3: Jackknife criterion

Drop in (x_i, y_i) $(i = 1, \ldots, n)$ the lth pair $(l = 1, \ldots, n)$. With the $n - 1$ remaining data pairs, the functions $f_j(x, \theta) \in F$ are fitted. Let $\hat{y}_l(j) = \hat{f}_j^{(l)}(x_l, \theta)$ be the value of the fitted function at x_l. Then

$$JK_j = \frac{1}{n}\sum_{i=1}^{n}[y_i - \hat{y}_i(j)]^2$$

is the value of the Jackknife criterion. The name is chosen analogously to the Jackknife estimate (Chapter 4).

K4: Modified Akaike criterion (Akaike, 1974)

With $S_j^2 = \frac{n-p_j}{n} s_j^2$ is

$$AIC_j = n\ln\left(S_j^2\right) + \frac{n\left(n+p_j\right)}{n-p_j-2}$$

K5: Schwarz criterion (Schwarz, 1978)

With $T_j^2 = \frac{n-p_j}{n} s_j^2$ is

$$SC_j = n\ln\left(T_j^2\right) + p_j\ln(n).$$

Rasch and van Wijk (1994) found the Jackknife version of the modified Akaike criterion to be the best one. In the simulation experiment, values have been generated with each function in F. To the values of the function error terms have been added. By the LS method each function was fitted to each generated data set. This was repeated 5000 times. In an 8×8 matrix it was shown how often data of a generating function (in the row of the matrix) was selected by one of the criteria (in the column) as best fitted. Of course a heavy main diagonal of the matrix is ideal for a criterion.

9.7 Exercises

9.1 Which of the regression functions below are linear, quasilinear or intrinsically non-linear?

a) $f(x, \theta) = \theta_1 + \theta_2 x + \theta_3 x^2$

b) $f(x,\theta) = \theta_1 \frac{1}{x} + \theta_2 x$

c) $f(x, \theta) = \theta_0 + \theta_1 x_1 + \theta_2 x_2$

d) $f(x,\theta) = \theta_1 x^{\theta_2}$

e) $f(x, \theta) = \theta_1 x_1 + \theta_2 x_2 + \theta_3 x_1 x_2$

f) $f(x,\theta) = \theta_1 + e^{\theta_2 x}$

g) $f(x,\theta) = \theta_1 + \frac{\theta_2 x}{\theta_3 x^2 + 1}$

9.2 Determine the non-linearity parameter(s) of the following regression functions:

a) $f(x, \theta) = \theta_1 + \sin(\theta_2 x + \theta_3)$

b) $f(x,\theta) = \frac{\theta_1 x}{\theta_2 + x}$

c) $f(x,\theta) = \theta_1 e^{\theta_2 x + \theta_3}$

d) $f(x,\theta) = \theta_1 + \theta_2 2^{\theta_3}$

e) $f(x,\theta) = \theta_1 + \theta_2 \frac{1}{x} + \theta_3\left(1 + e^{2\theta_4}\right)^2$

9.3 Fit the exponential regression function

$$y = f(x,\theta) = \alpha + \beta e^{\gamma x} + e,\ \theta = (\alpha,\beta,\gamma)^T,\ \gamma < 0$$

to the data below!

Time	0	1	2	3	4	5	6	7	8	9	10
Value	77.2	94.5	107.2	116.0	122.4	126.7	129.2	129.9	130.4	130.8	131.2

Calculate the estimates a, b, c for the parameters by the LS method. Give further an estimate for the variance σ^2.

9.4 For the no-load loss of a generator in dependency of the voltage measurements the data are:

X voltage	230	295	360	425	490	555	620
Loss L (kW)	64.0	66.0	69.5	74.0	80.8	91.0	103.5

Which of the regression models in Section 9.6 fits best?

Find estimates for the parameter of the best fitting function by the LS method and give an estimate for the error variance.

References

Akaike, H. (1974) A new look at the statistical model identification. *IEEE Trans. Autom. Control*, **19**, 716–723.

Atkinson, A. C. and Hunter, W. G. (1968) The design of experiments for parameter estimations. *Technometrics*, **10**, 271–289.

Barath, C. S., Rasch, D. and Szabo, T. (1996) Összefügges a kiserlet pontossaga es az ismetlesek szama között. *Allatenyesztes es takarmanyozas*, **45**, 359–371.

Bates, D. M. and Watts, D. G. (1988) *Nonlinear Regression Analysis and Its Applications*, John Wiley & Sons, Inc., New York.

Bauer, H. (2002) *Wahrscheinlichkeitstheorie*, de Gruyter, Berlin, New York.

Beale, E. M. L. (1960) Confidence regions in nonlinear regression. *J. R. Stat. Soc. Ser. B*, **22**, 41–88.

Bertalanffy, L. von (1929) Vorschlag zweier sehr allgemeiner biologischer Gesetze. *Biol. Zbl.*, **49**, 83–111.

Bliss, C. L. and James, A. T. (1968) Fitting on the rectangular hyperbola. *Biometrika*, **22**, 573–603.

Box, M. J. (1971) Bias in nonlinear estimation (with discussion). *J. R. Stat. Soc. Ser. B*, **46**, 171–201.

Box, G. E. P. and Draper, N. R. (1963) The choice of a second order rotatable design. *Biometrika*, **50**, 77–96.

Box, G. E. P. and Lucas, H. L. (1959) Design of experiments in nonlinear statistics. *Biometrics*, **46**, 77–96.

Bunke, H. (1973) Approximation of regression functions. *Math. Oper. Stat. Ser. Stat.*, **4**, 315–325.

Clarke, G. P. Y. (1980) Moments of the least squares estimators in a nonlinear regression model. *J. R. Stat. Soc. Ser. B*, **42**, 227–237.

Ermakoff, S. M. (1970) Ob optimalnich nesmeschtschennich planach regressionnich eksperimentov. *Trudy Nat. Inst. Steklowa*, **111**, 25–257.

Feller, W. (1961) *An Introduction to Probability Theory and Its Application*, John Wiley & Sons, Inc., New York.

Fleishman, A. J. (1978) A method for simulating non-normal distributions. *Psychometrika*, **43**, 521–532.

Gompertz, B. (1825) On the nature of the function expressive of the law of human mortality, and on a new method determining the value of life contingencies. *Philos. Trans. R. Soc. B*, London, **115**, 513–585.

Gretzebach, L. (1986) *Simulationsuntersuchungen zum t – Test für die Parameter der Funktion* $\alpha + \beta tanh(+ \delta x)$, Sektion Mathematik, Universität Rostock.

Halperin, M. (1963) Confidence interval estimation in nonlinear regression. *J. R. Stat. Soc. Ser. B*, **25**, 330–333.

Hartley, H. O. (1948) The estimation of non-linear parameters by "internal least squares". *Biometrika*, **35**, 32–48.

Hartley, H. O. (1961) The modified Gauss-Newton method for the fitting of nonlinear regression functions by least squares. *Technometrics*, **3**, 269–280.

Holland, P. W. (1973) Covariance stabilizing transformations. *Ann. Stat.*, **1**, 84–92.

Hotelling, H. (1927) Differential equations subject to error and population estimates. *J. Am. Stat. Assoc.*, **2**, 283–314.

Hougaard, P. (1982) Parametrizations of non-linear models. *J. R. Stat. Soc. Ser. B*, **44**, 244–282.

Hougaard, P. (1984) *Parameter Transformations in Multiparameter Nonlinear Regression Models*, Inst. Math. Statist. Univ. Copenhagen, Preprint 2/84.

Ivanov, A.V. and Zwanzig, S. (1983) An asymptotic expansion of the distribution of least squares estimators in the nonlinear regression model. *Math. Oper. Stat. Ser. Stat.*, **14**, 7–27.

Jennrich, R. J. (1969) Asymptotic properties of nonlinear least squares estimation. *Ann. Math. Stat.*, **40**, 633–643.

Johansen, S. (1984) *Functional Relations, Random Coefficients and Nonlinear Regression with Applications to Kinetic Data*, Springer, New York.

Karson, M. G., Manson, A. R. and Hader, R. G. (1969) Minimum bias estimation and experimental design for regression surface. *Technometrics*, **11**, 461–475.

Levenberg, K. (1944) A method for the solution of certain non-linear problems in least squares. *Q. Appl. Math.*, **2**, 164–168.

Malinvaud, E. (1970) The consistency of nonlinear regression. *Ann. Math. Stat.*, **41**, 56–64.

Mallows, C. L. (1973) Some comments on C_P. *Technometrics*, **15**, 661–675.

Maritz, J. S. (1962) Confidence regions for regression parameters. *Austr. J. Stat.*, **4**, 4–10.

Marquardt, D. W. (1963) An algorithm for least squares estimation of non-linear parameters. *SIAM J. Appl. Math.*, **11**, 431–441.

Marquardt, D. W. (1970) Generalized inverse, ridge regression, biased linear estimation and non-linear estimation. *Technometrics*, **13**, 591–612.

Melas, V. B. (2008) *Functional Approach to Optimal Experimental Design*, Springer, New York.

Morton, R. (1987) A symmetry of estimators in nonlinear regression. *Biometrika*, **74**, 679–685.

Nash, J. C. (1979) *Compact Numerical Methods for Computers – Linear Algebra and Function Minimization*, Adam Hilger Ltd., Bristol.

Paulo, M. J. and Rasch, D. (2002) Robustness and efficiency of D-optimal experimental designs in a growth problem. *Biom. J.*, **44**, 527–540.

Pazman, A. (1985) *Foundations of Optimum Experimental Design*, Reidel, Dordrecht.

Petersen, J. (1973) Linear minimax estimation for inadequate models. *Math. Oper. Stat. Ser. Stat.*, **4**, 463–471.

Rasch, D. (1990) Optimum experimental design in nonlinear regression. *Commun. Stat., – Theory Methods*, **19**, 4789–4806.

Rasch, D. (1993) The robustness against parameter variation of exact locally optimum experimental designs in growth models - a case study, *Technical Note 93-3*, Department of Mathematics, Wageningen Agricultural University.

Rasch, D. and Guiard, V. (2004) The robustness of parametric statistical methods. *Psychol. Sci.*, **46**, 175–208.

Rasch, D. and Melas, V. (Eds.) (2017) Advances in Statistics and Simulation, Statistical Theory and Practice, *Special Issue*, Taylor and Francis, Oxford, New York.

Rasch, D. and Schimke, E. (1983) Distribution of estimators in exponential regression – a simulation study. *Scand. J. Stat.*, **10**, 293–300.

Rasch, D. and van Wijk, H. (1994) *The evaluation of criteria for model selection in growth curve analysis, Proc. XVIIth Int. Biometric Conf.*, Hamilton, Ontario, Canada.

Rasch, D., Herrendörfer, G., Bock, J., Victor, N. and Guiard, V. (Eds.) (2008) *Verfahrensbibliothek Versuchsplanung und -auswertung, 2.* verbesserte Auflage in einem Band mit CD, R. Oldenbourg Verlag München, Wien.

Richards, F. J. (1959) A flexible growth function for empirical use. *J. Exp. Bot.*, **10**, 290–300.

Scharf, J. H. (1970) Innere Regressionsrechnung mit nichtlinearen Differentialgleichungen. *Biochem. Z.*, **12**, 228–241.

Schlettwein, K. (1987) *Beiträge zur Analyse von vier speziellen Wachstumsfunktionen*, Dipl.-Arbeit, Sektion Mathematik, Univ. Rostock.

Schmidt, W. H. (1979) Asymptotic results for estimations and testing variances in regression models. *Math. Oper. Stat. Ser. Stat.*, **10**, 209–236.

Schönfelder, E. (1987) *Eigenschaften der Schätzungen für Parameter in nichtlinearen Regressionsfunktionen – dargestellt am Beispiel der vierparametrigen Richards-Funktion*, Diss. Sektion Mathematik Univ. Rostock.

Schwarz, G. (1978) Estimating the dimension of a model. *Ann. Stat.*, **2**, 461–464.

Verhagen, A. W. M. (1960) Growth curves and their functional form. *Austr. J. Stat.*, **2**, 122–127.

Williams, E. J. (1962) Exact fiducial limits in nonlinear estimation. *J. R. Stat. Soc. Ser. B*, **24**, 125–139.

Wu, C. F. (1981) Asymptotic theory of nonlinear least squares estimation. *Ann. Stat.*, **9**, 501–513.

Yanagida, T. (2016) miscor: miscellaneous functions for the correlation coefficient. R package version 0.1-0. https://cran.r-project.org/web/packages/miscor. Accessed August 14, 2017.

Yanagida, T. (2017) Seqtest: an R package for sequential triangular tests, in Rasch, D. and Melas, V. (Eds.) Advances in Statistics and Simulation, Statistical Theory and Practice, *Special Issue*, Taylor and Francis, Oxford, New York.

10

Analysis of Covariance (ANCOVA)

10.1 Introduction

Analysis of covariance (ANCOVA) as a branch of applied statistics covers several objectives. In any case, the observations are influenced by at least two factors. At least one of these factors has several levels, by which the material is classified into classes. At least one further factor is a regressor in a regression model between different variables in the model and called a covariable or a covariate. One branch of the ANCOVA is to test whether the influence of the covariable is significant and as the case may be to eliminate it.

If the factor is qualitative (not numeric), this target can be achieved simply by blocking and using analysis of variance (ANOVA). Another branch of the ANCOVA is to estimate the parameters of the regression model within the classes of the classification factor.

If we have just one classification factor and one covariable, then we have four models of the ANCOVA:

Model I–I: Levels of the classification factor fixed and model I of regression
Model I–II: Levels of the classification factor fixed and model II of regression
Model II–I: Levels of the classification factor random and model I of regression
Model II–II: Levels of the classification factor random and model II of regression

In statistical (theoretical) textbooks, mainly model I–I was presented. However, in applications and in many examples, exclusively cases are found for which model I–II must be used. The results found for model I–I are used for model I–II. Real practical examples for model I–I can hardly be found. Graybill (1961) bypasses this difficulty by using a fictive numerical example.

Searle (1971, 2012) uses as levels of the classification factor three kinds of school education of fathers of a family and the number of children in the family as a covariable. The observed trait is the amount of expenditures. Here certainly

Mathematical Statistics, First Edition. Dieter Rasch and Dieter Schott.

a model of an incomplete two-way cross-classification could be used, and the question whether the covariable 'number of children' leads to model I or II of the regression analysis depends on data collection. Scheffé (1953) considered an introductory example with the classification factor 'kind of starch' and the covariable the thickness of the starch strata. This is an example where model I–II can be used although afterwards he discusses model I–I. However, Scheffé is one of the few recognising and discussing the two models. He gave a heuristic rationale of the application of the results derived for model I–I but applied for model I–II. The background for this is the applicability of methods of estimating and testing of model I of regression to model II of regression as described in Chapter 8.

In the text below exclusively model I–I as special case of model equation (4.1) is considered. Model II–II corresponds with problems treated in Chapter 6 (estimation of variance and covariance components).

10.2 General Model I–I of the Analysis of Covariance

We consider the following special case.

Definition 10.1 If $X\beta$ in (5.1) with the assumptions of Definition 5.1 can be written as

$$X\beta = W\alpha + Z\gamma$$

with

$$X = (W, Z),\ \ \beta^T = \left(\alpha^T, \gamma^T\right),$$
$$X : [X \times (a+1)] \text{ matrix of rank } \ p < a+1,$$
$$W : [N \times (t+1)] \text{ matrix of rank } \ r, 0 < r < t+1 < a+1,$$
$$W = (1_N, W^*),$$
$$Z : [N \times s] \text{ matrix of rank } 0 < s \le p < a+1,$$
$$\alpha = (\mu, \alpha_1, \dots, \alpha_t)^T,\ \ \gamma = (\gamma_1, \dots, \gamma_s)^T \ \ (t+s=a, r+s=p),$$

then the model equation

$$Y = W\alpha + Z\gamma + \boldsymbol{e},\ \ \Omega = R[W] \oplus \mathrm{R}[Z] \tag{10.1}$$

under the distributional assumption

$$Y:\ \ N\left(W\alpha + Z\gamma,\ \sigma^2 I_N\right),\ \ \boldsymbol{e} : N\left(0_N,\ \sigma^2 I_N\right) \tag{10.2}$$

is called a model I–I of the ANCOVA. Normality is only needed for testing and confidence estimation.

The columns of Z define the covariable.

First we will give an example.

Example 10.1 In a populations $G_1, \dots, G_a$ independent random vectors $\boldsymbol{Y}_1, \dots, \boldsymbol{Y}_a$ of size $n_1, \dots, n_a$ are available. Let $y_i = (y_{i1},\dots,y_{in_i})^T$. In G_i the $\boldsymbol{Y}_i$ are $N(\{\mu_i\}), \sigma^2 I_{n_i})$-distributed with $\{\mu_i\} = (\mu_{i1},\dots,\mu_{in_i})^T$.

Case (a). μ_{ij} can be written as

$$\mu_{ij} = \mu + \alpha_i + \gamma z_{ij} \quad (i = 1,\dots,a;\ j = 1,\dots,n_i).$$

z_{ij} are given values of a real (influence) variable Z, the covariable of the model. Case (b). μ_{ij} can be written as

$$\mu_{ij} = \mu + \alpha_i + \gamma_i z_{ij} \quad (i = 1,\dots,a;\ j = 1,\dots,n_i).$$

Then (10.1) has the special form:

Case (a):

$$\boldsymbol{y}_{ij} = \mu + \alpha_i + \gamma z_{ij} + \boldsymbol{e}_{ij} \quad (i = 1,\dots,a;\ j = 1,\dots,n_i) \tag{10.3}$$

Case (b):

$$\boldsymbol{y}_{ij} = \mu + \alpha_i + \gamma_i z_{ij} + \boldsymbol{e}_{ij} \quad (i = 1,\dots,a;\ j = 1,\dots,n_i). \tag{10.4}$$

In (10.3) and (10.4) (as special case of (10.1)),

$$W^{\mathrm{T}} = \begin{pmatrix} 1 & 1 & \dots & 1 & 1 & 1 & \dots & 1 & \dots & 1 & 1 & \dots & 1 \\ 1 & 1 & \dots & 1 & 0 & 0 & \dots & 0 & \dots & 0 & 0 & \dots & 0 \\ 0 & 0 & \dots & 0 & 1 & 1 & \dots & 1 & \dots & 0 & 0 & \dots & 0 \\ \vdots & \vdots & & \vdots & \vdots & \vdots & & \vdots & & \vdots & \vdots & & \vdots \\ \underbrace{0 \ \ 0}_{} & \dots & & 0 & 0 & 0 & \dots & 0 & \cdots & 1 & 1 & \dots & 1 \end{pmatrix}$$

with column blocks of sizes n_1, n_2, …, n_a,

$$\alpha = (\mu, \alpha_1,\dots,\alpha_a)^{\mathrm{T}},\quad \boldsymbol{Y}^{\mathrm{T}} = \left(\boldsymbol{Y}_1^{\mathrm{T}},\dots,\boldsymbol{Y}_a^{\mathrm{T}}\right) \text{ and } \boldsymbol{e} = (\boldsymbol{e}_{11},\dots,\boldsymbol{e}_{an_a})^{\mathrm{T}}$$

In (10.3) $Z^T = (z_{11},\ldots,z_{an_a})$ and γ is a scalar. In (10.4) $\gamma^T = (\gamma_1,\gamma_2,\ldots,\gamma_a)$ and

$$Z = \begin{pmatrix} z_{11} & 0 & \ldots & 0 \\ \vdots & \vdots & & \vdots \\ z_{1n_1} & 0 & \ldots & 0 \\ 0 & z_{21} & \ldots & 0 \\ \vdots & \vdots & & \vdots \\ 0 & z_{21} & \ldots & 0 \\ \vdots & \vdots & & \vdots \\ 0 & 0 & \ldots & z_{a1} \\ \vdots & \vdots & & \vdots \\ 0 & 0 & \ldots & z_{an_a} \end{pmatrix}.$$

Example 10.2 We consider the situation of Example 10.1 with a populations, but for $\boldsymbol{y}_{ij}$ we use the model equation

$$\mu_{ij} = \mu + \alpha_i + \gamma_1 z_{ij}^2 + \cdots + \gamma_s z_{ij}^s \quad (i = 1,\ldots,a;\ j = 1,\ldots,n_i),$$

so that (10.1) has the special form

$$\boldsymbol{y}_{ij} = \mu + \alpha_i + \gamma_1 z_{ij} + \gamma_1 z_{ij}^2 + \cdots + \gamma_s z_{ij}^s + \boldsymbol{e}_{ij}. \tag{10.5}$$

$\boldsymbol{Y}$, α and $\boldsymbol{e}$ are given in Example 10.1 with $\gamma^T = (\gamma_1,\gamma_2,\ldots,\gamma_a)$ and

$$Z = \begin{pmatrix} z_{11} & z_{11}^2 & \ldots & z_{11}^s \\ z_{12} & z_{12}^2 & \ldots & z_{12}^s \\ \vdots & \vdots & & \vdots \\ z_{an_a} & z_{an_a}^2 & \ldots & z_{an_a}^s \end{pmatrix}.$$

With Examples 10.1 and 10.2, all typical problems of the ANCOVA model I–I can be illustrated:

- Testing the hypothesis $H_0 : \alpha_1 = \cdots = \alpha_a$
- Testing the hypothesis $H_0 : \gamma = 0$ (Example 10.1. Case (a))
- Testing the hypothesis $H_0 : \gamma_1 = \cdots = \gamma_s$ (Example 10.1. Case (b))
- Testing the hypothesis $H_0 : \gamma_r = \gamma_{r+1} = \cdots = \gamma_s = 0$ for $2 \le r \le s-1$ (Example 10.2)
- Estimation of a γ or $\gamma_1, \ldots, \gamma_s$

Going back to the general case, we note that with W also X does not have full rank. Therefore $X^T X$ is singular, and the normal equations in Section 4.1

$$X^T X\beta^* = X^T Y$$

have no unique solution.

For model I–I of the ANCOVA due to

$$X = (W, Z),\ \beta^T = (\alpha^T, \gamma^T),$$

the normal equations have the form

$$\begin{pmatrix} W^T W & W^T Z \\ Z^T W & Z^T Z \end{pmatrix} \begin{pmatrix} \boldsymbol{\alpha}^* \\ \gamma^* \end{pmatrix} = \begin{pmatrix} W^T \boldsymbol{Y} \\ Z^T \boldsymbol{Y} \end{pmatrix}. \tag{10.6}$$

Let G_W be a generalised inverse of $W^T W$, and then from (10.6) we obtain

$$\boldsymbol{\alpha}^* = G_W\left(W^T \boldsymbol{Y} - W^T Z\gamma^*\right). \tag{10.7}$$

If $\boldsymbol{\alpha}^{**}$ is the solution of the normal equations of model equation (10.1) for $\gamma = 0_s$ (without covariable), then

$$\boldsymbol{\alpha}^* = \boldsymbol{\alpha}^{**} - G_W W^{\mathrm{T}} Z\boldsymbol{\gamma}^* \tag{10.8}$$

where $\boldsymbol{\alpha}^{**} = G_W W^T \boldsymbol{Y}$ is the solution of the normal equations for an ANOVA (model I) with model equation (5.1). The formula for $\boldsymbol{\alpha}^{**}$ is up to the notation identical with (5.3). If we apply $\boldsymbol{\alpha}^*$ in (10.6), we receive for $\boldsymbol{\gamma}^*$ formula (10.10). In the following theorem we show that $\boldsymbol{\gamma}^*$ is uniquely determined and BLUE of the estimable function γ.

Theorem 10.1 Let model equation (10.1) and the distributional assumption (10.2) be valid for $\boldsymbol{Y}$. Then γ is estimable and the solution $\boldsymbol{\gamma}^*$ of the normal equations (10.6) are unique (i.e. independent of the special choice of G_W) and BLUE *of* γ. We therefore write $\boldsymbol{\gamma}^* = \hat{\boldsymbol{\gamma}}$.

Proof: At first we show that $\boldsymbol{\gamma}^*$ is unique and write (10.6) detailed as

$$W^{\mathrm{T}} W\boldsymbol{\alpha}^* + W^{\mathrm{T}} Z\gamma^* = W^{\mathrm{T}} \boldsymbol{Y}$$
$$Z^{\mathrm{T}} W\boldsymbol{\alpha}^* + Z^{\mathrm{T}} Z\gamma^* = Z^{\mathrm{T}} \boldsymbol{Y}$$

or using (10.8) and $W^{\mathrm{T}} W G_W W^{\mathrm{T}} = W^T$ as

$$\left.\begin{aligned} &W^{\mathrm{T}} W\boldsymbol{\alpha}^{**} = W^{\mathrm{T}} \boldsymbol{Y} \\ &Z^{\mathrm{T}} W\boldsymbol{\alpha}^{**} - Z^{\mathrm{T}} W\ G_W W^{\mathrm{T}} Z\gamma^* + Z^{\mathrm{T}} Z\gamma^* = Z^{\mathrm{T}} \boldsymbol{Y} \end{aligned}\right\}. \tag{10.9}$$

Due to Theorem 4.13 the relation $\boldsymbol{\alpha}^{**} = G_W W^T \boldsymbol{Y}$ follows from the first equation of (10.9), and appointing this to the second equation of (10.9) leads to

$$Z^{\mathrm{T}} W\ G_W W^{\mathrm{T}} \boldsymbol{Y} - Z^{\mathrm{T}} W\ G_W W^{\mathrm{T}} Z\gamma^* + Z^{\mathrm{T}} Z\gamma^* = Z^{\mathrm{T}} \boldsymbol{Y}.$$

With the idempotent matrix $A = I_N - WG_W W^T$, this gives the solution

$$\gamma^* = \left(Z^{\mathrm{T}}AZ\right)^{-} Z^{\mathrm{T}}AY,$$

where $(Z^{\mathrm{T}}AZ)^{-}$ is a generalised inverse of $Z^{\mathrm{T}}AZ$. Because A is idempotent, we have $\mathrm{rk}(AZ) = \mathrm{rk}(Z^{\mathrm{T}}AZ)$. Further AZ has a full column rank, so that $Z^{\mathrm{T}}AZ$ is non-singular and $(Z^{\mathrm{T}}AZ)^{-} = (Z^{\mathrm{T}}AZ)^{-1}$. Therefore

$$\gamma^* = \hat{\gamma} = \left(Z^{\mathrm{T}}AZ\right)^{-1} Z^{\mathrm{T}}AY \tag{10.10}$$

is the unique solution component of (10.6).

From Lemma 4.1 we know that γ is estimable if it is a linear function of $E(\hat{\gamma})$. From (10.10) we get

$$E(\hat{\gamma}) = \left(Z^{\mathrm{T}}AZ\right)^{-1} Z^{\mathrm{T}}AE(Y) = \left(Z^{\mathrm{T}}AZ\right)^{-1} Z^{\mathrm{T}}A(W\alpha + Z\gamma).$$

Because $AW = (I_N - WG_W W^T)W = W - W = O$ we obtain $E(\hat{\gamma}) = \gamma$. Therefore $\hat{\gamma}$ is BLUE of γ, and this completes the proof.

Corollary 10.1 The estimator $\hat{\gamma}$ of γ in model equation (10.1) is a BLUE concerning γ in the model equation

$$Y = AZ\gamma + e \tag{10.11}$$

with $A = I_N - WG_W W^T$ where Y, W, Z and e fulfil the conditions of Definition 10.1.

This follows from the idempotence of A and the results of Chapter 8.

We now test the hypotheses

$$H_0 : M^{\mathrm{T}}\gamma = c, \quad \mathrm{rk}\left(M^{\mathrm{T}}\right) = v < s$$

and

$$H_0 : L^{\mathrm{T}}\alpha = R, \quad \mathrm{rk}\left(L^{\mathrm{T}}\right) = u < r.$$

Because γ is estimable,

$$H_0 : M^{\mathrm{T}}\gamma = c$$

is testable. For the second hypothesis only such matrices are admitted, for which the rows of $L^T\alpha$ are estimable. From (10.8) it follows that $L^T\alpha$ with α in (10.1) is estimable, if $L^T\alpha$ is in the corresponding model of the ANOVA (with $\gamma = 0$) estimable.

In Chapter 4 the F-test statistic (4.37) was used as a test statistic of the hypothesis $H_0 : \gamma = 0$. From the results of Chapter 4, we can now derive a test statistic for the general (nonlinear) hypothesis $H_0 : M^{\mathrm{T}}\gamma = c$, with a $(v \times s)$ matrix M with $v < s$ and of rank v. The hypothesis $H_0 : M^{\mathrm{T}}\gamma = c$ defines a subspace $\omega_\gamma \subset \Omega$ of dimension $s - v$.

We write $M^{\mathrm{T}}\gamma = c$ as a special case of the general hypothesis $K^T\beta = a$ in model (4.29). To test this hypothesis we obtain with $N-p$ and $q = \mathrm{rk}(K)$ and if $H_0: K^T\beta = a$ is true the F-distributed test statistic

$$\boldsymbol{F} = \frac{(K^{\mathrm{T}}\boldsymbol{\beta}^* - a)^{\mathrm{T}}[K^{\mathrm{T}}(X^{\mathrm{T}}X)^{-}K]^{-1}(K^{\mathrm{T}}\boldsymbol{\beta}^* - a)}{\boldsymbol{Y}^{\mathrm{T}}(I_N - X(X^{\mathrm{T}}X)^{-}X^{\mathrm{T}})\boldsymbol{Y}} \cdot \frac{N-p}{q}. \tag{10.12}$$

We now use

$$K^{\mathrm{T}} = \left(\mathrm{O}_{\nu,t+1}, M^{\mathrm{T}}\right), \quad \beta = \begin{pmatrix} \alpha \\ \gamma \end{pmatrix}, \quad X = (W, Z), \quad a = \begin{pmatrix} 0_{t+1} \\ c \end{pmatrix}$$

with a $(\nu \times s)$ matrix M^T of rank $\nu < s$. For $(X^TX)^{-}$ we write

$$\left(X^{\mathrm{T}}X\right)^{-} = \begin{pmatrix} W^{\mathrm{T}}W & W^{\mathrm{T}}Z \\ Z^{\mathrm{T}}W & Z^{\mathrm{T}}Z \end{pmatrix}^{-}, \tag{10.13}$$

and this becomes with $G_W = (W^TW)^{-}$ and with $D = Z^T(I_N - WG_WW^T)Z = Z^TAZ$

$$\left(X^{\mathrm{T}}X\right)^{-} = \begin{pmatrix} G_W + G_W W^{\mathrm{T}}ZD^{-1}Z^{\mathrm{T}}W\, G_W^{\mathrm{T}} & -G_W W^{\mathrm{T}}ZD^{-1} \\ -D^{-1}Z^{\mathrm{T}}W\, G_W^{\mathrm{T}} & D^{-1} \end{pmatrix}. \tag{10.14}$$

Therefore,

$$K^{\mathrm{T}}\left(X^{\mathrm{T}}X\right)^{-}K = M^{\mathrm{T}}\left(Z^{\mathrm{T}}AZ\right)^{-1}M$$

is non-singular (M has full row rank) and

$$X\left(X^{\mathrm{T}}X\right)^{-}X^{\mathrm{T}} = W\, G_W W^{\mathrm{T}} + AZ\left(Z^{\mathrm{T}}AZ\right)^{-1}Z^{\mathrm{T}}A. \tag{10.15}$$

That means that (10.12) for our special case becomes

$$\boldsymbol{F} = \frac{(M^{\mathrm{T}}\hat{\boldsymbol{\gamma}} - c)^{\mathrm{T}}\left[M^{\mathrm{T}}(Z^{\mathrm{T}}AZ)^{-1}M\right]^{-1}(M^{\mathrm{T}}\hat{\boldsymbol{\gamma}} - c)}{\boldsymbol{Y}^{\mathrm{T}}(I_N - X(X^{\mathrm{T}}X)^{-}X^{\mathrm{T}})\boldsymbol{Y}} \cdot \frac{N-r-s}{\nu}. \tag{10.16}$$

$\boldsymbol{F}$ is $F(N-r-s, \nu, \lambda)$-distributed with the non-centrality parameter

$$\lambda = \frac{1}{\sigma^2}\left(M^{\mathrm{T}}\gamma - c\right)^{\mathrm{T}}\left[M^{\mathrm{T}}\left(Z^{\mathrm{T}}AZ\right)^{-1}M\right]^{-1}\left(M^{\mathrm{T}}\gamma - c\right).$$

Analogously we obtain the test statistic for the special hypothesis $H_0: L^T\alpha = R$ as

$$\boldsymbol{F} = \frac{(L^{\mathrm{T}}\boldsymbol{\alpha}^* - R)^{\mathrm{T}}\left\{L^{\mathrm{T}}\left[G_W + G_W W^{\mathrm{T}}Z(Z^{\mathrm{T}}AZ)^{-1}Z^{\mathrm{T}}W\, G_W^{\mathrm{T}}\right]L\right\}^{-1}(L^{\mathrm{T}}\boldsymbol{\alpha}^* - R)}{\boldsymbol{Y}^{\mathrm{T}}(I_N - X(X^{\mathrm{T}}X)^{-}X^{\mathrm{T}})\boldsymbol{Y}} \cdot \frac{N-r-s}{u}, \tag{10.17}$$

if we put $K^T = (L^T, O_{v,s})$, where $\mathrm{rk}(L^T) = u < r$. The hypothesis $H_0 : L^T\alpha = R$ corresponds with a subspace $\omega_\alpha \subset \Omega$ of dimension $r{-}u$ and $\boldsymbol{F}$ in (10.17) is $F(N-r-s, u, \lambda)$-distributed with the non-centrality parameter

$$\lambda = \frac{1}{\sigma^2}\left(L^T\alpha - R\right)^T \left\{L^T\left[G_W + G_W W^T Z\left(Z^T A Z\right)^{-1} Z^T W G_W^T\right] L^T\right\}^{-1}\left(L^T\alpha - R\right).$$

For the special case $R = 0_v$, we write (10.12) as done in this section for model equation (10.1) where $\mathrm{rk}(L^T) = r-1$ and $\gamma = 0_s$. In (10.12) we replace X by W, K by L, $\boldsymbol{\beta}$ by $\boldsymbol{\alpha}^{**}$, a by 0_v, q by $r-1$ and p by r and receive

$$\boldsymbol{F} = \frac{\boldsymbol{\alpha}^{**T} L^T \left(L^T G_W L\right)^{-1} L^T \boldsymbol{\alpha}^{**}}{\boldsymbol{Y}^T\left(I_N - W\, G_W W^T\right)\boldsymbol{Y}} \cdot \frac{N-r}{r-1} \tag{10.12a}$$

as test statistic for the test of the null hypothesis $H_0 : L^T\alpha = 0$ in model (10.1), if $\gamma = 0$.

If we now consider the hypothesis

$$H_0 : L^T\left(\alpha + G_W W^T Z\gamma\right) = 0$$

for the general model (10.1) and observe

$$L^T\left(\alpha + G_W W^T Z\gamma\right) = K^T\begin{pmatrix}\alpha\\ \gamma\end{pmatrix} = K^T\beta = 0$$

with $K^T = L^T(I_{t+1}, G_W W^T Z)$ in (10.12) for $\alpha = 0_{a+1}$, we obtain a test statistic with numerator ***SS***

$$\boldsymbol{\beta}^{*T} K\left[K^T\left(X^T X\right)^{-1} K\right]^{-1} K^T \boldsymbol{\beta}^*,$$

and we show that this becomes $\boldsymbol{\alpha}^{**T} L\left(L^T G_W L\right)^{-1} L^T \boldsymbol{\alpha}^{**}$. This is true because $K^T\left(X^T X\right)^- = L^T G_W L$ after decomposing $\left(X^T X\right)^-$ as in (10.13) and because

$$\left(\boldsymbol{\alpha}^{*T}, \hat{\gamma}^T\right) K^T = \left(\boldsymbol{\alpha}^{*T}, \hat{\gamma}^T\right)\begin{pmatrix} I_{t+1} \\ Z^T W G_w\end{pmatrix} L = \left(\boldsymbol{\alpha}^{*T}, \hat{\gamma}^T\right)\left(Z^T W G_W\right) L = \boldsymbol{\alpha}^{**T} L.$$

By this we get a F-statistic for testing the hypothesis $H_0 : L^T\alpha = 0$ in (10.1) and we have $\gamma = 0_s$.

In (10.12a) we get the same numerator ***SS*** to test the hypothesis $H_0 : L^T\left(\alpha + G_W W^T Z\right) = 0$ for model (10.1).

We use this in Section 10.3.

10.3 Special Models of the Analysis of Covariance for the Simple Classification

The general formulae derived in Section 10.2 are now for special cases explicitly given.

Definition 10.2 Model equation (10.1) with the side conditions (10.2) is the model of the simple (one-way) classifications of the analysis of covariance if in (10.1)

$$W^T = \begin{pmatrix} 1 & 1 & \dots & 1 & 1 & 1 & \dots & 1 & \dots & 1 & 1 & \dots & 1 \\ 1 & 1 & \dots & 1 & 0 & 0 & \dots & 0 & \dots & 0 & 0 & \dots & 0 \\ 0 & 0 & \dots & 0 & 1 & 1 & \dots & 1 & \dots & 0 & 0 & \dots & 0 \\ & & \dots & & & & \dots & & \dots & & & \dots & \\ \vdots & \vdots & \dots & \vdots & \vdots & \vdots & \dots & \vdots & \dots & \vdots & \vdots & \dots & \vdots \\ & & \dots & & & & \dots & & \dots & & & \dots & \\ 0 & 0 & \dots & 0 & 0 & 0 & \dots & 0 & \dots & 1 & 1 & \dots & 0 \end{pmatrix}.$$

and $\alpha^T = (\mu, \alpha_1, \dots, \alpha_a)$ is chosen.

From Theorem 5.3 we know that the denominator of (10.12) is

$$\boldsymbol{Y}^{\mathrm{T}} A \boldsymbol{Y} = \boldsymbol{Y}^{\mathrm{T}}\left(I_N - W G_W W^{\mathrm{T}}\right)\boldsymbol{Y} = \sum_{i=1}^{a}\sum_{j=1}^{n_i} \boldsymbol{y}_{ij}^2 - \sum_{i=1}^{a}\frac{1}{n_i}\boldsymbol{Y}_{i}^2 = \boldsymbol{SS}_{res\,y}. \tag{10.18}$$

Analogously we write with $Z^T = (z_{11}, \dots, z_{an_a})$

$$Z^{\mathrm{T}} A Z = \sum_{i=1}^{a}\sum_{j=1}^{n_i} z_{ij}^2 - \sum_{i=1}^{a}\frac{1}{n_i} Z_{i\cdot}^2 = SS_{res\,z} \tag{10.19}$$

and

$$Z^{\mathrm{T}} A \boldsymbol{Y} = \sum_{i=1}^{a}\sum_{j=1}^{n_i} \boldsymbol{y}_{ij} z_{ij} - \sum_{i=1}^{a}\frac{1}{n_i}\boldsymbol{Y}_{i\cdot} Z_{i\cdot} = \boldsymbol{SP}_{res}. \tag{10.20}$$

Further is

$$SS_{total} = SS_T = Z^{\mathrm{T}} Z = Z^{\mathrm{T}} A Z + \sum_{i=1}^{a}\frac{1}{n_i} Z_{i\cdot}^2 - \frac{1}{N} Z_{\cdot\cdot}^2 = SS_Z + SS_{res\,z} \tag{10.21}$$

and

$$\boldsymbol{SP}_{total} = Z^{\mathrm{T}} \boldsymbol{Y} = Z^{\mathrm{T}} A \boldsymbol{Y} + \sum_{i=1}^{a}\frac{1}{n_i} Z_{i\cdot}\boldsymbol{Y}_{i\cdot} - \frac{1}{N} Z_{\cdot\cdot}\boldsymbol{Y}_{\cdot\cdot} = \boldsymbol{SP}_{between} + \boldsymbol{SP}_{res}. \tag{10.22}$$

Here $\boldsymbol{Y}^T = (\boldsymbol{Y}_1^T,\dots,\boldsymbol{Y}_a^T)$ is a vector with a independent random samples $\boldsymbol{Y}_i^T$ and the elements of the ith sample are $N(\mu+\alpha_i,\sigma^2)$-distributed.

Definition 10.2 is still rather general. Below we discuss special cases of Example 10.1. Let γ be a scalar and $Z^T = z^T = (z_{11},\dots, z_{1n_1},\dots,z_{a1},\dots z_{an_a})$ a row vector, such that (10.1) becomes (10.3). If $\gamma = (\gamma_1,\dots,\gamma_a)^T$ and

$$Z = \bigoplus_{i=1}^{a} Z_i \ \text{ with } \ Z_i = (z_{i1},\dots,z_{in_i})^{\mathrm{T}} \quad (i=1,\dots,a),$$

then (10.1) becomes (10.4). If $\gamma = (\gamma_{11},\dots,\gamma_{1n_1},\dots,\gamma_{a1},\dots\gamma_{an_a})^T$ and

$$Z = \bigoplus_{i=1}^{a}\left[\bigoplus_{j=1}^{a} z_{ij}\right],$$

then (10.1) becomes

$$\boldsymbol{y}_{ij} = \mu + \alpha_i + \gamma_{ij} z_{ij} + \boldsymbol{e}_{ij}. \tag{10.23}$$

If

$$Z^{\mathrm{T}} = \begin{pmatrix} u_{11} & \cdots & u_{1n_1} & \cdots & u_{a1} & \cdots & u_{an_a} \\ v_{11} & \cdots & v_{1n_1} & \cdots & v_{a1} & \cdots & v_{an_a} \end{pmatrix} \text{ and } \gamma = \begin{pmatrix} \gamma_1 \\ \gamma_2 \end{pmatrix},$$

(10.1) becomes

$$\boldsymbol{y}_{ij} = \mu + \alpha_i + \gamma_1 z_{ij} + \gamma_2 v_{ij} + \boldsymbol{e}_{ij}. \tag{10.24}$$

Finally we consider the case of Example 10.2. With Z and γ in Example 10.2, Equation (10.1) becomes (10.5). In applications mainly (10.3) and (10.4) are used, and these cases are discussed below.

10.3.1 One Covariable with Constant γ

In model equation (10.3) one covariable z occurs, and the factor is the same for all values z_{ij} of this covariable. The BLUE $\hat{\gamma}$ of γ is [see (10.10) and (10.18) to (10.20)] given by

$$\hat{\gamma} = \frac{\sum_{i=1}^{a}\sum_{j=1}^{n_i} \boldsymbol{y}_{ij} z_{ij} - \sum_{i=1}^{a} \frac{1}{n_i} \boldsymbol{Y}_{i.} Z_{i.}}{\sum_{i=1}^{a}\sum_{j=1}^{n_i} z_{ij}^2 - \sum_{i=1}^{a} \frac{1}{n_i} Z_{i.}^2} = \frac{\boldsymbol{SP}_{\text{res}}}{SS_{resz}}. \tag{10.25}$$

Formula (10.25) is the estimator of regression coefficients within classes. We shall test:

a) The null hypothesis $H_0 : \gamma = 0$
b) The null hypothesis $H_0 : \alpha_1 = \alpha_2 = \cdots = \alpha_a$
c) The null hypothesis $H_0 : \alpha_1 + \gamma\bar{z}_1 = \alpha_2 + \gamma\bar{z}_2 = \cdots = \alpha_a + \gamma\bar{z}_a$

a) To test the hypothesis $H_0: \gamma = 0$, we use the test (10.16) with $M = 1, c = 0$; the denominator is given by (10.18). If $\hat{\gamma}$ is taken from (10.25) and because $Z'AZ = \mathrm{SQ}_{Iz}$, formula (10.16) becomes

$$\boldsymbol{F} = \frac{\hat{\gamma} Z^{\mathrm{T}} A Z \hat{\gamma}(N-a-1)}{\boldsymbol{SS}_{res\,y}} = \frac{\boldsymbol{SP}^2_{res}(N-a-1)}{SS_{res\,z}\boldsymbol{SS}_{res\,y}} = \boldsymbol{t}^2. \tag{10.26}$$

$\boldsymbol{F}$ in (10.26) is $F(1, N-a-1, \lambda)$ with $\lambda = \frac{\gamma^2}{\sigma^2} SS_{res\,z}$. If H_0 is true, $\boldsymbol{t} = \sqrt{\boldsymbol{F}}$ is $t(N-a-1)$-distributed.

b) To test the hypothesis $H_0: \alpha_1 = \cdots = \alpha_a$, we use a special case of (10.17) with $r = 0_{a-1}$ and

$$L^{\mathrm{T}} = \begin{pmatrix} 1 & -1 & 0 & \cdots & 0 \\ 1 & 0 & -1 & \cdots & 0 \\ \vdots & \vdots & \vdots & & \vdots \\ 1 & 0 & 0 & \cdots & -1 \end{pmatrix}. \tag{10.27}$$

L^T is a $[(a-1) \times a]$ matrix of rank $a - 1$. Because $\left(W^{\mathrm{T}}\boldsymbol{Y}\right)^{\mathrm{T}} = (\bar{\boldsymbol{y}}_{..}, \bar{\boldsymbol{y}}_{1.}, \ldots, \bar{\boldsymbol{y}}_{a.})$ and correspondingly $\left(W^{\mathrm{T}}\boldsymbol{Z}\right)^{\mathrm{T}} = (\bar{z}_{..}, \bar{z}_{1.}, \ldots, \bar{z}_{a.})$ and by (10.8), we obtain

$$\boldsymbol{\alpha}^{**T} = (\bar{\boldsymbol{y}}_1 - \hat{\boldsymbol{\gamma}}\bar{\boldsymbol{z}}_{1.}, \ldots, \bar{\boldsymbol{y}}_{a.} - \hat{\boldsymbol{\gamma}}\bar{\boldsymbol{z}}_{a.}),$$

so that (10.17) becomes

$$\boldsymbol{F} = \frac{\boldsymbol{SS}_{total\,y} - \frac{\boldsymbol{SP}_{total}}{\boldsymbol{SS}_{total\,z}} - \left(\boldsymbol{SS}_{Zy} - \frac{\boldsymbol{SP}^2_{res}}{\boldsymbol{SS}_{res\,z}}\right)}{\boldsymbol{SS}_{Zy} - \frac{\boldsymbol{SP}^2_{res}}{\boldsymbol{SS}_{res\,z}}} \cdot \frac{N-a-1}{a-1}. \tag{10.28}$$

If the null hypothesis $H_0: \alpha_1 = \alpha_2 = \cdots = \alpha_a$ is true, $\boldsymbol{F}$ is $F(a-1, N-a-1)$-distributed.

c) To test the null hypothesis $H_0: \alpha_1 + \gamma\bar{z}_1 = \alpha_2 + \gamma\bar{z}_2 = \cdots = \alpha_a + \gamma\bar{z}_a$, we write H_0 with L^T from (10.27) as

$$H_0: L^{\mathrm{T}}\left(\alpha + G_W W^{\mathrm{T}} Z\gamma\right) = 0_{a-1}.$$

The test statistic of this hypothesis has the numerator $\boldsymbol{SS}$ given in (10.12a), and because $\mathrm{rk}(L^T) = a-1$, it is equal to

$$\boldsymbol{F} = \frac{\sum_{i=1}^{a} \frac{1}{n_i}\boldsymbol{Y}^2_{i.} - \frac{1}{N}\boldsymbol{Y}^2_{..}}{\sum_{i=1}^{a}\sum_{j=1}^{n_i} \boldsymbol{y}^2_{ij} - \sum_{i=1}^{a}\frac{1}{n_i}\boldsymbol{Y}^2_{i.}\frac{\boldsymbol{SP}^2_{res}}{\boldsymbol{SS}_{res\,z}}} \cdot \frac{N-a-1}{a-1} \tag{10.29}$$

with $\boldsymbol{SP}_{res}$ in (10.20) and $\boldsymbol{SS}_{resz}$ in (10.19).

Table 10.1 Analysis of covariance table for model equation (10.3) of the simple analysis of covariance.

Source of variation	*SS*	*df*	*MS*
Factor A	$\sum_{i=1}^{a}\frac{Y_{i\cdot}^2}{n_i}-\frac{Y_{\cdot\cdot}^2}{N}=SS_A$	$a-1$	$MS_A=\frac{SS_A}{a-1}$
γ	$\frac{SP_{res}^2}{SS_{res\,z}}$	1	$MS_\gamma=\frac{SP_{res}^2}{SS_{res\,z}}$
Residual	$SS_{res}=SS_{total}-SS_A-\frac{SP_{res}^2}{SS_{res\,z}}$	$N-a-1$	$MS_{\text{res}}=\frac{SS_{res}}{N-a-1}$
Total	$Y^TY-\frac{1}{N}Y_{\cdot\cdot}^2=SS_{total}$	$N-1$	

Table 10.2 *SS*, *df* and *MS* of model (10.1) of analysis of covariance.

Source of variation	*SS*	*df*	*MS*
Components of $\boldsymbol{\alpha}$	$Y^TWG_WW^TY-\frac{1}{N}Y_{\cdot\cdot}^2=SS_A$	$r-1$	$MS_A=\frac{SS_A}{r-1}$
Covariable	$Y^TAZ(Z^TAZ)^{-1}Z^TAY=SS_{cov}$	s	$MS_{\text{cov}}=\frac{SS_{cov}}{s}$
Residual	$Y^TY-Y^TWG_WW^TY-SS_{cov}=SS_{res}$	$N-r-s$	$MS_{\text{res}}=\frac{SS_{res}}{N-r-s}$
Total	$Y^TY-\frac{1}{N}Y_{\cdot\cdot}^2=SS_{total}$	$N-1$	

If $H_0:L^T(\alpha+G_WW^TZ\gamma)=0_{a-1}$, then F in (10.29) is centrally $F(a-1,N-a-1)$-distributed.

In the Example 10.3 we demonstrate how ANCOVA tables can be obtained by SPSS.

Table 10.1 is the ANOVA table for model equation (10.3), a special case of Table 10.2.

For our data set we obtain an analogue output.

10.3.2 A Covariable with Regression Coefficients γ_i Depending on the Levels of the Classification Factor

Similar to Section 10.3.1 the general formulae for special models may be simplified. We leave the derivation for the special cases to the reader.

For model equation (10.4) the null hypotheses below are of interest:

$$H_0 : \gamma_1 = \gamma_2 = \cdots = \gamma_a$$
$$H_0 : \gamma_1 = \gamma_2 = \cdots = \gamma_a = 0$$
$$H_0 : \alpha_1 = \alpha_2 = \cdots = \alpha_a$$

We write

$$SS_{res\,z,i} = \sum_{j=1}^{n_i} z_{ij}^2 - \frac{1}{n_i} Z_{i\cdot}^2, \quad \boldsymbol{SP}_{res,i} = \sum_{j=1}^{n_i} z_{ij}\boldsymbol{y}_{ij} - \frac{1}{n_i} Z_{i\cdot}\boldsymbol{Y}_{i\cdot}$$

The components γ_i of $(\gamma_1, \ldots, \gamma_a)^T$ with (10.10) are estimated by

$$\hat{\gamma}_i = \frac{\boldsymbol{SP}_{res,i}}{SS_{res\,z,i}}. \tag{10.30}$$

Because

$$\boldsymbol{SS}_{\text{res}} = \boldsymbol{SS}_{res\,y} - \sum_{i=1}^{a} \frac{\boldsymbol{SP}_{res,i}^2}{SS_{res\,z,i}}$$

is a quadratic form of rank $N-2a$,

$$\boldsymbol{F} = \frac{\sum_{i=1}^{a} \frac{\boldsymbol{SP}_{res,i}^2}{SS_{res\,z,i}} - \frac{\boldsymbol{SP}_{res}^2}{SS_{res\,z}}}{\boldsymbol{SS}_{\text{res}}} \cdot \frac{N-2a}{a-1} \tag{10.31}$$

under

$$H_0 : \gamma_1 = \gamma_2 = \cdots = \gamma_a$$

is $F(a-1, N-2a)$-distributed and can be used as test statistic if

$$H_0 : \gamma_1 = \gamma_2 = \cdots = \gamma_a$$

is true. If the hypothesis is not rejected, then

$$H_0 : \alpha_1 = \alpha_2 = \cdots = \alpha_a$$

can be tested with the test statistic (10.28), but the two tests are dependent!

10.3.3 A Numerical Example

Example 10.3 In Figure 10.1 we show laboratory data from four laboratories after (y) and before (z) some treatments have been applied as SPSS file.

We can either continue with

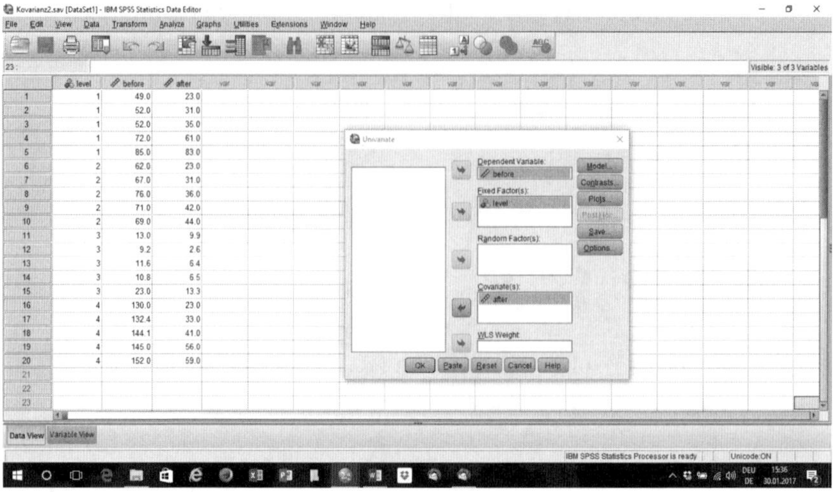

Figure 10.1 Laboratory data after (*y*) and before (*z*) some treatments from four laboratories. *Source:* Reproduced with permission of IBM.

Analyze
    General Linear Model
        Multivariate

(with before and after as two independent variables) or

Analyze
    General Linear Model
        Univariate

as we did in Figure 10.1 and will do in this example. In Figure 10.1 under 'options', we use those shown in the syntax in Figure 10.2a where the first results are also given. Further results are shown in Figure 10.2b and Figure 10.2c.

In Figure 10.2b we find the estimate of the regression coefficient under 'parameter estimates' in the row 'before' as $\hat{\gamma} = 1.45663$. In Figure 10.2c we find the estimated means and pairwise comparisons of the character 'after' for the four levels of the factor (labour).

Because the regression coefficients γ_i (i = 1, …, 4) differ significantly from each other, we can estimate them via the SPSS syntax:

```
UNIANOVA after BY level WITH before
  /METHOD=SSTYPE(1)
  /INTERCEPT=INCLUDE
  /PRINT=PARAMETER
  /CRITERIA=ALPHA(.05)
  /DESIGN=level level*before.
```

We obtain then the four estimates of the γ_i in Figure 10.3 using (10.30) as

$$\hat{\gamma}_1 = 1.56313$$
$$\hat{\gamma}_2 = 1.06604$$
$$\hat{\gamma}_3 = 0.66509$$

and

$$\hat{\gamma}_4 = 1.54462.$$

(a)

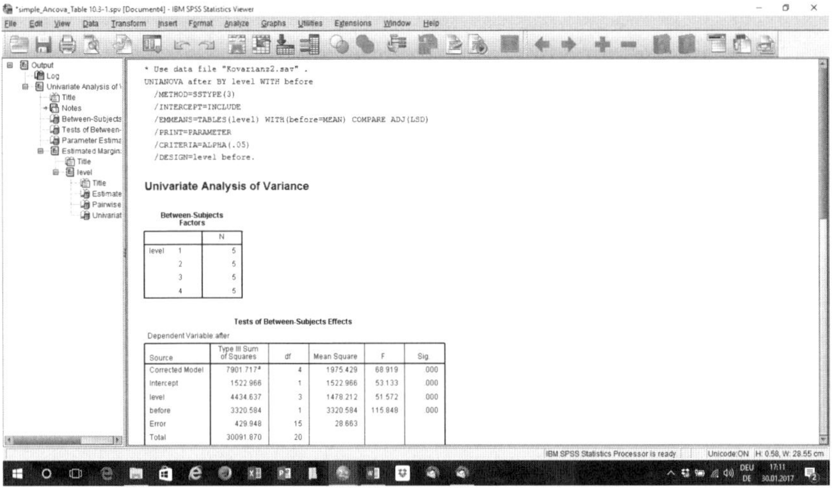

(b)

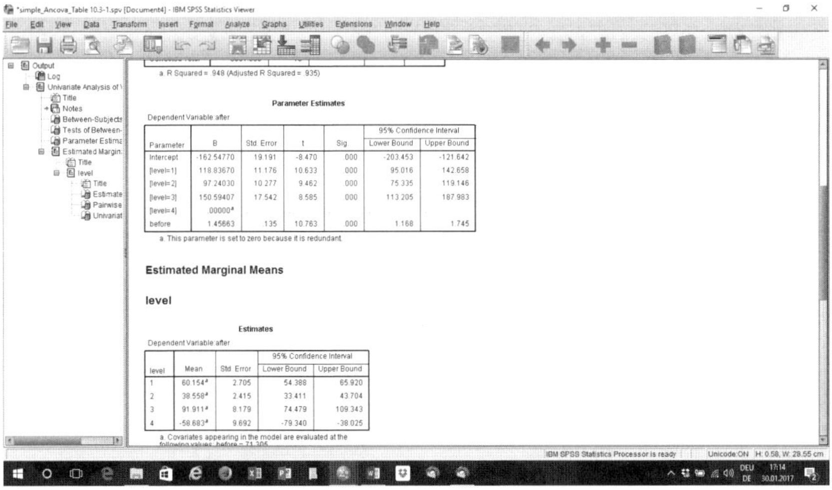

Figure 10.2 (a–c) ANCOVA for the laboratory data in Figure 10.1. *Source:* Reproduced with permission of IBM.

(c)

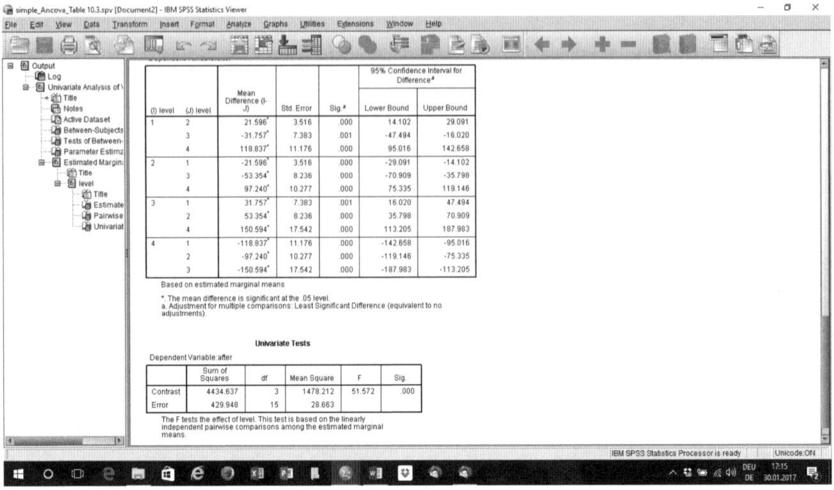

(I) level	(J) level	Mean Difference (I-J)	Std. Error	Sig.[a]	95% Confidence Interval for Difference[a]	
					Lower Bound	Upper Bound
1	2	21.596*	3.516	.000	14.102	29.091
	3	-31.757*	7.383	.001	-47.494	-16.020
	4	118.837*	11.176	.000	95.016	142.658
2	1	-21.596*	3.516	.000	-29.091	-14.102
	3	-53.354*	8.236	.000	-70.909	-35.798
	4	97.240*	10.277	.000	75.335	119.146
3	1	31.757*	7.383	.001	16.020	47.494
	2	53.354*	8.236	.000	35.798	70.909
	4	150.594*	17.542	.000	113.205	187.983
4	1	-118.837*	11.176	.000	-142.658	-95.016
	2	-97.240*	10.277	.000	-119.146	-75.335
	3	-150.594*	17.542	.000	-187.983	-113.205

Based on estimated marginal means

*. The mean difference is significant at the .05 level.
a. Adjustment for multiple comparisons: Least Significant Difference (equivalent to no adjustments).

Univariate Tests

Dependent Variable: after

	Sum of Squares	df	Mean Square	F	Sig.
Contrast	4434.637	3	1478.212	51.572	.000
Error	429.948	15	28.663		

The F tests the effect of level. This test is based on the linearly independent pairwise comparisons among the estimated marginal means.

Figure 10.2 (Continued)

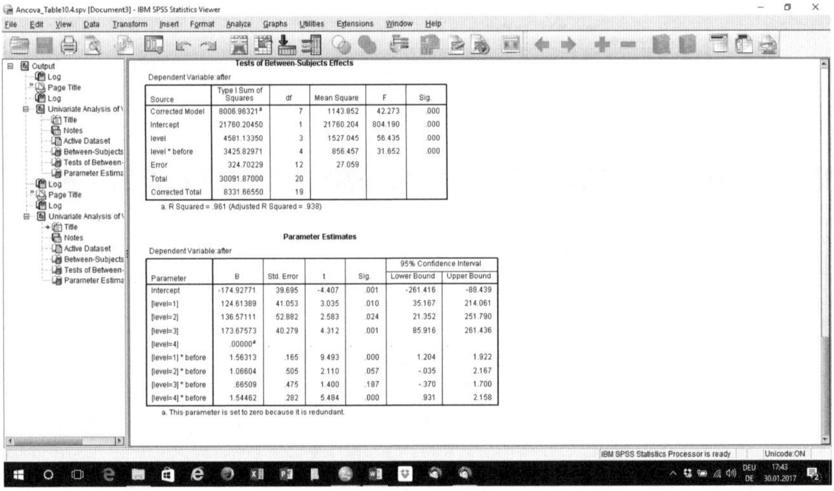

Tests of Between-Subjects Effects

Dependent Variable: after

Source	Type I Sum of Squares	df	Mean Square	F	Sig.
Corrected Model	8006.96321[a]	7	1143.852	42.273	.000
Intercept	21760.20450	1	21760.204	804.190	.000
level	4581.13350	3	1527.045	56.435	.000
level * before	3425.82971	4	856.457	31.652	.000
Error	324.70229	12	27.059		
Total	30091.87000	20			
Corrected Total	8331.66550	19			

a. R Squared = .961 (Adjusted R Squared = .938)

Parameter Estimates

Dependent Variable: after

Parameter	B	Std. Error	t	Sig.	95% Confidence Interval	
					Lower Bound	Upper Bound
Intercept	-174.92771	39.695	-4.407	.001	-261.416	-88.439
[level=1]	124.61389	41.053	3.035	.010	35.167	214.061
[level=2]	136.57111	52.882	2.583	.024	21.352	251.790
[level=3]	173.67573	40.279	4.312	.001	85.916	261.436
[level=4]	.00000[a]	.	.	.	.	.
[level=1] * before	1.56313	.165	9.493	.000	1.204	1.922
[level=2] * before	1.06604	.505	2.110	.057	-.035	2.167
[level=3] * before	.66509	.475	1.400	.187	-.370	1.700
[level=4] * before	1.54462	.282	5.484	.000	.931	2.158

a. This parameter is set to zero because it is redundant.

Figure 10.3 Results of ANCOVA for the laboratory data in Figure 10.1. *Source:* Reproduced with permission of IBM.

10.4 Exercises

10.1 Give a practical example for model equation (10.24).

10.2 Derive the test statistic (10.12).

References

Graybill, F.A. (1961) *An Introduction to Linear Statistical Methods*, McGraw Hill Book Comp., New York.

Scheffé, H. (1953) A method for judging all contrasts in the analysis of variance, *Biometrika*, **40**, 87–104.

Searle, S.R. (1971, 2012) *Linear Models*, John Wiley & Sons, Inc., New York.

11

Multiple Decision Problems

A multiple decision problem is given if a decision function can take on two or more values. A good overview about this is given in Gupta and Huang (1981). In this chapter, mainly the case of more than two decisions is discussed, we then speak about true multiple decision problems. Two-decision problems occur only in special cases (for $a = 2$). Statistical tests as typical statistical two-decision problems have a decision function with 'values' acceptation of H_0 and rejection of H_0.

In this chapter we assume that statements about $a \geq 2$ populations (distributions) from a set $G = \{P_1, \dots, P_a\}$ of populations must be made. These populations correspond with random variables $\boldsymbol{Y}_{\text{i}}$ with distribution functions $F(y, \theta_i), i = 1, \dots, a,$ where for each i at least one component θ_{ij} of $\theta_i^T = \left(\theta_{i1},\dots,\theta_{ip}\right) \in \Omega \subset R^p, p \geq 1$ is unknown. By a real-valued score function $g^*(\theta_i) = g_i^*$, the θ_i are mapped into R^1. For independent random samples $\boldsymbol{Y}_i^T = \left(\boldsymbol{y}_{i1},\dots,\boldsymbol{y}_{in_i}\right)$ with positive integer n_i, decisions about the g_i^* shall be made. We already have statistical tests for this. For instance, in simple analysis of variance model I with $N(\mu_i, \sigma^2)$ normally distributed $\boldsymbol{y}_i$, with $\theta_i^T = (\mu_i, \sigma^2)$ for $g^*(\theta_i) = g_i^* = \mu_i$, the null hypothesis

$$H_0 : \mu_1 = \mu_2 = \dots = \mu_a$$

has to be tested against

H_A: At least one pair $i, \ i'$ exists with $i \neq i' \ (i, i' = 1, \dots, a)$, so that $\mu_i \neq \mu_{i'}$

by the test statistic

$$\boldsymbol{F} = \frac{\boldsymbol{MS}_A}{\boldsymbol{MS}_{res}}$$

(Chapter 5). If $F(f_A, f_{res} \mid 1 - \alpha) < F$ with the degrees of freedom f_A and f_{res} of $\boldsymbol{MS}_A$ and $\boldsymbol{MS}_{res}$, respectively, H_0 is rejected, or otherwise accepted. This is an α-test of the solution of a statistical two-decision problems.

Mathematical Statistics, First Edition. Dieter Rasch and Dieter Schott.

Real multiple decision problems in this situation would be:

- Order the $g^*(\theta_i)$ according to magnitude.
- Select the $t < a$ largest (smallest) $g^*(\theta_i)$, $1 \le t \le a-1$.
- Decide which differences $g^*(\theta_i) - g^*(\theta_j)$, $(i \neq j; i, j = 1, \dots, a)$ are different from zero.
- Decide which differences $g^*(\theta_1) - g^*(\theta_i)$, $(i = 2, \dots, a)$ are different from zero.

The number of possible decisions differs in the four examples above, but for $a > 2$ it is always larger than two. The number of possible decisions equals $a!$, $\binom{a}{t}$, $2^{\binom{a}{2}}$ and 2^{a-1}, respectively.

11.1 Selection Procedures

To define selection procedures we first have to order the populations in G by magnitude. For this we need an order relation.

11.1.1 Basic Ideas

Definition 11.1 A population P_k is considered better than the population P_j $(j, k = 1, \dots, a, j \neq k)$, if $g_k^* = g^*(\theta_k) > g^*(\theta_j) = g_j^*$. P_k is considered not worse than P_j if $g_k^* \ge g_j^*$.

The values $g_1^*, \dots, g_a^*$ can be ordered as the a populations; if $g_{(i)}^*$ is the ith ordered (by magnitude) g^* value, then we have $g_{(1)}^* \le g_{(2)}^* \le \dots \le g_{(a)}^*$.

Next we renumber the populations by permuting the indices $1, \dots, a$. To avoid confusions between the original and the permuted indices, we denote the populations, the random variables, the parameters and the score functions afresh. The permutation $\begin{pmatrix} 1 & 2 & \dots & a \\ (1) & (2) & \dots & (a) \end{pmatrix}$ transforms the population P_j with its parameter θ_j and its random variable $\boldsymbol{y}_j$ belonging to $g_{(j)}^*$ into the population A_j, the random variable $\boldsymbol{x}_j$ with parameter η_j, respectively. We write further $g^*(\theta_{(i)}) = g(\eta_i) = g_i$ and by this get the rank order

$$g(\eta_1) \le g(\eta_2) \le \dots \le g(\eta_a), \tag{11.1}$$

and A_i is not worse than A_{i*} if $i \ge i^*$. But do not forget that the permutation is unknown, and we used it only to simplify writing.

Definition 11.2 If the set $G = \{A_1, \dots, A_a\} = \{P_1, \dots, P_a\}$ shall be partitioned in at least two subsets, so that in one of the subsets, the better elements of G following Definition 11.1 are contained, we have a selection problem.

A decision function (rule) performing such a partition is called selection rule or selection procedure.

Definition 11.3 If the elements of G are fixed (not randomly selected), we call this model I of selection. But if the elements have been randomly sampled from a larger universe, we call this model II of selection.

We restrict ourselves in this book to model I of selection. Model II occurs mainly in animal and plant breeding and is discussed within population genetics (see Rasch and Herrendörfer, 1990).

The theory of model I is about 65 years old (see Miescke and Rasch, 1996).

We consider the case that G shall be partitioned exactly into two subsets G_1 and G_2 so that $G = G_1 \cup G_2$, $G_1 \cap G_2 = \emptyset$, $G_1 = \{G_a, \dots, G_{a-t+1}\}$ and $G_2 = \{G_{a-t}, \dots, G_1\}$.

Problem 1 (Bechhofer, 1954).

For a given risk of wrong decision β with $\binom{a}{t}^{-1} < 1-\beta < 1$ and $d > 0$ from G, a subset $\boldsymbol{M}_B$ of size t has to be selected. Selection is made based on random samples $(\boldsymbol{x}_{i1},\dots,\boldsymbol{x}_{in_i})$ from A_i with $\boldsymbol{x}_i$ distributed components. Select $\boldsymbol{M}_B$ in such a way that the probability $P(CS)$ of a correct selection is

$$P(CS) = P_C = P(\boldsymbol{M}_B = G_1 | d(G_1, G_2) \geq d) \geq 1-\beta. \tag{11.2}$$

In (11.2) $d(g_{a-t+1}, g_{a-t})$ is the distance between A_{a-t+1} and A_{a-t}. The distance $d(G_1, G_2) = d(g_{a-t+1}, g_{a-t})$ between G_1 and G_2 equals at least to the value d, given in advance. A modified formulation is as follows:

Problem 1A Select a subset $\boldsymbol{M}_B$ of size t corresponding to Problem 1 in such a way that in place of (11.2),

$$P_C^* = P(\boldsymbol{M}_B \subset G_1^*) \geq 1-\beta \tag{11.3}$$

Here G_1^* is the set in G, containing all A_i with $g_i \geq g_{a-t+1} - d$.

The condition $\binom{a}{t}^{-1} < 1-\beta$ above is reasonable, because for $1-\beta \leq \binom{a}{t}^{-1}$ no real statistical problem exists. Without experimenting one could then denote any of the $\binom{a}{t}^{-1}$ subsets of size t by $\boldsymbol{M}_B$ and would with $\binom{a}{t}^{-1} \geq 1-\beta$ fulfil (11.2) and (11.3).

Problem 2 (Gupta, 1956).

For a given risk β of an incorrect decision with $\binom{a}{t}^{-1} < 1-\beta < 1$, select from G a subset $\boldsymbol{M}_G$ of random size $\boldsymbol{r}$ so that

$$P\{G_1 \subset \boldsymbol{M}_G\} \geq 1-\beta \tag{11.4}$$

By this an optimality criterion has to be considered. For instance, we could demand that one of the following properties holds:

- $E(\boldsymbol{r}) \Rightarrow \text{Min}$,
- $E(\boldsymbol{w}) \Rightarrow \text{Min}$, where $\boldsymbol{w}$ is the number of wrongly selected populations,
- The experimental costs are minimised.

In Problems 1 and 1A, selection is not named incorrect, as long as the distance between the worst of the t best populations and any non-best population does not exceed a value d fixed in advance. The region $[g_{a-t+1} - d, g_{a-t+1}]$ is called indifference zone, and Problem 1 often is called indifference zone formulation of the selection problem; Problem 2 however is called subset formulation of the selection problem.

One can ask, which of the two problem formulations for practical purposes should be used? Often experiments with a technologies, a varieties, medical treatments and others have the purpose to select the best of them (i.e. $t = 1$). If we then have a lot of candidates at the beginning (say about $a \approx 500$), such as in drug screening, then it is reasonable, at first, to reduce the number of candidates by a subset procedure down to let's say $r \leq 20$(or $r \leq 50$) and then in a second step to use an indifference zone procedure with $a = r$.

Before special cases are handled, we would say that in place of Problem 1, Problem 1A can always be used. There are advantages in application. The researcher could ask what can be said about the probability that we really selected the t best populations [(concerning $g(\eta)$] if $d(g_{a-t+1}, g_{a-t}) < d$. An answer to such a problem is not possible, but it is better to formulate Problem 1A, which can better be interpreted and where we now know at least with probability $1-\beta$ that we elected exactly that t populations not being more than d worse than A_{a-t+1}.

Guiard (1994) could show that the least favourable cases concerning the values of P_C and P_C^* for Problem 1 and 1A are identical. By this, the lower bounds $1-\beta$ in (11.2) and (11.3) are equal (for the same d). We call P_C^* the probability of a d-precise selection.

11.1.2 Indifference Zone Formulation for Expectations

In this section we discuss Problem 1A and restrict ourselves on univariate random variables $\boldsymbol{y}_i$ and $\boldsymbol{x}_i$. Further let be $g_{(i)}^* = g^*(\theta_{(i)}) = g(\eta_i) = E(\boldsymbol{x}_i) = \mu_i$.

Then $d(g_j, g_{a-t+1}) = |\mu_j - \mu_{a-t+1}|$ (and $d(g_{a-t+1}, g_{a-t}) = \mu_{a-t+1} - \mu_{a-t}$ in Problem 1). For the selection procedure, take from each of a populations a random sample $(\boldsymbol{x}_{i1},\ldots,\boldsymbol{x}_{in_i})$. These random samples are assumed to be stochastically independent; the components $\boldsymbol{x}_{ij}$ are assumed to be distributed like $\boldsymbol{x}_i$. Decisions will be based on estimators of μ.

Selection Rule 11.1 From the a independent random samples, the sample means $\bar{\boldsymbol{x}}_1,\ldots,\bar{\boldsymbol{x}}_a$ are calculated, and then we select the t populations with the t largest means into the set $\boldsymbol{M}_B$ (see Bechhofer, 1954).

Selection Rule 11.1 can be applied if only μ_i are an unknown component of η_i. If further components of η_i are unknown, we apply a multistage selection procedures (see Section 11.1.2.1).

11.1.2.1 Selection of Populations with Normal Distribution

We assume that the $\boldsymbol{x}_i$ introduced in Section 11.1.2 are $N\left(\mu_i,\sigma_i^2\right)$-distributed (i.e. $p = 2$). As mentioned above we renumber the $\boldsymbol{y}_i$ so that

$$\mu_i \le \mu_2 \le \cdots \le \mu_a \tag{11.5}$$

Let the σ_i^2 be known and equal to σ^2. Then we have

Theorem 11.1 (Bechhofer, 1954).

Under the assumptions of this section and if $n_i = n$ $(i = 1, \ldots, a)$,

$$P_0 = P\{\max(\bar{\boldsymbol{x}}_1,\ldots,\bar{\boldsymbol{x}}_{a-t}) < \min(\bar{\boldsymbol{x}}_{a-t+1},\ldots,\bar{\boldsymbol{x}}_a)\}, \tag{11.6}$$

and $\mu_{a-t+1} - \mu_{a-t} > d$, with $d^* = \dfrac{d\sqrt{n}}{\sigma}$

$$P_0 \ge t \int_{-\infty}^{\infty} [\Phi(z + d^*)]^{a-t}[1 - \Phi(z)]^{t-1}\varphi(z)\mathrm{d}z \tag{11.7}$$

always holds. If we apply Selection Rule 11.1, P_0 in (11.7) can be replaced by P_C.

Proof: P_0 is smallest if

$$\mu_1 = \cdots = \mu_{a-t} = \mu_{a-t+1} - d = \cdots = \mu_a - d \tag{11.8}$$

We now consider the t exclusive elements:

$$\max(\bar{\boldsymbol{x}}_{1\cdot},\ldots,\bar{\boldsymbol{x}}_{a-t}) < \bar{\boldsymbol{x}}_u < \min_{\substack{u \ne v \\ a-t+1 \le v \le a}} (\bar{\boldsymbol{x}}_v) \quad (u = a - t + 1,\ldots,a). \tag{11.9}$$

Under (11.8) all these events have the same probability so that for $\mu_{a-t+1} - \mu_{a-t} > d$ always $P_0 \ge t\, P_1$. With $f(\bar{\boldsymbol{x}}_{a-t+1})$, we write as density function of $\bar{\boldsymbol{x}}_{a-t+1}$

$$P_1 = \int_{-\infty}^{\infty} P\{\max(\bar{\boldsymbol{x}}_1,\ldots,\bar{\boldsymbol{x}}_{a-t}) < \bar{\boldsymbol{x}}_{a-t+1} < \min(\bar{\boldsymbol{x}}_{a-t+2\cdot},\ldots,\bar{\boldsymbol{x}}_a)\}\, f(\bar{\boldsymbol{x}}_{a-t+1})\mathrm{d}\,\bar{\boldsymbol{x}}_{a-t+1}.$$

If $\varphi(z)$ is as usual the density function of a $N(0,1)$ distribution, then with

$$A = \frac{\sqrt{n}}{\sigma}(\bar{\boldsymbol{x}}_{a-t+1} - \mu_{a-t}) \text{ and } B = \frac{\sqrt{n}}{\sigma}(\bar{\boldsymbol{x}}_{a-t+1} - \mu_{a-t+1})$$

we get

$$P_0 \geq tP_1 = t\int_{-\infty}^{\infty}\left[\int_{-\infty}^{A}\varphi(u)\mathrm{d}u\right]^{a-t}\left[\int_{B}^{\infty}\varphi(u)\mathrm{d}u\right]^{t-1}\frac{1}{\sigma}\sqrt{\frac{n}{2\pi}}\mathrm{e}^{-\frac{B^2}{2}}\mathrm{d}\,\bar{\boldsymbol{x}}_{a-t+1}.$$

Because

$$A - B = \frac{\sqrt{n}}{\sigma}(\mu_{a-t+1} - \mu_{a-t}) = \frac{\sqrt{n}}{\sigma}d$$

we complete the proof by using the distribution function Φ of the standard normal distribution.

For the often occurring special case $t = 1$ formula (11.7) becomes

$$P_0 \geq \int_{-\infty}^{\infty}[\Phi(z + d^*)]^{a-1}\varphi(z)\mathrm{d}z, \tag{11.10}$$

and this can be simplified following Theorem 11.2.

Theorem 11.2 Under the assumptions of Theorem 11.1 with $t = 1$ and with $\mu_a - \mu_j = d_{aj}$ $(j = 1, \dots, a - 1)$, we receive (without the condition $\mu_{a-t+1} - \mu_{a-t} > d$)

$$P_0 = P\{\max(\bar{\boldsymbol{x}}_1,\dots,\bar{\boldsymbol{x}}_{a-1.}) < \bar{\boldsymbol{x}}_{a.}\} = \frac{1}{\sqrt{a\pi^{a-1}}}\int_{-D_{a-1}}^{\infty}\cdots\int_{-D_1}^{\infty}\mathrm{e}^{-\frac{1}{2}t_v^T R^{-1} t_v}dt_v \tag{11.11}$$

with $\frac{d_{a,l}\sqrt{n}}{\sigma\sqrt{2}} = D_l$, $t_v = (t_1, \dots, t_{a-1})^T$, $R = (\varrho_{ij})$ and

$$\varrho_{ij} = \begin{cases} 1, & \text{if } i = j \\ \frac{1}{2}, & \text{if } i \neq j \end{cases} \quad (i, j = 1, \dots, a-1).$$

Proof: With $\boldsymbol{z}_i = \bar{\boldsymbol{x}}_{(a).} - \bar{\boldsymbol{x}}_{(i).}$ $(i = 1,\dots,a-1)$, P_0 becomes

$$P_0 = P\{\boldsymbol{z}_1 > 0,\dots,\boldsymbol{z}_{a-1} > 0\}.$$

Further $E(\boldsymbol{z}_1) = d_{a,i}$.

From the independency of the a random samples,

$$\mathrm{var}(\boldsymbol{z}_i) = \frac{2\sigma^2}{n}$$

and for $i \neq j$

$$\mathrm{cov}(\boldsymbol{z}_i, \boldsymbol{z}_j) = \mathrm{var}(\bar{\boldsymbol{x}}_a) = \frac{\sigma^2}{n}.$$

Because $\boldsymbol{x}_{ik}$ are independently $N(\mu_i, \sigma^2)$-distributed, $z = (\boldsymbol{z}_1, \dots, \boldsymbol{z}_{a-1})^T$ is $(a-1)$-dimensional normally distributed with

$$E(\boldsymbol{z}) = (d_{a,1},\dots,d_{a,a-1})^T = \Delta \text{ and } \operatorname{var}(\boldsymbol{z}) = \Sigma = \frac{\sigma^2}{n}(I_{a-1} + 1_{a-1,a-1}).$$

Therefore P_0 has the form

$$P_0 = \frac{|\Sigma|^{-\frac{1}{2}}}{\sqrt{(2\pi)^{a-1}}} \int_0^\infty \cdots \int_0^\infty e^{-\frac{1}{2}(z-\Delta)^T \Sigma^{-1}(z-\Delta)} dz.$$

From Lemma 6.1 it follows $|\Delta| = a\left(\frac{\sigma^2}{n}\right)^{a-1}$; and after the substitution $t = \sqrt{n}(z-\Delta)/(\sigma\sqrt{2})$, we obtain for P_0 Equation (11.11) with $R = (I_{a-1} + 1_{a-1,a-1})/2$.

But now $d_{a,a-1} \le d_{a,a-2} \le \cdots \le d_{a,1}$ and therefore

$$P_0 \ge \frac{1}{\sqrt{a\pi^{a-1}}} \int_z^\infty \cdots \int_z^\infty e^{-\frac{1}{2} t_v^T R^{-1} t_v} dt_v \tag{11.12}$$

with $z = -\sqrt{n} d_{a,a-1}/(\sigma\sqrt{2})$, that is, the least favourable case (equality sign in (11.12)) is that with $d_{a,a-1} = d_{a,1}$.

We now define the β-quantile $z(a-1,\beta) = -z(a-1, 1-\beta)$ of the $(a-1)$-dimensional normal distribution with expectation vector 0_{a-1} and covariance matrix R by

$$\beta = \frac{1}{\sqrt{a\pi^{a-1}}} \int_{-\infty}^{z(a-1,\beta)} \cdots \int_{-\infty}^{z(a-1,\beta)} e^{-\frac{1}{2} t_v^T R^{-1} t_v} dt_v. \tag{11.13}$$

Putting in (11.12),

$$z = z(a-1,\beta)/\sqrt{2} = -z(a-1, 1-\beta)/\sqrt{2},$$

gives $P_0 \ge 1-\beta$. If $d_{a,a-1} \ge d$ and if we choose n so that

$$n \ge \frac{2\sigma^2 z^2(a-1, 1-\beta)}{d^2}, \tag{11.14}$$

then Selection Rule 11.1 for $t = 1$ at least with probability $1-\beta$ gives a correct selection. Table 11.1 shows the values $z(a-1, 1-\beta)$ for $a = 2(1)39$.

Table 11.1 is not needed if we solve (11.14) by R. We use the OPDOE program with

```
>size.selection.bechhofer(a=.beta=…delta=…, sigma=…)
```

This program can also be used to calculate the smallest d from n, σ^2, β and a.

If $t > 1$, we determine n so that the right-hand side of (11.7) never takes a value smaller $1 - \beta$. Table 11.2 gives for some (a, t)-combinations the values $\sqrt{n}d/\sigma$. Later with given d, σ, the proper n can be calculated.

In the examples the populations are given in their original form $P_1, \dots, P_a$.

Example 11.1 Select from $a = 10$ given populations $P_1, \dots, P_{10}$ the $t = 4$ with the largest expectations!

Let us assume that we know from experiments in the past that the character investigated can be modelled by a $N(\mu_i, \sigma^2)$-distributed random variable with $\sigma^2 = 300$.

How many observations are needed in each of the ten populations to obtain for Problems 1 and 1A that $P_C \geq 0.95$ $\left(P_C^* \geq 0.95\right)$, respectively, if we choose $d = 22$?

In Table 11.2 we find for $1 - \beta = 0.95$, $a = 10$, $t = 4$

$$\sqrt{n}\frac{d}{\sigma} = 3.9184$$

so that

$$n = \left\lceil \frac{\sigma^2 \cdot 3.9184^2}{d^2} \right\rceil = \left\lceil \frac{300 \cdot 3.9184^2}{22^2} \right\rceil = \lceil 9.52 \rceil = 10.$$

By R of course we obtain the same value. We now observe 10 data per population and receive the sample means in Table 11.3.

Using Selection Rule 11.1 we have to select the populations P_1, P_2, P_3 and P_7.

Bechhofer for Selection Rule 11.1 showed that P_R is a maximal lower bound for the probability of a correct selection if we have normal distributions with known equal variances and use $n_i = n$ for fixed a, t and d.

If σ^2 is unknown for $t = 1$, a two-stage selection rule is proposed.

Selection Rule 11.2 Calculate from observations (realisations of $\boldsymbol{x}_{ij}$) $x_{ij}(i = 1, \dots, a; j = 1, \dots, n_0)$ from the a populations $A_1, \dots, A_a$ with $10 \leq n_0 \leq 30$ as in Table 5.2 the estimate $s_0^2 = MS_{res}$ with $f = a(n_0 - 1)$ degrees of freedom. For given d and $\beta = 0.05, 0.025$ or 0.01, respectively, we calculate further with the $(1 - \beta)$-quantile of the (a-1)–dimensional t-distribution with f degrees of freedom $t_v(a - 1, f, 1 - \beta)$ in Table 11.4 analogously to (11.14) the value

$$c = \frac{d}{t(a-1, f, 1-\beta)}. \tag{11.15}$$

Then we round s_0^2/c^2 up to the next integer (no rounding, if s_0^2/c^2 is already integer) and choose the maximum of n_0 and the rounded value as final sample size. If $n > n_0$, we take from each of a populations $n - n_0$ additional observations; otherwise n_0 is the final sample size. With n and n_0 we continue as in Selection Rule 11.1 for $t = 1$.

Table 11.1 Quantiles $z(a-1, 1-\beta)$ of the $(a-1)$-dimensional standardised normal distribution with correlation ½.

	β				
$a-1$	**0.01**	**0.025**	**0.05**	**0.10**	**0.25**
1	2.326	1.960	1.645	1.282	0.675
2	2.558	2.212	1.916	1.577	1.014
3	2.685	2.350	2.064	1.735	1.189
4	2.772	2.442	2.160	1.838	1.306
5	2.837	2.511	2.233	1.916	1.391
6	2.889	2.567	2.290	1.978	1.458
7	2.933	2.613	2.340	2.029	1.514
8	2.970	2.652	2.381	2.072	1.560
9	3.002	2.686	2.417	2.109	1.601
10	3.031	2.716	2.448	2.142	1.636
11	3.057	2.743	2.477	2.172	1.667
12	3.079	2.768	2.502	2.180	1.696
13	3.100	2.790	2.525	2.222	1.724
14	3.120	2.810	2.546	2.244	1.745
15	3.138	2.829	2.565	2.264	1.767
16	3.154	2.846	2.583	2.283	1.787
17	3.170	2.863	2.600	2.301	1.805
18	3.185	2.878	2.616	2.317	1.823
19	3.198	2.892	2.631	2.332	1.839
20	3.211	2.906	2.645	2.347	1.854
21	3.223	2.918	2.658	2.361	1.869
22	3.235	2.930	2.671	2.374	1.883
23	3.246	2.942	2.683	2.386	1.896
24	3.257	2.953	2.694	2.392	1.908
25	3.268	2.964	2.705	2.409	1.920
26	3.276	2.973	2.715	2.420	1.931
27	3.286	2.983	2.725	2.430	1.942
28	3.295	2.993	2.735	2.440	1.953
29	3.303	3.001	2.744	2.450	1.963
30	3.312	3.010	2.753	2.459	1.972
31	3.319	3.018	2.761	2.467	1.982
32	3.327	3.026	2.770	2.476	1.990

(*Continued*)

Table 11.1 (Continued)

$a-1$	β 0.01	0.025	0.05	0.10	0.25
33	3.335	3.034	2.777	2.484	1.999
34	3.342	3.041	2.785	2.492	2.007
35	3.349	3.048	2.792	2.500	2.015
36	3.355	3.055	2.800	2.507	2.023
37	3.362	3.062	2.807	2.514	2.031
38	3.368	3.069	2.813	2.521	2.038
39	3.374	3.075	2.820	2.528	2.045

Table 11.2 Values $\sqrt{n}\frac{d}{\sigma}$ for the selection of the t best of a populations with normal distribution with probability of a correct selection at least equal to $1-\beta$ (Bechhofer, 1954).

$1-\beta$	$a=2$ $t=1$	$a=3$ $t=1$	$a=4$ $t=1$	$a=4$ $t=2$	$a=5$ $t=1$
0.99	3.2900	3.6173	3.7970	3.9323	3.9196
0.98	2.9045	3.2533	3.4432	3.5893	3.5722
0.97	2.6598	3.0232	3.2198	3.3734	3.3529
0.96	2.4759	2.8504	3.0522	3.2117	3.1885
0.95	2.3262	2.7101	2.9162	3.0808	3.0552
0.94	2.1988	2.5909	2.8007	2.9698	2.9419
0.93	2.0871	2.4865	2.6996	2.8728	2.8428
0.92	1.9871	2.3931	2.6092	2.7861	2.7542
0.91	1.8961	2.3082	2.5271	2.7075	2.6737
0.90	1.8124	2.2302	2.4516	2.6353	2.5997
0.88	1.6617	2.0899	2.3159	2.5057	2.4668
0.86	1.5278	1.9655	2.1956	2.3910	2.3489
0.84	1.4064	1.8527	2.0867	2.2873	2.2423
0.82	1.2945	1.7490	1.9865	2.1921	2.1441
0.80	1.1902	1.6524	1.8932	2.1035	2.0528
0.75	0.9539	1.4338	1.6822	1.9038	1.8463
0.70	0.7416	1.2380	1.4933	1.7253	1.6614
0.65	0.5449	1.0568	1.3186	1.5609	1.4905
0.60	0.3583	0.8852	1.1532	1.4055	1.3287
0.55	0.1777	0.7194	0.9936	1.2559	1.1726

Table 11.2 (Continued)

$1-\beta$	$a=5$ $t=2$	$a=6$ $t=1$	$a=6$ $t=2$	$a=6$ $t=3$	$a=7$ $t=1$
0.99	4.1058	4.0121	4.2244	4.2760	4.0861
0.98	3.7728	3.6692	3.8977	3.9530	3.7466
0.97	3.5635	3.4528	3.6925	3.7504	3.5324
0.96	3.4071	3.2906	3.5393	3.5992	3.3719
0.95	3.2805	3.1591	3.4154	3.4769	3.2417
0.94	3.1732	3.0474	3.3104	3.3735	3.1311
0.93	3.0795	2.9496	3.2187	3.2831	3.0344
0.92	2.9959	2.8623	3.1370	3.2026	2.9479
0.91	2.9201	2.7829	3.0628	3.1296	2.8694
0.90	2.8505	2.7100	2.9948	3.0627	2.7972
0.88	2.7257	2.5789	2.8729	2.9427	2.6676
0.86	2.6153	2.4627	2.7651	2.8368	2.5527
0.84	2.5156	2.3576	2.6677	2.7411	2.4486
0.82	2.4241	2.2609	2.5784	2.6535	2.3530
0.80	2.3391	2.1709	2.4955	2.5720	2.2639
0.75	2.1474	1.9674	2.3086	2.3887	2.0626
0.70	1.9765	1.7852	2.1421	2.2256	1.8824
0.65	1.8191	1.6168	1.9888	2.0756	1.7159
0.60	1.6706	1.4575	1.8443	1.9342	1.5583
0.55	1.5277	1.3037	1.7054	1.7985	1.4062

$1-\beta$	$a=7$ $t=2$	$a=7$ $t=3$	$a=8$ $t=1$	$a=8$ $t=2$	$a=8$ $t=3$
0.99	4.3140	4.3926	4.1475	4.3858	4.4807
0.98	3.9917	4.0758	3.8107	4.0669	4.1683
0.97	3.7895	3.8773	3.5982	3.8668	3.9728
0.96	3.6385	3.7293	3.4390	3.7175	3.8270
0.95	3.5164	3.6097	3.3099	3.5968	3.7093
0.94	3.4130	3.5086	3.2002	3.4946	3.6097
0.93	3.3228	3.4203	3.1043	3.4054	3.5229
0.92	3.2423	3.3417	3.0186	3.3258	3.4456
0.91	3.1693	3.2704	2.9407	3.2537	3.3755
0.90	3.1024	3.2051	2.8691	3.1876	3.3113

(Continued)

Table 11.2 (Continued)

$1-\beta$	$a=7$ $t=2$	$a=7$ $t=3$	$a=8$ $t=1$	$a=8$ $t=2$	$a=8$ $t=3$
0.88	2.9824	3.0880	2.7406	3.0691	3.1963
0.86	2.8764	2.9847	2.6266	2.9644	3.0948
0.84	2.7806	2.8915	2.5235	2.8698	3.0032
0.82	2.6929	2.8061	2.4286	2.7832	2.9194
0.80	2.6113	2.7269	2.3403	2.7027	2.8416
0.75	2.4277	2.5485	2.1407	2.5215	2.6666
0.70	2.2641	2.3899	1.9621	2.3601	2.5111
0.65	2.1137	2.2442	1.7970	2.2116	2.3683
0.60	1.9719	2.1071	1.6407	2.0718	2.2340
0.55	1.8355	1.9754	1.4899	1.9374	2.1051

$1-\beta$	$a=8$ $t=4$	$a=9$ $t=1$	$a=9$ $t=2$	$a=9$ $t=3$	$a=9$ $t=4$
0.99	4.5078	4.1999	4.4455	4.5513	4.5950
0.98	4.1972	3.8653	4.1292	4.2423	4.2888
0.97	4.0029	3.6543	3.9308	4.0489	4.0974
0.96	3.8581	3.4961	3.7829	3.9048	3.9548
0.95	3.7412	3.3679	3.6633	3.7885	3.8398
0.94	3.6424	3.2590	3.5620	3.6902	3.7426
0.93	3.5562	3.1637	3.4736	3.6045	3.6579
0.92	3.4794	3.0785	3.3948	3.5280	3.5825
0.91	3.4099	3.0012	3.3234	3.4589	3.5142
0.90	3.3462	2.9301	3.2579	3.3955	3.4516
0.88	3.2322	2.8024	3.1405	3.2820	3.3395
0.86	3.1316	2.6893	3.0368	3.1818	3.2408
0.84	3.0408	2.5868	2.9433	3.0915	3.1518
0.82	2.9577	2.4926	2.8575	3.0088	3.0703
0.80	2.8807	2.4049	2.7778	2.9321	2.9947
0.75	2.7074	2.2067	2.5984	2.7596	2.8249
0.70	2.5535	2.0293	2.4387	2.6064	2.6741
0.65	2.4122	1.8653	2.2919	2.4658	2.5359
0.60	2.2794	1.7102	2.1535	2.3335	2.4059
0.55	2.1520	1.5604	2.0206	2.2066	2.2814

Table 11.2 (Continued)

$1-\beta$	$a=10$ $t=1$	$a=10$ $t=2$	$a=10$ $t=3$	$a=10$ $t=4$	$a=10$ $t=5$
0.99	4.2456	4.4964	4.6100	4.6648	4.6814
0.98	3.9128	4.1823	4.3037	4.3619	4.3796
0.97	3.7030	3.9854	4.1120	4.1727	4.1911
0.96	3.5457	3.8385	3.9693	4.0319	4.0509
0.95	3.4182	3.7198	3.8541	3.9184	3.9378
0.94	3.3099	3.6193	3.7567	3.8224	3.8422
0.93	3.2152	3.5316	3.6718	3.7387	3.7589
0.92	3.1305	3.4534	3.5962	3.6643	3.6848
0.91	3.0536	3.3826	3.5277	3.5969	3.6177
0.90	2.9829	3.3176	3.4650	3.5351	3.5563
0.88	2.8560	3.2011	3.3526	3.4246	3.4463
0.86	2.7434	3.0983	3.2535	3.3272	3.3494
0.84	2.6418	3.0055	3.1642	3.2395	3.2621
0.82	2.5479	2.9203	3.0824	3.1591	3.1822
0.80	2.4608	2.8413	3.0065	3.0847	3.1082
0.75	2.2637	2.6635	2.8360	2.9174	2.9419
0.70	2.0873	2.5051	2.6845	2.7690	2.7944
0.65	1.9242	2.3595	2.5456	2.6330	2.6592
0.60	1.7700	2.2224	2.4149	2.5052	2.5322
0.55	1.6210	2.0907	2.2896	2.3827	2.4106

$1-\beta$	$a=11$ $t=2$	$a=11$ $t=3$	$a=11$ $t=4$	$a=11$ $t=5$	$a=12$ $t=3$
0.99	4.5408	4.6602	4.7229	4.7506	4.7039
0.98	4.2286	4.3560	4.4227	4.4522	4.4016
0.97	4.0329	4.1658	4.2353	4.2660	4.2126
0.96	3.8869	4.0242	4.0958	4.1274	4.0719
0.95	3.7689	3.9099	3.9834	4.0158	3.9584
0.94	3.6691	3.8133	3.8883	3.9214	3.8624
0.93	3.5819	3.7291	3.8055	3.8392	3.7788
0.92	3.5042	3.6541	3.7318	3.7661	3.7043
0.91	3.4338	3.5862	3.6652	3.6999	3.6369
0.90	3.3693	3.5239	3.6041	3.6393	3.5751

(Continued)

Table 11.2 (Continued)

$1-\beta$	$a=11$ $t=2$	$a=11$ $t=3$	$a=11$ $t=4$	$a=11$ $t=5$	$a=12$ $t=3$
0.88	3.2536	3.4126	3.4948	3.5309	3.4645
0.86	3.1514	3.3143	3.3984	3.4354	3.3670
0.84	3.0592	3.2258	3.3117	3.3494	3.2791
0.82	2.9747	3.1447	3.2323	3.2707	3.1986
0.80	2.8963	3.0695	3.1587	3.1978	3.1240
0.75	2.7196	2.9006	2.9934	3.0341	2.9563
0.70	2.5624	2.7505	2.8468	2.8890	2.8075
0.65	2.4179	2.6129	2.7125	2.7560	2.6709
0.60	2.2818	2.4835	2.5863	2.6312	2.5426
0.55	2.1510	2.3594	2.4654	2.5117	2.4196

$1-\beta$	$a=12$ $t=4$	$a=12$ $t=5$	$a=13$ $t=4$	$a=13$ $t=5$	$a=14$ $t=5$
0.99	4.7725	4.8083	4.8158	4.8576	4.9005
0.98	4.4746	4.5126	4.5197	4.5641	4.6089
0.97	4.2886	4.3281	4.3350	4.3810	4.4271
0.96	4.1502	4.1909	4.1975	4.2449	4.2919
0.95	4.0387	4.0803	4.0867	4.1353	4.1831
0.94	3.9444	3.9870	3.9932	4.4027	4.0911
0.93	3.8623	3.9057	3.9117	3.9521	4.0111
0.92	3.7893	3.8333	3.8391	3.8904	3.9399
0.91	3.7232	3.7678	3.7735	3.8255	3.8756
0.90	3.6626	3.7079	3.7134	3.7661	3.8166
0.88	3.5543	3.6007	3.6059	3.6599	3.7113
0.86	3.4588	3.5063	3.5111	3.5664	3.6185
0.84	3.3729	3.4213	3.4259	3.4822	3.5350
0.82	3.2942	3.3435	3.3478	3.4052	3.4586
0.80	3.2213	3.2715	3.2755	3.3339	3.3879
0.75	3.0577	3.1098	3.1132	3.1739	3.2292
0.70	2.9125	2.9666	2.9693	3.0321	3.0887
0.65	2.7796	2.8354	2.8374	2.9023	2.9600
0.60	2.6547	2.7122	2.7137	2.7805	2.8394
0.55	2.5352	2.5944	2.5952	2.6640	2.7240

Table 11.3 Sample means of Example 11.1.

Population	P_1	P_2	P_3	P_4	P_5	P_6	P_7	P_8	P_9	P_{10}
$\bar{y}_{i\cdot}$	138.6	132.2	138.4	122.7	130.6	131.0	139.2	131.7	128.0	122.5

Table 11.4 Quantiles $t(a-1, f, 1-\beta)$ of the $(a-1)$-dimensional t-distribution with correlation 1/2.

$\beta = 0.05$	$a-1$								
f	1	2	3	4	5	6	7	8	9
5	2.02	2.44	2.68	2.85	2.98	3.08	3.16	3.24	3.30
6	1.94	2.34	2.56	2.71	2.83	2.92	3.00	3.07	3.12
7	1.89	2.27	2.48	2.62	2.73	2.82	2.89	2.95	3.01
8	1.86	2.22	2.42	2.55	2.66	2.74	2.81	2.87	2.92
9	1.83	2.18	2.37	2.50	2.60	2.68	2.75	2.81	2.86
10	1.81	2.15	2.34	2.47	2.56	2.64	2.70	2.76	2.81
11	1.80	2.13	2.31	2.44	2.53	2.60	2.67	2.72	2.77
12	1.78	2.11	2.29	2.41	2.50	2.58	2.64	2.69	2.74
13	1.77	2.09	2.27	2.39	2.48	2.55	2.61	2.66	2.71
14	1.76	2.08	2.25	2.37	2.46	2.53	2.59	2.64	2.69
15	1.75	2.07	2.24	2.36	2.44	2.51	2.57	2.62	2.67
16	1.75	2.06	2.23	2.34	2.43	2.50	2.56	2.61	2.65
17	1.74	2.05	2.22	2.33	2.42	2.49	2.54	2.59	2.64
18	1.73	2.04	2.21	2.32	2.41	2.48	2.53	2.58	2.62
19	1.73	2.03	2.20	2.31	2.40	2.47	2.52	2.57	2.61
20	1.72	2.03	2.19	2.30	2.39	2.46	2.51	2.56	2.60
24	1.71	2.01	2.17	2.28	2.36	2.43	2.48	2.53	2.57
30	1.70	1.99	2.15	2.25	2.33	2.40	2.45	2.50	2.54
40	1.68	1.97	2.13	2.23	2.31	2.37	2.42	2.47	2.51
60	1.67	1.95	2.10	2.21	2.28	2.35	2.39	2.44	2.48
120	1.66	1.93	2.08	2.18	2.26	2.32	2.37	2.41	2.45
∞	1.64	1.92	2.06	2.16	2.23	2.29	2.34	2.38	2.42
$\beta = 0.025$	$a-1$								
f	1	2	3	4	5	6	7	8	9
5	2.57	3.03	3.39	3.66	3.88	4.06	4.22	4.36	4.49
6	2.45	2.86	3.18	3.41	3.60	3.75	3.88	4.00	4.11

(Continued)

Table 11.4 (Continued)

$\beta = 0.025$	$a-1$								
f	1	2	3	4	5	6	7	8	9
7	2.36	2.75	3.04	3.24	3.41	3.54	3.66	3.76	3.86
8	2.31	2.67	2.94	3.13	3.28	3.40	3.51	3.60	3.68
9	2.26	2.61	2.86	3.04	3.18	3.29	3.39	3.48	3.55
10	2.23	2.57	2.81	2.97	3.11	3.21	3.31	3.39	3.46
11	2.20	2.53	2.76	2.92	3.05	3.15	3.24	3.31	3.38
12	2.18	2.50	2.72	2.88	3.00	3.10	3.18	3.25	3.32
13	2.16	2.48	2.69	2.84	2.96	3.06	3.14	3.21	3.27
14	2.14	2.46	2.67	2.81	2.93	3.02	3.10	3.17	3.23
15	2.13	2.44	2.64	2.79	2.90	2.99	3.07	3.13	3.19
16	2.12	2.42	2.63	2.77	2.88	2.96	3.04	3.10	3.16
17	2.11	2.41	2.61	2.75	2.85	2.94	3.01	3.08	3.13
18	2.10	2.40	2.59	2.73	2.84	2.92	2.99	3.05	3.11
19	2.09	2.39	2.58	2.72	2.82	2.90	2.97	3.04	3.09
20	2.09	2.38	2.57	2.70	2.81	2.89	2.96	3.02	3.07
24	2.06	2.35	2.53	2.66	2.76	2.84	2.91	2.96	3.01
30	2.04	2.32	2.50	2.62	2.72	2.79	2.86	2.91	2.96
40	2.02	2.29	2.47	2.58	2.67	2.75	2.81	2.86	2.90
60	2.00	2.27	2.43	2.55	2.63	2.70	2.76	2.81	2.85
120	1.98	2.24	2.40	2.51	2.59	2.66	2.71	2.76	2.80
∞	1.96	2.21	2.37	2.47	2.55	2.62	2.67	2.71	2.75

$\beta = 0.01$	$a-1$								
f	1	2	3	4	5	6	7	8	9
5	3.37	3.90	4.21	4.43	4.60	4.73	4.85	4.94	5.03
6	3.14	3.61	3.88	4.07	4.21	4.33	4.43	4.51	4.59
7	3.00	3.42	3.66	3.83	3.96	4.07	4.15	4.23	4.30
8	2.90	3.29	3.51	3.67	3.79	3.88	3.96	4.03	4.09
9	2.82	3.19	3.40	3.55	3.66	3.75	3.82	3.89	3.94
10	2.76	3.11	3.31	3.45	3.56	3.64	3.71	3.78	3.83
11	2.72	3.06	3.25	3.38	3.48	3.56	3.63	3.69	3.74
12	2.68	3.01	3.19	3.32	3.42	3.50	3.56	3.62	3.67
13	2.65	2.97	3.15	3.27	3.37	3.44	3.51	3.56	3.61
14	2.62	2.94	3.11	3.23	3.32	3.40	3.46	3.51	3.56

Table 11.4 (Continued)

$\beta = 0.01$	$a-1$								
f	1	2	3	4	5	6	7	8	9
15	2.60	2.91	3.08	3.20	3.29	3.36	3.42	3.47	3.52
16	2.58	2.88	3.05	3.17	3.26	3.33	3.39	3.44	3.48
17	2.57	2.86	3.03	3.14	3.23	3.30	3.36	3.41	3.45
18	2.55	2.84	3.01	3.12	3.21	3.27	3.33	3.38	3.42
19	2.54	2.83	2.99	3.10	3.18	3.25	3.31	3.36	3.40
20	2.53	2.81	2.97	3.08	3.17	3.23	3.29	3.34	3.38
24	2.49	2.77	2.92	3.03	3.11	3.17	3.22	3.27	3.31
30	2.46	2.72	2.87	2.97	3.05	3.11	3.16	3.21	3.24
40	2.42	2.68	2.82	2.92	2.99	3.05	3.10	3.14	3.18
60	2.39	2.64	2.78	2.87	2.94	3.00	3.04	3.08	3.12
120	2.36	2.60	2.73	2.82	2.89	2.94	2.99	3.03	3.06
∞	2.33	2.56	2.68	2.77	2.84	2.89	2.93	2.97	3.00

11.1.2.2 Approximate Solutions for Non-normal Distributions and $t = 1$

Let the random variables $\boldsymbol{x}_i$ be distributed in populations A_i with the distribution function $F(x_i; \mu_i, \eta_{i2}, \dots, \eta_{ip})$. The distribution of the $\boldsymbol{x}_i$ may be such that for the purposes of a practical investigation, it can adequately be characterised by the expectation μ_i and the standard deviation $\sigma(\mu_i)$, and we have

$$F\left(x_i;\mu_i,\eta_{i2},\dots,\eta_{ip}\right) \approx G(x_i;\mu_i,\sigma(\mu_i))$$

For a random samples of size n, the sample means $\bar{\boldsymbol{x}}_i$ are approximately normally distributed with expectation μ_i and variance $\frac{\sigma^2(\mu_i)}{n}$. For $n \geq 30$ the approximation is in most cases sufficient for practical purposes. For $t = 1$ from (11.10), by taking into account the variance homogeneity with

$$\gamma = \frac{\sigma\left(\mu_{(a)} - d\right)}{\sigma\left(\mu_{(a)}\right)},$$

we obtain

$$P_0 \geq \int_{-\infty}^{\infty} \left(\Phi\left\{\frac{1}{\gamma}\left[y + \frac{d\sqrt{n}}{\sigma\left(\mu_{(a)}\right)}\right]\right\}\right)^{a-1} \varphi(y)\mathrm{d}y. \tag{11.16}$$

Table 11.5 Approximative values of $d\sqrt{n}/\sigma(\mu)$ for the selection of the population with the largest expectations from a populations for the given minimum probability $1-\beta$ of a correct selection with $\gamma = \frac{\sigma\left(\mu_{(a)}-d\right)}{\sigma\left(\mu_{(a)}\right)}$.

$1-\beta = 0.90$	γ								
a	0.6	0.7	0.8	0.9	1	1.1	1.2	1.3	1.4
2	1.495	1.564	1.641	1.724	1.812	1.905	2.002	2.102	2.205
3	1.770	1.877	1.990	2.108	2.230	2.357	2.487	2.620	2.757
4	1.914	2.041	2.173	2.310	2.452	2.597	2.745	2.896	3.050
5	2.010	2.150	2.296	2.446	2.600	2.757	2.918	3.081	3.247
6	2.081	2.231	2.387	2.547	2.710	2.877	3.047	3.219	3.394
7	2.136	2.295	2.459	2.626	2.797	2.971	3.149	3.329	3.511
8	2.182	2.348	2.518	2.692	2.869	3.050	3.233	3.419	3.607
9	2.221	2.393	2.568	2.747	2.930	3.116	3.304	3.496	3.689
10	2.255	2.431	2.611	2.796	2.983	3.173	3.366	3.562	3.760
					$1-\beta = 0.95$				
2	1.918	2.008	2.106	2.213	2.326	2.445	2.569	2.698	2.830
3	2.178	2.300	2.430	2.567	3.710	2.858	3.011	3.169	3.329
4	2.315	2.456	2.603	2.757	2.916	3.081	3.249	3.422	3.599
5	2.407	2.560	2.719	2.885	3.055	3.231	3.410	3.594	3.781
6	2.475	2.637	2.806	2.980	3.159	3.343	3.531	3.723	3.918
7	2.528	2.699	2.875	3.056	3.242	3.432	3.621	3.825	4.027
8	2.572	2.749	2.931	3.118	3.310	3.506	3.706	3.910	4.117
9	2.610	2.792	2.979	3.171	3.368	3.569	3.774	3.982	4.194
10	2.642	2.829	3.021	3.217	3.418	3.623	3.832	4.045	4.260
					$1-\beta = 0.99$				
2	2.713	2.840	2.979	3.130	3.290	3.458	3.634	3.816	4.002
3	2.945	3.097	3.261	3.435	3.617	3.808	4.005	4.209	4.418
4	3.070	3.237	3.415	3.602	3.797	4.000	4.210	4.426	4.647
5	3.155	3.332	3.519	3.715	3.920	4.131	4.350	4.574	4.804
6	3.218	3.403	3.598	3.801	4.012	4.231	4.455	4.686	4.922
7	3.268	3.460	3.660	3.869	4.086	4.310	4.540	4.776	5.017
8	3.309	3.506	3.712	3.926	4.147	4.376	4.611	4.851	5.096
9	3.344	3.546	3.756	3.974	4.200	4.432	4.671	4.915	5.164
10	3.375	3.581	3.795	4.017	4.246	4.481	4.723	4.971	5.223

The selection rule below is a modification from Chambers and Jarratt (1964) of Selection Rule 11.2.

Selection Rule 11.2a Take from each population A_i a random sample of size n_0 $(10 \le n_0 \le 30)$ and determine the maximum sample mean $\bar{x}_{(a)}^{(0)}$, and use it as estimate of μ_a. Determine the sample sizes n per population with $\sigma\left(\bar{x}_{(a)}\right)$ in place of $\sigma(\mu_a)$ so that the integral in (11.16) is not below $1-\beta$. Then observe (if $n > n_0$) $n - n_0$ further values from each population. We then say that the population with the largest sample mean calculated with n observations is best.

In Selection Rule 11.2a it was assumed that the function $\sigma(\mu)$ is known. If $\bar{\boldsymbol{x}}$ is $B(n, p)$-distributed, we have $\sigma(\mu) = \sqrt{\mu(1-\mu)}$; if $\boldsymbol{x}$ is $P(\lambda)$-distributed, we have $\sigma(\mu) = \sqrt{\mu}$. If $\sigma(\mu)$ is unknown, we estimate it by regression of s on $\bar{\boldsymbol{x}}$. But for non-normal continuous distributions, we also can use the method described in Section 11.1.2.1 because it is robust against non-normality as shown in Domröse and Rasch (1987).

The values $\dfrac{d\sqrt{n}}{\sigma\left(\mu_{(a)}\right)}$ needed in (11.16) can be found in Table 11.5.

11.1.3 Selection of a Subset Containing the Best Population with Given Probability

We discuss now Problem 2 of Section 11.1.1 for $t = 1$, $\boldsymbol{Y} = \boldsymbol{y}_1 = \boldsymbol{y}$ and $\Omega = R^1$. Let $\boldsymbol{y}_i$ in P_i be continuously distributed with distribution function $F(y, \theta)$ and density function $f(y, \theta)$. Let F and f be known but the θ_i of the P_i be unknown. We assume that $g^*(\theta) = \theta$.

In Problem 2 we have to find a (non-empty) subset $(A_{i_1}, \dots, A_{i_r}) = \boldsymbol{M}_G$ of the populations $A_1, \dots, A_a$ so that the probability of a correct selection $P(CS)$ that the best population (with parameter $\vartheta_{(a)}$) is in the subset is at least $1-\beta$. Again as in Section 11.1.2, we assume $\frac{1}{a} < 1-\beta < 1$. If for more than one P_i the parameter is $\theta_{(a)} = \eta_a$, any of them is called the best.

The following selection rule (class of selection rules) stems from Gupta and Panchapakesan (1970, 1979).

Selection Rule 11.3 First we select a proper estimator $\hat{\boldsymbol{\eta}}$ of the unknown parameters η (and θ). With $H(\hat{\eta}, \eta)$ and $h(\hat{\eta}, \eta)$, we denote the distribution function and the density function of $\hat{\eta}_i$, respectively. We assume that for $\eta > \eta'$, always $H(\hat{\eta}, \eta) \le H(\hat{\eta}, \eta')$ and for at least one $\hat{\eta}$, we have $H(\hat{\eta}, \eta) < H(\hat{\eta}, \eta')$.

Further let $d_{u,v}(x)$ be a real differentiable function with parameters $u \ge 1$, $v \ge 0$, so that for each x from the domain of definitions Ω of $H(x, \eta)$, the conditions below are fulfilled:

$\cdots d_{u,v}(x) \geq x,$

$\cdots d_{1,0}(x) = x,$

$\cdots d_{u,v}(x)$ is continuous in u and v,

and at least one of the relations

$$\lim_{v \to \infty} d_{u,v}(x) = \infty \quad \text{for given } u,$$

$$\lim_{u \to \infty} d_{u,v}(x) = \infty \quad \text{for given } v \text{ and } x \neq 0.$$

is valid.

Then M_G contains all populations A_i, for which

$$d_{u,v}(\hat{\eta}_1) \geq \eta_a.$$

Analogous to (11.10) is by Selection Rule 11.3

$$P(CS) \geq \int_{\Omega} \{H[d_{u,v}(\hat{\eta}),\eta_a]\}^{a-1} h(\hat{\eta},\eta_a) \mathrm{d}\hat{\eta}. \tag{11.17}$$

We put

$$\int_{\Omega} \{H[d_{u,v}(\hat{\eta}),\eta]\}^{i} h(\hat{\eta},\eta_a) \mathrm{d}\hat{\eta} = I(\eta,u,v,i+1), \tag{11.18}$$

so that (11.17) can be written as $P(CS) \geq I(\eta_a, u, v, a)$. For I in (11.18) it follows from the conditions of Selection Rule 11.3:

$$\left.\begin{aligned} & I(\eta,u,v,a) \geq \frac{1}{a}, \\ & I(\eta,1,0,a) = \frac{1}{a}, \\ & \text{either} \quad \lim_{v \to \infty} I(\eta,u,v,a) = 1 \quad \text{for fixed } u \\ & \text{or} \quad \lim_{u \to \infty} I(\eta,u,v,a) = 1 \quad \text{for fixed } v \\ & (\text{or both}). \end{aligned}\right\} \tag{11.19}$$

From (11.19) it follows that u and v are chosen appropriately so that $P(CS) > 1 - \beta$ can be fulfilled for each β. This leads to

Theorem 11.3 For a continuous random variable $\boldsymbol{\eta}$ *with* $H(\hat{\eta},\eta) \geq H(\hat{\eta},\eta')$ for $\eta < \eta' \in \Omega = R^1$ and $t = 1$, Problem 2 of Section 11.1.1 is solvable with Selection Rule 11.3 for all β with $\frac{1}{a} < \beta < 1$.

Gupta and Panchapakesan (1970) proved the theorem below under the assumption that $(\eta_i \leq \eta_j)$

$$\frac{\partial}{\partial\eta_i}H[d_{u,v}(\hat{\eta}),\eta_i]h\left(\hat{\eta},\eta_j\right)-\frac{\partial}{\partial\hat{\eta}}d_{u,v}(\hat{\eta})\frac{\partial}{\partial\eta_i}H(\hat{\eta},\eta_i)h\left[d_{u,v}(\hat{\eta}),\eta_j\right]\geq 0. \quad (11.20)$$

Theorem 11.4 (Gupta and Panchapakesan).
By Selection Rule 11.3 and with the assumptions of Theorem 11.3 and (11.20), the supremum of the expectations $E(\boldsymbol{r})$ and $E(\boldsymbol{w})$ is taken for $\eta_1 = \cdots = \eta_a$. Here $\boldsymbol{w}$ is the number of those A_i in $\boldsymbol{M}_G$ obtained by Selection Rule 11.3, not having the largest parameter η_a.

Therefore $\eta_1 = \cdots = \eta_a$ is the least favourable parameter constellation for Problem 2.

We now consider the special case that θ is a location parameter with $\Omega = (-\infty, \infty)$. Then essential simplifications appear, because $H(\hat{\eta},\eta) = G(\hat{\eta}-\eta)$ $(-\infty < \eta < \infty)$. Then (11.20) with

$$d^*_{u,v}(\hat{\eta}) = \frac{\partial d_{u,v}(\hat{\eta})}{\partial\hat{\eta}}$$

becomes

$$d^*_{u,v}(\hat{\eta})h(\hat{\eta},\eta_i)h\left[d_{u,v}(\hat{\eta}),\eta_j\right]-h(\hat{\eta},\eta_i)h[d_{u,v}(\hat{\eta}),\eta_i]\geq 0.$$

If the distribution of $\hat{\boldsymbol{\eta}}$ has a monotone likelihood ratio in $\hat{\eta}$, then (11.20) is fulfilled. An appropriate choice of $d_{u,v}(\hat{\eta})$ is $d(\hat{\eta}) = \hat{\eta} + d\,(u = 1, v = d)$ with $\hat{\boldsymbol{\eta}} = \bar{\boldsymbol{x}}$ and $\eta = \mu$, so that by Selection Rule 11.3, all the A_i are put in $\boldsymbol{M}_G$, for which

$$\bar{x}_{i\cdot} \geq \bar{x}_{(a)\cdot} - d \quad \left(\bar{x}_{(a)\cdot}\text{ largest sample mean}\right). \quad (11.21)$$

We have to choose d so that

$$\int_{-\infty}^{\infty}[H(\bar{x}+d,\mu)]^{a-1}h(\bar{x},\mu)\mathrm{d}\bar{x} = 1-\beta. \quad (11.22)$$

Another important special case is that θ is a scale parameter and $H(\hat{\eta},\eta) = G\left(\frac{\hat{\eta}}{\eta}\right)$. Then $\Omega = [0, \infty)$ and $\eta \geq 0$, and (11.20) with $\hat{\eta} = s^2, \eta = \sigma^2$ becomes

$$s^2 d^*_{u,v}(s^2)h(s^2,\sigma_i^2)h\left[d_{u,v}(s^2),\sigma_j^2\right]-d_{u,v}(s^2)h\left[d_{u,v}(s^2),\sigma_i^2\right]h\left(s^2,\sigma_j^2\right)\geq 0.$$

If the distribution of $\boldsymbol{y}$ has a monotone likelihood ratio and

$$s^2 d^*_{u,v}(s^2) \geq d_{u,v}(s^2) \geq 0,$$

then (11.20) is fulfilled. Therefore

$$d_{u,v}\left(s^2\right) = us^2 \quad (u > 1)$$

is a possible (and often used) choice of $d_{u,v}(s^2)$.

11.1.3.1 Selection of the Normal Distribution with the Largest Expectation

The most important special case is that $\boldsymbol{x}$ is $N(\mu, \sigma^2)$-distributed; σ^2 may be unknown. From n observations from each of the $A_i (i = 1, \ldots, a)$, the sample means $\bar{x}_i.$ are calculated. The likelihood ratio of the normal distribution and that of the t-distribution are monotone for known as well as for unknown σ^2. Therefore a selection rule, 'Choose for $\boldsymbol{M}_G$ all A_i, with $\bar{x}_i \geq \bar{x}_{(a)} - d$', can be used. $\bar{x}_{(a)}$ is the largest sample mean. We start with the case where σ^2 is known. We put $d = D\sigma/\sqrt{n}$ with a D (analogous to (11.22)) so that

$$1 - \beta = \int_{-\infty}^{\infty} [\Phi(u + D)]^{a-1} \varphi(u) \mathrm{d}u \tag{11.23}$$

where Φ and φ are the distribution function and the density function of the standardised normal distribution, respectively. If σ^2 is unknown, we write approximately $d \approx Ds/\sqrt{n}$, where s^2 is an estimate of σ^2, based on f degrees of freedom. (11.23) is replaced by

$$1 - \beta = \int_0^{\infty} \int_{-\infty}^{\infty} [\Phi(u + Dy)]^{a-1} \varphi(u) h_f(y) \mathrm{d}u \mathrm{d}y, \tag{11.24}$$

where $h_f(y)$ is the density function of $\sqrt{\chi_f^2/f}$ and χ_f^2 is $CS(f)$-distributed.

In Table 11.2 the values $D = d$ fulfilling (11.23) are given in dependency of α and β for $t = 1$. If the experimenter selected the values d, α and β, then n can be calculated by (11.14).

For independent random samples from a populations with normal distribution and known variance, Problem 1 for $t = 1$ is solved by Selection Rule 11.1, and Problem 2 is solved by Selection Rule 11.3, leading to the same sample size.

11.1.3.2 Selection of the Normal Distribution with Smallest Variance

Let the random variable $\boldsymbol{x}$ in P_i be $N\left(\mu_i, \sigma_i^2\right)$-distributed. From n observations from each population $P_i (i = 1, \ldots, a)$ with known μ_i

$$q_i = \frac{1}{n} \sum_{j=1}^{n} \left(x_{ij} - \mu_i\right)^2$$

and with unknown μ_i,

$$q_i = \frac{1}{n-1}\sum_{j=1}^{n}\left(x_{ij}-\bar{x}_i\right)^2, \quad (i=1,\ldots,a)$$

are calculated. q_i will be used to select the population with the smallest variance; each $\boldsymbol{q}_i$ has the same number f of degrees of freedom (if μ_i is known we have $f=n$, and if μ_i is unknown then $f = n - 1$).

We use $d_{u,v}(q) = zq$ and the selection for the smallest variance follows from Selection Rule 11.4.

Selection Rule 11.4. Put into $\boldsymbol{M}_G$ all A_i, for which

$$s_i^2 \le \frac{s_{(1)}^2}{z^*} \quad \left(z^{-1} = z^* \le 1\right),$$

$s_{(1)}^2$ is the smallest sample variance. $z^* = z(f, a, \beta)$ depends on the degrees of freedom f, on the number a of populations and on $1 - \beta$.

For z^* we choose the largest number, so that the right-hand side of (11.17) equals $1 - \beta$. We have to calculate $P(CS)$ for the least favourable case given by $s_{(2)}^2 = \cdots = s_{(a)}^2$ (monotonicity of the likelihood ratio is given). We denote the estimates of σ_i^2 as usual by s_i^2 and formulate

Theorem 11.5 Let the $\boldsymbol{y}_i$ in a populations be $N\left(\mu_i, \sigma_i^2\right)$-distributed. There may be independent estimators $\boldsymbol{s}_i^2$ of σ_i^2 with f degrees of freedom each. Select from the a populations a subset $\boldsymbol{N}_G$ so that it contains the smallest variance σ_1^2 at least with probability $1 - \beta$. Using Selection Rule 11.3 with an appropriate chosen $z^* = z(f, a, \beta)$, the probability of a correct selection $P(CS)$ then is

$$P(CS) \ge \int_0^\infty \left[1 - G_f(z^* v)\right]^{a-1} g_f(v)\mathrm{d}v. \tag{11.25}$$

In (11.25) G_f and g_f are the distribution function and the density function of the central χ^2-distribution with f degrees of freedom, respectively.

Proof: If s_i^2 is the estimate of σ_i^2, we have (because $z^* < 1$)

$$P(CS) = P\left\{\boldsymbol{s}_1^2 \le \frac{1}{z^*}\min\left(\boldsymbol{s}_2^2,\ldots,\boldsymbol{s}_a^2\right)\right\}$$

$$= P\left\{\frac{z^* f\,\boldsymbol{s}_1^2}{\sigma_2^2} \le \frac{f\,\boldsymbol{s}_2^2}{\sigma_2^2},\ldots,\frac{z^* f\,\boldsymbol{s}_1^2}{\sigma_a^2} \le \frac{f\,\boldsymbol{s}_a^2}{\sigma_a^2}\right\}$$

$$= \prod_{j=2}^{a} P\left\{\frac{f\,\boldsymbol{s}_j^2}{\sigma_j^2} \ge \frac{z^*\,\sigma_1^2}{\sigma_j^2}\cdot\frac{f\,\boldsymbol{s}_1^2}{\sigma_1^2}\right\},$$

and further

$$P(CS) = \int_0^\infty g_f(\nu)\prod_{j=2}^{a}\left[1-G_f\left(\frac{z^*\sigma_1^2}{\sigma_j^2}\nu\right)\right]d\nu. \tag{11.26}$$

If $\sigma_1^2 = \sigma_2^2 = \cdots = \sigma_a^2$, then $P(CS)$ is minimum and this completes the proof.

Table 11.6 shows the values $z^* = z(f, a, \beta)$ that make the right-hand side of (11.25) equal to $1-\beta$. Approximately $z^* = z(f, a, \beta)$ can also obtained from Table 11.2 using

$$\sqrt{n}\frac{d}{\sigma} = \sqrt{\frac{1}{2}(f-1)\ln\frac{1}{z^*}}.$$

Table 11.6 Values of $10^4 z = 10^4 z(f, a, \beta)$, for which the right-hand side of (11.25) equals $1-\beta$.

	$a-1$									
f	1	2	3	4	5	6	7	8	9	10
					$1-\beta = 0.75$					
2	3333	1667	1111	0833	0667	0556	0476	0417	0310	0333
4	4844	3168	2494	2112	1860	1678	1540	1530	1340	1264
6	5611	4040	3369	2973	2704	2505	2350	2225	2121	2033
8	6099	4628	3978	3587	3317	3116	2957	2828	2720	2627
10	6446	5060	4434	4054	3788	3588	3430	3301	3192	3098
12	6711	5395	4794	4424	4165	3968	3813	3684	3576	3483
14	6921	5667	5087	4728	4475	4283	4130	4004	3898	3806
16	7094	5892	5332	4984	4737	4550	4400	4276	4171	4081
18	7239	6084	5542	5203	4963	4779	4633	4511	4408	4319
20	7364	6250	5724	5394	5160	4980	4837	4718	4616	4529
22	7472	6395	5883	5562	5333	5158	5017	4900	4801	4715
24	7568	6523	6026	5712	5488	5317	5179	5064	4967	4882
26	7653	6635	6153	5847	5628	5460	5325	5212	5117	5034
28	7729	6742	6268	5969	5754	5590	5457	5347	5253	5171
30	7798	6836	6373	6080	5870	5708	5578	5470	5377	5297
32	7861	6922	6469	6182	5976	5817	5689	5583	5492	5413
34	7919	7001	6558	6276	6074	5918	5792	5687	5598	5520

Table 11.6 (Continued)

	$a-1$									
f	1	2	3	4	5	6	7	8	9	10
36	7972	7074	6640	6363	6164	6011	5887	5784	5696	5619
38	8021	7142	6715	6444	6248	6098	5976	5874	5788	5712
40	8067	7205	6786	6519	6327	6178	6058	5958	5873	5799
42	8109	7264	6852	6590	6400	6254	6136	6038	5952	5880
44	8149	7319	6914	6656	6470	6326	6209	6112	6029	5957
46	8186	7371	6973	6718	6534	6393	6278	6182	6100	6029
48	8221	7420	7028	6777	6596	6456	6343	6248	6167	6097
50	8254	7466	7080	6832	6654	6516	6404	6311	6231	6162
	$1-\beta=0.90$									
2	1111	0556	0370	0278	0222	0185	0159	0139	0123	0111
4	2435	1630	1297	1106	0979	0886	0816	0759	0713	0674
6	3274	2417	2039	1813	1657	1541	1450	1377	1315	1263
8	3862	3002	2610	2370	2202	2076	1976	1894	1826	1766
10	4306	3457	3062	2818	2645	2515	2410	2325	2252	2190
12	4657	3825	3433	3188	3014	2881	2775	2688	2613	2549
14	4944	4132	3744	3501	3327	3194	3087	2999	2924	2859
16	5186	4392	4011	3770	3597	3464	3358	3270	3194	3129
18	5394	4618	4243	4004	3833	3702	3596	3508	3433	3368
20	5575	4816	4447	4112	4043	3913	3808	3720	3646	3581
22	5734	4992	4629	4397	4230	4101	3997	3911	3837	3772
24	5876	5149	4792	4564	4399	4272	4169	4083	4010	3946
26	6004	5291	4940	4715	4553	4427	4325	4240	4168	4104
28	6119	5420	5076	4854	4693	4569	4468	4384	4312	4250
30	6225	5539	5199	4981	4822	4700	4600	4517	4446	4384
32	6322	5648	5314	5098	4942	4820	4722	4640	4570	4508
34	6411	5749	5419	5207	5052	4933	4836	4754	4684	4624
36	6493	5842	5518	5308	5156	5037	4941	4861	4792	4732
38	6570	5929	5609	5402	5252	5135	5040	4960	4892	4833
40	6642	6011	5695	5491	5342	5227	5133	5054	4987	4928
42	6709	6087	5776	5574	5427	5313	5220	5142	5076	5017
44	6772	6159	5852	5653	5508	5394	5303	5226	5160	5102
46	6831	6227	5924	5727	5583	5472	5381	5304	5239	5182

(*Continued*)

Table 11.6 (Continued)

					$1-\beta=0.90$					
48	6887	6291	5992	5797	5655	5544	5454	5379	5314	5258
50	6940	6352	6056	5863	5723	5614	5525	5450	5386	5330
					$1-\beta=0.95$					
2	0526	0263	0175	0132	0105	0088	0075	0066	0058	0053
4	1565	1062	0851	0728	0646	0586	0540	0504	0473	0448
6	2334	1749	1486	1327	1217	1134	1069	1017	0972	0935
8	2909	2293	2007	1830	1706	1612	1573	1476	1424	1379
10	3358	2732	2436	2250	2119	2018	1938	1872	1815	1767
12	3722	3096	2796	2606	2470	2366	2283	2214	2155	2104
14	4026	3405	3103	2911	2774	2668	2583	2512	2452	2399
16	4285	3671	3370	3178	3039	2933	2847	2775	2714	2661
18	4510	3903	3604	3413	3274	3168	3081	3009	2947	2894
20	4708	4109	3813	3622	3484	3378	3291	3219	3157	3104
22	4883	4294	4000	3811	3674	3568	3481	3409	3348	3294
24	5041	4460	4170	3982	3846	3740	3654	3582	3521	3467
26	5184	4611	4324	4138	4003	3898	3812	3741	3680	3626
28	5313	4749	4465	4281	4147	4043	3958	3887	3826	3773
30	5432	4876	4595	4413	4280	4177	4093	4022	3962	3909
32	5542	4993	4716	4536	4404	4302	4218	4148	4088	4036
34	5643	5102	4828	4649	4519	4418	4335	4265	4206	4154
36	5737	5203	4932	4756	4627	4526	4444	4375	4316	4264
38	5825	5298	5030	4855	4728	4628	4546	4478	4419	4368
40	5907	5387	5122	4949	4822	4724	4643	4575	4517	4466
42	5984	5470	5208	5037	4912	4814	4734	4667	4609	4558
44	6057	5549	5290	5120	4996	4899	4820	4753	4696	4646
46	6126	5624	5367	5199	5076	4980	4901	4835	4778	4729
48	6190	5694	5440	5274	5152	5057	4979	4913	4857	4808
50	6252	5761	5510	5345	5224	5130	5053	4988	4932	4883
					$1-\beta=0.99$					
2	0101	0051	0034	0025	0020	0017	0014	0013	0011	0010
4	0626	0434	0351	0302	0269	0245	0226	0211	0199	0189
6	1181	0907	0779	0701	0646	0605	0572	0545	0522	0503
8	1659	1339	1186	1089	1024	0968	0926	0891	0862	0837
10	2062	1717	1548	1440	1362	1303	1255	1215	1181	1152

Table 11.6 (Continued)

					$1-\beta=0.99$					
12	2407	2046	1867	1752	1668	1604	1552	1508	1472	1439
14	2704	2334	2149	2029	1942	1874	1820	1774	1734	1700
16	2966	2590	2401	2278	2188	2118	2061	2014	1973	1937
18	3197	2819	2627	2501	2410	2338	2280	2232	2190	2153
20	3404	3025	2831	2704	2612	2539	2480	2431	2388	2351
22	3591	3212	3017	2890	2796	2723	2663	2613	2570	2532
24	3761	3382	3188	3060	2966	2892	2832	2782	2738	2700
26	3916	3539	3344	3216	3122	3048	2988	2937	2894	2855
28	4059	3684	3490	3362	3268	3194	3133	3082	3038	3000
30	4191	3818	3635	3497	3403	3329	3268	3217	3173	3135
32	4314	3943	3750	3623	3529	3455	3395	3344	3300	3261
34	4428	4060	3868	3741	3648	3574	3513	3462	3418	3380
36	4535	4169	3979	3852	3759	3685	3625	3574	3530	3492
38	4636	4272	4089	3957	3864	3791	3730	3680	3636	3598
40	4730	4369	4181	4056	3963	3890	3830	3780	3736	3698
42	4819	4461	4274	4149	4057	3984	3925	3874	3831	3793
44	4903	4548	4362	4238	4146	4074	4014	3964	3921	3883
46	4983	4630	4445	4322	4231	4159	4100	4050	4007	3969
48	5059	4709	4525	4402	4312	4240	4181	4132	4089	4051
50	5131	4784	4601	4479	4389	4318	4259	4210	4167	4129

11.2 Multiple Comparisons

We know from Chapter 3 that in a statistical test concerning a parameter $\theta \in \Omega$, a null hypothesis $H_0 : \theta \in \omega$ is tested against an alternative hypothesis $H_A : \theta \in \Omega \backslash \omega$, and one has to decide between H_0 and H_A. If however the parameter space Ω is partitioned into more than two disjoint subsets $\omega_1,\ldots,\omega_r$, $\cup_{i=1}^{r} \omega_i = \Omega$,, we can call one of the hypotheses $H_i : \theta \in \omega_i$ null hypothesis. For instance, we can accept one (null) hypothesis $(H_1 : \theta \in \omega_1)$ or reject $(H_2 : \theta \in \omega_2)$ or make no statement $(H_3 : \theta \in \omega_3)$ (ω_3 is then an indifference zone).

Real multiple decision problems (with more than two decisions) are present. If results of some tests are considered simultaneously, their risks must be mutually evaluated. Of course we cannot give a full overview about methods available in this field. For more details, see Miller (1981) and Hsu (1996).

We restrict ourselves in this section on hypotheses about expectations of normal distributions. The set of populations $G = (P_1, \dots, P_a)$ can, for instance, be interpreted as levels in a simple analysis of variance model I or as level combinations in higher classifications.

Let the random variable $\boldsymbol{y}_i$ in P_i be $N(\mu_i, \sigma^2)$-distributed. We assume that from each populations P_i, an independent random sample $\boldsymbol{Y}_i^T = (\boldsymbol{y}_{i1},\dots,\boldsymbol{y}_{in_i})$ $(i = 1,\dots,a)$ of size n_i is obtained. Concerning the μ_i we consider several problems.

Problem 3 The null hypothesis

$$H_0 : \mu_1 = \mu_2 = \cdots = \mu_a$$

has to be tested against the alternative hypothesis

$$H_A : \text{there exists at least one pair } (i,j) \text{ with } i \neq j \text{ for that } \mu_i \neq \mu_j$$

with a given first kind risk α_e.

Problem 4 Each of the $\binom{a}{2}$ null hypotheses

$$H_{0ij} : \mu_i = \mu_j \quad (i \neq j; i, j = 1,\dots,a)$$

has to be tested against the alternative hypothesis

$$H_{Aij} : \mu_i \neq \mu_j,$$

where the first kind risk α_{ij} is given. Often we choose $\alpha_{ij} = \alpha$. If we perform $\binom{a}{2}$ t-tests, then we speak about the multiple t-procedure.

If for each $i \neq j$ the null hypothesis H_{0ij} is correct, then H_0 in Problem 3 is also correct. Therefore one is often interested in the probability $1 - \alpha_e$, that none of the $\binom{a}{2}$ null hypotheses H_{0ij} is wrongly rejected. We call the α_{ij} error probabilities per comparison (comparison-wise risk of the first kind) and α_e error probability per experiment (global error probability or experiment-wise risk of the first kind).

Problem 5 One of the populations (w.l.o.g. P_a) is prominent (a standard method, a control treatment and so on). Each of the $a - 1$ null hypotheses

$$H_{0i} : \mu_i = \mu_a \quad (i = 1,\dots,a-1)$$

has to be tested against the alternative hypothesis

$$H_{Ai} : \mu_i \neq \mu_a.$$

Again the first kind risk α_i is given in advance. Often we choose $\alpha_i = \alpha$. As in Problem 4 we like to know the probability $1 - \alpha_e$, that none of the $a - 1$ null hypotheses is wrongly rejected; again we call α_e the experiment-wise risk of the first kind.

If we use the term experiment-wise risk of the first kind in Problems 4 and 5, we must know that it is no first kind risk of a test. Instead α_e is the probability that at least one of the $\binom{a}{2}$ and $a - 1$ null hypotheses, respectively, is wrongly rejected. Let us consider all possible pairs (null hypothesis–alternative hypothesis) of Problem 4 or 5. Then we have a multiple decision problem with more than two possible decisions if $a > 2$.

In general α_e and α cannot be converted into each other. In Table 11.7 it is shown how α_e increases if the number of pairs of hypotheses k in Problem 4 or 5 is increasing. For calculating the values of Table 11.7, the asymptotic (for known σ^2) relations for k orthogonal contrast

$$\alpha_e = 1 - (1 - \alpha)^k, \tag{11.27}$$

$$\alpha = 1 - (1 - \alpha_e)^{1/k}, \tag{11.28}$$

have been used. (11.27) and (11.28) follow from elementary rules of probability theory, because we can assign to the independent contrasts independent F-tests (transformed z-tests) with $f_1 = 1$, $f_2 = \infty$ degrees of freedom.

Definition 11.4 A (linear) contrast L_r is a linear function

$$L_r = \sum_{i=1}^{a} c_{ri}\mu_i \text{ with the condition} \sum_{i=1}^{a} c_{ri} = 0.$$

Two linear contrasts L_u and L_v are called orthogonal if $\sum_{i=1}^{a} c_{ui}c_{vi} = 0$.

Before we solve Problems 3–5, we first construct confidence intervals for differences of expectation as well as for linear contrasts in these expectations. With these confidence intervals, the problems can be handled.

Most of the tables mentioned below can also be found in Rasch et al. (2008).

Program Hint

If data are present in an SPSS file, most of the methods below (and more) can be applied by the commands

Analyze

  Compare Means

    One-Way ANOVA

or for higher classifications by

Analyze
General Linear Model
Univariate

Table 11.7 Asymptotic relation between comparison-wise risk of the first kind (α) and experiment-wise risk of the first kind (α_e) for k orthogonal contrasts.

k	$10^4\alpha_e$ for $\alpha = 0.05$	$10^5\alpha$ for $\alpha_e = 0.05$	k	$10^4\alpha_e$ for $\alpha = 0.05$	$10^5\alpha$ for $\alpha_e = 0.05$
1	500	5000	15	5367	341
2	975	2532	20	6415	256
3	1426	1695	30	7854	171
4	1855	1274	50	9231	103
5	2262	1021	80	9835	64
6	2649	851	100	9941	51
7	3017	730	200	9999.6	26
8	3366	639	500	10000	10
9	3698	568	1000	10000	5
10	4013	512	5000	10000	1
12	4596	427			

If we use packages like R, SPSS or SAS, no tables of the quantiles are needed, because the packages give the correct significance value.

11.2.1 Confidence Intervals for All Contrasts: Scheffé's Method

As we know from Chapter 5, Problem 3 is solved by the F-test of the one-way analysis of variance.

Using the notation of Chapter 4, Problem 3 is with $\beta^T = (\mu, \alpha_1, \dots, \alpha_a)$ and X from Example 4.1 a special case of

$$H_0 : X\beta \in \omega \subset \Omega \text{ with } \dim(\omega) = 1, \quad \dim(\Omega) = a, \quad \Omega = R[X],$$

against

$$H_A : X\beta \notin \omega.$$

We reformulate Problem 1 as a problem to construct confidence intervals. If H_0 is correct, then all linear contrasts in the $\mu_i = \mu + \alpha_i$ equal zero. Conversely it follows that all linear contrasts vanish, the validity of H_0 (see Section 4.1.4).

Therefore confidence intervals $\boldsymbol{K}_r$ for all linear contrasts L_r can be constructed in such a way that the probability that $L_r \in \boldsymbol{K}_r$ for all r is at least $1 - \alpha_e$. We then reject H_0 with a first kind risk α_e if at least one of the $\boldsymbol{K}_r$ does not cover L_r.

The method proposed in Scheffé (1953) allows the calculation of simultaneous confidence intervals for all linear contrasts for β in Equation (5.1), lying in a subspace $\omega \subset \Omega$ where Ω is the rank space $R[X]$ of X in (4.1). The confidence coefficient $1 - \alpha_e$ is the probability that all linear contrasts in ω lie in the corresponding confidence interval. This confidence interval can easily be derived from Theorems 4.6 and 4.9 together with Example 4.4.

Theorem 11.6 We use model I of the analysis of variance in Definition 4.1. Further, let $k_i^T\beta(i=1,\dots,q)$ with $k_i^T = (k_{i1},\dots,k_{i,k+1})$ be estimable functions such that with the matrix $K = (k_1, \dots, k_q)^T = X^T T$ by $K^T\beta = 0$, a null hypothesis is given. For all vectors $c \in R[K]$ with $\text{rk}(K) = q$ and $\text{rk}(X) = \dim(\Omega) = p$

$$\left\{\left[c^T\boldsymbol{\beta}^* - \boldsymbol{G}, c^T\boldsymbol{\beta}^* + \boldsymbol{G}\right]\right\} \tag{11.29}$$

a class of simultaneous confidence intervals for the $c^T\beta$ with confidence coefficient $1 - \alpha_e$ is defined, if we put

$$\boldsymbol{G}^2 = qF(q, N-p|1-\alpha_e)\boldsymbol{s}^2 c^T\left(X^TX\right)^- c$$

with $\boldsymbol{\beta}^*$ from (5.3) and

$$\boldsymbol{s}^2 = \frac{1}{n-p}\boldsymbol{Y}^T\left[I_N - X\left(X^TX\right)^- X^T\right]\boldsymbol{Y}.$$

Proof: We apply Theorem 11.9 and Formula (11.23) and put $\theta = X\beta$ (by (5.1)). Then with $T^TX = K^T$ by

$$\left(\boldsymbol{\beta}^{*T}X^TT - \beta^TX^TT\right)\left[T^TX\left(X^TX\right)^-X^TT\right]^{-1}\left(T^TX\boldsymbol{\beta}^* - T^TX\beta\right)$$
$$\le q\boldsymbol{s}^2F(q, N-p|1-\alpha_e),$$

a confidence interval with confidence coefficient $1 - \alpha_e$ for $K^T\beta$ is given. Therefore, all (estimable) linear combinations $c^T\beta$ lie with probability $1 - \alpha_e$ in the interval given by (11.29).

Example 11.2 We use Scheffé's method to test the null hypothesis of Problem 3 for the one-way analysis of variance in Example 5.1. We have $q = a - 1$, $p = a = \text{rk}(X)$, $\beta^T = (\mu, \alpha_1, \dots, \alpha_a)$, and considering all linear contrasts L_r in the μ_i, (11.29) becomes

$$\left[\hat{\boldsymbol{L}}_r - \boldsymbol{s}\sqrt{(a-1)F(a-1, N-a|1-\alpha_e)}\sqrt{\sum_{i=1}^{a}\frac{c_{ri}^2}{n_i}},\right.$$
$$\left.\hat{\boldsymbol{L}}_r + \boldsymbol{s}\sqrt{(a-1)F(a-1, N-a|1-\alpha_e)}\sqrt{\sum_{i=1}^{a}\frac{c_{ri}^2}{n_i}},\right] \tag{11.30}$$

and (11.30) contains all L_r with $\sum_{i=1}^{a} c_{ri} = 0$ with probability $1 - \alpha_e$. Here is $s = \sqrt{MS_{res}}$ (from Table 5.2).

From Lemma 5.1 it follows that all differences $\mu_i - \mu_{i'}$ and the linear contrast in $\mu + \alpha_i$ are estimable. Using (11.30) to construct confidence intervals for all $\binom{a}{2}$ differences of expectations only, then the confidence interval in (11.30) is too large and contains the differences of expectations with a probability $\geq 1 - \alpha_e$. We say in such cases that the confidence intervals and the corresponding tests are conservative.

Example 11.3 We consider a two-way cross-classification with model equation (5.13), that is, we assume interactions and put $n_{ij} = n$ for all $i = 1, \ldots, a$; $j = 1, \ldots, b$. We denote $\mu + \alpha_i (i = 1, \ldots, a)$ as row-means and $\mu + \beta_j$ $(j = 1, \ldots, b)$ as column 'means'. If the null hypothesis in Problem 3 has to be tested against the corresponding alternative hypothesis for the row-means, we obtain from (11.29) the confidence interval with confidence coefficient $1 - \alpha_e$

$$\left.\begin{array}{l}[\hat{\boldsymbol{L}}_{zr} - \boldsymbol{A}, \hat{\boldsymbol{L}}_{zr} + \boldsymbol{A}] \\ \text{with } \boldsymbol{A} = \boldsymbol{s}\sqrt{(a-1)F[a-1, ab(n-1)|1-\alpha_e]}\sqrt{\dfrac{1}{bn}\displaystyle\sum_{i=1}^{a} c_{ri^2}}\end{array}\right\} \tag{11.31}$$

for an arbitrary (but fixed) linear contrast in the row-means:

$$L_{zr} = \sum_{i=1}^{a} l_{ri}(\mu + \alpha_i).$$

If

$$L_{sr} = \sum_{j=1}^{b} l_{sj}\left(\mu + \beta_j\right)$$

correspondingly is a linear contrast in the column-means, so it is analogous to (11.31),

$$\left.\begin{array}{l}[\hat{\boldsymbol{L}}_{sr} - \boldsymbol{B}, \hat{\boldsymbol{L}}_{sr} + \boldsymbol{B},] \\ \text{with } \boldsymbol{B} = \boldsymbol{s}\sqrt{(b-1)F[b-1, ab(n-1)|1-\alpha_e]}\sqrt{\dfrac{1}{an}\displaystyle\sum_{j=1}^{b} c_{sj}^2}\end{array}\right\} \tag{11.32}$$

a confidence interval with confidence coefficient $1-\alpha_e$ for an arbitrary (but fixed) linear contrast L_{sr}.

For estimable functions in ω_{ij}, we analogously can construct confidence intervals.

In (11.31) and (11.32) $s^2 = MS_{res}$ is given in Table 5.13. From Lemma 5.1 it follows that all differences and linear contrasts between the row 'means' or between the column 'means' are estimable functions and (11.31) and (11.32) can be used. If only differences of the row 'means' or column 'means' are of interest, remarks at the end of Example 11.2 are again valid.

From Theorem 11.6 it follows that by Scheffé's method, simultaneous confidence intervals for all linear combinations $c^T\beta$ with $c \in R[K]$ can be constructed.

Of course, if only differences in expectations are of interest, confidence intervals with Scheffé's method have a too large expected width, and the power of the corresponding tests is too small. In those cases, Scheffé's method will not be applied. To show this, we consider an example for this and competing methods.

Example 11.4 A (pseudo-)random number generator has generated ten samples of size five each. The values of the samples 1–8 are realisations of an $N(50, 64)$ normally distributed random variables; the two other samples differ only in expectations. We have $\mu_9 = 52$ and $\mu_{10} = 56$, respectively. The generated samples are shown in Table 11.8.

Differences between means are given in Table 11.9.

Table 11.8 Simulated observations of Example 11.4.

	Number of sample									
y_{ij}	1	2	3	4	5	6	7	8	9	10
y_{i1}	63.4	49.6	50.3	55.5	62.5	30.7	56.7	64.5	44.4	55.7
y_{i2}	46.7	48.4	52.8	36.1	45.8	48.6	46.2	42.2	38.2	64.7
y_{i3}	59.1	49.3	52.5	54.0	52.8	45.8	41.9	49.6	64.8	61.8
y_{i4}	60.7	48.3	58.6	55.9	44.9	44.9	55.8	48.9	43.7	38.9
y_{i5}	54.9	51.5	48.0	52.9	51.3	52.9	48.9	40.7	61.3	61.8
$\bar{y}_{i.}$	56.96	49.42	52.44	50.88	51.46	44.58	49.90	49.18	50.48	56.58

Table 11.9 Differences $\left(\bar{y}_{i.} - \bar{y}_{j.}\right)$ between means of Example 11.4.

j \ i	2	3	4	5	6	7	8	9	10
1	7.54	4.52	6.08	5.50	12.38	7.06	7.78	6.48	0.38
2		−3.02	−1.46	−2.04	4.84	−0.48	0.24	−1.06	−7.16
3			1.56	0.98	7.86	2.54	3.26	1.96	−4.14
4				−0.58	6.30	0.98	1.70	0.40	−5.70
5					6.88	1.56	2.28	0.98	−5.12
6						−5.32	4.60	−5.90	−12.00
7							0.72	−0.58	−6.68
8								−1.30	−7.40
9									−6.10

The analysis was done with SPSS via

Analyze
Compare Means
One-Way ANOVA

and define sample as factor and the observations as dependent as in Figure 11.1

If we continue in Figure 11.1 with ok, we obtain the results of a one-way analysis of variance with ten samples as factor levels in Figure 11.2.

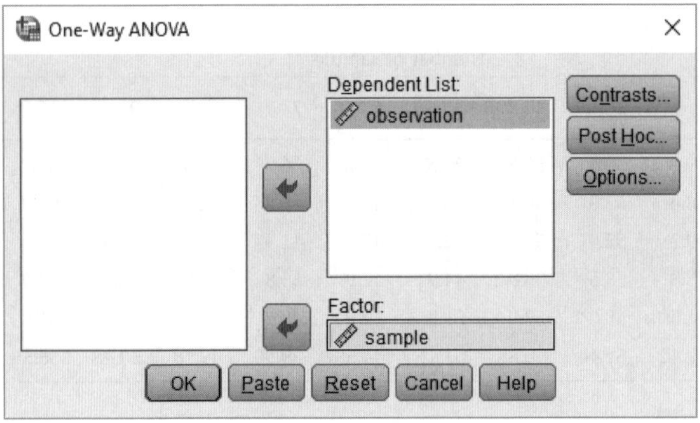

Figure 11.1 Program start of Example 11.4 in SPSS. *Source:* Reproduced with permission of IBM.

Because $F = 1.041 < F(9.40|0.95)$ (with $\alpha_e = 0.05$), H_0 in Problem 11.1 is accepted.

If in Figure 11.1 we press 'post hoc', we can select Scheffé's method in Figure 11.3 (page 542).

If we continue and press 'OK', we obtain Table 11.10.

ANOVA

Observation

	Sum of squares	*df*	Mean square	*F*	Sig.
Between groups	585.549	9	65.061	1.041	.426
Within groups	2500.784	40	62.520		
Total	3086.333	49			

Figure 11.2 Output of ANOVA of Example 11.4. *Source:* Reproduced with permission of IBM.

Table 11.10 Confidence intervals by Scheffé's method of Example 11.4 (shortened).

Multiple comparisons
Dependent variable: observation
Scheffe

					95% Confidence interval	
(I) sample	**(J) sample**	**Mean difference (I–J)**	**Std. error**	**Sig.**	**Lower bound**	**Upper bound**
1	2	7.5400	5.0008	0.983	−14.325	29.405
	3	4.5200	5.0008	1.000	−17.345	26.385
	4	6.0800	5.0008	0.997	−15.785	27.945
	5	5.5000	5.0008	0.998	−16.365	27.365
	6	12.3800	5.0008	0.721	−9.485	34.245
	7	7.0600	5.0008	0.990	−14.805	28.925
	8	7.7800	5.0008	0.980	−14.085	29.645
	9	6.4800	5.0008	0.994	−15.385	28.345
	10	0.3800	5.0008	1.000	−21.485	22.245
2	1	−7.5400	5.0008	0.983	−29.405	14.325
	3	−3.0200	5.0008	1.000	−24.885	18.845
	4	−1.4600	5.0008	1.000	−23.325	20.405
	5	−2.0400	5.0008	1.000	−23.905	19.825
	6	4.8400	5.0008	0.999	−17.025	26.705
	7	−.4800	5.0008	1.000	−22.345	21.385
	8	0.2400	5.0008	1.000	−21.625	22.105
	9	−1.0600	5.0008	1.000	−22.925	20.805
	10	−7.1600	5.0008	0.988	−29.025	14.705

Source: Reproduced with permission of IBM.

11.2.2 Confidence Intervals for Given Contrasts: Bonferroni's and Dunn's Method

Confidence intervals by Scheffé's method are not appropriate if confidence intervals for k special but not for all contrasts are wanted. Sometimes shorter intervals can be obtained using the Bonferroni inequality.

Theorem 11.7 If the k components $\boldsymbol{x}_i$ of k-dimensional random variables $\boldsymbol{x}^T = (\boldsymbol{x}_1, \dots, \boldsymbol{x}_k)$ with distribution function $F(x_1, \dots, x_k)$ have the same marginal distribution functions $F(x)$, then the Bonferroni inequality

$$1 - F(x_1,\dots,x_k) \le \sum_{i=1}^{k} [1 - F(x_i)]. \tag{11.33}$$

is valid.

Proof: Given k events $A_1, A_2, \dots, A_k$ of the probability space $(A, \mathfrak{B}_A, P)$, that is, let $A_i \in \mathfrak{B}_A (i = 1,\dots,k)$. Then by mathematical induction it follows from inclusion and exclusion that

$$P\left(\bigcup_{i=1}^{k} A_i\right) \le \sum_{i=1}^{k} P(A_i).$$

If $A_i = \{\boldsymbol{x}_i < x_i\}$, then because $\overline{\cap_{i=1}^{k} A_i} = \cup_{i=1}^{k} \bar{A}_i$ (11.33) follows.

If k special linear contrast $L_r = \sum_{j=1}^{a} c_{rj}\mu_j$ $(r = 1, \dots, k)$ is given, then the estimator $\hat{\boldsymbol{L}}_r = \sum c_{rj}\bar{\boldsymbol{y}}_j$ is for each r $N(L_r, k_r\sigma^2)$-distributed with $k_r = \sum_{i=1}^{a} \frac{c_{ri}^2}{n_i}$. Then

$$\boldsymbol{t}_r = \frac{\hat{\boldsymbol{L}}_r - L_r}{\sqrt{k_r}\boldsymbol{s}} \quad (r = 1,\dots,k) \tag{11.34}$$

with $\boldsymbol{s} = \sqrt{\boldsymbol{MS}_{res}}$ are components of a k-dimensional random variable. The marginal distributions are central t-distributions with $\nu = \sum_{i=1}^{a}(n_i - 1)$ degrees of freedom and the density $f(t, \nu)$.

The Bonferroni inequality allows us to find a lower bound of the probability that all t_r-values $(r = 1, \dots, k)$ lie between $-w$ and w $(w > 0)$. Due to the symmetry of t-distribution and Theorem 11.7, we get

$$P = P\{-w \le \boldsymbol{t}_r < w | r = 1,\dots,k\} \ge 1 - 2k\int_{w}^{\infty} f(t,\nu)\mathrm{d}t. \tag{11.35}$$

We select w so that the right-hand side of (11.35) equals $(1 - \alpha_e)$ and obtain simultaneous $(1 - \alpha_e)$ confidence intervals for the L_r as

$$\left[\hat{\boldsymbol{L}}_r - w\sqrt{k_r}\boldsymbol{s}, \hat{\boldsymbol{L}}_r + w\sqrt{k_r}\boldsymbol{s}\right].$$

Table 11.11 $\left(1-\frac{0.05}{2k}\right)$-Quantiles of the central *t*-distribution with *f* degrees of freedom.

	k								
f	2	3	4	5	6	7	8	9	10
5	3.163	3.534	3.810	4.032	4.219	4.382	4.526	4.655	5.773
6	2.969	3.287	3.521	3.707	3.863	3.997	4.115	4.221	4.317
7	2.841	3.128	3.335	3.499	3.636	3.753	3.855	3.947	4.029
8	2.752	3.016	3.206	3.355	3.479	3.584	3.677	3.759	3.832
9	2.685	2.933	3.111	3.250	3.364	3.462	3.547	3.622	3.690
10	2.634	2.870	3.038	3.169	3.277	3.368	3.448	3.518	3.581
11	2.593	2.820	2.981	3.106	3.208	3.295	3.370	3.437	3.497
12	2.560	2.779	2.934	3.055	3.153	3.236	3.308	3.371	3.428
15	2.490	2.694	2.837	2.947	3.036	3.112	3.177	3.235	3.286
20	2.423	2.613	2.744	2.845	2.927	2.996	3.055	3.107	3.153
30	2.360	2.536	2.657	2.750	2.825	2.887	2.941	2.988	3.030
40	2.329	2.499	2.616	2.704	2.776	2.836	2.887	2.931	2.971
50	2.311	2.477	2.591	2.678	2.747	2.805	2.855	2.898	2.937
60	2.299	2.463	2.575	2.660	2.729	2.786	2.834	2.877	2.915
80	2.284	2.445	2.555	2.639	2.705	2.761	2.809	2.850	2.887
100	2.276	2.435	2.544	2.626	2.692	2.747	2.793	2.834	2.871
∞	2.241	2.394	2.498	2.579	2.638	2.690	2.734	2.773	2.807

This means we determine w so that

$$\int_{w}^{\infty} f(t,\nu)\mathrm{d}t = \frac{\alpha_e}{2k} = \alpha \tag{11.36}$$

and the Bonferroni inequality (11.33) has the form

$$P > 1-\alpha_e \geq 1-2k\alpha. \tag{11.37}$$

For $\alpha_e = 0.05$ these w-values $w(k,f,0.95)$ for some k and f are given in Table 11.11.

Dunn (1961) published a table with cases for which his method was better than Scheffé's method. If, among the k contrasts, all $\binom{a}{2}$ differences of the expectations can be found, then Ury and Wiggins (1971, 1974) gave a modification and corresponding tables (but see Rodger, 1973).

11.2.3 Confidence Intervals for All Contrasts for $n_i = n$: Tukey's Method

Definition 11.5 If $Y = (y_1, \dots, y_a)^T$ is a random sample with independent components and $N(\mu, \sigma^2)$-distributed and if $\nu s^2/\sigma^2$ is independent of Y $CS(\nu)$-distributed, we call the random variable

$$q_{a,\nu} = \frac{w}{s}$$

the studentised range of Y, if $w = \max_{i=1,\dots,a}(y_i) - \min_{i=1,\dots,a}(y_i)$ is the range of Y.

The augmented studentised range is the random variable

$$q^*_{a,\nu} = \frac{1}{s}[\max(w, \max\{|y_i - \mu|\})].$$

Tukey's method is based on the distribution of $q_{a,\nu}$. We can show that the distribution function of the studentised range $q_{a,\nu}$ is given by

$$\frac{2a}{\Gamma\left(\frac{\nu}{2}\right)}\left(\frac{\nu}{2}\right)^{\frac{\nu}{2}}\int_0^\infty\int_{-\infty}^\infty \varphi(z)[\Phi(z) - \Phi(z - q_{a,\nu}x)]^{a-1}x^{\nu-1}e^{-\frac{\nu x^2}{2}}dzdx. \qquad (11.38)$$

In (11.38) $x = s/\sigma$, $\varphi(z)$ is the density function and $\Phi(z)$ the distribution function of the standardised normal distribution.

We denote by $q(a, \nu|1 - \alpha_\nu)$ the $(1 - \alpha)$-quantile of the distribution function of $q_{a,\nu}$ in (11.38), which depends on the number a of components of Y and the degrees of freedom of s^2 in Definition 11.5.

Tukey's method (1953) to construct confidence intervals for the differences $\mu_i - \mu_{i'}$ between the expectations of a independent $N(\mu_i, \sigma^2)$-distributed random variables $y_i (i = 1, \dots, a)$ is based on the equivalence of the probabilities

$$P\left\{\frac{1}{s}[(y_i - y_{ik}) - (\mu_i - \mu_k)] \le K \text{ for all } i \ne k; i,k = 1,\dots,a\right\}$$

and

$$P\left\{\frac{1}{s}\max_{i,k}[y_i - \mu_i - (y_k - \mu_k)] \le K \ (i,k = 1,\dots,a)\right\}.$$

This equivalence is a consequence of the fact that the validity of the inequality in the second term is necessary and sufficient for the validity of the inequality in the first term. The maximum of a set of random variables is understood as its largest-order statistic, and

$$\max_{\substack{i,k \\ i,k=1,\dots,a}} [y_i - \mu_i - (y_k - \mu_k)]$$

is the range w of $N(0, \sigma^2)$-distributed random variables; if y_i are independent of each other, it is $N(\mu_i, \sigma^2)$-distributed. From this it follows Theorem 11.8.

Theorem 11.8 If $y_1, \dots, y_a$ are independently $N(\mu_i, \sigma_i^2)$-distributed random variables $(i = 1, \dots, a)$ with $\sigma_i^2 = \sigma^2$ and $\boldsymbol{s}^2/\sigma^2$ is independent of the $\boldsymbol{y}_i$ $(i = 1, \dots, a)$ $CS(f)$-distributed, then

$$P\{|(\boldsymbol{y}_i - \boldsymbol{y}_{ik}) - (\mu_i - \mu_k)| \le q(a, f|1 - \alpha_e)\boldsymbol{s} \ \ (i, k = 1, \dots, a)\} = 1 \ \ \alpha_e. \tag{11.39}$$

Therefore by (11.39) a class of simultaneous confidence intervals with confidence coefficient $1 - \alpha_e$ is given.

The results of Theorem 11.8 are shown in two examples.

Example 11.5 Tukey's method is used to construct confidence intervals for differences in expectations and to test the first problem in one-way analysis of variance (see Example 5.1). For this we have to assume $n_i = n$. $\bar{\boldsymbol{y}}_{1\cdot}, \dots, \bar{\boldsymbol{y}}_{a}$ are the means of the observations $\boldsymbol{y}_{ij}$ of the a factor levels. For the differences $\mu + \alpha_i - (\mu + \alpha_{i'}) = \alpha_i - \alpha_{i'}$, simultaneous confidence intervals can be constructed with (11.39). For $i = 1, \dots, n$ it holds $\operatorname{var}(\bar{\boldsymbol{y}}_{i\cdot}) = \frac{1}{n}\sigma^2$. We estimate σ^2 by $\boldsymbol{MS}_{res} = \boldsymbol{s}^2$ in Table 4.2 $(n_i = n)$, that is,

$$\boldsymbol{s}^2 = \frac{1}{a(n-1)}\left[\sum_{i=1}^{a}\sum_{j=1}^{n}\boldsymbol{y}_{ij}^2 - \frac{1}{n}\sum_{i=1}^{a}\boldsymbol{Y}_{i\cdot}^2\right].$$

Now $\frac{a(n-1)}{\sigma^2}\boldsymbol{s}^2$ is $CS[a(n-1)]$-distributed and independent of the $\boldsymbol{y}_{i.} - \boldsymbol{y}_{i'\cdot}$. From Theorem 11.8 with $f = a(n-1)$ and $\frac{\sigma^2}{n}$ for σ^2, we obtain the class of simultaneous confidence intervals with confidence coefficient $1 - \alpha_e$ for $\mu_i - \mu_{i'}$:

$$\left[\bar{\boldsymbol{y}}_{i\cdot} - \bar{\boldsymbol{y}}_{k\cdot} - q[a, a(n-1)|1-\alpha_e]\frac{\boldsymbol{s}}{\sqrt{n}}, \bar{\boldsymbol{y}}_{i\cdot} - \bar{\boldsymbol{y}}_{k\cdot} + q[a, a(n-1)|1-\alpha_e]\frac{\boldsymbol{s}}{\sqrt{n}}\right]$$
$$(i \ne k; i, k = 1, \dots, a). \tag{11.40}$$

Example 11.6 Analogously to Example 11.3 we consider the two-way cross-classification with model equation (5.13) and construct simultaneous confidence intervals for the differences between the row 'means' and column 'means' introduced in Example 11.3. Again, let $n = n_{ij}$ for all (i, j). Of course, this is a limitation of the method.

For the row 'means' we have $\operatorname{var}(\bar{\boldsymbol{y}}_{i\cdot\cdot}) = \sigma^2/(bn)$ and for the column 'means' $\operatorname{var}\left(\bar{\boldsymbol{y}}_{\cdot j\cdot}\right) = \sigma^2/(an)$. With $\boldsymbol{s}^2$ from Table 5.13 and $f = ab(n-1)$ from Theorem 11.8, it follows the class of simultaneous intervals for $\mu + \alpha_i - (\mu + \alpha_k) = \alpha_i - \alpha_k$ with confidence coefficient $(1 - \alpha_e)$:

$$\left[\bar{\boldsymbol{y}}_{i\cdot\cdot} - \bar{\boldsymbol{y}}_{k\cdot\cdot} - q[a, ab(n-1)|1-\alpha_e]\frac{\boldsymbol{s}}{\sqrt{bn}}, \bar{\boldsymbol{y}}_{i\cdot\cdot} - \bar{\boldsymbol{y}}_{k\cdot\cdot} + q[a, ab(n-1)|1-\alpha_e]\frac{\boldsymbol{s}}{\sqrt{bn}}\right]$$
$$(i \ne k; i, k = 1, \dots, a) \tag{11.41}$$

and analogous to the column 'means', the class of confidence intervals

$$\left[\bar{\boldsymbol{y}}_{\cdot j\cdot} - \bar{\boldsymbol{y}}_{\cdot k\cdot} - q[b, ab(n-1)|1-\alpha_e]\frac{s}{\sqrt{an}}, \bar{\boldsymbol{y}}_{\cdot j\cdot} - \bar{\boldsymbol{y}}_{\cdot k\cdot} + q[b, ab(n-1)|1-\alpha_e]\frac{s}{\sqrt{an}}\right]$$

$$(j \neq k; j, k = 1, \ldots, b). \tag{11.42}$$

We can show that for any contrast $L = \sum_{i=1}^{a} c_i \mu_i$, c_i real, in generalisation of (11.39) with

$$\hat{\boldsymbol{L}} = \sum_{i=1}^{a} c_i \bar{\boldsymbol{y}}_{i\cdot}$$

the relation

$$1-\alpha_e = P\left\{\hat{\boldsymbol{L}} - \frac{s}{\sqrt{n}} q(a, f|1-\alpha_e)\frac{1}{2}\sum_{i=1}^{a}|c_i| < L < \hat{\boldsymbol{L}} + \frac{s}{\sqrt{n}} q(a, f|1-\alpha_e)\frac{1}{2}\sum_{i=1}^{a}|c_i|\right\} \tag{11.43}$$

holds for all L.

If only the set of $\binom{a}{2}$ differences in expectations $\mu_i - \mu_j$ $(i \neq j; i, j = 1, \ldots, a)$ is considered, Tukey's method gives smaller simultaneous confidence intervals as the Scheffé's method. Tukey's method is then preferable if $n_i = n$ is given. We continue with Example 11.4 by using now in Figure 11.3 the button 'Tukey'. Analogously to Table 11.10 we receive Table 11.12.

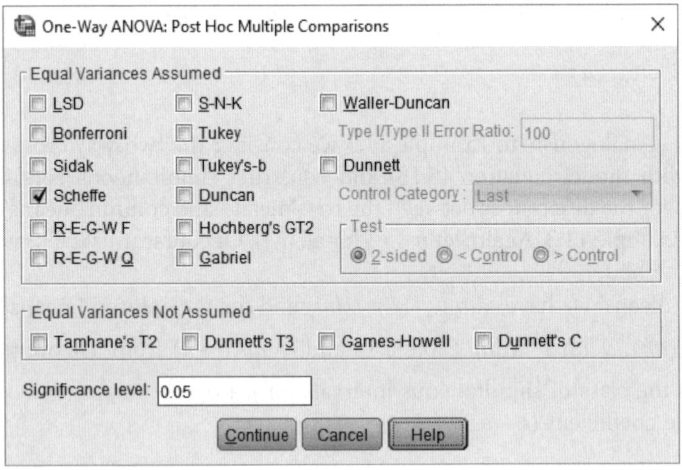

Figure 11.3 Post hoc for multiple comparisons of means in SPSS. *Source:* Reproduced with permission of IBM.

Table 11.12 Confidence intervals by Tukey's method of Example 11.4 (shortened).

Multiple comparisons Dependent variable Tukey HSD						
					95% Confidence interval	
(I) sample	(J) sample	Mean difference (I–J)	Std. error	Sig.	Lower bound	Upper bound
1	2	7.54000	5.00078	0.881	−9.2017	24.2817
	3	4.52000	5.00078	0.995	−12.2217	21.2617
	4	6.08000	5.00078	0.965	−10.6617	22.8217
	5	5.50000	5.00078	0.982	−11.2417	22.2417
	6	12.38000	5.00078	0.312	−4.3617	29.1217
	7	7.06000	5.00078	0.916	−9.6817	23.8017
	8	7.78000	5.00078	0.861	−8.9617	24.5217
	9	6.48000	5.00078	0.949	−10.2617	23.2217
	10	0.38000	5.00078	1.000	−16.3617	17.1217
2	1	−7.54000	5.00078	.881	−24.2817	9.2017
	3	−3.02000	5.00078	1.000	−19.7617	13.7217
	4	−1.46000	5.00078	1.000	−18.2017	15.2817
	5	−2.04000	5.00078	1.000	−18.7817	14.7017
	6	4.84000	5.00078	0.993	−11.9017	21.5817
	7	−0.48000	5.00078	1.000	−17.2217	16.2617
	8	0.24000	5.00078	1.000	−16.5017	16.9817
	9	−1.06000	5.00078	1.000	−17.8017	15.6817
	10	−7.16000	5.00078	0.910	−23.9017	9.5817

Source: Reproduced with permission of IBM.

11.2.4 Confidence Intervals for All Contrasts: Generalised Tukey's Method

Spjøtvoll and Stoline (1973) generalised Tukey's method of Section 11.2.3 without assuming $n_i = n$.

Definition 11.6 Given a independent $N(\mu, \sigma^2)$-distributed random variables $\boldsymbol{y}_i$ $(i = 1, \dots, a)$, let $\boldsymbol{s}^2$ be an estimator of σ^2 independent of the $\boldsymbol{y}_i$ with ν degrees of freedom. Then the random variable

$$\frac{\max\limits_{1 \le i \le a}\{|\boldsymbol{y}_i - \mu|\}}{\boldsymbol{s}} = \boldsymbol{q}^{**}(a, \nu)$$

is called the augmented studentised range of $\boldsymbol{y}_i$ with ν degrees of freedom.

Theorem 11.9 (Spjøtvoll and Stoline)

If all conditions of Theorem 11.8 are fulfilled, all linear contrasts $L = \sum_{i=1}^{a} c_i \mu_i$ simultaneously are covered with probability $1 - \alpha_e$ by intervals

$$\left[\hat{\boldsymbol{L}} - \frac{1}{2}\sum |c_i| q^*(a,f|1-\alpha_e)\boldsymbol{s}, \hat{\boldsymbol{L}} + \frac{1}{2}\sum |c_i| q^*(a,f|1-\alpha_e)\boldsymbol{s}\right]. \tag{11.44}$$

In (11.44) $\hat{\boldsymbol{L}} = \sum_{i=1}^{a} c_i \boldsymbol{y}_{i.}$, and $q^*(a,f|1-\alpha)$ is the $(1-\alpha)$-quantile of the distribution of the augmented studentised range $\boldsymbol{q}^{**}(a,f)$ corresponding to Definition 11.6.

The proof is given in Spjøtvoll and Stoline (1973). It is based on the transition to random variables $\boldsymbol{x}_i = \frac{1}{\sigma_i}\boldsymbol{y}_i$, having the same variance and restoring the problem to that handled in Section 11.2.3. Spjøtvoll and Stoline approximate the quantiles $q^*(a,f|1-\alpha_e)$ by the quantiles $q(a,f|1-\alpha_e)$ of the studentised range, but Stoline (1978) gave tables of $q^*(a,f|1-\alpha_e)$.

The generalised Tukey's method gives as well shorter as also larger confidence intervals than the Scheffé's method, depending on the degree of unbalancedness.

A further generalisation of Tukey's method can be found in Hochberg (1974) and Hochberg and Tamhane (1987).

Theorem 11.10 Theorem 11.9 is still valid if (11.44) is replaced by

$$\left[\hat{\boldsymbol{L}} - \boldsymbol{s}\sum \frac{|c_i|}{\sqrt{2n_i}} q^{**}\left(\binom{a}{2}, f|1-\alpha_e\right), \hat{\boldsymbol{L}} + \boldsymbol{s}\sum \frac{|c_i|}{\sqrt{2n_i}} q^{**}\left(\binom{a}{2}, f|1-\alpha_e\right)\right]. \tag{11.45}$$

Here $q^{**}\left[\binom{a}{2}, f|1-\alpha_e\right]$ is the quantile of the distribution of $\boldsymbol{q}^{**}\left(\binom{a}{2}, f\right)$ in Definition 11.6. These quantiles are given in Stoline and Ury (1979).

11.2.5 Confidence Intervals for the Differences of Treatments with a Control: Dunnett's Method

Sometimes $a - 1$ treatments have to be compared with a standard procedure called control.

Then simultaneous $1 - \alpha_e$ confidence intervals for the $a - 1$ differences

$$\mu_i - \mu_a \quad (i = 1, \dots, a-1)$$

shall be constructed. (After renumbering μ_a is always the expectation of the control.)

We consider a independent $N(\mu_i, \sigma^2)$-distributed random variables $\boldsymbol{y}_i$ independent of a $CS(f)$-distributed random variable $\frac{f\boldsymbol{s}^2}{\sigma^2}$.

Dunnett (1955) derived the distribution of

$$\left(\frac{\boldsymbol{y}_1 - \boldsymbol{y}_a}{\boldsymbol{s}}, \ldots, \frac{\boldsymbol{y}_{a-1} - \boldsymbol{y}_a}{\boldsymbol{s}}\right)$$

Dunnett (1964) and Bechhofer and Dunnett (1988) present the quantiles $d(a-1, f|1-\alpha_e)$of the distribution of

$$\boldsymbol{d} = \frac{\max\limits_{1 \le i \le a-1} \left[|\boldsymbol{y}_i - \boldsymbol{y}_a - (\mu_i - \mu_a)| \right]}{\boldsymbol{s}\sqrt{2}}.$$

We see that $\boldsymbol{d} \le d(a-1, f|1-\alpha_e)$ is necessary and sufficient for

$$\frac{1}{\boldsymbol{s}\sqrt{2}} |\boldsymbol{y}_i - \boldsymbol{y}_a - (\mu_i - \mu_a)| \le d(a-1, f|1-\alpha_e).$$

For all i by

$$\left[\boldsymbol{y}_i - \boldsymbol{y}_a - d(a-1, f|1-\alpha_e)\boldsymbol{s}\sqrt{2}, \boldsymbol{y}_i - \boldsymbol{y}_a + d(a-1, f|1-\alpha_e)\boldsymbol{s}\sqrt{2}\right], \tag{11.46}$$

a class of confidence intervals is given, covering all differences $\mu_i - \mu_a$ with probability $1 - \alpha_e$.

For the one-way classification with the notation of Example 11.5, we receive for equal subclass numbers n the class of confidence intervals:

$$\left[\bar{\boldsymbol{y}}_{i\cdot} - \bar{\boldsymbol{y}}_{a\cdot} - d(a-1, a(n-1)|1-\alpha_e)\boldsymbol{s}\frac{\sqrt{2n}}{n}, \bar{\boldsymbol{y}}_{i\cdot} - \bar{\boldsymbol{y}}_{a\cdot} + d(a-1, a(n-1)|1-\alpha_e)\boldsymbol{s}\frac{\sqrt{2n}}{n}\right]$$
$$(i = 1, \ldots, a-1). \tag{11.47}$$

For the two-way cross-classification (model (5.13)) with the notation of Example 11.6, we receive for equal subclass numbers the class of confidence intervals for the row 'means':

$$\left[\bar{\boldsymbol{y}}_{i\cdot\cdot} - \bar{\boldsymbol{y}}_{a\cdot\cdot} - d(a-1, ab(n-1)|1-\alpha_e)\boldsymbol{s}\frac{\sqrt{2n}}{bn}, \; \bar{\boldsymbol{y}}_{i\cdot\cdot} - \bar{\boldsymbol{y}}_{a\cdot\cdot} + d(a-1, ab(n-1)|1-\alpha_e)\boldsymbol{s}\frac{\sqrt{2n}}{bn}\right]$$
$$(i = 1, \ldots, a-1) \tag{11.48}$$

and for the column 'means'

$$\left[\bar{\boldsymbol{y}}_{\cdot i\cdot} - \bar{\boldsymbol{y}}_{\cdot b\cdot} - d(b-1, ab(n-1)|1-\alpha_e)\boldsymbol{s}\frac{\sqrt{2n}}{an}, \bar{\boldsymbol{y}}_{\cdot i\cdot} - \bar{\boldsymbol{y}}_{\cdot b\cdot} + d(b-1, ab(n-1)|1-\alpha_e)\boldsymbol{s}\frac{\sqrt{2n}}{an}\right]$$
$$(i = 1, \ldots, b-1). \tag{11.49}$$

We continue with Example 11.4 using now in Figure 11.3 'Dunnett' and receive in Table 11.13 analogous to Tables 11.10 and 11.12 taking control as the last (10-th) sample.

Table 11.13 Risk of the first kind multiple comparisons.

Dependent variable: observation						
Dunnett *t* (2-sided) a						
					95% Confidence interval	
(I) sample	(J) sample	Mean difference (I–J)	Std. error	Sig.	Lower bound	Upper bound
1	10	0.38000	5.00078	1.000	−13.6810	14.4410
2	10	−7.16000	5.00078	0.629	−21.2210	6.9010
3	10	−4.14000	5.00078	0.964	−18.2010	9.9210
4	10	−5.70000	5.00078	0.829	−19.7610	8.3610
5	10	−5.12000	5.00078	0.893	−19.1810	8.9410
6	10	−12.00000	5.00078	0.125	−26.0610	2.0610
7	10	−6.68000	5.00078	0.698	−20.7410	7.3810
8	10	−7.40000	5.00078	0.594	−21.4610	6.6610
9	10	−6.10000	5.00078	0.778	−20.1610	7.9610
Dunnett *t*-tests treat one group as a control and compare all other groups against it						

Source: Reproduced with permission of IBM.

11.2.6 Multiple Comparisons and Confidence Intervals

We now discuss the problems given at the start of Section 11.2. Let P_i be the a levels of the factor in a one-way analysis of variance model I. Independent of random samples $\boldsymbol{Y}_i$, we have for the components the model equation

$$\boldsymbol{y}_{ij} = \mu_i + \boldsymbol{e}_{ij} \quad (i = 1,\dots,a; j = 1,\dots,n_i) \tag{11.50}$$

with error terms $\boldsymbol{e}_{ij}$ that are $N(0, \sigma^2)$-distributed.

$\boldsymbol{s}^2 = \boldsymbol{MS}_{res}$ in Table 4.2 is a of the a sample 'means' $\bar{\boldsymbol{y}}_{i.}$ independent estimator of σ^2. The degrees of freedom of $\boldsymbol{MS}_{res}$ are

$$\sum_{i=1}^{a} (n_i - 1) = N - a,$$

and $\frac{1}{\sigma^2}(N-a)\boldsymbol{s}^2$ is $CS(N-a)$-distributed.

Problem 3 can be handled by the F-test; H_0 is rejected, if

$$\frac{\boldsymbol{MS}_A}{\boldsymbol{MS}_{res}} > F(a-1, N-a \mid 1-\alpha_e). \tag{11.51}$$

Problems 4 and 5 are solved by the methods of construction of confidence intervals (see Problem 4 and Problem 5 below).

If the $\binom{a}{2}$ and $a - 1$ null hypotheses of Problems 4 and 5, respectively, have to be tested in particular and not in total so that for each pair of hypotheses (H_{0ij}, H_{Aij}) the first kind risk is $\alpha_{ij} = \alpha$, then we use the multiple t-procedure or reject H_{0ij}, if

$$|t_{ij}| = \frac{|\bar{y}_{i\cdot} - \bar{y}_{j\cdot}|}{s}\sqrt{\frac{n_i n_j}{n_i + n_j}} > t\left(N - a\middle| 1 - \frac{\alpha}{2}\right). \tag{11.52}$$

The risk α must be understood for each of $\binom{a}{2}$ and $a - 1$ single comparisons, respectively, and we call it therefore a comparison-wise first kind risk. In Problem 5 the result is always $i = a$ and $j = 1, \ldots, a - 1$; in Problem 4 we have $i \neq j$; $i, j = 1, \ldots, a$.

The minimal size of the experiment $N = \sum_{i=1}^{a} n_i$ comes out if $n_i = n$; $i = 1, \ldots, a$. The value n depends on the comparison-wise first kind risk α and on the comparison-wise second kind risk β and the effect size δ analogous to Section 3.4.2.1.1 as

$$n = \left\lceil \left[t\left(a(n-1)\middle|1 - \frac{\alpha}{2}\right) + t(a(n-1)|1 - \beta) \right]^2 \frac{2\sigma^2}{\delta^2} \right\rceil$$

Example 11.7 We plan pairwise comparisons for $a = 8$ factor levels and use $\alpha = 0.05$; $\beta = 0.1$; and $\delta = \sigma$. We start with $n = \infty$ degrees of freedom and calculate iteratively

$$n_1 = \left\lceil 2[t(\infty|0.975) + t(\infty|0.9)]^2 \right\rceil = \left\lceil 2(1.96 + 1.2816)^2 \right\rceil = 22$$

and in the second step

$$n_2 = \left\lceil 2[t(168|0.975) + t(168|0.9)]^2 \right\rceil = \left\lceil 2(1.9748 + 1.2864)^2 \right\rceil = 22.$$

Therefore $n = 22$.

However, if the first kind risk $\alpha_{ij} = \alpha$ shall be chosen so that that the probability that at least one of the null hypotheses H_{0ij} is wrongly rejected, we proceed as follows.

Problem 4 If all $n_i = n$, we use the Tukey procedure. For all pairs $\mu_i - \mu_j$ ($i \neq j$; $i, j = 1, \ldots, a$), we calculate a confidence interval by (11.40).

If the corresponding realised confidence interval covers 0, H_{0ij} is rejected. In other words we reject H_{0ij} if

$$\frac{\left|\bar{y}_{i\cdot}-\bar{y}_{j\cdot}\right|\sqrt{n}}{s}>q(a,a(n-1)|1-\alpha_e) \tag{11.53}$$

For unequal n_i we calculate in place of (11.53) a confidence interval with $M_{ij} = \min\left(\sqrt{n_i},\sqrt{n_j}\right)$ and $f = N - a$ by

$$\left[\bar{y}_{i\cdot}-\bar{y}_{j\cdot}-q^*(a,f|1-\alpha_e)\frac{s}{M_{ij}},\bar{y}_{i\cdot}-\bar{y}_{j\cdot}+q^*(a,f|1-\alpha_e)\frac{s}{M_{ij}}\right] \tag{11.54}$$

and continue analogously, that is, we reject H_{0ij}, if

$$\frac{\left|\bar{y}_{i\cdot}-\bar{y}_{j\cdot}\right|M_{ij}}{s}>q^*(a,(N-a)|1-\alpha_e) \tag{11.55}$$

The minimal size of the experiment $N = \sum_{i=1}^{a} n_i$ comes out if $n_i = n\,; i = 1\,,\ldots,a$.

With experiment-wise error probability α_e and comparison-wise risk of the second kind β, we receive

$$n = \left\lceil 2\left[\frac{q(a,N-a|1-\alpha_e)}{\sqrt{2}}+t(a(n-1)|1-\beta)\right]^2\right\rceil$$

Alternatively to (11.55) with $R_{ij} = \sqrt{\frac{1}{n_i}+\frac{1}{n_j}}$ in place of (11.54), we can use

$$\left[\bar{y}_{i\cdot}-\bar{y}_{j\cdot}-q^{**}\left(\binom{a}{2},f|1-\alpha_e\right)R_{ij}s,\bar{y}_{i\cdot}-\bar{y}_{j\cdot}+q^{**}\left(\binom{a}{2},f|1-\alpha_e\right)R_{ij}s\right] \tag{11.56}$$

and reject H_{0ij}, if

$$\frac{\left|\bar{y}_{i\cdot}-\bar{y}_{j\cdot}\right|}{sR_{ij}}>q^{**}\left(\binom{a}{2},N-a|1-\alpha_e\right) \tag{11.57}$$

(Hochberg procedure).

Problem 5 If $n_i = n$, we use the Dunnett procedure, based on confidence intervals of Dunnett's method. Then H_{0i} is rejected if

$$\sqrt{n}\frac{|\bar{y}_{i\cdot}-\bar{y}_{a\cdot}|}{\sqrt{2}s}>d(a-1,a(n-1)|1-\alpha_e) \tag{11.58}$$

If n_i are not equal, we use a method also proposed by Dunnett (1964) with modified quantiles.

To determine the minimal sample sizes in multiple comparisons for the pairs of hypotheses (H_{0ij}, H_{Aij}) and (H_{0i}, H_{Ai}), respectively, with α or α_e, an upper bound β_0 for the second kind risks β_{ij} and β_i and $|\mu_i - \mu_j| > \Delta_{ij}$ given in advance, we use the R-commands in the program OPDOE:

`>size.multiple_t.test.` for `Problem 1 ;`
`>size.multiple_t.test.comp_standard.`

or

`>sizes.dunnett.exp_wise.` for comparisons with a control.

11.2.7 Which Multiple Comparison Shall Be Used?

To answer the question, which of the different multiple comparison procedures shall be used, we must at first decide whether Problem 3, 4 or 5 shall be solved (all corresponding assumptions for the procedures must be fulfilled).

Because Problem 3 is a two-decision problem, we use the *F*-test here; α_e and $1 - \beta_e$ are probabilities to be understood experiment-wise.

If Problem 4 is to be solved, we at first have to decide whether the first kind risk shall be comparison-wise α for each test separately or whether it has to be understood as probability α_e (called the experiment-wise first kind risk) that none of the null hypotheses H_{0ij} is wrongly rejected. If the first kind risk is comparison-wise, we use the multiple *t*-procedure or otherwise the Tukey procedure for $n_i = n$ and either the Spjøtvoll–Stoline procedure or the Hochberg procedure for unequal subclass numbers. Ury (1976) argued to use the Spjøtvoll–Stoline procedure mainly if small differences between the n_i occur. If n_i differs strongly he recommends to use the Hochberg procedure.

For Problem 5 with a comparison-wise α, the multiple *t*-procedure must be used; otherwise the Dunnett procedure is recommended.

If all or many linear contrasts shall be tested, the Scheffé, the Spjøtvoll–Stoline or the Hochberg procedure is recommended. Occasionally the Dunn procedure leads to useful intervals and tests. The Bonferroni and the Scheffé procedure can be used even if the random variables are correlated.

11.3 A Numerical Example

We demonstrate the methods of this chapter by Example 11.4 and solve the problems below.

Problem (a) Test the null hypotheses H_{0ij} against H_{Aij}. For Problem 5 the sample 10 corresponds with the control. We use $\alpha_e = 0.05$; $\alpha = 0.05$.

Problem (b) Construct simultaneous (1 – 0.05) confidence intervals for the contrasts

$$L_1 = 9\mu_{10} - \sum_{i=1}^{9} \mu_i, \quad L_2 = 3\mu_1 - \mu_2 - \mu_3 - \mu_4,$$

$$L_2 = 5\mu_1 - 3\mu_2 - 2\mu_3, \quad L_4 = 25\mu_1 - 15\mu_2 - 8\mu_3 - 2\mu_4$$

and for the differences in expectations $\mu_i - \mu_{10}$ $(i = 4, \dots, 9)$, denoted by L_5 bis L_{10}.

Problem (c) Select the population with the largest expectation and select the 2, 3, 4 and 5 populations with the largest expectations.

Problem (a). Due to (11.52), H_{0ij} is rejected, if

$$\left|\bar{y}_{i\cdot} - \bar{y}_{j\cdot}\right| > s \cdot t(40|0.975)\sqrt{\frac{2}{5}} = 10.107,$$

where s is obtained from Figure 11.2 as $s = \sqrt{62.52} = 7.907$ and from Table D.3 $t(40|0.975) = 2.0211$. We see in Table 11.9 that for Problem 4 with the multiple t-procedure, the null hypothesis $H_{01,6}$ is wrongly but $H_{06,10}$ is correctly rejected. Among the 43 accepted null hypotheses, 16 have been wrongly accepted. The reason for this high percentage is the small subclass number to hold a reasonable risk of the second kind. In Section 11.2.6 we presented a formula for the needed subclass number, and this gives for $\alpha = 0.05$, $\beta = 0.2$ and $\delta = \sigma$ the value $n = 16$, but we simulated in Example 11.4 only a subclass number $n = 5$.

For Problem 5 H_{06} is wrongly rejected, but the other eight null hypotheses are wrongly accepted. Because the subclass numbers are equal, the Tukey procedure can be applied.

Because $q(10, 40|0.95) = 4.735$, all those H_{0ij} of Problem 4 have to be rejected, if

$$\left|\bar{y}_{i\cdot} - \bar{y}_{j\cdot}\right| > \frac{7.907}{\sqrt{5}} \cdot 4.735 = 16.744$$

and this is not the case for any pair (i, j).

For Problem 5 $H_{0i}(i = 1, \dots, 9)$ is rejected, if $[q(9, 40|0.95) = 2.81]$

$$\left|\bar{y}_{i\cdot} - \bar{y}_{j\cdot}\right| > \frac{\sqrt{2} \cdot 7.907}{\sqrt{5}} \cdot 2.81 = 14.05$$

and neither is this the case for any pair (i, j).

With a subclass number 5 in Example 11.4, many incorrect decisions have been made. The subclass number for the α- and α_e-values above are chosen so that a difference $|\mu_i - \mu_j| > 8$ with probability of at most $\beta = 0.1$ is not detected. This is calculated with OPDOE of R.

Problem 3 We receive from the command in Section 5.2.2.2 (delta = δ/σ)

```
> size.anova(model="a",a=10, alpha=0.05, beta=0.1,
+delta=1, case="minimin")
n
9
> size.anova(model="a",a=10, alpha=0.05, beta=0.1,
+delta=1, case="maximin")
n
21
```

and choose $n = 15$.

Problem 4 Multiple t-procedure: The output of R shows $n = 22$ observations per population.

For the Tukey procedure, we receive $n = 36$.

Problem 5 Multiple t-procedure: We obtain $n_{10} = n_0 = 45$ and $n_i = 14$ $(i < 10)$.

Dunnett procedure: We obtain $n_{10} = n_0 = 63$ and $n_i = 23$ $(i < 10)$.

Problem (b). The estimates of the contrast we calculate from the means $\bar{y}_{i\cdot}$ in Table 11.8:

$$\hat{L}_1 = 53.92,\quad \hat{L}_2 = 18.14,\quad \hat{L}_3 = 31.66,\quad \hat{L}_4 = 161.42.$$

Scheffé's method: Using (11.30), we need for each contrast $\sqrt{\sum_{i=1}^{a} \frac{c_{ri}^2}{n_i}} = w_r$.

We obtain

$$w_1 = 4.2426,\quad w_2 = 1.5492,\quad w_3 = 2.7568,\quad w_4 = 13.5499.$$

For all contrasts $L_{4+i} = \mu_{3+i} - \mu_{10}$ $(i = 1, \dots, 6)$ results, $w_{4+i} = 0.6325$. From $F(9, 40|0.95) = 2.1240$ results

$$s\sqrt{(a-1)F(a-1, N-a|0{,}95)} = 34.5709.$$

The confidence intervals of the contrast have the bounds $\hat{L}_r \pm D_r^s$ $(r = 1,\dots,10, s\ \text{stands for Scheffé})$ with

$$D_1^s = 146.67,\quad D_2^s = 54.56,\quad D_3^s = 95.30,\quad D_4^s = 468.43,$$
$$D_{4+i}^s = 21.86\quad (i = 1,\dots,6).$$

All these confidence intervals cover 0, and therefore none of the hypotheses $H_{0r}: L_r = 0$ $(r = 1, \dots, 10)$ is rejected.

Dunn's method: Using (11.36), the bounds of the confidence intervals are $\hat{L}_r \pm D_r^D$ with $D_r^D = w_r \cdot s \cdot w$, where w can be found in Table 11.11. Because the number of contrasts equals 10, we read for f = 40, $w = 2.97$ Calculating confidence intervals for L_1, L_2, L_3, L_4 only, then $w = 2.62$. For the complete set of contrasts, we have

$$D_1^D = 99.63, \quad D_2^D = 36.38, \quad D_3^D = 64.74, \quad D_4^D = 318.20,$$
$$D_{4+i}^D = 14.85 \quad (i = 1,\ldots,6).$$

Simultaneous 0.95 confidence intervals for L_1 bis L_4 only give

$$D_1^{D^*} = 87.89, \quad D_2^{D^*} = 32.09, \quad D_3^{D^*} = 57.11, \quad D_4^{D^*} = 280.70.$$

Tukey's method: Using (11.43), simultaneous 0.95 confidence intervals are $L_r \pm D_r^T$ $(r = 1,\ldots,10)$ with

$$D_r^T = \frac{s}{2\cdot\sqrt{5}} q(10, 40|0.95) \sum_{i=1}^{a} |c_i|.$$

Because $q(10, 40|0.95) = 4.735$, we receive

$$D_1^T = 150.69, \quad D_2^T = 50.23, \quad D_3^T = 83.72, \quad D_4^T = 418.59,$$
$$D_{4+i}^T = 16.74 \ (i = 1,\ldots,6).$$

Confidence intervals with an individual confidence coefficient 0.95 by the multiple t-method of course are shorter but not comparable.

From Table 11.14 we see that only the method of DUNN is uniformly better than the others.

Problem (c). Using Selection Rule 11.1, then

$$\bar{y}_{(10)\cdot} = \bar{y}_{1\cdot} = 56.96 = \max_{1 \le i \le 10} (\bar{y}_{i\cdot}).$$

We wrongly call population 1 the best one. This happens due to the fact that $n = 5$ is too small. Also other rules lead to wrong conclusions.

Finally we compare the minimal sample sizes of the methods used for Example 11.4.

Method	n for $\alpha = 0.05$, $\alpha_e = 0.05$ $\beta = 0.05$; $d = \sigma$	Remarks
Selection rule 11.1 ($t = 1$)	12	
Tukey's procedure	40	45 comparisons
Dunnett's procedure	27 average	9 comparisons
multiple t procedure	17.1 average	α comparisonwise
F – test	15	1 test

Average means, for instance, $\bar{n} = (9n_1 + n_{10})$.

Table 11.14 Half widths of simultaneous confidence intervals for the contrasts of Example 11.4.

Method	L_1	L_2	L_3	L_4	$L_5, \ldots, L_{10}$
Scheffé	146.67	54.56	95.30	468.43	21.86
Dunn	99.63	36.38	64.74	318.20	14.85
Tukey	150.69	50.23	83.72	418.59	16.74

11.4 Exercises

11.1 Corresponding to Problem 1 and 1A, calculate the sample sizes of eight populations to obtain

$P_C \geq 0.99 \quad (P_{C^*} \geq 0.99)$

for the selection of the $t = 1, 2, 3, 4$ best populations if $\frac{d}{\sigma} = 0.1;$ $0.2; 0.5$ and 1.

11.2 Calculate the minimal experimental size concerning the multiple t-test for five groups and the comparison-wise risks $\alpha = 0.05$ and $\beta = 0.05$; 0.1 and 0.2 for $\delta = \sigma$ and $\delta = 0.5\ \sigma$.

11.3 Calculate the minimal experimental size concerning the Tukey test for $a = 3, 4, 5, 10, 20$ groups and the experiment-wise risks $\alpha = 0.05$ and 0.10 as well as the comparison-wise risks $\beta = 0.05$; 0.1 and 0.2 for $\delta = \sigma$ and $\delta = 0.5\ \sigma$.

References

Bechhofer, R. E. (1954) A single sample multiple decision procedure for ranking means of normal populations with known variances. *Ann. Math. Stat.*, **25**, 16–39.

Bechhofer, R. E. and Dunnett, Ch. W (1988) Percentage points of multivariate Student t – distribution in selected tables in mathematical statistics, in Odeh, R. E., Davenport, J. M. and Pearson, N. S. (Ed.), *Selected Tables in Mathematical Statistics*, vol. **11**, American Mathematical Society, Providence, RI.

Chambers, M. L. and Jarratt, P. (1964) Use of double sampling for selecting best populations. *Biometrika*, **51**, 49–64.

Domröse, H. and Rasch, D. (1987) Robustness of selection procedures. *Biom. J*, **5**, 541–553.

Dunn, G. J. (1961) Multiple comparisons among means. *J. Am. Stat. Assoc.*, **56**, 52–64.

Dunnett, C. W. (1955) A multiple comparison procedure for comparing several treatments with a control. *J. Am. Stat. Assoc.*, **50**, 1096–1121.

Dunnett, C. W. (1964) New tables for multiple comparisons with a control. *Biometrics*, **20**, 482–491.

Guiard, V. (1994) Different definitions of Δ-correct selection for the indifference zone formulation, in Mieske, K. J. and Rasch, D. (Eds.) (1996) Special Issue on 40 Years of Statistical Selection Theory J. Stat. Plann. Inference 54, pp. 176–199.

Gupta, S. S. (1956) *On a Decision Rule for a Problem in Ranking Means*, Mim. Ser., No 150, Univ. North Carolina, Chapel Hill.

Gupta, S. S. and Huang, D. Y. (1981) *Multiple Statistical Decision Theory: Recent Developments*, Springer, New York.

Gupta, S. S. and Panchapakesan, S. (1970) *On a Class of Subset Selection Procedures*, Mim. Ser. 225, Purdue Univ., West Lafayette.

Gupta, S. S. and Panchapakesan, S. (1979) *Multiple Decision Procedures: Theory and Methodology of Selecting and Ranking Populations*, John Wiley & Sons, Inc., New York.

Hochberg, Y. (1974) Some generalization of the *T*- method in simultaneous inference. *J. Multivar. Anal..*, **4**, 224–234.

Hochberg, Y. and Tamhane, A. C. (1987) *Multiple Comparison Procedures*, John Wiley & Sons, Inc., New York.

Hsu, J. C. (1996) *Multiple Comparisons*, Chapman & Hall, London.

Miescke, K. J. and Rasch, D. (Eds.) (1996) Special Issue on 40 Years of Statistical Selection Theory, Part I: *J. Stat. Plann. Inference* 54, number 2; Part II. *J. Stat. Plann. Inference* 54, number 3.

Miller, R. G. (1981) *Simultaneous Statistical Inference*, Springer Verlag, New York.

Rasch, D. and Herrendörfer, G. (1990) *Handbuch der Populationsgenetik und Züchtungsmethodik*, Deutscher Landwirtschaftsverlag, Berlin.

Rasch, D., Herrendörfer, G., Bock, J., Victor, N. and Guiard, V. (Eds.) (2008) *Verfahrensbibliothek Versuchsplanung und - auswertung,* 2. verbesserte Auflage in einem Band mit CD, R. Oldenbourg Verlag München, Wien.

Rodger, R. S. (1973) Confidence intervals for multiple comparisons and the misuse of the Bonferroni inequality. *Br. J. Math. Stat. Psychol.*, **26**, 58–60.

Scheffé, H. (1953) A method for judging all contrasts in the analysis of variance. *Biometrika*, **40**, 87–104.

Spjøtvoll, E. and Stoline, M. R. (1973) An extension of the *T*-method of multiple comparisons to include the cases with unequal sample size. *J. Am. Stat. Assoc.*, **68**, 975–978.

Stoline, M. R. (1978) Tables of the studentized augmented range and applications to problems of multiple comparison. *J. Am. Stat. Assoc.*, **73**, 656–660.

Stoline, M. R. and Ury, H. K. (1979) Tables of studentized maximum modules and an applications to problems of multiple comparisons. *Technometrics*, **21**, 87–93.

Tukey, J. W. (1953) Multiple comparisons. *J. Am. Stat. Assoc.*, **48**, 624–625.

Ury, H. K. (1976) A comparison of four procedures for multiple comparisons among means (pairwise contrasts) for arbitrary sample size. *Technometrics*, **18**, 89–97.
Ury, H. K. and Wiggens, A. D. (1971) Large sample and other multiple comparisons among means. *Br. J. Math. Stat. Psychol.*, **24**, 174–194.
Ury, H. K. and Wiggens, A. D. (1974) Use of the Bonferroni inequality for comparisons among means with posthoc contrasts. *Br. J. Math. Stat. Psychol.*, **27**, 176–178.

12

Experimental Designs

Experimental designs originated in the early years of the 20th century mainly in agricultural field experimentation in connection with open land variety testing. The centre was Rothamsted Experimental Station near London, where Sir Ronald Aylmer Fisher was the head of the statistical department (since 1919). There he wrote one of the first books about statistical design of experiments (Fisher, 1935), a book that was fundamental, and promoted statistical technique and application. The mathematical justification of the methods was not stressed and proofs were often barely sketched or omitted. In this book Fisher also outlined the Lady tasting tea, which is now a famous design of a statistical randomised experiment that uses Fisher's exact test and is the original exposition of Fisher's notion of a null hypothesis.

Because soil fertility in trial fields varies enormously, a field is partitioned into so-called blocks and each block subdivided in plots. It is expected that the soil within the blocks is relatively homogeneous so that yield differences of the varieties planted at the plots of one block are due only to the varieties and not due to soil differences. To ensure homogeneity of soil within blocks, the blocks must not be too large. On the other hand, the plots must be large enough so that harvesting (mainly with machines) is possible. Consequently, only a limited number of plots within the blocks are possible, and only a limited number of varieties within the blocks can be tested. If all varieties can be tested in each block, we speak about a complete block design. The number of varieties is often larger than the number of plots in a block. Therefore incomplete block designs were developed, chief among them completely balanced incomplete block designs, ensuring that all yield differences of varieties can be estimated with equal variance using models of the analysis of variance.

If two disturbing influences occur in two directions (as humidity from east to west and soil fertility from north to south), then the so-called row–column designs (RCDs) are in use, especially Latin squares (LS).

Mathematical Statistics, First Edition. Dieter Rasch and Dieter Schott.

The experimental designs originally developed in agriculture soon were used in medicine, in engineering or more generally in all empirical sciences. Varieties (v) were generalized to treatments (t), and plots to experimental units. But even today the number v of treatments or the letter y (from yield) in the models of the analysis of variance recalls this to the agricultural origin. Sometimes in technical books or papers, t is used in place of v.

Later experimental designs have been handled not only within statistics but also within the combinatorics and many were published in statistical journals (as *Biometrika*) as well as in combinatorial ones (as *Journal of Combinatorial Designs*).

12.1 Introduction

Experimental designs are an important part in the planning (designing) of experiments. The main principles are (the three Rs)

1) Replication
2) Randomisation
3) Reduction of the influence of noisy factors (blocking)

Statements in the empirical sciences can almost never be derived based on an experiment with only one measurement. Because we often use the variance as a measure of variability of the observed character and then we need at least two observations (replications) to estimate it (in statistics the term replication mainly means one measurement; thus, two measurements are two replications and not one measurement and one replication).

Therefore, two replications are the lower bound for the number the replications. The sample size (the number the replications) has to be chosen and was already discussed at several places in previous chapters.

Experimental designs are used mainly for the reduction of possible influences of known nuisance factors.

This is the main topic of this chapter, but initially we consider the situation where the nuisance factors are not known or not graspable. In this case, we try to solve the problem by randomisation here understood as the unrestricted random assignment of the experimental units to the treatments (not vice versa!). Randomisation is (as shown in Chapter 1) also understood as the random selection of experimental units from a universe. But in this chapter in designing experiments, we assume that experimental units and blocks are already randomly selected.

Randomisation is used to keep the probability of some bias by some unknown nuisance factors as small as possible. It shall ensure that statistical models as base for planning and analysing represent the situation of an experiment adequately and the analysis with statistical methods is justified.

We distinguish between pure and restricted forms of randomisation experimental designs. We at first assume that the experimental material is unstructured, which means there is no blocking.

This is the simplest form of an experimental design. If in an experimental design exactly n_i experimental units are randomly allocated to the ith of v treatments ($\Sigma n_i = N$), we call this a complete or unrestricted randomisation, and we call the experimental design a simple or a completely randomised experimental design. Such designs were used in the previous chapters.

In this chapter we define experimental designs as model independent (i.e. independent of the models for the analysis, e.g. for the analysis of variance) and consider experiments with N experimental units, numbered from 1 to N, and these numbers are used as names of the units. In an experiment the effects of p treatment factors $A^{(1)}, \ldots, A^{(p)}$ have to be estimated or tested, and the effects of q nuisance factors $B^{(1)}$, …, $B^{(q)}$ must be eliminated. The possible values of a factor are called factor levels (levels).

N and p are positive integers, and q is nonnegative and an integer. An experiment is always the combination of an experimental design with a rule of randomisation.

Definition 12.1 The assignment of a given number $N > 2$ of experimental units to the levels $A_i^{(h)}$ $(i = 1,\ldots,v_h,\ h = 1,\ldots,p)$ of $p \geq 1$ treatment factors $A^{(1)}$, …, $A^{(p)}$ and the levels $B_j^{(c)}$ $(j = 1,\ldots,b_c,\ c = 1,\ldots,q)$ of $q \geq 0$ nuisance factors (block factors) $B^{(1)}$, …, $B^{(q)}$ is called a p-factorial experimental design with q block factors. If $p = 1$ the one-factorial experimental design is called a simple experimental design. If $p > 0$ we speak about a factorial experiment. If $q = 0$ we speak about a completely randomised or a simple experimental design.

Simple experimental designs are, for instance, the base of the methods in Chapters 2 and 3, the randomisation in these experimental designs means that N experimental units are randomly assigned (e.g. by random number generators) to the v level combinations of some treatment factors or the v levels of one treatment factor.

To illustrate the assignment rules of Definition 12.1, we use matrices U_h and Z_c combining to the matrix

$$Z = \left(U_1,\ldots,U_p, Z_1,,\ldots,Z_q\right) \tag{12.1}$$

The elements of the submatrices U_h and Z_c are defined as follows:

$$u_{lk}^{(h)} = \begin{cases} 1, \text{ if the } l-\text{th experimental unit is assigned to the } k-\text{th level of } A^{(h)} \\ 0, \text{ otherwise} \end{cases}$$

and

$$z_{lk}^{(c)} = \begin{cases} 1, \text{ if the } l-\text{th experimental unit is assigned to the } k-\text{th Level of } B^{(c)} \\ 0, \text{ otherwise} \end{cases}$$

We obtain

$$U_h^T 1_N = r^{(h)},\ r^{(h)^T} = \left(r_1^{(h)},\ldots,r_{v_h}^{(h)}\right);\ Z_c^T 1_N = k^{(c)},\ k^{(c)^T} = \left(k_1^{(c)},\ldots,k_{b_c}^{(c)}\right). \tag{12.2}$$

We consider mainly one-factorial experimental designs and write then for $A^{(1)} = A$ and for $v_1 = v$ and further

$$r^T = r^{(1)^T} = \left(r_1^{(h)},\ldots,r_{v_h}^{(h)}\right) = (r_1,\ldots,r_v)$$

with $r_i \geq 1$ and $N = \sum_{i=1}^{v} r_i \geq v+1$.

Example 12.1 We consider the structure of Example 5.12.

		Forage crop	
		Green rye	Lucerne
Storage	Glass	8.39	9.44
		7.68	10.12
		9.46	8.79
		8.12	8.89
	Sack	5.42	5.56
		6.21	4.78
		4.98	6.18
		6.04	5.91

Here is $N = 16$, $q=0$ and $p = 2$. In the first column are the elements 1–8 and in the second column, the elements 9–16 (numbered from above).

The factors are $A^{(1)}$ and $A^{(2)}$ and further

$$U_1^T = \begin{pmatrix} 1&1&1&1&1&1&1&1&0&0&0&0&0&0&0&0 \\ 0&0&0&0&0&0&0&0&1&1&1&1&1&1&1&1 \end{pmatrix}$$

and

$$U_2^T = \begin{pmatrix} 1&1&1&1&0&0&0&0&1&1&1&1&0&0&0&0 \\ 0&0&0&0&1&1&1&1&0&0&0&0&1&1&1&1 \end{pmatrix}.$$

Besides $v_1 = 2$, $v_2 = 2$ and ${r^{(h)}}^T = (8,\ 8)$ with $h = 1,2$.

Definition 12.2 A one-factorial experimental design is K-balanced of order t, if a given operator K transfers the matrix $Z = (U_1, Z_1)$ in (12.1) into a $(v \times v)$-matrix with identical elements in the main diagonal and exactly t different elements outside the main diagonal.

12.2 Block Designs

Block designs are experimental designs to eliminate one disturbance variable. In case of a quantitative nuisance factor, we also can use the analysis of covariance if the type of the dependency and the underlying function are known (for instance linear or quadratic). The parameters are estimated from the observed values of the character and the nuisance factor as already shown in Chapter 10. A general (i.e. also for qualitative nuisance factors) applicable method is blocking or stratification by the levels of the nuisance factor. We restrict ourselves to one treatment factor. This is no loss of generality. If we have several treatment factors, we consider all level combinations of these treatment factors as treatments of some new factor.

As already said, a block design helps to eliminate the effects of a disturbance variable, that is, the matrix Z in (12.1) contains just one matrix Z_1 and is of the form $Z = (U_1, \ldots, U_p, Z_1) = (Z_0, Z_1)$ with $Z_0 = (U_1, \ldots, U_p)$. We form

$$Z^T Z = \begin{pmatrix} Z_0^T Z_0 & Z_0^T Z_1 \\ Z_1^T Z_0 & Z_1^T Z_1 \end{pmatrix}.$$

$Z_0^T Z_0$ is a diagonal matrix and is for one treatment factor of the form $U_1^T U_1 = \mathrm{diag}(r_1, \ldots, r_v)$; $Z_1^T Z_1 = \mathrm{diag}(k_1, \ldots, k_b)$ is also a diagonal matrix. Now we have

$$Z^T Z = \begin{pmatrix} U_1^T U_1 & U_1^T Z_1 \\ Z_1^T U_1 & Z_1^T Z_1 \end{pmatrix} = \begin{pmatrix} \mathrm{diag}(r_1, \ldots, r_v) & U_1^T Z_1 \\ Z_1^T U_1 & \mathrm{diag}(k_1, \ldots, k_b) \end{pmatrix}.$$

The submatrix $U_1^T Z_1 = \mathcal{N}$ is called incidence matrix. By this $Z^T Z$ has the form

$$Z^T Z = \begin{pmatrix} \mathrm{diag}(r_1, \ldots, r_v) & \mathcal{N} \\ \mathcal{N}^T & \mathrm{diag}(k_1, \ldots, k_b) \end{pmatrix}.$$

A block design has therefore a finite incidence structure built up from an incidence matrix. A finite set $\{1, 2, \ldots, v\}$ of v elements (treatments) and a finite set $\{B_1, B_2, \ldots, B_b\}$ of b sets, called blocks, are the levels of the nuisance and block factor. The elements of the incidence matrix $\mathcal{N} = (n_{ij})$ lead to the two diagonal matrices because $\mathcal{N}1_b = \text{diag}(r_1,\ldots,r_v)$ and $\mathcal{N}^T 1_v = \text{diag}(k_1,\ldots,k_b)$.

Definition 12.3 The elements of the incidence matrix $\mathcal{N} = (n_{ij})$ with v rows and b columns show how often the ith treatment (representing the ith row) occurs in the jth block (representing the jth column). If all n_{ij} are either 0 or 1, the incidence matrix and the corresponding block design are called binary. The b column sums k_j of the incidence matrix are the elements of $\text{diag}(k_1, \ldots, k_b)$ and are called block sizes. The v row sums r_i of the incidence matrix are the elements of $\text{diag}(r_1, \ldots, r_v)$ and are called replications. A block design is complete, if the elements of the incidence matrix are all positive ($n_{ij} \geq 1$). A block design is incomplete, if the incidence matrix contains at least one zero. Blocks are called incomplete, if in the corresponding column of the incidence matrix, there is at least one zero.

In block designs the randomisation has to be done as follows: the experimental units in each block are randomly assigned to the treatments, occurring in this block. This randomisation is done separately for each block. In the complete randomisation in Section 12.1, we only have to replace N by the block size k and v by the number of plots in the block.

In complete block designs with v plots per block, where each of them is assigned to exactly one of the v treatments, the randomisation is completed. If $k < v$, (incomplete block designs) the abstract blocks, obtained by the mathematical construction have to be randomly assigned to the real blocks using the method in Section 12.1 for $N = b$, if b is the number of blocks.

For incomplete binary block designs in place of the incidence matrix often a shorter writing is in use. Each block is represented by a bracket including the symbols (numbers) of the treatments, contained in the block.

Example 12.2 A block design with $v = 4$ treatments and $b = 6$ blocks may have the incidence matrix:

$$\begin{pmatrix} 1 & 0 & 1 & 0 & 0 & 0 \\ 0 & 1 & 0 & 1 & 1 & 1 \\ 1 & 0 & 1 & 0 & 0 & 0 \\ 0 & 1 & 0 & 1 & 1 & 1 \end{pmatrix}.$$

Because zeros occur, we have an incomplete block design. This can now be written as

$$\{(1,3), (2,4), (1,3), (2,4), (2,4), (2,4)\}.$$

The first bracket represents block 1 with treatments 1 and 3 corresponding to the fact that in column 1 (representing the first block) in row 1 and 3 occurs a one.

Definition 12.4 (Tocher, 1952)
A block design for which in Definition 12.2 the operator K maps the matrix Z into the matrix, $\mathcal{N}\mathcal{N}^T$ is called $\mathcal{N}\mathcal{N}^T$ -balanced. A $\mathcal{N}\mathcal{N}^T$-balanced incomplete block design of order t is called partially balanced with t association classes.

Definition 12.5 A block design with a symmetric incidence matrix is a symmetric block design. If all treatments in a block design occur equally often (the number of replications is $r_i = r$ for all i), it is called equireplicate. If the number of plots in a block design is the same in each block ($k_j = k$, for all j), it is called proper.

It can easily be seen that the sum of the replications r_i as well as the sum of all block sizes k_j equals the number N of the experimental units of a block design. Therefore, for each block design we have

$$\sum_{i=1}^{v} r_i = \sum_{j=1}^{b} k_j = N, \tag{12.3}$$

especially for equireplicate and proper block designs ($r_i = r$ and $k_j = k$); this gives

$$vr = bk. \tag{12.4}$$

In symmetric block designs are $b = v$ and $r_i = k_i$ $(i = 1,\ldots, v)$.

Definition 12.6

a) An incomplete block design is connected, if for each pair (A_k, A_l) of treatments $A_1, \ldots, A_v$, there exists a chain of treatments starting with A_k and ending with A_l, in which each of two adjacent treatments in this chain occur in at least one block. Otherwise, the block design is disconnected.
b) Alternatively we say: a block design with incidence matrix $\mathcal{N}$ is disconnected if, by permuting its rows and columns in a suitable way, $\mathcal{N}$ can be transformed into a matrix M that can be written as the direct sum of at least two matrices. Otherwise, it is connected.

Both parts of Definition 12.6 are equivalent; the proof is left to the reader.

Both parts of this definition are very abstract and their meaning is perhaps unclear. But the feature 'connected' is very important for the analysis. Disconnected block designs, for instance, cannot be analyzed as a whole by the analysis of variance (side conditions not fulfilled), but rather as two or more independent experimental designs.

Example 12.2 – Continued In the design of Example 12.2, the first and the second treatments occur together in none of the six blocks. There is no chain of treatments as in Definition 12.6, and therefore the design is a disconnected block design. This can be seen if the blocks and the treatments are renumbered, which means that the columns and the rows of the incidence matrix are properly interchanged. We interchange the blocks 2 and 3 and the treatments 1 and 4. Thus, in the incidence matrix the columns 2 and 3 and the rows 1 and 4 are interchanged. The result is the matrix:

$$\begin{pmatrix} 0 & 0 & 1 & 1 & 1 & 1 \\ 0 & 0 & 1 & 1 & 1 & 1 \\ 1 & 1 & 0 & 0 & 0 & 0 \\ 1 & 1 & 0 & 0 & 0 & 0 \end{pmatrix}.$$

And this is the direct sum of two matrices, and that means that we have two designs with two separate subsets of treatments. In the first design there are two treatments (1 and 2) in four blocks, while in the second design there are two further treatments (3 and 4) in two other blocks.

In the rest of these chapters, we only consider complete (and by this by definition connected) or connected incomplete block designs. Further, we restrict ourselves to proper and equireplicate block designs.

Definition 12.7 Let $\mathcal{N}_i; i = 1,2$ be the incidence matrices of two block designs with the parameters v_i, b_i, k_i, r_i. The Kronecker product $\mathcal{N} = \mathcal{N}_1 \otimes \mathcal{N}_2$ is the incidence matrix of a Kronecker product design with the parameters $v = v_1 v_2, b = b_1 b_2, k = k_1 k_2, r = r_1 r_2$.

Theorem 12.1 If the $\mathcal{N}_i$, $i = 1,2$ in Definition 12.7 are binary, then the Kronecker product design with the incidence matrix $\mathcal{N} = \mathcal{N}_1 \otimes \mathcal{N}_2$ is also binary, and we have $\mathcal{N}\mathcal{N}^T = \mathcal{N}_1\mathcal{N}_1^T \otimes \mathcal{N}_2\mathcal{N}_2^T$. If the Kronecker product designs with the incidence matrices $\mathcal{N}_i; i = 1,2$ are both $\mathcal{N}_i\mathcal{N}_i^T$-balanced of order t_i, so are $\mathcal{N} = \mathcal{N}_1 \otimes \mathcal{N}_2$ and $\mathcal{N}^* = \mathcal{N}_2 \otimes \mathcal{N}_1$ incidence matrices of $\mathcal{N}\mathcal{N}^T$-balanced and $\mathcal{N}^*\mathcal{N}^{*T}$-balanced block designs, respectively, of order $t^* \le (t_1 + 1)(t_2 + 1) - 1$.

Proof: The first part of the theorem is a consequence of the definition of Kronecker product designs. Because the design is $\mathcal{N}_i\mathcal{N}_i^T$-balanced of order t_i, these matrices have exactly $t_i + 1$ (with the main diagonal element) different elements. Because $\mathcal{N}\mathcal{N}^T = \mathcal{N}_1\mathcal{N}_1^T \otimes \mathcal{N}_2\mathcal{N}_2^T$ in $\mathcal{N}\mathcal{N}^T$(or $\mathcal{N}^*\mathcal{N}^{*T}$) all $(t_1 + 1)(t_2 + 1)$ products will be found. But all elements in the main diagonal of $\mathcal{N}\mathcal{N}^T$ are equal, so that maximal $(t_1 + 1)(t_2 + 1) - 1$ different values in $\mathcal{N}\mathcal{N}^T$ (or $\mathcal{N}^*\mathcal{N}^{*T}$) exist.

12.2.1 Completely Balanced Incomplete Block Designs (BIBD)

Definition 12.8 A (completely) balanced incomplete block design (BIBD) is a proper and equireplicate incomplete blocks design with the additional property that each pair of treatments occurs in equally many, say, in λ, blocks. Following

Definition 12.4 it is a $\mathcal{N}\mathcal{N}^T$-balanced incomplete Block design with $t = 1$, a BIBD with v treatments with r replications in b Blocks of size $k < v$, is called a $B(v, k, \lambda)$-design. A BIBD for a pair (v, k) is called elementary, if it cannot be decomposed in at least two BIBD for this pair (v, k). A BIBD for a pair (v, k) is a smallest BIBD for this pair (v, k), if r (and by this also b and λ) is minimal.

In $B(v, k, \lambda)$ only three of the five parameters v, b, k, r, λ of a BIBD occur. But this is sufficient, because exactly three of the five parameters can be fixed, while the two others are automatically fixed.

This can be seen as follows. The number of possible pairs of treatments in the design is $\binom{v}{2} = \frac{v(v-1)}{2}$. However, in each of the b blocks exactly $\binom{k}{2} = \frac{k(k-1)}{2}$ pairs of treatments exist so that

$$\lambda v(v-1) = bk(k-1)$$

if each of the $\binom{v}{2}$ pairs of treatments occurs λ -times in the experiment. From Formula (12.4) we replace bk with vr and after division by v we obtain

$$\lambda(v-1) = r(k-1). \tag{12.5}$$

The Equations (12.4) and (12.5) are necessary conditions for the existence of a BIBD. These necessary conditions reduce the set of possible quintuple of integers v, b, r, k, λ on a subset of integers, for which the conditions (12.4) and (12.5) are fulfilled. If we characterize a BIBD by three of these parameters, like $\{v, k, \lambda\}$, the other parameters can be calculated via (12.4) and (12.5).

The necessary conditions are not always sufficient for the existence of a BIBD. To show this we give a counter example.

Example 12.3 We show that the conditions that are necessary for the existence of a BIBD must not be sufficient. The values

$$v = 16, r = 3, b = 8, k = 6, \lambda = 1$$

give $16 \cdot 3 = 8 \cdot 6$ and $1 \cdot 15 = 3 \cdot 5$ by (12.4) and (12.5), but no BIBD with these parameters exists.

Besides (12.4) and (12.5) there is a further necessary condition, Fisher's inequality

$$b \geq v. \tag{12.6}$$

This inequality is not fulfilled in Example 12.3. But even if (12.4), (12.5) and (12.6) are valid, a BIBD does not always exist. Examples for that are

$$v = 22,\ k = 8,\ b = 33,\ r = 12,\ \lambda = 4$$

and

$$v = 34,\ k = 12,\ r = 12,\ b = 34,\ \lambda = 4.$$

The smallest existing BIBD for

$$v = 22,\ k = 8 \text{ and } v = 34 \text{ and } k = 12$$

has the parameters

v = 22, k = 8, b = 66, r =24, λ=8 and
v = 34, k = 12, r = 18, b = 51, λ = 6, respectively.

A BIBD (a so-called unreduced or trivial BIBD) for any positive integer v and $k < v$ can always be constructed by writing down all possible k-tuples from v elements. Then $b = \binom{v}{k}$, $r = \binom{v-1}{k-1}$ and $\lambda = \binom{v-2}{k-2}$.

Often a smaller BIBD (with fewer blocks than the trivial one) can be found as a subset of a trivial BIBD. A case where such a reduction is not possible, is that with $v = 8$ and $k = 3$. This is the only one case for $v \le 25$ and $2 < k < v - 1$ where no smaller BIBD than the trivial one exists. Rasch et al. (2016) formulate and support the following conjecture.

Conjecture:
The cases v = 8 and k = 3 are the only cases for $k > 2$ and $k < v - 2$ where the trivial BIBD is elementary.

This conjecture is even now neither confirmed nor disproved. But the following theorem is proved.

Theorem 12.2 The conjecture above is true, if at least one of the following conditions is fulfilled:

a) $v < 26$, $2 < k < v - 1$
b) $k < 6$
c) for $v > 8$ and k a BIBD exists with $b = v(v-1)$.

Proof: For (a) and (b) the theorem is proved constructively that for all parameter combinations there exists a non-trivial BIBD.

If there exists a BIBD with $b = v(v - 1)$, then for each $k \le v/2$ we write

$$\binom{v}{k} = \frac{v}{1}\frac{(v-1)}{2}\frac{(v-2)}{3}\cdots\frac{(v-k+1)}{k} = v(v-1)\frac{(v-2)}{6}\frac{(v-3)}{4}\cdots\frac{(v-k+1)}{k}.$$

All factors of $\frac{(v-2)}{6}\frac{(v-3)}{4}\cdots\frac{(v-k+1)}{k}$ are larger than 1, so that $v(v-1) < \binom{v}{k}$.

This completes the proof.

The block designs with $b = v(v-1)$ blocks often are not the smallest. One reason is that in some designs each block occurs w times. Removing $w-1$ copies of each block leads to a BIBD with $\frac{v(v-1)}{w}$ blocks.

In the meantime F. Teuscher (2017, Constructing a BIBD with $v = 26$, $k = 11$, $b = 130$, $r = 55$, $\lambda = 22$, personal communication) showed that the conjecture is true for $v = 26$, $k = 11$, because he constructed a design with $v = 26$, $k = 11$, $b = 130$, $r = 55$, $\lambda = 22$.

In constructing BIBD we can restrict ourselves to $k \le \frac{v}{2}$ due to the definition below.

Definition 12.9 A complementary block design for a given BIBD for a pair (v,k) is a block design for $(v,v-k)$ with the same number of blocks, so that each block of the complementary block design contains just those treatments not occurring in the corresponding block of the original BIBD.

We receive (parameters of the complementary design are indicated with *)

$$v^* = v,\ b^* = b,\ k^* = v - k,\ r^* = b - r.$$

The incidence matrix of the complementary design is $\mathcal{N}^* = 1_{vb} - \mathcal{N}$, and this adds up to

$$\begin{aligned}\mathcal{N}^*\mathcal{N}^{*T} &= (1_{vb} - \mathcal{N})(1_{vb} - \mathcal{N})^T = b1_{vv} - r1_{vv} - r1_{vv} + \mathcal{N}\mathcal{N}^T \\ &= (r-\lambda)I_v + (b - 2r + \lambda)1_{vv}.\end{aligned}$$

That means that the complementary block design of a BIBD is also a BIBD, with $\lambda^* = b - 2r + \lambda$

Theorem 12.3 The complementary block design to a given BIBD for a pair (v,k) is a BIBD for $(v,v-k)$ with the parameters v^*, b^*, k^*, r^*, λ^* and $v^* = v$, $b^* = b$, $k^* = v - k$, $r^* = b - r$, $\lambda^* = b - 2r + \lambda$.

From this it follows that a BIBD cannot be complementary to a block design that is not a BIBD.

Of course smallest (v,k) – BIBD are elementary, but not all elementary BIBD are smallest, as we will show in Example 12.4.

In applications the number v of treatments and the block size k are often given, and we like to find the smallest BIBD for a pair (v,k). This is possible with the R-programme in OPDOE (Rasch et al., 2011) for $v \le 25$.

If $k = 1$ each of the v elements define a block of a degenerated BIBD with $v = b$, $r = 1$ and $\lambda = 0$. These BIBD are trivial and elementary. The same is true for its complementary BIBD with $v = b$, $r = k = v - 1$ and $\lambda = v - 2$. Here in each block another treatment is missing. Even for k=2 all BIBD and their complementary BIBD are trivial as well as elementary. That is why in the future, we will confine ourselves to $3 \le k \le v/2$.

Definition 12.10 A BIBD is said to be α- resolvable or *α-RBIBD*, if its blocks can be arranged to form $l \ge 2$ classes in such a way that each block occurs exactly α times in each class. We write $RB(v, k, \lambda)$. A *α-RBIBD* is said to be affine α – resolvable, if each pair of blocks from a given class has exactly $\alpha = q_1$ treatments in common and pairs of blocks from different classes have q_2 treatments in common, a 1-resolvable BIBD is called resolvable or a *RBIBD*.

For affine α- resolvable *RBIBD* we have $b = v + r - 1$ and $\alpha = \frac{k^2}{v}$.

Example 12.4 The BIBD with $v = 9$, $k = 3$, $\lambda = 1$ and $b = 12$ is affine 1-resolvable in four classes (the columns of the scheme)

$$\begin{Bmatrix} (1,2,3) & (1,4,7) & (1,5,9) & (1,6,8) \\ (4,5,6) & (2,5,8) & (2,6,7) & (3,5,7) \\ (7,8,9) & (3,6,9) & (3,4,8) & (2,4,9) \end{Bmatrix}$$

because $\alpha = \frac{3^2}{9} = 1$.

Definition 12.11 If $\mathcal{N}$ is the incidence matrix of a BIBD, then $\mathcal{N}^T$ is the incidence matrix of the dual BIBD, obtained by interchanging rows and columns in the incidence matrix of a BIBD.

The parameters v^*, b^*, r^*, k^* and λ^* of the dual BIBD of *a* BIBD with parameters v, b, r, k, and λ are $v^* = b$, $b^* = v$, $r^* = k$, $k^* = r$ and $\lambda = \lambda^*$.

Example 12.5 For $v = 7$ and $k = 3$ the trivial BIBD is given by the following:

(1,2,3)	(1,3,6)	(1,6,7)	(2,4,7)	*(3,5,6)*
(1,2,4)	***(1,3,7)***	(2,3,4)	(2,5,6)	(3,5,7)
(1,2,5)	(1,4,5)	***(2,3,5)***	(2,5,7)	(3,6,7)
(1,2,6)	(1,4,6)	(2,3,6)	***(2,6,7)***	(4,5,6)
(1,2,7)	(1,4,7)	(2,3,7)	(3,4,5)	***(4,5,7)***
(1,3,4)	***(1,5,6)***	(2,4,5)	***(3,4,6)***	*(4,6,7)*
(1,3,5)	*(1,5,7)*	(2,4,6)	(3,4,7)	(5,6,7)

An elementary BIBD has the parameters $b = 7$, $r = 3$, $\lambda = 1$ and the blocks {(1,2,4); (1,3,7), (1,5,6), (2,3,5), (2,6,7), (4,5,7), (3,4,6)} – in bold print above. The incidence matrix is

$$\begin{pmatrix} 1 & 1 & 1 & 0 & 0 & 0 & 0 \\ 1 & 0 & 0 & 1 & 1 & 0 & 0 \\ 0 & 1 & 0 & 1 & 0 & 0 & 1 \\ 1 & 0 & 0 & 0 & 0 & 1 & 1 \\ 0 & 0 & 1 & 1 & 0 & 1 & 0 \\ 0 & 0 & 1 & 0 & 1 & 0 & 1 \\ 0 & 1 & 0 & 0 & 1 & 1 & 0 \end{pmatrix}.$$

The complementary BIBD is

$$\{(1,2,3,6),(1,3,4,5),(1,4,6,7),(1,2,5,7),(2,4,5,6),(2,3,4,7),(3,5,6,7)\}.$$

A further elementary BIBD with parameters $b = 7$, $r = 3$, $\lambda = 1$ is the septuplet printed italic (but not bold) in the trivial BIBD. It is isomorph to the BIBD with the italic and bold printed blocks. The set of the residual 21 of the 35 blocks cannot be split up into smaller BIBD; they also build an elementary BIBD, but of course not the smallest.

To show that there are no further BIBD with 7 blocks (and by thus no BIBD with 14 blocks) within the residual 21 blocks, we consider one of the 21 residual blocks, namely, (1,2,3). Because $r = 3$ we need two further blocks with a 1, where (1,4), (1,5), (1,6) and (1,7) are contained. The only possibility is (1,4,5) and (1,6,7), and other possibilities are already in a block of the two elementary designs or contradict $\lambda = 1$. The block design we are looking for must start with (1,2,3), (1,4,5) and (1,6,7). Now we need two further blocks with a 2 with the pairs (2,4), (2,5), 2,6), and (2,7). Possibilities are (2,4,6) with (2,5,7) or (2,4,7) with (2,5,6).

It means we have two possibilities for the first five blocks:

(1,2,3)	or	(1,2,3)
(1,4,5)		(1,4,5)
(1,6,7)		(1,6,7)
(2,4,6)		(2,4,7)
(2,5,7)		(2,5,6)

Now we need two blocks with a 3 in each to add them to the five blocks. The blocks (3,6,7) and (3,4,5) are not permissible; the pairs 4,5 and 6,7 are already present in the first five blocks, (3,4,7) is in the first quintuple permissible, but the needed partner (3,5,6) is already in a block of the two elementary designs. Therefore, the first quintuple must be withdrawn. In the second quintuple we could continue with (3, 5, 7), but here the needed partner (3, 4, 6) is also gone. Therefore the remaining 21 blocks build an elementary BIBD.

The dual BIBD of the bold printed elementary design above has the incidence matrix

$$\begin{pmatrix} 1 & 1 & 0 & 1 & 0 & 0 & 0 \\ 1 & 0 & 1 & 0 & 0 & 0 & 1 \\ 1 & 0 & 0 & 0 & 1 & 1 & 0 \\ 0 & 1 & 1 & 0 & 1 & 0 & 0 \\ 0 & 1 & 0 & 0 & 0 & 1 & 1 \\ 0 & 0 & 0 & 1 & 1 & 0 & 1 \\ 0 & 0 & 1 & 1 & 0 & 1 & 0 \end{pmatrix}.$$

The corresponding BIBD is

$$\{(1,2,3); (1,4,5), (1,6,7), (2,4,7), (2,5,6), (3,4,6)\ (3,5,7)\}$$

and of course elementary as well.

In the following we give some results where the necessary conditions (12.4), (12.5) and (12.6) are sufficient.

Theorem 12.4 (Hanani, 1961, 1975; Abel and Greig, 1998; Abel et al., 2001) The necessary conditions (12.4) to (12.6) are sufficient, if

$k = 3$ and $k = 4$ for all $v \geq 4$ and for all λ
$k = 5$ with exception of $v = 15$ and $\lambda = 2$
$k = 6$ for all $v \geq 7$ and $\lambda > 1$ with exception of $v = 21$ and $\lambda = 2$
$k = 7$ for all $v \geq 7$ and $\lambda = 0, 6, 7, 12, 18, 24, 30, 35, 36 \pmod{(42)}$ and all $\lambda > 30$ not divisible by 2 or 3
$k = 8$ for $\lambda = 1$, with 38 possible exceptions for v, namely, the values

113, 169, 176, 225, 281, 337, 393, 624, 736, 785, 1065, 1121, 1128, 1177, 1233, 1240, 1296, 1345, 1401, 1408, 1457, 1464, 1513, 1520, 1569, 1576, 1737, 1793, 1905, 1961, 2185, 2241, 2577, 2913, 3305, 3417, 3473, 3753.

From these 38 values of v exist $(v, 8, 2)$-BIBD with exception of $v = 393$, but for $\lambda = 2$ there are further values of v: 29, 36, 365, 477, 484, 533, 540, 589 for which the existence is not clear. The necessary conditions are sufficient for all $\lambda > 5$ and for $\lambda = 4$ if $v \neq 22$.

Because the proof of this theorem is enormous, we refer to the original literature. For $\lambda = 4$ and $v = 22$ there exists no BIBD, the smallest BIBD for $v = 22$ and $k = 8$ given by the R-programme OPDOE is that for $\lambda = 8$, $b = 66$ and $r = 24$.

Theorem 12.5 (Theorem 1.2, Abel et al. (2002a, 2002b, 2004))
The necessary conditions for the existence of a $(v, k = 9, \lambda)$-BIBD in the following cases are sufficient:

a) For $\lambda = 2$ (necessary conditions: $v \equiv 1,9 \pmod{36}$) with the possible exception of v = 189, 253, 505, 765, 837, 1197, 1837 and 1845
b) For $\lambda = 3$ (necessary conditions: $v \equiv 1,9 \pmod{24}$) with the possible exception of v = 177, 345 and 385
c) For $\lambda = 4$ (necessary conditions: $v \equiv 1,9 \pmod{18}$) with the possible exception v = 315, 459 and 783
d) For $\lambda = 6$ (necessary conditions: $v \equiv 1,9 \pmod{12}$) with the possible exception v=213
e) For $\lambda = 8$ (necessary conditions: $v \equiv 0,1 \pmod 9$)
f) For $\lambda = 9$ (necessary conditions: $v \equiv 1 \pmod 8$)
g) For $\lambda = 12$ (necessary conditions: $v \equiv 1,3 \pmod 6$ with $v \geq 9$)
h) For λ = 18, 24, 36, 72 and all further values of λ, not being divisor of 72

The proof is given in Abel et al. (2002a, 2002b, 2004), where it was stated that the possible exceptions could not be definite shown as exceptions, for all other block designs the existence was shown. Cases not yet clear are given in Tables 12.1 and 12.2.

Hanani (1989) showed that the necessary conditions (12.4), (12.5) and (12.6) are sufficient for the existence of a BIBD with k = 7 and λ =3 and λ =21 with the possible exception for the values λ = 3 and v = 323, 351, 407, 519, 525, 575, 665.

Sun (2012) showed that if the number of treatments is a prime power, in many cases the necessary conditions are sufficient for the existence of a BIBD.

Table 12.1 Values of v in not yet constructed $(v, k = 9, \lambda)$-BIBD with $\lambda = 1$.

145 153 217 225 289 297 361 369 505 793 865 873 945 1017 1081 1305 1441 1513 1585 1593
1665 1729 1809 1881 1945 1953 2025 2233 2241 2305 2385 2449 2457 2665 2737 2745 2881 2889 2961 3025 3097 3105 3241 3321 3385 3393 3601 3745 3753 3817 4033 4257 4321 4393 4401 4465 4473 4825 4833 4897 4905 5401 5473 5481 6049 6129 6625 6705 6769 6777 6913 7345 7353 7425 9505 10017 10665 12529 12537 13185 13753 13833 13969 14113 14473 14553 14625 14689 15049 15057 16497.

Table 12.2 $(v, k = 9, \lambda)$-BIBD with $\lambda > 1$ not yet constructed.

(177,9,3) (189,9,2) (213,9,6) (253,9,2) (315,9,4) (345,9,3) (385,9,3) (459,9,4) (505,9,2) (765,9,2) (783,9,4) (837,9,2) (1197,9,2) (1837,9,2) (1845,9,2)

Results for the existence of symmetric BIBD contains the following theorem:

Theorem 12.6 (Bruck-Ryser-Chowla-Theorem, Mohan et al. (2004))
If the parameters v; k; λ of a BIBD fulfil the existence condition (12.5) for $k = r$, so for the existence of a symmetric BIBD, it is necessary that either

a) v is even and $k - \lambda$ a is a square number or
b) v is odd and $z^2 = (k-\lambda)x^2 + (-1)^{\frac{v-1}{2}}\lambda y^2$ has a non-trivial integer solution x; y; z.

Some authors published tables of BIBD; the first for $r \le 10$ stems from Fisher and Yates (1963). For $11 \le r \le 15$ we find a table in Rao (1961) and for $16 \le r \le 20$ in Sprott (1962). Takeuchi (1962) gives further tables for $v \le 100$, $k \le 30$, $\lambda \le 14$. The parameter combinations of further tables are given in Raghavarao (1971) for $v \le 100$, $k \le 15$, $\lambda \le 15$; in Collins (1976) $v \le 50$, $k \le 23$, $\lambda \le 11$; in Mathon and Rosa (2006) for $r \le 41$ and in Mohan et al. (2004) for $v \le 111$ $k \le 55$ $\lambda \le 30$ (Colbourn and Dinitz, 2006).

12.2.2 Construction Methods of BIBD

In this section we show the multiplicity of methods of the construction of BIBD, but these are not exhaustive. Further methods are, for instance, given in Abel et al. (2004) or in Rasch et al. (2011). In the latter R-programme, methods are described using difference sets and difference families, not described here.

Definition 12.12 Let p be a prime. Then for an integer h put s = p. Each ordered set $X = (x_0,\ldots,x_n)$ of $n + 1$ elements x_i of a Galois field $GF(s)$ is a point of a (finite) projective geometry $PG(n,s)$. Two sets $Y = (y_0,\ldots,y_n)$ and $X = (x_0,\ldots,x_n)$ with $y_i = qx_i (i = 0,\ldots,n)$ and an element q of the $GF(s)$ unequal 0 represent the same point. The elements $x_i (i = 0,\ldots,n)$ of X are coordinates of X. All points of a $PG(n,s)$, fulfilling the n-m linear independent homogeneous equations $\sum_{i=0}^{n} a_{ji}x_i = 0; j = 1,\ldots,n-m; a_{ji} \in GF(s)$, create an m-dimensional subspace of the $PG(n,s)$. Subspaces with $x_0 = 0$ are subspaces in the infinite. In a $PG(n,s)$ there are $Q_n = \dfrac{s^{n+1}-1}{s-1}$ different points and $Q_m = \dfrac{s^{m+1}-1}{s-1}$ points in each m-dimensional subspace. The number of m-dimensional subspaces of a $PG(n,s)$ is

$$\varphi(n,m,s) = \frac{(s^{n+1}-1)(s^n-1)\ldots(s^{n-m+1}-1)}{(s^{m+1}-1)(s^m-1)\ldots(s-1)}, (m \ge 0; n \ge m). \tag{12.7}$$

The number of different m-dimensional subspaces of a $PG(n,s)$, having no point in common, is

$$\varphi(n,m,s)\frac{s^{m+1}-1}{s^{n+1}-1} (= \varphi(n-1,m-1,s) \text{ if } m \ge 1).$$

The number of different m-dimensional subspaces of a $PG(n,s)$ with two different points in common is

$$\varphi(n,m,s)\frac{(s^{m+1}-1)(s^m-1)}{(s^{n+1}-1)(s^n-1)}\ (=\varphi(n-2,m-2,s)\,,\text{if } m\geq 2).$$

Method 12.1 We construct a $PG(n,s)$ and consider their points as the v treatments and for each m, the m-dimensional subspace as a block. This gives a BIBD with

$$v=\frac{s^{n+1}-1}{s-1},$$

$$b=\varphi(n,m,s),$$

$$r=\frac{s^{m+1}-1}{s^{n+1}-1}\varphi(n,m,s),$$

$$k=\frac{s^{m+1}-1}{s-1},$$

$$\lambda=\frac{(s^{m+1}-1)\cdot(s^m-1)}{(s^{n+1}-1)\cdot(s^n-1)}\varphi(n,m,s)$$

where $\varphi(n, m, s)$ is defined in Definition 12.12.

Example 12.6 We construct a $PG(3,2)$ with s $=p = 2$; $h = 1$ and $n = 3$. The GF (2) is {0, 1}, a minimal function we do not need, because $h = 1$. The 15 elements (treatments) of the $PG(3,2)$ are all possible combinations of (0;1)-values in $X=(x_0,\ldots,x_3)$ with the exception of (0,0,0,0):

$$\{(1,0,0,0), (0,1,0,0), (0,0,1,0), (0,0,0,1), (1,1,0,0), (1,0,1,0), (1,0,0,1), (0,1,1,0),$$
$$(0,1,0,1), (0,0,1,1), (1,1,1,0), (1,1,0,1), (1,0,1,1), (0,1,1,1), (1,1,1,1)\}.$$

With m =2 the equation ($n - m = 1$) for the two dimensional subspaces is $a_0+a_1x_1+a_2x_2+a_3x_3=0$ with all combinations of coefficients of the $GF(2)$ (except (0,0,0,0)). These are just the same quadruple as the 15 points above. We create now a (15×15) matrix with rows defined by the treatments and columns defined by the subspaces (blocks). In each cell of the matrix, we insert a 1 if the point lies in the block and a 0 otherwise. We consider the first block defined by $a_0 = 0$. All points with a_0 at the first place are in that block. These are the points 2, 3, 4, 8, 9, 10 and 14. The second equation is $x_1 = 0$. In that block are all points with a_0 as the second entry. These are the points 1, 3, 4, 6, 7, 10 and 13. So we continue with all 15 blocks and receive the symmetric BIBD with $v = b = 15$, $r = k = 7$ and $\lambda = 3$.

Block	Treatments						
1	1	2	4	5	8	10	15
2	2	3	5	6	9	11	1
3	3	4	6	7	10	12	2
4	4	5	7	8	11	13	3
5	5	6	8	9	12	14	4
6	6	7	9	10	13	15	5
7	7	8	10	11	14	1	6
8	8	9	11	12	15	2	7
9	9	10	12	13	1	3	8
10	10	11	13	14	2	4	9
11	11	12	14	15	3	5	10
12	12	13	15	1	4	6	11
13	13	14	1	2	5	7	12
14	14	15	2	3	6	8	13
15	15	1	3	4	7	9	14

Definition 12.13 Let p be a prime. Then for an integer h is $s = p^h$. Each ordered set $X = (x_1,\dots,x_n)$ of n elements x_i of a $GF(s)$ is a point of a (finite) Euclidean geometry $EG(n,s)$. Two sets $Y = (y_0,\dots,y_n)$ and $X = (x_0,\dots,x_n)$ with $y_i = x_i (i = 0,\dots,n)$ represent the same point. The elements $x_i (i = 1,\dots,n)$ of X are coordinates of X. All points of a $EG(n,s)$, fulfilling the $n - m$ linear independent equations $\sum_{i=1}^{n} a_{ji}x_i = 0; j = 1,\dots,n-m; a_{ji} \in GF(s)$ and $x_0 = 1$, create an m-dimensional subspace of the $EG(n,s)$.

In $EG(n,s)$ there are s^n different points and s^m points in each m-dimensional subspace. The number of m-dimensional subspaces of an $EG(n,s)$ passing through one fixed point is

$$\varphi(n-1, m-1, s).$$

The number of different m-dimensional subspaces of an $EG(n,s)$ with two different points in common is

$$\varphi(n-2, m-2, s).$$

Method 12.2 We can construct an $EG(n,s)$ and consider its points as v treatments and for each m, the m-dimensional subspaces as block. This gives a BIBD with

$$
\begin{aligned}
v &= s^n, \\
b &= \varphi(n,m,s) - \varphi(n-1,m,s), \\
r &= \frac{s^{m+1}-1}{s^{n+1}-1}\varphi(n,m,s), \\
k &= s^m, \\
\lambda &= \frac{(s^{m+1}-1)\cdot(s^m-1)}{(s^{n+1}-1)\cdot(s^n-1)}\varphi(n,m,s)
\end{aligned}
$$

Example 12.7 We construct an EG(3,2) with s $=p = 2$; $h = 1$, $n = 3$ and $m = 2$. The parameters of the block design are

$$
\begin{aligned}
v &= 2^3 = 8, \\
b &= \varphi(3,2,2) - \varphi(2,2,2) = 15 - 1 = 14, \\
r &= \frac{s^3-1}{s^4-1}\cdot 15 = 7, \\
k &= s^2 = 4, \\
\lambda &= \frac{(s^3-1)\cdot(s^2-1)}{(s^4-1)\cdot(s^3-1)}\cdot 15 = 3
\end{aligned}
$$

and the block design is

Block	Treatments			
1	1	3	5	7
2	1	2	5	6
3	1	4	5	8
4	1	2	3	4
5	1	3	6	8
6	1	2	7	8
7	1	4	6	7
8	2	4	6	8
9	3	4	7	8
10	2	3	6	7
11	5	6	7	8
12	2	4	5	7
13	3	4	5	6
14	2	3	5	8

Method 12.3 If $\mathcal{N}$ is the incidence matrix of a BIBD with parameters

$$v = b = 4l + 3, r = k = 2l + 1 \text{ and } \lambda = l; (l = 1, 2, \ldots)$$

and if $\widetilde{\mathcal{N}}$ is the incidence matrix of the complementary BIBD, then the matrix $\mathcal{N}^* = \begin{pmatrix} \mathcal{N} & \widetilde{\mathcal{N}} \\ 0_v^T & 1_v^T \end{pmatrix}$ is the incidence matrix of a BIBD ($4l + 4$, $8l + 6$, $4l + 3$, $2l + 2$, $2l + 1$).

Example 12.8 Let $l = 1$ then

$$\mathcal{N} = \begin{pmatrix} 1 & 1 & 1 & 0 & 0 & 0 & 0 \\ 1 & 0 & 0 & 1 & 1 & 0 & 0 \\ 0 & 1 & 0 & 1 & 0 & 1 & 0 \\ 1 & 0 & 0 & 0 & 0 & 1 & 1 \\ 0 & 0 & 1 & 1 & 0 & 0 & 1 \\ 0 & 0 & 1 & 0 & 1 & 1 & 0 \\ 0 & 1 & 0 & 0 & 1 & 0 & 1 \end{pmatrix} \text{ and } \widetilde{\mathcal{N}} = \begin{pmatrix} 0 & 0 & 0 & 1 & 1 & 1 & 1 \\ 0 & 1 & 1 & 0 & 0 & 1 & 1 \\ 1 & 0 & 1 & 0 & 1 & 0 & 1 \\ 0 & 1 & 1 & 1 & 1 & 0 & 0 \\ 1 & 1 & 0 & 0 & 1 & 1 & 0 \\ 1 & 1 & 0 & 1 & 0 & 0 & 1 \\ 1 & 0 & 1 & 1 & 0 & 1 & 0 \end{pmatrix}.$$

This results in

$$\mathcal{N}^* = \begin{pmatrix} 1 & 1 & 1 & 0 & 0 & 0 & 0 & 0 & 0 & 0 & 1 & 1 & 1 & 1 \\ 1 & 0 & 0 & 1 & 1 & 0 & 0 & 0 & 1 & 1 & 0 & 0 & 1 & 1 \\ 0 & 1 & 0 & 1 & 0 & 1 & 0 & 1 & 0 & 1 & 0 & 1 & 0 & 1 \\ 1 & 0 & 0 & 0 & 0 & 1 & 1 & 0 & 1 & 1 & 1 & 1 & 0 & 0 \\ 0 & 0 & 1 & 1 & 0 & 0 & 1 & 1 & 1 & 0 & 0 & 1 & 1 & 0 \\ 0 & 0 & 1 & 0 & 1 & 1 & 0 & 1 & 1 & 0 & 1 & 0 & 0 & 1 \\ 0 & 1 & 0 & 0 & 1 & 0 & 1 & 1 & 0 & 1 & 1 & 0 & 1 & 0 \\ 1 & 1 & 1 & 1 & 1 & 1 & 1 & 0 & 0 & 0 & 0 & 0 & 0 & 0 \end{pmatrix}.$$

This is the incidence matrix of a BIBD with $v = 8$, $b = 14$, $r = 7$, $k = 4$ and $\lambda = 3$, and it is isomorphic with that in Example 12.7.

As we have seen, different methods can lead to the same block design. We now need the minimal functions of a $GF(p^h)$ as presented in Table 12.3. A minimal function $P(x)$ can be used to generate the elements of a $GF(p^h)$. We need the function

$$f(x) = a_0 + a_1 x + \cdots + a_{h-1} x^{h-1}$$

Table 12.3 Minimal functions $P(x)$ of a $GF(p^h)$.

p	h	P(x)	p	h	P(x)	p	h	P(x)
2	2	x^2+x+1	5	2	x^2+2x+3	11	2	x^2+x+7
	3	x^3+x^2+1		3	x^3+x^2+2		3	x^3+x^2+3
	4	x^4+x^3+1		4	$x^4+x^3+2x^2+2$		4	x^4+4x^3+2
	5	x^5+x^3+1		5	x^5+x^2+2		5	$x^5+x^3+x^2+9$
	6	x^6+x^5+1		6	x^6+x^5+2	13	2	x^2+x+2
3	2	x^2+x+2	7	2	x^2+x+3		3	x^3+x^2+2
	3	x^3+2x+1		3	x^3+x^2+x+2		4	$x^4+x^3+3x^2+2$
	4	x^4+x+2		4	$x^4+x^3+x^2+3$	17	2	x^2+x+3
	5	x^5+2x^4+1		5	x^5+x^4+4		3	x^3+x+3
	6	x^6+x^5+2		6	$x^6+x^5+x^4+3$		4	x^4+4x^2+x+3

with integer coefficients $a_i\,(i=0,\dots,h-1)$ as the elements of a $GF(p)$. The function

$$F(x)=f(x)+pq(x)+P(x)Q(x) \tag{12.8}$$

with the minimal function $P(x)$ and certain polynomials $q(x)$ and $Q(x)$ creates a class of functions, the residues modulo p and $P(x)$. We write

$$F(x)\equiv f(x)(\mathrm{mod}\,p;P(x)). \tag{12.9}$$

If p and $P(x)$ are fixed and $f(x)$ is variable $F(x)$ generates just p^h classes (functions) representing a $GF(p^h)$ iff p is prime and $P(x)$ is a minimal function of $GF(p^h)$.

Method 12.4 If $v=p^m$, where p is prime and m a natural number with the elements of a GF $\{a_0=0;a_1=1;\dots,a_{v-1}\}$, we construct $v-1$ LS (see Section 12.3) $A_l=\left(a_{ij}^{(l)}\right)$; $l=1,\dots,v-1$ as follows: $A_1=\left(a_{ij}^{(1)}\right)$ is the addition table of a group, the elements of $A_1=\left(a_{ij}^{(1)}\right)$; $t=2,\dots,v-1$ are $a_{ij}^t=a_{ij}^1\cdot a_t$. We construct the $v(v-1)$ matrix $A=(A_1,\dots,A_{v-1})$. With the desired block size k, we choose k different elements from the GF. Each column of A defines one block of the BIBD; its elements are just the row numbers of A of the k selected elements of the *GF*. If each block occurs $w\ge 2$ times, we delete $w-1$ copies. To find out, whether blocks occur more than once, we order the elements in the blocks lexicographically. The parameters of the original BIBD are

$$v=p^m;\,b=v(v-1);\,r=k(v-1);\,k;\,\lambda=k(k-1).$$

The reduced BIBD then has the parameters

$$v^* = v,\ b^* = \frac{b}{w},\ r^* = \frac{r}{w},\ k^* = k,\ \lambda^* = \frac{\lambda}{w}.$$

Example 12.9 We try to construct a BIBD with $v = 9$. For $v = 9 = 3^2$ is $p = 3$; $m = 2$. The minimal function is $x^2 + x + 2$ and $f(x) = \alpha_0 + \alpha_1 x$ with coefficients $\alpha_i; i = 0, 1$ from $GF(3) = \{0, 1, 2\}$. The function $F(x) \equiv f(x) (\bmod 3; x^2 + x + 2)$ gives the nine elements of $GF(9)$ for all values of $f(x)$:

α_0	α_1	f(x) = F(x)
0	0	$a_0 = 0$
0	1	$a_2 = x$
0	2	$a_3 = 2x$
1	0	$a_1 = 1$
1	1	$a_4 = 1 + x$
1	2	$a_5 = x^2 = 1 + 2x$
2	0	$a_6 = 2$
2	1	$a_7 = 2 + x$
2	2	$a_8 = 2 + 2x$

The addition table of $GF(9)$ is a LS:

$$\begin{pmatrix}
0 & 1 & x & 2x & 1+x & 1+2x & 2 & 2+x & 2+2x \\
1 & 2 & 1+x & 1+2x & 2+x & 2+2x & 0 & x & 2x \\
x & 1+x & 2x & 0 & 1+2x & 1 & 2+x & 2+2x & 2 \\
2x & 1+2x & 0 & x & 1 & 1+x & 2+2x & 2 & 2+x \\
1+x & 2+x & 1+2x & 1 & 2+2x & 2 & x & 2x & 0 \\
1+2x & 2+2x & 1 & 1+x & 2 & 2+x & 2x & 0 & x \\
2 & 0 & 2+x & 2+2x & x & 2x & 1 & 1+x & 1+2x \\
2+x & x & 2+2x & 2 & 2x & 0 & 1+x & 1+2x & 1 \\
2+2x & 2x & 2 & 2+x & 0 & x & 1+2x & 1 & 1+x
\end{pmatrix}$$

The seven other matrices are (at first we multiply with $a_2 = x$) the following:

$$\begin{pmatrix}
0 & x & 1+2x & 2+x & 1 & 2+2x & 2x & 1+x & 2\\
x & 2x & 1 & 2+2x & 1+x & 2 & 0 & 1+2x & 2+x\\
1+2x & 1 & 2+x & 0 & 2+2x & x & 1+x & 2 & 2x\\
2+x & 2+2x & 0 & 1+2x & x & 1 & 2 & 2x & 1+x\\
1 & 1+x & 2+2x & x & 2 & 2x & 1+2x & 2+x & 0\\
2+2x & 2 & x & 1 & 2x & 1+x & 2+x & 0 & 2+2x\\
2x & 0 & 1+x & 2 & 1+2x & 2+x & x & 1 & 2+2x\\
1+x & 1+2x & 2 & 2x & 2+x & 0 & 1 & 2+2x & x\\
2 & 2+x & 2x & 1+x & 0 & 2+2x & 2+2x & x & 1
\end{pmatrix}$$

$$\begin{pmatrix}
0 & 2x & 2+x & 1+2x & 2 & 1+x & x & 2+2x & 1\\
2x & x & 2 & 1+x & 2+2x & 1 & 0 & 2+x & 1+2x\\
2+x & 2 & 1+2x & 0 & 1+x & 2x & 2+2x & 1 & x\\
1+2x & 1+x & 0 & 2+x & 2x & 2 & 1 & x & 2+2x\\
2 & 2+2x & 1+x & 2x & 1 & x & 2+x & 1+2x & 0\\
1+x & 1 & 2x & 2 & x & 2+2x & 1+2x & 0 & 2+x\\
x & 0 & 2+2x & 1 & 2+x & 1+2x & 2x & 2 & 1+x\\
2+2x & 2+x & 1 & x & 1+2x & 0 & 2 & 1+x & 2x\\
1 & 1+2x & x & 2+2x & 0 & 2+x & 1+x & 2x & 2
\end{pmatrix}$$

$$\begin{pmatrix}
0 & 1+x & 1 & 2 & 2+x & x & 2+2x & 2x & 1+2x\\
1+x & 2+2x & 2+x & x & 2x & 1+2x & 0 & 1 & 2\\
1 & 2+x & 2 & 0 & x & 1+x & 2x & 1+2x & 2+2x\\
2 & x & 0 & 1 & 1+x & 2+x & 1+2x & 2+2x & 2x\\
2+x & 2x & x & 1+x & 1+2x & 2+2x & 1 & 2 & 0\\
x & 1+2x & 1+x & 2+x & 2+2x & 2x & 2 & 0 & 1\\
2+2x & 0 & 2x & 1+2x & 1 & 2 & 1+x & 2+x & x\\
2x & 1 & 1+2x & 2+2x & 2 & 0 & 2+x & x & 1+x\\
1+2x & 2 & 2+2x & 2x & 0 & 1 & x & 1+x & 2+x
\end{pmatrix}$$

$$\begin{pmatrix}
0 & 1+2x & 2+2x & 1+x & x & 2 & 2+x & 1 & 2x \\
1+2x & 2+x & x & 2 & 1 & 2x & 0 & 2+2x & 1+x \\
2+2x & x & 1+x & 0 & 2 & 1+2x & 1 & 2x & 2+x \\
1+x & 2 & 0 & 2+2x & 1+2x & x & 2x & 2+x & 1 \\
x & 1 & 2 & 1+2x & 2x & 2+x & 2+2x & 1+x & 0 \\
2 & 2x & 1+2x & x & 2+x & 1 & 1+x & 0 & 2+2x \\
2+x & 0 & 1 & 2x & 2+2x & 1+x & 1+2x & x & 2 \\
1 & 2+2x & 2x & 2+x & 1+x & 0 & x & 2 & 1+2x \\
2x & 1+x & 2+x & 1 & 0 & 2+2x & 2 & 1+2x & x
\end{pmatrix}$$

$$\begin{pmatrix}
0 & 2 & 2x & x & 2+2x & 2+x & 1 & 1+2x & 1+x \\
2 & 1 & 2+2x & 2+x & 1+2x & 1+x & 0 & 2x & x \\
2x & 2+2x & x & 0 & 2+x & 2 & 1+2x & 1+x & 1 \\
x & 2+x & 0 & 2x & 2 & 2+2x & 1+x & 1 & 1+2x \\
2+2x & 1+2x & 2+x & 2 & 1+x & 1 & 2x & x & 0 \\
2+x & 1+x & 2 & 2+2x & 1 & 1+2x & x & 0 & 2x \\
1 & 0 & 1+2x & 1+x & 2x & x & 2 & 2+2x & 2+x \\
1+2x & 2x & 1+x & 1 & x & 0 & 2+2x & 2+x & 2 \\
1+x & x & 1 & 1+2x & 0 & 2x & 2+x & 2 & 2+2x
\end{pmatrix}$$

$$\begin{pmatrix}
0 & 2+x & 1+x & 2+2x & 2x & 1 & 1+2x & 2 & x \\
2+x & 1+2x & 2x & 1 & 2 & x & 0 & 1+x & 2+2x \\
1+x & 2x & 2+2x & 0 & 1 & 2+x & 2 & x & 1+2x \\
2+2x & 1 & 0 & 1+x & 2+x & 2x & x & 1+2x & 2 \\
2x & 2 & 1 & 2+x & x & 1+2x & 1+x & 2+2x & 0 \\
1 & x & 2+x & 2x & 1+2x & 2 & 2+2x & 0 & 1+x \\
1+2x & 0 & 2 & x & 1+x & 2+2x & 2+x & 2x & 1 \\
2 & 1+x & x & 1+2x & 2+2x & 0 & 2x & 1 & 2+x \\
x & 2+2x & 1+2x & 2 & 0 & 1+x & 1 & 2+x & 2x
\end{pmatrix}$$

$$\begin{pmatrix} 0 & 2+2x & 2 & 1 & 1+2x & 2x & 1+x & x & 2+x \\ 2+2x & 1+x & 1+2x & 2x & x & 2+x & 0 & 2 & 1 \\ 2 & 1+2x & 1 & 0 & 2x & 2+2x & x & 2+x & 1+x \\ 1 & 2x & 0 & 2 & 2+2x & 1+2x & 2+x & 1+x & x \\ 1+2x & x & 2x & 2+2x & 2+x & 1+x & 2 & 1 & 0 \\ 2x & 2+x & 2+2x & 1+2x & 1+x & x & 1 & 0 & 2 \\ 1+x & 0 & x & 2+x & 2 & 1 & 2+2x & 1+2x & 2x \\ x & 2 & 2+x & 1+x & 1 & 0 & 1+2x & 2x & 2+2x \\ 2+x & 1 & 1+x & x & 0 & 2 & 2x & 2+2x & 1+2x \end{pmatrix}.$$

We now choose the four elements 0; 1;2;x and get the blocks (1,2,3,7) [from the first row of the addition table] and

(1,2,7,8); (1,4,6,9); (3,4,5,8); (4,6,7,9); (3,5,8,9); (1,2,5,7); (2,4,6,9); (3,5,6,8).

From the next matrix we get

(1,2,5,9); (1,3,6,7); (2,4,6,8); (3,5,6,7); (1,4,5,9); (2,3,4,8); (2,3,7,8); (3,6,7,9); (1,5,8,9).

We continue in this way and get

(1,5,7,9); (2,3,6,7); (2,4,8,9); (3,6,7,8); (1,5,6,9); (2,4,5,8); (1,2,4,8); (3,4,6,7); (1,3,5,9);
(1,3,4,6); (4,7,8,9); (1,3,4,5); (1,2,3,4); (3,7,8,9); (1,7,8,9); (2,5,6,9); (2,5,6,8); (2,5,6,7);
(1,5,6,8); (3,4,5,7); (2,4,5,7); (2,3,6,9); (1,2,3,9); (1,4,6,8); (2,3,8,9); (1,6,7,8); (4,5,7,9);
(1,2,4,7); (1,2,7,9); (3,4,6,9); (1,3,5,8); (4,6,8,9); (3,5,7,8); (1,2,6,7); (4,5,6,9); (2,3,5,8);
(1,6,8,9); (4,5,6,7); (4,5,7,8); (2,3,7,9); (2,3,5,9); (1,2,6,8); (2,3,4,9); (1,3,6,8); (1,4,5,7);
(1,3,4,8); (5,7,8,9); (1,3,4,7); (1,3,4,9); (2,7,8,9); (6,7,8,9); (2,3,5,6); (1,2,5,6); (2,4,5,6).

All blocks are different, which means $w = 1$ and $v = 9; b = 72; r = 32; k = 4; \lambda = 12$.

We know from Theorem 12.3 that for $k = 4$, a BIBD exists with parameter fulfilling the necessary conditions [$v = 9; b = 18; r = 8; k = 4; \lambda = 3$]. This shows that Method 12.4 even for $w = 1$ does not necessarily lead to a smallest BIBD. We recommend, therefore, to use this method only if no other method for the pair (v,k) is available.

Method 12.5 A BIBD with parameters $v = s^2, b = s(s+1), k = s$ can be partitioned into $s + 1$ groups with s blocks each. The blocks of the groups 2 to $s + 1$ are $(s - 1)$ times included into the BIBD to be constructed. The blocks from group 1 occur once. Finally this so obtained set is complemented by all

$(s-1)$-tuples from the blocks of group 1 complemented by the treatment $v+1$. The so constructed BIBD has the parameters

$$v = s^2 + 1; k = s; b = s(s^2 + 1); r = s^2; \lambda = s - 1.$$

Example 12.10 We construct a BIBD with parameters $v = 10; k = 3; b = 30; r = 9; \lambda = 2$. The BIBD with the parameters $v = 9; k = 3; b = 12; r = 4; \lambda = 1$ $(s = 3)$ is written in four groups:

$$\text{Group 1: } \begin{Bmatrix} (1,2,6) \\ (3,4,5) \\ (7,8,9) \end{Bmatrix}; \text{ Group 2: } \begin{Bmatrix} (1,3,7) \\ (2,4,9) \\ (5,6,8) \end{Bmatrix}; \text{ Group 3: } \begin{Bmatrix} (1,4,8) \\ (2,5,7) \\ (3,6,9) \end{Bmatrix}.;$$

$$\text{Group 4: } \begin{Bmatrix} (1,5,9) \\ (2,3,8) \\ (4,6,7) \end{Bmatrix}.$$

The blocks of groups 2–4 are used twice for the BIBD to be constructed, and group 1 is used once giving all together 21 blocks. The nine pairs (1,2), (1,6), (2,6), (3,4), (3,5), (4,5), (7,8, (7,9) and (8,9) from the blocks of group 1 are complemented by the treatment 10 giving nine more blocks and finally the design with $v = 10; k = 3; b = 30; r = 9; \lambda = 2$.

In this BIBD some (but not all) blocks occur repeatedly.

Definition 12.14 A square matrix H_n of order n with elements -1 and $+1$ is a Hadamard matrix, if $H_n H_n^T = nI_n$.

A necessary condition for the existence of a Hadamard matrix for $n > 2$ is $n \equiv 0 \pmod 4$. This necessary condition is sufficient for all $n<201$.

Trivially $H_1 = (1); H_2 = \begin{pmatrix} 1 & 1 \\ 1 & -1 \end{pmatrix}$.

Each Hadamard matrix w.l.o.g. can be written in a normal form in which the first row and the first column contains only the elements $+1$ and the Kronecker product $H_{n_1} \otimes H_{n_2} = H_{n_1 n_2}$ of two Hadamard matrices $H_{n_1}; H_{n2}$ is a Hadamard matrix of order $n_1 \cdot n_2$.

Method 12.6 Let H be a Hadamard matrix of order $n = 4\,t$ in normal form and B be the matrix gained from H by deleting the first row and the first column. In B we replace the elements -1 by 0 and receive the incidence matrix of a BIBD with $v = b = 4t - 1;\ r = k = 2t - 1,\ \lambda = t - 1$.

Example 12.11 A BIBD with $v = b = 15;\ r = k = 7,\ \lambda = 3$ is obtained from a Hadamard matrix of order 16 $(t = 4)$ in normal form

$$\begin{pmatrix}
1 & 1 & 1 & 1 & 1 & 1 & 1 & 1 & 1 & 1 & 1 & 1 & 1 & 1 & 1 & 1 \\
1 & -1 & 1 & -1 & 1 & -1 & 1 & -1 & 1 & -1 & 1 & -1 & 1 & -1 & 1 & -1 \\
1 & 1 & -1 & -1 & 1 & 1 & -1 & -1 & 1 & 1 & -1 & -1 & 1 & 1 & -1 & -1 \\
1 & -1 & -1 & 1 & 1 & -1 & -1 & 1 & 1 & -1 & -1 & 1 & 1 & -1 & -1 & 1 \\
1 & 1 & 1 & 1 & -1 & -1 & -1 & -1 & 1 & 1 & 1 & 1 & -1 & -1 & -1 & -1 \\
1 & -1 & 1 & -1 & -1 & 1 & -1 & 1 & 1 & -1 & 1 & -1 & -1 & 1 & -1 & 1 \\
1 & 1 & -1 & -1 & -1 & -1 & 1 & 1 & 1 & 1 & -1 & -1 & -1 & -1 & 1 & 1 \\
1 & -1 & -1 & 1 & -1 & 1 & 1 & -1 & 1 & -1 & -1 & 1 & -1 & 1 & 1 & -1 \\
1 & 1 & 1 & 1 & 1 & 1 & 1 & 1 & -1 & -1 & -1 & -1 & -1 & -1 & -1 & -1 \\
1 & -1 & 1 & -1 & 1 & -1 & 1 & -1 & -1 & 1 & -1 & 1 & -1 & 1 & -1 & 1 \\
1 & 1 & -1 & -1 & 1 & 1 & -1 & -1 & -1 & -1 & 1 & 1 & -1 & -1 & 1 & 1 \\
1 & -1 & -1 & 1 & 1 & -1 & -1 & 1 & -1 & 1 & 1 & -1 & -1 & 1 & 1 & -1 \\
1 & 1 & 1 & 1 & -1 & -1 & -1 & -1 & -1 & -1 & -1 & -1 & 1 & 1 & 1 & 1 \\
1 & -1 & 1 & -1 & -1 & 1 & -1 & 1 & -1 & 1 & -1 & 1 & 1 & -1 & 1 & -1 \\
1 & 1 & -1 & -1 & -1 & -1 & 1 & 1 & -1 & -1 & 1 & 1 & 1 & 1 & -1 & -1 \\
1 & -1 & -1 & 1 & -1 & 1 & 1 & -1 & -1 & 1 & 1 & -1 & 1 & -1 & -1 & 1
\end{pmatrix}$$

We delete the first row and the first column and replace −1 by 0:

$$\begin{pmatrix}
0 & 1 & 0 & 1 & 0 & 1 & 0 & 1 & 0 & 1 & 0 & 1 & 0 & 1 & 0 \\
1 & 0 & 0 & 1 & 1 & 0 & 0 & 1 & 1 & 0 & 0 & 1 & 1 & 0 & 0 \\
0 & 0 & 1 & 1 & 0 & 0 & 1 & 1 & 0 & 0 & 1 & 1 & 0 & 0 & 1 \\
1 & 1 & 1 & 0 & 0 & 0 & 0 & 1 & 1 & 1 & 1 & 0 & 0 & 0 & 0 \\
0 & 1 & 0 & 0 & 1 & 0 & 1 & 1 & 0 & 1 & 0 & 0 & 1 & 0 & 1 \\
1 & 0 & 0 & 0 & 0 & 1 & 1 & 1 & 1 & 0 & 0 & 0 & 0 & 1 & 1 \\
0 & 0 & 1 & 0 & 1 & 1 & 0 & 1 & 0 & 0 & 1 & 0 & 1 & 1 & 0 \\
1 & 1 & 1 & 1 & 1 & 1 & 1 & 0 & 0 & 0 & 0 & 0 & 0 & 0 & 0 \\
0 & 1 & 0 & 1 & 0 & 1 & 0 & 0 & 1 & 0 & 1 & 0 & 1 & 0 & 1 \\
1 & 0 & 0 & 1 & 1 & 0 & 0 & 0 & 0 & 1 & 1 & 0 & 0 & 1 & 1 \\
0 & 0 & 1 & 1 & 0 & 0 & 1 & 0 & 1 & 1 & 0 & 0 & 1 & 1 & 0 \\
1 & 1 & 1 & 0 & 0 & 0 & 0 & 0 & 0 & 0 & 1 & 1 & 1 & 1 & 1 \\
0 & 1 & 0 & 0 & 1 & 0 & 1 & 0 & 1 & 0 & 1 & 1 & 0 & 1 & 0 \\
1 & 0 & 0 & 0 & 0 & 1 & 1 & 0 & 0 & 1 & 1 & 1 & 1 & 0 & 0 \\
0 & 0 & 1 & 0 & 1 & 1 & 0 & 0 & 1 & 1 & 0 & 1 & 0 & 0 & 1
\end{pmatrix}$$

This is the incidence matrix of the design:

$$(2,4,6,8,10,12,14);(1,4,5,8,9,12,13);(3,4,7,8,11,12,15);(1,2,3,8,9,10,11);(2,5,7,8,10,13,15);$$
$$(1,6,7,8,9,14,15);(3,5,6,8,11,13,14);(1,2,3,4,5,6,7);(2,4,6,9,11,13,15):(1,4,5,10,11,14,15);$$
$$(3,4,7,9,10,13,14);(1,2,3,12,13,14,15);(2,5,7,9,11,12,14);(1,6,7,10,11,12,13);(3,5,6,9,10,12,15).$$

We find that all pairs occur three times and each element seven times.

Method 12.7 Let $v = p^n = m(\lambda - 1) + 1$, p *a* prime, $m \geq 1$ and x a primitive element of $GF(v)$. The blocks $\left(0, x^i, x^{i+m}, x^{i+2m}, \ldots, x^{i+(\lambda-2)m}\right), i = 0, \ldots, m-1$ are so-called initial blocks. From these initial blocks we construct a BIBD with v; $b = mv$; $k = \lambda$; $r = mk$, λ by adding modulo p after increasing all elements by 1.

Example 12.12 We construct a BIBD for $v = p = 29 = 7 \cdot 4 + 1$, with $m = 7$, $\lambda = 5$. The initial blocks are $\left(0, x^i, x^{i+7}, x^{i+14}, x^{i+21}\right), i = 0, \ldots, 6$, a primitive element of $GF(29)$ is $x = 2$. We receive a BIBD with $b = 203$ blocks, $k = \lambda = 5$ and $r = 35$. The initial block for $i = 0$ is, for instance, $(0, 1, 2^7 = 128 = 12, 2^{14} = 28, 2^{21} = 17)$. Adding 1 to these treatments, we obtain the next of the 29 blocks of this initial block, namely, (1,2,13,0,18). Adding 1 to all treatments, results in the first two blocks, that is, (1,2,13,18,29) and (1,2,3,14,19). Thus all 203 blocks can be generated.

Method 12.8 In a symmetric BIBD with parameters $v = b$, $k = r$, λ, we delete one block and then delete from all other blocks all the elements occurring in the deleted block. By this we obtain a BIBD with parameters

$$v^* = v - k, b^* = v - 1, k^* = k - \lambda, r^* = k, \lambda^* = \lambda.$$

If particularly $v = b = 4t - 1$; $r = k = 2t - 1$, $\lambda = t - 1$, we get a BIBD with

$$v^* = 2t, b^* = 4t - 2, k^* = t, r^* = 2t - 1, \lambda^* = t - 1.$$

This BIBD is said to be a residual design to the initial BIBD.

Example 12.13 We start with the symmetric BIBD of Example 12.6 with $v = b = 15$, $r = k = 7$ and $\lambda = 3$ and delete its first block and then in the other blocks all treatments occurring in the deleted block (bold print in the scheme):

Block	Treatments						
1	**1**	**2**	**4**	**5**	**8**	**10**	**15**
2	**2**	3	**5**	6	9	11	**1**
3	3	**4**	6	7	**10**	12	**2**
4	**4**	**5**	7	**8**	11	13	3
5	**5**	6	**8**	9	12	14	**4**
6	6	7	9	**10**	13	**15**	**5**
7	7	**8**	**10**	11	14	**1**	6

Block	Treatments						
8	**8**	9	11	12	**15**	**2**	7
9	9	**10**	12	13	**1**	3	**8**
10	**10**	11	13	14	**2**	**4**	9
11	11	12	14	**15**	3	**5**	**10**
12	12	13	**15**	**1**	**4**	6	11
13	13	14	**1**	**2**	**5**	7	12
14	14	**15**	**2**	3	6	**8**	13
15	**15**	**1**	3	**4**	7	9	14.

We rename the remaining eight treatments with 3 in 1, 6 in 2, 7 in 3, 9 in 4, 11 in 5, 12 in 6, 13 in 7, 14 in 8 and obtain the BIBD:

Block	Treatments			
1	1	2	4	5
2	1	2	3	6
3	3	5	7	1
4	2	4	6	8
5	2	3	4	7
6	3	5	8	2
7	4	5	6	3
8	4	6	7	1
9	5	7	8	4
10	5	6	8	1
11	6	7	2	5
12	7	8	3	6
13	8	1	2	7
14	1	3	4	8.

Method 12.9 From a symmetric BIBD with parameters $v = b$, $k = r$, λ, we delete one block and in the remaining blocks we drop the treatments, not contained in this block. We obtain a BIBD with parameters $v^* = k, b^* = v-1, k^* = \lambda$, $r^* = k-1, \lambda^* = \lambda - 1$.

Example 12.14 We choose the design of Example 12.13 with $v = b = 15$, $r = k = 7$ and $\lambda = 3$ and delete its first block and then in the other blocks all treatments not occurring in the deleted block and rename as in Example 12.13.

We obtain the blocks (1,2,4), (2,3,6), (3,4,5), (3,4,5), (4,6,7), (1,5,6), (2,5,7), (1,5,6), (2,3,6), (4,6,7), (1,3,7), (1,2,4), (2,5,7) and (1,3,7).

In this BIBD each block occurs twice. We reduce it to a BIBD with 7 blocks dropping one of these blocks and obtain the BIBD:

(1,2,4), (1,3,7), (1,5,6), (2,3,6), (2,5,7), (3,4,5), (4,6,7) with $v = 7$, $b = 7$, $k = r = 3$ and $\lambda = 1$.

12.2.3 Partially Balanced Incomplete Block Designs

Partially balanced incomplete block designs are of less practical interest than completely balanced ones. They do not allow estimating all treatment differences with equal precision.

Definition 12.15 We consider v treatments $1,2,\ldots,v$, an association scheme with m classes fulfils the conditions:

1) Two given treatments are either first, second,…or mth associates.
2) Each treatment w in $\{1,2,\ldots,v\}$has n_i ith associates ($i = 1,\ldots, m$); the number n_i does not depend on w.
3) If the treatments w and z are ith associates, then the number of treatments that are jth associates of w and lth associates of z is p_{jl}^i independent of w and z.

   We write this in form of the matrices:

$$P_1 = \begin{pmatrix} p_{11}^1 & p_{12}^1 \\ p_{21}^1 & p_{22}^1 \end{pmatrix} \text{ and } P_2 = \begin{pmatrix} p_{11}^2 & p_{12}^2 \\ p_{21}^2 & p_{22}^2 \end{pmatrix}.$$

The numbers v, n_i and p_{jl}^i are the parameters of the association scheme.

Definition 12.16 An incomplete proper and equireplicate block design with v treatments in b blocks with $k < v$ elements each is a partially balanced incomplete block design PBIBD; if in the case that the treatments w and z are ith associates, they occur together in exactly λ_i blocks independent of the pair w and z.

For a PBIBD beside (12.4), we have

$$\sum_{i=1}^{m} n_i = v - 1 \tag{12.10}$$

and in place of (12.5)

$$\sum_{i=1}^{m} n_i \lambda_i = r(k-1). \tag{12.11}$$

A BIBD is a special case of a PBIBD with $m = 1$. Then (12.10) and (12.11) become (12.5). Of special interest are PBIBD(2) with two association classes. A part of the treatment pairs then occurs together in exactly λ_1 and all the rest in exactly λ_2 blocks. We give the following:

Example 12.15 We show a PBIBD with $m = 2$ and the parameters

$$v = 8, k = 3, b = 16, r = 6, \lambda_1 = 2,\ \lambda_2 = 1, n_1 = 5, n_2 = 2.$$

Block	Treatments		
1	1	2	4
2	2	3	5
3	3	4	6
4	4	5	7
5	5	6	8
6	6	7	1
7	7	8	2
8	8	1	3
9	1	2	5
10	2	3	6
11	3	4	7
13	5	6	1
14	6	7	2
15	7	8	3
16	8	1	4

The pair (1,2) occurs twice, but the pair (1,7) only once; the five first associates of 1 are 2,4,5,6,8 and the two second associates are 3 and 7, the pairs with 1, so as the pair (1,2), where the partner is a first associates of 1, occur twice, the pair (1,7) however only once. The pairs with 1 such as pair (1,7) with a partner that is second associates of 1 occurs once. The PBIBD(2) with $v = 8$, $k = 3$ has only 16 blocks, the BIBD has 56.

In Rasch et al. (2008) PBIBD(2) are given and close now the topic of construction methods (with one exception) and define only some special cases with an example.

Example 12.16 Let

$\mathcal{N}_1 = \mathcal{N}_2 = \begin{pmatrix} 1 & 1 & 0 \\ 1 & 0 & 1 \\ 0 & 1 & 1 \end{pmatrix}$ be incidence matrices of two (identical) BIBD with parameters $v = 3$, $b = 3$, $k = 2$, $r = 2$ and $\lambda = 1$. Then the incidence matrix of the Kronecker product design is

$$\mathcal{N} = \mathcal{N}_1 \otimes \mathcal{N}_2 = \begin{pmatrix} 1 & 1 & 0 & 1 & 1 & 0 & 0 & 0 & 0 \\ 1 & 0 & 1 & 1 & 0 & 1 & 0 & 0 & 0 \\ 0 & 1 & 1 & 0 & 1 & 1 & 0 & 0 & 0 \\ 1 & 1 & 0 & 0 & 0 & 0 & 1 & 1 & 0 \\ 1 & 0 & 1 & 0 & 0 & 0 & 1 & 0 & 1 \\ 0 & 1 & 1 & 0 & 0 & 0 & 0 & 1 & 1 \\ 0 & 0 & 0 & 1 & 1 & 0 & 1 & 1 & 0 \\ 0 & 0 & 0 & 1 & 0 & 1 & 1 & 0 & 1 \\ 0 & 0 & 0 & 0 & 1 & 1 & 0 & 1 & 1 \end{pmatrix}.$$

This matrix is symmetric and the product $\mathcal{N}\mathcal{N}$ equals

$$\mathcal{N}\mathcal{N} = \begin{pmatrix} 4 & 2 & 2 & 2 & 1 & 1 & 2 & 1 & 1 \\ 2 & 4 & 2 & 1 & 2 & 1 & 1 & 2 & 1 \\ 2 & 2 & 4 & 1 & 1 & 2 & 1 & 1 & 2 \\ 2 & 1 & 1 & 4 & 2 & 2 & 2 & 1 & 1 \\ 1 & 2 & 1 & 2 & 4 & 2 & 1 & 2 & 1 \\ 1 & 1 & 2 & 2 & 2 & 4 & 1 & 1 & 2 \\ 2 & 1 & 1 & 2 & 1 & 1 & 4 & 2 & 2 \\ 1 & 2 & 1 & 1 & 2 & 1 & 2 & 4 & 2 \\ 1 & 1 & 2 & 1 & 1 & 2 & 2 & 2 & 4 \end{pmatrix}.$$

This matrix corresponds to a PBIBD(2) with parameters $v = 9, b = 9, k = 4, r = 4, \lambda_1 = 1, \lambda_2 = 2$, and the necessary conditions (12.10) and (12.11) are fulfilled.

We consider now some subgroups of PBIBD(2).

Definition 12.17 A PBIBD(2) is said to be divisible, if $v = qw$ and the treatments can be arranged into q groups of w elements each so that pairs of treatments in the same group occur in λ_1 blocks, and pairs of treatments not from the same group occur in λ_2 blocks.

Example 12.17 A block design with $v = 6, b = 4, k = 3, r = 2$ and the blocks (1,3,5), (1,4,6) and (2,3,6) is a divisible PBIBD(2) with $q = 3, \lambda_1 = 0, \lambda_2 = 1$ and the three groups $[1, 2]$; $[3, 4]$; $[5, 6]$.

Definition 12.18 A PBIBD(2) is said to be simple, if one of the $\lambda_i (i = 1, 2)$ equals zero.

As we can see the classes of PBIBD(2) in Definitions 12.17 and 12.18 can contain the same design; the design of Example 12.17 is simple.

In the PBIBD(2) with the blocks (1,2,3), (4,5,6), (7,8,9), (1,4,7), (2,5,8), (3,6,9), (1,5,9), (2,6,7) and (3,4,8), each of the $v = 9$ treatments occur in three blocks of $(r = 3)$; the $b = 9$ blocks are of size $k = 3$. Pairs of treatments occur either once $(\lambda_1 = 1)$ or not at all $(\lambda_2 = 0)$ together in a block. Therefore the design is a simple PBIBD(2).

Definition 12.19 A PBIBD(2) is said to be a triangular design, if $v = \frac{u(u-1)}{2}$, and the treatments can be arranged in an upper triangular matrix of a square $(u \times u)$ matrix in such a way that after the triangular matrix is transformed into a 'symmetric matrix' without a main diagonal by reflection and if two treatments in the same row or column occur λ_1 - times and two treatments not in the same row or column occur λ_2 - times in the same block.

Triangular design exists for $v \geq 6$ only.

Example 12.18 The blocks (1,2,7,8,10), (1,3,5,9,10), (1,4,6,8,9), (2,3,6,7,9), (2,4,5,6,10) and (3,4,5,7,8) are from a triangular design with parameters $v = 10$, $b = 6$, $k = 5$, $r = 3$, $\lambda_1 = 1, \lambda_2 = 2$ and $u = 5$. Arranging the treatments as

	1	2	3	4
1		5	6	7
2	5		8	9
3	6	8		10
4	7		10	

pairs of treatments in the same row or column occur in one block, all others in two blocks.

Definition 12.20 A PBIBD(2) is said to be cyclic, if $v \geq 5$, the PBIBD(2) is not divisible and $v = 4t + 1$; $n_1 = n_2 = 2t$.

For cyclic designs the association matrices are $P_1 = \begin{pmatrix} t-1 & t \\ t & t \end{pmatrix}$ and $P_2 = \begin{pmatrix} t & t \\ t & t-1 \end{pmatrix}$.

Example 12.19 We choose $t = 3$, so that $v = 13$. The associations matrices are $P_1 = \begin{pmatrix} 2 & 3 \\ 3 & 3 \end{pmatrix}$ and $P_2 = \begin{pmatrix} 3 & 3 \\ 3 & 2 \end{pmatrix}$. Further $n_1 = n_2 = 6$. The condition (12.11) reads

$$n_1 \lambda_1 + n_2 \lambda_2 = 6(\lambda_1 + \lambda_2) = r(k-1).$$

We get solutions ($\lambda_1 = \lambda_2$ is impossible, it gives a BIBD) as the following:

$$(\lambda_1 + \lambda_2) = 1,\ r = k = 3;$$
$$(\lambda_1 + \lambda_2) = 5,\ r = k = 6$$
$$(\lambda_1 + \lambda_2) = 7,\ r = k = 7.$$

Each solution defines a cyclic PBIBD(2). Next we give the design for $\lambda_1 = 1$, $\lambda_2 = 0$ and $r = k = 3$. The 13 blocks are the following:

(1,3,9), (1,6,8), (1,7,12), (2,4,10), (2,7,9), (2,8,13), (3,5,11), (3,8,10), (4,6,12), (4,9,11), (5,7,13), (5,10,12), (6,11,13)

The 39 pairs

$$1,3;1,6;1,7;1,8;1,9;1,12;2,4,\ 2,7;2,8;2,9;2,10;2,13;3,5;3,8\,;3,9;$$
$$3,10;3,11;4,6;4,9;4,10;4,11;4,12;5,7;5,10;5,11;5,12;5,13;6,8;$$
$$6,11;6,12;6,13;7,9;7,12;7,13;8,10;8,13;9,11;10;12;11,13$$

are first associates and occur once in the design; the other 39 do not occur.

12.3 Row–Column Designs

We consider now some RCD. These experimental designs are used to eliminate two nuisance factors in two directions written as rows and columns.

The name RCD stems from the fact that the design can be characterized by a matrix so that its r rows correspond to the levels of one and its c columns correspond to the levels of the other nuisance factor. The elements represent the treatments. Construction and analysis depend on the special type of an RCD. The most important RCD are shown below:

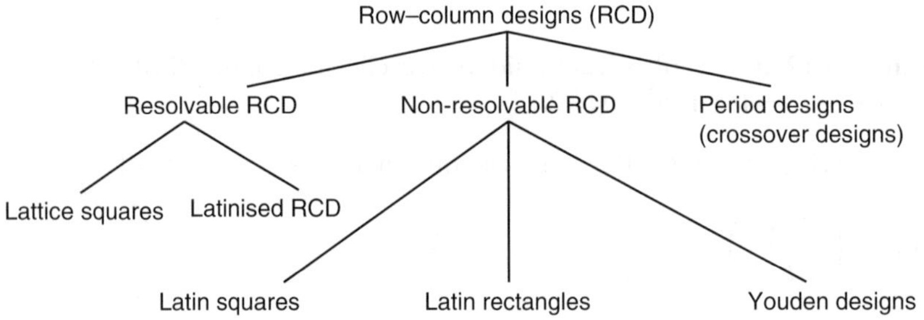

Definition 12.21 Resolvable RCD are experimental designs, with an arrangement of v treatments in t matrices with r rows and c columns in such a way that $v = rc$ and all v treatments occur in each matrix. The matrices are not

understood as levels of a third nuisance factor; they are t replications with changed order of the treatments in the matrices.

An important group of resolvable RCD are lattice squares with $r = c$ and consequently v a square number, if both the row blocks and the column blocks build a BIBD, they are called balanced.

Another group is the Latinised RCD. An experimental design constructed from t replications of a resolvable RCD with tr rows and $c > t$ columns is said to be columnwise Latinised, if not treatment occurs more often than once in a column. Analogous rowwise Latinised RCD are defined.

Example 12.20 A balanced lattice square with $r = c = 3$, $v = 9$, $t = 3$ is given by the replications 1–4 in the schema.

1			2			3			4		
1	2	3	1	4	7	1	6	8	1	9	5
4	5	6	2	5	8	9	2	4	6	2	7
7	8	9	3	6	9	5	7	3	8	4	3

Non-resolvable RCD are the LS and Latin rectangles (LR) and the Youden designs (YD). First we will consider LS.

Definition 12.22 If in an experiment with v treatments a square matrix of order v is given so that each of the v treatments $A_1, \ldots, A_v$ occurs exactly once in each row and in each column, we say it is an LS of side v. If the treatments are in natural order, then an LS where the A_i in the first row and in the first column are arranged in this natural order is called an LS in standard form. If A_i are arranged in this natural order only in the first row, we have an LS in semi-standard form. In the LS treatments often are represented by letters A, B, C,....

Example 12.21 An LS of side seven is

D	E	A	B	C	G	F
B	D	E	F	A	B	C
A	B	C	D	E	F	G
E	C	B	G	F	D	A
C	G	F	E	B	A	D
F	A	G	C	B	E	D
G	F	D	A	C	B	E

Each complete Sudoku scheme is a LS of side nine with additional conditions.

Randomisation of LS means that we first must determine the set M of possible standardised LS of a given side. From this set we randomly select one element. Some of these LS are isomorphic, which means that one results from the other by permutation of rows, columns and treatments. Otherwise, they belong to different classes. Some elements of different classes are conjugate, which means that one results from the other by interchanging rows and columns. For instance, for v = 6 exist M = 9408 different standardised LS in 22 classes. In ten of these classes per class are two standardised LS conjugated. But for the randomisation we simply can randomly select one of the 9408 standardised LS. Then we randomly assign the levels of the two nuisance factors to the rows and columns, respectively, and the treatments to the numbers 1 to n.

Special LS are used in many applications. The next definition is given in Freeman (1979).

Definition 12.23 Complete LS are LS with all ordered pairs of experimental units occurring next to each other once in each row and column. Quasi-complete LS are LS with all unordered pairs of experimental units occurring twice in each row and column.

Bailey (1984) gave methods for the construction of quasi-complete LS and discussed randomisation problems. She could show that randomisation in a subset is valid while in the whole set is not.

Definition 12.24 Two LS the side v with $A = (a_{ij})$ and $B = b_{ij}$ $(i,j = 1,\dots,v,$ $a_{ij} \in \{1,\dots,v\}, b_{ij} \in \{1,\dots,v\})$ are orthogonal, if each combination (f,g) $(f,g \in \{1,\dots,v\})$ occurs exactly once among the v^2 pairs (a_{ij}, b_{ij}). A set of $m > 2$ LS of the same side is called a set of mutually orthogonal LS (MOLS), if all pairs of this set are orthogonal.

There exist maximal $v - 1$ MOLS. It is not fully clear how many MOLS exist. Wilson (1974) showed that the maximal number of MOLS is ≥ 6 as long as $v \geq 90$ and for large v it is $\geq v^{\frac{1}{17}} - 2$.

Up to v = 13 we have the following:

v	3	4	5	6	7	8	9	10	11	12	13
Number of MOLS	2	3	4	1	6	7	8	≥ 2	10	≥ 5	12

The case v = 6 was investigated by Leonard Euler (1782). Tsarina Catherine the Great set Euler the task to arrange six regiments consisting of six officers each of different ranks in a 6 × 6 square so that no rank or regiment will be repeated in any row or column. That means one has to construct two orthogonal LS of side six. Euler conjectured that this is impossible. This conjecture was proved by Tarry (1900, 1901). However, Euler's more general conjecture that no orthogonal LS of side v = 4t +2 exist was disproved. Bose and Shrikhande (1960) showed that two orthogonal LS of side 10 exist.

Definition 12.25 An RCD with v treatments and two disturbance variables with r and c levels, respectively, is said to be a LR, if $2 \le r \le v; 2 \le c \le v$, and the design can be written as $(r \times c)$ matrix with v different elements $(1, \ldots, v)$ in such a way, which in each row or column, each of the v elements occurs at most once.

Special cases are the LS and the YD.

Definition 12.26 A YD is an RCD that is generated from a LS by dropping at least one column so that the rows form a BIBD. Therefore a YD has exactly v rows and $c < v$ columns.

If one column is dropped, a YD certainly results. If more columns are dropped, balance must be checked.

Definition 12.27 A groups period design (GPD) is an experimental design, in which the experimental units are investigated in successive periods, the groups correspond to the rows and the periods to the columns of an RCD.

GPD was first used as feeding experiments with animals. The groups of animals were fed differently in the periods of observation. Generally is a GPD a RCD with the experimental units as rows and the periods of observation as columns. More about this can be found in Johnson (2010) and Raghavarao and Padgett (2014).

12.4 Factorial Designs

Factorial designs are only shortly defined here to complete this chapter. Originally, the idea of such designs was developed in Fisher (1935). A general description can be found in Mukerjee and Wu (2006). Factorial designs play a fundamental role in efficient experimentation with multiple input variables and is used in various fields of application, including engineering, agriculture and life sciences. The factors are not applied and observed one after the other but at the same time. This can spare time and costs. Fractional factorial designs are described in Gunst and Mason (2009).

Definition 12.28 An experiment with $p \ge 2$ (treatment) factors $F_i (i = 1,\ldots,p)$ arranged so that these p factors occur at the same time with different levels in this experiment is said to be a factorial experiment or a factorial design with p factors. If $s_i \ge 2$ are the number of the levels of the ith factor $(i = 1,\ldots,p)$, the factorial experiment is called an $(s_1, s_2, \ldots, s_p)$ factorial design. Experiments with $s_1 = s_2 = \ldots = s_p = s$ are symmetric, all other experiments are asymmetric. Symmetric experiments with s levels of p factors are n-experiments. If not all factor level combinations occur in a factorial design but some conditions are fulfilled, we speak about fractional factorial designs.

If in a factorial design for N experimental objects it is counted how many of these objects belong to the factor level combinations, the result is a contingency table.

If the factors are qualitative and the observed character is quantitative, we can analyse the design by ANOVA.

If the factors are quantitative and the observed character is quantitative, we can analyse the design by the regression analysis.

More details can be found in Rasch et al. (2011) and in Rasch et al. (2008).

12.5 Programs for Construction of Experimental Designs

By the R-programme OPDOE in CRAN completely and partially balanced block designs and fractional factorial designs can be constructed. The following is an example of the construction of a BIBD with $v = b = 15$, $k = r = 7$, $\lambda = 3$:

The command is

```
> make,BIBD(s=2,n=3,m=2,method=3)
```

As the result we obtain

```
Balanced Incomplete Block Design: BIBD(15,15,7,7,3)
(1, 2, 3, 4, 5, 6, 7) (1, 2, 3, 8, 9,10,11)
(1, 2, 3,12,13,14,15) (1, 4, 5, 8, 9,12,13)
(1, 4, 5,10,11,14,15) (1, 6, 7, 8, 9,14,15)
(1, 6, 7,10,11,12,13) (2, 4, 6, 8,10,12,14)
(2, 4, 6, 9,11,13,15) (2, 5, 7, 8,10,13,15)
(2, 5, 7, 9,11,12,14) (3, 4, 7, 8,11,12,15)
(3, 4, 7, 9,10,13,14) (3, 5, 6, 8,11,13,14)
(3, 5, 6, 9,10,12,15),
```

The method 3 of the program is the method 1 in this chapter.

12.6 Exercises

12.1 Randomise the trivial BIBD:

(1,2,3)	(1,3,6)	(1,6,7)	(2,4,7)	(3,5,6)
(1,2,4)	**(1,3,7)**	(2,3,4)	(2,5,6)	(3,5,7)
(1,2,5)	(1,4,5)	**(2,3,5)**	(2,5,7)	(3,6,7)
(1,2,6)	(1,4,6)	(2,3,6)	**(2,6,7)**	(4,5,6) .
(1,2,7)	(1,4,7)	(2,3,7)	(3,4,5)	**(4,5,7)**
(1,3,4)	**(1,5,6)**	(2,4,5)	**(3,4,6)**	(4,6,7)
(1,3,5)	(1,5,7)	(2,4,6)	(3,4,7)	(5,6,7)

12.2 Construct the dual BIBD to the BIBD with the parameters $b = 7$, $r = 3$, $\lambda = 1$ and the incidence matrix

$$\begin{pmatrix} 1 & 1 & 1 & 0 & 0 & 0 & 0 \\ 1 & 0 & 0 & 1 & 1 & 0 & 0 \\ 0 & 1 & 0 & 1 & 0 & 0 & 1 \\ 1 & 0 & 0 & 0 & 0 & 1 & 1 \\ 0 & 0 & 1 & 1 & 0 & 1 & 0 \\ 0 & 0 & 1 & 0 & 1 & 0 & 1 \\ 0 & 1 & 0 & 0 & 1 & 1 & 0 \end{pmatrix}$$

Write the generated BIBD in bracket form.

12.3 Give the parameters of a BIBD constructed by a $PG(3,4)$.

12.4 Give the parameters of a BIBD constructed by a $EG(3,4)$.

12.5 Construct a BIBD by Method 12.3 with $\lambda = 2$.

12.6 Give the parameters of a BIBD constructed by Method 12.4 with $m = 4$.

12.7 Show the equivalence of a) and b) in Definition 12.6.

12.8 Transform the LS of Example 12.21 by interchanging the rows in a semi-standardised LS.

12.9 In the LS of Example 12.21, drop the two last columns and check whether the result is a YD.

References

Abel, R. J. R. and Greig, M. (1998) Balanced incomplete block designs with block size 7. *J. Des. Codes Crypt.*, **13**, 5–30.

Abel, R. J. R., Bluskov, I. and Greig, M. (2001) Balanced incomplete block designs with block size 8. *J. Comb. Des.*, **9**, 233–268.

Abel, R. J. R., Bluskov, I. and Greig, M. (2002a) Balanced incomplete block designs with block size 9 and λ = 2,4,8. *Des. Codes Crypt.*, **26**, 33–59.

Abel, R. J. R., Bluskov, I. and Greig, M. (2002b) Balanced incomplete block designs with block size 9, *III. Aust. J. Combin.*, **30**, 57–73.

Abel, R. J. R., Bluskov, I. and Greig, M. (2004) Balanced incomplete block designs with block size 9, *II. Discret. Math.*, **279**, 5–32.

Bailey, R. A. (1984) Quasi-complete Latin squares: construction and randomisation. *J. R. Stat. Soc. B.*, **46**, 323–334.

Bose, R. C. and Shrikhande, S. S. (1960) On the construction of sets of mutually orthogonal Latin squares and the falsity of a conjecture of Euler. *Trans. Am. Math. Soc.*, **95**, 191–209.

Colbourn, C. J. and Dinitz, J. H. (2006) *The CRC Handbook of Combinatorial Designs*. Chapman and Hall, Boca Raton.

Collins, J. R. (1976) Constructing BIBDs with a computer. *Ars Comb.*, **1**, 187–231.

Euler, L. (1782) Recherches sur une nouvelle espece de quarres magiques, Verhandelingen Zeeuwsch Genootschap der Wetenschappen te Vlissingen 9, Middelburg, 85–239. Reprinted *Communicationes arithmeticae* 2, 1849, 202–361.

Fisher, R. A. (1935) *The Design of Experiments*, Oliver & Boyd, Edinburgh.

Fisher, R. A. and Yates, F. (1963) *Statistical Tables for Biological, Agricultural and Medical Research*, 6th edition, Oliver & Boyd, Edinburgh and London.

Freeman, G. H. (1979) Complete Latin squares and related experimental designs. *J. R. Stat. Soc. B.*, **41**, 253–262.

Gunst, R. F. and Mason, R. L. (2009) *Fractional Factorial Design*. John Wiley & Sons, Inc., New York.

Hanani, H. (1961) The existence and construction of balanced incomplete block designs. *Ann. Math. Stat.*, **32**, 361–386.

Hanani, H. (1975) Balanced incomplete block designs and related designs. *Discrete Mathem.*, **11**, 275–289.

Hanani, H. (1989) BIBD's with block-size seven. *Discret. Math.*, **77**, 89–96.

Johnson, D. E. (2010) Crossover experiments. *Comput. Stat.*, **2**, 620–625.

Mathon, R. and Rosa, A. (2006) 2-(v; k;λ) designs of small order, in Colbourn, C. J. and Dinitz, J. H. (Eds.), *Handbook of Combinatorial Designs*, 2nd edition, Chapman & Hall/CRC, Boca Raton, pp. 25–58.

Mohan, R. N., Kageyama, S. and Nair, M. N. (2004) On a characterization of symmetric balanced incomplete block designs. *Discussiones Mathematicae, Probability and Statistics*, **24**, 41–58.

Mukerjee, R. and Wu, C. F. J (2006) *A Modern Theory of Factorial Design*, Springer, New York.

Raghavarao, D. (1971) *Constructions and Combinatorial Problems in Design of Experiments*, John Wiley & Sons, Inc., New York.

Raghavarao, D. and Padgett, L. (2014) *Repeated Measurements and Cross-Over Designs*, John Wiley & Sons, Inc., New York.

Rao, C. R. (1961) A study of BIB designs with replications 11 to 15. *Sankhya Ser. A*, **23**, 117–129.

Rasch, D., Herrendörfer, G., Bock, J., Victor, N. and Guiard, V. (Eds.) (2008) *Verfahrensbibliothek Versuchsplanung und - auswertung*. 2. verbesserte Auflage in einem Band mit CD, R. Oldenbourg Verlag, München.

Rasch, D., Pilz, J., Verdooren, R.L. and Gebhardt, A. (2011) *Optimal Experimental Design with R*, Chapman and Hall, Boca Raton.

Rasch, D., Teuscher, F. and Verdooren, R. L. (2016) A conjecture about BIBDs. *Commun. Stat. Simul. Comput.*, **45**, 1526–1537.

Sprott, D. A. (1962) A list of BIB designs with r = 16 to 20. *Sankhya Ser. A.*, **24**, 203–204.

Sun, H. M. (2012) On the existence of simple BIBDs with number of elements a prime power. *J. Comb. Des.*, **21**, 47–59.

Takeuchi, K. (1962) A table of difference sets generating balanced incomplete block designs. *Rev. Int. Stat. Inst.*, **30**, 361–366.

Tarry, G. (1900) Le Probléme de 36 Officiers. *Compte Rendu de l'Association Française pour l'Avancement de Science Naturel, Secrétariat de l'Association*, **1**, 122–123.

Tarry, G. (1901) Le Probléme de 36 Officiers. *Compte Rendu de l'Association Française pour l'Avancement de Science Naturel, Secrétariat de l'Association*, **2**, 170–203.

Tocher, K. D. (1952) The design and analysis of block experiments. *J. R. Stat. Soc. B.*, **14**, 45–91.

Wilson, R. M. (1974) Concerning the number of mutually orthogonal Latin squares. *Discret. Math.*, **9**, 181–198.

Appendix A: Symbolism

Partially we distinguish in notation from other mathematical disciplines. We do not use capital letters as in probability theory to denote random variables but denote them by bold printing. We do this not only to distinguish between a random variable $\boldsymbol{F}$ with F-distribution and its realisation F but mainly because linear models are important in this book. In a mixed model in the two-way cross-classification of the analysis of variance with a fixed factor A and a random factor $\boldsymbol{B}$, the model equation with capital letters is written as

$$Y_{ijk} = \mu + a_i + B_j + (aB)_{ij} + E_{ijk}.$$

This looks strange and is unusual. We use instead

$$\boldsymbol{y}_{ijk} = \mu + a_i + \boldsymbol{b}_j + (\boldsymbol{ab})_{ij} + \boldsymbol{e}_{ijk}.$$

Functions are never written without an argument to avoid confusion. So is $p(y)$ often a probability function but p a probability. Further is $f(y)$ a density function but f the symbol for degrees of freedom.

Sense	Symbol
Rounding-up function	$\lceil x \rceil$ = smallest integer $\geq x$
Binomial distribution with parameters n, p	$B(n,p)$
Chi-squared (χ^2) distribution with f degrees of freedom	$CS\ (f)$
Determinant of the matrix A	$\lvert A \rvert$, det(A)
Diagonal matrix of order n	$\mathrm{diag}(a_1, \ldots, a_n)$
Direct product of the sets A and B	$A \otimes B$
Direct sum of the sets A and B	$A \oplus B$
Identity matrix of order n	I_n
$(n \times m)$ matrix with only zeros	$O_{n,m}$

(Continued)

Mathematical Statistics, First Edition. Dieter Rasch and Dieter Schott.

Sense	Symbol
$(n \times m)$ matrix with only ones	1_{nm}
Euclidean space of dimension n and 1, respectively (real axis). Positive real axis	R^n; $R^1 = R$; R^+
y is distributed as	$y \sim$
Indicator function	If A is a set and $x \in A$, then $I_A(x) = \begin{cases} 1, \text{ if } x \in A \\ 0, \text{ if } x \notin A \end{cases}$
Interval on the x-axis	
Open	$(a,b) : a < x < b$
Half open	$[a,b) : a \le x < b$, $(a,b] : a < x \le b$
Closed	$[a,b] : a \le x \le b$
ith-order statistic	$\boldsymbol{y}_{(i)}$
Cardinality (number) of elements in S	$card(S)$; $\|S\|$
Constant in formulae	const.
Kronecker product of matrices $\mathcal{N}_1$ and $\mathcal{N}_2$	$\mathcal{N} = \mathcal{N}_1 \otimes \mathcal{N}_2$
Empty set	$\emptyset$
Multivariate normal distribution with expectation vector μ and covariance matrix Σ	$N(\mu,\Sigma)$
Normal distribution with expectation μ and variance σ^2	$N(\mu,\sigma^2)$
Null vector with n *elements*	0_n
Vector with n ones	1_n
Parameter space	Ω
Poisson distribution with parameter λ	$P(\lambda)$
P-quantile of the $N(0, 1)$ distribution	$z(P)$ or z_P (see Table D.3 last line)
P-quantile of the χ^2 distribution with f degrees of freedom	$CS(f\|P)$ (see Table D.4)
P-quantile of the t-distribution with f degrees of freedom	$t(f\|P)$ (see Table D.3)
P-quantile of the F-distribution with f_1 and f_2 degrees of freedom	$F(f_1,f_2 \| P) = F_P(f_1,f_2)$ (see Table D.5)
Rank of matrix A	$\text{rk}(A)$
Rank space of matrix A	$R[A]$
Standard normal distribution with Expectation 0; variance 1	$N(0,1)$
Trace of matrix A	$\text{tr}(A)$
Transposed vector of Y	Y^T
Vector (column vector)	Y
Distribution function of a $N(0,1)$ distribution	$\Phi(x)$
Density function of a $N(0,1)$ distribution	$\varphi(x)$
Random variable (bold print)	$\boldsymbol{y}$, $\boldsymbol{Y}$

Appendix B: Abbreviations

ASN	average sample number
BAN	best asymptotic normal (estimator)
BIBD	balanced incomplete block design
BLUE	best linear unbiased estimator
BLUP	best linear unbiased prediction
BQUE	best quadratic unbiased estimator
df	degrees of freedom
iff	if and only if
LS	Latin square
LSE	least squares estimator
LSM	least squares method
LVUE	locally variance-optimal unbiased estimator
MINQUE	minimum quadratics norm estimator
ML	maximum likelihood
MLE	maximum likelihood estimator
MS	mean squares
MSD	mean square deviation
PBIBD	partially balanced incomplete block design
RCD	row–column design

Mathematical Statistics, First Edition. Dieter Rasch and Dieter Schott.

REML	restricted maximum likelihood
SLRT	sequential likelihood ratio test
SS	sum of squares
UMP	uniformly most powerful (test)
UMPU	uniformly most powerful unbiased (test)
UVUE	uniformly variance-optimal unbiased estimator
W.l.o.g.	without loss of generality
YD	Youden design

Appendix C: Probability and Density Functions

Bernoulli distribution	$p(y,p)=p^y(1-p)^{1-y},\ 0<p<1,\ y=0,1$
Beta distribution	$f(y.\theta)=\dfrac{1}{B(a,b)}y^{a-1}(1-y)^{b-1}$ $0<y<1; 0<a,b<\infty$
Binomial distribution	$p(y,p)=\binom{n}{y}p^y(1-p)^{n-y}; 0<p<1; y=0,\ldots,n$
Exponential family	$f(y,\theta)=h(y)e^{\sum_{i=1}^{k}\eta_i(\theta)\cdot T_i(y)-B(\theta)}$
Exponential family in canonical form	$f(y,\eta)=h(y)e^{\sum_{i=1}^{k}\eta_i\cdot T_i(y)-A(\eta)}$
Exponential distribution	$f(y,\lambda)=\lambda e^{-\lambda y}; \lambda\in R^+; y\ge 0$
Geometrical distribution	$p(y,p)=p(1-p)^{y-1};\ y=1,2,\ldots; 0<p<1$
Uniform distribution in (a,b)	$f(y,a,b)=\dfrac{1}{b-a},\ a<b,\ a\le y\le b$
Hypergeometric distribution	$p(y,M,N,n)=\dfrac{\binom{M}{y}\binom{N-M}{n-y}}{\binom{N}{n}}, n\in\{1,\ldots,N\}$ $y\in\{0,\ldots,N\}; M\le N$ integer
Negative binomial distribution	$p(y,p,r)=\binom{y-1}{r-1}p^r(1-p)^{y-r}$ $0<p<1, y\ge r,\ r\in\{0,1,\ldots\}$

Mathematical Statistics, First Edition. Dieter Rasch and Dieter Schott.

Normal distribution	$f(y,\mu,\sigma^2) = \frac{1}{\sigma\sqrt{2\pi}}e^{-\frac{(y-\mu)^2}{2\sigma^2}}$; $-\infty < \mu, y < \infty, \sigma > 0$; see Table D.1
Pareto distribution	$f(y,\theta) = \frac{\theta a^{\theta}}{y^{\theta+1}}, y > a > 0, \theta \in \Omega = R^{+}$
Poisson distribution	$p(y, \lambda) = \frac{\lambda^y}{y!}e^{-\lambda}, \lambda > 0 \quad y = 0, 1, 2, \ldots$
Weibull distribution	$f(y,\theta) = \theta a(\theta y)^{a-1}e^{-(\theta y)^a}, a \geq 0, y \geq 0$ $\theta \in \Omega = R^{+}$

Appendix D: Tables

Mathematical Statistics, First Edition. Dieter Rasch and Dieter Schott.
© 2018 John Wiley & Sons Ltd. Published 2018 by John Wiley & Sons Ltd.

Table D.1 Density function $\varphi(z)$ of the standard normal distribution

z	0.00	0.01	0.02	0.03	0.04	0.05	0.06	0.07	0.08	0.09
0.0	.39894	.39892	.39886	.39876	.39862	.39844	.39822	.39797	.39767	.39733
0.1	.39695	.39654	.39608	.39559	.39505	.39448	.39387	.39322	.39253	.39181
0.2	.39104	.39024	.38940	.38853	.38762	.38667	.38568	.38466	.38361	.38251
0.3	.38139	.38023	.37903	.37780	.37654	.37524	.37391	.37255	.37115	.36973
0.4	.36827	.36678	.36526	.36371	.36213	.36053	.35889	.35723	.35553	.35381
0.5	.35207	.35029	.34849	.34667	.34482	.34294	.34105	.33912	.33718	.33521
0.6	.33322	.33121	.32918	.32713	.32506	.32297	.32086	.31874	.31659	.31443
0.7	.31225	.31006	.30785	.30563	.30339	.30114	.29887	.29659	.29431	.29200
0.8	.28969	.28737	.28504	.28269	.28034	.27798	.27562	.27324	.27086	.26848
0.9	.26609	.26369	.26129	.25888	.25647	.25406	.25164	.24923	.24681	.24439
1.0	.24197	.23955	.23713	.23471	.23230	.22988	.22747	.22506	.22265	.22025
1.1	.21785	.21546	.21307	.21069	.20831	.20594	.20357	.20121	.19886	.19652
1.2	.19419	.19186	.18954	.18724	.18494	.18265	.18037	.17810	.17585	.17360
1.3	.17137	.16915	.16694	.16474	.16256	.16038	.15822	.15608	.15395	.15183
1.4	.14973	.14764	.14556	.14350	.14146	.13943	.13742	.13542	.13344	.13147
1.5	.12952	.12758	.12566	.12376	.12188	.12001	.11816	.11632	.11450	.11270
1.6	.11092	.10915	.10741	.10567	.10396	.10226	.10059	.09893	.09728	.09566
1.7	.09405	.09246	.09089	.08933	.08780	.08628	.08478	.08329	.08183	.08038
1.8	.07895	.07754	.07614	.07477	.07341	.07206	.07074	.06943	.06814	.06687
1.9	.06562	.06438	.06316	.06195	.06077	.05959	.05844	.05730	.05618	.05508

2.0	.05399	.05292	.05186	.05082	.04980	.04879	.04780	.04682	.04586	.04491
2.1	.04398	.04307	.04217	.04128	.04041	.03955	.03871	.03788	.03706	.03626
2.2	.03547	.03470	.03394	.03319	.03246	.03174	.03103	.03034	.02965	.02898
2.3	.02833	.02768	.02705	.02643	.02582	.02522	.02463	.02406	.02349	.02294
2.4	.02239	.02186	.02134	.02083	.02033	.01984	.01936	.01888	.01842	.01797
2.5	.01753	.01709	.01667	.01625	.01585	.01545	.01506	.01468	.01431	.01394
2.6	.01358	.01323	.01289	.01256	.01223	.01191	.01160	.01130	.01100	.01071
2.7	.01042	.01014	.00987	.00961	.00935	.00909	.00885	.00861	.00837	.00814
2.8	.00792	.00770	.00748	.00727	.00707	.00687	.00668	.00649	.00631	.00613
2.9	.00595	.00578	.00562	.00545	.00530	.00514	.00499	.00485	.00470	.00457
3.0	.00443	.00327	.00238	.00172	.00123	.00087	.00061	.00042	.00029	.00020
4.0	.00013	.00009	.00006	.00004	.00002	.00002	.00001	.00001	–	–

Table D.2 Distribution function $\Phi(z), z \geq 0$ of the standard normal distribution (the values of $\Phi(z), z < 0$ are $1-\Phi(z), z \geq 0$).

z	0.00	0.01	0.02	0.03	0.04	0.05	0.06	0.07	0.08	0.09
0.0	.500000	.503989	.507978	.511967	.515953	.519939	.523922	.527903	.531881	.535856
0.1	.539828	.543795	.547758	.551717	.555670	.559618	.563559	.567495	.571424	.575345
0.2	.579260	.583166	.587064	.590954	.594835	.598706	.602568	.606420	.610261	.614092
0.3	.617911	.621719	.625516	.629300	.633072	.636831	.640576	.644309	.648027	.651732
0.4	.655422	.659097	.662757	.666402	.670031	.673645	.677242	.680822	.684386	.687933
0.5	.691462	.694974	.698468	.701944	.705401	.708840	.712260	.715661	.719043	.722405
0.6	.725747	.729069	.732371	.735653	.738914	.742154	.745373	.748571	.751748	.754903
0.7	.758036	.761148	.764238	.767305	.770350	.773373	.776373	.779350	.782305	.785236
0.8	.788145	.791030	.793892	.796731	.799546	.802337	.805106	.807850	.810570	.813267
0.9	.815940	.818589	.821214	.823814	.826391	.828944	.831472	.833977	.836457	.838913
1.0	.841345	.843752	.846136	.848495	.850830	.853141	.855428	.857690	.859929	.862143
1.1	.864334	.866500	.868643	.870762	.872857	.874928	.876976	.878999	.881000	.882977
1.2	.884930	.886860	.888767	.890651	.892512	.894350	.896165	.897958	.899727	.901475
1.3	.903199	.904902	.906582	.908241	.909877	.911492	.913085	.914656	.916207	.917736
1.4	.919243	.920730	.922196	.923641	.925066	.926471	.927855	.929219	.930563	.931888
1.5	.933193	.934478	.935744	.936992	.938220	.939429	.940620	.941792	.942947	.944083
1.6	.945201	.946301	.947384	.948449	.949497	.950529	.951543	.952540	.953521	.954486
1.7	.955435	.956367	.957284	.958185	.959071	.959941	.960796	.961636	.962462	.963273
1.8	.964070	.964852	.965621	.966375	.967116	.967843	.968557	.969258	.969946	.970621
1.9	.971284	.971933	.972571	.973197	.973810	.974412	.975002	.975581	.976148	.976705

2.0	.977250	.977784	.978308	.978822	.979325	.979818	.980301	.980774	.981237	.981691
2.1	.982136	.982571	.982997	.983414	.983823	.984222	.984614	.984997	.985371	.985738
2.2	.986097	.986447	.986791	.987126	.987455	.987776	.988089	.988396	.988696	.988989
2.3	.989276	.989556	.989830	.990097	.990358	.990613	.990863	.991106	.991344	.991576
2.4	.991802	.992024	.992240	.992451	.992656	.992857	.993053	.993244	.993431	.993613
2.5	.993790	.993963	.994132	.994297	.994457	.994614	.994766	.994915	.995060	.995201
2.6	.995339	.995473	.995603	.995731	.995855	.995975	.996093	.996207	.996319	.996427
2.7	.996533	.996636	.996736	.996833	.996928	.997020	.997110	.997197	.997282	.997365
2.8	.997445	.997523	.997599	.997673	.997744	.997814	.997882	.997948	.998012	.998074
2.9	.998134	.998193	.998250	.998305	.998359	.998411	.998462	.998511	.998559	.998605
3.0	.998650	.999032	.999313	.999517	.999663	.999767	.999841	.999892	.999928	.999952

Table D.3 *P*-quantiles of the *t*-distribution with *df* degrees of freedom (for $df = \infty$ *P*-quantiles of the standard normal distribution).

	P								
df	0.60	0.70	0.80	0.85	0.90	0.95	0.975	0.99	0.995
1	0.3249	0.7265	1.3764	1.9626	3.0777	6.3138	12.7062	31.8205	63.6567
2	0.2887	0.6172	1.0607	1.3862	1.8856	2.9200	4.3027	6.9646	9.9248
3	0.2767	0.5844	0.9785	1.2498	1.6377	2.3534	3.1824	4.5407	5.8409
4	0.2707	0.5686	0.9410	1.1896	1.5332	2.1318	2.7764	3.7469	4.6041
5	0.2672	0.5594	0.9195	1.1558	1.4759	2.0150	2.5706	3.3649	4.0321
6	0.2648	0.5534	0.9057	1.1342	1.4398	1.9432	2.4469	3.1427	3.7074
7	0.2632	0.5491	0.8960	1.1192	1.4149	1.8946	2.3646	2.9980	3.4995
8	0.2619	0.5459	0.8889	1.1081	1.3968	1.8595	2.3060	2.8965	3.3554
9	0.2610	0.5435	0.8834	1.0997	1.3830	1.8331	2.2622	2.8214	3.2498
10	0.2602	0.5415	0.8791	1.0931	1.3722	1.8125	2.2281	2.7638	3.1693
11	0.2596	0.5399	0.8755	1.0877	1.3634	1.7959	2.2010	2.7181	3.1058
12	0.2590	0.5386	0.8726	1.0832	1.3562	1.7823	2.1788	2.6810	3.0545
13	0.2586	0.5375	0.8702	1.0795	1.3502	1.7709	2.1604	2.6503	3.0123
14	0.2582	0.5366	0.8681	1.0763	1.3450	1.7613	2.1448	2.6245	2.9768
15	0.2579	0.5357	0.8662	1.0735	1.3406	1.7531	2.1314	2.6025	2.9467
16	0.2576	0.5350	0.8647	1.0711	1.3368	1.7459	2.1199	2.5835	2.9208
17	0.2573	0.5344	0.8633	1.0690	1.3334	1.7396	2.1098	2.5669	2.8982
18	0.2571	0.5338	0.8620	1.0672	1.3304	1.7341	2.1009	2.5524	2.8784
19	0.2569	0.5333	0.8610	1.0655	1.3277	1.7291	2.0930	2.5395	2.8609

20	0.2567	0.5329	0.8600	1.0640	1.3253	1.7247	2.0860	2.5280	2.8453
21	0.2566	0.5325	0.8591	1.0627	1.3232	1.7207	2.0796	2.5176	2.8314
22	0.2564	0.5321	0.8583	1.0614	1.3212	1.7171	2.0739	2.5083	2.8188
23	0.2563	0.5317	0.8575	1.0603	1.3195	1.7139	2.0687	2.4999	2.8073
24	0.2562	0.5314	0.8569	1.0593	1.3178	1.7109	2.0639	2.4922	2.7969
25	0.2561	0.5312	0.8562	1.0584	1.3163	1.7081	2.0595	2.4851	2.7874
26	0.2560	0.5309	0.8557	1.0575	1.3150	1.7056	2.0555	2.4786	2.7787
27	0.2559	0.5306	0.8551	1.0567	1.3137	1.7033	2.0518	2.4727	2.7707
28	0.2558	0.5304	0.8546	1.0560	1.3125	1.7011	2.0484	2.4671	2.7633
29	0.2557	0.5302	0.8542	1.0553	1.3114	1.6991	2.0452	2.4620	2.7564
30	0.2556	0.5300	0.8538	1.0547	1.3104	1.6973	2.0423	2.4573	2.7500
40	0.2550	0.5286	0.8507	1.0500	1.3031	1.6839	2.0211	2.4233	2.7045
50	0.2547	0.5278	0.8489	1.0473	1.2987	1.6759	2.0086	2.4033	2.6778
60	0.2545	0.5272	0.8477	1.0455	1.2958	1.6706	2.0003	2.3901	2.6603
70	0.2543	0.5268	0.8468	1.0442	1.2938	1.6669	1.9944	2.3808	2.6479
80	0.2542	0.5265	0.8461	1.0432	1.2922	1.6641	1.9901	2.3739	2.6387
90	0.2541	0.5263	0.8456	1.0424	1.2910	1.6620	1.9867	2.3685	2.6316
100	0.2540	0.5261	0.8452	1.0418	1.2901	1.6602	1.9840	2.3642	2.6259
300	0.2536	0.5250	0.8428	1.0382	1.2844	1.6499	1.9679	2.3451	2.5923
500	0.2535	0.5247	0.8423	1.0375	1.2832	1.6479	1.9647	2.3338	2.5857
∞	0.2533	0.5244	0.8416	1.0364	1.2816	1.6449	1.9600	2.3263	2.5758

Table D.4 *P*-quantiles *CS* (*df*, *P*) of the χ^2 distribution.

	P												
df	0.005	0.010	0.025	0.050	0.100	0.250	0.500	0.750	0.900	0.950	0.975	0.990	0.995
1	$3927 \cdot 10^{-8}$	$1571 \cdot 10^{-7}$	$9821 \cdot 10^{-7}$	$3932 \cdot 10^{-6}$	0.01579	0.1015	0.4549	1.323	2.706	3.841	5.024	6.635	7.879
2	0.01003	0.02010	0.05064	0.1026	0.2107	0.5754	1.386	2.773	4.605	5.991	7.378	9.210	1.60
3	0.07172	0.1148	0.2158	0.3518	0.5844	1.213	2.366	4.108	6.251	7.815	9.348	11.34	12.84
4	0.2070	0.2971	0.4844	0.7107	1.064	1.923	3.357	5.385	7.779	9.488	11.14	13.28	14.86
5	0.4117	0.5543	0.8312	1.145	1.610	2.675	4.351	6.626	9.236	11.07	12.83	15.09	16.75
6	0.6757	0.8721	1.237	1.635	2.204	3.455	5.348	7.841	10.64	12.59	14.45	16.81	18.55
7	0.9893	1.239	1.690	2.167	2.833	4.255	6.346	9.037	12.02	14.07	16.01	18.48	2.28
8	1.344	1.646	2.180	2.733	3.490	5.071	7.344	10.22	13.36	15.51	17.53	2.09	21.96
9	1.735	2.088	2.700	3.325	4.168	5.899	8.343	11.39	14.68	16.92	19.02	21.67	23.59
10	2.156	2.558	3.247	3.940	4.865	6.737	9.342	12.55	15.99	18.21	2.48	23.21	25.19
11	2.603	3.053	3.816	4.575	5.578	7.584	10.34	13.70	17.28	19.68	21.92	24.72	26.76
12	3.074	3.571	4.404	5.226	6.304	8.438	11.34	14.85	18.55	21.03	23.34	26.22	28.30
13	3.565	4.107	5.009	5.892	7.042	9.299	12.34	15.98	19.81	22.36	24.74	27.69	29.82
14	4.075	4.660	5.629	6.571	7.790	10.17	13.34	17.12	21.06	23.68	26.12	29.14	31.32
15	4.601	5.229	6.262	7.261	8.547	11.04	14.34	18.25	22.31	25.00	27.49	3.58	32.80
16	5.142	5.812	6.908	7.962	9.312	11.91	15.34	19.37	23.54	26.30	28.85	32.00	34.27
17	5.697	6.408	7.564	8.672	10.09	12.79	16.34	2.49	24.77	27.59	3.19	33.41	35.72
18	6.265	7.015	8.231	9.390	10.86	13.68	17.34	21.60	25.99	28.87	31.53	34.81	37.16
19	6.844	7.633	8.907	10.12	11.65	14.56	18.34	22.72	27.20	3.14	32.85	36.19	38.58

20	7.434	8.260	9.591	10.85	12.44	15.45	19.34	23.83	28.41	31.41	34.17	37.57	4.00
21	8.034	8.897	10.28	11.59	13.24	16.34	20.34	24.93	29.62	32.67	35.48	38.93	41.40
22	8.643	9.542	10.98	12.34	14.04	17.24	21.34	26.04	8.81	33.92	36.78	40.22	42.80
23	9.260	10.20	11.69	13.09	14.85	18.14	22.34	27.14	32.01	35.17	38.08	41.64	44.18
24	9.886	10.86	12.40	13.85	15.66	19.04	23.34	28.24	33.20	36.42	39.36	42.98	45.56
25	10.52	11.52	13.12	14.61	16.47	19.94	24.34	29.34	34.38	37.65	40.65	44.31	46.93
26	11.16	12.20	1384	15.38	17.29	2.84	25.34	30.43	35.56	38.89	41.92	45.64	48.29
27	11.81	12.88	14.57	16.15	18.11	21.75	26.34	31.53	36.74	4.11	43.19	46.96	49.64
28	12.46	13.56	15.31	16.93	18.94	22.06	27.34	32.62	37.92	41.34	44.46	48.28	50.99
29	13.12	14.26	16.05	17.71	19.77	23.57	28.34	33.71	39.09	42.56	45.72	49.59	52.34
30	13.79	14.95	16.79	18.49	20.60	24.48	29.34	34.80	40.26	43.77	46.98	50.89	53.67
40	20.71	22.16	24.43	26.51	29.05	33.66	39.34	45.62	51.80	55.76	59.34	63.69	66.77
50	27.99	29.71	32.36	34.76	37.69	42.94	49.33	56.33	63.17	67.50	71.42	76.15	79.49
60	35.53	37.48	40.48	43.19	46.46	52.29	59.33	66.98	74.40	79.08	83.30	88.38	91.95
70	43.28	45.44	48.76	51.74	55.33	61.70	69.33	77.58	85.53	90.53	95.02	10.42	104.22
80	51.17	53.54	57.15	60.39	64.28	71.14	79.33	88.13	96.58	101.88	106.63	112.33	116.32
90	59.20	61.75	65.65	69.13	73.29	80.62	89.33	98.65	107.56	113.14	118.14	124.12	128.30
100	67.33	70.06	74.22	77.93	82.36	90.13	99.33	109.14	118.50	124.34	129.56	135.81	140.17

Table D.5 *95 %* quantiles of the F-distribution with f_1 and f_2 degrees of freedom.

f_2 \ f_1	1	2	3	4	5	6	7	8	9
1	161.4	199.5	215.7	224.6	230.2	234.0	236.8	238.9	240.5
2	18.51	19.00	19.16	19.25	19.30	19.33	19.35	19.37	19.38
3	10.13	9.55	9.28	9.12	9.01	8.94	8.89	8.85	8.81
4	7.71	6.94	6.59	6.39	6.26	6.16	6.09	6.04	6.00
5	6.61	5.79	5.41	5.19	5.05	4.95	4.88	4.82	4.77
6	5.99	5.14	4.76	4.53	4.39	4.28	4.21	4.15	4.10
7	5.59	4.74	4.35	4.12	3.97	3.87	3.79	3.73	3.68
8	5.32	4.46	4.07	3.84	3.69	3.58	3.50	3.44	3.39
9	5.12	4.26	3.86	3.63	3.48	3.37	3.29	3.23	3.18
10	4.96	4.10	3.71	3.48	3.33	3.22	3.14	3.07	3.02
11	4.84	3.98	3.59	3.36	3.20	3.09	3.01	2.95	2.90
12	4.75	3.89	3.49	3.27	3.11	3.00	2.91	2.85	2.80
13	4.67	3.81	3.41	3.18	3.03	2.92	2.83	2.77	2.71
14	4.60	3.74	3.34	3.11	2.96	2.85	2.76	2.70	2.65
15	4.54	3.68	3.29	3.06	2.90	2.79	2.71	2.64	2.59
16	4.49	3.63	3.24	3.01	2.85	2.74	2.66	2.59	2.54
17	4.45	3.59	3.20	2.96	2.81	2.70	2.61	2.55	2.49
18	4.41	3.55	3.16	2.93	2.77	2.66	2.58	2.51	2.46
19	4.38	3.52	3.13	2.90	2.74	2.63	2.54	2.48	2.42
20	4.35	3.49	3.10	2.87	2.71	2.60	2.51	2.45	2.39
21	4.32	3.47	3.07	2.84	2.68	2.57	2.49	2.42	2.37
22	4.30	3.44	3.05	2.82	2.66	2.55	2.46	2.40	2.34
23	4.28	3.42	3.03	2.80	2.64	2.53	2.44	2.37	2.32
24	4.26	3.40	3.01	2.78	2.62	2.51	2.42	2.36	2.30
25	4.24	3.39	2.99	2.76	2.60	2.49	2.40	2.34	2.28
26	4.23	3.37	2.98	2.74	2.59	2.47	2.39	2.32	2.27
27	4.21	3.35	2.96	2.73	2.57	2.46	2.37	2.31	2.25
28	4.20	3.34	2.95	2.71	2.56	2.45	2.36	2.29	2.24
29	4.18	3.33	2.93	2.70	2.55	2.43	2.35	2.28	2.22
30	4.17	3.32	2.92	2.69	2.53	2.42	2.33	2.27	2.21
40	4.08	3.23	2.84	2.61	2.45	2.34	2.25	2.18	2.12
60	4.00	3.15	2.76	2.53	2.37	2.25	2.17	2.10	2.04
120	3.92	3.07	2.68	2.45	2.29	2.17	2.09	2.02	1.96
∞	3.84	3.00	2.60	2.37	2.21	2.10	2.01	1.94	1.88

Table D.5 (Continued)

f_2 \ f_1	10	12	15	20	24	30	40	60	120	∞
1	241.9	243.9	245.9	248.0	249.1	250.1	251.1	252.2	253.3	254.
2	19.40	19.41	19.43	19.45	19.45	19.46	19.47	19.48	19.49	19.5
3	8.79	8.74	8.70	8.66	8.64	8.62	8.59	8.57	8.55	8.5
4	5.96	5.91	5.86	5.80	5.77	5.75	5.72	5.69	5.66	5.6
5	4.74	4.68	4.62	4.56	4.53	4.50	4.46	4.43	4.40	4.3
6	4.06	4.00	3.94	3.87	3.84	3.81	3.77	3.74	3.70	3.6
7	3.64	3.57	3.51	3.44	3.41	3.38	3.34	3.30	3.27	3.2
8	3.35	3.28	3.22	3.15	3.12	3.08	3.04	3.01	2.97	2.9
9	3.14	3.07	3.01	2.94	2.90	2.86	2.83	2.79	2.75	2.7
10	2.98	2.91	2.85	2.77	2.74	2.70	2.66	2.62	2.58	2.5
11	2.85	2.79	2.72	2.65	2.61	2.57	2.53	2.49	2.45	2.4
12	2.75	2.69	2.62	2.54	2.51	2.47	2.43	2.38	2.34	2.3
13	2.67	2.60	2.53	2.46	2.42	2.38	2.34	2.30	2.25	2.2
14	2.60	2.53	2.46	2.39	2.35	2.31	2.27	2.22	2.18	2.1
15	2.54	2.48	2.40	2.33	2.29	2.25	2.20	2.16	2.11	2.0
16	2.49	2.42	2.35	2.28	2.24	2.19	2.15	2.11	2.06	2.0
17	2.45	2.38	2.31	2.23	2.19	2.15	2.10	2.06	2.01	1.9
18	2.41	2.34	2.27	2.19	2.15	2.11	2.06	2.02	1.97	1.9
19	2.38	2.31	2.23	2.16	2.11	2.07	2.03	1.98	1.93	1.8
20	2.35	2.28	2.20	2.12	2.08	2.04	1.99	1.95	1.90	1.8
21	2.32	2.25	2.18	2.10	2.05	2.01	1.96	1.92	1.87	1.8
22	2.30	2.23	2.15	2.07	2.03	1.98	1.94	1.89	1.84	1.7
23	2.27	2.20	2.13	2.05	2.01	1.96	1.91	1.86	1.81	1.7
24	2.25	2.18	2.11	2.03	1.98	1.94	1.89	1.84	1.79	1.7
25	2.24	2.16	2.09	2.01	1.96	1.92	1.87	1.82	1.77	1.71
26	2.22	2.15	2.07	1.99	1.95	1.90	1.85	1.80	1.75	1.6
27	2.20	2.13	2.06	1.97	1.93	1.88	1.84	1.79	1.73	1.6
28	2.19	2.12	2.04	1.96	1.91	1.87	1.82	1.77	1.71	1.65
29	2.18	2.10	2.03	1.94	1.90	1.85	1.81	1.75	1.70	1.64
30	2.16	2.09	2.01	1.93	1.89	1.84	1.79	1.74	1.68	1.62
40	2.08	2.00	1.92	1.84	1.79	1.74	1.69	1.64	1.58	1.51
60	1.99	1.92	1.84	1.75	1.70	1.65	1.59	1.53	1.47	1.39
120	1.91	1.83	1.75	1.66	1.61	1.55	1.50	1.43	1.35	1.25
∞	1.83	1.75	1.67	1.57	1.52	1.46	1.39	1.32	1.22	1.00

Solutions and Hints for Exercises

Chapter 1

Exercise 1.1

The sample is not random because inhabitants without entry in the telephone book cannot be selected.

Exercise 1.2

It is recommended to use SPSS avoiding long-winded calculations by hand. Write the 81 different quadruples of the numbers 1, 2, 3 due to the random sampling with replacement into the columns y_1, y_2, y_3, y_4 of a SPSS data sheet (Statistics Data Editor). In the command sequence 'Transform – Compute Variable', denote the effect variable by 'Mean' and form $(y_1 + y_2 + y_3 + y_4)/4$ using the command MEAN = MEAN(y1,y2,y3,y4). See also the SPSS syntax below. Now the mean values occur in column 5 of the data sheet. Analogously create the variable s2 = VARIANCE(y1,y2,y3,y4) and in column 6 of the data sheet s2 is given.

After performing of the command sequence 'Analyze – Descriptive Statistics – Descriptive', the mean value and the variance of the population are calculated from the means (set under options) of MEAN and s2. The value of the VARIANCE of the variable MEAN must be multiplied by $(N-1)/N$ to get the population variance of the sample mean (2/3)/4, because from the population of $N = 81$ samples of size 4 SPSS calculates a sample variance with denominator $N-1$. The corresponding graphical representations are obtained via 'Graphs – Legacy Dialogs – Bar'.

Mathematical Statistics, First Edition. Dieter Rasch and Dieter Schott.

SPSS output

Descriptive statistics

	N	Minimum	Maximum	Mean	Std. deviation	Variance
Mean	81	1.00	3.00	2.0000	.41079	.16875
Valid N (listwise)	81					

Descriptive statistics

	N	Minimum	Maximum	Mean
s2	81	.00	1.33	.6667
Valid N (listwise)	81			

Population mean = 3 and population variance s2 = .6667 are obtained.

Remark

The population variance is $\sigma^2 = 2/3$. The population variance of a sample mean of size 4 with replacement is $\sigma^2/4$. From the population of $N = 81$ sample means of possible different samples of size 4, the package SPSS calculates $S^2/4$ with $S^2 = \sigma^2/(N-1)$.

Exercise 1.3

The conditional distributions are as follows:

a) $P(\boldsymbol{Y} = Y|M(\boldsymbol{Y}) = t) = \dfrac{t!}{n^t \prod_{i=1}^{n} y_i!} I_{\{Y:\Sigma y_i = t\}}$,

b) $f(Y|M) = \dfrac{I_{\{y_{(1)} = \min y_i, y_{(n)} = \max y_i\}}}{n(n-1)\left(y_{(n)} - y_{(1)}\right)^{n-2}}$,

c) $f(Y|M) = \dfrac{I_{\{0 < y_{(1)} = \min y_i \le \max y_i = y_{(n)}\}}}{n\, y_{(n)}^{n-1}}$,

d) $M(Y)$ is gamma distributed.

Exercise 1.4

The sufficient statistics are:

a) $\boldsymbol{M} = \ln \prod_{i=1}^{n} y_i$,

b) $\boldsymbol{M} = \sum_{i=1}^{n} y_i^a$,

c) $\boldsymbol{M} = \ln \prod_{i=1}^{n} y_i$.

Exercise 1.5

The minimal sufficient statistics are:

a) $\boldsymbol{M} = \sum_{i=1}^{n} \boldsymbol{y}_i$,

b) $\boldsymbol{M} = \left(\boldsymbol{y}_{(1)}, \ldots, \boldsymbol{y}_{(n)}\right)$,

c) i) $\boldsymbol{M} = \sum_{i=1}^{n} \boldsymbol{y}_i$,

ii) $\boldsymbol{M} = \left(\boldsymbol{y}_{(1)}, \ldots, \boldsymbol{y}_{(n)}\right)$,

d) i) $\boldsymbol{M} = \Pi_{i=1}^{n} \boldsymbol{y}_i$,

ii) $\boldsymbol{M} = \Pi_{i=1}^{n} (1 - \boldsymbol{y}_i)$.

Exercise 1.6

a) Apply the uniqueness theorem for power series (concerning the fact that the coefficients of power series are uniquely determined).

b) If $\int_{\theta_1}^{\theta_2} h(y)dy = 0$ holds for each interval $(\theta_1, \theta_2) \subset R^1$, then the integrable function $h(y)$ has to be almost everywhere identical to 0.

Exercise 1.7

The statistic $y_{(n)}$ has the density function $f(t) = \frac{n}{\theta^n} t^{n-1}, 0 < t < \theta$. The family of these distributions $(\theta \in R^+)$ is complete. This can be proven as in Exercise 1.6 (b).

Exercise 1.8

Let $h(y)$ be an arbitrary discrete function with $E(h(\boldsymbol{y})) = 0$ for all $\theta \in (0,1)$. For $\theta = 0$ it follows $h(0) = 0$, and putting $y = k$ further,

$$\sum_{k=1}^{\infty} h(k)\theta^{k-1} = -\frac{h(-1)}{(1-\theta)^2} = -h(-1)\sum_{k=1}^{\infty} k\theta^{k-1}, \quad \theta \in (0,1).$$

Because of the uniqueness theorem for power series (compare Exercise 1.6), we get $h(k) = -kh(-1), \ k = 1, 2, \ldots$

If $h(y)$ is bounded, then $h(-1) = 0$ and therefore $h(y) \equiv 0$. On the other hand, for $h(-1) = -1$, the function

$$h(y) = \begin{cases} y & \text{for } y = -1, 0, 1, \ldots \\ 0 & \text{else} \end{cases}$$

is an unbiased estimator of zero.

Exercise 1.9

We obtain for the Fisher information the expressions

a) $I(\theta) = B''(\theta) - \frac{\eta''(\theta)B'(\theta)}{\eta'(\theta)}$,

b) i) $I(p) = \frac{n}{p(1-p)}$,

ii) $I(\lambda) = \frac{1}{\lambda}$,

iii) $I(\theta) = \frac{1}{\theta^2}$,

iv) $I(\sigma) = \frac{2}{\sigma^2}$.

Exercise 1.10

a) This follows by considering $E\left[\frac{\partial}{\partial\theta_i}L(\mathbf{y},\theta)\right] = 0, \quad i = 1,2,\ldots,p.$

b) This can be shown analogously as $I(\theta) = -E\left[\frac{\partial^2}{\partial\theta^2}\ln L(\mathbf{y},\theta)\right]$ following the derivation after Definition 1.10.

Exercise 1.11

Let be $\mathbf{M} = M(\mathbf{Y})$.

a) $E(\mathbf{M}) = e^{-\theta}, \quad \text{var}(\mathbf{M}) = e^{-\theta}\left(1 - e^{-\theta}\right), \quad I(\theta) = \frac{1}{\theta}, \quad \text{var}(\mathbf{M}) > \theta\, e^{-2\theta}.$

b) $n\bar{y}$ is $P(n\theta)$ – distributed (Poisson).

$$E(\mathbf{M}) = e^{-\theta}, \quad \text{var}(\mathbf{M}) = e^{-2\theta}\left(e^{\theta/n} - 1\right), \quad I(\theta) = \frac{n}{\theta}, \quad \text{var}(\mathbf{M}) > \frac{\theta\, e^{-2\theta}}{n}.$$

c) $E(\mathbf{M}) = \frac{1}{\theta}, \quad \text{var}(\mathbf{M}) = \frac{1}{n\theta^2}, \quad I(\theta) = \frac{n}{\theta^2}, \quad \text{var}(\mathbf{M}) > \frac{\left(\frac{dg}{d\theta}\right)^2}{nI(\theta)} = \frac{1}{n^2\theta^2} \quad (n > 1)$

with $g(\theta) = E(\mathbf{M}) = \frac{1}{\theta}$.

Exercise 1.12

a)

i	1	2	3	4	5	6	7	8	9
$R(d_i(\mathbf{y}),\ \theta_1)$	0	7	3.5	3	10	6.5	1.5	8.5	5
$R(d_i(\mathbf{y}),\ \theta_2)$	12	7.6	9.6	5.4	1	3	8.4	4	6

where $R(d(\mathbf{y}),\ \theta) = L(d(0),\ \theta)\ p_\theta(0)\ +\ L(d(1),\ \theta)\ p_\theta(1)$.

b)

i	1	2	3	4	5	6	7	8	9
$\max_{j=1,2}\{R(d_i,\theta_j)\}$	12	7.6	9.6	5.4	10	6.5	8.4	8.5	6

$\min_i \max_j \{R(d_i,\theta_j)\} = R(d_4,\theta_2) = 5.4$, minimax decision function $d_M = d_4(\boldsymbol{y})$.

c) $r(d_i,\pi) = E[R(d_i(\boldsymbol{y}),\theta)] = R(d_i(\boldsymbol{y}),\theta_1)\pi(\theta_1) + R(d_1(\boldsymbol{y}),\theta_2)\pi(\theta_2)$.

i	1	2	3	4	5	6	7	8	9
$r\ (d_i,\ \pi)$	9.6	7.48	8.38	4.92	2.8	3.7	7.02	4.9	5.8

$\min\{r(d_i,\pi)\} = 2.8$, Bayesian decision function $d_B = d_5(\boldsymbol{y})$.

Exercise 1.13

a) $$R(d_{r,s}(\bar{y}),\theta) = \begin{cases} c\Phi((\theta-r)\sqrt{n}) + b\Phi((\theta-s)\sqrt{n}) & \text{for } \theta<0, \\ b[1-\Phi(\sqrt{n}s)+\Phi(\sqrt{n}r)] & \text{for } \theta=0, \\ b\Phi(\sqrt{n}(r-\theta)) + c\Phi((s-\theta)\sqrt{n}) & \text{for } \theta>0. \end{cases}$$

b) i) $$R(d_{-1,1}(\bar{y}),\theta) = \begin{cases} \Phi(\theta+1)+\Phi(\theta-1) & \text{for } \theta<0, \\ 2\Phi(-1) & \text{for } \theta=0, \\ 2-\Phi(\theta+1)-\Phi(\theta-1) & \text{for } \theta>0. \end{cases}$$

ii) $$R(d_{-1,2}(\bar{y}),\theta) = \begin{cases} \Phi(\theta+1)+\Phi(\theta-2) & \text{for } \theta<0, \\ \Phi(-2)+\Phi(-1) & \text{for } \theta=0, \\ 2-\Phi(\theta+1)-\Phi(\theta-2) & \text{for } \theta>0. \end{cases}$$

$d_{-1,1}(\bar{y})$ is for $\theta>0$ 'better' than $d_{-1,2}(\bar{y})$ (in the meaning of a smaller risk).

Chapter 2

Exercise 2.1

a) $\sum_{k=-1}^{3} P(\boldsymbol{y}=k) = 1$,

b) $$U(\boldsymbol{y}) = \begin{cases} a & \text{for} & \boldsymbol{y}=-1, \\ 0 & \text{for} & \boldsymbol{y}=0,3, \\ -2a & \text{for} & \boldsymbol{y}=1, \\ 2a & \text{for} & \boldsymbol{y}=2, \end{cases} \quad a\in R^1,$$

c) i) $S_0(y) = \begin{cases} 0 & \text{for } y = -1,0 \\ 1 & \text{for } y = 1,3, \quad E(S_0) = p, \\ 2 & \text{for } y = 2, \end{cases}$

$$\tilde{S}_1(y) = \begin{cases} -a & \text{for} \quad y = -1, \\ 0 & \text{for} \quad y = 0, \\ 1+2a & \text{for} \quad y = 1, \qquad \text{with } a = -\dfrac{1+p}{3}, \\ 2+2a & \text{for} \quad y = 2, \\ 1 & \text{for} \quad y = 3, \end{cases}$$

ii) $S_0(y) = \begin{cases} \dfrac{1}{2} & \text{for} \quad y = -1, \\ 0 & \text{for } y = 0,1,2,3, \end{cases} \qquad E(S_0(y)) = p(1-p),$

$$\tilde{S}_2(y) = \begin{cases} \dfrac{1}{2} - a & \text{for} \quad y = -1, \\ 0 & \text{for} \quad y = 0,3, \quad \text{with } a = \dfrac{1}{6}, \\ 2a & \text{for} \quad y = 1, \\ -2a & \text{for} \quad y = 2, \end{cases}$$

d) $\tilde{S}_1(y)$ is only a LVUE, and $\tilde{S}_2(y)$ is even a UVUE with $E\left[\tilde{S}_2(y)\ U(y)\right] = 0$.

Exercise 2.2

First observe that $M(Y) = \sum y_i$ is completely sufficient.

a) $\hat{\psi}(Y) = \dfrac{1}{Nn} M(Y) = \dfrac{1}{N} \bar{Y}$ with $E(\hat{\psi}(Y)) = p$ is a UVUE according to Theorem 2.4.

b) $S(Y) = \dfrac{1}{nN} M(Y)$ is completely sufficient; therefore also $\hat{\psi}(Y) = \dfrac{nN}{Nn-1} S(Y)(1 - S(Y))$ with $E(\hat{\psi}(Y)) = p(1-p)$. Because of Theorem 2.4, the function $\hat{\psi}(Y)$ is a UVUE.

Exercise 2.3

a) $\dfrac{1}{\sigma^2} \sum_{i=1}^{n} (y_i - \bar{y})^2$ is distributed as $CS(n-1)$ and has therefore the expectation $n-1$. This implies the assertion.

b) It is $S(Y) = \dfrac{n}{n-1}(1-\bar{y})\bar{y}$. Hence, the assertion follows.

Exercise 2.4

a) $S_{ML}(\boldsymbol{Y}) = \boldsymbol{y}_{(n)},\ S_M(\boldsymbol{Y}) = 2\bar{\boldsymbol{y}},$

b) $S_{ML}(\boldsymbol{Y}) = \frac{1}{2}\boldsymbol{y}_{(n)},\ S_M(\boldsymbol{Y}) = \frac{2}{3}\bar{\boldsymbol{y}},$

c) $S_{ML}(Y)$ each value of the interval $\left[y_{(n)} - 1, y_{(1)}\right]$ (MLE not uniquely determined!), $S_M(\boldsymbol{Y}) = \bar{\boldsymbol{y}} - \frac{1}{2}.$

Exercise 2.5

a) $\widetilde{S}_{ML}(\boldsymbol{Y}) = \left(1 + \frac{1}{n}\right),\ S_{ML}(\boldsymbol{Y}) = \left(1 + \frac{1}{n}\right)\boldsymbol{y}_{(n)},\ \widetilde{S}_M(\boldsymbol{Y}) = S_M(\boldsymbol{Y}).$

b) $\widetilde{S}_{ML}(\boldsymbol{Y})$ is complete sufficient and unbiased, that is, a UVUE.

$$S_0(\boldsymbol{Y}) = \widetilde{S}_{ML}(\boldsymbol{Y}).$$

$$E_0\left(\widetilde{S}_{ML}(\boldsymbol{Y})\right) = 1,\ E_0\left(\widetilde{S}_M(\boldsymbol{Y})\right) = \frac{3}{n+2}.$$

Exercise 2.6

a) $\hat{\boldsymbol{a}} = \bar{\boldsymbol{x}},\ \hat{\boldsymbol{b}} = \bar{\boldsymbol{y}},\ \hat{\boldsymbol{c}} = \bar{\boldsymbol{z}},$

b) $\hat{\boldsymbol{a}}^* = \bar{\boldsymbol{x}}\left(1 - \lambda_a^2\right) + \lambda_a^2(\bar{\boldsymbol{z}} - \bar{\boldsymbol{y}}),\ \hat{\boldsymbol{b}}^* = \bar{\boldsymbol{y}}\left(1 - \lambda_b^2\right) + \lambda_b^2(\bar{\boldsymbol{z}} - \bar{\boldsymbol{x}}),$

$\hat{\boldsymbol{c}}^* = \bar{\boldsymbol{z}}\left(1 - \lambda_c^2\right) + \lambda_c^2(\bar{\boldsymbol{x}} + \bar{\boldsymbol{y}})$

with $\sigma^2 = \sigma_a^2 + \sigma_b^2 + \sigma_c^2$ and $\lambda_a^2 = \frac{\sigma_a^2}{\sigma^2},\ \lambda_b^2 = \frac{\sigma_b^2}{\sigma^2},\ \lambda_c^2 = \frac{\sigma_c^2}{\sigma^2}.$

c) $E(\hat{\boldsymbol{a}}) = a,\ E\left(\hat{\boldsymbol{b}}\right) = b,\ E(\hat{\boldsymbol{c}}) = c,\ \text{var}(\hat{\boldsymbol{a}}) = \frac{\sigma_a^2}{n},\ \text{var}\left(\hat{\boldsymbol{b}}\right) = \frac{\sigma_b^2}{n},\ \text{var}(\hat{\boldsymbol{c}}) = \frac{\sigma_c^2}{n},$

$E(\hat{\boldsymbol{a}}^*) = a,\ E\left(\hat{\boldsymbol{b}}^*\right) = b,\ E(\hat{\boldsymbol{c}}^*) = c,$

$\text{var}(\hat{\boldsymbol{a}}^*) = \frac{1}{n}\sigma_a^2\left(1 - \lambda_a^2\right),\ \text{var}\left(\hat{\boldsymbol{b}}^*\right) = \frac{1}{n}\sigma_b^2\left(1 - \lambda_b^2\right),\ \text{var}(\hat{\boldsymbol{c}}^*) = \frac{1}{n}\sigma_c^2\left(1 - \lambda_c^2\right).$

Exercise 2.7

The problems

$$L(\bar{Y}) = \frac{1}{(2\pi)^{n/2}\sigma^n}\exp\left\{-\frac{1}{2\sigma^2}\sum_{i=1}^{n}(y_i - f_i(x_i,\theta))^2\right\} \to \max$$

and

$$\sum_{i=1}^{n}(y_i - f_i(x_i,\theta))^2 \to \min$$

are equivalent.

Exercise 2.8

a) $\hat{\theta}_M = \bar{y}$,

b) $\hat{\alpha}_M = \bar{y} - \hat{\beta}_M \bar{x}$, $\hat{\beta}_M = \dfrac{\sum_{i=1}^{n} x_i y_i - n\bar{x}\bar{y}}{\sum_{i=1}^{n} x_i^2 - n\bar{x}^2}$.

Exercise 2.9

$$v_n\left(\theta, y_{(n)}\right) = -\frac{\theta}{n+1}.$$

Exercise 2.10

Using the Taylor expansion for $\bar{x}/\bar{y}$ at (η, μ), we obtain

$$E(\hat{g}) = \frac{\eta}{\mu} + \frac{\sigma^2\eta}{n\mu^3} - \frac{\varrho\sigma\tau}{n\mu^2} + O\left(\frac{1}{n^2}\right) = \frac{\eta}{\mu} + O\left(\frac{1}{n}\right),$$

$$E(J[\hat{g}]) = \frac{\eta}{\mu} + O\left(\frac{1}{n(n-1)}\right) = \frac{\eta}{\mu} + O\left(\frac{1}{n^2}\right).$$

Exercise 2.11

a) By (2.33) we get $E\left(y_{(j)}\right) = \mu + \left(\dfrac{2j}{n+1} - 1\right)\alpha$, $j = 1,\ldots,n$.

b) The assertion follows using the result of (a).

Exercise 2.12

a) $I(\alpha) = \dfrac{1}{\alpha^2}$, $e(S(\boldsymbol{Y})) = \dfrac{\alpha^2}{n\,\mathrm{var}(S(\boldsymbol{Y}))}$ because of (2.44).

b) $n\hat{\alpha}_{ML}(\boldsymbol{Y}) = n/\bar{y}$ is gamma distributed with the parameters n, α such that $\tilde{\alpha}(\boldsymbol{Y}) = \dfrac{n-1}{n\bar{y}}$ with $\mathrm{var}(\tilde{\alpha}(\boldsymbol{Y})) = \dfrac{\alpha^2}{n-2}$ is fulfilled. Finally, we find $e(\bar{\alpha}\boldsymbol{Y}) = 1 - \dfrac{2}{n}$.

Exercise 2.13

It is

$$E(\hat{\alpha}_{ML}(\boldsymbol{Y})) = E\left(\frac{1}{\bar{y}}\right) = \frac{n}{n-1}\alpha \to \alpha \quad \text{for } n \to \infty.$$

The consistency can be proven using the Chebyshev inequality $P\{|\boldsymbol{Y} - E(\boldsymbol{Y})| \geq \varepsilon\} \leq \dfrac{1}{\varepsilon^2}\mathrm{var}(\boldsymbol{Y})$.

Exercise 2.14

It is

$$\hat{\boldsymbol{\theta}}_{ML} = \sqrt{1 + \frac{1}{n}\sum_{i=1}^{n} \boldsymbol{y}_i^2} - 1.$$

Taking the (weak) law of large numbers into account, we arrive at

$$\frac{1}{n}\sum_{i=1}^{n} \boldsymbol{y}_i^2 \xrightarrow{P} E\left(\boldsymbol{y}_i^2\right) = 2\theta + \theta^2 \quad \text{for } n \to \infty.$$

Chapter 3

Exercise 3.1

a) $E(k_1(\boldsymbol{y})|H_0) = E(k_2(\boldsymbol{y})|H_0) = \alpha,\ \pi_{k_1}(H_A) = \pi_{k_2}(H_A) = 1$,
b) The test

$$k_1(y) = \begin{cases} \alpha & \text{for } L(y|H_A) = cL(y|H_0) \\ 1 & \text{for } L(y|H_A) > cL(y|H_0) \end{cases},$$

is randomised for $c = 0$. The test $k_2(y)$ cannot be represented in the form (3.5).

Exercise 3.2

We put

a) $A = 1 + \dfrac{\ln c_\alpha - n\ln\dfrac{1-p_1}{1-p_0}}{n\ln\dfrac{p_1}{p_0}}$,

i) $k(Y) = \begin{cases} 1 & \text{for } \bar{y} > A, \\ \gamma(Y) & \text{for } \bar{y} = A, \\ 0 & \text{for } \bar{y} < A, \end{cases}$ if $p_1 > p_0$,

ii) $k(Y) = \begin{cases} 1 & \text{for } \bar{y} < A, \\ \gamma(Y) & \text{for } \bar{y} = A, \\ 0 & \text{for } \bar{y} > A; \end{cases}$ if $p_1 < p_0$,

b) $c_\alpha = 1.8,\ \gamma(y) = 0.1,\ \beta = 0.91.$

Exercise 3.3

$$k(Y)=\begin{cases}1 & \text{for } T>4,\\ 0.413 & \text{for } T=4,\\ 0 & \text{for } T<4\end{cases} \quad \text{with } T=\sum_{i=1}^{10} y_i,\ \beta=0.0214.$$

Exercise 3.4

$$\lambda_0<\lambda_1: k(Y)=\begin{cases}1 & \text{for } 2n\bar{y}\lambda_0<CS(2n|\alpha),\\ 0 & \text{for } 2n\bar{y}\lambda_0>CS(2n|\alpha);\end{cases}$$

$$\lambda_0>\lambda_1: k(Y)=\begin{cases}1 & \text{for } 2n\bar{y}\lambda_0>CS(2n|1-\alpha),\\ 0 & \text{for } 2n\bar{y}\lambda_0<CS(2n|1-\alpha).\end{cases}$$

Exercise 3.5

Use

$$k(Y)=k^*(M)=\begin{cases}1 & \text{for } M<M_\alpha,\\ \gamma_\alpha & \text{for } M=M_\alpha,\\ 0 & \text{for } M>M_\alpha,\end{cases}$$

instead of (3.24) and α instead of $1-\alpha$ in the inequality for M_0^α as well as M_α instead of $M_{1-\alpha}$. The proof is analogous to that of Theorem 3.8.

Exercise 3.6

a) $k(Y)=\begin{cases}1 & \text{for } 2\lambda_0 n\bar{y}>CS(2n|\alpha),\\ 0 & \text{for } 2\lambda_0 n\bar{y}<CS(2n|\alpha).\end{cases}$

b) $\pi(\lambda)=F_{\gamma^2_{(2n)}}\left(\frac{\lambda}{\lambda_0}CS(2n|\alpha)\right)$, where $F_{\gamma^2_{(2n)}}$ is the distribution function of $CS(2n)$.

c) H_0 is accepted.

Exercise 3.7

a) $\pi(\theta)=1-\min\left(1,\left(\frac{c}{\theta}\right)^n\right)$,

b) $c=\sqrt[n]{0.95}$,

c) $c=0.4987$,

d) $\beta=0.02;\ n=9$.

Exercise 3.8

a) The existence follows from Theorem 3.8. Putting $M(Y)=\sum_{i=1}^{n} y_i^2$ we get

$$k(Y)\approx\begin{cases}1 & \text{for } M(Y)>2\sqrt{n}(z_{1-\alpha}+\sqrt{n}),\\ 0 & \text{for } M(Y)<2\sqrt{n}(z_{1-\alpha}+\sqrt{n}).\end{cases}$$

b) The power function is $\pi(\theta)\approx 1-\Phi\left(\frac{1}{\theta^2}\left(z_{1-\alpha}+\sqrt{n}\right)-\sqrt{n}\right)$.

Exercise 3.9

a) The existence follows from Theorem 3.11.
b) $e^{-\lambda_0 c_1}-e^{-\lambda_0 c_2}=1-\alpha,\quad c_1 e^{-\lambda_0 c_1}-c_2 e^{-\lambda_0 c_2}=0,$
c) $k^*(y)=\begin{cases}1 & \text{for } y<0.00253 \text{ or } y>0.3689,\\ 0 & \text{else.}\end{cases}$

$\pi(10,1)=0.04936<\alpha=0.05$.

Exercise 3.10

a) $k(Y)=\begin{cases}1 & \text{for } h_n \lesseqgtr np\mp\sqrt{np_0(1-p_0)}z_{1-\frac{\alpha}{2}},\\ 0 & \text{else.}\end{cases}$

b) $H_0: p=\frac{1}{2}$ has to be rejected.

c) $H_0: p=\frac{1}{6}$ has to be accepted.

Exercise 3.11

a) $H_0:\mu=3.5$ has to be rejected.
b) The probability is 0.68.
c) $\delta\geq 0.065$.
d) $H_0:\mu=3.5$ has to be rejected.

Exercise 3.12

Acception of $H_0:\mu\leq 9.5$.
Rejection of $H_0:\sigma^2\leq 6.25$.

Exercise 3.13

In both cases H_0 is accepted.

Exercise 3.14

Sample size $n=15$.

Exercise 3.15

a) $K_1 = \left[\frac{y_{(n)}}{\sqrt[n]{1-\alpha_1}}, \frac{y_{(n)}}{\sqrt[n]{\alpha_2}}\right]$, $K_2 = \left[y_{(n)}, \frac{y_{(n)}}{\sqrt[n]{\alpha}}\right]$, $K_3 = \left[\frac{y_{(n)}}{\sqrt[n]{1-\alpha}}, +\infty\right)$,

b) $l_1 = \frac{n\theta_0}{n+1}\left(\frac{1}{\sqrt[n]{\alpha_2}} - \frac{1}{\sqrt[n]{1-\alpha_1}}\right)$, $l_2 = \frac{n\theta_0}{n+1}\left(\frac{1}{\sqrt[n]{\alpha}} - 1\right)$, $l_3 = \infty$, K_2 has the smallest mean length.

c) $$W_1(\theta|\theta_0) = \begin{cases} 0 & \text{for } \theta \le 0 \text{ or } \theta \ge \frac{\theta_0}{\sqrt[n]{\alpha_2}}, \\ \left(\frac{\theta}{\theta_0}\right)^n (1-\alpha_1) & \text{for } 0 \le \theta \le \frac{\theta_0}{\sqrt[n]{1-\alpha_1}}, \\ 1-\left(\frac{\theta}{\theta_0}\right)^n \alpha_2 & \text{for } \frac{\theta_0}{\sqrt[n]{1-\alpha_1}} \le \theta \le \frac{\theta_0}{\sqrt[n]{\alpha_2}}, \end{cases}$$

$$W_2(\theta|\theta_0) = \begin{cases} 0 & \text{for } \theta \le 0 \text{ or } \theta \ge \frac{\theta_0}{\sqrt[n]{\alpha}}, \\ \left(\frac{\theta}{\theta_0}\right)^n (1-\alpha) & \text{for } 0 \le \theta \le \theta_0, \\ 1-\left(\frac{\theta}{\theta_0}\right)^n \alpha & \text{for } \theta_0 \le \theta \le \frac{\theta_0}{\sqrt[n]{\alpha}}, \end{cases}$$

$$W_3(\theta|\theta_0) = \begin{cases} 0 & \text{for } \theta \le 0, \\ \left(\frac{\theta}{\theta_0}\right)^n (1-\alpha) & \text{for } 0 \le \theta \le \frac{\theta_0}{\sqrt[n]{1-\alpha}}, \\ 1 & \text{for } \theta \ge \frac{\theta_0}{\sqrt[n]{1-\alpha}}. \end{cases}$$

Only K_2 is unbiased.

Exercise 3.16

a) $K_L = \left[\frac{CS(2n|\alpha)}{2n\bar{y}}; +\infty\right)$, $K_R = \left[0; \frac{CS(2n|1-\alpha)}{2n\bar{y}}\right]$,

b) $K_L = [0.0065; +\infty)$, $K_R = [0; 0.0189]$.

Exercise 3.17

Confidence interval

$$\left[\frac{s_1^2}{s_2^2 F\left(n_1-1, n_2-1 \middle| 1-\frac{\alpha}{2}\right)}; \frac{s_1^2 F\left(n_1-1, n_2-1 \middle| 1-\frac{\alpha}{2}\right)}{s_2^2}\right].$$

Exercise 3.18

a) $\pi(p) = 1-(1-p)^{n_0}$,

b) $E(\boldsymbol{n}|p) = \frac{1}{p}(1-(1-p)^{n_0})$,

c) $\alpha = 0.0956,\ \ \beta = 0.3487,\ \ E(\boldsymbol{n}|p_0) = 9.56,\ \ E(\boldsymbol{n}|p_1) = 6.51$.

Exercise 3.19

Sample sizes
a) $n = 139$,
b) $n = 45$.

Exercise 3.20

a) Use the test statistic of the Welch test, namely, $\boldsymbol{t}^* = \dfrac{\bar{y}_1 - \bar{y}_2}{\sqrt{\dfrac{s_1^2}{n_1} + \dfrac{s_2^2}{n_2}}}$.

b) (b1) $n_1 = 206;\ n_2 = 103$,
(b2) $n_1 = 64;\ n_2 = 32$.

Exercise 3.21

Proceed analogously to Section 3.4.2.1.

Chapter 4

Exercise 4.1

The equation $C^T b = 0_p$ ($b \in R^n$) defines the null space of the $(p \times n)$ matrix C^T. Since the columns of C are supposed to be an orthonormal basis of the p-dimensional linear subspace Ω, it is the rank space (range) of C and the null space of C^T is its $(n-p)$-dimensional orthogonal complement.

Exercise 4.2

We know that the $(n \times p)$ matrix X has the rank $p > 0$. The second derivative of $||Y - X\beta^2||$ according to β is equal to $2X^TX$, and it is therefore positive definite.

Exercise 4.3

We put $B = X - XGX^TX$ and obtain

$$B^TB = \left(X - XGX^TX\right)^T\left(X - XGX^TX\right) = X^TXG^TX^TXGX^TX - X^TXG^TX^TX = O.$$

This implies the assertion.

Exercise 4.4

Because of $X^T(I_n - XGX^T) = X^T - X^TXGX^T$, we can continue as in Exercise 4.3.

Exercise 4.5

Obviously the matrix

$$A = \begin{pmatrix} \frac{1}{n} & \cdots & \frac{1}{n} \\ \vdots & & \vdots \\ \frac{1}{n} & \cdots & \frac{1}{n} \end{pmatrix}$$

is symmetric, idempotent ($A^2 = A$) and of rank 1.

Chapter 5

Exercise 5.1

Since a normal distribution is supposed, we have only to show that the covariances vanish. We demonstrate this briefly for $\text{cov}(\bar{y}_{..}, \bar{y}_{i.} - \bar{y}_{..})$; the other cases follow analogously.

$$\text{cov}(\bar{y}_{..}, \bar{y}_{i.} - \bar{y}_{..}) = \text{cov}\left(\frac{\sum_{i=1}^{a}\sum_{j=1}^{b} y_{ij}}{ab}, \frac{\sum_{j=1}^{b} y_{ij}}{b} - \frac{\sum_{i=1}^{a}\sum_{j=1}^{b} y_{ij}}{ab}\right)$$

$$= \text{cov}\left(\frac{\sum_{i=1}^{a}\sum_{j=1}^{b} y_{ij}}{ab}, \frac{a\sum_{j=1}^{b} y_{ij}}{ab} - \frac{\sum_{i=1}^{a}\sum_{j=1}^{b} y_{ij}}{ab}\right) = 0.$$

Exercise 5.2

The data are fed in as follows (see Figure S1).

Then call 'Analyze – General Linear Model – Univariate' and work on the menu window as follows (see Figure S2).

Now pressing 'ok' supplies the result in Table S1. (The original output was adapted to the text in the book. More information you can find in the SPSS help under managing a pivot table.)

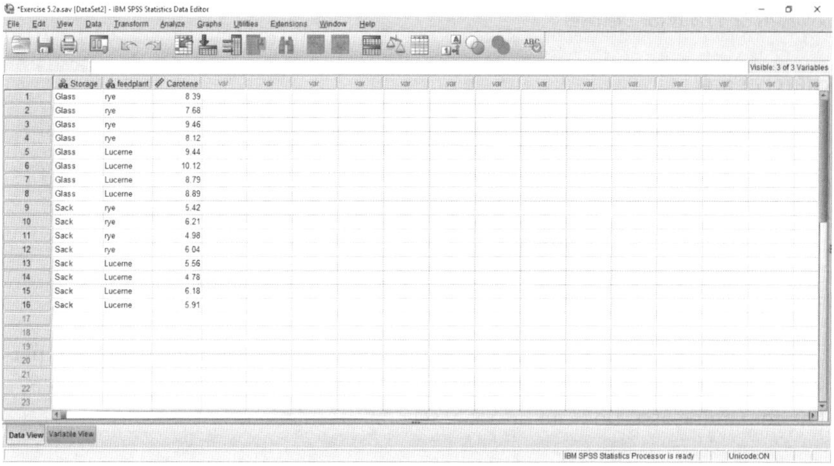

Figure S1 Data of Exercise 5.2. *Source:* Reproduced with permission of IBM.

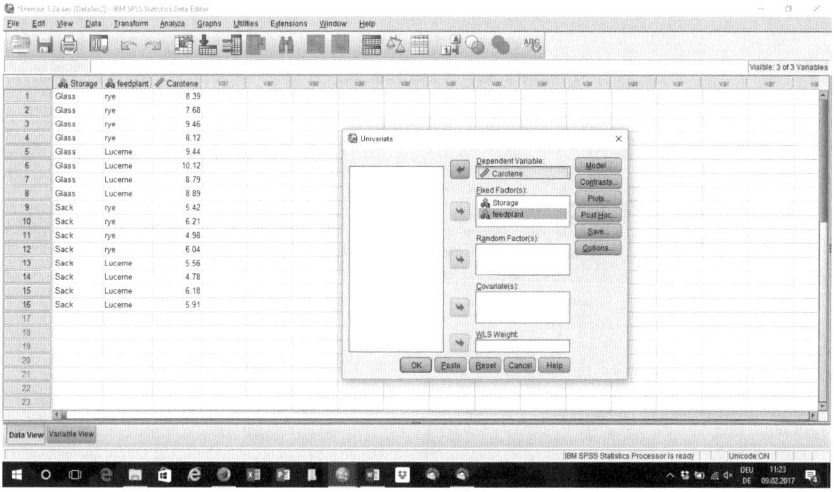

Figure S2 Data of Exercise 5.2 and menu. *Source:* Reproduced with permission of IBM.

Table S1 Tests of between-subject effects of Exercise 5.2

Dependent variable: carotene					
Source	**Type III sum of squares**	**df**	**Mean square**	**F**	**Sig.**
Storage	41.635	1	41.635	101.696	.000
feedplant	.710	1	.710	1.734	.213
Storage feedplant	.907	1	.907	2.216	.162
Error	4.913	12	.409		
Total	888.730	16			

Source: Reproduced with permission of IBM.

Exercise 5.3

Proceed analogously to Exercise 5.2.

Exercise 5.4

See the solution of Exercise 4.3.

Exercise 5.5

Obviously all three matrices are symmetric. The idempotence can be shown by performing the products $(B_2 - B_3(B_2 - B_3))$, $(B_1 - B_2)(B_1 - B_2)$ and $(I_n - B_1)(I_n - B_1)$. The remaining assertion can be easily seen by calculation.

Exercise 5.6

Use the brand new R-packages via https://cran.r-project.org/ and download for your computer (see Figure S3).

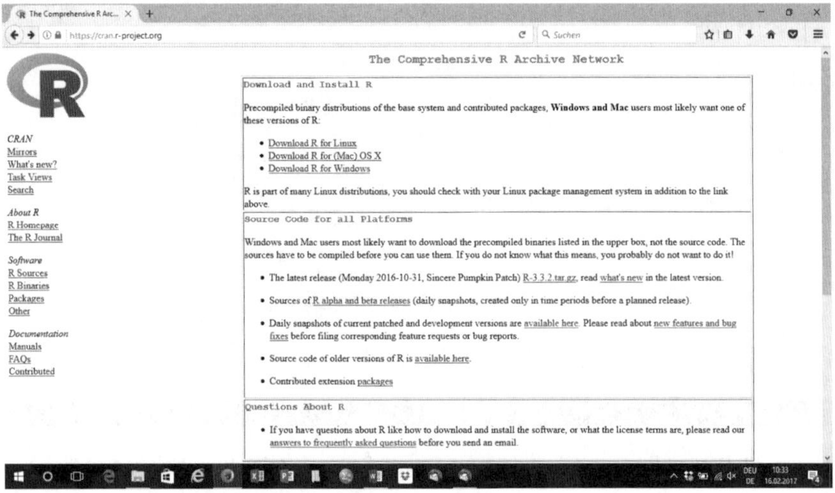

Figure S3 Start of the R-program.

Further, activate at the left-hand side 'Packages' and list of available packages by name. Then a list with R-packages appears, which contains also OPDOE (see Figure S4).

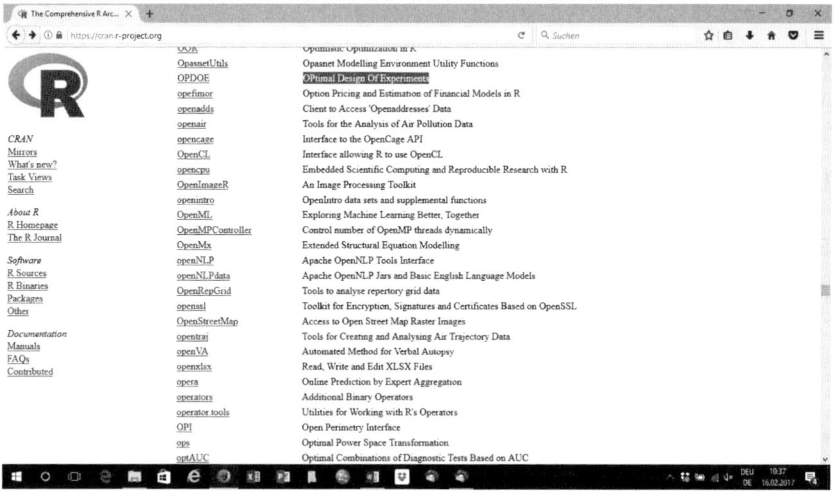

Figure S4 List of program packages in R.

Exercise 5.7

Maximin	40
Minimin	14

Exercise 5.8

Maximin	9
Minimin	4

Exercise 5.9

Factor *A*

Maximin	9
Minimin	4

Factor *B*

Maximin	51
Minimin	5

Exercise 5.10

Maximin	48
Minimin	5

Exercise 5.11

Maximin	7
Minimin	3

Chapter 6

Exercise 6.1

First we put the data in a SPSS data sheet (statistics data editor) and choose 'Analyze – General Linear Model – Univariate'. Then we get with our special data (see Figure S5).

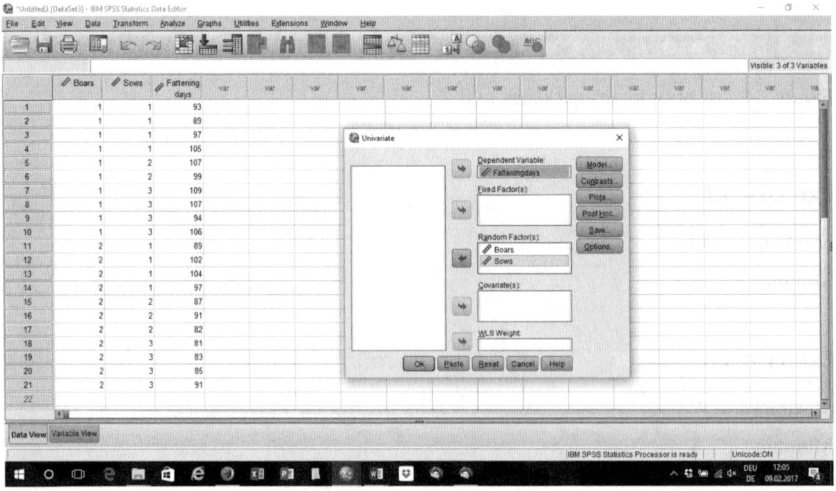

Figure S5 Data of Exercise 6.1 with menu. *Source:* Reproduced with permission of IBM.

Now we continue with 'Paste' and modify the command sequence as described in Chapter 5 in the part concerning the nested classification. We push on the button 'Execute' and obtain the following results (see Figure S6).

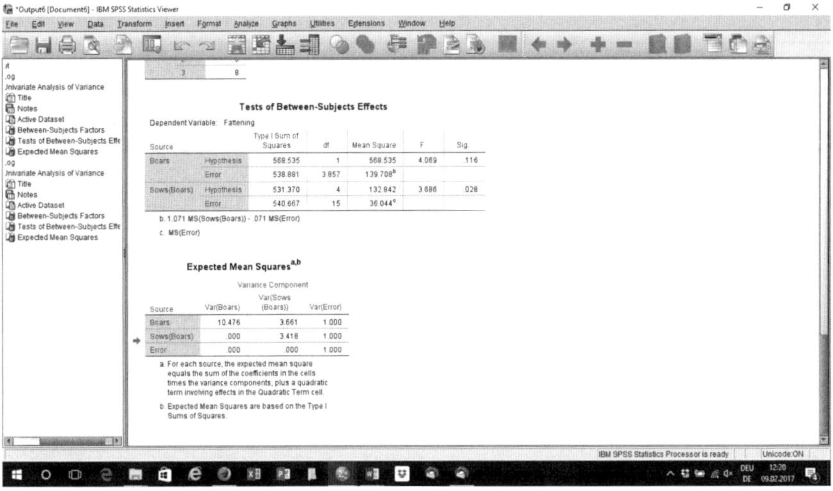

Figure S6 SPSS-ANOVA table of Exercise 6.1. *Source:* Reproduced with permission of IBM.

Estimate the variance components via

Analyze

General Linear Model

Variance Components

Exercise 6.2

First we find $a_2 = 83.67$.

Choosing $a_2 = 83$, the expression

$$A(200{,}83) = \left\{\frac{2}{82}\left(0.5+\frac{0.5}{200}\right)^2 + \frac{4}{117}0.5^2 + 2\frac{0.5}{200}\left[1-\frac{0.5}{200}199\right]\right\}\text{const}$$

turns out to be greater than the corresponding expression

$$A(200{,}84) = \left\{\frac{2}{83}\left(0.5+\frac{0.5}{200}\right)^2 + \frac{4}{116}0.5^2 + 2\frac{0.5}{200}\left[1-\frac{0.5}{200}199\right]\right\}\text{const}$$

for $a_2 = 84$. Now look for the optimal solution starting with the pairs

$$(a = 83, n = 2); (a = 83, n = 3); (a = 84, n = 2); (a = 83, n = 3).$$

Exercise 6.3

The completed data table is given as follows (see Table S2).

Table S2 Data of Exercise 6.3.

Sire									
B_1	B_2	B_3	B_4	B_5	B_6	B_7	B_8	B_9	B_{10}
120	152	130	149	110	157	119	150	144	159
155	144	138	107	142	107	158	135	112	105
131	147	123	143	124	146	140	150	123	103
130	103	135	133	109	133	108	125	121	105
140	131	138	139	154	104	138	104	132	144
140	102	152	102	135	119	154	150	144	129
142	102	159	103	118	107	156	140	132	119
146	150	128	110	116	138	145	103	129	100
130	159	137	103	150	147	150	132	103	115
152	132	144	138	148	152	124	128	140	146
115	102	154	122.70	138	124	100	122	106	108
146	160	139.82	122.70	115	142	135.64	154	152	119

If you like you can calculate the variance components by hand using the described method of analysis of variance. Alternatively you can use statistical software as SPSS or R.

Chapter 7

Exercise 7.1

Use the completed data table of Exercise 6.3. In the solution of Exercise 6.3, it is Table S2.

The random division into two classes can be realised with pseudo-random numbers that are uniformly distributed in the interval (0,1). A sire is assigned to class 1 if the result is less than 0.5, or otherwise to class 2. If in one of the two classes are 6 sires, then the remaining sires are put into the other class. We have a mixed model of twofold nested classification with the fixed factor 'Location' and the random factor 'Sire'.

Exercise 7.2

We recommend using SPSS for the solution. Observe the necessary syntax modification described in Chapter 5 for nested classification.

Exercise 7.3

It suffices to estimate the variance components of the factor 'Sire' using the method of analysis of variance by hand. Of course you can do it also with SPSS.

Chapter 8

Exercise 8.1

The partial derivatives of S according to β_0 and β_1 are

$$\frac{\partial S}{\partial \beta_0} = -2\sum_{i=1}^{n}(y_i - \beta_0 - \beta_1 x_i),$$

$$\frac{\partial S}{\partial \beta_1} = -2\sum_{i=1}^{n}x_i(y_i - \beta_0 - \beta_1 x_i).$$

If these derivatives are put to 0, we get the simultaneous equations

$$\sum_{i=1}^{n} y_i - nb_0 - b_1\sum_{i=0}^{n} x_i = 0,$$

$$\sum_{i=1}^{n} x_i y_i - b_0\sum_{i=0}^{n} x_i - b_1\sum_{i=0}^{n} x_i^2 = 0.$$

The first equation supplies (8.10) (if we replace the realisations by random variables). If we put $\boldsymbol{b}_0 = \bar{\boldsymbol{y}} - \boldsymbol{b}_1\bar{x}$ into the second equation and use random variables instead of realisations, the Equation (8.9) is obtained after rearrangement.

Exercise 8.2

Because of $\boldsymbol{b} = \hat{\boldsymbol{\beta}} = (X^T X)^{-1} X^T \boldsymbol{Y}$ (see Theorem 8.1), we get

$$E(\boldsymbol{b}) = E\left(\left(X^T X\right)^{-1} X^T \boldsymbol{Y}\right) = \left(X^T X\right)^{-1} X^T E(\boldsymbol{Y}) = \left(X^T X\right)^{-1} X^T X\beta = \beta$$

and as special cases $E(\boldsymbol{b}_0) = \beta_0$ and $E(\boldsymbol{b}_1) = \beta_1$. Further, it is

$$\text{var}(\boldsymbol{b}) = \left(X^T X\right)^{-1} X^T \text{var}(\boldsymbol{Y}) X \left(X^T X\right)^{-1}.$$

Considering now $\text{var}(\boldsymbol{Y}) = \sigma^2 I_n$, we find $\text{var}(\boldsymbol{b}) = \sigma^2 (X^T X)^{-1}$.

In our special case it is

$$X = \begin{pmatrix} 1 & x_1 \\ \vdots & \vdots \\ 1 & x_n \end{pmatrix}$$

and therefore

$$X^T X = \begin{pmatrix} n & \sum_{i=1}^{n} x_i \\ \sum_{i=1}^{n} x_i & \sum_{i=1}^{n} x_i^2 \end{pmatrix}$$

as well as

$$(X^T X)^{-1} = \frac{1}{n\sum_{i=1}^{n} x_i^2 - \left(\sum_{i=1}^{n} x_i\right)^2} \begin{pmatrix} n\sum_{i=1}^{n} x_i^2 & -\sum_{i=1}^{n} x_i \\ -\sum_{i=1}^{n} x_i & n \end{pmatrix}$$

$$= \frac{1}{n\sum_{i=1}^{n} (x_i - \bar{x})^2} \begin{pmatrix} \sum_{i=1}^{n} x_i^2 & -\frac{1}{n}\sum_{i=1}^{n} x_i \\ -\frac{1}{n}\sum_{i=1}^{n} x_i & 1 \end{pmatrix}.$$

This implies (8.14) and (8.15).

Exercise 8.3

Substituting $x_1 = \cos(2x)$, $x_2 = \ln(6x)$, the case is traced back to a twofold linear regression. In $\boldsymbol{b} = \hat{\boldsymbol{\beta}} = (X^T X)^{-1} X^T \boldsymbol{Y}$, we have now to put

$$X = \begin{pmatrix} 1 & \cos(2x_1) & \ln(6x_1) \\ 1 & \cos(2x_2) & \ln(6x_2) \\ \vdots & \vdots & \vdots \\ 1 & \cos(2x_{n-1}) & \ln(6x_{n-1}) \\ 1 & \cos(2x_n) & \ln(6x_n) \end{pmatrix}.$$

Exercise 8.4

After feeding the data of Example 8.3 concerning the storage in glass in a SPSS data sheet (see Figure S7), we select ‘Analyze – Regression – Linear’ and fill the appearing box correspondingly.

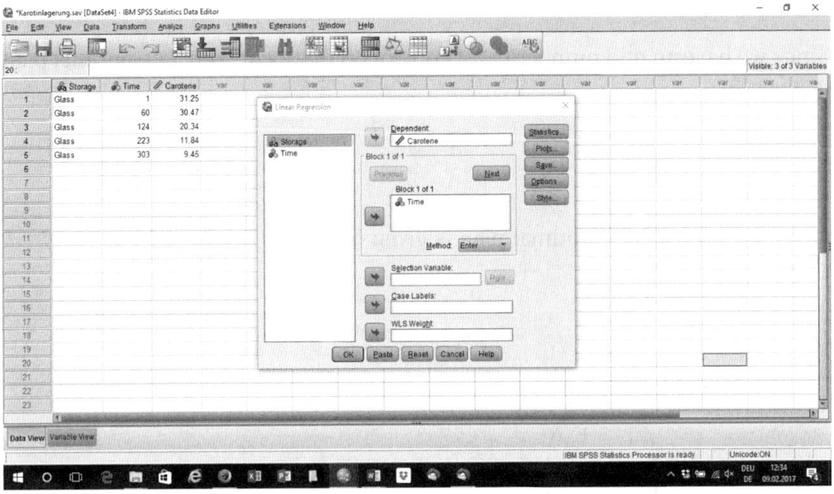

Figure S7 Data of Exercise 8.4 with menu. *Source:* Reproduced with permission of IBM.

Under 'Statistics' we request the covariance matrix of the estimations. Then the result is presented after pressing the button 'ok'. There we deleted the correlation coefficients, since we dealt with model I (see Figure S8).

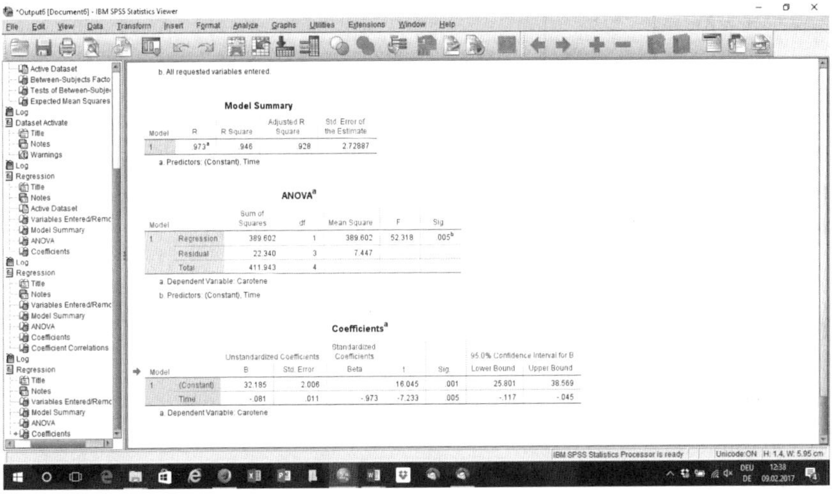

Figure S8 SPSS-output of Exercise 8.4. *Source:* Reproduced with permission of IBM.

Exercise 8.5

Since we have an odd number (5) of control points, a concrete D – optimal plan is given by

$$\begin{pmatrix} 1 & 303 \\ 3 & 2 \end{pmatrix},$$

and the concrete G – optimal plan is given by

$$\begin{pmatrix} 1 & 152 & 303 \\ 2 & 1 & 2 \end{pmatrix}.$$

For the D – optimal plan it is

$$X = X_D = \begin{pmatrix} 1 & 1 \\ 1 & 1 \\ 1 & 1 \\ 1 & 303 \\ 1 & 303 \end{pmatrix},$$

and for the G – optimal plan it is

$$X = X_G = \begin{pmatrix} 1 & 1 \\ 1 & 1 \\ 1 & 152 \\ 1 & 303 \\ 1 & 303 \end{pmatrix}.$$

Therefore the determinant $\left|X_G^T X_G\right| = 456\,010$ of the G – optimal plan is smaller than the corresponding determinant $\left|X_D^T X_D\right| = 547\,224$ of the D – optimal plan, which maximises $|X^T X|$ for $n = 5$ in the interval [1; 303].

Chapter 9

Exercise 9.1

a) Quasilinear
b) Quasilinear
c) Linear
d) Intrinsically non-linear
e) Quasilinear
f) Intrinsically non-linear
g) Intrinsically non-linear

Exercise 9.2

The non-linearity parameters are
a) θ_2, θ_3
b) θ_2
c) θ_2, θ_3
d) θ_3
e) θ_4

Exercise 9.3

The normal equations for the given $n = 11$ points serving to determine a, b, c are non-linear in c. If the first two equations are solved for a and b and if the corresponding values are put into the third equation, then a non-linear equation $g(c) = 0$ for c follows, which has to be iteratively solved.

If one of the usual iterative methods is used initialised, for example, with $c_0 = -0.5$, then after a few iterations a value $c \approx -0.406$ is obtained. If you want to check the quality of iterates c_k, you can calculate the values $f(c_k)$, which should lie nearby 0. If c is replaced in the solution formulas for the two other parameters a and b by its approximate value -0.406, then $a = 132.96$ and $b = -56.43$ is obtained. Hence, the estimated regression function is

$$f^*(x, \theta^*) = 132.96 - 56.43\,\mathrm{e}^{-0.406x}.$$

The estimate for the variance can be calculated using the formula

$$s^2 = \frac{1}{n-3} \sum_{i=1}^{n} (y_i - a - b\mathrm{e}^{cx_i})^2$$

The result is $s^2 = 0.761$.

In SPSS we edit the data in a data matrix and program the exponential function as shown in Figure S9.

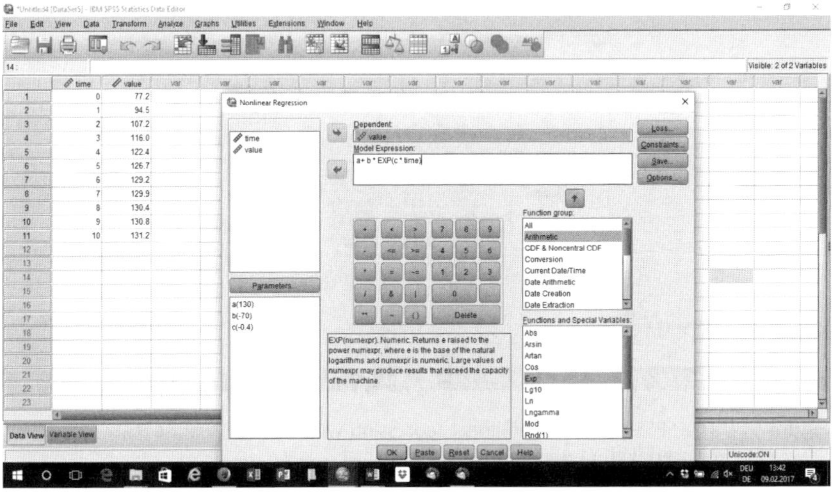

Figure S9 Menu for 'Non-linear Regression' in SPSS for Exercise 9.3. *Source:* Reproduced with permission of IBM.

The result is as follows (see Figure S10).

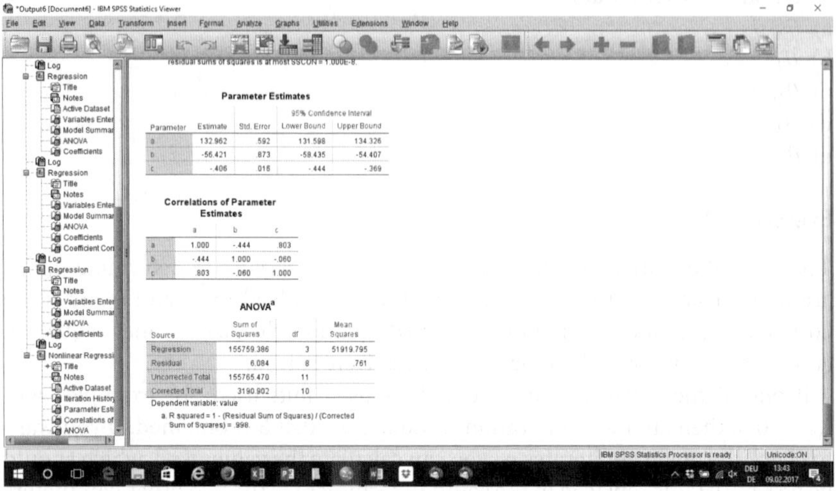

Figure S10 SPS-output 'Non-linear Regression'. *Source:* Reproduced with permission of IBM.

Exercise 9.4

First, we select a model using the criterion of residual variance. The best fit is reached for the arc tan (4) function.

We now use initial values in SPSS – Non-linear Regression as in Figure S11.

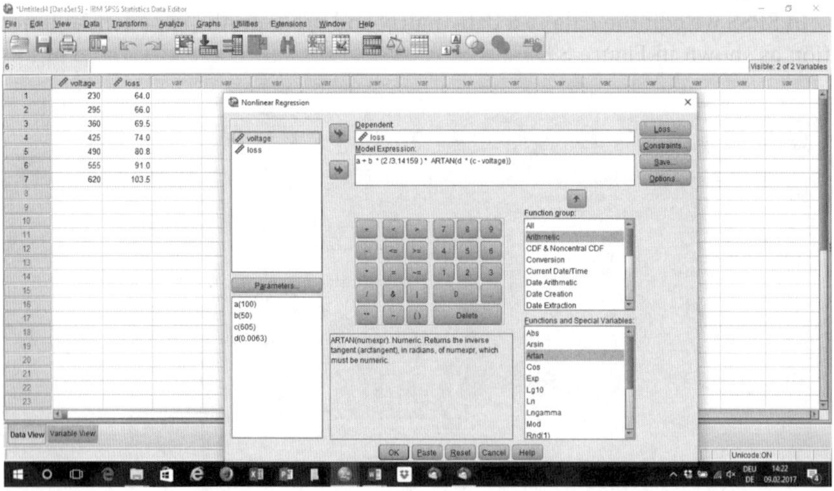

Figure S11 The SPSS-program non-linear regression with data of Exercise 9.4. *Source:* Reproduced with permission of IBM.

We obtain the results given in Figure S12.

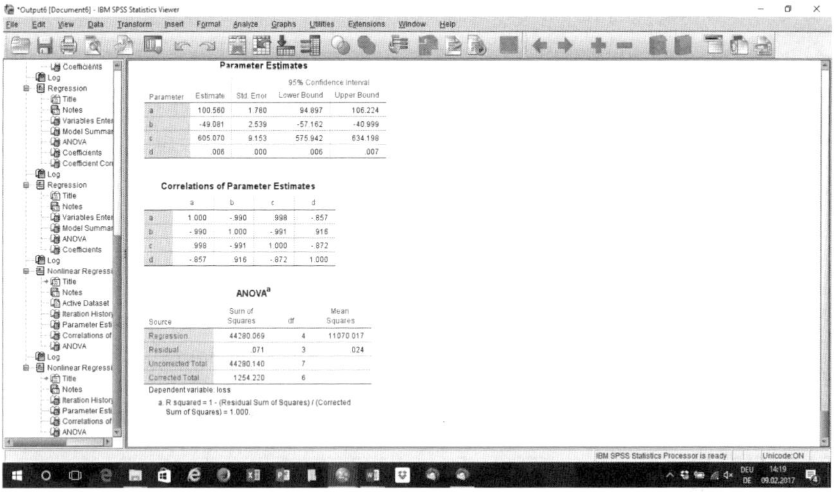

Figure S12 Menu for 'Non-linear Regression' in SPSS for Exercise 9.4. *Source:* Reproduced with permission of IBM.

Chapter 10

Exercise 10.1

Since the storage type is a fixed factor and the instants of time were prescribed by the experimenter, a model I–I of the form

$$\mu_{ij} = \mu + \alpha_i + \gamma z_{ij};\ i = 1, 2;\ j = 1,\ldots,5$$

is given with the main effects α_1 and α_2 for the both storage types and the contents z_{ij} of carotene.

Exercise 10.2

After the command sequence 'Analyze – General Linear Model – Univariate', the fixed factor and the covariate is entered, as Figure S13 shows.

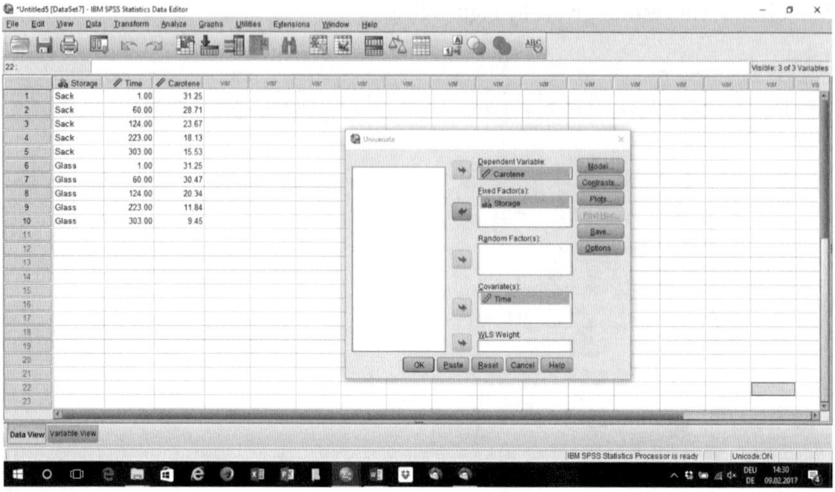

Figure S13 Data and menu in SPSS for Exercise 10.2. *Source:* Reproduced with permission of IBM.

Pressing 'OK' the result follows (see Figure S14).

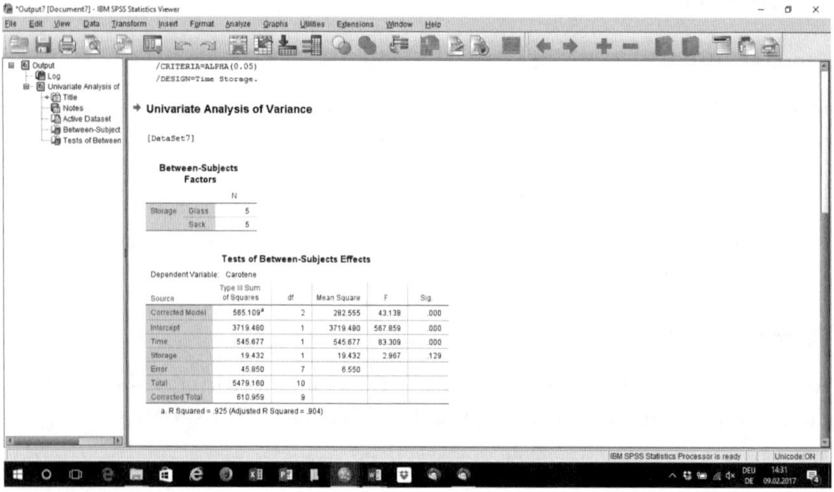

Figure S14 SPSS-output for the analysis of covariance for Exercise 10.2. *Source:* Reproduced with permission of IBM.

Chapter 11

The results are given in this chapter in form of tables (see Table S3, Table S4 and Table S5).

Exercise 11.1

Table S3 Sample sizes for Exercise 11.1.

		t			
d/σ		1	2	3	4
	0.1	1721	1654	1738	1762
	0.2	431	414	435	441
	0.5	69	67	70	71
	1	18	17	18	18

Exercise 11.2

Table S4 Sample sizes for Exercise 11.2 ($\alpha = 0.05$).

		β		
		0.05	0.1	0.2
d	0.5	105	85	64
	1	27	22	17

Exercise 11.3

Table S5 Sample sizes for Exercise 11.3 ($\beta = 0.05$, $d = \sigma$)

		a				
		3	4	5	10	20
α	0.05	28	31	33	40	47
	0.1	23	27	29	36	43

The remaining sample sizes for other values of β and d are omitted here.

Chapter 12

Exercise 12.1

Without computer use you can encode the 35 blocks into the numbers from 1 to 35, write these numbers down on corresponding sheets of paper, lay down these sheets into a bowl and draw these sheets without replacement by random. The block belonging to the first drawn number gets the first place. The randomisation within the blocks can be realised by throwing the dice. For each treatment the dice is once thrown: 1 or 4 means position 1; 2 or 5 means position 2; and finally 3 or 6 supplies position 3. This can lead to a repeated rearrangement within the blocks.

Exercise 12.2

The dual balanced incomplete block design (BIBD) has the parameters $k = r = 4$ and $\lambda = 2$. The design is

$$(1,2,4,6); (1,2,5,7); (1,3,4,7); (1,3,5,6); (2,3,4,5); (2,3,6,7); (4,5,6,7).$$

Exercise 12.3

We choose $m = 2$ and obtain

$$v = 85, \quad b = 85, \quad r = 21, \quad k = 21, \quad \lambda = 5.$$

Exercise 12.4

We choose $m = 1$, which supplies

$$v = 64, \quad b = 336, \quad r = 21, \quad k = 4, \quad \lambda = 1.$$

Exercise 12.5

Analogously to Example 12.3, a BIBD is obtained with the parameters

$$v = 12, \quad b = 22, \quad r = 11, \quad k = 6, \quad \lambda = 5.$$

Exercise 12.6

The parameters of the original BIBD are

$$v = 8, \quad b = 56, \quad r = 21, \quad k = 3, \quad \lambda = 6.$$

Exercise 12.7

In the LS (Latin square)

D	*E*	*A*	*B*	*C*	*G*	*F*
B	*D*	*E*	*F*	*A*	*B*	*C*
A	*B*	*C*	*D*	*E*	*F*	*G*
E	*C*	*B*	*G*	*F*	*D*	*A*
C	*G*	*F*	*E*	*B*	*A*	*D*
F	*A*	*G*	*C*	*B*	*E*	*D*
G	*F*	*D*	*A*	*C*	*B*	*E*

the columns have to be exchanged such that in the first row the sequence *A*,*B*,*C*, *D*,*E*,*F*,*G* appears.

Exercise 12.8

If we cancel in the LS of Exercise 12.7 the last both columns, then we get the design

D	*E*	*A*	*B*	*C*
B	*D*	*E*	*F*	*A*
A	*B*	*C*	*D*	*E*
E	*C*	*B*	*G*	*F*
C	*G*	*F*	*E*	*B*
F	*A*	*G*	*C*	*B*
G	*F*	*D*	*A*	*C*

This is no Youden design, since, for example, the pair (*A*,*B*) occurs four times, while the pair (*A*,*E*) occurs only three times.

Index Mathematical Statistics

Mathematical Statistics, First Edition. Dieter Rasch and Dieter Schott.

d

f

m

n

o

p

t

U

V